Presented to the

Ironside Memorial Library

by


Alumni

Miss Joanne Rankin

Computational

Methods of

Linear Algebra

**A SERIES OF BOOKS IN MATHEMATICS**

Editors: *R. A. Rosenbaum*   *G. Philip Johnson*

**D. K. FADDEEV** *and* **V. N. FADDEEVA**

# Computational
# Methods of
# Linear Algebra

*Translated by* **ROBERT C. WILLIAMS**

W. H. FREEMAN AND COMPANY
SAN FRANCISCO AND LONDON

*Originally published June 14, 1960, by State Publishing House for Physico-Mathematical Literature, Moscow B-71, Leninskii Prospekt, 15. E. Sokolova Printing House No. 2, Leningrad, Izmailovskii Prospekt, 29.*

26313

# PREFACE TO THE RUSSIAN EDITION

The present book is devoted to an exposition of computational methods for solving the basic problems of linear algebra.

These problems are: the solution of a system of linear equations, the inversion of a matrix, and the solution of the complete and special eigenvalue problem.

The great number of numerical methods for solving these problems (which have appeared mainly in recent years) has made it necessary for the authors to try to systematize them and present them from certain general points of view. Despite this, the authors have tried to present this material without going beyond the concepts of linear algebra, insofar as this is possible. Thus, for example, the authors have consciously excluded the use of the theory of continued fractions, replacing it with the theory of orthogonal polynomials, taking orthogonality in the linear-algebraic sense.

The important question of the influence of approximation errors on the results of computations is hardly touched on in the book.

The first chapter of the book is of an introductory nature. The remaining eight chapters are devoted to the presentation of computational methods. The material of these chapters was partially covered in V. N. Faddeeva's book which appeared in 1950 under the same title.

A bibliography on computational methods of linear algebra and on questions of estimating and evaluating the eigenvalues of a matrix has been added at the end of the book. I. A. Lifshitz and R. S. Aleksandrova gave the authors essential aid in compiling the book. The authors extend to them their gratitude.

The manuscript of the book was read by V. N. Kublanovskaya, who made a series of valuable comments. The authors extend to her their deep gratitude. The authors thank also the editor of the book, G. P. Akilov, and all his associates who displayed such interest in their work.

# CONTENTS

*Chapter III*

## Iterative Methods for Solving Systems of Linear Equations

*Chapter IV*

## The Complete Eigenvalue Problem

*Chapter VIII*

## Iterative Methods for Solving the Complete Eigenvalue Problem

*Chapter I X*

## Universal Algorithms

# BASIC MATERIAL FROM LINEAR ALGEBRA

## 1. MATRICES

**1. Definitions.** An aggregate of numbers, generally complex, which is arranged in the form of a rectangular table containing $n$ rows and $m$ columns is called a *rectangular matrix*.

Such a matrix is denoted by:

$$\begin{bmatrix} a_{11} & a_{12} & \cdots & a_{1m} \\ a_{21} & a_{22} & \cdots & a_{2m} \\ \cdot & \cdot & \cdot & \cdot \\ a_{n1} & a_{n2} & \cdots & a_{nm} \end{bmatrix} \tag{1}$$

or in the abbreviated form:

$$A = (a_{ij}); \quad i = 1, 2, \cdots, n; \quad j = 1, 2, \cdots, m .$$

Two matrices are called *equal* if their corresponding elements are equal.

A matrix consisting of one row is called simply a *row*; a matrix consisting of one column is called a *column*; a matrix $A = (a)$ consisting of one number is identified with the number. If the number $n$ of the rows of a matrix is equal to the number of its columns, then the matrix is called *square*. In this case the number $n$ is called the *order* of the matrix.

Among square matrices an important role is played by the so-called *diagonal matrices*, that is, matrices which have elements different from zero only along the main diagonal. Diagonal matrices are denoted by $[\alpha_1, \alpha_2, \cdots, \alpha_n]$, so that

$$[\alpha_1, \alpha_2, \cdots, \alpha_n] = \begin{bmatrix} \alpha_1 & 0 & \cdots & 0 \\ 0 & \alpha_2 & \cdots & 0 \\ \cdot & \cdot & \cdot & \cdot \\ 0 & 0 & \cdots & \alpha_n \end{bmatrix} . \tag{2}$$

Moreover, if all the numbers $\alpha_i$ are equal to each other, the matrix is called *scalar*:

$$[\alpha] = \begin{bmatrix} \alpha & 0 & \cdots & 0 \\ 0 & \alpha & \cdots & 0 \\ & & \cdots & \\ 0 & 0 & \cdots & \alpha \end{bmatrix} \tag{3}$$

and in the case where $\alpha = 1$, a *unit matrix*:

$$E = \begin{bmatrix} 1 & 0 & \cdots & 0 \\ 0 & 1 & \cdots & 0 \\ & & \cdots & \\ 0 & 0 & \cdots & 1 \end{bmatrix} = (\delta_{ij}) , \tag{4}$$

where $\delta_{ij}$ is the so-called Kronecker symbol, i.e. $\delta_{ij} = 0$ for $i \neq j$, $\delta_{ii} = 1$.

Finally, a matrix, all of whose elements are equal to zero, is called a *null matrix*. We shall denote it by the symbol 0.

If in the matrix

$$A = \begin{bmatrix} a_{11} & a_{12} & \cdots & a_{1m} \\ a_{21} & a_{22} & \cdots & a_{2m} \\ & & \cdots & \\ a_{n1} & a_{n2} & \cdots & a_{nm} \end{bmatrix}$$

we interchange rows and columns, we obtain the so-called *transposed matrix*

$$A' = \begin{bmatrix} a_{11} & a_{21} & \cdots & a_{n1} \\ a_{12} & a_{22} & \cdots & a_{n2} \\ & & \cdots & \\ a_{1m} & a_{2m} & \cdots & a_{nm} \end{bmatrix} . \tag{5}$$

A square matrix $A$ is equal to its transpose $A'$ if and only if it is *symmetric*, that is, if $a_{ij} = a_{ji}$.

It is obvious that a matrix transposed from a row matrix will be a column matrix containing the same elements. We shall often use this fact to obtain a convenient notation for columns. Thus, instead of the column

$$\begin{bmatrix} 4 \\ 2 \\ 3 \\ 5 \end{bmatrix}$$

we shall write $(4, 2, 3, 5)'$.

By replacing the elements of a matrix with conjugate complex numbers we obtain the so-called *complex-conjugate matrix* $\bar{A}$. If the elements of the matrix $A$ are real, then $\bar{A} = A$.

The matrix $A^* = \bar{A}'$, the complex-conjugate of the transposed matrix, is called the matrix *conjugate to* the matrix $A$, or the conjugate of $A$. It is obvious that

$$(A^*)^* = A$$

If the matrix A is real, then the matrix conjugate to it coincides with the transposed matrix.

The determinant whose elements are the elements of a square matrix is exactly the same arrangement is called the *determinant of the square matrix.* The determinant of the matrix $A$ will be denoted by $|A|$.

The determinant obtained by eliminating a certain number of rows, and the same number of columns, from a square matrix is called *a minor* of the matrix. That is, a minor of order $k$ of a matrix $A$ is a determinant of that order formed from elements located at the intersections of $k$ rows and $k$ columns of the matrix, the order of appearance being unchanged.

The maximal order of any non-zero minor of a matrix $A$ is called the *rank* of the matrix. In other words, the rank of a matrix will be a number $r$ such that among the minors of the matrix there exists a minor of order $r$, not equal to zero, and all minors of order $r + 1$, or higher, are either equal to zero or cannot be formed.

**2. Multiplication of a matrix by a number; addition of matrices.** The *product of a matrix* $A = (a_{ji})$ *by a number* $\alpha$ is a matrix whose elements are obtained from the elements of the matrix by multiplying each of them by the number $\alpha$:

$$\alpha A = \begin{bmatrix} \alpha a_{11} & \alpha a_{12} & \cdots & \alpha a_{1m} \\ \cdot & \cdot & \cdot & \cdot \\ \alpha a_{n1} & \alpha a_{n2} & \cdots & \alpha a_{nm} \end{bmatrix}, \tag{6}$$

The *sum* of two rectangular matrices $A = (a_{ij})$ and $B = (b_{ij})$ having the same number of rows as well as columns is a matrix $C$ whose elements are equal to the sums of corresponding elements of the matrices $A$ and $B$, that is:

$$A + B = \begin{bmatrix} a_{11} + b_{11} & a_{12} + b_{12} & \cdots & a_{1m} + b_{1m} \\ a_{21} + b_{21} & a_{22} + b_{22} & \cdots & a_{2m} + b_{2m} \\ \cdot & \cdot & \cdot & \cdot \\ a_{n1} + b_{n1} & a_{n2} + b_{n2} & \cdots & a_{nm} + b_{nm} \end{bmatrix}. \tag{7}$$

It is easy to see that the operations introduced above induce the matrix properties:

1. $A + (B + C) = (A + B) + C$ ;
2. $A + B \qquad = B + A$ ;
3. $A + O \qquad = A$ ;
4. $(\alpha + \beta)A \qquad = \alpha A + \beta A$ ;
5. $\alpha(A + B) \qquad = \alpha A + \alpha B$ ;
6. $1 \cdot A \qquad = A$ ;
7. $\alpha(\beta A) \qquad = \alpha \beta A$ .

Here $A, B$, and $C$ are matrices, $\alpha$ and $\beta$ are numbers.

**3. Multiplication of matrices.** Multiplication of two matrices $A$ and $B$ is defined only under the assumption that the number of columns of matrix $A$ is equal to the number of rows of matrix $B$. Under this assumption the elements of the product $C$ are defined in the following way: the element of the $i$th row and $j$th column of the matrix $C$ is equal to the sum of the products of elements of the $i$th row of matrix $A$ with corresponding elements of the $j$th column of matrix $B$.  Thus,

$$AB = \begin{bmatrix} a_{11} & a_{12} & \cdots & a_{1m} \\ a_{21} & a_{22} & \cdots & a_{2m} \\ \vdots & & & \vdots \\ a_{n1} & a_{n1} & \cdots & a_{nm} \end{bmatrix} \begin{bmatrix} b_{11} & b_{12} & \cdots & b_{1p} \\ b_{21} & b_{22} & \cdots & q_{2p} \\ \vdots & & & \vdots \\ b_{m1} & b_{m2} & \cdots & b_{mp} \end{bmatrix}$$

$$= \begin{bmatrix} c_{11} & c_{12} & \cdots & c_{1p} \\ c_{21} & c_{22} & \cdots & c_{2p} \\ \vdots & & & \vdots \\ c_{n1} & c_{n2} & \cdots & c_{np} \end{bmatrix}. \tag{8}$$

where

$$c_{ij} = a_{i1}b_{1j} + a_{i2}b_{2j} + \cdots + a_{im}b_{mj} \qquad (i = 1, \cdots, j = 1, \cdots, p). \tag{9}$$

Note that the product of two rectangular matrices is again a rectangular matrix, the number of rows of which is equal to the number of rows of the first matrix, and the number of columns equal to the number of columns of the second matrix.  Thus, for example, the product of a square matrix with a matrix which consists of one column is again a matrix of one column.

In general, the commutative law does not hold for multiplication of matrices.  It is easy to see that the question of equality of the matrix products $AB$ and $BA$ is meaningful only if $A$ and $B$ are square matrices of the same order.  Both matrices $AB$ and $BA$ can exist simultaneously only if the number of rows of one matrix is equal to the number of columns of the other and the number of columns of the first matrix is equal to the number of rows of the second.  But under these conditions, both matrices $AB$ and $BA$ will both be square, although of different orders if $A$ and $B$ are not square.  But even for square matrices of the same order it is generally true that $AB \neq BA$.

For example,

$$\begin{bmatrix} 1 & 2 \\ 3 & 4 \end{bmatrix} \begin{bmatrix} 1 & 1 \\ -3 & 1 \end{bmatrix} = \begin{bmatrix} -5 & 3 \\ -9 & 7 \end{bmatrix},$$

$$\begin{bmatrix} 1 & 1 \\ -3 & 1 \end{bmatrix} \begin{bmatrix} 1 & 2 \\ 3 & 4 \end{bmatrix} = \begin{bmatrix} 4 & 6 \\ 0 & -2 \end{bmatrix}.$$

In particular cases multiplication may be commutative.  Thus, for example, scalar matrices are commutative with all square matrices of the same order, since

$$\begin{bmatrix} \alpha & 0 & \cdots & 0 \\ 0 & \alpha & \cdots & 0 \\ \vdots & & & \vdots \\ 0 & 0 & \cdots & \alpha \end{bmatrix} \begin{bmatrix} a_{11} & a_{12} & \cdots & a_{1n} \\ a_{21} & a_{22} & \cdots & a_{2n} \\ \vdots & & & \vdots \\ a_{n1} & a_{n2} & \cdots & a_{nn} \end{bmatrix}$$

$$= \begin{bmatrix} a_{11} & a_{12} & \cdots & a_{1n} \\ a_{21} & a_{22} & \cdots & a_{2n} \\ \vdots & & & \vdots \\ a_{n1} & a_{n2} & \cdots & a_{nn} \end{bmatrix} \begin{bmatrix} \alpha & 0 & \cdots & 0 \\ 0 & \alpha & \cdots & 0 \\ \vdots & & & \vdots \\ 0 & 0 & \cdots & \alpha \end{bmatrix} = \begin{bmatrix} \alpha a_{11} & \alpha a_{12} & \cdots & \alpha a_{1n} \\ \alpha a_{21} & \alpha a_{22} & \cdots & \alpha a_{2n} \\ \vdots & & & \vdots \\ \alpha a_{n1} & \alpha a_{n2} & \cdots & \alpha a_{nn} \end{bmatrix}$$

The special role of the unit matrix in the multiplication of matrices follows from the above formula, namely: among all square matrices of a given order the unit matrix plays the same role as the number one among numbers. In fact,

$$AE = EA = A$$

It may be shown that multiplication of matrices is associative. That is, if the matrix products $AB$ and $(AB)C$ exist, then $BC$ and $A(BC)$ exist, and

$$A(BC) = (AB)C .$$

The element of the $i$th row and $j$th column of $(AB)C$ is equal to

$$\sum_\beta [\sum_\alpha a_{i\alpha}b_{\alpha\beta}]c_{\beta j} = \sum_\alpha \sum_\beta a_{i\alpha}b_{\alpha\beta}c_{\beta j} ,$$

and the element of the $i$th row and $j$th column of $A(BC)$ is equal to

$$\sum_\alpha a_{i\alpha}[\sum_\beta b_{\alpha\beta}c_{\beta j}] = \sum_\alpha \sum_\beta a_{i\alpha}b_{\alpha\beta}c_{\beta j} .$$

Thus corresponding elements of the matrices $(AB)C$ and $A(BC)$ are equal; consequently the matrices themselves are also equal.

The product of matrices also possesses the properties:

$$\alpha(AB) = (\alpha A)B = A(\alpha B) ,$$

$$(A + B)C = AC + BC ,$$

$$C(A + B) = CA + CB ,$$

where $A, B,$ and $C$ are matrices and $\alpha$ is a number.

The following rule for transposing a product holds:

$$(AB)' = B'A'. \tag{10}$$

The element of the $i$th row and the $j$th column of the matrix $(AB)'$ is equal to the element of the $j$th row and the $i$th column of the matrix $AB$, that is, equal to

$$a_{j_1}b_{1i} + a_{j_2}b_{2i} + \cdots + a_{jm}b_{mi} .$$

Obviously the last expression is equal to the sum of the products of elements of the $i$th row of the matrix $B'$ with corresponding elements of the $j$th column of the matrix $A'$, that is, equal to the element of the $i$th row and the $j$th column of the matrix $B'A'$.

It is likewise clear that

$$\overline{AB} = \overline{A}\,\overline{B}$$

and

$$(AB)^* = B^*A^*. \tag{11}$$

As was said above, the matrix $AB$ will be square if the number $n$ of the rows of matrix $A$ is equal to the number of columns of matrix $B$. We denote by $m$ the number of columns of matrix $A$ (it is also equal to the number of rows of matrix $B$, since only under this condition is the product $AB$ meaningful). From the theory of determinants it is known that, for $n > m$, the determinant of matrix $AB$ is equal to zero; for $n \leq m$ it is equal to the sum of the products of all minors of order $n$ contained in matrix $A$ with the corresponding minors contained in matrix $B$. More precisely, if

$$A = \begin{bmatrix} a_{11} & \cdots & a_{1m} \\ \cdot & \cdot \cdot \cdot \cdot & \cdot \\ a_{n1} & \cdots & a_{nm} \end{bmatrix},$$

$$B = \begin{bmatrix} b_{11} & \cdots & b_{1n} \\ \cdot & \cdot \cdot \cdot \cdot & \cdot \\ b_{m1} & \cdots & b_{mn} \end{bmatrix}$$

and $n \leq m$, then

$$|AB| = \sum_{i_1 < i_2 < \cdots < i_n} \begin{vmatrix} a_{1i_1} & \cdots & a_{1i_n} \\ \cdot & \cdot \cdot \cdot \cdot & \cdot \\ a_{ni_1} & \cdots & a_{ni_n} \end{vmatrix} \cdot \begin{vmatrix} b_{i_11} & \cdots & b_{i_1n} \\ \cdot & \cdot \cdot \cdot \cdot & \cdot \\ b_{i_n1} & \cdots & b_{i_nn} \end{vmatrix}.$$

In particular, for $m = n$

$$|AB| = \begin{vmatrix} a_{11} & \cdots & a_{1n} \\ \cdot & \cdot \cdot \cdot \cdot & \cdot \\ a_{n1} & \cdots & a_{nn} \end{vmatrix} \cdot \begin{vmatrix} b_{11} & \cdots & b_{1n} \\ \cdot & \cdot \cdot \cdot \cdot & \cdot \\ b_{n1} & \cdots & b_{nn} \end{vmatrix},$$

that is, *the determinant of the product of two square matrices is equal to the product of the determinants of the matrices which are being multiplied.*

**4. Partitioning of matrices into cells.** Often it is useful to reduce the operations on matrices of higher orders to operations on matrices of lower orders. Such a reduction may be realized by partitioning given matrices into so-called cells. That is, we consider every matrix as made up of several matrices of a lower order. This may usually be done in one of several ways. For example,

$$\begin{bmatrix} a_{11} & a_{12} & a_{13} & a_{14} \\ a_{21} & a_{22} & a_{23} & a_{24} \\ a_{31} & a_{32} & a_{33} & a_{34} \end{bmatrix} = \left[ \begin{array}{c|ccc} a_{11} & a_{12} & a_{12} & a_{14} \\ \hline a_{21} & a_{22} & a_{23} & a_{24} \\ a_{31} & a_{32} & a_{33} & a_{34} \end{array} \right] = \left[ \begin{array}{cc|cc} a_{11} & a_{12} & a_{13} & a_{14} \\ a_{21} & a_{22} & a_{23} & a_{24} \\ a_{31} & a_{32} & a_{33} & a_{34} \end{array} \right].$$

The matrices into which a given matrix is partitioned are called **cells.** When partitioning into cells it is assumed that the horizontal and vertical dividing lines cut across the whole matrix.

We shall not consider the general case of partitioning matrices into cells but only such partitions of square matrices for which the diagonal cells are square.

The basic operations on matrices with diagonal cells of the same order are naturally connected with the operations on the cells themselves.

That is, if

$$A = \begin{bmatrix} A_{11} & A_{12} & \cdots & A_{1k} \\ A_{21} & A_{22} & \cdots & A_{2k} \\ \cdot & \cdot \cdot \cdot \cdot & \cdot \\ A_{k1} & A_{k2} & \cdots & A_{kk} \end{bmatrix}$$

and

$$B = \begin{bmatrix} B_{11} & B_{12} & \cdots & B_{1k} \\ B_{21} & B_{22} & \cdots & B_{2k} \\ \cdot & \cdot & \cdot & \cdot & \cdot & \cdot & \cdot & \cdot \\ B_{k1} & B_{k2} & \cdots & B_{kk} \end{bmatrix},$$

where $A_{ii}$ and $B_{ii}$ are square matrices of the same order, then

$$A + B = \begin{bmatrix} A_{11} + B_{11} & A_{12} + B_{12} & \cdots & A_{1k} + B_{1k} \\ A_{21} + B_{21} & A_{22} + B_{22} & \cdots & A_{2k} + B_{2k} \\ \cdot & \cdot & \cdot & \cdot & \cdot & \cdot & \cdot & \cdot & \cdot \\ A_{k1} + B_{k1} & A_{k2} + B_{k2} & \cdots & A_{kk} + B_{kk} \end{bmatrix} \qquad (12)$$

and

$$AB = \begin{bmatrix} C_{11} & C_{12} & \cdots & C_{1k} \\ C_{21} & C_{22} & \cdots & C_{2k} \\ \cdot & \cdot & \cdot & \cdot & \cdot & \cdot \\ C_{k1} & C_{k2} & \cdots & C_{kk} \end{bmatrix}, \qquad (13)$$

where

$$C_{ij} = A_{i1}B_{1j} + A_{i2}B_{2i} + \cdots + A_{ik}B_{kj} \qquad i, j = 1, \cdots, k .$$

All the products $A_{il}B_{lj}$ are meaningful, since the number of columns of matrix $A_{il}$ is always equal to the number of rows of matrix $B_{lj}$. The sum $A_{i1}B_{1j} + \cdots + A_{ik}B_{kj}$ is meaningful, since all elements of the matrix have the same structure. Furthermore, let $c_{\alpha\beta}$ be some element of the cell $C_{ij}$.

$$c_{\alpha\beta} = (a_{\alpha 1}b_{1\beta} + \cdots + a_{\alpha s_1}b_{s_1\beta})$$
$$+ \cdot \quad \cdot \quad \cdot \quad \cdot \quad \cdot \quad \cdot \quad \cdot \quad \cdot \quad \cdot$$
$$+ (a_{\alpha s_{k-1}+1}b_{s_{k-1}+1\beta} + \cdots + a_{\alpha s_k}b_{s_k\beta}) .$$

Here $s_1, s_2 - s_1, \cdots, s_k - s_{k-1}$ denote the orders of the matrices $A_{11}, A_{22}, \cdots,$ $A_{kk}$. It is clear that the elements in parentheses, whose sum composes $c_{\alpha\beta}$, are elements of the matrices $A_{i1}B_{1k}, \cdots, A_{ik}B_{kj}$, occupying in these matrices the same position that $c_{\alpha\beta}$ occupies in the matrix $C_{ij}$. Consequently,

$$C_{ij} = A_{i1}B_{1j} + \cdots + A_{ik}B_{kj} .$$

Formulas (12) and (13) show that operations with matrices which have been partitioned into cells of the type indicated are carried out just as if there were numbers in place of the cells.

*Bordered matrices* are an important special case of partitioned matrices. Consider a square matrix $A_{n-1}$ of order $n - 1$:

$$A_{n-1} = \begin{bmatrix} a_{11} & a_{12} & \cdots & a_{1,n-1} \\ a_{21} & a_{22} & \cdots & a_{2,n-1} \\ \cdot & \cdot & \cdot & \cdot & \cdot & \cdot & \cdot \\ a_{n-1,1} & a_{n-1,2} & \cdots & a_{n-1,n-1} \end{bmatrix}.$$

We shall form a matrix $A_n$ of the $n$th order by appending to the matrix $A_{n-1}$ a row $v_{n-1} = (a_{n1}\cdots a_{n,n-1})$, a column $u_{n-1} = (a_{1n}a_{2n}\cdots a_{n-1,n})$, and a number $a_{nn}$:

$$A_n = \left\{ \begin{array}{cc} A_{n-1} & \begin{array}{c} a_{1n} \\ a_{2n} \\ \vdots \\ a_{n-1,n} \end{array} \\ a_{n1} \cdots a_{n,n-1} \, a_{nn} \end{array} \right\} = \left[ \begin{array}{cc} A_{n-1} & u_{n-1} \\ v_{n-1} & a_{nn} \end{array} \right]. \tag{14}$$

We shall say that the matrix $A_n$ has been obtained by bordering the matrix $A_{n-1}$. The matrix $A_n$ is naturally divisible into cells.

Operations on bordered matrices are carried out according to the general rules for operations on partitioned matrices.

Let

$$A = \left[ \begin{array}{cc} M & u \\ v & a \end{array} \right],$$

$$B = \left[ \begin{array}{cc} P & y \\ x & b \end{array} \right]$$

be two bordered matrices of order $n$. The meaning of the notation $M, v,$ $u, a$ and $P, x, y, b$ is the same as in the definition.

Then the following equalities are valid:

$$\alpha A = \left[ \begin{array}{cc} \alpha M & \alpha u \\ \alpha v & \alpha a \end{array} \right],$$

$$A + B = \left[ \begin{array}{cc} M + P & u + y \\ v + x & \alpha + b \end{array} \right], \tag{15}$$

$$AB = \left[ \begin{array}{cc} MP + ux & My + ub \\ vP + \alpha x & vy + ab \end{array} \right].$$

Here $MP$ and $ux$ are matrices of the $(n-1)$th order; $My$ and $ub$ are columns consisting of the $(n-1)$th element; $vP$ and $ax$ are corresponding rows; $vy + ab$ is a number.

**5. Quasi-diagonal matrices.** We shall consider once more a special case of partitioned matrices, namely, the so-called *quasi-diagonal matrices*. By a quasi-diagonal matrix is meant a square matrix having square cells along its main diagonal with the remaining elements equal to zero. For example the matrix of 7th order

$$\left[ \begin{array}{cc|ccc|cc} \alpha_{11} & a_{12} & 0 & 0 & 0 & 0 & 0 \\ a_{21} & a_{22} & 0 & 0 & 0 & 0 & 0 \\ \hline 0 & 0 & b_{11} & b_{12} & b_{13} & 0 & 0 \\ 0 & 0 & b_{21} & b_{22} & b_{23} & 0 & 0 \\ 0 & 0 & b_{31} & b_{32} & b_{33} & 0 & 0 \\ \hline 0 & 0 & 0 & 0 & 0 & c_{11} & c_{12} \\ 0 & 0 & 0 & 0 & 0 & c_{21} & c_{22} \end{array} \right]$$

is quasi-diagonal. Obviously the cells of this matrix are the matrices

$$A = \begin{bmatrix} a_{11} & a_{12} \\ a_{21} & a_{22} \end{bmatrix};$$

$$B = \begin{bmatrix} b_{11} & b_{12} & b_{13} \\ b_{21} & b_{22} & b_{23} \\ b_{31} & b_{32} & b_{33} \end{bmatrix}$$

$$C = \begin{bmatrix} c_{11} & c_{12} \\ c_{21} & c_{22} \end{bmatrix}$$

and six null matrices.

If the structure of two quasi-diagonal matrices is the same, then the product of these matrices will again be a quasi-diagonal matrix of the same structure, the diagonal cells of which are equal to the products of the corresponding cells of the matrices being multiplied.

By the well-known theorem of Laplace the determinant of a quasi-diagonal matrix is equal to the product of the determinants of the diagonal cells.

**6. Inverse and adjoint matrices.** A square matrix $A = (a_{ij})$ is called *non-singular* or non-degenerate if its determinant is not equal to zero; in the contrary case the matrix is called *singular*.

We shall now introduce the important concept of the *inverse matrix*. We shall call the matrix $B$ *inverse* to the square matrix $A$ if

$$AB = E \tag{16}$$

where $E$ is the unit matrix of that order.

We shall prove that a necessary and sufficient condition for the existence of an inverse matrix is the non-singularity of the matrix $A$.

The necessity follows immediately from the theorem on the determinant of the product of two matrices. If $AB = E$, then $|A| \, |B| = 1$ and consequently $|A| \neq 0$.

We shall now assume $|A| \neq 0$. To construct an inverse matrix we shall first consider the so-called *adjoint matrix*, that is, the matrix

$$C = \begin{bmatrix} A_{11} & A_{21} & \cdots & A_{n1} \\ A_{12} & A_{22} & \cdots & A_{n2} \\ \cdot & \cdot & \cdot \cdot \cdot & \cdot \\ A_{1n} & A_{2n} & \cdots & A_{nn} \end{bmatrix}. \tag{17}$$

Here $A_{ij}$ is the algebraic cofactor of the element $a_{ij}$ in the determinant of the matrix $A$.[†]

We shall prove that the adjoint matrix possesses the following property:

$$AC = |A| E \tag{18}$$

In fact, computing any element of the matrix $AC$ by the rules for multiplying matrices we find that it is equal to

$$a_{i1} A_{j1} + a_{i2} A_{j2} + \cdots + a_{in} A_{jn},$$

---

[†] The algebraic cofactor is $(-1)^{i+j}$ times the minor obtained by striking out the row and column containing $a_{ij}$.

that is, equal to zero for $i \neq j$ and to the determinant of the matrix $A$ for $i = j$, on the basis of a well-known theorem on the expansion of a determinant.

In the same way one may verify the equality

$$CA = |A|E \qquad (18')$$

The adjoint matrix is meaningful for any square matrix $A$. From the equality $AC = |A|E$ it follows that the matrix

$$B = \frac{1}{|A|}C \qquad (19)$$

is inverse to the non-singular matrix $A$.

In fact,

$$AB = A\frac{1}{|A|}C = \frac{1}{|A|}AC = E \,.$$

The matrix constructed also possesses the property

$$BA = E \,, \qquad (20)$$

which follows from the equality $CA = |A|E$.

We shall prove finally the uniqueness of the inverse matrix. We assume that there exists a matrix $X$ such that $AX = E$. Multiplying this equality by $B$ from the left we get $X = B$. If it is assumed that $YA = E$, then on multiplying by $B$ from the right we get $Y = B$.

The matrix which is inverse to the matrix $A$ is denoted by $A^{-1}$. It is obvious that $|A^{-1}| = |A|^{-1}$. Furthermore, $(A^{-1})^{-1} = A$. This follows directly from the equality $A^{-1}A = E$.

It is readily shown that the matrix which is the inverse of the product of two matrices is equal to the product of the inverted matrices taken in reverse order, that is,

$$(A_1A_2)^{-1} = A_2^{-1}A_1^{-1}. \qquad (21)$$

In fact,

$$A_1A_2A_2^{-1}A_1^{-1} = A_1A_1^{-1} = E \,.$$

Equality (19) makes it possible to compute the inverse matrix. However the computation of the adjoint matrix is so laborious that the equality cited is important only theoretically.

Finding the numerical solution of an inverse matrix is one of the most important computational problems of linear algebra and we shall return to it often in the following chapters.

**7. Polynomial functions of matrices.** We shall now define the positive integral power of a square matrix, assuming

$$A^n = \underbrace{A \cdots A}_{n \text{ times}} \,. \qquad (22)$$

Because of the associative law it is not necessary to use parentheses in this product. From the definition it is clear that

$$\left.\begin{array}{l} A^n A^m = A^{n+m} \\ (A^n)^m = A^{nm} \end{array}\right\} \tag{23}$$

It is clear that *powers of the same matrix are commutative.*
An expression of the form

$$\alpha_0 A^n + \alpha_1 A^{n-1} + \cdots + \alpha_n E ,$$

where $\alpha_0, \alpha_1, \cdots, \alpha_n$ are complex numbers, is called a *polynomial in a matrix* or a *matrix polynomial.* A matrix polynomial may be made up by substitution of a matrix $A$ for the variable $t$ in the algebraic polynomial

$$f(t) = \alpha_0 t^n + \alpha_1 t^{n-1} + \cdots + \alpha_n \tag{24}$$

It is important to note that the rules of operations on matrix polynomials do not differ from the rules of operations on algebraic polynomials.
That is, if

$$\left.\begin{array}{l} \rho(t) = \phi(t) \pm \chi(t) \\ \omega(t) = \phi(t)\chi(t) \\ \\ \rho(A) = \phi(A) \pm \chi(A) \\ \omega(A) = \phi(A)\chi(A) . \end{array}\right\} \tag{25}$$

then

## 8. The characteristic polynomial. The Cayley-Hamilton theorem. The minimum polynomial. The equation

$$\begin{vmatrix} a_{11} - t & a_{12} & \cdots & a_{1n} \\ a_{21} & a_{22} - t & \cdots & a_{2n} \\ \cdots & \cdots & \cdots & \cdots \\ a_{n1} & a_{n2} & \cdots & a_{nn} - t \end{vmatrix} = 0 . \tag{26}$$

is called the *secular* or *characteristic* equation of the matrix $A - (a_{ij})$. The left member of this equation, which may be written in the abbreviated form $|A - tE|$, is called the characteristic polynomial of the matrix. Secular equations are of considerable importance in applied mathematics.

Direct computation of the characteristic polynomial often presents considerable technical difficulties. In fact, if

$$\varphi(t) = |A - tE| = (-1)^n [t^n - p_1 t^{n-1} - p_2 t^{n-2} - \cdots - p_n] , \tag{27}$$

then

$$\begin{array}{l} p_1 = a_{11} + a_{22} + \cdots + a_{nn} \\ p_n = (-1)^{n-1} |A| , \end{array} \tag{28}$$

and [the remaining coefficient $p_k$ are sums of all the principal minors of order $k$ in matrix $A$, (i.e. for elements resting on the principal diagonal), these taken with the sign $(-1)^{k-1}$. The number of such minors is equal to the number of combinations of $n$ elements taken $k$ at a time.

The roots $\lambda_i$ of the characteristic polynomial are called the *eigenvalues* or *characteristic numbers* of the matrix. From the well-known theorem of Vieta, which connects the roots of an equation with its coefficients, it follows that

$$\left. \begin{aligned} \lambda_1 + \lambda_2 + \cdots + \lambda_n &= p_1 = a_{11} + a_{22} + \cdots + a_{nn} \\ \lambda_1 \lambda_2 \cdots \lambda_n &= (-1)^{n-1} p_n = |A| \, . \end{aligned} \right\} \qquad (29)$$

The quantity $p_1 = a_{11} + \cdots + a_{nn}$ is called the *spur* (or *trace*) of the matrix $A$ and is denoted by $Sp\ A$, (or by some writers, $Tr\ A$).

Numerical determination of the roots of the characteristic polynomial is one of the most important problems of linear algebra. Convenient methods for determining the coefficients and roots of the characteristic polynomial will be investigated later.

The leading coefficient of the characteristic polynomial is equal to $(-1)^n$. Sometimes instead of the characteristic polynomial one considers the normalized characteristic polynomial which differs from the ordinary one by the factor $(-1)^n$. The leading coefficient of a normalized characteristic polynomial is equal to one.

For any matrix the following remarkable relationship, known as the Cayley-Hamilton theorem, holds: *if $\varphi(t)$ is the characteristic polynomial of the matrix $A$, then $\varphi(A) = 0$, that is, a matrix is a zero of its own characteristic polynomial.*

For a proof we shall consider the matrix $B$, adjoint of the matrix $A - tE$. Since every algebraic cofactor in the determinant $|A - tE|$ is a polynomial in $t$ of degree not exceeding $n - 1$, the adjoint matrix may be represented as

$$B = B_{n-1} + B_{n-2}t + \cdots + B_0 t^{n-1},$$

where $B_{n-1}, \cdots, B_0$ are certain matrices which do not involve $t$. From a basic property of adjoint matrices we have:

$$(B_{n-1} + B_{n-2}t + \cdots + B_0 t^{n-1})(A - tE) = |A - tE| E$$
$$= (-1)^n (t^n - p_n t^{n-1} - \cdots - p_n) E \, .$$

This equation is equivalent to the system of equations

$$\begin{aligned} B_{n-1}A \quad\quad\quad &= (-1)^{n+1} p_n E \, , \\ B_{n-2}A - B_{n-1} &= (-1)^{n+1} p_{n-1} E \, , \\ \cdot\ \cdot\ \cdot\ \cdot\ \cdot\ \cdot\ \cdot\ &\cdot\ \cdot\ \cdot\ \cdot\ \cdot\ \cdot\ \cdot\ \cdot\ , \\ B_0 A - B_1 \quad\ &= (-1)^{n+1} p_1 E \, , \\ - B_0 \quad\quad\quad\ &= (-1)^n E \, . \end{aligned}$$

If we post-multiply[†] these equalities by $E, A, A^2, \cdots, A^{n-1}, A^n$, and add, we obtain on the left the null matrix and on the right

---

† Post-multiplication of a matrix $B$ by a matrix $A$ means the product $BA$; premultiplication of $B$ by $A$ means $AB$.

$$(-1)^n[-p_nE - p_{n-1}A - p_{n-1}A^2 - \cdots + A^n] = \varphi(A) . \qquad (30)$$

Thus it has been proved that $\varphi(A) = 0$.

The Cayley-Hamilton relationship shows that given any matrix there exists a polynomial which has the matrix as a zero. It is obvious that such a polynomial is not unique, since if $\psi(t)$ possesses this property, then every polynomial divisible by $\psi(t)$ also possesses it. The polynomial of least degree which possesses the property that the matrix $A$ is a zero of the polynomial is called the *minimum polynomial* of the matrix.

We shall prove that the characteristic polynomial is divisible by the minimum polynomial.

Let $q(t)$ and $r(t)$ be the quotient and the remainder obtained when the characteristic polynomial $\varphi(t)$ is divided by the minimum polynomial $\psi(t)$. Then

$$\varphi(t) = \psi(t)q(t) + r(t) ,$$

where the degree of $r(t)$ is less than the degree of $\psi(t)$.

Substituting $A$ for $t$ in this equation, we get $r(A) = \varphi(A) - \psi(A)q(A) = 0$. Thus the matrix $A$ is a zero of the polynomial $r(t)$. From this it follows $r(t) \equiv 0$, since otherwise $\psi(t)$ could not be the minimum polynomial. Consequently $\varphi(t)$ is in fact divisible by $\psi(t)$.

In exactly the same way it can be proved that any polynomial $\omega(t)$ having the matrix $A$ as a zero, that is, which satisfies the condition $\omega(A) = 0$, is divisible by the minimum polynomial.

**9. Similar matrices.** The matrix B is said to be *similar* to the matrix $A$ if there exists a non-singular matrix $C$, such that $B = C^{-1}AC$. In this case one says that matrix $B$ is obtained from matrix $A$ by means of a *similarity transformation*.

A similarity transformation possesses the following properties:

$$C^{-1}A_1C + C^{-1}A_2C + \cdots + C^{-1}A_nC = C^{-1}(A_1 + A_2 + \cdots + A_n)C$$
$$C^{-1}A_1C \cdot C^{-1}A_2C \cdots C^{-1}A_nC = C^{-1}(A_1A_2 \cdots A_n)C . \qquad (31)$$

In particular, $(C^{-1}AC)^n = C^{-1}A^nC$. Furthermore,

$$f(C^{-1}AC) = C^{-1}f(A)C$$

for any polynomial $f(t)$.

From the last property it follows immediately that *the minimum polynomials of similar matrices are identical.*

We shall show that *similar matrices also have the same characteristic polynomials.*

In fact,

$$|B - tE| = |C^{-1}AC - tE| = |C^{-1}AC - tC^{-1}EC|$$
$$= |C|^{-1}|A - tE||C| = |A - tE| .$$

**10.  Eigenvalues of a polynomial in matrices.**  Let $A$ be a matrix with eigenvalues $\lambda_1, \cdots, \lambda_n$, (these are not necessarily distinct) and let $f(t) = a_0 + a_1 t + \cdots + a_m t^m$ be a given polynomial.  We shall show that the eigenvalues of the matrix $f(A)$ are the numbers $f(\lambda_1), \cdots, f(\lambda_n)$.

First we compute the determinant of the matrix $f(A)$.  With this in mind we factor the polynomial $f(t)$ into linear factors

$$f(t) = a_m(t - \mu_1) \cdots (t - \mu_m),$$

where $\mu_1, \cdots, \mu_m$ are zeros of the polynomial $f(t)$.

Then

$$f(A) = a_m(A - \mu_1 E) \cdots (A - \mu_m E)$$

and consequently,

$$|f(A)| = a_m^n |A - \mu_1 E| \cdots |A - \mu_m E| = a_m^n \varphi(\mu_1) \cdots \varphi(\mu_m),$$

where $\varphi(t)$ is the characteristic polynomial of the matrix $A$.  But

$$\varphi(t) = (\lambda_1 - t) \cdots (\lambda_n - t).$$

Therefore

$$|f(A)| = a_m^n \prod_{i=1}^{n} \prod_{j=1}^{m} (\lambda_i - \mu_j) = \prod_{i=1}^{n} \left( a_m \prod_{j=1}^{m} (\lambda_i - \mu_j) \right) = \prod_{i=1}^{n} f(\lambda_i).$$

The equation

$$|f(A)| = f(\lambda_1) \cdots f(\lambda_n)$$

is an identity involving the coefficients of the polynomial $f(t)$.  Applying this identity to the polynomial $f(t) - u$ we obtain

$$|f(A) - uE| = (f(\lambda_1) - u) \cdots (f(\lambda_n) - u).$$

This means that the eigenvalues of the matrix $f(A)$ are the numbers $f(\lambda_1)$, $\cdots, f(\lambda_n)$.

In particular we observe that the eigenvalues of the matrix $A^k$ are $\lambda_1^k, \cdots, \lambda_n^k$.

**11.  Elementary transformations.**  It is often convenient to perform the following operations on matrices:

a)  multiplying the elements of some row by a number;

b′)  adding to the elements of some row numbers which are proportional to the corresponding elements of some previous row;

b″)  adding to the elements of some row numbers which are proportional to the elements of some subsequent row.

These transformations may also be performed on columns.  Transformations of the type indicated will be called *elementary transformations* of matrices.

It is easily verified that each elementary transformation on rows is equivalent to pre-multiplying the matrix by a certain non-singular matrix of a special type. In fact, operation a) is equivalent to pre-multiplying the matrix by the matrix

$$
\begin{bmatrix}
1 & & & & & & \\
& \cdot & & & & & \\
& & \cdot & & & & \\
& & & 1 & & & \\
& & & & \alpha & & \\
& & & & 1 & & \\
& & & & & \cdot & \\
& & & & & & \cdot \\
& & & & & & & 1
\end{bmatrix}, \tag{32}
$$

operation b') is equivalent to pre-multiplication by the matrix

$$
\begin{bmatrix}
1 & & & & & & \\
& \cdot & & & & & \\
& & \cdot & & & & \\
& & & 1 & & & \\
& & & \cdot & \cdot & & \\
& & & \cdot & & \cdot & \\
& & & \cdot & & & \cdot \\
& & & \alpha & & 1 & \\
& & & & & & \cdot \\
& & & & & & & 1
\end{bmatrix}, \tag{33}
$$

operation b') is equivalent to pre-multiplication by the matrix

$$
\begin{bmatrix}
1 & & & & & & \\
& \cdot & & & & & \\
& & \cdot & & & & \\
& & & 1 & \cdot \cdot \cdot & \alpha & \\
& & & & \cdot & \cdot & \\
& & & & & \cdot & \cdot \\
& & & & & \cdot \cdot & \\
& & & & & 1 & \\
& & & & & & \cdot \\
& & & & & & & 1
\end{bmatrix}. \tag{34}
$$

For example,

$$\begin{bmatrix} 1 & 0 & 0 \\ 0 & \alpha & 0 \\ 0 & 0 & 1 \end{bmatrix} \begin{bmatrix} a & b & c \\ x & y & z \\ u & v & w \end{bmatrix} = \begin{bmatrix} a & b & c \\ \alpha x & \alpha y & \alpha z \\ u & v & w \end{bmatrix}$$

$$\begin{bmatrix} 1 & 0 & 0 \\ 0 & 1 & 0 \\ 0 & \alpha & 1 \end{bmatrix} \begin{bmatrix} a & b & c \\ x & y & z \\ u & v & w \end{bmatrix} = \begin{bmatrix} a & b & c \\ x & y & z \\ u + \alpha x & v + \alpha y & w + \alpha z \end{bmatrix}$$

$$\begin{bmatrix} 1 & 0 & 0 \\ 0 & 1 & \alpha \\ 0 & 0 & 1 \end{bmatrix} \begin{bmatrix} a & b & c \\ x & y & z \\ u & v & w \end{bmatrix} = \begin{bmatrix} a & b & c \\ x + \alpha u & y + \alpha v & z + \alpha w \\ u & v & w \end{bmatrix}.$$

Operations a), b''), b') on columns are carried out by post-multiplication by the same matrices.

Later we shall often have occasion to make transformations of the form a) and b') on matrices.

The result of several successive transformations of this type is equivalent to pre-multiplying a matrix by some *"left triangular" matrix*, which is a matrix of form

$$\begin{bmatrix} \gamma_{11} & 0 & \cdots & 0 \\ \gamma_{21} & \gamma_{22} & \cdots & 0 \\ \cdot & \cdot & \cdot & \cdot \\ \gamma_{n1} & \gamma_{n2} & \cdots & \gamma_{nn} \end{bmatrix} \tag{35}$$

with non-zero diagonal elements $\gamma_{ii}$.

Actually each separate transformation of the form a) and b') is equivalent to pre-multiplication by some triangular matrix of the type indicated, and the product of two or more triangular matrices of the same structure (e.g. both left triangular) is again a triangular matrix.

Analogously, the result of several transformations of the form a) and b'') is equivalent to pre-multiplication by a *"right triangular" matrix*, which is, a matrix of form

$$\begin{bmatrix} b_{11} & b_{12} & \cdots & b_{1n} \\ 0 & b_{22} & \cdots & b_{2n} \\ \cdot & \cdot & \cdot & \cdot \\ 0 & 0 & \cdots & b_{nn} \end{bmatrix}. \tag{36}$$

Furthermore, the result of successive transformations of the form a) and b') on columns is equivalent to post-multiplication by a right triangular matrix; the result of successive transformations of the form a) and b'') on columns is equivalent to post-multiplication by a left triangular matrix.

We observe further that the result of the several transformations of form b') and b'') on columns, such that the elements of the $i$th column (which itself remains unchanged) are added to each of the other columns, is equivalent to post-multiplication by a matrix of the form:

$$\begin{bmatrix} 1 & 0 & \cdots & \cdots & \cdots & \cdots & \cdots & \cdots & 0 \\ 0 & 1 & \cdots & \cdots & \cdots & \cdots & \cdots & \cdots & 0 \\ \cdot & \cdot & \cdot & \cdot & \cdot & \cdot & \cdot & \cdot & \cdot \\ m_{i1} & m_{i2} & \cdots & m_{i,\,i-1} & 1 & m_{i,\,i+1} & \cdots & m_{in} \\ \cdot & \cdot & \cdot & \cdot & \cdot & \cdot & \cdot & \cdot & \cdot \\ 0 & 0 & \cdots & \cdots & \cdots & \cdots & \cdots & \cdots & 1 \end{bmatrix} \tag{37}$$

**12. Expression of a matrix as the product of two triangular matrices.** Triangular matrices possess various convenient properties: the determinant of a triangular matrix is equal to the product of the elements of the principal diagonal; the product of two triangular matrices of the same structure is again a triangular matrix of the same structure; a non-singular triangular matrix is easily inverted and its inverse matrix has an identical structure, etc.

The following theorem is therefore of interest.

THEOREM 1.1.  *The matrix[†]*

$$A = \begin{bmatrix} a_{11} & a_{12} & \cdots & a_{1n} \\ a_{21} & a_{22} & \cdots & a_{2n} \\ \cdot & \cdot & \cdot & \cdot \\ a_{n1} & a_{n2} & \cdots & a_{nn} \end{bmatrix}$$

*may be represented as the product of a left and a right triangular matrix, provided that*

$$a_{11} \neq 0 , \begin{vmatrix} a_{11} & a_{12} \\ a_{21} & a_{22} \end{vmatrix} \neq 0 , \cdots , | A | \neq 0 .$$

We shall use mathematical induction to carry out the proof. For $n = 1$ the statement is obvious: $(a_{11}) = (b_{11})(c_{11})$, where one of the (non-zero) factors may be taken arbitrarily. Assume that the theorem is true for matrices of order $(n - 1)$. We shall show that it is true for matrices of order $n$.

We partition matrix $A$ into cells in the following way:

$$A = \begin{bmatrix} a_{11} & \cdots & a_{1n} \\ a_{21} & \cdots & a_{2n} \\ \cdot & \cdot & \cdot \\ a_{n1} & \cdots & a_{nn} \end{bmatrix}$$

$$A = \begin{bmatrix} A_{n\,1} & \begin{matrix} a_{1n} \\ \vdots \\ a_{n-1,n} \end{matrix} \\ a_{n1}\ a_{n2}\ \cdots\ a_{n,n-1}\ a_{nn} \end{bmatrix} = \begin{bmatrix} A_{n-1} & u \\ v & a_{nn} \end{bmatrix} ,$$

We shall look for a factorization $A = CB$ of the matrix $A$ into the product of the matrices $B$ and $C$ of the type required, partitioning these matrices into cells as we did with $A$:

$$C = \begin{bmatrix} C_{n-1} & 0 \\ x & c_{nn} \end{bmatrix} ,$$

$$B = \begin{bmatrix} B_{n-1} & y \\ 0 & b_{nn} \end{bmatrix} .$$

According to the rule for multiplying partitioned matrices:

---

[†] Here, as elsewhere, we assume that the elements belong to a field: as a matter of fact the real or the complex number field.

$$CB = \begin{bmatrix} C_{n-1} & 0 \\ x & c_{nn} \end{bmatrix} \begin{bmatrix} B_{n-1} & y \\ 0 & b_{nn} \end{bmatrix}$$

$$= \begin{bmatrix} C_{n-1}B_{n-1} & C_{n-1}y \\ xB_{n-1} & xy + c_{nn}b_{nn} \end{bmatrix} = A .$$

The proposed goal will be achieved if we define the matrices $C_{n-1}, B_{n-1}$, $x, y$ and the numbers $c_{nn}, b_{nn}$, such that

$$C_{n-1}B_{n-1} = A_{n-1}$$
$$C_{n-1}y = u$$
$$xB_{n-1} = v$$
$$xy + c_{nn}b_{nn} = a_{nn}$$

The first condition is satisfied because of the induction hypothesis. From the assumption that $|A_{n-1}| \neq 0$ it follows that $|C_{n-1}| \neq 0$ and $|B_{n-1}| \neq 0$. Furthermore, $y$ and $x$ are easily determined from the formulas

$$y = C_{n-1}^{-1}u ,$$
$$x = vB_{n-1}^{-1} .$$

Thus it remains to determine only the diagonal elements $c_{nn}$ and $b_{nn}$ from the equation

$$c_{nn}b_{nn} = a_{nn} - xy$$

It is obvious that this is always possible, since one may assign to one of the numbers $c_{nn}$ or $b_{nn}$ an arbitrary, non-zero, value; the second number is then easily determined. The theorem is thus proved.

From the proof it follows that the representation of a matrix as the product of two triangular matrices will be unique if a priori we assign fixed values to the diagonal elements of one of the triangular matrices.

It is convenient to consider, for example, that $b_{ii} = 1$, $i = 1, 2, \cdots, n$. If one then selects diagonal elements from the rows of matrix $C$, then we arrive at the expansion

$$A = \tilde{C}\varLambda B ,$$

where $\varLambda = [\alpha_1, \cdots, \alpha_n]$, $\alpha_i$ are the diagonal elements of matrix $C$, and $\tilde{c}_{ii} = 1$. It is easily verified that

$$\alpha_i = \frac{|A_i|}{|A_{i-1}|} \quad (i = 1, \cdots, n) .$$

In fact from this construction it follows that

$$|A_i| = |C_i| \cdot |B_i| = c_{ii} |C_{i-1}| b_{ii} |B_{i-1}| = \alpha_i |C_{i-1}| |B_{i-1}| = \alpha_i |A_{i-1}| .$$

It is not hard to obtain formulas for expressing a matrix as the product of two triangular matrices.

We shall denote by $\beta_{ik}, i \leq k$, that minor of matrix $A$ formed of the first $i$ rows, the first $i-1$ columns, and the $k$th column. Correspondingly, we shall denote by $\gamma_{ik}$ the minor of matrix $A$ formed of the first $i$ columns, the first $i - 1$ rows, and the $k$th row. Using this notation $\beta_{kk} = \gamma_{kk} = |A_k|$.

We shall denote by $\bar{\beta}_{ik}$, $i \leq k$, the algebraic cofactor of the $i$th element of the last row of matrix $A_k$, by $\bar{\gamma}_{ki}$ the algebraic cofactor of the $i$th element of the last column of matrix $A_k$. Using this notation $\bar{\beta}_{kk} = \bar{\gamma}_{kk} = |A_{k-1}|$.

We construct the following matrices:

$$B_1 = \begin{bmatrix} \beta_{11} & \beta_{12} & \cdots & \beta_{1n} \\ 0 & \beta_{22} & \cdots & \beta_{2n} \\ & \cdot & \cdot & \\ 0 & 0 & \cdots & \beta_{nn} \end{bmatrix}, \quad C_1 = \begin{bmatrix} \gamma_{11} & 0 & \cdots & 0 \\ \gamma_{21} & \gamma_{22} & \cdots & 0 \\ & \cdot & \cdot & \\ \gamma_{n1} & \gamma_{n2} & \cdots & \gamma_{nn} \end{bmatrix}$$

$$B_2 = \begin{bmatrix} \bar{\beta}_{11} & \bar{\beta}_{12} & \cdots & \bar{\beta}_{1n} \\ 0 & \bar{\beta}_{22} & \cdots & \bar{\beta}_{2n} \\ & \cdot & \cdot & \\ 0 & 0 & \cdots & \bar{\beta}_{nn} \end{bmatrix}, \quad C_2 = \begin{bmatrix} \bar{\gamma}_{11} & 0 & \cdots & 0 \\ \bar{\gamma}_{21} & \bar{\gamma}_{22} & \cdots & 0 \\ & \cdot & \cdot & \\ \bar{\gamma}_{n1} & \bar{\gamma}_{n2} & \cdots & \bar{\gamma}_{nn} \end{bmatrix}.$$

From the elementary properties of determinants it follows that

$$\begin{aligned} AB_2 &= C_1 ; \\ C_2A &= B_1 . \end{aligned} \tag{38}$$

We introduce for consideration the diagonal matrices

$$S_1 = [\beta_{11}, \beta_{22}, \cdots, \beta_{nn}] = [\gamma_{11}, \gamma_{22}, \cdots, \gamma_{nn}] = [\,|A_1|, |A_2|, \cdots, |A_n|\,]$$

and

$$S_2 = [\bar{\beta}_{11}, \bar{\beta}_{22}, \cdots, \bar{\beta}_{nn}] = [\bar{\gamma}_{11}, \bar{\gamma}_{22}, \cdots, \bar{\gamma}_{nn}] = [1, |A_1|, \cdots, |A_{n-1}|\,] .$$

We introduce the notation

$$\begin{aligned} \tilde{B} &= S_1^{-1}B_1 , \\ \tilde{C} &= C_1S_1^{-1} , \\ S &= S_1S_2^{-1} = S_2^{-1}S_1 , \\ B_0 &= B_2S_2^{-1} , \\ C_0 &= S_2^{-1}C . \end{aligned}$$

It is obvious that matrices $\tilde{B}$, $\tilde{C}$, $B_0$ and $C_0$ have units as diagonal elements. From equalities (*38*) it follows that

$$\begin{aligned} AB_0 &= S\tilde{C} ; \\ C_0A &= S\tilde{B} , \end{aligned}$$

from which

$$A = \tilde{C}SB_0^{-1} = C_0^{-1}S\tilde{B} .$$

The matrices $\tilde{C}$ and $C_0^{-1}$ are left triangular matrices with unit diagonal elements; matrices $S\tilde{B}$ and $SB_0^{-1}$ are right triangular matrices. From the uniqueness of the representation of a matrix as the product of triangular matrices with prescribed diagonal elements for one of the factors, we conclude that $\tilde{C} = C_0^{-1}$ and $SB_0^{-1} = S\tilde{B}$, from which $\tilde{B} = B_0^{-1}$.

Thus

$$A = \tilde{C} S \tilde{B} . \tag{39}$$

It is clear that

$$\tilde{C} = \begin{bmatrix} 1 & & & \\ \dfrac{\gamma_{21}}{\gamma_{11}} & 1 & & 0 \\ \cdot & \cdot & \cdot & \cdot & \cdot & \cdot \\ \dfrac{\gamma_{n1}}{\gamma_{11}} & \dfrac{\gamma_{n2}}{\gamma_{22}} & \cdots & 1 \end{bmatrix}$$

$$\tilde{B} = \begin{bmatrix} 1 & \dfrac{\beta_{12}}{\beta_{11}} & \cdots & \dfrac{\beta_{1n}}{\beta_{11}} \\ & 1 & \cdots & \dfrac{\beta_{2n}}{\beta_{22}} \\ & & \cdot & \vdots \\ 0 & & & 1 \end{bmatrix} \tag{40}$$

**13. Expansion of a partitioned matrix into the product of two quasi-diagonal matrices.** A partitioned matrix with square diagonal cells is called *right quasi-triangular* if it has the form

$$\begin{bmatrix} B_{11} & B_{12} & \cdots & B_{1n} \\ 0 & B_{22} & \cdots & B_{2n} \\ \cdot & \cdot & \cdot & \cdot & \cdot & \cdot \\ 0 & 0 & \cdots & B_{nn} \end{bmatrix},$$

and *left quasi-triangular* if it has the form

$$\begin{bmatrix} C_{11} & 0 & \cdots & 0 \\ C_{21} & C_{22} & \cdots & 0 \\ \cdot & \cdot & \cdot & \cdot & \cdot & \cdot \\ C_{n1} & C_{n2} & \cdots & C_{nn} \end{bmatrix}.$$

It is obvious that the determinant of a quasi-triangular matrix is equal to the product of the determinants of its diagonal cells.

The following theorem holds.

THEOREM 1.2. *If the matrices*

$$A_1 = A_{11} ,$$

$$A_2 = \begin{bmatrix} A_{11} & A_{12} \\ A_{21} & A_{22} \end{bmatrix},$$

$$\cdots,$$

$$A_{n-1} = \begin{bmatrix} A_{11} & \cdots & A_{1n-1} \\ \cdot & \cdot & \cdot & \cdot & \cdot \\ A_{n-11} & \cdots & A_{n-1n-n} \end{bmatrix}$$

*are non-singular, then the matrix*

$$A = \begin{bmatrix} A_{11} & A_{12} & \cdots & A_{1n} \\ A_{21} & A_{22} & \cdots & A_{2n} \\ \cdot & \cdot & \cdot & \cdot & \cdot & \cdot \\ A_{n1} & A_{n1} & \cdots & A_{nn} \end{bmatrix}$$

*may be represented as a product of left and right quasi-triangular matrices, where the diagonal cells of one of them may be any chosen non-singular matrices of the suitable orders.*

The proof is carried out analogously to the proof of theorem 1.1. We assume that, for the matrix $A_{n-1}$, a known factorization $A_{n-1} = C_{n-1}B_{n-1}$ has already been obtained. Since $|A_{n-1}| \neq 0$, then $|C_{n-1}| \neq 0$ and $|B_{n-1}| \neq 0$. We assume further that

$$A = \begin{bmatrix} A_{n-1} & U \\ V & A_{nn} \end{bmatrix}$$

and we seek quasi-triangular matrices $B$ and $C$ in the form

$$B = \begin{bmatrix} B_{n-1} & Y \\ 0 & B_{nn} \end{bmatrix} \quad \text{and} \quad C = \begin{bmatrix} C_{n-1} & 0 \\ X & C_{nn} \end{bmatrix}.$$

The matrix equation $CB = A$ may be broken down into the equations

$$C_{n-1}B_{n-1} = A_{n-1} ,$$
$$C_{n-1}Y = U ,$$
$$XB_{n-1} = V ,$$
$$XY + C_{nn}B_{nn} = A_{nn} .$$

The first of these equations is satisfied automatically; from the second and third we get $Y = C_{n-1}^{-1}U$, $X = VB_{n-1}^{-1}$; from the last we obtain

$$B_{nn} = C_{nn}^{-1}(A_{nn} - XY) \quad \text{(or } C_{nn} = (A_{nn} - XY)B_{nn}^{-1}) ,$$

taking any non-singular matrix for $C_{nn}$ (or $B_{nn}$).

**14. Matrix notation for a system of linear equations.** We consider a system of $n$ linear equations with $n$ unknowns:

$$
\begin{aligned}
a_{11}x_1 + a_{12}x_2 + \cdots + a_{1n}x_n &= f_1 , \\
a_{21}x_1 + a_{22}x_2 + \cdots + a_{2n}x_n &= f_2 , \\
&\cdots \cdots \\
a_{n1}x_1 + a_{n2}x_2 + \cdots + a_{nn}x_n &= f_n .
\end{aligned}
\tag{41}
$$

Utilizing matrix notation, and the rules for multiplying matrices, we may write system (41) as a single matrix equation

$$Ax = f . \tag{41'}$$

Here $A$ denotes the matrix of coefficients of the system, $f$ is a column of constant terms, and $x$ is a column matrix whose elements are the unknowns.

If the matrix $A$ is non-singular we immediately obtain a solution for the system by pre-multiplying (41') by $A^{-1}$. That is,

$$x = A^{-1}f = \frac{1}{|A|}Bf , \tag{42}$$

where $B$ is the matrix adjoint to $A$.

We shall show that the last formula is the matrix notation for the well-known formulas of Cramer

$$x_i = \frac{|A_i|}{|A|} \, ,$$                                (43)

where $A_i$ is the matrix which is obtained from matrix $A$ by replacing the elements of the $i$th column with the constants of column $f$.

Actually, the matrix equation (42) is equivalent to $n$ equations

$$x_i = \frac{A_{1i}f_1 + A_{2i}f_2 + \cdots + A_{ni}f_n}{|A|} \quad i = 1, \cdots, n \, .$$

Since $A_{ki}$ are the algebraic cofactors of the element $a_{ki}$ in the determinant of matrix $A$, we find that

$$A_{1i}f_1 + A_{2i}f_2 + \cdots + A_{ni}f_n = |A_i| \, ,$$

which proves our statement.

**15. Matrix notation for quadratic forms.** In many areas of mathematics an essential role is played by *quadratic forms*, that is, homogeneous polynomials of the second degree in several variables. It is clear that a quadratic form is made up of terms of two kinds: squared variables and pairwise products of variables, each such term having some numerical coefficient. We shall divide up each term containing a pairwise product of variables into two equal terms, writing the product of the variables in both possible orders. That is we may write the quadratic form in the following format:

$$\begin{aligned} F(x_1, x_2, \cdots, x_n) = a_{11}x_1^2 &+ a_{12}x_1x_2 + \cdots + a_{1n}x_1x_n \\ &+ a_{21}x_2x_1 + \quad a_{22}x_2^2 + \cdots + a_{2n}x_2x_n \\ &\cdots\cdots\cdots\cdots\cdots\cdots\cdots\cdots\cdots \\ &+ a_{n1}x_nx_1 + a_{n2}x_nx_2 + \cdots + a_{nn}x_n^2 \, . \end{aligned}$$                (44)

where $a_{ij} = a_{ji}$ for $i \neq i$.

The matrix

$$A = \begin{bmatrix} a_{11} & a_{12} & \cdots & a_{1n} \\ a_{21} & a_{22} & \cdots & a_{2n} \\ \cdots & \cdots & \cdots & \cdots \\ a_{n1} & a_{n2} & \cdots & a_{nn} \end{bmatrix}$$

is called the matrix of the quadratic form. Because of the manner in which it was composed, $A$ is a symmetric matrix; i.e. $a_{ij} = a_{ji}$. Thus, every quadratic form is associated in a natural way with a unique symmetric matrix and, conversely, every symmetric matrix may be associated with a certain quadratic form.

A quadratic form may be written in an abbreviated matrix notation. In fact,

$$F(x_1, x_2, \cdots, x_n) = x_1(a_{11}x_1 + a_{12}x_2 + \cdots + a_{1n}x_n)$$
$$+ x_2(a_{21}x_1 + a_{22}x_2 + \cdots + a_{1n}x_n)$$
$$\cdot \ \cdot \ \cdot \ \cdot \ \cdot \ \cdot \ \cdot \ \cdot \ \cdot \ \cdot \ \cdot \ \cdot$$
$$+ x_n(a_{n1}x_1 + a_{n2}x_2 + \cdots + a_{nn}x_n)$$

$$= (x_1, x_2, \cdots, x_n) \begin{bmatrix} a_{11}x_1 + a_{12}x_2 + \cdots + a_{1n}x_n \\ a_{21}x_1 + a_{22}x_2 + \cdots + a_{2n}x_n \\ \cdot \ \cdot \ \cdot \ \cdot \ \cdot \ \cdot \ \cdot \ \cdot \\ a_{n1}x_1 + a_{n2}x_2 + \cdots + a_{nn}x_n \end{bmatrix} \qquad (45)$$

$$= (x_1, x_2, \cdots, x_n) \begin{bmatrix} a_{11} & a_{12} & \cdots & a_{1n} \\ a_{21} & a_{22} & \cdots & a_{2n} \\ \cdot & \cdot & \cdot & \cdot \\ a_{n1} & a_{n2} & \cdots & a_{nn} \end{bmatrix} \begin{bmatrix} x_1 \\ x_2 \\ \vdots \\ x_n \end{bmatrix} = x'Ax,$$

$$x' = (x_1, x_2, \cdots, x_n).$$

A real quadratic form is called *positive definite* if, for real values of the variables, its value is always positive, with the obvious exception for $(x_1 = x_2 = \cdots = x_n = 0)$.

The form $x_1^2 + x_2^2 + \cdots + x_n^2$ is one example of a positive definite form.

The term "positive definite" is also extended to symmetric matrices. A real symmetric matrix $A = (a_{ij})$ is called *positive definite* if the quadratic form $F(x_1, x_2, \cdots, x_n) = \sum_{i,\,j=1}^{n} a_{ij} x_i x_j$ is positive definite. Thus, for example, the unit matrix is positive definite, since the quadratic form corresponding to it is positive definite. Moreover, it is obvious that a quasi-diagonal matrix which is composed of positive definite cells is positive definite.

The positive definite concept may be extended also to complex matrices of a special type—the so-called Hermitian matrices which are related to Hermitian forms. These forms will be considered later.

We consider how the coefficients of a quadratic form change under a linear transformation of variables. Assume that

$$x_1 = b_{11}y_1 + b_{12}y_2 + \cdots + b_{1n}y_n,$$
$$x_2 = b_{21}y_1 + b_{22}y_2 + \cdots + b_{2n}y_n,$$
$$\cdot \ \cdot \ \cdot \ \cdot \ \cdot \ \cdot \ \cdot \ \cdot \ \cdot \ \cdot \ \cdot \ \cdot \ \cdot,$$
$$x_n = b_{n1}y_1 + b_{n2}y_2 + \cdots + b_{nn}y_n$$

or, in matrix notation, $x = By$. Then

$$x'Ax = (By)'A(By) = y'B'ABy = y'Cy,$$

where $C = B'AB$. It is readily seen that the matrix $C$ is symmetric. That is, $C' = (B'AB)' = B'A'(B')' = B'AB = C$. since $A' = A, (B')' = B$.

Thus under a linear transformation of variables, associated with a matrix $B$, the quadratic form becomes again a quadratic form, while its coefficient matrix is replaced by the matrix $B'AB$.

We remark that for a linear transformation of variables of a form associated with a non-singular matrix $B$, a positive definite quadratic form remains positive definite. In fact, if one assumes that, for a certain system of values $y_0$, the transformed quadratic form has a negative or zero value, then the initial quadratic form has the same value for the system of values for the variables $x_0 = By_0$, which is possible only for $x_0 = 0$ (and consequently for $y_0 = 0$).

**16. Gauss transformations.** The solution of a system of linear equations

$$Ax = f$$

with a non-degenerate matrix $A$ may always be reduced to the solution of a system with a symmetric, and in fact a positive definite matrix. This information is based on the following theorem.

THEOREM 1.3. *If $A$ is a non-degenerate matrix, then the matrices $A'A$ and $AA'$ are positive definite.*

*Proof.* If in a quadratic form associated with a unit matrix $E$ one replaces the variables with the matrix $A$ (or correspondingly, with the matrix $A'$), then one obtains a quadratic form with the matrix $A'EA = A'A$ or correspondingly, $AEA' = AA'$). Since $E$ is positive definite, the matrices $A'A$ and $AA'$ will also be positive definite.

If one pre-multiplies the system of equations

$$Ax = f$$

by the matrix $A'$, one obtains the equivalent system

$$A'Ax = A'f$$

with the positive definite matrix $A'A$. We shall call such a transformation a *first (left) Gauss transformation.*

A *second (right) Gauss transformation* consists in considering, instead of the system

$$Ax = f,$$

the auxiliary system

$$AA'y = f$$

with the positive definite matrix $AA'$. Having found a solution $y$ of the auxiliary system, we then find a solution of the initial system from the formula

$$x = A'y$$

## 2. MATRICES OF SPECIAL TYPE

**1. Symmetric matrices.** Symmetric matrices possess a number of remarkable properties which we shall consider later on. At the moment we shall observe only that the product of two symmetric matrices will not always be a symmetric matrix. More precisely, the product of symmetric matrices is a symmetric matrix if and only if they are commutative. In fact, $(AB)' = B'A' = BA$, so that $AB = (AB)'$ only if $BA = AB$.

**2. Orthogonal matrices.** A real matrix is called *orthogonal* if the sum of the squares of the elements of each column is equal to one, and if the sum of the products of corresponding elements of two different columns is equal to zero. The orthogonality of a matrix may be characterized by a single matrix equation, namely,

$$A'A = E. \tag{1}$$

The diagonal elements of the matrix $A'A$ are the sums of the squares of the elements of the columns of matrix $A$, and the non-diagonal elements of the columns of matrix $A$, and the non-diagonal elements are equal to the sums of the products of corresponding element columns. Orthogonal matrices possess the following properties:

1. The unit matrix is orthogonal.
2. If $A$ is orthogonal, then $A^{-1} = A'$. This follows from the equality $A'A = E$.
3. If $A$ is orthogonal, then $A'$ is also orthogonal. In other words, in satisfying the conditions of orthogonality for the columns of a matrix $A$ one satisfies the same conditions for the rows of matrix $A$. In fact.

$$(A')'A' = AA' = AA^{-1} = E.$$

4. The product of two orthogonal matrices is an orthogonal matrix. In fact, if $A$ and $B$ are orthogonal, then

$$(AB)'AB = B'A'AB = B'EB = E.$$

5. The determinant of an orthogonal matrix equals 1 or $-1$. From $A'A = E$ it follows that

$$|A'A| = |A'| \cdot |A| = |A|^2 = 1.$$

This last result determines a natural separation of orthogonal matrices into two classes: those called **properly orthogonal** (with determinant $+1$), and those called **improperly orthogonal** (with determinant $-1$).

To the class of orthogonal matrices belong so-called *elementary rotation matrices* these are of form:

$$T_{ij} = \begin{bmatrix} 1 & & & & & & \\ & \ddots & & & & & \\ & & \ddots & & & & \\ & & & c & \cdots & -s & \\ & & & \vdots & & \vdots & \\ & & & s & \cdots & c & \\ & & & & & & \ddots \\ & & & & & & & 1 \end{bmatrix} \tag{2}$$

where $c^2 + s^2 = 1$. This last relationship shows that there exists an angle $\varphi$, such that $c = \cos \varphi$, $s = \sin \varphi$.

Elementary rotation matrices differ from the unit matrix only in four elements, located at the intersection of two rows and two columns with the numbers $i$ and $j, i < j$. These four elements form the matrix

$$\begin{bmatrix} c, & -s \\ s, & c \end{bmatrix} = \begin{bmatrix} \cos \varphi, & -\sin \varphi \\ \sin \varphi, & \cos \varphi \end{bmatrix}$$

which coincides with the matrix for transforming Cartesian coordinates in the plane by rotating the axis through the angle $\varphi$.

Later on, elementary rotation matrices will often be employed as auxiliary matrices for transforming a given matrix $A$ by means of a sequence of multiplications from the left or right (or from both sides) by these matrices.

It is obvious that left multiplication of the matrix $A = (a_{\alpha\beta})$ by the matrix $T_{ij}$ changes only the $i$th and $j$th rows of matrix $A$; for the matrix $A^{(1)} = T_{ij} A$ we shall have

$$\begin{aligned} a_{i\beta}^{(1)} &= c a_{i\beta} - s a_{j\beta} \\ a_{j\beta}^{(1)} &= s a_{i\beta} + c a_{j\beta} \end{aligned} \qquad (\beta = 1, 2, \cdots, n) . \qquad (3)$$

Correspondingly, multiplying matrix $A = (a_{\alpha\beta})$ on the right by matrix $T_{ij}$ changes only the $i$th and $j$th columns, according to the formulas

$$\begin{aligned} a_{\alpha i}^{(1)} &= c a_{\alpha i} + s a_{\alpha j} \\ a_{\alpha j}^{(1)} &= - s a_{\alpha j} + c a_{\alpha j} \end{aligned} \qquad (\alpha = 1, 2, \cdots, n) . \qquad (4)$$

It is clear that if at least one of the two elements $a_{i\beta}$ and $a_{j\beta}$ is different from zero, then one may choose $c$ and $s$, such that for the matrix $A^{(1)} = T_{ij} A$ the element $a_{j\beta}^{(1)}$ turns out to be equal to zero. For this, one must take

$$s = - \frac{a_{j\beta}}{\sqrt{a_{i\beta}^2 + a_{j\beta}^2}} ,$$

$$c = \frac{a_{i\beta}}{\sqrt{a_{i\beta}^2 + a_{j\beta}^2}} . \qquad (5)$$

From such a choice of $s$ and $c$ we get

$$a_{i\beta}^{(1)} = \sqrt{a_{i\beta}^2 + a_{j\beta}^2} > 0 ,$$

$$a_{j\beta}^{(1)} = = 0 .$$

THEOREM 2.1. *Every real, non-degenerate matrix may be transformed by means of a sequence of pre-multiplications by elementary rotation matrices into a right triangular matrix, all of whose diagonal elements—except possibly the last—will be positive.*

*Proof.* Let

$$A = \begin{bmatrix} a_{11} & a_{12} & \cdots & a_{1n} \\ a_{21} & a_{22} & \cdots & a_{2n} \\ \cdot & \cdot & \cdots & \cdot \\ a_{n1} & a_{n2} & \cdots & a_{nn} \end{bmatrix}$$

be a non-degenerate real matrix.

We assume first that $a_{11} \neq 0$. We pre-multiply matrix $A$ successively by the matrices $T_{12}, T_{13}, \cdots, T_{1n}$, choosing them so that one successively reduces all elements of the first column, except the upper one, to zero. We shall make the initial element positive at the first step and then preserve its sign.

If we have $a_{11} = 0$, we begin the sequence of transformations with premultiplication by $T_{1j_0}$, where $j_0$ is the least number for which $a_{j1} \neq 0$. Since the matrix $A$ is non-degenerate, at least one element of the first column is different from zero, so that such a number $j_0$ may be found.

After such transformations we arrive at the matrix

$$A^{(1)} = T_{1n}T_{1n-1} \cdots T_{12}A = \begin{bmatrix} a_{11}^{(1)} & a_{12}^{(1)} & \cdots & a_{1n}^{(1)} \\ 0 & a_{22}^{(1)} & \cdots & a_{2n}^{(1)} \\ \cdot & \cdot & \cdot & \cdot \\ 0 & a_{n2}^{(1)} & \cdots & a_{nn}^{(1)} \end{bmatrix},$$

where $a_{11}^{(1)} > 0$.

From the non-degeneracy of the matrix $A^{(1)}$ at least one of the elements $a_{22}^{(1)}, \cdots, a_{2n}^{(1)}$ is different from zero. We now select elementary rotation matrices $T_{23}, T_{24}, \cdots, T_{2n}$ such that after the succession of multiplications by these matrices all the elements of the second column below the diagonal element becomes positive. We then proceed to make the sub-diagonal elements of the third column zero, etc. At the end of this process we obtain the matrix

$$A^{(n-1)} = T_{n-1n} \cdots T_{12}A = \begin{bmatrix} a_{11}^{(1)} & a_{12}^{(1)} & \cdots & a_{1n}^{(1)} \\ 0 & a_{22}^{(2)} & \cdots & a_{2n}^{(2)} \\ \cdot & \cdot & \cdot & \cdot \\ 0 & 0 & \cdots & a_{nn}^{(n-1)} \end{bmatrix},$$

in which all the diagonal elements, except possibly the last, $a_{nn}^{(n-1)}$, are positive. The sign of $a_{nn}^{(n-1)}$ coincides, obviously, with the sign of the determinant of matrix $A$. The theorem is proved.

We note that the usual number of matrix multiplications for obtaining the matrix $A^{(n-1)}$ will not exceed the number of subdiagonal elements, that is, $n(n-1)/2$.

COROLLARY 1. *Every non-degenerate real matrix is the product of a properly orthogonal matrix by a right triangular one.*

In fact, $A = PA^{(n-1)}$, where $P = (T_{n-1n} \cdots T_{12})^{-1}$ is a properly orthogonal matrix.

COROLLARY 2. *Every properly orthogonal matrix is a product of elementary rotation matrices.*

Let $A$ be a properly orthogonal matrix. Then

$$A = T_{12}^{-1} \cdots T_{n-1n}^{-1}A^{(n-1)},$$

where all the diagonal elements of the matrix $A^{(n-1)}$ are positive. It is readily seen that the matrix $A^{(n-1)}$ is the unit matrix. The sum of the squares of the elements of the first column is equal to one, from which it follows that $a_{11}^{(1)} = 1$. Since the sum of the products of elements of the first column by the corresponding elements of any other column is equal to zero, we conclude that $a_{12}^{(1)} = \cdots = a_{1n}^{(1)} = 0$. Furthermore, the sum of the squares of elements of the second column is equal to one. Consequently $a_{22}^{(2)} = 1$, and so forth. Thus we conclude that all the non-diagonal elements of matrix $A^{(n-1)}$ are equal to zero and all the diagonal elements are equal to one.

Symmetric matrices, like orthogonal matrices, come under the more general class of so-called normal matrices. A real matrix is called *normal* if it commutes with its own transpose, that is, if

$$A'A = AA'$$

**3. Hermitian matrices.** A matrix with complex elements is called *Hermitian* if

$$a_{ij} = \bar{a}_{ji}, \tag{6}$$

or, in matrix terminology,

$$A = A^*.$$

From this definition it follows that the diagonal elements of a Hermitian matrix are real. Real symmetric matrices are a special case of Hermitian matrices. Many properties of real symmetric matrices are preserved almost unchanged for Hermitian matrices. In particular, the product of two Hermitian matrices will be a Hermitian matrix if and only if the matrices commute.

Furthermore, for any matrix with complex elements the matrix $A^*A$ will be Hermitian.

**4. Unitary matrices.** A matrix with complex elements is called *unitary* if the sum of the squares of the moduli of the elements of each column is equal to one, and also the sum of the products of the elements of one column by numbers conjugate to the corrresponding elements of another column is equal to zero. Unitary matrices may be characterized by the matrix equation

$$A^*A = E. \tag{7}$$

Orthogonal matrices are obviously a special case of unitary matrices.

Properties 1–4 of orthogonal matrices also hold for unitary matrices. The determinant of a unitary matrix is a complex number whose modulus is equal to one. Unitary matrices as well as Hermitian matrices belong to the more general class of complex normal matrices which are characterized by the fact that each commutes with its conjugate matrix.

The class of unitary matrices contains so-called **elementary unitary matrices**; these are of form

$$\begin{bmatrix} 1 & & & & & \\ & \cdot & & & & \\ & & ce^{i\varphi_1}\cdots -se^{i\varphi_2} & & & \\ & & \cdot \quad \cdot & & & \\ & & se^{i\varphi_3}\cdots \quad ce^{i\varphi_4} & & & \\ & & & & \cdot & \\ & & & & & 1 \end{bmatrix} \tag{8}$$

with $c > 0, s > 0, c^2 + s^2 = 1, \varphi_1 - \varphi_2 = \varphi_3 - \varphi_4$. The determinant of such a matrix is equal to $e^{i(\varphi_1+\varphi_4)}$. It is equal to one if and only if $\varphi_4 = -\varphi_1$ and, consequently, $\varphi_3 = -\varphi_2$ (modulo $2\pi$).

Given two complex numbers $a$ and $b$, not both equal to zero, it is always possible to choose $c, s, \varphi_1$ and $\varphi_2$ such that

$$ace^{i\varphi_1} - bse^{i\varphi_2} > 0 ,$$
$$ase^{-i\varphi_2} + bce^{-i\varphi_1} = 0 .$$

For this it is sufficient to take

$$c = \frac{|b|}{\sqrt{|a|^2 + |b|^2}} ;$$

$$s = \frac{|a|}{\sqrt{|a|^2 + |b|^2}} ; \tag{9}$$

$$\varphi_1 = -\arg a; \varphi_2 = \pi - \arg b .$$

This notation allows us to prove the theorem:

THEOREM 2.2. *Every non-degenerate complex matrix A may be pre-multiplied by a succession of elementary unitary matrices, with determinants equal to one, so as to be transformed into a right triangular matrix, all of whose diagonal elements, except possibly the last, are positive.*

It is clear that the argument of this last element coincides with the argument of the determinant of matrix $A$.

From the theorem arise the following corollaries:

COROLLARY 1. *Every non-degenerate complex matrix may be represented as the product of a unitary matrix with determinant equal to one and a right triangular matrix, all of whose diagonal elements except possibly the last are positive.*

COROLLARY 2. *Every unitary matrix with determinant equal to one is the product of elementary unitary matrices with determinant equal to one.*

### 5. Tridiagonal matrices. A *tridiagonal matrix* is a matrix of the form

$$\begin{bmatrix} a_1 & b_1 & 0 & \cdots & 0 & 0 \\ c_1 & a_2 & b_2 & \cdots & 0 & 0 \\ 0 & c_2 & a_3 & \cdots & 0 & 0 \\ \cdot & \cdot & \cdot & \cdots & \cdot & \cdot \\ 0 & 0 & 0 & \cdots & a_{n-1} & b_{n-1} \\ 0 & 0 & 0 & \cdots & c_{n-1} & a_n \end{bmatrix}. \tag{10}$$

A real tridiagonal matrix is called *Jacobian* if $b_i c_i > 0$ for $i = 1, 2, \cdots,$ $n - 1$. Every symmetric tridiagonal matrix with non-zero non-diagonal elements will automatically be Jacobian.

Tridiagonal matrices are distinguished by the fact that their characteristic polynomials may be computed by simple recurrence formulas. Let $\varphi_k(t)$ be the normalized characteristic polynomial of a truncated matrix, that is, the matrix

$$A_k = \begin{bmatrix} a_1 & b_1 & & \\ c_1 & a_2 & b_2 & \\ \cdot & \cdot & \cdot & \cdot & \cdot & \cdot \\ & & c_{k-1} & a_k \end{bmatrix}.$$

Then

$$\varphi_k(t) = (-1)^k \, | A_k - tE |$$

$$= (-1)^k \begin{vmatrix} a_1 - t & b_1 & & \\ c_1 & a_2 - t & b_2 & \\ \cdot & \cdot & \cdot & \cdot & \cdot & \cdot & \cdot & \cdot \\ & & a_{k-1} - t & b_{k-1} \\ & & c_{k-1} & a_k - t \end{vmatrix}.$$

Evaluating this determinant according to the elements of the last column we obtain, for $k \geq 3$.

$$\varphi_k(t) = -(a_k - t)\varphi_{k-1}(t) - (-1)^k b_{k-1} \begin{vmatrix} a_1 - t & b_1 & & \\ & \cdot & & \\ & & \cdot & \\ & & a_{k-2} - t & b_{k-2} \\ & & 0 & c_{k-1} \end{vmatrix} \tag{11}$$

$$= (t - a_k)\varphi_{k-1}(t) - b_{k-1}c_{k-1}\varphi_{k-2}(t) .$$

This last formula also remains true for $k = 2$ if one assumes $\varphi_0 = 1$. It is clear that $\varphi_1(t) = t - a_1$. Letting $k = 2, 3, \cdots, n$, we determine respectively the polynomials $\varphi_2(t), \varphi_3(t), \cdots, \varphi_n(t)$, where $\varphi_n(t)$ is the characteristic polynomial of the given tridiagonal matrix.

For a Jacobian matrix the sequence of polynomials $\varphi_0, \varphi_1(t), \cdots, \varphi_n(t)$ is the sequence of Sturm functions.[†] Since the leading coefficients of all these polynomials are positive, then according to Sturm's theorem all the roots of these polynomials are real, while the roots of two adjacent polynomials are distinct. Thus all eigenvalues of a Jacobian matrix are real and distinct.

---

[†] A. G. Kurosh, *A Course on Higher Algebra*, 1949, p. 269. Govt. Pub. House, the USSR.

We shall describe one method for solving a system of linear equations with tridiagonal matrix. Let

$$
\begin{aligned}
a_1 x_1 + b_1 x_2 &= f_1 , \\
c_1 x_1 + a_2 x_2 + b_2 x_3 &= f_2 , \\
c_2 x_2 + a_3 x_3 + b_3 x_4 &= f_3 , \\
\cdots\cdots\cdots\cdots\cdots\cdots\cdots\cdots &, \\
c_{n-2} x_{n-2} + a_{n-1} x_{n-1} + b_{n-1} x_n &= f_{n-1} , \\
c_{n-1} x_{n-1} + a_n x_n &= f_n .
\end{aligned}
$$

We assume that $b_i \neq 0$, $i = 1, 2, \cdots, n-1$. If this condition is not satisfied, then a system with fewer unknowns is determined.

We shall eliminate the last equation and find two solutions $x^{(0)}$ and $x^{(1)}$ of the remaining system of $n-1$ equations, letting $x_1^{(0)} = 0$ and $x_1^{(1)} = 1$. For this it is necessary to solve the system twice for a triangular matrix. It is obvious that the column $x^{(0)} + t(x^{(1)} - x^{(0)})$ will give a solution for the system for all $t$. We shall find $t$ such that the last equation, rejected earlier, is also satisfied.

For this it is necessary to solve the equation

$$
c_{n-1} x_{n-1}^{(0)} + c_{n-1} t(x_{n-1}^{(1)} - x_{n-1}^{(0)}) + a_n x_n^{(0)} + a_n t(x_n^{(1)} - x_n^{(0)}) = f_n ,
$$

from which

$$
t = \frac{r_n^{(0)}}{r_n^{(0)} - r_n^{(1)}} ,
$$

where $r_n^{(0)}$ and $r_n^{(1)}$ are the remainders of the last equation when the solutions $x^{(0)}$ and $x^{(1)}$ are substituted.

**6. Almost triangular matrices.** A matrix is called (right) *almost triangular* if it has the form

$$
\begin{vmatrix}
a_{11} & a_{12} & a_{13} & \cdots & a_{1n-1} & a_{1n} \\
a_{21} & a_{22} & a_{23} & \cdots & a_{2n-1} & a_{2n} \\
0 & a_{32} & a_{33} & \cdots & a_{3n-1} & a_{3n} \\
\cdot & \cdot & \cdot & \cdots & \cdot & \cdot \\
0 & 0 & 0 & & a_{nn-1} & a_{nn}
\end{vmatrix} . \tag{12}
$$

The determinant of an almost triangular matrix is connected by a simple recurrence relationship with its principal minors. That is, if we assume $\Delta_0 = 1$ and write

$$
\Delta_k =
\begin{vmatrix}
a_{11} & a_{12} & \cdots & a_{1k} \\
a_{21} & a_{22} & \cdots & a_{2k} \\
\cdot & \cdot & \cdots & \cdot \\
0 & 0 & \cdots & a_{kk}
\end{vmatrix}
\qquad (k = 1, 2, \cdots, n) ,
$$

then

$$
\begin{aligned}
\Delta_k = a_{kk}\Delta_{k-1} - a_{kk-1}a_{k-1k}\Delta_{k-2} + a_{kk-1}a_{k-1k-2}a_{k-2k}\Delta_{k-3} \\
+ \cdots + (-1)^{k-1} a_{kk-1}a_{k-1k-2}\cdots a_{21}a_{1k}\Delta_0 .
\end{aligned} \tag{13}
$$

It is easy to convince oneself of this by expanding the determinant $\varDelta_k$ according to the elements of the last column.

Application of this formula for computing a determinant demands approximately $n^3/6$ multiplications, which, as we shall see later, is slightly less than the number of operations needed to compute a determinant in the general case.

Formula (13) may be applied in constructing by recurrence the characteristic polynomial of an almost triangular matrix, namely,

$$\varphi_k(t) = (t - a_{kk})\varphi_{k-1}(t) - a_{kk-1}a_{k-1k}\varphi_{k-2}(t)$$
$$- a_{kk-1}a_{k-1k-2}a_{k-2k}\varphi_{k-3}(t) - \cdots - a_{kk-1}a_{k-1k-2}\cdots a_{21}a_{1k} .$$

$$(14)$$

Here

$$\varphi_k(t) = \begin{vmatrix} t - a_{11} & - a_{12} & \cdots & - a_{1k} \\ - a_{21} & t - a_{22} & \cdots & - a_{2k} \\ \cdots & \cdots & \cdots & \cdots \\ 0 & 0 & \cdots & t - a_{kk} \end{vmatrix} \qquad (k = 1, 2, \cdots, n) .$$

## 3.   AXIOMS OF LINEAR SPACE

As is well-known, the methods of analytic geometry make it possible to compare one of the most important geometric objects, a vector in a space, with a triple of real numbers — the projections of a vector on selected coordinate axes. Such a comparison makes it possible, on the one hand, to investigate the properties of geometric objects by algebraic means and, on the other hand, to interpret algebraic problems geometrically. For example, the complete solution of a system of three linear equations in three unknowns is of interest as a problem concerning the intersection of three planes in space.

This circumstance makes the introduction of geometric terminology into linear algebra both intelligible and useful.

It is clear that to treat linear algebra geometrically the concept of a vector must be throughly reviewed. This will be effected by introducing so-called linear spaces. We shall introduce this concept axiomatically.

A *linear space* is a collection of mathematical (or physical) objects for which two operations are defined: addition, and multiplication by all real and complex numbers; these operations satisfy the following conditions (axioms):

1.   $X + Y = Y + X$ (commutative law);
2.   $(X + Y) + Z = X + (Y + Z$ (associative law);
3.   there exists an element "**0**" such that $X + 0 = X$ for all $X$;
4.   for all $X$ there exists a negative element "$- X$" such that $X + (-X)$ $= 0$;
5.   $1 \cdot X = X$;
6.   $(a + b)X = aX + bX$;

7.  $a(X + Y) = aX + aY$;
8.  $a(bX) = abX$ .

The elements of a linear space are called *vectors*.

From the above axioms it is easy to deduce the uniqueness of the zero element, the uniqueness of the negative element, and the equalities $0X = a0 = 0$, $(-X) = (-1)X$. We shall not stop to prove these statements.

A space is called *real* if multiplication of its vectors is defined only for real numbers and *complex* if multiplication is defined for complex numbers.

A space is called *finite-dimensional* if the following axiom is satisfied:

9.  There exists a finite number of vectors $X_1, \cdots, X_N$ such that every vector in the space may be represented in the form

$$c_1 X_1 + \cdots + c_N X_N .$$

The dimension of a finite-dimensional space is the least number of vectors satisfying the condition of axiom 9. If axiom 9 is not satisfied, then the space is called infinite-dimensional. The study of infinite-dimensional spaces leads beyond the limits of linear algebra, and infinite-dimensional spaces, with additional limitations, are investigated in one of the most important mathematical disciplines—functional analysis.

In complex as well as real linear space one may introduce the concept of *scalar product* in the following way. To each pair of vectors $X, Y$ there corresponds a number $(X, Y)$ (real in a real space, complex in a complex space); such that the following axioms are satified:

10.  $(X, X) > 0$ for $X \neq 0$; $(X, X) = 0$ for $X = 0$;
11.  $(X, Y) = (\overline{Y, X})$;
12.  $(X_1 + X_2, Y) = (X_1, Y) + (X_2, Y)$;
13.  $(aX, Y) = a(X, Y)$ .

For a real space the eleventh axiom takes on a simpler form, namely, $(X, Y) = (Y, X)$.

A real linear space with the scalar product defined in such a manner is called *Euclidean*; a complex one is called *unitary*.

The number $\sqrt{(X, X)}$ is called the *length* of a vector and is denoted by $|X|$. The scalar product of two vectors satisfies the following important inequality:

$$|(X, Y)| \leq |X| \cdot |Y|, \tag{1}$$

which is called the Cauchy-Bunyakovskii inequality. We shall prove it. For $X = 0$ the inequality is obvious. Let $X \neq 0$. We introduce the vector $Z = Y - \alpha X$, where $\alpha = (Y, X)/(X, X)$, and compute the square of its length. Remembering that

$$(Z, X) = (Y, X) - \alpha(X, X) = 0 ,$$

we have

$$|Z|^2 = (Z, Z) = (Z, Y - \alpha X) = (Z, Y) = (Y - (\alpha X, Y) = (Y, Y) - \alpha(X, Y)$$

$$= (Y, Y) - \frac{(Y, X)(X, Y)}{(X, X)} = \frac{|X|^2 \cdot |Y|^2 - |(X, Y)|^2}{|X|^2} .$$

Consequently $|X|^2 \cdot |Y|^2 - |(X, Y)|^2 = |X|^2 \cdot |Z|^2 \geq 0$, from which there follows directly the inequality (1).

For a pair of non-zero vectors $X$ and $Y$ in Euclidean space the concept of an angle arises naturally from the formula

$$\cos \varphi = \frac{(X, Y)}{|X| \cdot |Y|} . \tag{2}$$

This definition is always meaningful, since the number $(X, Y)/|X| \cdot |Y|$ in absolute value does not exceed one, on the basis of the Cauchy-Bunyakovskii inequality.

An important example of a linear space is the so-called *arithmetic* space. The vectors of this space are ordered sets of $n$ real (complex) numbers which are called components. The operations of addition and multiplication by a real (complex) number are defined in terms of components. These operations for vectors of an arithmetic space do not differ in any way from the same operations for rows (i.e. one-row matrices). Consequently all the formal laws for these operations, listed on p. 3 for arbitrary matrices, are also true for vectors, so that axioms 1–8 for a linear space turn out to be satisfied. The role of the null vector is played by the vector whose components are all equal to zero. The negative vector $- X$ for the vector is $(- 1)X$.

Axiom 9 is satisfied since every vector in an arithmetic space may be represented in the form

$$X = x_1 e_1 + \cdots x_n e_n ,$$

where $e_i$ is the vector whose $i$th component is equal to 1 and the rest of whose elements are equal to zero. Later we shall show that there does not exist a system of vectors which satisfies the condition of axiom 9 and which consists of less than $n$ vectors. An arithmetic space whose vectors are formed from $n$ numbers turns out to be an $n$-dimensional linear space.

A space consisting of vectors with real components is a real linear space; a space consisting of vectors with complex components is a complex space.

The scalar product is introduced by the formula

$$(X, Y) = x_1 \bar{y}_1 + \cdots + x_n \bar{y}_n ,$$

which in a real space assumes the simpler form

$$(X, Y) = x_1 y_1 + \cdots + x_n y_n .$$

It is easily verified that the scalar product satisfies axioms 10–13.

The length of a vector is equal to $\sqrt{|x_1|^2 + \cdots + |x_n|^2}$.

The scalar product of vectors in an arithmetic space may be expressed in terms of matrices. That is,

$$(\boldsymbol{X}, \boldsymbol{Y}) = x_1 \bar{y}_1 + \cdots + x_n \bar{y}_n$$

$$= (\bar{y}_1, \cdots, \bar{y}_n) \begin{bmatrix} x_1 \\ \cdot \\ \cdot \\ \cdot \\ x_n \end{bmatrix} = (x_1, \cdots, x_n) \begin{bmatrix} \bar{y}_1 \\ \cdot \\ \cdot \\ \cdot \\ \bar{y}_n \end{bmatrix},$$

where $(\bar{y}_1, \cdots, \bar{y}_n)$ is a row of numbers conjugate to the components of the vector $\boldsymbol{Y}$ and $(x_1, \cdots, x_n)$ is a row whose elements are equal to the components of the vector $\boldsymbol{X}$.

Other examples of linear finite-dimensional spaces are: a set of all the square matrices of a given order; the set of polynomials in one variable, each of whose degrees do not exceed a given number; the set of solutions of a linear homogeneous differential equation; etc.

In some of these spaces it is natural to introduce the scalar product. Thus in the space of polynomials of finite degree with real coefficients the scalar product of polynomials $f_1(t)$ and $f_2(t)$ may be defined as $\int_a^b f_1(t) f_2(t)\, dt$. It is easy to see that in this case axioms 10–13 will be satisfied.

In linear spaces which arise in connection with the study of applied problems the scalar product is usually introduced in some way so that it is closely connected with the specific features of elements of the space under consideration.

A linear space in which the operation of scalar product is not introduced is called an *affine* space; in general, those properties of spaces which are not connected with the concept of scalar product are called affine properties.

We begin our systematic study with a description of affine properties. In general we shall not specify what kind of space is being considered—real or complex—in view of the complete parallelism in formulating and proving the results.

## 4. BASES AND COORDINATES

**1. Linear dependence.** The vector $\boldsymbol{Y} = c_1 \boldsymbol{X}_1 + c_2 \boldsymbol{X}_2 + \cdots + c_m \boldsymbol{X}_m$ is called a *linear combination* of the vectors $\boldsymbol{X}_1, \boldsymbol{X}_2, \cdots, \boldsymbol{X}_m$.

It is easy to see that if the vectors $\boldsymbol{Y}_1, \boldsymbol{Y}_2, \cdots, \boldsymbol{Y}_k$ are linear combinations of the vectors $\boldsymbol{X}_1, \boldsymbol{X}_2 \cdots, \boldsymbol{X}_m$, then every linear combination $\gamma_1 \boldsymbol{Y}_1 + \gamma_2 \boldsymbol{Y}_2 + \cdots + \gamma_k \boldsymbol{Y}_k$ is also a linear combination of the vectors $\boldsymbol{X}_1, \boldsymbol{X}_2, \cdots, \boldsymbol{X}_m$.

The vectors $\boldsymbol{X}_1, \boldsymbol{X}_2, \cdots, \boldsymbol{X}_m$ are said to be *linearly dependent* if there exist numbers $c_1, c_2, \cdots, c_m$, not equal to zero simultaneously, such that the following equation holds:

$$c_1 \boldsymbol{X}_1 + c_2 \boldsymbol{X}_2 + \cdots + c_m \boldsymbol{X}_m = 0 \tag{1}$$

If this equality holds only when all the constants $c_1, c_2, \cdots, c_m$ are equal to zero, then the vectors $\boldsymbol{X}_1, \boldsymbol{X}_2, \cdots, \boldsymbol{X}_m$ are said to be *linearly independent*.

If the vectors $\boldsymbol{X}_1, \boldsymbol{X}_2, \cdots, \boldsymbol{X}_m$ are linearly dependent, then at least one of them is a linear combination of the others. In fact if $c_m \neq 0$, for example, then from (1) we get

$$X_m = -\frac{c_1}{c_m}X_1 - \frac{c_2}{c_m}X_2 - \cdots - \frac{c_{m-1}}{c_m}X_{m-1} \,.  \tag{2}$$

**THEOREM 4.1.** *If the vectors $Y_1, Y_2, \cdots, Y_k$ are linear combinations of the vectors $X_1, X_2, \cdots, X_m$ and $k > m$, then they are linearly dependent.*

We shall carry out the proof using mathematical induction.

For $m = 1$ the theorem is obvious. We assume that the theorem is true on the assumption that the number of vectors being combined is equal to $m - 1$ and under this assumption we shall prove it for a combination of $m$ vectors. Let

$$Y_1 = c_{11}X_1 + \cdots + c_{1m}X_m$$
$$\cdots \cdots \cdots \cdots \cdots$$
$$Y_k = c_{k1}X_1 + \cdots + c_{km}X_m \,.$$

The cases may be considered:

1. *All the coefficients $c_{11}, c_{21}, \cdots, c_{k1}$ are equal to zero.* Then $Y_1, Y_2, \cdots,$ $Y_m$ are in fact linear combinations of only $m - 1$ vectors $X_2, \cdots, X_m$. In view of the induction hypothesis $Y_1, \cdots, Y_k$ will be linearly dependent.

2. *At least one coefficient of $X_1$ is different from zero.* Without loss of generality we may assume that $c_{11} \neq 0$.

Consider the system of vectors

$$Y_2' = Y_2 - \frac{c_{21}}{c_{11}}Y_1 \,,$$
$$\cdots \cdots \cdots \cdots \cdots$$
$$Y_k' = Y_k - \frac{c_{k1}}{c_{11}}Y_1 \,.$$

These vectors are obviously linear combinations of the vectors $X_2, \cdots,$ $X_m$ and the number of them, $k - 1$, is more than $m - 1$. In view of the induction hypothesis they are linearly dependent, that is, one may find numbers $\gamma_2, \cdots, \gamma_k$, not simultaneously equal to zero, such that

$$\gamma_2 Y_2' + \cdots + \gamma_k Y_k' = 0 \,.$$

Substituting for $Y_2', \cdots, Y_k'$ in terms of $Y_1, \cdots, Y_k$ we get

$$\gamma_1 Y_1 + \gamma_2 Y_2 + \cdots + \gamma_k Y_k = 0 \,,$$

where $\gamma_1 = -(c_{21}/c_{11})\gamma_2 - \cdots - (c_{k1}/c_{11})\gamma_k$. The numbers $\gamma_1, \gamma_2, \cdots, \gamma_k$ are not equal to zero simultaneously and consequently $Y_1, Y_2, \cdots, Y_k$ are linearly dependent.

**2. Basis of a space.** A system of linearly independent vectors is called a *basis* of a space if every vector of the space is a linear combination of this system.

For example, in an arithmetic space the vectors $e_1, \cdots, e_n$ form a basis. They are linearly independent, since the components of a vector $c_1 e_1 + \cdots + c_n e_n$ are $c_1, \cdots, c_n$, and from the equality $c_1 e_1 + \cdots + c_n e_n = \mathbf{0}$ it follows that $c_1 = c_2 = \cdots = c_n = 0$. Moreover, for every vector $X$ with components $x_1, \cdots, x_n$ we have

$$X = x_1 e_1 + \cdots + x_n e_n .$$

We shall call the basis $e_1, \cdots, e_n$ a *natural basis* for a arithmetic space.

We shall now show that in a general, linear, finite-dimensional space there always exists a basis. Let $n$ be the dimension of the space. In view of the definition of dimension for a space there exists a system of $n$ vectors $U_1, \cdots, U_n$ such that all vectors of the space are linear combinations of them and there does not exist a system with a smaller number of vectors which possesses the same property.

We shall show that the vectors $U_1, \cdots, U_n$ form a basis of the space; for this it is sufficient to establish their linear independence. But this is almost obvious. In fact, if one lets the vectors $U_1, \cdots, U_n$ be linearly dependent, then at least one of them, for example $U_n$, will be a linear combination of the remaining $U_1, \cdots, U_{n-1}$. Then all the vectors of the space will turn out to be linear combinations of the vectors $U_1, \cdots, U_{n-1}$, which contradicts the fact that $n$ is the dimension of the space.

As we shall see below, the basis of a space is not unique and the choice of a basis is quite arbitrary.

THEOREM 4.2. *In n-dimensional space there do not exist more than n linearly independent vectors.*

*Proof.* In $n$-dimensional space there will exist a basis consisting of $n$ vectors $U_1, \cdots, U_n$. Every system of $m > n$ vectors will consist of linearly dependent vectors, on the basis of theorem 4.1.

COROLLARY. *The number of vectors in a basis of a space does not depend on the choice of a basis and coincides with the dimension of the space.*

In fact, the basis of an $n$-dimensional space may not consist of less than $n$ vectors, by the definition of dimension, nor of more than $n$ vectors, in view of theorem 4.2.

From theorem 4.2 it follows that the dimension of an arithmetic space which contains $n$-component vectors is actually equal to $n$, since there exists in this space a natural basis $e_1, \cdots, e_n$ consisting of exactly $n$ vectors.

The following theorem shows that the choice of a basis is arbitrary.

THEOREM 4.3. *Any system of n linearly independent vectors forms a basis for an n-dimensional space.*

*Proof.* Let $U_1, \cdots, U_n$ be a system of $n$ linearly independent vectors and let $X$ be any vector of the space. From theorem 4.1 the vectors $U_1, \cdots,$ $U_n$, $X$ will be linearly dependent, since each of them is a linear combination of basis vectors, the number of which is equal to $n$. Consequently one may find the numbers $c_0, c_1, \cdots, c_n$, not equal to zero simultaneously, such that $c_0 X + c_1 U_1 + \cdots + c_n U_n = 0$. In this case, $c_0 \neq 0$, since if $c_0 = 0$, then $c_1 U_1 + \cdots + c_n U_n = 0$, which contradicts the linear indepence of the vectors $U_1, \cdots, U_n$. Consequently,

$$X = -\frac{c_1}{c_0} U_1 - \cdots - \frac{c_n}{c_0} U_n .$$

Thus we have proved that any vector of a space is a linear combination of the vectors $U_1, \cdots, U_n$, from which it follows that the vectors $U_1, \cdots,$ $U_n$ form a basis.

This theorem permits us to use the following construction for setting up a basis. We take the vectors $U_1$ arbitrarily, except that it is different from zero. We take the vector $U_2$ arbitrarily, but not equal to a linear combination of the vector $U_1$ (such a vector may be found for $n > 1$). Moreover, we take for the vector $U_3$ an arbitrary vector which is not a linear combination of the first two, and so forth. In view of the definition of dimension this construction permits us to set up a system of $n$ vectors $U_1, U_2, \cdots, U_n$ which will be linearly independent on the basis of this construction. It also follows from this construction that one may extend any system of linearly independent vectors to a basis of a space.

**3. Vector coordinates** Let $U_1, \cdots, U_n$ be any basis of a space. Then every vector $X$ is a linear combination of the vectors $U_1, \cdots, U_n$, namely,

$$X = x_1 U_1 + \cdots + x_n U_n .  \tag{3}$$

The coefficients in this expansion are simply determined by the vector $X$, since if

$$X = x_1 U_1 + \cdots + x_n U_n = x_1' U_1 + \cdots + x_n' U_n ,$$

then

$$(x_1 - x_1') U_1 + \cdots + (x_n - x_n') U_n = 0 ,$$

and consequently

$$x_1 - x_1' = \cdots = x_n - x_n' = 0$$

because of the linear independence of the vectors $U_1, \cdots, U_n$.

The coefficients $x_1, \cdots, x_n$ are called the coordinates of the vector $X$ in the basis $U_1 \cdots, U_n$.

In arithmetic space the components of the vector $x_1, \cdots, x_n$ are obviously its coordinates in a natural basis.

The introduction of coordinates makes it possible to associate with each vector $X$ of a general, linear, $n$-dimensional space a column $X = (x_1, \cdots,$ $x_n)'$ of its coordinates in the selected basis $U_1, \cdots, U_n$.

Every column $X = (x_1, \cdots, x_n)'$ turns out to correspond to some vector, namely, the vector $X = x_i U_n$. If to a vector $X$ there corresponds a column $X$, then to the vector $cX$ will correspond the column $cX$; if to the vectors $X$ and $Y$ there correspond the columns $X$ and $Y$, then to the vector $X + Y$ will correspond the column $X + Y$. This correspondence is obviously one-to-one.

Thus each choice of a basis determines a representation of the vectors of an $n$-dimensional space as columns containing their coordinates. Each such representation is one-to-one. The product of a number and a vector is represented as the product of the same number and a column which represents the vector. A sum of vectors is represented as the sum to columns. In other words, these representations are *isomorphic*; operations on vectors correspond to the similar operations on their representative columns.

For an arithmetic space, in particular, a natural basis generates a representation of vectors as columns of their components. Such a representation will be called natural. However other choices of a basis give other representations of the vectors for an arithmetic space as columns.

From the preceding argument it is clear that a general linear space of dimension $n$ is isomorphic to an arithmetic space of the same dimension.

In computational problems of linear algebra sets of unknowns and sets of numbers being introduced into the original data which are being determined should be combined in columns. It is advisable to consider them as vectors in an arithmetic space in their natural representation; in solving problems one should then pass on to other representations connected in some way with the specific problem.

**4. Transformation of coordinates.** We shall consider how the coordinates of a vector change when the basis is changed. Let $U_1, U_2, \cdots, U_n$ and $U'_1, U'_2, \cdots, U'_n$ be two bases and let

$$
\begin{aligned}
U'_1 &= a_{11} U_1 + a_{21} U_2 + \cdots + a_{n1} U_n , \\
U'_2 &= a_{12} U_1 + a_{22} U_2 + \cdots + a_{n2} U_n , \\
&\ \cdot \cdot \cdot \cdot \cdot \cdot \cdot \cdot \cdot \cdot \cdot \cdot \cdot \cdot \cdot \cdot , \\
U'_n &= a_{1n} U_1 + a_{2n} U_2 + \cdots + a_{nn} U_n .
\end{aligned}
\tag{4}
$$

We associate with a transformation of coordinates a matrix whose columns consist of the coordinates of the vectors $U'_1, U'_2, \cdots, U'_n$ in the basis $U_1, U_2, \cdots, U_n$, that is, the matrix

$$
A = \begin{bmatrix}
a_{11} & a_{12} & \cdots & a_{1n} \\
a_{21} & a_{22} & \cdots & a_{2n} \\
\cdot & \cdot & \cdot & \cdot \\
a_{n1} & a_{n2} & \cdots & a_{nn}
\end{bmatrix}
\tag{5}
$$

Matrix $A$ is non-singular, since it has an inverse matrix, by means of which the vectors $U_1, U_2, \cdots, U_n$ are expressed as the vectors $U'_1, U'_2, \cdots, U'_n$.

We shall denote by $x_1, x_2, \cdots, x_n$ the coordinates of the vector $X$ in the basis $U_1, U_2, \cdots, U_n$, by $x'_1, x'_2, \cdots, x'_n$ the coordinates in the basis $U'_1, U'_2, \cdots, U'_n$. We shall find the dependence between the old and new coordinates. We have

$$
\begin{aligned}
X &= x_1 U_1 + x_2 U_2 + \cdots + x_n U_n \\
&= x'_1 U'_1 + x'_2 U'_2 + \cdots + x'_n U'_n \\
&= x'_1(a_{11} U_1 + a_{21} U_2 + \cdots + a_{n1} U_n) \\
&\quad + x'_2(a_{12} U_1 + a_{22} U_2 + \cdots + a_{n2} U_n) \\
&\quad + \cdots \cdots \cdots \cdots \cdots \\
&\quad + x'_n(a_{1n} U_1 + a_{2n} U_2 + \cdots + a_{nn} U_n) \\
&= (a_{11}x'_1 + a_{12}x'_2 + \cdots + a_{1n}x'_n) U_1 \\
&\quad + (a_{21}x'_1 + a_{22}x'_2 + \cdots + a_{2n}x'_n) U_2 \\
&\quad + \cdots \cdots \cdots \cdots \cdots \\
&\quad + (a_{n1}x'_1 + a_{n2}x'_2 + \cdots + a_{nn}x'_n) U_n \,.
\end{aligned}
$$

From here, in view of the linear independence of the vectors $U_1, U_2, \cdots, U_n$,

$$
\begin{aligned}
x_1 &= a_{11}x'_1 + a_{12}x'_2 + \cdots + a_{1n}x'_n \,, \\
x_2 &= a_{21}x'_1 + a_{22}x'_2 + \cdots + a_{2n}x'_n \,, \\
&\cdots \cdots \cdots \cdots \cdots \cdots \cdots \\
x_n &= a_{n1}x'_1 + a_{n2}x'_2 + \cdots + a_{nn}x'_n .
\end{aligned} \tag{6}
$$

The last equalities may be written in the matrix form

$$X = AX',$$

where

$$
X = \begin{bmatrix} x_1 \\ x_2 \\ \cdot \\ \cdot \\ \cdot \\ x_n \end{bmatrix} \quad \text{and} \quad X' = \begin{bmatrix} x'_1 \\ x'_2 \\ \cdot \\ \cdot \\ \cdot \\ x'_n \end{bmatrix}
$$

are columns which contain the coordinates of the vector $X$ in the bases $U_1, U_2, \cdots, U_n$ and $U'_1, U'_2, \cdots, U'_n$.

## 5. SUBSPACES

1. **Definition of subspace, dimension, basis.** A *subspace* of a space $R$ is a set of vectors $X \in R$ such that any linear combination of vectors in the set is again a vector in the set. It is obvious that the set consisting of the null vector and also the whole space are subspaces in the sense of this definition. We shall call them *trivial subspaces*.

It is clear that a collection of vectors which forms a subspace satisfies axioms 1–8 for linear space so that a subspace, considered by itself, is a linear space. It is not hard to convince oneself that a subspace of an $n$-dimensional space will be finite-dimensional and its dimension will not exceed the number $n$. In fact, in a subspace (as in a whole space) there may exist no more than $n$ linearly independent vectors.

Each subspace obviously has its own basis, while the number of vectors of the basis (the dimension of the subspace) is not greater than $n$.

If the dimension of a subspace is equal to $n$, then it coincides with the whole space. In fact, the basis $U_1, \cdots, U_n$ of a subspace consists of linearly independent vectors equal in number to the dimension of the whole space; consequently it is also a basis of the whole space.

THEOREM 5.1. *Any basis* $U_1, \cdots, U_m$ *of a subspace may be extended to a basis of the whole space.*

*Proof.* One may take for the first $m$ basis vectors the vectors $U_1, \cdots, U_m$, since they are linearly independent and, as we saw above, any system of linearly independent vectors may be extended to a basis of the whole space.

**2. Subspaces spanned by a given system of vectors.** Given a set of linearly independent or even linearly dependent vectors $X_1, \cdots, X_m$, then it is obvious that the set of all possible linear combinations of them forms a subspace. Such a subspace is called the subspace *spanned* by the system of vectors $X_1, \cdots, X_m$.

THEOREM 5.2. *The dimension of subspace spanned by a given system of vectors* $X_1, \cdots, X_m$ *is equal to the rank of the matrix which contains the coordinates of these vectors with respect to any basis.*

*Proof.* Let

$$\Xi = \begin{bmatrix} x_{11} & \cdots & x_{1m} \\ x_{21} & \cdots & x_{2m} \\ \cdot & \cdot \cdot \cdot \cdot & \cdot \\ x_{n1} & \cdots & x_{nm} \end{bmatrix}$$

be a matrix whose columns are, respectively, the coordinates of the given vectors $X_1, \cdots, X_m$ with respect to some basis. Let the rank of this matrix be equal to $r$. Then, from the definition of rank there exists a non-zero minor of order $r$ of matrix $A$ and all minors of order $r + 1$ or higher are either equal to zero or may not be formed. Without loss of generality (if necessary one may change the numbering of the given vectors $X_1, \cdots, X_m$ and of the basis vectors, which is equivalent to changing the numbering of the columns and rows of matrix $\Xi$) one may assume that the non-zero minor is

$$\delta = \begin{bmatrix} x_{11}, & \cdots, & x_{1r} \\ \cdot & \cdot \cdot \cdot \cdot & \cdot \\ x_{r1}, & \cdots, & x_{rr} \end{bmatrix}.$$

We shall prove that the vectors $X_1, \cdots, X_r$ form a basis of the subspace $P$ spanned by the vectors $X_1, \cdots, X_m$. First of all we establish the linear independence of the vectors $X_1, \cdots, X_r$. Let

$$c_1 X_1 + \cdots + c_r X_r = 0 \, .$$

Writing this equality in coordinates we get

$$
\begin{aligned}
c_1 x_{11} &+ \cdots + c_r x_{1r} = 0 \, , \\
&\cdots\cdots\cdots\cdots\cdots \\
c_1 x_{r1} &+ \cdots + c_r x_{rr} = 0 \, , \\
c_1 x_{r+11} &+ \cdots + c_r x_{r+1r} = 0 \, , \\
&\cdots\cdots\cdots\cdots\cdots \\
c_1 x_{n1} &+ \cdots + c_r x_{nr} = 0 \, ,
\end{aligned}
$$

The first $r$ equalities represent a system of $r$ linear, homogeneous equations with respect to $c_1, \cdots, c_r$, the determinant of whose coefficients is equal to $\delta \neq 0$. Consequently this system has a unique solution, namely $c_1 = \cdots = c_r = 0$. The linear independence of the vectors $X_1, \cdots, X_r$ is established.

We shall now show that all the given vectors $X_1, \cdots, X_r, X_{r+1}, \cdots, X_m$ are linear combinations of $X_1, \cdots, X_r$. For the vectors $X_1, \cdots, X_r$ this is trivial, so that it is necessary to consider only the vectors $X_s, s = r + 1, \cdots, m$.

We consider the determinant

$$
\Delta_s =
\begin{vmatrix}
x_{11} & \cdots & x_{1r} & x_{1s} \\
\cdot & \cdot\ \cdot\ \cdot & \cdot & \cdot \\
x_{r1} & \cdots & x_{rr} & x_{rs} \\
z_1 & \cdots & z_r & z
\end{vmatrix} ,
$$

where $z_1, \cdots, z_r, z$ are certain numbers whose values are immaterial for us. We shall denote by $M_1, M_2, \cdots, M_r, M$ the algebraic cofactors of the elements of the last row in the determinant $\Delta_s$.

We shall consider the vector $Y = M_1 X_1 + \cdots + M_r X_r + M X_s$. Its coordinates will be

$$
\begin{aligned}
y_1 &= M_1 x_{11} + \cdots + M_r x_{1r} + M x_{1s} \, , \\
&\cdots\cdots\cdots\cdots\cdots\cdots\cdots \\
y_r &= M_1 x_{r1} + \cdots + M_r x_{rr} + M x_{rs} \, , \\
y_{r+1} &= M_1 x_{r+11} + \cdots + M_r x_{r+1r} + M x_{r+1s} \, , \\
&\cdots\cdots\cdots\cdots\cdots\cdots\cdots \\
y_n &= M_1 x_{n1} + \cdots + M_r x_{nr} + M x_{ns} \, .
\end{aligned}
$$

The first $r$ coordinates $y_1, \cdots, y_r$ are equal to zero, since they represent the sums of the products of the algebraic cofactors of elements of the last row of the determinant $\Delta_s$ and the corresponding elements of the other rows. The remaining coordinates $y_{r+1}, \cdots, y_n$ are also equal to zero. In fact,

$$
y_{r+1} =
\begin{vmatrix}
x_{11} & \cdots & x_{1r} & x_{1s} \\
\cdot & \cdot\ \cdot\ \cdot & \cdot & \cdot \\
x_{r1} & \cdots & x_{rr} & x_{rs} \\
x_{r+11} & \cdots & x_{r+1r} & x_{r+1s}
\end{vmatrix} ,
$$

$$
\cdots\cdots\cdots\cdots\cdots\cdots\cdots ,
$$

$$
y =
\begin{vmatrix}
x_{11} & \cdots & x_{1r} & x_{1s} \\
\cdot & \cdot\ \cdot\ \cdot & \cdot & \cdot \\
x_{r1} & \cdots & x_{rr} & x_{rs} \\
x_{n1} & \cdots & x_{nr} & x_{ns}
\end{vmatrix}
$$

and they are equal to zero as minors of the $(r + 1)$th order contained in the matrix $\mathit{\Xi}$ of rank $r$. Consequently $M_1 X_1 + \cdots + M_r X_r + M X_s = 0$. Since $M = \delta \neq 0$, then

$$X_s = -\frac{M_1}{M} X_1 - \cdots - \frac{M_r}{M} X_r .$$

Thus we have proved that the vectors $X_1, \cdots, X_r$ are linearly independent and that all vectors $X_1, \cdots, X_m$ are linear combinations of them. Consequently the vectors $X_1, \cdots, X_r$ form a basis of the subspace $P$, since any linear combination of the vectors $X_1, \cdots, X_m$ is also a linear combination of the vectors $X_1, \cdots, X_r$. Thus it has been proved that the dimension of the subspace is equal to $r$.

In terms of the theory of matrices theorem 5.2 may be reformulated in the following manner: *the maximum number of linearly independent columns of a matrix as well as the maximum number of linearly independent rows coincides with the rank of the matrix.*

3. **Relative bases.** Let $P$ be a subspace of dimension $m$ in an $n$-dimension space $R$. The vectors $V_1, \cdots, V_k$ are called *linearly independent relative to $P$* if no linear combination of them (except the null one) belongs to $P$, in other words, if from $c_1 V_1 + \cdots + c_k V_k \in P$ it follows that $c_1 = \cdots = c_k = 0$.

The system of vectors $V_1, \cdots, V_k$ is said to be a *basis of $R$ relative to $P$* if the vectors $V_1, \cdots, V_k$ are linearly independent relative to $P$ and if every vector of $R$ is represented as the sum of a certain vector of $P$ and the linear combination of vectors $V_1, \cdots, V_k$.

THEOREM 5.3. *Let $U_1, \cdots, U_m$ be a basis of the subspace $P$. The vectors $V_1, \cdots, V_k$ are linearly independent relative to the subspace $P$ if and only if the vectors $U_1, \cdots, U_m, V_1, \cdots, V_k$ are linearly independent.*

*Proof.* Let $V_1, \cdots, V_k$ be linearly independent relative to $P$ and let $c_1 V_1 + \cdots + c_k V_k + d_1 U_1 + \cdots + d_m U_m = 0$. Then $c_1 V_1 + \cdots + c_k V_k = -d_1 U_1 - \cdots - d_m U_m \in P$. Consequently $c_1 = \cdots = c_k = 0$ from the definition of linear independence relative to $P$. Therefore $d_1 U_1 + \cdots + d_m U_m = 0$, from which $d_1 = \cdots = d_m = 0$. Thus the vectors $V_1, \cdots, V_k, U_1, \cdots, U_m$ are linearly independent, so that the necessity of the condition formulated has been proved.

We shall now assume that the vectors $V_1, \cdots, V_k, U_1, \cdots, U_m$ are linearly independent. Let $Y = c_1 V_1 + \cdots + c_k V_k \in P$. Then $Y = d_1 U_1 + \cdots + d_m U_m$, from which $c_1 V_1 + \cdots + c_k V_k - d_1 U_1 - \cdots - d_m U_m = 0$. In view of the linear independence of the vectors $V_1, \cdots, V_k, U_1, \cdots, U_m$ all the coefficients in the last equality are equal to zero. In particular, $c_1 = \cdots = c_k = 0$, so that $V_1, \cdots, V_k$ are linearly independent relative to $P$. Thus the sufficiency of the condition is also proved.

THEOREM 5.4. *In order that the vectors $V_1, \cdots, V_k$ form a basis of the space $R$ relative to the subspace $P$ it is necessary and sufficient that the vectors*

$V_1, \cdots, V_k, U_1, \cdots, U_m$ form a basis of the space $R$. Here the vectors $U_1$, $\cdots, U_m$ form a basis of $P$.

*Proof.* Let $V_1, \cdots, V_k$ be a basis of $R$ relative to $P$. Then in view of theorem 5.3 the vectors $V_1, \cdots, V_k, U_1, \cdots, U_m$ are linearly independent. Furthermore, for any vector $X$ in $R$ we have

$$X = c_1 V_1 + \cdots + c_k V_k + Y,$$

where $Y \in P$, and therefore

$$X = c_1 V_1 + \cdots + c_k V_k + d_1 U_1 + \cdots + d_m U_m.$$

The necessity has been proved.

Let now the vectors $V_1, \cdots, V_k, U_1, \cdots, U_m$ form a basis of $R$. Then from theorem 5.3 the vectors $V_1, \cdots, V_k$ will be linearly independent relative to $P$. Furthermore, for any vector $X \in R$ we have $X = c_1 V_1 + \cdots + c_k V_k + d_1 U_1 + \cdots + d_m U_m = c_1 V_1 + \cdots + c_k V_k + Y$, where $Y \in P$. The sufficiency has been proved.

COROLLARY 1. *A relative basis always exists and the number of vectors forming it is equal to the difference of the dimensions of $R$ and $P$.*

In fact, as we saw before, any basis $U_1, \cdots, U_m$ of a subspace $P$ may be extended to a basis of the space $R$. The set of extended vectors $V_1, \cdots, V_k$ is a basis of $R$ relative to $P$ and the the number of them is equal to difference between the dimensions of $R$ and $P$.

COROLLARY 2. *Let $P_k \supset P_{k-1} \supset \cdots \supset P_1$ be a decreasing sequence of subspaces. Then the union of the basis of $P_1$, the basis of $P_2$ relative to $P_1$, $\cdots$, the basis of $P_k$ relative to $P_{k-1}$ forms a basis of $P_k$.*

THEOREM 5.5. *Any system of vectors $V_1, \cdots, V_s$, linearly independent relative to $P$, may be extended to a basis of $R$ relative to $P$.*

*Proof.* The vectors $U_1, \cdots, U_m, V_1, \cdots, V_s$ are linearly independent in view of theorem 5.3. This system may be extended to the system $U_1, \cdots, U_m, V_1, \cdots, V_s, V_{s+1}, \cdots, V_k$, forming a basis of $R$. Then the vectors $V_1, \cdots, V_k$ also form a basis of $R$ relative to $P$ on the basis of theorem 5.4.

**4. Vector sum and the intersection of subspaces.** Let $P$ and $Q$ be two subspaces of the space $R$. The *vector sum* of the subspaces $P$ and $Q$ is the set of all vectors $Z = X + Y$, where $X \in P$, $Y \in Q$. It is obvious that a vector sum of two subspaces is again a subspace. It may be characterized as the smallest subspace containing the subspaces $P$ and $Q$. We shall denote the vector sum of two subspaces $P$ and $Q$ by $(P, Q)$.

The *intersection* of two subspaces $P$ and $Q$ is the set of all vectors belonging both to subspace $P$ and to subspace $Q$. It is clear that the intersection of two subspaces is again a subspace. It may be characterized as the largest subspace contained in $P$ and $Q$. The intersection of subspaces $P$ and $Q$ is denoted by $P \cap Q$.

THEOREM 5.6. *Let* $s$ *be the dimension of* $(P, Q)$ *and* $t$ *be the dimension of* $P \cap Q$. *Then* $s + t = p + q$, *where* $p$ *is the dimension of* $P$ *and* $q$ *is the dimension of* $Q$.

*Proof.* It is clear that $t \le p \le s, t \le q \le s$. Let $U_1, \cdots, U_t$ be a basis of $P \cap Q$. We include it in the bases $U_1, \cdots, U_t, V_1, \cdots, V_{p-t}$, and $U_1, \cdots, U_t, W_1, \cdots, W_{q-t}$ of subspaces $P$ and $Q$. We shall prove that the vectors $U_1, \cdots, U_t, V_1, \cdots, V_{p-t}, W_1, \cdots, W_{q-t}$ form a basis of $(P, Q)$. We first establish their linear independence. Let

$$c_1 U_1 + \cdots + c_t U_t + d_1 V_1 + \cdots + d_{p-t} V_{p-t} + d'_1 W_1 + \cdots + d'_{q-t} W_{q-t} = 0 . \quad (1)$$

We assume

$$Z = c_1 U_1 + \cdots + c_t U_t + d_1 V_1 + \cdots + d_{p-t} V_{p-t} .$$

It is clear that $Z \in P$. On the other hand, we conclude from equality (1) that

$$Z = - d'_1 W_1 - \cdots - d'_{q-t} W_{q-t} ,$$

from which $Z \in Q$. Consequently $Z \in P \cap Q$ and therefore

$$Z = \nu_1 U_1 + \cdots + \nu_t U_t$$

for some $\nu_1, \cdots, \nu_t$. Setting the second and third representations of the vector $Z$ equal to each other we obtain

$$\nu_1 U_1 + \cdots + \nu_t U_t + d'_1 W_1 + \cdots + d'_{q-t} W_{q-t} = 0 ,$$

from which we conclude that $\nu_1 = \cdots = \nu_t = 0$, $d'_1 = \cdots = d'_{q-t} = 0$ in view of the linear independence of the vectors $U_1, \cdots, U_t, W_1, \cdots, W_{q-t}$. Equation (1) is then

$$c_1 U_1 + \cdots + c_t U_t + d_1 V_1 + \cdots + d_{p-t} V_{p-t} = 0 ,$$

from which $c_1 = \cdots = c_t = 0$, $d_1 = \cdots = d_{p-t} = 0$.

Thus all coefficients in equality (1) turn out to be zero and consequently the vectors $U_1, \cdots, U_t, V_1, V_{p-t}, \cdots, W_1, \cdots, W_{q-t}$ are linearly independent. It remains to prove that any vector in $(P, Q)$ is a linear combination of them. Let $Z \in (P, Q)$. Then $Z = X + Y$ where $X \in P, Y \in Q$. Writing $X$ and $Y$ in terms of basis vectors of the subspaces $P$ and $Q$ we get

$$X = c_1 U, + \cdots + c_t U_t + d_1 V, , + \cdots + d_{p-t} V_{p-t} ,$$
$$Y = c'_1 U, , + \cdots + c'_t U_t + d'_1 W_1, + \cdots + d'_{q-t} W_{q-t} ,$$

from which

$$Z = (c_1 + c'_1) U_1 + \cdots + (c_t + c'_t) U_t + d_1 V_1,$$
$$+ \cdots + d_{p-t} V_{p-t} + d'_1 W_1 + \cdots + d'_{q-t} W_{q-t} .$$

Thus we have proved that the vectors $U_1, \cdots, U_t, V_1, \cdots, V_{p-t}, W_1, \cdots, W_{q-t}$ form a basis of the subspace $(P, Q)$. Thus the dimension $s$ of the subspace $(P, Q)$ is equal to $t + p - t + q - t = p + q - t$, from which it follows that

$$s + t = p + q .$$

**5. Direct sums.** If every vector $X$ of a space $R$ may be represented as a sum of vectors $Y, , \cdots, Y_k$ of subspaces $P_1, \cdots, P_k$, then it is said that $R$ is the *vector sum* of the subspaces $P_1, \cdots, P_k$. If the representation

$$X = Y_1, + \cdots + Y_k, \ Y_i \in P_i \ (i = 1, \cdots, k)$$

is unique, then $R$ is the *direct sum* of subspaces $P_1, \cdots, P_k$.

THEOREM 5.7.   *For the space $R$ to be a direct sum of its subspaces $P_1$, $\cdots, P_k$ it is necessary and sufficient that the union of the bases of the subspaces form a basis of the whole space.*

*Proof.* Let $R$ be a direct sum of the subspaces $P_1, \cdots, P_k$ and let the vectors $U_1, \cdots, U_{s_1}; \cdots; U_{s_{k-1}+1}, \cdots, U_{s_k}$ form the bases of these subspaces. Then for any vector in $R$ we have

$$X = Y_1 + \cdots + Y_k ,$$

where $Y_i \in P_i$, and therefore

$$X = c_1 U_1 + \cdots + c_{s_1} U_{s_1} + \cdots + c_{s_{k-1}+1} U_{s_{k-1}+1} + \cdots + c_{s_k} U_{s_k} .$$

It remains to prove the linear independence of the vectors $U_1, \cdots, U_{s_k}$. Let

$$c_1 U_1 + \cdots + c_{s_1} U_{s_1} + \cdots + c_{s_{k-1}+1} U_{s_{k-1}+1} + \cdots + c_{s_k} U_{s_k} = 0 .$$

We introduce the notation:

$$c_1 U_1 + \cdots + c_{s_1} U_{s_1} = Y_1 ,$$
$$\cdots \cdots \cdots \cdots \cdots ,$$
$$c_{s_{k-1}+1} U_{s_{k-1}+1} + \cdots + c_{s_k} U_{s_k} = Y_k .$$

Then $Y_i \in P_i$ and $0 = Y_1 + \cdots + Y_k$. But all the subspaces $P_i$ contain the null vector and $0 = 0 + \cdots + 0$. In view of the natural decomposition of the vectors of $R$ according to the subspaces $P_1, \cdots, P_k$, we conclude that $Y_1 = \cdots = Y_k = 0$. Consequently all the coefficients $c_1, \cdots, c_{s_1}, \cdots, c_{s_{k-1}+1}$, $\cdots, c_{s_k}$ are equal to zero. The linear independence of the vectors $U_1, \cdots$, $U_{s_k}$ has been proved. Thus the necessity of the condition has been proved.

We now assume that the vectors $U_1, \cdots, U_{s_1}; \cdots; U_{s_{k-1}+1}, \cdots, U_{s_k}$ which form the bases of subspaces $P_1, \cdots, P_k$ form a basis of $R$. Then for any vector $X \in R$ we have

$$X = c_1 U_1 + \cdots + c_{s_1} U_{s_1} + \cdots + c_{s_{k-1}+1} U_{s_{k-1}+1} + \cdots + c_{s_k} U_{s_k} = Y_1 + \cdots + Y_k$$

where

$$Y_1 = c_1 U_1 + \cdots + c_{s_1} U_{s_1} \in P_1 ,$$
$$\cdots \cdots \cdots \cdots \cdots \cdots \cdots ,$$
$$Y_k = c_{s_{k-1}+1} U_{s_{k-1}+1} + \cdots + c_{s_k} U_{s_k} \in P_k .$$

This representation will be unique, since if

$$X = Y_1' + \cdots + Y_k'$$

for $Y_i' \in P_i$, then

$$X = c_1' U_1 + \cdots + c_{s_1}' U_{s_1} + \cdots + c_{s_{k-1}+1}' U_{s_{k-1}+1} + \cdots + c_{s_k}' U_{s_k} ,$$

where

$$Y_1' = c_1' U_1 + \cdots + c_1' U_{s_1} ,$$
$$\cdots \cdots \cdots \cdots \cdots ,$$
$$Y_k' = c_{s_{k-1}+1}' U_{s_{k-1}+1} + \cdots + c_{s_k}' U_{s_k} .$$

In view of the natural decomposition of the vector $X$ according to the basis vectors, we conclude that $c_1 = c_1', \cdots, c_{s_1} = c_{s_1}', \cdots, c_{s_{k-1}+1} = c_{s_{k-1}+1}', \cdots, c_{s_k} = c_{s_k}'$ and therefore $Y_1 = Y_1', \cdots, Y_k = Y_k'$. The sufficiency part of the theorem is also proved.

THEOREM 5.8. *If a space $R$ is the vector sum of the subspaces $P_1, \cdots, P_k$ and the dimension of $R$ is equal to the sum of the dimensions of $P_1, \cdots, P_k$, then $R$ is the direct sum of $P_1, \cdots, P_k$.*

*Proof.* The vectors of the space $R$ are linear combinations of the basis vectors of all the subspaces $P_1, \cdots, P_k$. Consequently the dimension of $R$ will not exceed the sum of the dimensions of the subspaces $P_1, \cdots, P_k$; it will be equal to this sum only if the set of all basis vectors of all $P_i$ is linearly independent. But in this case, in view of theorem 5.7, the space $R$ is a direct sum of the subspaces $P_1, \cdots, P_k$.

From the last theorem it follows, in particular, that a vector sum of two subspaces will be a direct sum if and only if the intersection of these subspaces has dimension 0, that is, contains only the null vector. This last statement is easily and directly proved.

## 6. LINEAR OPERATORS

1. **Function of a vector argument.** The function of a vector argument with a range of values $\Omega$ is a law of correspondence for each vector of the space $R$ (or of one of its subsets) with an element of $\Omega$.

If the range of values $\Omega$ is a collection of numbers, then the function of a vector argument is called a *functional*; if the range of values $\Omega$ is a collection of vectors of the same space, then the function of a vector argument is called a *transformation* or an *operator*.

The scalar product $(X, Y_0)$ for a fixed vector $Y_0$, the length of a vector $X$, and a quadratic form of coordinates of a vector in some basis may serve as examples of functionals. In general, if a basis is chosen for a space then any function of $n$ variables (namely of the $n$ coordinates of a variable vector in this given basis) will be a functional. It is obvious that, given a change of basis, a function of the coordinates of a vector which determines a functional must undergo a corresponding transformation of variables.

A functional $\Phi$ is said to be *linear* if

$$\Phi(c_1 X_1 + c_2 X_2) = c_1 \Phi(X_1) + c_2 \Phi(X_2) . \tag{1}$$

It is clear that a more general form of linear functional is

$$\Phi(X) = \Phi(x_1 U_1 + \cdots + x_n U_n) = a_1 x_1 + \cdots + a_n x_1 ,$$

where $a_1, \cdots, a_n$ are numbers, namely $a_1 = \Phi(U_1), \cdots, a_n = \Phi(U_n)$; here $U_1,$ $\cdots, U_n$ is the selected basis and $x_1, \cdots, x_n$ are the coordinates of the vector $X$ in this basis. Quadratic and certain other functionals will play an important role later on.

**2. Linear operators.** An operator is called *linear* if it satisfies the following conditions of linearity.

1. $A(\alpha X) = \alpha AX$ for any complex number $\alpha$.
2. $A(X_1 + X_2) = AX_1 + AX_2$.

Here $AX$ denotes the result of applying the operator $A$ to the vector $X$.

We shall define operations on linear operators. We shall call the *product AB* of linear operators $A$ and $B$ that operator $C$ which consists of applying first the linear operator $B$ and then linear operator $A$.

The product of linear operators $A$ and $B$, as it is easy to see, is again a linear operator.

In fact,

$$C(X_1 + X_2) = A(B(X_1 + X_2)) = A(BX_1 + BX_2)$$
$$= A(BX_1) + A(BX_2) = CX_1 + CX_2 .$$

It follows directly from the definition that multiplication for operators is associative.

The operator $E$ associating to each vector $X$ the same vector is called the *unit operator*. It is clear that the unit operator is linear and $EA = AE = A$ for any operator $A$.

The *sum $A + B$* of linear operators $A$ and $B$ will be an operator $C$ which associates to the vector $X$ a vector $AX + BX$.

The operator $0$ which maps all the vectors of a space onto the null vector is called the *null operator*. It is clear that the null operator is linear and $A + 0 = 0 + A = A$ for any operator.

The *product $\alpha A$* of a linear operator $A$ *by a number* $\alpha$ is an operator which associates with the vector $X$ the vector $\alpha(AX)$.

It is obvious that the sum of linear operators and also the product of an operator by a number are linear operators. The given definitions for operations permit us to define in a natural way the operator $A^k$ as the successive employment of $A, k$ times, and to define a polynomial operator according to the format

$$f(A) = a_0 A^n + a_1 A^{n-1} + \cdots + a_{n-1} A + a_n E ,$$

where

$$f(t) = a_0 t^n + a_1 t^{n-1} + \cdots + a_{n-1} t + a_n .$$

In the sequel we shall deal only with linear operators and therefore the qualifying word 'linear' will not be explicity mentioned.

**3. Representation of an operator by a matrix.** We select in a space $R$ some basis $U_1, U_2, \cdots, U_n$. The operator $A$ associates to these vectors of the basis the vectors $AU_1, AU_2, \cdots, AU_n$.

Let the vectors $AU_1, AU_2, \cdots, AU_n$ be given by their coordinates in the basis $U_1, U_2, \cdots, U_n$, that is, let

$$AU_1 = a_{11}U_1 + a_{21}U_2 + \cdots + a_{n1}U_n ,$$
$$AU_2 = a_{12}U_1 + a_{22}U_2 + \cdots + a_{n2}U_n ,$$
$$\cdots\cdots\cdots\cdots\cdots\cdots\cdots\cdots , \qquad (2)$$
$$AU_n = a_{1n}U_1 + a_{2n}U_2 + \cdots + a_{nn}U_n .$$

We consider the matrix $A$ whose columns consist of the coordinates of the vectors $AU_1, AU_2, \cdots, AU_n$;

$$A = \begin{bmatrix} a_{11} & a_{12} & \cdots & a_{1n} \\ a_{21} & a_{22} & \cdots & a_{2n} \\ \cdot & \cdot & \cdots & \cdot \\ a_{n1} & a_{n2} & \cdots & a_{nn} \end{bmatrix} . \qquad (3)$$

We shall show that matrix $A$ completely determines the operator $A$.[†]

If the matrix $A$ is known for the operator $A$, then the coefficient matrix for operator $A$ with respect to the basis vectors $U_1, U_2, \cdots, U_n$ is also known, and, in view of the linearity of the operator, it is easy to determine its effect on any vector. Namely, if $X = x_1U_1 + \cdots + x_nU_n$, then

$$AX = x_1AU_1 + \cdots + x_nAU_n .$$

From here the coordinates of the transformed vector are easily found. Namely,

$$Y = AX = \sum_{k=1}^{n} y_k U_k = \sum_{i=1}^{n} x_i AU_i = \sum_{i=1}^{n} \sum_{k=1}^{n} a_{ki}x_i U_k ,$$

from which

$$y_k = \sum_{i=1}^{n} a_{ki}x_i$$

or in matrix notation,

$$Y = AX \qquad (4)$$

where $Y$ and $X$ are column matrices whose elements are the coordinates of the vectors $Y$ and $X$.

Conversely, an arbitrary matrix $A$ may be associated with a certain operator.

Moreover a transformation given by the formula

$$Y = AX ,$$

where $Y$ and $X$, as before, are columns exhibiting the coordinates of the vectors $Y$ and $X$, is linear for any matrix $A$.

---

[†] We note that the matrix of coefficients in equations (2) forms a matrix which is the transpose of the one which we associate with the operator.

The established one-to-one relationship between operators and matrices is preserved under operations on operators. That is, the matrix of a sum of operators is equal to the sum of the matrices of the operators; the matrix of the product of operators is equal to the product of the corresponding matrices.

In brief, the set of operators over an $n$-dimensional space is isomorphic to the set of matrices of order $n$, and such an isomorphism may be realized by associating to each operator the matrix which corresponds to it relative to a certain fixed basis of the space.

In those discussions in which the basis of a space is fixed beforehand, the identification of an operator with its corresponding matrix is meaningful, as is the identification of a vector with the column matrix made up of its coordinates, as we have done above. Using such identifications the result of applying an operator to a vector coincides with the result of multiplying the operator matrix and the column matrix.

**4. The relation between matrices of an operator in different bases.** We shall now consider how the matrix of an operator changes when the basis of a space is changed.

We assume that we have passed from the basis $U_1, U_2, \cdots, U_n$ to the basis $U'_1, U'_2, \cdots, U'_n$. The coordinates of any vector of the space will then change according to the formula

$$X = CX'$$

where $X$ is the column matrix comprising the coordinates of the vector $X$ in the basis $U_1, \cdots, U_n$, $X'$ is the column of coordinates relative to the basis $U'_1, U'_2, \cdots, U'_n$, and $C$ is the transformation matrix.

We now consider an operator $A$ to which corresponds a matrix $A$ in the basis $U_1, \cdots, U_n$ and a matrix $B$ in the basis $U'_1, \cdots, U'_n$.

Let $Y = AX$ and let $Y$ and $Y'$ be the column matrices whose elements are the coordinates of the vector $Y$ in the bases $U_1, \cdots, U_n$ and $U'_1, \cdots, U'_n$ respectively.

Then

$$Y = AX,$$
$$Y' = BX'.$$

But $X = CX'$, $Y = CY'$ and therefore

$$CY' = ACX',$$

for which

$$Y' = C^{-1}ACX'$$

and consequently

$$B = C^{-1}AC.$$

Thus similar matrices correspond to one and the same operator in different bases. The matrix by which a similarity transformation is realized coincides with the matrix for transforming the coordinates.

**5. Rank of an operator.** The set $AP$ of vectors $AX$, where $A$ is a given operator and $X$ is a vector variable whose domain is a certain subspace $P$ of an $n$-dimensional space $R$, forms a subspace. In fact, if $Y_1 \in AP$ and $Y_2 \in AP$, then $Y_1 = AX_1$, $Y_2 = AX_2$, where $X_1$ and $X_2$ are certain vectors of $P$, and consequently $c_1 Y_1 + c_2 Y_2 = A(c_1 X_1 + c_2 X_2) \in AP$, since $c_1 X_1 + c_2 X_2 \in P$.

In particular, $AR$ is a subspace of $R$. This subspace is called the *image* (space) of operator $A$. The dimension of this subspace is called the *rank* of the operator $A$. It is obvious that $AR$ is a subspace spanned by the vectors $AU_1, \cdots, AU_n$, where $U_1, \cdots, U_n$ is a basis of $R$. Therefore, in accordance with theorem 5.2, the rank of an operator is equal to the rank of the matrix which corresponds to the operator in the basis $U_1, \cdots, U_n$.

We observe that since the dimension of the subspace $AR$ does not depend on the choice of basis the ranks of all matrices which correspond to the operator $A$ in the different bases are equal to each other. Consequently, the *ranks of similar matrices are equal.*

The image $AR$ coincides with the whole space if and only if the rank of the operator $A$ is equal to $n$, that is, if the determinant of its matrix is not equal to zero. In this case the operator is called *non-degenerate*. An operator whose rank is less than the dimension of the space is said to be *degenerate*.

The set $Q$ of vectors $Y \in R$, such that $AY = 0$, is also a subspace. In fact if $Y_1 \in Q$, and $Y_2 \in Q$, then $AY_1 = AY_2 = 0$ and $A(c_1 Y_1 + c_2 Y_2) = c_1 AY_1 + c_2 AY_2 = 0$ and consequently $c_1 Y_1 + c_2 Y_2 \in Q$. The subspace $Q$ is called the *kernel* of the operator $A$.

THEOREM 6.1. *The sum of the dimensions of the kernel and image of an operator equals the dimension of the whole space.*

*Proof.* Let $U_1, \cdots, U_m$ be a basis of the kernel $Q$ of an operator $A$. We shall add to it the basis of a space $R$ with vectors $V_1, \cdots, V_{n-m}$. We shall show that the vectors $AV_1, \cdots, AV_{n-m}$ form the basis of the image $AR$ of the operator $A$. We first prove the linear independence of these vectors. Let

$$c_1 AV_1 + \cdots + c_{n-m} AV_{n-m} = 0 .$$

Then $A(c_1 V_1 + \cdots + c_{n-m} V_{n-m}) = 0$, that is, $c_1 V_1 + \cdots + c_{n-m} V_{n-m} \in Q$. But this is possible only for $c_1 = \cdots = c_{n-m} = 0$, since the vectors $V_1, \cdots, V_{n-m}$ are linearly independent with respect to $Q$.

Now let $Y \in AR$. Then $Y = AX$. We expand $X$ into the vectors of the chosen basis:

$$X = c_1 U_1 + \cdots + c_m U_m + d_1 V_1 + \cdots + d_{n-m} V_{n-m} .$$

Consequently

$$Y = d_1 AV_1 + \cdots + d_{n-m} AV_{n-m} ,$$

since $AU_1 = \cdots = AU_m = 0$. Thus we have proved that the dimension of the image $AR$ of an operator $A$ is equal to $n - m$, where $m$ is the dimension of the kernel. The theorem has been proved.

From this theorem it follows that the kernel consists of the null vector alone if and only if the dimension of $AR$ is equal to $n$, that is, if the operator is non-degenerate. We observe that if the kernel and image of an operator have a null intersection, then the whole space is their direct sum. However this condition is not often satisfied.

In terms of matrix theory the theorem may be formulated in the following way. *The maximum number of linearly independent solutions of a system of n linear, homogeneous equations in n unknowns is equal to $n - r$, where $r$ is the rank of the matrix composed of the coefficients of the system.*

Let the following system be given:

$$
\begin{aligned}
a_{11}y_1 + \cdots + a_{1n}y_n &= 0 , \\
\cdots \cdots \cdots \cdots \cdots &, \\
a_{n1}y_1 + \cdots + a_{nn}y_n &= 0 .
\end{aligned}
\tag{6}
$$

This system is equivalent to the vector equation

$$AY = 0 ,$$

where $A$ is an operator with the matrix

$$
A = \begin{bmatrix} a_{11} & \cdots & a_{1n} \\ \cdot & \cdots & \cdot \\ a_{n1} & \cdots & a_{nn} \end{bmatrix}
$$

and $Y$ is a vector with coordinates $y_1, \cdots, y_n$, Therefore every solution of system (6) is a vector of the kernel $Q$ of operator $A$ and conversely, the coordinates of any vector $Y \in Q$ form a solution for system (6), so that the maximum number of linearly independent solutions is equal to the dimension of $Q$. By theorem 6.1 the dimension of $Q$ is equal to $n - r$, where $r$ is the dimension of the image of operator $A$, that is, the rank of matrix $A$.

**6. The inverse operator.** As we have seen, the image of a non-degenerate operator is the whole space so that a non-degenerate operator maps the space into itself. Such a mapping is one-to-one. In fact if $AX = Z$ and $AY = Z$ then $A(X - Y) = 0$, from which it follows that $X = Y$, since the kernel of a non-degenerate operator consists of only the null vector. Therefore for each non-degenerate operator $A$ there exists an inverse operator $A^{-1}$ which associates to each vector $Z \in R$ a uniquely determined vector $X$ such that $AX = Z$. The linearity of the operator $A^{-1}$ is obvious.

From the definition of an inverse operator it follows that $A^{-1}A = AA^{-1} = E$.

In any given basis, mutually inverse operators $A$ and $A^{-1}$ correspond to mutually inverse matrices.

**7. Eigenvectors and eigenvalues of an operator.** An *eigenvalue* (or characteristic number) of an operator $A$ is a number $\lambda$ such that for some non-zero vector $X$ the following equality holds:

$$AX = \lambda X. \tag{7}$$

Any non-zero vector $X$ which satisfies equation (7) is called an *eigenvector* of the operator $A$ corresponding (or belonging) to the eigenvalue $\lambda$.

The *spectrum* of an operator is the set of all its eigenvalues.

Eigenvectors and eigenvalues of an operator are found in the following way.

Let an operator $A$ be represented in some basis by the matrix $A = (a_{ik})$; let the coordinates of an eigenvector in this basis be $x_1, \cdots, x_n$. Then the coordinates of the vector $AX$ will be

$$\sum_{k=1}^{n} a_{1k}x_k, \cdots, \sum_{k=1}^{n} a_{nk}x_k,$$

and therefore the coordinates $x_1, \cdots, x_n$ and the eigenvalue $\lambda$ determine the system of equations:

$$\begin{aligned}
a_{11}x_1 + a_{12}x_2 + \cdots + a_{1n}x_n &= \lambda x_1, \\
a_{21}x_1 + a_{22}x_2 + \cdots + a_{2n}x_n &= \lambda x_2, \\
&\cdots\cdots\cdots, \\
a_{n1}x_1 + a_{n2}x_2 + \cdots + a_{nn}x_n &= \lambda x_n,
\end{aligned} \tag{8}$$

or

$$\begin{aligned}
(a_{11} - \lambda)x_1 + a_{12}x_2 + \cdots + a_{1n}x_n &= 0, \\
a_{21}x_1 + (a_{22} - \lambda)x_2 + \cdots + a_{2n}x_n &= 0, \\
&\cdots\cdots\cdots, \\
a_{n1}x_1 + a_{n2}x_2 + \cdots + (a_{nn} - \lambda)x_n &= 0.
\end{aligned} \tag{9}$$

This system of equations, homogeneous in $x_1, x_2, \cdots, x_n$, will have a non-zero solution if and only if

$$\begin{vmatrix}
a_{11} - \lambda & a_{12} & \cdots & a_{1n} \\
a_{21} & a_{22} - \lambda & \cdots & a_{2n} \\
\cdots & \cdots & \cdots & \cdots \\
a_{n1} & a_{n2} & \cdots & a_{nn} - \lambda
\end{vmatrix} = 0, \tag{10}$$

that is, if $\lambda$ is a root of the characteristic polynomial of the matrix. Thus the following theorem holds:

THEOREM 6.2. *The eigenvalues of an operator coincide with the roots of the characteristic polynomial of the matrix which represents the operator.*

We recall that matrices which represent the same operator in different bases are similar to each other and consequently their characteristic polynomials coincide. Thus it is reasonable to call the characteristic polynomial of any matrix which represents an operator the *characteristic polynomial of the operator.*

If $\varphi(t)$ is the characteristic polynomial of an operator $A$ and $A$ is the matrix representing it, then by the Cayley-Hamilton relationship $\varphi(A) = 0$.

Consequently $\varphi(A) = 0$, since $\varphi(A)$ is represented by the null matrix $\varphi(A)$.

There is also associated with an operator a minimum polynomial, defined as the polynomial of least degree which vanishes for the operator. It is clear that the minimum polynomial of an operator is also the minimum polynomial for the matrix which corresponds to the operator in an arbitrary basis. The characteristic polynomial is divisible by the minimum polynomial.

Therefore every root of the minimum polynomial is a root of the characteristic polynomial. The converse is also true. Namely, every eigenvalue of an operator, that is, every root of the characteristic polynomial, is also a root of the minimum polynomial of the operator.

In fact, let $\lambda$ be an eigenvalue of an operator, let $X$ be the eigenvector corresponding to it, and let $\psi(t)$ be the minimum polynomial of the operator. By the theorem of Bezout $\psi(t) = p(t)(t - \lambda) + \psi(\lambda)$. Consequently $\psi(A)X = p(A)(A - \lambda E)X + \psi(\lambda)X$. But $(A - \lambda E)X = 0$ and $\psi(A)X = 0$. Therefore $\psi(\lambda)X = 0$ and $\psi(\lambda) = 0$.

Thus the roots of the characteristic and minimum polynomials of an operator coincide, and hence can differ only in multiplicity.

From the so-called fundamental theorem of higher algebra we know that every polynomial equation has at least one root. Consequently an operator has at least one eigenvalue, which may be complex even if the matrix of the operator is real.

For every eigenvalue, corresponding eigenvectors may be determined from system (9) by substituting the appropriate numerical value for the latter $\lambda$. There are infinitely many eigenvectors corresponding to the eigenvalue $\lambda$ and they form a subspace of the space $R$.

In fact all the eigenvectors corresponding to the eigenvalue $\lambda$ form the kernel of the operator $A - \lambda E$. The dimension of this subspace, that is, the number of linearly independent eigenvectors which correspond to the eigenvalue $\lambda$, is equal to $l = n - r$, where $r$ is the rank of the operator $A - \lambda E$.

We shall show that $l$ does not exceed the multiplicity $k$ of the number $\lambda$ as a root of the characteristic polynomial of the operator. In fact, let $X_1$, $\cdots$, $X_l$ be linearly independent eigenvectors corresponding to the eigenvalue $\lambda$. We shall construct a basis $X_1, \cdots, X_n$ of the space, taking the vectors $X_1, \cdots, X_l$ as the first $l$ vectors. In this basis the operator is represented as a matrix whose first $l$ columns have the form:

$$\begin{matrix} \lambda & 0 & \cdots & 0 \\ 0 & \lambda & \cdots & 0 \\ & \cdot & \cdot & \\ 0 & 0 & \cdots & \lambda \\ & \cdot & \cdot & \\ 0 & 0 & \cdots & 0, \end{matrix}$$

since $AX_1 = \lambda X_1, \cdots, AX_l = \lambda X_l$. The characteristic polynomial of this matrix is divisible by $(t - \lambda)^l$ and consequently $\lambda$ has multiplicity $k$, not less than $l$, that is $l \le k$. It would be natural to assume that $l = k$, that is, that there corresponds $k$ linearly independent eigenvectors to the multiple roots of the characteristic polynomial. However in fact this is not so. That is, the number of linearly independent vectors may be less than the multiplicity of a given eigenvalue.

We emphasize this with an example. Consider an operator with matrix

$$A = \begin{bmatrix} 3 & 1 \\ 0 & 3 \end{bmatrix}$$

Then $|A - tE| = (t - 3)^2$ and consequently $\lambda = 3$ is a double root of the characteristic polynomial.

The system of equations for determining the coordinates of an eigenvector of the operator $A$ will be

$$3x_1 + x_2 = 3x_1 ,$$
$$3x_2 = 3x_2 .$$

from which $x_2 = 0$; therefore all eigenvectors of this operator are $(x_1, 0) = x_1(1, 0)$. Thus there corresponds to the double eigenvalue in the example only one linearly independent eigenvector, so that here $l$ is strictly less than $k$.

It is important to note that if $k = 1$, that is, if $\lambda$ is a simple root of the characteristic polynomial, then $l = k = 1$. In fact, $l \leq 1$ and $l > 0$, since there exists at least one eigenvector belonging to the eigenvalue.

**8. Eigenvectors of a matrix.** An *eigenvector of a matrix* A is a non-zero column (vector) which satisfies the condition

$$AX = \lambda X ,  \tag{11}$$

where $\lambda$ is some number. It is clear that an eigenvector of matrix $A$ is the column formed of the coordinates of an eigenvector of the operator $A$, to which the given matrix $A$ is associated for a given basis.

We observe that if an eigenvalue of a real matrix is complex, the coordinates of the eigenvector will also be complex. A vector whose coordinates are (complex) conjugate to the coordinates of an eigenvector of a real matrix is also an eigenvector of this matrix and belongs to the conjugate complex eigenvalue.

To be convinced of this it is sufficient to replace all numbers in the equation $AX = \lambda X$ by their conjugates.

It was established above that similar matrices have the same characteristic polynomials and consequently the same spectra of eigenvalues.

We have explained the geometric reason for this circumstance, namely, that it is possible to consider similar matrices as matrices of the same operator relative to different bases. Therefore "eigenvectors" of similar matrices will be columns containing of the coordinates of the eigenvectors of the given operator in different bases and consequently will be connected by the relationship

$$X' = C^{-1}X ,  \tag{12}$$

where $C$ is the matrix for transforming the coordinates.

This relation may also be verified formally: if

$$AX = \lambda X, \text{ then } (C^{-1}AC)(C^{-1}X) = \lambda C^{-1}X.$$

### 9. Eigenvectors of a triangular matrix. Let

$$B = \begin{vmatrix} b_{11} & b_{12} & \cdots & b_{1n} \\ & b_{22} & \cdots & b_{2n} \\ & & \cdot & \cdot \\ & 0 & & \cdot \\ & & & \cdot \\ & & & b_{nn} \end{vmatrix}$$

be a right triangular matrix whose diagonal elements are pairwise distinct. It is obvious that these diagonal elements will be eigenvalues of the matrix $B$. We shall find the eigenvectors corresponding to them. Let $X_i = (x_{1i}, \cdots, x_{ni})'$ be an eigenvector belonging to the eigenvalue $b_{ii}$. To determine the components of the vector $X_i$ we first equate components, in the equation

$$BX_i = b_{ii}X_i, \tag{13}$$

beginning with the $(i + 1)$th. This gives

$$(b_{i+1i+1} - b_{ii})x_{i+1i} + b_{i+1i+2}x_{i+2i} + \cdots + b_{i+1n}x_{ni} = 0,$$
$$(b_{i+2i+2} - b_{ii})x_{i+2i} + \cdots + b_{i+2n}x_{ni} = 0,$$
$$\cdots\cdots\cdots\cdots\cdots\cdots,$$
$$(b_{nn} - b_{ii})x_{ni} = 0,$$

from which we find that $x_{ni} = 0, x_{n-1i} = 0, \cdots, x_{i+1i} = 0$.
Equating the $i$th components in equation (13) gives the identity

$$(b_{ii} - b_{ii})x_{ii} = 0,$$

which shows that it is possible to take component $x_{ii}$ arbitrarily. For simplicity let us take $x_{ii} = 1$. Then the first $i - 1$ components of the vector $x_i$ are determined from the triangular system

$$(b_{11} - b_{ii})x_{1i} + b_{12}x_{2i} + \cdots + b_{1i-1}x_{i-1i} = - b_{1i},$$
$$(b_{22} - b_{ii})x_{2i} + \cdots + b_{2i-1}x_{i-1i} = - b_{2i},$$
$$\cdots\cdots\cdots\cdots\cdots\cdots\cdots\cdots,$$
$$(b_{i-1i-1} - b_{ii})x_{i-1i} = - b_{i-1i}.$$

Thus the eigenvector $X_i$ belonging to the eigenvalue $b_{ii}$ has all its components, beginning with the $i + 1$th, equal to zero. Therefore the matrix $\varXi$ whose columns are made up of the components of the eigenvectors $X_1, \cdots, X_n$ will be a right triangular matrix.

In a similar way it may be established that the components of the eigenvectors of a left triangular matrix with pairwise distinct diagonal elements form a left triangular matrix.

### 10. Transforming the matrix of an operator to diagonal form. We shall consider the question of the conditions which an operator must satisfy so that the space has a basis consisting of its eigenvectors. This circumstance will not always hold, as the example on p. 55 shows.

A basis consisting of eigenvectors is distinguished by the fact that in it the matrix of an operator has the diagonal form $[\lambda_1, \lambda_2, \cdots, \lambda_n]$.  In fact if $X_1$, $\cdots$, $X_n$ is a basis consisting of the eigenvectors of the operator $A$ and $\lambda_1$, $\lambda_2, \cdots, \lambda_n$ are the corresponding eigenvalues (not necessarily all distinct), then $AX_1 = \lambda_1 X_1 \cdots$, $AX_n = \lambda_n X_n$, so that in this basis the $i$th coordinate of the vector $AX_i$ is equal to $\lambda_i$, and all remaining coordinates are equal to zero.  Consequently the matrix of operator $A$ in the basis $X_1, \cdots, X_n$ will be

$$\begin{bmatrix} \lambda_1 & 0 & \cdots & 0 \\ 0 & \lambda_2 & \cdots & 0 \\ & \cdot & \cdot & \\ 0 & 0 & \cdots & \lambda_n \end{bmatrix}.$$

Conversely, if the operator $A$ has a diagonal matrix in some basis $U_1, \cdots$, $U_n$, then the vectors of this basis are linearly independent eigenvectors of the operator $A$.  In fact, in the column made up of the coordinates of the vector $AU_i$ only the $i$th coordinate will be non-zero and therefore $AU_i = \lambda_i U_i$.

THEOREM 6.3. *The eigenvectors corresponding to pairwise distinct eigenvalues are linearly independent.*

*Proof.*  Let $\lambda_1, \cdots, \lambda_s$ be the pairwise distinct eigenvalues of the operator $A$ and let $X_1, \cdots, X_s$ be eigenvectors corresponding to them.  Let us assume that they are linearly dependent.  Without loss of generality we assume that the vectors $X_1, \cdots, X_j$, where $j < s$, are linearly independent and that the vectors $X_{j+1}, \cdots, X_s$ are linear combinations of them.  In particular, let

$$X_s = \sum_{i=1}^{j} c_i X_i$$

Then

$$AX_s = \sum_{i=1}^{j} c_i AX_i = \sum_{i=1}^{j} c_i \lambda_i X_i .$$

On the other hand,

$$AX_s = \lambda_s X_s = \sum_{i=1}^{j} c_i \lambda_s X_i ,$$

It follows that

$$\sum_{i=1}^{j} (\lambda_s - \lambda_i) c_i X_i = 0 .$$

In view of the linear independence of the vectors $X_i$ all the coefficients $(\lambda_s - \lambda_i) c_i = 0$ and, since from the assumption $\lambda_s \neq \lambda_i$ for $i = 1, 2, \cdots, j$, all $c_i$ are equal to zero and therefore $X_s = 0$; this contradicts the fact that $X_s$ is an eigenvector.  Thus the vectors $X_1, \cdots, X_s$ are linearly independent.

THEOREM 6.4. *If the characteristic polynomial of an operator has only simple roots, then there exists a basis in the space which consists of the eigenvectors of the operator.*

*Proof.* From the theorem's condition the operator has $n$ different eigenvalues, where $n$ is the dimension of the space. The eigenvectors $X_1, \cdots, X_n$ corresponding to them are linearly independent, from theorem 6.3, which means that they form a basis.

In terms of matrix theory this theorem may be rephrased in the following manner. *If the eigenvalues of a matrix $A$ are pairwise distinct, then there exists a non-singular matrix $C$ such that $C^{-1}AC = \Lambda$, where $\Lambda = [\lambda_1, \cdots, \lambda_n]$.*

Consider the operator $A$ with matrix $A$ relative to some basis. For the operator $A$ there exists a basis of eigenvectors. In this basis there corresponds to operator $A$ a diagonal matrix $\Lambda = [\lambda_1, \cdots, \lambda_n]$, where $\Lambda = C^{-1}AC$. Here $C$ is the matrix for transforming coordinates when passing from the initial basis to a basis consisting of the eigenvectors. Consequently its columns contain the coordinates of the eigenvectors with respect to their initial basis.

THEOREM 6.5. *For a basis consisting of the eigenvectors of operator $A$ to exist it is necessary and sufficient that there correspond to each eigenvalue as many linearly independent vectors as the multiplicity of that eigenvalue as a root of the characteristic equation.*

*Proof.* Let $\lambda_1, \cdots, \lambda_s$ be all the distinct eigenvalues of an operator $A$ and let $k_1, \cdots, k_s$ be their multiplicities as zeros of the characteristic polynomial, where $k_1 + \cdots + k_s = n$. We shall denote by $l_i$ the number of linearly independent eigenvectors associated with the eigenvalue $\lambda_i$. Let $X_{i1}, \cdots, X_{il_i}$ be these eigenvectors. We shall show that the vectors

$$X_{11}, \cdots, X_{1l_1}, \cdots, X_{s1}, \cdots, X_{sl_s}$$

are linearly independent.

We assume that

$$c_{11}X_{11} + \cdots + c_{1l_1}X_{1l_1} + \cdots + c_{s1}X_{s1} + \cdots + c_{sl_s}X_{sl_s} = 0 \qquad (14)$$

We write

$$Y_1 = c_{11}X_1 + \cdots + c_{1l_1}X_{1l_1}; \cdots; Y_s = c_{s1}X_{s1} + \cdots + c_{sl_s}X_{sl_s} \qquad (14')$$

Then

$$Y_1 + \cdots + Y_s = 0, \qquad (14'')$$

Each of the vectors $Y_i$ is either a null vector or an eigenvector belonging to the eigenvalue $\lambda_i$. In view of theorem 6.3 equality $(14'')$ is possible only if all the vectors $Y_1, \cdots, Y_s$ are equal to the null vector. But then, from $(14')$ and the linear independence of the vectors $X_{i1}, \cdots, X_{il_i}$ for each $i$, we conclude that

$$c_{11} = \cdots = c_{1l_1} = \cdots = c_{s1} = c_{sl_s} = 0,$$

so that equality (14) is possible only with zero coefficients. Thus the linear independence of the vectors $X_{11}, \cdots, X_{1l_1}, \cdots, X_{s1}, \cdots, X_{sl_s}$ is proved. The maximum number of linearly independent vectors corresponding to all eigenvalues is equal to $l_1 + \cdots + l_s$. Therefore to construct a basis of eigenvectors it is necessary and sufficient that $l_1 + \cdots + l_s = n$, which will happen only if all $l_i = k_i$.

## 7. THE JORDAN CANONICAL FORM

In the preceding paragraph we saw that if a matrix does not have multiple eigenvalues, then it may always be changed to a diagonal form by means of a similarity transformation. However if multiple values exist, then a transformation to the diagonal form may not be possible. This circumstance is noteworthy in the sense that the space of matrices which have multiple eigenvalues has lower dimensionality than the space of all matrices of that order. Nevertheless, investigation of the structure of such matrices is of great interest in practical as well as theoretical application. In computational mathematics, in circumstances when matrix elements are given only by approximation, the sharp line between cases of simple and multiple eigenvalues is erased, since for small deformations of matrix elements one may always go from a matrix with multiple eigenvalues to a matrix with simple eigenvalues. For this reason in computational algebra the investigation of matrices with multiple eigenvalues is important primarily for a correct orientation in the structure of matrices which have very close but distinct eigenvalues. It is necessary to cope with such matrices very frequently in practice.

The present paragraph is devoted to a study of the structure of matrices which do not go over to diagonal form and, in particular, to establishing a certain very simple canonical form, more general than the diagonal, to which a arbitrary matrix may be changed by means of a similarity transformation.

**1. Invariant subspaces.** Let $A$ be an operator in an $n$-dimensional space $R$. A subspace $P$ of the space $R$ is said to be *invariant* with respect to the operator $A$ if the vectors of $P$ are transformed by the operator again into vectors of $P$; that is, from $X \in P$ it follows that $AX \in P$ (or, in abbreviated notation: $AP \subset P$).

From the given definition it follows that if $P$ is an invariant subspace for $A$, then it will also be invariant for the operator $f(A)$, where $f(t)$ is any polynomial. In fact if $X \in P$ and $P$ is invariant, then $AX \in P$, $A^2X \in P$, $\cdots$. and consequently $f(A)X \in P$.

In particular we observe that a subspace invariant with respect to the operator $A$ is also invariant with respect to the operator $A - \mu E$ for any number $\mu$. The converse is also true: if a subspace is invariant with respect to an operator $A - \mu E$, then it is invariant with respect to $A$, since $A = A - \mu E + \mu E$.

It is obvious that the whole space and the space consisting of the null vector are invariant subspaces. As non-trivial examples of invariant subspaces we have, for example, subspaces spanned by one or more eigenvectors of the operator $A$. In fact, let $X_1, \cdots, X_k$ be eigenvectors of $A$ and $P$ a subspace spanned by them. Then any vector $X$ belonging to $P$ may be written in the form

$$X = c_1 X_1 + \cdots + c_k X_k ,$$

and therefore $AX = c_1 A X_1 + \cdots + c_k A X_k = c_1 \lambda_1 X_1 + \cdots + c_k \lambda_k X_k$. (Among the numbers $\lambda_1, \cdots, \lambda_k$ some may be equal to each other). In the case where all the eigenvalues of the operator $A$ are different the indicated subspaces, as we shall see later, exhaust all invariant subspaces of the operator.

Another important type of invariant subspaces is the cyclic subspace. To define this concept we shall consider the following construction. Let a vector $X_0$ be given. We shall construct a system of vectors $X_0, A X_0, A^2 X_0, \cdots$. It is clear that in this system of vectors at some time we shall encounter, for the first time, the vector $A^q X_0$ which is a linear combination of the preceding $X_0, A X_0, \cdots, A^{q-1} X_0$.

A *cyclic subspace* $Q$, generated by the vector $X_0$, is the subspace spanned by the vectors $X_0, A X_0, \cdots, A^{q-1} X_0$. Since the vectors $X_0, A X_0, \cdots, A^{q-1} X_0$ are linearly independent, they form basis of the cyclic subspace $Q$ and therefore the dimension of $Q$ is equal to the exponent of degree $q$.

We shall now prove that the cyclic subspace generated by the vector $X_0$ is the smallest invariant subspace which contains $X_0$, that is, that it is itself invariant and that it is contained in every invariant subspace containing $X_0$.

In fact let $A^q X_0 = \gamma_0 X_0 + \cdots + \gamma_{q-1} A^{q-1} X_0$ and let $Y \in Q$. Then

$$Y = c_0 X_0 + c_1 A X_0 + \cdots + c_{q-2} A^{q-2} X_0 + c_{q-1} A^{q-1} X_0;$$

$$AY = c_0 A X_0 + c_1 A^2 X_0 + \cdots + c_{q-2} A^{q-1} X_0 + c_{q-1} A^q X_0$$

$$= c_0 A X_0 + c_1 A^2 X_0 + \cdots + c_{q-2} A^{q-1} X_0$$

$$+ c_{q-1}(\gamma_0 X_0 + \gamma_1 A X_0 + \cdots + \gamma_{q-1} A^{q-1} X_0)$$

$$= c_0' X_0 + c_1' A X_0 + \cdots + c_{q-1}' A^{q-1} X_0 \in Q .$$

Thus the invariance of $Q$ is proved.

Furthermore, let $Q'$ be any invariant subspace containing $X_0$. Then $X_0 \in Q', A X_0 \in Q', \cdots, A^{q-1} X_0 \in Q'$ and consequently $Q \subset Q'$ which proves the minimality of $Q$ among invariant subspaces containing $X_0$.

We observe for later reference that any subspace cyclic with respect to an operator $A$ and generated by the vector $X_0$ will also be cyclic with respect to the operator $A - \mu E$, for any numerical value of $\mu$. In fact any subspace invariant with respect to $A$ will also be invariant for $A - \mu E$, and conversely; consequently minimum invariant subspaces containing $X_0$ should coincide.

2. **Minimum annihilating polynomial of the vector $X_0$.** The dimension $q$ of a cyclic subspace is also connected with the following important concept. Let $A$ be a given operator. We shall call a polynomial $\chi(t)$ an annihilating polynomial for the vector $X_0$ if $\chi(A)X_0 = 0$. Among the annihilating polynomials for the vector $X_0$ there exists a polynomial $\theta(t)$ of least degree called the *minimum annihilating polynomial* for the vector $X_0$.

The degree of the minimum annihilating polynomial is equal to the dimension of the cyclic subspace generated by the vector $X_0$.

In fact let $q$ be the dimension of the cyclic subspace generated by the vector $X_0$, and let $A^q X_0 = \gamma_0 X_0 + \cdots + \gamma_{q-1} A^{q-1} X_0$.

Placing $\theta(t) = t^q - \gamma_{q-1} t^{q-1} - \cdots - \gamma_0$ we get $\theta(A)X_0 = 0$; that is, $\theta(t)$ is a polynomial annihilating the vector $X_0$. On the other hand, if $\chi(t)$ is a polynomial of lower degree than $q$, then $\chi(A)X_0 \neq 0$, because of the linear independence of the vectors $X_0, AX_0, \cdots, A^{q-1} X_0$. Consequently the polynomial of degree $q, \theta(t)$, is the minimum annihilating polynomial for the vector $X_0$.

It is easy to prove that any polynomial which annihilates $X_0$ is divisible by the minimum annihilating polynomial for $X_0$. In fact let $\chi(A)X_0 = 0$. Dividing the polynomial $\chi(t)$ by the polynomial $\theta(t)$ we get $\chi(t) = p(t)\theta(t) + r(t)$, where the remainder $r(t)$ has degree less than $q$. Consequently, $0 = \chi(A)X_0 = p(A)\theta(A)X_0 + r(A)X_0 = r(A)X_0$, from which $r(t) = 0$, since otherwise the polynomial $\theta(t)$ would not be the minimum annihilating polynomial for $X_0$. In particular the minimum polynomial of the operator (and consequently its characteristic polynomial) is divisible by the minimum annihilating polynomial for $X_0$. Therefore the dimension of any cyclic subspace will not exceed the degree of the operator's minimum polynomial.

THEOREM 7.1. *If the minimum annihilating polynomial $\theta(t)$ for a vector $X$ is formed as the product of relatively prime pairs of factors*

$$\theta(t) = \theta_1(t)\theta_2(t)\cdots\theta_s(t) ,$$

*then the vector $X$ may be represented as the sum of the vectors $X_1, X_2, \cdots, X_s$ which are annihilated respectively by the polynomials $\theta_1(t), \theta_2(t), \cdots, \theta_s(t)$. The elements $X_1, X_2, \cdots, X_s$ may be taken as belonging to any invariant subspace which contains $X$.*

*Proof.* It is obvious that it is sufficient to prove the theorem for $s = 2$, since the passage to the general case may be shown by mathematical induction. Since $\theta_1(t)$ and $\theta_2(t)$ are relatively prime, two polynomials $p_1(t)$ and $p_2(t)$ may be found such that $\theta_1(t)p_1(t) + \theta_2(t)p_2(t) = 1$.[†] This equality implies the operator equation

$$\theta_1(A)p_1(A) + \theta_2(A)p_2(A) = E ,$$

and therefore the following vector equation is true:

$$X = \theta_1(A)p_1(A)X + \theta_2(A)p_2(A)X .$$

---

[†] A. G. Kurosh, *A Course on Higher Algebra*, 1940, p. 175.

we let

$$X_1 = \theta_2(A)p_2(A)X, \ X_2 = \theta_2(A)p_1(A)X \ .$$

Then

$$X = X_1 + X_2$$

while

$$\theta_1(A)X_1 = \theta_1(A)\theta_2(A)p_2(A)X = \theta(A)p_2(A)X = p_2(A)\theta(A)X = p_2(A)0 = 0 \ ,$$

and analogously $\theta_2(A)X_2 = 0$.

The constructed vectors $X_1$ and $X_2$ belong to any invariant subspace containing $X$, since if $X \in P$ and $P$ is invariant, then

$$X_1 = \theta_2(A)p_2(A)X \in P, \ X_2 = \theta_1(A)p_1(A)X \in P \ ,$$

*Note.* It is not hard to show that the polynomials $\theta_1(t), \cdots, \theta_s(t)$ will be minimum annihilating polynomials for the vectors $X_1, \cdots, X_s$ respectively.

**3. Induced operators.** Let an operator $A$ act in an $n$-dimensional space $R$ and let $P$ be an invariant subspace for this operator. Then the operator $A$ associates to each vector of $P$ a vector of $P$, that is, it determines a ₍certain transformation in the subspace $P$. It is obvious that this transformation is a linear operator defined in $P$. This operator is called an operator *induced by the operator $A$ in the subspace $P$*. An induced operator differs from the operator $A$ only in its ·domain of definition.

Let $P$ be an invariant subspace relative to an operator $A$, let $U_1, \cdots, U_m$ be a basis of $P$, and let $U_1, \cdots, U_m, V_1, \cdots, V_{n-m}$ be a basis of the whole space. We shall consider the form which the matrix of the operator $A$ must have in this basis. Since the vectors $AU_1, \cdots, AU_m$ belong to $P$, that is, are linear combinations only of the vectors $U_1, \cdots, U_m$, then their coordinates in the given basis, beginning with the $m + 1$th, are equal to zero. Consequently the matrix of the operator $A$ takes the form:

$$\begin{bmatrix} a_{11} & \cdots & a_{1m} & a_{1m+1} & \cdots & a_{1n} \\ \cdot & \cdot & \cdot & \cdot & \cdot & \cdot \\ a_{m1} & \cdots & a_{mm} & a_{mm+1} & \cdots & a_{mn} \\ 0 & \cdots & 0 & a_{m+1m+1} & \cdots & a_{m+1n} \\ \cdot & \cdot & \cdot & \cdot & \cdot & \cdot \\ 0 & \cdots & 0 & a_{nm+1} & \cdots & a_{nn} \end{bmatrix}$$

or the shortened form

$$\begin{bmatrix} A_P & B \\ 0 & \tilde{A}_P \end{bmatrix}.$$

Here $A_P$ is a square matrix of order $m$, $\tilde{A}_P$ is a square matrix of order $n - m$, $B$ is a rectangular matrix with $m$ rows and $n - m$ columns, 0 is a null rectangular matrix. It is clear that $A_P$ is the matrix of the operator induced in $P$.

The matrix of the operator is even simpler if the space $R$ is a direct sum of two invariant subspaces. In fact let $R = P_1 + P_2$. We shall take for a basis of $R$ the union of the bases of $P_1$ and $P_2$. In this basis the matrix of the operator $A$ obviously has the form

$$\begin{bmatrix} A_{P_1} & 0 \\ 0 & A_{P_2} \end{bmatrix}.$$

where $A_{P_1}$ and $A_{P_2}$ are matrices of operators induced by the operator $A$ in $P_1$ and $P_2$. If the space is decomposed into the direct sum of $k$ invariant subspaces, then in the basis consisting of the union of the bases of these subspaces the matrix of the operator $A$ will have the quasi-diagonal form

$$\begin{bmatrix} A_{P_1} & & & & 0 \\ & A_{P_2} & & & \\ & & \cdot & & \\ & & & \cdot & \\ 0 & & & & A_{P_k} \end{bmatrix}, \tag{1}.$$

where $A_{P_1}, A_{P_2}, \cdots, A_{P_k}$ are matrices of the operators induced in $P_1, P_2, \cdots, P_k$.

From the above theory, there arises the following theorem.

THEOREM 7.2. *If a space $R$ is a direct sum of subspaces $P_1, P_2, \cdots, P_k$ invariant relative to an operator $A$, then the characteristic polynomial of the operator $A$ is equal to the product of the characteristic polynomials of operators $A_1, A_2, \cdots, A_k$ induced by operator $A$ in the subspaces $P_1, P_2, \cdots, P_k$.*

For the proof it is sufficient to take out $t$ from the elements of the principal diagonal of matrix (1) and to use the theorem which states that the determinant of a quasi-diagonal matrix is the product of the determinants formed from its diagonal cells.

If in expanding the space into a direct sum of invariant subspaces one introduces one-dimensional invariant subspaces (i.e. subspace spanned by eigenvectors), then the corresponding diagonal cells of the first order, namely, the diagonal elements of the matrix. It is obvious that these diagonal elements will be the eigenvalues of the matrix.

Later on we shall frequently replace the expression "operator induced by the operator $A$ in $P$" by the expression "operator $A$ in $P$".

**4. Root subspaces.**[†] Among invariant subspaces a very important role is

---

[†] The "root vectors" mentioned hereafter are also called *principal vectors* by some authors (see, for example, papers by G. E. Forsythe in the References). The so-called "root subspace" formed by such vectors is called also *the invariant subspace of $A$ relative to $\mu$*, and the concept called "height" of $X$ has also been designated in English as *the degree of the relative minimum polynomial* of $X$. However, since this terminology is by no means standardized in the literature in English, there would appear to be no point in departing from the verbatim terminology of the original text—except where the English terminology does have rather universal acceptance.

This convention will be followed in the subsequent material of the translation. Indulgence of the reader is asked if subsequently the translation errors in the use of this convention; however, readers literate enough in this subject matter to recognize such errant terminology are hardly likely to be misled or confused by it. (*Editor.*)

played by so-called root subspaces. A *root vector* for an operator $A$, corresponding to a number $\mu$, is a vector $X$ such that $(A - \mu E)^m X = 0$ for some whole number $m > 0$. It is clear that the set of root vectors satisfying a given number $\mu$ forms a subspace. In fact if $(A - \mu E)^{m_1} X_1 = 0$ and $(A - \mu E)^{m_2} X_2 = 0$, then $(A - \mu E)^m (c_1 X_1 + c_2 X_2) = 0$, where $m = \max (m_1, m_2)$. This subspace is called the *root subspace* corresponding to the number $\mu$. We shall show that it is invariant. In fact if $(A - \mu E)^m X = 0$, then $(A - \mu E)^m A X = A(A - \mu E)^m X = 0$.

The concept of a root vector is a more general concept of an eigenvector, namely, each eigenvector $X$ belonging to an eigenvalue $\lambda$ is also a root vector for the same number $\lambda$, since $(A - \lambda E)X = 0$.

The *height* of a non-zero root vector is the least number among the exponents $m$ such that $(A - \mu E)^m X = 0$. In other words, the height of a root vector is a number $k$ such that $(A - \mu E)^k X = 0$, but $(A - \mu E)^{k-1} X \neq 0$. The height of a null vector is equal to zero by definition. Eigenvectors are root vectors with height one.

The polynomial $(t - \mu)^k$ is the minimum annihilating polynomial for a root vector of height $k$. In fact, $(A - \mu E)^k X = 0$ and consequently the minimum annihilating polynomial for $X$ is a divisor of $(t - \mu)^k$. But divisors of $(t - \mu)^k$ are only polynomials $(t - \mu)^j$ for $j \leq k$. However the polynomials $(t - \mu)^j$ for $j < k$ do not annihilate the vector $X$, since $(A - \mu E)^j X \neq 0$.

THEOREM 7.3.   *In order for a non-zero root vector to exist, for some number $\mu$, it is necessary and sufficient that the number $\mu$ be an eigenvalue of the operator $A$. In this case the height of the root vector will not exceed the multiplicity m of the number $\mu$ as a root of the minimum polynomial. There will exist root vectors of height m.*

*Proof.* If $\mu$ is an eigenvalue, then there exist non-zero root vectors corresponding to it, for example, eigenvectors. Conversely, if for a number $\mu$ there exists a non-zero root vector $X$ of height $k$, then $Z = (A - \mu E)^{k-1} X \neq 0$ and $(A - \mu E)Z = (A - \mu E)^k X = 0$, so that $Z$ is an eigenvector corresponding to the number $\mu$ and consequently $\mu$ is an eigenvalue. The minimum annihilating polynomial $(t - \mu)^k$ for a vector $X$ is a divisor of the minimum polynomial for the operator $A$. Therefore the height $k$ of the vector $X$ will not exceed the multiplicity $m$ of the number $\mu$ as a root of the minimum polynomial.

It remains to prove the last statement of the theorem. Let $\phi(t) = (t - \mu)^m f(t)$ be the minimum polynomial of an operator $A$. We shall choose a vector $U$ so that it is not annihilated by the operator $(A - \mu E)^{m-1} f(A)$. Such a vector $U$ may be found, since otherwise $\phi(t)$ would not be the minimum polynomial for $A$.

We let $X = f(A)U$. Then $(A - \mu E)^{m-1} X = (A - \mu E)^{m-1} f(A)U \neq 0$. But $(A - \mu E)^m X = (A - \mu E)^m f(A)U = \phi(A)U = 0$. Thus $X$ is a root vector of height $m$ for the number $\mu$.

**5. Properties of an operator induced in a root subspace.** Let $A$ be an operator in the space $R$, $\lambda$ its eigenvalue with multiplicity $m$ as a root of the minimum polynomial, and $P$ the root subspace corresponding to this eigenvalue. Let $AP$ be the operator induced by the operator $A$ in the subspace $P$.

THEOREM 7.4. *The minimum polynomial of an operator $AP$ is equal to $(t - \lambda)^m$, the characteristic polynomial of an operator $AP$ is equal to $(t - \lambda)^p$, where $p$ is the dimension of the space $P$.*

*Proof.* The operator $(A - \lambda E)^m$ annihilates all vectors of the subspace $P$ and the operator $(A - \lambda E)^{m-1}$ annihilates only some vectors of $P$. Consequently $(AP - \lambda E)^m = 0$ and $(AP - \lambda E)^{m-1} \neq 0$. From here it follows that $(t - \lambda)^m$ is the minimum polynomial of the operator $AP$. Furthermore, every eigenvalue of the operator is a root of the minimum polynomial. Consequently the operator $AP$ has a unique eigenvalue $\lambda$ and therefore the characteristic polynomial of the operator $AP$ equals $(t - \lambda)^p$. The exponent $p$ is equal to the dimension of the subspace $P$, since the degree of the characteristic polynomial of any operator is equal to the dimension of the space in which it is defined. We shall show below that $p$ is equal to the multiplicity of the eigenvalue as a root of the characteristic polynomial of the operator $A$.

**6. Linear independence of root vectors**

THEOREM 7.5. *The non-zero root vectors corresponding to the pairwise distinct eigenvalues of operator $A$ are linearly independent.*

*Proof.* Let $X_1, \cdots, X_s$ be non-zero root vectors of an operator $A$ corresponding to the eigenvalues $\lambda_1, \cdots, \lambda_s$ where $\lambda_i \neq \lambda_j$ for $i \neq j$. Let $k_1, \cdots, k_s$ be the heights of the vectors $X_1, \cdots, X_s$. We shall denote by $f_i(t)$ the polynomial

$$(t - \lambda_1)^{k_1} \cdots (t - \lambda_i)^{k_i - 1} \cdots (t - \lambda_s)^{k_s}.$$

We shall prove that in the relationship

$$c_1 X_1 + \cdots + c_i X_i + \cdots + c_s X_s = 0$$

all the coefficients must be equal to zero. Applying the operator $f_i(A)$ to both sides of the equation we get

$$c_1 f_i(A) X_1 + \cdots + c_i f_i(A) X_i + \cdots + c_s f_i(A) X_s = 0 . \tag{2}$$

It is clear that $f_i(A) X_j = 0$ for $i \neq j$, since the polynomial $f_i(t)$ is divisible by the polynomials $(t - \lambda_j)^{k_j}$, $j \neq i$, which annihilate the vectors $X_j$ respectively.

Furthemore, $f_i(A) X_i \neq 0$, since the polynomial $f_i(t)$ is not divisible by the polynomial $(t - \lambda_i)^{k_i}$, which is the minimum annihilating polynomial for the vector $X_i$. Thus equation (2) becomes

$$c_i f_i(A) X_i = 0 \ ,$$

while $f_i(A) X_i \neq 0$.  Consequently $c_i = 0$ for all $i = 1, 2, \cdots, s$.  Thus the linear independence of the vectors $X_1, \cdots, X_s$ is proved.

### 7.  Expansion of a space into a direct sum of root subspaces

THEOREM 7.6.  *A space $R$ is a direct sum of all the root subspaces of an operator $A$.*

*Proof.*  The vector sum $R'$ of all root subspaces is a direct sum in view of the linear independence of the root vectors proved above; these vectors correspond to the pairwise distinct eigenvalues, that is, those which belong to pairwise distinct root subspaces.  It remains to prove only that $R'$ coincides with the whole space $R$, that is, that any vector $X$ of $R$ may be expanded as the sum of root vectors $X_i$ for $i = 1, \cdots, s$.  We shall prove this.  Let the polynomial $\theta(t)$ be the minimum annihilating polynomial for the vector $X$.  We expand it into linear factor.

$$\theta(t) = (t - \lambda_1)^{k_1} \cdots (t - \lambda_s)^{k_s}, \ \lambda_i \neq \lambda_j \ ,$$

The factors $(t - \lambda_1)^{k_1}, \cdots, (t - \lambda_s)^{k_s}$ are relatively prime.  Consequently we use theorem 7.1 to obtain the expansion

$$X = X_1 + \cdots + X_s \ ,$$

where the vectors $X_1, \cdots, X_s$ will be annihilated, respectively, by the polynomials $(t - \lambda_1)^{k_1}, \cdots, (t - \lambda_s)^{k_s}$.  Therefore the vectors $X_1, \cdots, X_s$ are root vectors.  The theorem is proved.

We observe that if the vector $X$ belongs to some invariant subspace, then the vectors $X_1, \cdots, X_s$ belong to the same subspace.  This follows from the above theorem.

The $X_i$ of the vector $X$ which correspond to the root subspaces will be called *projections* on these subspaces.

The following corollary arises from the theorem.  *The dimension of a root subspace corresponding to an eigenvalue $\lambda_i$ is equal to the multiplicity of $\lambda_i$ as a root of the characteristic polynomial of the operator $A$.*

In fact, the characteristic polynomial $\varphi(t)$ of the operator $A$ (in view of what was said earlier in par. 2) is the product of characteristic polynomials of the operators induced by the operator $A$ on the root subspaces $P_1, \cdots, P_s$; each polynomial may be written as $(\lambda_i - t)^{p_i}$ where $p_i$ is the dimension of the corresponding root subspace.  Thus

$$\varphi(t) = (\lambda_1 - t)^{p_1} \cdots (\lambda_s - t)^{p_s} \ ,$$

from which it follows that the dimensions $p_1, \cdots, p_s$ are the multiplicities of the eigenvalues in the characteristic polynomial of the operator $A$.

### 8.  Canonical basis of a root subspace.

We shall study more closely the structure of a particular root subspace for an operator $A$.  In order to simplify the notation we shall denote the root subspace by the letter $P$ and the corresponding eigenvalue by $\lambda$, ignoring any indices.

A root subspace $P$ is easily broken down into "stages". By the stage of height $j$ we shall mean the set of all vectors of height $j$. Stages are not subspaces, since in particular they do not contain the null vector. However a set of vectors whose heights do not exceed a given number $j$ do form a subspace. In fact if the heights of the vectors $X_1$ and $X_2$ do not exceed $j$, then $(A - \lambda E)^j X_1 = (A - \lambda E)^j X_2 = 0$ and consequently $(A - \lambda E)^j (c_1 X_1 + c_2 X_2) = 0$, that is, the height of the vector $c_1 X_1 + c_2 X_2$ does not exceed $j$. We shall denote such a subspace by $P^{(j)}$. It is obvious that $P^{(j)}$ is invariant. Furthermore, $P^{(1)} \subset P^{(2)} \subset \cdots \subset P^{(m)} = P$.

Along with "horizontal" invariant subspaces $P^{(j)}$ we shall consider invariant subspaces of a completely different type, the so-called "vertical" ones. If $X_0$ is a root vector of height $j \geq 1$, then the vector $X_1 = (A - \lambda E)X_0$ will have height $j - 1$. We shall say that the vector $X_1$ **lies under** the vector $X_0$. The set of vector $X_0, X_1, \cdots, X_{j-1}$ such that

$$X_1 = (A - \lambda E)X_0 \, ,$$
$$X_2 = (A - \lambda E)X_1 \, ,$$
$$\cdots \cdots \cdots \cdots ,$$
$$X_{j-1} = (A - \lambda E)X_{j-2}$$

is called a "tower". It is clear that $(A - \lambda E)X_{j-1} = 0$. The height of a tower (i.e. the number of its elements) is equal to the height of its "upper" generating vector $X_0$. We shall show that the vectors which form a tower are linearly independent. In fact let

$$c_0 X_0 + c_1 X_1 + \cdots + c_{j-1} X_{j-1} = 0 \, .$$

Applying successively to this equality the operators $(A - \lambda E)$, $(A - \lambda E)^2$, $\cdots$, $(A - \lambda E)^{j-1}$, we obtain

$$c_0 X_1 + c_1 X_2 + \cdots + c_{j-2} X_{j-1} = 0 \, ,$$
$$c_0 X_2 + \cdots + c_{j-3} X_{j-1} = 0 \, ,$$
$$\cdots \cdots \cdots \cdots \cdots ,$$
$$c_0 X_{j-1} = 0 \, ,$$

from which we conclude that $c_0 = 0, c_1 = 0, \cdots, c_{j-1} = 0$.

A subspace spanned by a tower will have dimension $j$. It will be invariant and cyclic for the operator $A - \lambda E$ and consequently will also be invariant and cyclic for the operator $A$.

We shall show that in a subspace $P$ there exists a basis which is obtained by joining several towers having no elements in common. Such a basis will be called a *canonical basis* of a root subspace.

It is clear that each choice of a canonical basis determines an expansion of a subspace $P$ into a direct sum of invariant cyclic subspace. Namely, subspaces spanned by basis vectors contained in separate towers will be such subspaces.

The following lemma precedes our existence proof for a canonical basis.

LEMMA. *If the vectors $Z_1, \cdots, Z_s$ belong to $P^{(j+1)}$ and are linearly independent relative to $P^{(j)}$, then the vectors $(A - \lambda E)Z_1, \cdots, (A - \lambda E)Z_s$ belong to $P^{(j)}$ and are linearly independent relative to $P^{(j-1)}$.*

*Proof.* The first statement of the lemma is obvious. We now assume that

$$c_1(A - \lambda E)Z_1 + \cdots + c_s(A - \lambda E)Z_s = V \in P^{(j-1)} .$$

This means that

$$(A - \lambda E)^{j-1} V = (A - \lambda E)^j (c_1 Z_1 + \cdots + c_s Z_s) = 0 .$$

Consequently

$$c_1 Z_1 + \cdots + c_s Z_s \in P^{(j)} ,$$

which is possible only for $c_1 = \cdots = c_s = 0$, in view of the linear independence of the vectors $Z_1, \cdots, Z_s$ relative to $P^{(j)}$. Thus the lemma is proved.

We now proceed to prove the existence of a canonical basis.

Let $P^{(1)} \subset P^{(2)} \subset \cdots \subset P^{(m)} = P$, where as before $P^{(j)}$ is the set of all vectors whose heights do not exceed $j$. We choose an arbitrary basis $X_{11}$, $\cdots, X_{1k_1}$ of the subspace $P^{(m)}$ relative to $P^{(m-1)}$. The number $k_1$ is the difference of the dimensions of $P^{(m)}$ and $P^{(m-1)}$. Then, from the lemma, the vectors $(A - \lambda E)X_{11}, \cdots, (A - \lambda E)X_{1k_1}$ belong to $P^{(m-1)}$ and are linearly independent relative to $P^{(m-2)}$. Consequently they may be included in the basis of $P^{(m-1)}$ relative to $P^{(m-2)}$. Let $(A - \lambda E)X_{11}, \cdots, (A - \lambda E)X_{1k_1}; X_{21}$, $\cdots, X_{2k_2}$, be a basis of $P^{(m-1)}$ relative to $P^{(m-2)}$. In view of the lemma the vectors $(A - \lambda E)^2 X_{11}, \cdots, (A - \lambda E)^2 X_{1k_1}, (A - \lambda E)X_{21}, \cdots, (A - \lambda E)X_{2k2}$ belong to $P^{(m-2)}$ and are linearly independent relative to $P^{(m-3)}$. We shall add them, with vectors $X_{31}, \cdots, X_{3k_3}$, to a basis of $P^{(m-2)}$ relative to $P^{(m-3)}$, applying to the constructed basis the operator $A - \lambda E$; we then add the resultant system of vectors to the basis $P^{(m-3)}$ relative to $P^{(m-4)}$ and so forth. At the $m$th step we come to the basis $(A - \lambda E)^{m-1} X_{11}, \cdots, (A - \lambda E)^{m-1} X_{1k_1}, (A - \lambda E)^{m-2} X_{21}, \cdots, (A - \lambda E)^{m-2} X_{2k_2}, \cdots, X_{m1}, \cdots, X_{mk_m}$ of the subspace $P^{(1)}$. The union of all these relative bases is a basis of $P$ and this basis will obviously be canonical.

The construction of a canonical basis is presented in visual form in the scheme on page 69. In this scheme $B = A - \lambda E$.

From the construction it is clear that the choice of a canonical basis is not a simple task. However it is readily seen that a canonical basis may indeed be constructed by means of this method. Therefore the structure of any canonical basis (the number of towers of a given height) will be the same.

**9. Canonical basis of a space and the Jordan canonical form for the matrix of an operator.** A canonical basis of a space $R$ in which an operator $A$ is applied is a basis obtained by joining the canonical bases of all root subspaces for that operator. A canonical basis is easily separated

| | $X_{11}$ | $\cdots$ | $X_{1k_1}$ | $X_{21}$ | $\cdots$ | $X_{2k_2}$ | $\vdots$ | $X_{m-11}$ | $\cdots$ | $X_{m-1k_{m-1}}$ | $X_{m1}$ | $\cdots$ | $X_{mk_m}$ |
|---|---|---|---|---|---|---|---|---|---|---|---|---|---|
| basis of $P^{(m)}$ W.R.T. $P^{(m-1)}$ | $X_{11}$ | $\cdots$ | $X_{1k_1}$ | | | | | | | | | | |
| basis of $P^{(m-1)}$ W.R.T. $P^{(m-2)}$ | $BX_{11}$ | $\cdots$ | $BX_{1k_1}$ | $X_{21}$ | $\cdots$ | $X_{2k_2}$ | | | | | | | |
| $\cdots\cdots\cdots\cdots$ | $\vdots$ | $\vdots$ | $\vdots$ | $\vdots$ | $\vdots$ | $\vdots$ | $\vdots$ | | | | | | |
| basis of $P^{(2)}$ W.R.T. $P^{(1)}$ | $B^{m-2}X_{11}$ | $\cdots$ | $B^{m-2}X_{1k_1}$ | $B^{m-3}X_{21}$ | $\cdots$ | $B^{m-3}X_{2k_2}$ | | $X_{m-11}$ | $\cdots$ | $X_{m-1k_{m-1}}$ | | | |
| basis of $P^{(1)}$ | $B^{m-1}X_{11}$ | $\cdots$ | $B^{m-1}X_{1k_1}$ | $B^{m-2}X_{21}$ | $\cdots$ | $B^{m-2}X_{2k_2}$ | $\cdots$ | $BX_{m-11}$ | $\cdots$ | $BX_{m-1k_{m-1}}$ | $X_{m1}$ | $\cdots$ | $X_{mk_m}$ |

into towers and correspondingly the whole space is broken down into a direct sum of invariant cyclic subspaces spanned by vectors from different towers. Therefore the matrix of an operator in a canonical basis will be quasi-diagonal, consisting of "boxes" satisfying the separate towers. We shall consider the form of these boxes. Let $Q$ be one of the invariant subspaces spanned by a tower of height $j$ which contains the vectors $X_0$, $X_1 = (A - \lambda_i E)X_0$, $X_2 = (A - \lambda_i E)X_1$, $\cdots$, $X_{j-1} = (A - \lambda_i E)X_{j-2}$. It is clear that

$$
\begin{aligned}
AX_0 &= \lambda_i X_0 + X_1 \, , \\
AX_1 &= \lambda_i X_1 + X_2 \, , \\
&\cdots\cdots\cdots\cdots \, , \\
AX_{j-2} &= \lambda_i X_{j-2} + X_{j-1} \, , \\
AX_{j-1} &= \lambda_i X_{j-1} \, .
\end{aligned}
$$

Thus to an operator $A$ in a subspace $Q$ in the given basis $X_0, \cdots, X_{j-1}$ there corresponds a matrix

$$
\begin{bmatrix}
\lambda_i & 0 & \cdots & 0 & 0 \\
1 & \lambda_i & \cdots & 0 & 0 \\
\cdot & \cdot & \cdot & \cdot & \cdot \\
0 & 0 & \cdots & \lambda_i & 0 \\
0 & 0 & \cdots & 1 & \lambda_i
\end{bmatrix}.
$$

Such a matrix is called a *Jordan canonical box*.

In the whole space there will correspond to the operator $A$ a quasi-diagonal matrix consisting of Jordan canonical boxes, that is, a matrix of the form

$$
\begin{bmatrix}
\lambda_1 & 0 & \cdots & 0 & 0 & & & & & & & \\
1 & \lambda_1 & \cdots & 0 & 0 & & & & & & & \\
\cdot & \cdot & \cdot & \cdot & \cdot & \cdot & & & & & & \\
0 & 0 & \cdots & \lambda_1 & 0 & & & & & & & \\
0 & 0 & \cdots & 1 & \lambda_1 & & & & & & & \\
& & & & & \cdot & & & & & & \\
& & & & & & \cdot & & & & & \\
& & & & & & & \cdot & & & & \\
& & & & & & & & \lambda_s & 0 & \cdots & 0 & 0 \\
& & & & & & & & 1 & \lambda_s & \cdots & 0 & 0 \\
& & & & & & & & \cdot & \cdot & \cdot & \cdot & \cdot \\
& & & & & & & & 0 & 0 & \cdots & \lambda_s & 0 \\
& & & & & & & & 0 & 0 & \cdots & 1 & \lambda_s
\end{bmatrix}
$$

The number of Jordan boxes is equal to the number of "towers" and consequently to the number of first "stages" of these towers, that is, to the number of linearly independent eigenvectors of the operator $A$. The number of Jordan boxes containing one and the same eigenvalue $\lambda_i$ is equal to the number of towers into which the basis of the root subspace corresponding to $\lambda_i$ is divided; that is, it is equal to the number of linearly independent eigenvectors belonging to the eigenvalue $\lambda_i$. The maximum order of the Jordan boxes containing $\lambda_i$ is equal to the multiplicity of the eigenvalue $\lambda_i$

as a root of the minimum polynomial. The sum of the orders of all cano-
nical boxes containing $\lambda_i$ is equal to the multiplicity of $\lambda_i$ as a root of the
characteristic polynomial.

Let an operator $A$ be given by the matrix $A$ corresponding to it in a
certain basis $U_1, \cdots, U_n$. Let $V_1, \cdots, V_n$ be a canonical basis for the operator $A$. In the latter basis there corresponds to the operator $A$ a Jordan
canonical matrix $J$. If we denote by the letter $C$ the matrix for transforming the coordinates from the basis $U_1, \cdots, U_n$ to the basis $V_1, \cdots, V_n$, then
$J = C^{-1}AC$, that is, $J$ is obtained from $A$ by a similarity transformation.
This transformation is called a *reduction of the matrix to the Jordan canonical form*. Thus, knowing the canonical basis gives us the transformation
matrix $C$ as well as the canonical matrix $J$. It is not hard to see, conversely, that if the canonical matrix $J$ equals $C^{-1}AC$, then the basis $V_1, \cdots,$
$V_n$ (connected with the initial basis by the coordinate transformation of
matrix $C$) will be a canonical basis for the operator $A$.

Computing a canonical basis for an operator given by a matrix is quite
complex. But it is often important to determine only the canonical form
for a given matrix $A$ without computing the transitional matrix $C$, that is,
without computing a canonical basis for the corresponding operator. This
turns out to be possible in more than one way. One such method is connected with a detailed study of the matrix $A - tE$.

We shall denote by $D_i(t)$ the greatest common divisor of all minors of
order $i$ in the determinant $|A - tE|$. In particular $D_n(t)$ coincides with
the characteristic polynomial. One may prove that all $D_i(t)$ similar to $D_n(t)$
are common divisors for the class of similar matrices. Furthermore, it is
possible to prove that $D_i(t)$ is divisible by $D_{i-1}(t)$. We write

$$\frac{D_i(t)}{D_{i-1}(t)} = E_i(t) .$$

It is obvious that $D_n(t) = \prod\limits_{i=1}^{n} E_i(t).$

Moreover it turns out that $E_n(t) = \dfrac{D_n(t)}{D_{n-1}(t)}$ is the minimum polynomial
for the matrix.

We expand $E_i(t)$ into linear factors. Then

$$E_i(t) = \prod_{j=1}^{s} (\lambda_j - t)^{m_{ij}} .$$

Here $s$ denotes the number of different eigenvalues, $\sum\limits_{i=1}^{n} m_{ij} = n_j$, and $\sum\limits_{j=1}^{s} \sum\limits_{i=1}^{n} m_{ij}$
$= n$. It is obvious that only a few of the exponents $m_{ij}$ are different from
zero.

The binomials $(\lambda_i - t)^{m_{ij}}$ are called *elementary divisors of* the matrix $A$.
By knowing the elementary divisors we can construct a canonical form.
Namely, Jordan boxes are constructed from the numbers $\lambda_j$ and the orders
of these boxes are equal to the exponents $m_{ij}$. The number of boxes

containing $\lambda_j$ is equal to the number of non-zero exponents $m_{ij}$.

In the case where the elementary divisors are linear, that is, if all the non-zero exponents $m_{ij}$ are equal to one, the Jordan boxes degenerate into the diagonal elements and the canonical form turns out to be simply the diagonal form where, naturally, one and the same eigenvalue will come in as a diagonal element as many times as its multiplicity as a root of the characteristic polynomial.

The converse is also true, namely: If a matrix transforms to the diagonal form, its elementary divisors will be linear. Thus matrices with different eigenvalues possess linear elementary divisors.

If all elementary divisors $(\lambda_j - t)^{m_{ij}}$ are relatively prime (which holds if and only if $D_{n-1}(t) = 1$), then each eigenvalue enters only one canonical box, while the order of the box is equal to the multiplicity of the corresponding eigenvalue. In this case (and only then) the minimum and characteristic polynomials of a matrix coincide.

## 8. STRUCTURE OF INVARIANT SUBSPACES

### 1. Structure of invariant subspaces of the general type.

THEOREM 8.1. *Any invariant subspace is the direct sum of the intersection of this subspace with the root subspaces.*

*Proof.* Let $T$ be some invariant subspace. We denote by $T_i$ the intersection of $T$ with the root subspace $P_i$. It is obvious that the subspace $T_i$ is invariant. We shall show that the subspace is broken down into a direct sum of the subspaces $T_1, \cdots, T_s$ where $s$ is the number of different eigenvalues of the operator. The inclusion $T_1 + \cdots + T_s \subset T$ is trivial, since all $T_i \subset T$. The sum $T_1 + \cdots + T_s$ is direct, since $T_i$ is in the subspace $P_i$. We shall now prove the converse inclusion $T \subset T_1 + T_2 + \cdots + T_s$. Let $X \in T$. We factor $X$ into the root subspaces:

$$X = X_1 + X_2 + \cdots + X_s .$$

Then, as we have seen,[†] $X_i \in T$ and consequently $X_i \in T \cap P_i = T_i$. Thus the inclusion we desired is proved.

We note that certain of the $T_i$ may also be null subspaces.

From the theorem it follows that every invariant subspace is the direct sum of invariant subspaces contained in the root subspaces.

### 2. Structure of a cyclic subspace.

THEOREM 8.2. *Let $X_0$ be an arbitrary vector of a space $R$ and $Q$ a cyclic subspace for an operator $A$, generated by the vector $X_0$. Then $Q$ is a direct sum of cyclic subspaces generated by the projections of the vector $X_0$ on the root subspaces.*

---

† Refer to theorem 7.6.

*Proof.* From theorem 8.1, $Q$ is the direct sum of the subspaces $Q_i = Q \cap P_i$. We expand the vector $X_0$ according to the root subspaces:

$$X_0 = X_1 + \cdots + X_s \, ;$$
$$X_i \in P_i \, .$$

Then cyclic subspaces $Q_i'$ generated by the vectors $X_i$ enter correspondingly into $Q_i$ and their direct sum $Q'$ enters into $Q$. But $Q'$ is obviously invariant. Consequently $Q' = Q$ and all $Q_i' = Q_i$ for $i = 1, 2, \cdots, s$.

Let the height of the vector $X_i$ be equal to $j_i$. Then the vectors

$$X_i, (A - \lambda_i E)X_i, \cdots, (A - \lambda_i E)^{j_i - 1} X_i$$

are linearly independent and they generate an invariant subspace which coincides with $Q_i$ because of the minimality of $Q_i$. Consequently these vectors form a basis of $Q_i$ and therefore the dimension of $Q_i$ coincides with the height $j_i$ of the vector $X_i$. From here the following theorems arise directly.

THEOREM 8.3.   *The dimension of a cyclic subspace of an operator* $A$ *generated by the vector* $X_0$ *is equal to the sum of the heights of the projections of the generating vector* $X_0$ *on the root subspaces.*

THEOREM 8.4.   *The minimum annihilating polynomial for* $X_0$ *is equal to* $(t - \lambda_1)^{j_1} \cdots (t - \lambda_s)^{j_s}$.

This polynomial obviously annihilates all projections $X_1, \cdots, X_s$ of the vector $X_0$. Conversely, if $\theta(A)X_0 = 0$, then $\theta(A)X_1 = \cdots = \theta(A)X_s = 0$. But the polynomial of least degree which annihilates $X_i$ is $(t - \lambda_i)^{j_i}$, since $(A - \lambda_i E)^{j_i - 1} X_i \neq 0$ .

From theorem 8.4 it follows that for any divisor $g(t)$ of the minimum polynomial of an operator $A$ there may be found a vector $X_0$ for which this divisor will be the minimum annihilating polynomial (Luzin-Khlodovskii theorem),[†] since in a subspace $P_i$ there exist vectors with any height from 0 to $m_i$ inclusive.

## 9. ORTHOGONALITY OF VECTORS AND SUBSPACES

The present paragraph and paragraph 10 are devoted to describing the properties of Euclidean and unitary spaces. In view of the complete parallelism of the theory we shall present the facts and their proofs in terms of $n$-dimensional unitary space, formulating, where necessary, certain specifics of Euclidean space.

**1. Orthogonal systems of vectors.** Two vectors of a space are said to be mutually *orthogonal* if their scalar product is equal to zero. A system of vectors is said to be an orthogonal set if any two distinct vectors of the

[†] M. N. Luzin (1), (2), I. N. Khlodovsvski (1).

system are mutually orthogonal. When speaking of an orthogonal system
we shall always assume that all vectors of the system are different from
zero.

THEOREM 9.1.  *Vectors which form an orthogonal system are linearly
independent.*

*Proof.* Let $X_1, \cdots, X_k$ be an orthogonal system and let

$$c_1X_1 + \cdots + c_iX_i + \cdots + c_kX_k = 0 .$$

From the properties of the scalar product we have

$$0 = (c_1X_1 + \cdots + c_iX_i + \cdots + c_kX_k, X_i) =$$
$$= c_1(X_1, X_i) + \cdots + c_i(X_i, X_i) + \cdots + c_k(X_k, X_i) = c_i |X_i|^2$$

and since $|X_i|^2 > 0, c_i = 0$ for any $i = 1, 2, \cdots, k$. Thus the only possible
values for $c_1, c_2, \cdots, c_k$ in the equation $c_1X_1 + \cdots + c_kX_k = 0$ are $c_1 = c_2 = \cdots$
$= c_k = 0$, that is, the vectors $X_1, \cdots, X_k$ are linearly independent.  This
implies, first, that the number of vectors forming an orthogonal system does
not exceed $n$ and, second, that any orthogonal system of $n$ vectors forms
a basis of the space.  Such a basis is called an *orthogonal basis*.  If moreover, the lengths of all the vectors of an orthogonal basis are equal to one,
then the basis is called *an orthonormal basis*.

In an arithmetic space, in which the scalar product is introduced by the
formula $(X, Y) = \sum_{i=1}^{n} x_i\bar{y}_i$, the natural basis $e_1, \cdots, e_n$ is orthonormal.

It is possible to pass from any system of linearly independent vectors
$X_1, \cdots, X_k$ to an orthogonal system of vectors $Y_1, \cdots, Y_k$ by means of the
so-called *orthogonalization process*.  This process is described in the following theorem.

THEOREM 9.2.  *Let $X_1, \cdots, X_k$ be linearly independent.  Then it is possible
to construct an orthogonal system of vectors $Y_1, \cdots, Y_k$, which is connected
with the initial one by the relationships*:

$$Y_1 = X_1 ,$$
$$Y_2 = X_2 + \alpha_{21}X_1 ,$$
$$\cdots \cdots \cdots ,$$
$$Y_k = X_k + \alpha_{k1}X_1 + \cdots + \alpha_{kk-1}X_{k-1} .$$
\hfill (1)

For a proof we use the following inductive construction.  We let $Y_1 = X_1$.
We assume further that the vectors $Y_1, \cdots, Y_{m-1}$ have already been constructed and are different from zero.  We shall look for $Y_m$ in the form

$$Y_m = X_m - \gamma_{m1}Y_1 - \cdots - \gamma_{mm-1}Y_{m-1} .$$
\hfill (2)

We choose the coefficients $\gamma_{m1}, \cdots, \gamma_{mm-1}$ so that $(Y_m, Y_j) = 0$ for $j = 1, \cdots,$
$m - 1$.  This is easy to do since

$$(Y_m, Y_j) = (X_m, Y_j) - \gamma_{mj}(Y_j, Y_j) .$$

But $(Y_j, Y_j) \neq 0$, since $Y_j \neq 0$ in view of the induction hypothesis, and consequently it suffices to take

$$\gamma_{mj} = \frac{(X_m, Y_j)}{(Y_j, Y_j)} . \tag{3}$$

Substituting in equation (2), in place of $Y_1, \cdots, Y_{m-1}$, their expressions in $X_1, \cdots, X_{m-1}$, we finally obtain

$$Y_m = X_m + \alpha_{m1}X_1 + \cdots + \alpha_{mm-1}X_{m-1} .$$

It remains to verify that $Y_m \neq 0$. But this is obvious, since otherwise the vector $X_m$ would be a linear combination of the vectors $X_1, \cdots, X_{m-1}$, which would contradict the condition of the theorem.

*Note.* The linear independence of the vectors $X_1, \cdots, X_k$ was used in the proof only to establish that every constructed vector was different from zero. Therefore if the orthogonalization process is applied to a system of linearly dependent vectors, then along the way the construction of a null vector will turn out to be necessary. This will happen for the first time in the $r$th step if the vectors $X_1, \cdots, X_{r-1}$ are linearly independent and the vector $X_r$ is a linear combination of them. Therefore the orthogonalization process may be applied to verify the linear independence, or else to establish the linear dependence, of a given system of vectors.

It is easy to pass from an orthogonal system of vectors to an orthonormal system: we divide each vector of the system by its length. (This process is called "normalization").

This procedure makes the choice of an orthonormal basis quite arbitrary. In fact, it is possible to pass from any basis to an orthonormal one by means of orthogonalization and normalization.

The scalar product of two vectors is expressed very simply in terms of the coordinates of these vectors in any orthonormal basis.

In fact if $U_1, \cdots, U_n$ is an orthonormal basis and

$$X = \xi_1 U_1 + \cdots + \xi_n U_n, \quad Y = \eta_1 U_1 + \cdots + \eta_n U_n ,$$

then

$$(X, Y) = (\xi_1 U_1 + \cdots + \xi_n U_n, \eta_1 U_1 + \cdots + \eta_n U_n) =$$
$$= \sum_{i=1}^{n} \sum_{j=1}^{n} (\xi_i U_i, \eta_j U_j) = \sum_{i=1}^{n} \sum_{j=1}^{n} \xi_i \bar{\eta}_j (U_i, U_j) = \sum_{i=1}^{n} \xi_i \bar{\eta}_i . \tag{4}$$

Thus the scalar product is expressed in terms of the coordinates of the vectors in any orthonormal basis by employing precisely the same formula that expresses the scalar product of two vectors in an arithmetic space. Thus the correspondence to each vector of the column of its coordinates in an orthonormal basis maps a general linear space onto an arithmetic space (complex for a unitary, real for a Euclidean space); this correspondence is an isomorphism not only with respect to the operations of addition and multiplication by a number but also with respect to the operation of scalar multiplication.

**2. Coordinate transformations under a change of orthonormal basis.** Let $U_1, \cdots, U_n$ and $U_1', \cdots, U_n'$ be two orthonormal bases. We shall show that the coordinate transformation matrix for passing from the first to the second basis will be a unitary matrix if $R$ is a unitary space and an orthogonal matrix if $R$ is a Euclidean space.

Let the columns of the matrix

$$A = \begin{bmatrix} a_{11} & \cdots & a_{1n} \\ \cdot & \cdot \cdot \cdot \cdot & \cdot \\ a_{n1} & \cdots & a_{nn} \end{bmatrix}$$

be the coordinates of the vectors $U_1', \cdots, U_n'$ in the basis $U_1, \cdots, U_n$. Since $(U_i', U_j') = \delta_{ij}$, where $\delta_{ij}$ is the Kronecker symbol,[†] and since the basis $U_1, \cdots, U_n$ is orthonormal, the following relationships holds:

$$\sum_{k=1}^{n} a_{ki} \bar{a}_{kj} = \delta_{ij}$$

(for unitary space) and

$$\sum_{k=1}^{n} a_{ki} a_{kj} = \delta_{ij}$$

(for Euclidean space).

This says that the matrix $A$ is unitary (for a unitary space) or orthogonal (for a Euclidean space).

**3. Orthogonally complementary subspaces.** A vector $Z$ is said to be orthogonal to a subspace $P$ if it is orthogonal to every vector which belongs to $P$. This circumstance is written in the form $Z \perp P$.

Two spaces are called *mutually orthogonal* if each vector of one is orthogonal to every vector of the other. Mutually orthogonal subspaces have only the null vector in common. In fact, a vector $X$ belonging to two mutually orthogonal subspaces satisfies the condition $(X, X) = 0$, from which it follows that $X$ is the null vector.

The set of all vectors of $R$ orthogonal to all the vectors of a subspace $P$ form, obviously, a subspace $Q$ which is said to be an *orthogonally complementary* subspace to $P$ or its *orthogonal complement*. One may construct such a subspace, for example, in the following way. Let the subspace $P$ have dimension $p$. We shall take a basis of $P$ and adjoin it to a basis of the whole space. We apply the orthogonalization process to the resultant basis. Let $U_1, \cdots, U_p, U_{p+1}, \cdots, U_n$ be the orthogonal basis so-obtained. It is obvious that the vectors $U_1, \cdots, U_p$ form an orthogonal basis of the subspace $P$. We shall show that the subspace spanned by the vectors $U_{p+1}, \cdots, U_n$ will be the desired orthogonally complementary subspace $Q$. In fact if $Z \in Q$, then $Z$ is orthogonal to all $U_1, \cdots, U_p$ and consequently in the expansion

$$Z = c_1 U_1 + \cdots + c_p U_p + c_{p+1} U_{p+1} + \cdots + c_n U_n$$

[†] $\delta_{ij} = \begin{cases} 0 & \text{if } i \neq j \\ 1 & \text{if } i = j \end{cases}$.

the coefficients $c_1, \cdots, c_p$ are equal to zero. Conversely, if

$$Z = c_{p+1} U_{p+1} + \cdots + c_n U_n ,$$

then $Z$ is orthogonal to $U_1, \cdots, U_p$ and consequently $Z$ is orthogonal to any linear combination of them, that is, to any vector belonging to $P$.

From the preceding construction it follows that the dimension $q$ of the orthogonal complement is equal to $n - p$. It is easy to prove that the orthogonal complement of an orthogonal complement is the initial subspace. In fact let $Q$ be the orthogonal complement to $P$, and let $P_1$ be the orthogonal complement to $Q$. From the definition it is clear that $P \subset P_1$. But the dimensions of $P$ and $P_1$ are the same. Consequently $P_1 = P$.

Moreover, from the same construction it follows that the whole space is a direct sum of a subspace and its orthogonal complement. In fact, the union of the bases of these subspaces is a basis of the whole space. Consequently any vector $X$ of a space $R$ is simply represented in the form

$$X = Y + Z ,$$

where $Y \in P, Z \perp P$. The vector $Y$ is called the *orthogonal projection* of the vector $X$ on the subspace $P$.

**4. Dual bases.** Let $U_1, \cdots, U_n$ be a certain basis in the space $R$. A basis $V_1, \cdots, V_n$ is called *dual* with respect to $U_1, \cdots, U_n$ if

$$(U_i, V_i) = 1 ,$$
$$(U_i, V_j) = 0 \qquad \text{(for } i \neq j) . \tag{5}$$

For each basis there exists a dual basis. In fact, let $Q_1$ be a subspace spanned by the vectors $U_2, \cdots, U_n$ and let $S_1$ be the orthogonal complement to $Q_1$. It is obvious that the dimension of $Q_1$ is equal to $n - 1$, the dimension of $S_1$ equal to 1. Let $Z \neq 0$ by any vector of $S_1$. Then $(Z, U_2) = \cdots = (Z, U_n) = 0$ and consequently $(Z, U_1) = a_1 \neq 0$. Therefore the vector $V_1 = (1/a_1)Z$ satisfies condition (5). All the remaining vectors of the dual basis are constructed in an analogous manner. The uniqueness of the dual basis follows directly from the definition.

It is obvious that a basis will be dual with itself if and only if it is orthonormal.

If the basis $V_1, \cdots, V_n$ is dual to the basis $U_1, \cdots, U_n$, then the basis $U_1, \cdots, U_n$ is dual to the basis $V_1, \cdots, V_n$, that is, the duality relationship for bases is symmetric. Therefore it makes sense to talk about a pair of *mutually dual bases* (or *biorthogonal* bases).

Introducing the dual basis makes it possible to represent the coordinates of a vector as scalar products. That is, if the coordinates of a vector $X$ in a basis $U_1, \cdots, U_n$ are numbers $x_1, \cdots, x_n$, then

$$x_i = (X, V_i) .$$

In fact, $(X, V_i) = (x_1 U_1 + \cdots + x_n U_n, V_i) = x_i$. Also

$$x_i' = (X, U_i)$$

where $x'_1, \cdots, x'_n$ are the coordinates of the vector $X$ in the basis $V_1, \cdots,$ $V_n$. In tensor algebra the coordinates $x_1, \cdots, x_n$ are called *contravariant* with respect to the basis $U_1, \cdots, U_n$ and the coordinates $x'_1, \cdots, x'_n$ are called *covariant* with respect to the same basis. Simultaneous use of these and other coordinates turns out to be very convenient. Thus, for example, the scalar product of two vectors is expressed by their coordinates with respect to any (non-orthogonal) basis by the following simple formula:

$$(X, Y) = \sum_{i=1}^{n} x_i \bar{y}'_i = \sum_{i=1}^{n} x'_i \bar{y}_i .$$

We shall now outline an inductive method for constructing a basis dual to a given one by a procedure which resembles the orthogonalization process. Let $U_1, \cdots, U_n$ be a basis for which a dual is to be constructed and let $V_1^{(0)}, \cdots, V_n^{(0)}$ be some other basis. We assume that the determinants

$$\Delta_1 = (V_1^{(0)}, U_1),$$

$$\Delta_2 = \begin{vmatrix} (V_1^{(0)}, U_1) & (V_2^{(0)}, U_1) \\ (V_1^{(0)}, U_2) & (V_2^{(0)}, U_2) \end{vmatrix}$$

$$\cdots \cdots \cdots \cdots \cdots \cdots$$

$$\Delta_n = \begin{vmatrix} (V_1^{(0)}, U_1) \cdots (V_n^{(0)}, U_1) \\ \cdots \cdots \cdots \cdots \cdots \\ (V_1^{(0)}, U_n) \cdots (V_n^{(0)}, U_n) \end{vmatrix} ,$$

are different from zero.

We construct the sequential systems of vectors

$$\{V_1^{(1)}, \cdots, V_n^{(1)}\}, \{V_1^{(2)}, \cdots, V_n^{(2)}\}, \cdots, \{V_1^{(n)}, \cdots, V_n^{(n)}\}$$

so that the $k$th system satisfies the first $k$ groups of biorthogonality conditions

$$(V_i^{(k)}, U_j) = \delta_{ij} \text{ for } i = 1, 2, \cdots, n; j = 1, \cdots, k .$$

We shall take

$$V_1^{(1)} = \frac{1}{(V_1^{(0)}, U_1)} V_1^{(0)},$$

$$V_i^{(1)} = V_i^{(0)} - (V_i^{(0)}, U_1) V_1^{(1)} \text{ for } i > 1 .$$

It is clear that $(V_1^{(1)}, U_1) = 1$ and $(V_i^{(1)}, U_1) = 0$ for $i > 1$.

We observe that

$$(V_2^{(1)}, U_2) = (V_2^{(0)}, U_2) - (V_2^{(0)}, U_1) \frac{(V_1^{(0)}, U_2)}{(V_1^{(0)}, U_1)} = \frac{\Delta_2}{\Delta_1} \neq 0 .$$

We shall assume that we have already constructed vectors $V_1^{(k-1)}, V_2^{(k-1)},$ $\cdots, V_n^{(k-1)}$ which satisfy the given conditions and that we have been convinced that $(V_k^{(k-1)}, U_k) = (\Delta_k/\Delta_{k-1}) \neq 0$. We let

$$V_k^{(k)} = \frac{1}{(V_k^{(k-1)}, U_k)} V_k^{(k-1)}$$

and

$$V_i^{(k)} = V_i^{(k-1)} - (V_i^{(k-1)}, U_k) V_k^{(k)}, \quad i \neq k.$$

Then

$$(V_k^{(k)}, U_k) = 1,$$
$$(V_i^{(k)}, U_k) = 0 \quad (\text{for } i \neq k)$$

and

$$(V_i^{(k)}, U_j) = (V_i^{(k-1)}, U_j) - \frac{(V_i^{(k-1)}, U_k)}{(V_k^{(k-1)}, U_k)} (V_k^{(k-1)}, U_j)$$

$$= (V_i^{(k-1)}, U_j) = \delta_{ij} \quad (\text{for } j \leq k - 1).$$

It remains to convince ourselves that $(V_{k+1}^{(k)}, U_{k+1}) = \Delta_{k+1}/\Delta_k$ and therefore that $(V_{k+1}^{(k)}, U_{k+1}) \neq 0$. With this in mind, we consider the sequence of matrices

$$A^{(k)} = \begin{bmatrix} (V_1^{(k)}, U_1) & \cdots & (V_1^{(k)}, U_n) \\ \cdots & \cdots & \cdots \\ (V_n^{(k)}, U_1) & \cdots & (V_n^{(k)}, U_n) \end{bmatrix}$$

$$= \begin{bmatrix} 1 & 0 & \cdots & 0 & (V_1^{(k)}, U_{k+1}) & \cdots & (V_1^{(k)}, U_n) \\ \cdots & \cdots & \cdots & \cdots & \cdots & \cdots & \cdots \\ 0 & 0 & \cdots & 1 & (V_k^{(k)}, U_{k+1}) & \cdots & (V_k^{(k)}, U_n) \\ 0 & 0 & \cdots & 0 & (V_{k+1}^{(k)}, U_{k+1}) & \cdots & (V_{k+1}^{(k)}, U_n) \\ \cdots & \cdots & \cdots & \cdots & \cdots & \cdots & \cdots \end{bmatrix}.$$

The matrix $A^{(k)}$ is obtained from the matrix $A^{(k-1)}$ by dividing the $k$th row by the number $(V_k^{(k-1)}, U_k) = \Delta_k/\Delta_{k-1}$ and by adding the $k$th row (with an appropriate multiplier) to each of the remaining rows. In this way all the principal minors of the matrix $A^{(k)}$, beginning with a minor of order $k$, are equal to the products of the corresponding minors of the matrix $A^{(k-1)}$ by $\Delta_{k-1}/\Delta_k$.

Analogously, all the principal minors of the matrix $A^{(k-1)}$, beginning with the minor of order $k-1$, are equal to the corresponding minors of the matrix $A^{(k-2)}$ multiplied by $\Delta_{k-2}/\Delta_{k-1}$, and so forth. Applying this to the principal minor of order $k+1$ of the matrix $A^{(k)}$, which is obviously equal to $(V_{k+1}^{(k)}, U_{k+1})$, we obtain

$$(V_{k+1}^{(k)}, U_{k+1}) = \frac{\Delta_{k-1}}{\Delta_k} \cdot \frac{\Delta_{k-2}}{\Delta_{k-1}} \cdots \frac{1}{\Delta_1} \Delta_{k+1} = \frac{\Delta_{k+1}}{\Delta_k}.$$

Computing a basis $V_1, \cdots, V_n$ dual to a basis $U_1, \cdots, U_n$ is essentially equivalent to inverting a matrix $A$ consisting of the coordinates of the vectors $U_1, \cdots, U_n$ with respect to some orthonormal basis. In fact if the matrix $B$ consists of the coordinates of the vector $V_1, \cdots, V_n$ in the same basis, then

$$B^* A = E,$$

which follows directly from the multiplication rule for matrices and the orthogonality conditions.

## 10.  LINEAR OPERATORS IN UNITARY AND EUCLIDEAN SPACE

**1.  Conjugate operators.**  Let $A$ be an operator defined in a unitary space. The operator $A^*$ is said to be *conjugate to* $A$ if for any two vectors $X$ and $Y$ the equality

$$(AX, Y) = (X, A^*Y)$$

is satisfied.

We shall prove the existence and uniqueness of the conjugate operator. Let there correspond to an operator $A$ in a certain orthonormal basis the matrix

$$A = \begin{bmatrix} a_{11} & a_{12} & \cdots & a_{1n} \\ a_{21} & a_{22} & \cdots & a_{2n} \\ \cdot & \cdot & \cdots & \cdot \\ a_{n1} & a_{n2} & \cdots & a_{nn} \end{bmatrix},$$

let there correspond to the vectors $X$ and $Y$ in the same basis the columns of coordinates $(x_1, x_2, \cdots, x_n)'$ and $(y_1, y_2, \cdots, y_n)'$. Then to the vector $AX$ will correspond the column matrix

$$\begin{bmatrix} a_{11}x_1 + \cdots + a_{1n}x_n \\ a_{21}x_1 + \cdots + a_{2n}x_n \\ \cdot \quad \cdot \quad \cdot \quad \cdot \quad \cdot \\ a_{n1}x_1 + \cdots + a_{nn}x_n \end{bmatrix}.$$

Consequently,

$$\begin{aligned}
(AX, Y) &= a_{11}x_1\bar{y}_1 + \cdots + a_{1n}x_n\bar{y}_1 \\
&\quad + a_{21}x_1\bar{y}_2 + \cdots + a_{2n}x_n\bar{y}_2 \\
&\quad + \cdot \quad \cdot \quad \cdot \quad \cdot \quad \cdot \quad \cdot \quad \cdot \quad \cdot \\
&\quad + a_{n1}x_1\bar{y}_n + \cdots + a_{nn}x_n\bar{y}_n \\
&= x_1(a_{11}\bar{y}_1 + a_{21}\bar{y}_2 + \cdots + a_{n1}\bar{y}_n) \\
&\quad + \cdot \quad \cdot \quad \cdot \quad \cdot \quad \cdot \quad \cdot \quad \cdot \quad \cdot \\
&\quad + x_n(a_{1n}\bar{y}_1 + a_{2n}\bar{y}_2 + \cdots + a_{nn}\bar{y}_n) \\
&= (X, A^*Y),
\end{aligned}$$

where $A^*$ is an operator having in the same basis the matrix

$$A^* = \begin{bmatrix} \bar{a}_{11} & \bar{a}_{21} & \cdots & \bar{a}_{n1} \\ \bar{a}_{12} & \bar{a}_{22} & \cdots & \bar{a}_{n2} \\ \cdot & \cdot & \cdots & \cdot \\ \bar{a}_{1n} & \bar{a}_{2n} & \cdots & \bar{a}_{nn} \end{bmatrix},$$

conjugate to the matrix $A$.

Thus we may take as a conjugate operator an operator having in some orthonormal basis a matrix conjugate to the matrix of the initial operator in the same basis.

We shall now prove the uniqueness of the conjugate operator. Let $A_1^*$ and $A_2^*$ be two operators conjugate to an operator $A$. Then $(AX, Y) = (X, A_1^*Y) = (X, A_2^*Y)$, from which $(X, (A_1^* - A_2^*)Y) = 0$. Consequently the vector $(A_1^* - A_2^*)Y$ is orthogonal to every vector $X$ of the space $R$ and therefore $(A_1^* - A_2^*)Y = 0$ for any $Y$. From this we conclude that $A_1^* - A_2^*$ is the null operator, that is, $A_1^* = A_2^*$.

From the definition and from the uniqueness of the conjugate operator it is clear that $(A^*)^* = A$.

In Euclidean space there will correspond to a conjugate operator (with respect to an orthonormal basis) a matrix which is the transpose of the matrix of operator $A$ (with respect to the same basis), since operators in Euclidean space are written as real matrices.

**2. Properties of eigenvalues and eigenvectors for operators in unitary space.** We shall consider certain relationships between the eigenvalues and eigenvectors of mutually conjugate operators $A$ and $A^*$.

First of all we observe that the characteristic polynomials of these operators have conjugate complex coefficients and therefore the eigenvalues of the operator $A^*$ will be conjugate to the eigenvalues of the operator $A$. This follows from the fact that the matrix $A^*$ of an operator $A^*$, in an orthonormal basis, is conjugate to the matrix $A$ of an operator $A$ in the same basis. In the case where the matrix of an operator $A$ is real, in some orthonormal basis, then the characteristic polynomials and the eigenvalues of the operators $A$ and $A^*$ coincide.

The relationships among eigenvectors are characterized by the following theorem.

THEOREM 10.1. *If $X_1$ is an eigenvector of an operator $A$ corresponding to an eigenvalue $\lambda_i$ and $Y_j$ is an eigenvector of an operator $A^*$ corresponding to an eigenvalue $\bar\lambda_j$, then $(X_i, Y_j) = 0$ for $\lambda_i \neq \lambda_j$.*

*Proof.* We shall calculate $(AX_i, Y_j)$ in two ways. On the one hand

$$(AX_i, Y_j) = (\lambda_i X_i, Y_j) = \lambda_i(X_i, Y_j).$$

On the other hand

$$(AX_i, Y_j) = (X_i, A^*Y_j) = (X_i, \bar\lambda_j Y_j) = \lambda_j(X_i, Y_j).$$

Therefore

$$(\lambda_i - \lambda_j)(X_i, Y_j) = 0$$

and since $\lambda_i - \lambda_j \neq 0$, it follows that

$$(X_i, Y_j) = 0,$$

which was to be proved.

From this it follows that if all the eigenvalues $\lambda_1, \cdots, \lambda_n$ of an operator $A$ are distinct, then for the eigenvectors $X_1, \cdots, X_n$ of operator $A$ and $Y_1, \cdots, Y_n$ of operator $A^*$, $n^2 - n$ orthogonality relationships will hold, namely, $(X_i, Y_j) = 0$, if $i \neq j$. We shall now show that by choosing the vectors $Y_1, \cdots, Y_n$ in a convenient way we can normalize the vectors $X_1, \cdots, X_n$ so that $(X_i, Y_i) = 1$.

First of all, we show that $(X_i, Y_i) = \alpha_i \neq 0$. In fact if $(X_i, Y_i) = 0$, then the vector $Y_i$ would be orthogonal to all the eigenvectors $X_1, \cdots, X_i, \cdots, X_n$ of the operator $A$ and consequently to all the vectors of the space $R$ as well, which would mean that $Y_i$ was a null vector. Taking, in place of the vectors $X_1, \cdots, X_n$, the vectors $(1/\alpha_1)X_1 \cdots, (1/\alpha_n)X_n$ we obtain the desired normalization, since

$$\left(\frac{1}{\alpha_i}X_i, \, Y_i \right) = \frac{1}{\alpha_i}(X_i, \, Y_i) = 1 \, .$$

Thus after the vector systems $X_1, \cdots, X_n$ and $Y_1, \cdots, Y_n$ have been normalized they form mutually dual bases of the space.

**3. Two sets of orthogonality relationships for the eigenvectors of a matrix.** Let A be a matrix whose eigenvalues $\lambda_1, \cdots, \lambda_n$ are distinct and let $X_1, \cdots, X_n$ be their corresponding eigenvectors (columns). As we have seen, the conjugate matrix $A^*$ has for its eigenvalues complex numbers conjugate to $\lambda_1, \cdots, \lambda_n$. Let $Y_1, \cdots, Y_n$ be the eigenvectors of the matrix $A^*$, normalized in accordance with the preceding notation. We form the matrices

$$X = \begin{bmatrix} x_{11} & x_{12} & \cdots & x_{1n} \\ x_{21} & x_{22} & \cdots & x_{2n} \\ \cdot & \cdot & \cdot & \cdot \\ x_{n1} & x_{n2} & \cdots & x_{nn} \end{bmatrix} \text{ and } Y = \begin{bmatrix} y_{11} & y_{12} & \cdots & y_{1n} \\ y_{21} & y_{22} & \cdots & y_{2n} \\ \cdot & \cdot & \cdot & \cdot \\ y_{n1} & y_{n2} & \cdots & y_{nn} \end{bmatrix}$$

whose columns contain the components of the eigenvectors of the matrix $A$ and the matrix $A^*$, respectively. The orthogonality and normalization relationships depicted above have in coordinate notation the form

$$x_{1i}\bar{y}_{1j} + x_{2i}\bar{y}_{2j} + \cdots + x_{ni}\bar{y}_{nj} = \begin{cases} 0 & (i \neq j) \\ 1 & (i = j) \end{cases},$$

which is equivalent to the matrix equality

$$Y^*X = E \, ,$$

where $Y^*$ is the matrix conjugate to matrix $Y$. We note that the $i$th row of the matrix $Y^*$ consists of the components of the eigenvector of matrix $A'$ (the transpose of $A$) which corresponds to the eigenvalue $\lambda_i$. From the equality $Y^*X = E$ it follows that

$$XY^* = E \, ,$$

which gives a *second set of orthogonality relationships*

$$x_{i1}\bar{y}_{j1} + x_{i2}\bar{y}_{j2} + \cdots + x_{in}\bar{y}_{jn} = \begin{cases} 0 & (i \neq j) \\ 1 & (i = j) \end{cases} .$$

**4. Orthogonality properties of root vectors**

THEOREM 10.2. *Any root vector of an operator which corresponds to an eigenvalue $\lambda$ is orthogonal to any root vector of the operator $A^*$ which corresponds to an eigenvalue $\mu \neq \bar{\lambda}$.*

*Proof.* Let $X_0$ be a root vector of an operator $A$ of height $m$ corresponding to an eigenvalue $\lambda$ and let $Y_0$ be a root vector of the conjugate operator $A^*$ of height k corresponding to an eigenvalue $\mu \neq \bar{\lambda}$. We construct the sequence of root vectors

$$X_1 = (A - \lambda E)X_0 ,$$
$$X_2 = (A - \lambda E)X_1 ,$$
$$\cdots \cdots \cdots ,$$
$$X_{m-1} = (A - \lambda E)X_{m-2} ,$$
$$(A - \lambda E)X_{m-1} = 0 ,$$

and

$$Y_1 = (A^* - \mu E)Y_0 ,$$
$$Y_2 = (A^* - \mu E)Y_1 ,$$
$$\cdots \cdots \cdots ,$$
$$Y_{k-1} = (A^* - \mu E)Y_{k-2} ,$$
$$(A^* - \mu E)Y_{k-1} = 0 .$$

The vectors $X_{m-1}$ and $Y_{k-1}$ will be eigenvectors for the operators $A$ and $A^*$, respectively, belonging to the eigenvalues $\lambda$ and $\mu$. Consequently $X_{m-1}$ and $Y_{k-1}$ are orthogonal. The proof of the theorem is carried out by induction. Let it given that $(X_{m-i}, Y_{k-j}) = 0$ for $i + j < l, l \geq 3$. We shall prove that $(X_{m-i}, Y_{k-j}) = 0$ for $i + j = l$ as well. In fact

$$\lambda(X_{m-i}, Y_{k-j}) = (\lambda X_{m-i}, Y_{k-j}) = (AX_{m-i} - X_{m-i+1}, X_{k-j})$$
$$= (AX_{m-i}, Y_{k-j}) - (X_{m-i+1}, X_{k-j}) = (X_{m-i}, A^*Y_{k-j})$$
$$- (X_{m-i+1}, Y_{k-j}) = (X_{m-i}, \mu Y_{k-j})$$
$$+ (X_{m-i}, Y_{k-j+1}) - (X_{m-i+1}, Y_{k-j})$$
$$= \mu(X_{m-i}, Y_{k-j}) ,$$

since $(X_{m-i}, Y_{k-j+1})=0$ and $(X_{m-i+1}, Y_{k-j}) = 0$ on the basis of the induction hypothesis. From here $(\lambda - \bar{\mu})(X_{m-i}, Y_{k-j}) = 0$ and consequently $(X_{m-i}, X_{k-j}) = 0$, since $\lambda \neq \bar{\mu}$.

## 5. Dual bases for canonical bases

THEOREM 10.3. *A basis of an operator $A$ which is dual to a canonical basis is canonical for the conjugate operator $A^*$ with "inverted towers"; more precisely, if $X_0, \ldots, X_{m-1}$ is a "tower" taken from a canonical basis for the operator $A$ so that*

$$AX_0 = \lambda X_0 + X_1 ,$$
$$AX_1 = \lambda X_1 + X_2 ,$$
$$\cdots \cdots \cdots ,$$
$$AX_{m-2} = \lambda X_{m-2} + X_{m-1} ,$$
$$AX_{m-1} = \lambda X_{m-1} ,$$

*then the dual vectors $Y_0, \cdots, Y_{m-1}$ satisfy the relationships*

$$A^* Y_{m-1} = \bar{\lambda} Y_{m-1} + Y_{m-2},$$
$$A^* Y_{m-2} = \bar{\lambda} Y_{m-2} + Y_{m-3},$$
$$\cdots \cdots \cdots \cdots \cdots,$$
$$A^* Y_0 = \bar{\lambda} Y_1 + Y_0,$$
$$A^* Y_0 = \bar{\lambda} Y_0.$$

*Proof.* First we observe that if to an operator $A$ in a certain basis $U_1, \cdots, U_n$ there corresponds a matrix $A$, then to the the conjugate operator $A^*$ in the dual basis $V_1, \cdots, V_n$ there corresponds a matrix $A^*$ conjugate to the matrix $A$.

Let $A = (a_{ij})$ for $i, j = 1, 2, \cdots, n$. Then $A U_j = \sum_{i=1}^{n} a_{ij} U_i$, from which it follows that $a_{ij} = (A U_j, V_i)$. Furthermore, let $(B = (b_{ij}) (i - 1, \cdots, n; j = 1, \cdots, n)$ be a matrix corresponding to the operator $A^*$ in the basis $V_1, \cdots, V_n$. Then $b_{ij} = (A^* V_j, U_i)$. But $(A^* V_j, U_i) = (V_j, A U_i) = \overline{A U_i, V_j)} = \bar{a}_{ji}$. This also proves that $B = A^*$.

In a canonical basis there corresponds to an operator $A$ a Jordan canonical matrix, and there corresponds to a tower $X_0, \cdots, X_{m-1}$ a Jordan canonical box

$$\begin{bmatrix} \lambda & & & & \\ 1 & \lambda & & & \\ & \cdot & \cdot & & \\ & & \cdot & \cdot & \\ & & & 1 & \lambda \end{bmatrix}.$$

In a dual basis there corresponds to an operator $A^*$ the conjugate matrix. In particular, there will correspond to the vectors $Y_0, \cdots, Y_{m-1}$ the box

$$\begin{bmatrix} \bar{\lambda} & 1 & & & \\ & \bar{\lambda} & \cdot & & \\ & & \cdot & \cdot & \\ & & & \cdot & 1 \\ & & & & \bar{\lambda} \end{bmatrix}.$$

This means that

$$A^* Y_{m-1} = \bar{\lambda} Y_{m-1} + Y_{m-2},$$
$$A^* Y_{m-2} = \bar{\lambda} Y_{m-2} + Y_{m-3},$$
$$\cdots \cdots \cdots \cdots \cdots,$$
$$A^* Y_1 = \bar{\lambda} Y_1 + Y_0,$$
$$A^* Y_0 = \bar{\lambda} Y_0,$$

which was to be proved.

## 11. SELF-CONJUGATE OPERATORS

In the present paragraph we shall assume not a unitary space but a Euclidean space, since the results of the theory given here, as distinct from the general theory of eigenvalues, do not require an extension to complex space.

The results also carry over to a unitary space with hardly any changes. We shall confine ourselves to formulating only the relevant theorems.

**1. Definition.** An operator $A$ is called *self-conjugate* if, for any two vectors $X$ and $Y$ belonging to $R$, the following equation holds:

$$(AX, Y) = (X, AY),$$

that is, the operator $A$ coincides with its conjugate, The matrix of a self-conjugate operator with respect to some orthonormal basis is real (Euclidean space) and symmetric, since it must coincide with its transpose.

Any linear combination of self-conjugate operators is a self-conjugate operator. Furthermore, if a self-conjugate operator $A$ is non-degenerate, then its inverse operator $A^{-1}$ is self-conjugate. In fact, letting $A^{-1}X = U$, $A^{-1}Y = V$ we have $(A^{-1}X, Y) = (U, AV) = (AU, V) = (X, A^{-1}Y)$. The converse is also obvious; if to an operator $A$ in some orthonormal basis there corresponds a symmetric matrix, then the operator is self-conjugate. In fact, in this case, to the operators $A$ and $A^*$ there corresponds one and the same matrix, and consequently they coincide.

Thus if one selects an orthonormal basis for a space, then self-conjugate operators are found to be in a natural one-to-one correspondence with symmetric matrices. Similarly, a close connection exists between self-conjugate operators and quadratic forms: namely, a scalar product $(AX, X)$ expressed by the coordinates of the vector $X$ is none other than a quadratic form, with a matrix which coincides with the matrix $A$ of the operator in the orthonormal basis. In fact

$$(AX, X) = \sum_{i=1}^{n} \sum_{j=1}^{n} a_{ij}x_ix_j \quad (a_{ij} = a_{ji}).$$

**2. A property of invariant subspaces of a self-conjugate operator**

THEOREM 11.1. *If $P$ is an invariant subspace of a self-conjugate operator $A$, then the orthogonally complementary subspace $Q$ is also an invariant subspace.*

*Proof.* Let $X \in Q$. This means that $(X, Y) = 0$ for any $Y \in P$. But then for any $Y \in P$ $(AX, Y) = (X, AY) = 0$ (since $AY \in P$ in view of the invariance of $P$). From here $AX \in Q$, which is what we were to prove.

**3. Constructing a system of mutually orthogonal eigenvectors of a self-conjugate operator.** The eigenvectors and eigenvalues of a self-conjugate operator possess a set of extremal properties from which there directly arise such important properties as: all the eigenvalues are real; there exists an orthogonal basis consisting of eigenvectors. At the base of these extremal properties lies the following theorem.

THEOREM 11.2. *There exists a maximum for the quotient $(AX, X)/(X, X)$*

*for X coursing the whole space (excluding the null vector). Any vector for which this maximum is attained is an eigenvector of the operator A and the value of the maximum is the corresponding eigenvalue.*

*Proof.*  Since

$$\frac{(AX, X)}{(X, X)} = \frac{(AX, X)}{|X|^2} = \left( A\frac{X}{|X|}, \frac{X}{|X|} \right),$$

then the maximum being considered is equal to the maximum for $(AX, X)$ with $X$ varying over the "unit sphere", that is, so that the magnitude of $X$ is equal to one. The existence of a maximum for $(AX, X)$ follows directly from the Weierstrass theorem on achieving a least upper bound for a continuous function on a bounded, closed set. Let $\lambda_1 = \max_{X \in R} (AX, X)/(X, X)$ $= (AX_1, X_1)$ where $X_1$ is a vector of length one which realizes the maximum. Then for any vector $X \in R$

$$(AX, X) \le \lambda_1(X, X).$$

We let $Y = AX_1 - \lambda_1 X_1$ and we shall prove that $Y = 0$. First of all we shall show that $Y$ is orthogonal to $X_1$. In fact $(Y, X_1) = (AX_1 - \lambda_1 X_1, X_1)$ $= (AX_1, X_1) - \lambda_1(X_1, X_1) = 0$ from the definition of $\lambda_1$.

Let $X = X_1 + \varepsilon Y$ where $\varepsilon$ is a positive real number. We have

$$(A(X_1 + \varepsilon Y), X_1 + \varepsilon Y) \le \lambda_1(X_1 + \varepsilon Y, X_1 + \varepsilon Y),$$

from which

$$(AX_1, X_1) + \varepsilon(AX_1, Y) + \varepsilon(AY, X_1) + \varepsilon^2(AY, Y)$$
$$\le \lambda_1(X_1, X_1) + 2\lambda_1\varepsilon(X_1, Y) + \lambda_1\varepsilon^2(Y, Y).$$

Taking into consideration that $(AX_1, X_1) = \lambda_1$; $(AX_1, Y) = (AY, X_1)$; $(X_1, X_1) = 1$ and $(X_1, Y) = 0$ we get

$$2\varepsilon(AX_1, Y) + \varepsilon^2(AY, Y) \le \varepsilon^2\lambda_1(Y, Y).$$

Dividing both members by $\varepsilon$, and letting $\varepsilon$ tend to zero, we get

$$(AX_1, Y) \le 0$$

Moreover,

$$(Y, Y) = (AX_1 - \lambda_1 X_1, Y) = (AX_1, Y) - \lambda_1(X_1, Y) = (AX_1, Y) \le 0,$$

From here $Y = 0$, that is, $AX_1 = \lambda_1 X_1$.

*Note.*  The quotient $(AX, X)/(X, X)$ is often called the *Rayleigh quotient*.

THEOREM 11.3.  *Let A be a self-conjugate operator. Then there exists a system of pairwise orthogonal eigenvectors of A which form a basis of the space. All the eigenvalues of A are real.*

*Proof.*  Let $X_1$ be a normalized eigenvector of the operator $A$ which gives a maximum for $(AX, X)$ on the unit sphere. We shall denote by $P_1$ the one dimensional space spanned by $X_1$ and by $Q_1$ its orthogonal complement.

It is obvious that $P_1$, and consequently $Q_1$, will be invariant subspaces. The dimension of $Q_1$ is equal to $n - 1$. The operator $A$ considered for $Q_1$ will obviously be self-conjugate and consequently there will be found for it a normalized eigenvector $X_2$ which realizes $\max\limits_{X \in Q_1;\ |X|=1} (AX, X)$. It is easy to see that its corresponding eigenvalue $\lambda_2$ will not be greater than $\lambda_1$. According to the construction $(X_1, X_2) = 0$.

Let $P_2$ be the subspace spanned by $X_1$ and $X_2$ and let $Q_2$ be its orthogonal complement (of dimension $n - 2$). The subspace $P_2$, and consequently also the subspace $Q_2$, will be invariant. We shall consider the operator $A$ on $Q_2$. One finds for this self-conjugate operator an eigenvector $X_3$ which realizes $\max\limits_{|X|=1;\ X \in Q_2} (AX, X)$; its corresponding eigenvalue $\lambda_3$ will not be greater than $\lambda_2$. It is obvious that $(X_1, X_3) = 0$ and $(X_2, X_3) = 0$. Continuing this process we arrive at a system of pairwise orthogonal eigenvectors $X_1, \cdots, X_n$. Their corresponding eigenvalues will satisfy the inequalities $\lambda_1 \geq \lambda_2 \geq \cdots \geq \lambda_n$. We shall show that in this sequence every eigenvalue is respected as many times as its multiplicity as a root of the characteristic polynomial. In fact there corresponds to the operator $A$ in the basis $X_1, \cdots, X_n$ the diagonal matrix

$$\begin{bmatrix} \lambda_1 & & & & \\ & \lambda_2 & & & \\ & & \cdot & & \\ & & & \cdot & \\ & & & & \cdot \\ & & & & & \lambda_n \end{bmatrix},$$

and consequently the characteristic polynomial of $A$ is

$$(\lambda_1 - t)(\lambda_2 - t)\cdots(\lambda_n - t) .$$

From the given construction it follows that all the eigenvalues $\lambda_1, \cdots, \lambda_n$ are real and also that for each repeated eigenvalue there exist as many pairwise orthogonal (and consequently linearly independent) eigenvectors as the multiplicity of this eigenvalue as a root of the characteristic polynomial.

In algebraic form theorem 11.3 may be formulated in the following way: *for any symmetric matrix $A$ there exists an orthogonal matrix $P$ such that $P'AP$ is diagonal.* In fact it is possible to consider the matrix $A$ as the matrix of a certain self-conjugate operator $A$ in an orthonormal basis. In the basis formed from the normalized eigenvectors the matrix $\Lambda$ of an operator $A$ will be diagonal. Consequently $\Lambda = P^{-1}AP$, where $P$ is the matrix formed from the components of the normalized eigenvectors. The matrix $P$ is orthogonal so that $P^{-1} = P'$ and $\Lambda = P'AP$.

The diagonal elements of the matrix $\Lambda$ are eigenvalues of the matrix $A$. A direct corollary of this extremal construction is the following theorem.

THEOREM 11.4. *Let $\lambda_1, \lambda_2, \cdots, \lambda_n$ be the eigenvalues of a self-conjugate operator $A$ and let $X_1, \cdots, X_n$ be the pairwise orthogonal eigenvectors belonging* *to them.* **Then**

$$\lambda_1 = \max_{X \neq 0} \frac{(AX, X)}{(X, X)},$$

$$\lambda_2 = \max_{X \perp X_1} \frac{(AX, X)}{(X, X)},$$

$$\cdots \cdots \cdots \cdots,$$

$$\lambda_k = \max_{\substack{X \perp X_1 \\ \cdots \\ X \perp X_{k-1}}} \frac{(AX, X)}{(X, X)}.$$

It is obvious that it is possible to change the form of this construction of the eigenvalues and their associated eigenvectors by computing, instead of the successive maximums of the quotient $(AX, X)/(X, X)$, the sequence of minimums of this quotient. Under such a construction the eigenvalues are defined in increasing order. Therefore the following theorem holds.

THEOREM 11.5. *Let $\lambda_1 \geq \lambda_2 \geq \cdots \geq \lambda_n$ be the eigenvalues of a self-conjugate operator $A$ and let $X_1, \cdots, X_n$ be their corresponding pairwise orthogonal eigenvectors.* **Then**

$$\lambda_n = \min_{X \neq 0} \frac{(AX, X)}{(X, X)},$$

$$\lambda_{n-1} = \min_{X \perp X_n} \frac{(AX, X)}{(X, X)},$$

$$\cdots \cdots \cdots \cdots,$$

$$\lambda_k = \min_{\substack{X \perp X_n \\ \cdots \\ X \perp X_{k+1}}} \frac{(AX, X)}{(X, X)}.$$

**4. Invariant subspaces of a self-conjugate operator.** Let $\lambda_1 > \lambda_2 > \cdots > \lambda_s$ be the eigenvalues of a self-conjugate operator $A$ with multiplicities $n_1$, $n_2, \cdots, n_s$. As was pointed out, it is possible in a space to construct a basis composed of the eigenvectors of the operator $A$, while the number of basis eigenvectors corresponding to an eigenvalue $\lambda_i$ is equal to its multiplicity $n_i$. We denote by $P_i$ the subspace spanned by the basis eigenvector corresponding to the value $\lambda_i$. The subspaces $P_1, \cdots, P_s$ are mutually orthogonal in view of the construction and the whole space $R$ is the direct sum of the subspaces $P_1, \cdots, P_s$. It is obvious that every vector of a subspace $P_i$ is an eigenvector of the operator $A$ corresponding to the eigenvalue $\lambda_i$. Conversely, every eigenvector belonging to the eigenvalue $\lambda_i$ belongs to $P_i$, that is, $P_i$ is the set of all eigenvectors belonging to $\lambda_i$. In fact let $AX = \lambda_i X$. We decompose the vector $X$ into subspaces of $P_i$

$$X = X_1 + \cdots + X_i + \cdots + X_s, X_j \in P_j.$$

Then

$$AX = \lambda_i X = \lambda_i X_1 + \cdots + \lambda_i X_i + \cdots + \lambda_i X_s.$$

On the other hand,

$$AX = AX_1 + \cdots + AX_i + \cdots + AX_s = \lambda_1 X_1 + \cdots + \lambda_i X_i + \cdots + \lambda_s X_s.$$

In view of the uniqueness of the decomposition of a vector via subspaces of $P_i$ we have

$$\lambda_i X_j = \lambda_j X_j$$

for all $j$; consequently $X_j = 0$ for $j \neq i$, that is, $X = X_i \in P_i$. From here arises the following important property of the eigenvectors of a self-conjugate operator $A$.

*The eigenvectors of a self-conjugate operator $A$ belonging to different eigenvalues are orthogonal. In fact they belong to mutually orthogonal subspaces.*

This property of eigenvectors is easily and directly proved without using the extremal construction, since if $X_i$ and $X_j$ are eigenvectors belonging to $\lambda_i$ and $\lambda_j$ where $\lambda_i \neq \lambda_j$, then

$$(\lambda_i - \lambda_j)(X_i, X_j) = (\lambda_i X_i, X_j) - (X_i, \lambda_j X_j),$$
$$= (AX_i, X_j) - (X_i, AX_j),$$
$$= (AX_i, X_j) - (AX_i, X_j) = 0,$$

from which $(X_i, X_j) = 0$.

*Note.* It is obvious that the subspaces $P_i$ are root subspaces for the self-conjugate operator $A$. Thus all root vectors for a self-conjugate operator are simply eigenvectors. Therefore the minimum polynomial for a self-conjugate operator has only simple roots.

From the expansion of the space $R$ into a direct sum of the subspaces $P_1, \cdots, P_s$ and from the extremal properties of eigenvalues of a self-conjugate operator (theorem 11.4 and 11.5) follows directly the validity of the following theorems.

THEOREM 11.6. *If $\lambda_1 > \cdots > \lambda_s$ are pairwise distinct eigenvalues of the operator $A$ and $P_1, \cdots, P_s$ are their corresponding subspaces of eigenvectors, then*

$$\lambda_1 = \max_{X \neq 0} \frac{(AX, X)}{(X, X)},$$

$$\lambda_2 = \max_{X \perp P_1} \frac{(AX, X)}{(X, X)},$$

$$\cdots \cdots \cdots,$$

$$\lambda_k = \max_{X \perp P_1 + \cdots + P_{k-1}} \frac{(AX, X)}{(X, X)}.$$

THEOREM 11.7. *If $\lambda_1 > \cdots > \lambda_s$ are pairwise distinct eigenvalues of the operator $A$ and $P_1, \cdots, P_s$ are their corresponding subspaces of eigenvectors, then*

$$\lambda_s = \min_{X \neq 0} \frac{(AX, X)}{(X, X)},$$

$$\lambda_{s-1} = \min_{X \perp P_s} \frac{(AX, X)}{(X, X)},$$

$$\cdots \cdots \cdots,$$

$$\lambda_k = \min_{X \perp P_s + \cdots + P_{k+1}} \frac{(AX, X)}{(X, X)}.$$

The construction of invariant subspaces for a self-conjugate operator is simpler than in the general case. In fact let $T$ be any invariant subspace. The operator $\tilde{A}$ induced by the operator $A$ on $T$ will obviously be self-conjugate and consequently in the space $T$ there exists a basis consisting of the eigenvectors of the operator $\tilde{A}$. But every eigenvector of the operator $\tilde{A}$ is at the same time an eigenvector of the operator $A$. *Thus any invariant subspace of a self-conjugate operator is spanned by some system of eigenvectors.* Moreover a subspace, similar to the whole space, is representable as a direct sum of subspaces $T_i$ consisting of eigenvectors belonging to eigenvalues $\lambda_i$. It is clear that $T_i = T \cap P_i$.

If $X \in T$ and $X = X_1 + \cdots + X_s$ is the representation of $X$ via subspaces of $P_i$, then all $X_i \in T_i$. In fact $T = T \cap P_1 + \cdots + T \cap P_s$ and consequently the expansion $X = X_1' + \cdots + X_s'$ also holds, where $X_i' \in T \cap P_i$. In view of the uniqueness of the representation of the vector $X$ by subspaces of $P_i$ all the projections $X_i'$ coincide with $X_i$. But $X_i'$ belongs to $T$.

The construction of cyclic subspaces is described by the following theorem.

THEOREM 11.8. *Let $X_0$ be an arbitrary vector of a space $R$. Then a cyclic subspace for a self-conjugate operator $A$ generated by the vector $X_0$ is a subspace spanned by non-zero projections of $X_0$ on the subspaces $P_1, \cdots, P_s$.*

This theorem is a direct corollary of the more general theorem 8.2. However we shall carry out an independent proof. Let $Q$ be a cyclic subspace generated by the vector $X_0$ and let

$$X_0 = X_{k_1} + \cdots + X_{k_j},$$
$$X_{k_i} \in P_{k_i}$$

be the representation of $X_0$ by vectors of the subspaces $P_1, \cdots, P_s$; in this expansion we have only non-zero projections. In view of what was said above, all $X_{k_1}, \cdots, X_{k_j}$ belong to $Q$ and consequently the subspace $Q'$ spanned by them is contained in $Q$. But $Q'$ is invariant, since it is spanned by the eigenvectors $X_{k_1}, \cdots, X_{k_j}$ and $Q'$ contains $X_0$. Consequently $Q \subset Q'$ and therefore $Q = Q'$.

**5. Positive definite operators.** Among self-conjugate operators a particularly important role is played by positive definite operators. A self-conjugate operator $A$ is said to be *positive definite* if for any vector $X$ different from zero $(AX, X) > 0$.

From the definition it follows that the kernel of a positive definite operator contains only the null vector, so that a positive definite operator is not degenerate and consequently there exists an inverse operator $A^{-1}$.

The matrix of a positive definite operator in an orthonormal basis is positive definite and, conversely, an operator with a positive definite matrix in an orthonormal basis is positive definite. In fact $(AX, X)$ is the quadratic form composed of the coordinates of a vector in an orthonormal basis; the matrix of the quadratic form coincides with the matrix corresponding to the operator in the same basis.

We shall note a series of interesting properties of positive definite operators.

THEOREM 11.9. *If $A$ and $B$ are positive definite operators, then the operator $c_1A + c_2B$ is positive definite for $c_1 > 0, c_2 > 0$. In other words, the positive definite operators form a convex cone in the space of the operators.*

*Proof.* We have $((c_1A + c_2B)X, X) = c_1(AX, X) + c_2(BX, X) > 0$ for $X \neq 0$.

THEOREM 11.10. *If an operator $A$ is positive definite, then the operator $A^{-1}$ is also positive definite.*

In fact

$$(A^{-1}X, X) = (A^{-1}X, AA^{-1}X) = (AY, Y) > 0 .$$

Here $Y = A^{-1}X$.

THEOREM 11.11. *All the eigenvalues of a positive definite operator are positive.*

If $\lambda$ is an eigenvalue of a positive definite operator $A$ and $X$ is its corresponding eigenvector, then $AX = \lambda X$ and

$$\lambda = \frac{(AX, X)}{(X, X)} > 0 .$$

THEOREM 11.12. *If $A$ is a positive definite operator and $C$ is any non-degenerate operator, then $C^*AC$ is positive definite.*

In fact $(C^*ACX, X) = (ACX, CX) > 0$, since for $X \neq 0$ the vector $CX \neq 0$.

In particular it follows from this for any non-degenerate operator $C$ the operator $C^*C$ is positive definite.

THEOREM 11.13. *For any positive definite operator $A$ there exists a positive definite "square root" $A^{\frac{1}{2}}$, that is, a positive definite operator $B$ such that $B^2 = A$.*

In fact let $\lambda_1, \cdots, \lambda_n$ be eigenvalues of an operator $A$ and let $X_1, \cdots, X_n$ be pairwise orthogonal eigenvectors belonging to them. Then $AX_i = \lambda_iX_i$ $(i = 1, 2, \cdots, n)$. We shall determine the operator $B$, letting $BX_i = \sqrt{\lambda_i}X_i$ and extending the operator linearly to the whole space. Then $B^2X_i = \lambda_iX_i$ $= AX_i$, that is, the operators $B^2$ and $A$ coincide in the basis $X_1, \cdots, X_n$. Consequently in view of the linearity they coincide in the whole space as well. Thus $B^2 = A$. It is obvious that the operator $B$ is positive definite.

THEOREM 11.14. *If $A$ is a self-conjugate operator and $B$ is a positive definite operator, then all the eigenvalues of the operator $BA$ (and consequently of $B^{-1}A$) are real.*

$BA = B^{\frac{1}{2}}(B^{\frac{1}{2}}AB^{\frac{1}{2}})B^{-\frac{1}{2}}$.    Therefore the eigenvalues of the operator $BA$ are equal to the eigenvalues of the operator $B^{\frac{1}{2}}AB^{\frac{1}{2}}$, which is obviously self-conjugate.

THEOREM 11.15.  *If $A$ and $B$ are positive definite operators, then all the eigenvalues of the operator $BA$ are positive.*

In this case the operator $B^{\frac{1}{2}}AB^{\frac{1}{2}} = (B^{\frac{1}{2}})^*AB^{\frac{1}{2}}$ is positive definite on the basis of theorem 11.12.

THEOREM 11.16.  *If $B$ is a positive definite operator, and if $A$ is a self-conjugate operator, and all the eigenvalues of the operator $BA$ are positive, then $A$ is positive definite.*

In fact if all eigenvalues of the operator $BA$ are positive, then all eigenvalues of the operator $B^{-\frac{1}{2}}(BA)B^{\frac{1}{2}} = B^{\frac{1}{2}}AB^{\frac{1}{2}}$ are also positive, that is, the self-conjugate operator $B^{\frac{1}{2}}AB^{\frac{1}{2}}$ is positive.  On the basis of theorem 11.12 operator $A$ will also be positive definite.

THEOREM 11.17.  *If $A$ and $B$ are commutative positive definite operators, then $AB$ is a positive definite operator.*

In this case $AB$ is self-conjugate, and its positive definite quality follows from the fact that its eigenvalues are positive.

It is obvious that all the properties described in theorems 11.9—11.17 hold true for positive definite matrices as well.

Every positive definite operator determines a certain metric for the space which satisfies all the axioms for the usual Euclidean metric.  Namely, the *A-scalar product* $(X, Y)_A$ of two vectors $X$ and $Y$ is the scalar product $(AX, Y)$.  Under such a definition all four axioms of a Euclidean metric are satisfied:

1.  $(X, X)_A > 0, X \neq 0$
2.  $(X, Y)_A = (Y, X)_A$
3.  $(aX, Y)_A = a(X, Y)_A$
4.  $(X_1 + X_2, Y)_A = (X_1, Y)_A + (X_2, Y)_A$

The third and fourth axioms are satisfied in view of the linearity of the operator $A$, the second in view of the self-conjugate, and the first in view of the positive definiteness.

In this metric the role of the length of a vector $X$ is played by the *A-length*, that is, the value $\sqrt{(AX, X)}$.  *A-orthogonal* or *conjugate* (relative to $A$) are vectors $X$ and $Y$ for which $(AX, Y) = 0$.  All theorems proved in section 9 are obviously carried over to a space with an *A*-metric.  In particular the following theorems are valid.

THEOREM 11.18.  $(AX, Y)^2 \le (AX, X)(AY, Y)$.

THEOREM 11.19.  *Pairwise A-orthogonal vectors are linearly independent.*

THEOREM 11.20.  *Let* $X_1, \cdots, X_k$ *be a given system of linearly independent vectors. Then it is possible to construct an A-orthogonal system of vectors connected with the initial one by the relationships*:

$$S_1 = X_1 ,$$
$$S_2 = X_2 + \alpha_{21}X_1 ,$$
$$\cdots \cdots \cdots ,$$
$$S_k = X_k + \alpha_{k1}X_1 + \cdots + \alpha_{kk-1}X_{k-1} .$$

In view of the fact that in pursuing certain numerical methods the $A$-orthogonalization process will be applied, and since numerical realization of these methods will be carried out according to the formulas of this process, we shall carry out a proof of the theorem, although it is almost a verbatim repetition of the proof of theorem 9.2.

*Proof.* We shall carry out the construction by induction. Let the vectors $S_1, \cdots, S_{m-1}$ be already constructed, all different from zero. The vector $S_m$ will be sought in the form

$$S_m = X_m - \gamma_{m1}S_1 - \cdots - \gamma_{mm-1}S_{m-1} . \qquad (1)$$

The coefficients $\gamma_{mj}$ will be determined from the condition of $A$-orthogonality of the vectors $S_1, \cdots, S_{m-1}$ to the vector $S_m$. In view of the $A$-orthogonality of the system of vectors $S_1, \cdots, S_{m-1}$ and since $S_j \ne 0$, $j = 1, \cdots, m-1$, we have

$$\gamma_{mj} = \frac{(X_m, AS_j)}{(S_j, AS_j)} = \frac{(S_j, AX_m)}{(S_j, AS_j)} \qquad (2)$$

Substituting for $S_1, \cdots, S_{m-1}$ in equation (1) their expressions in terms of $X_1, \cdots, X_{m-1}$, we obtain

$$S_m = X_m + \alpha_{m1}X_1 + \cdots + \alpha_{mm-1}X_{m-1} .$$

From this it follows that the vector $S_m$ is not the null vector, since otherwise the vectors $X_1, \cdots, X_m$ would be linearly dependent, which contradicts the condition of the theorem. The basis for induction is given by the trivial case $m = 1$.

Thus a system of $A$-orthogonal (conjugate) vectors is determined by the recurrence relationships

$$S_1 = X_1, \cdots, S_m = X_m - \gamma_{m1}S_1 - \cdots - \gamma_{mm-1}S_{m-1}$$

in which the coefficients are determined by formula (2).

In practical computations it is more convenient to use several other formulas for the coefficients. Namely, from the equation

$$X_j = S_j + \gamma_{j1}S_1 + \cdots + \gamma_{jj-1}S_{j-1}$$

it follows that

$$AX_j = AS_j + \gamma_{j1}AS_1 + \cdots + \gamma_{jj-1}AS_{j-1} .$$

It follows that

$$(S_m , AX_j) = 0 \quad (\text{for } j < m) ,$$
$$(S_j , AX_j) = (S_j , AS_j) .$$

Thus

$$\gamma_{mj} = \frac{(S_j , AX_m)}{(S_j , AX_j)} .$$

**6. Self-conjugate operators in unitary space.** For unitary space the definition of a self-conjugate operator coincides with the definition given above for Euclidean space. Namely, an operator $A$ is said to be self-conjugate if it coincides with its conjugate, that is, if

$$(AX, Y) = (X, AY)$$

for any $X$ and $Y$.

The matrix of a self-conjugate operator with respect to any orthogonal basis is Hermitian, since it coincides with its conjugate. For any vector $X$ the scalar product $(AX, X)$ is real. In fact, $(AX, X) = (X, AX) = \overline{(AX, X)}$. This identification means that the theory of self-conjugate operators in unitary space coincides formally with the similar theory for Euclidean space: namely, one carries out, with certain changes, the construction for setting up in sequence a complete system of pairwise orthogonal eigenvectors with the help of extremal considerations. From this construction it follows that *all eigenvalues of a self-conjugate operator are real and the matrix of a self-conjugate operator may be changed to diagonal form. In algebraic terms this means that any Hermitian matrix may be transformed to diagonal form, using a unitary matrix for the similarity transformation.*

Moreover, all extremal properties of eigenvalues are preserved, as is the structure of invariant subspaces. The scalar product $(AX, X)$, expressed in terms of the coordinates of $X$ in an orthonormal basis, is a Hermitian form of these coordinates with a matrix equal to the matrix of the operator $A$ in the same basis.

A self-conjugate operator is called *positive definite* if for any vector $X$ different from zero $(AX, X) > 0$. All eigenvalues of a positive definite operator are positive.

**7. Normal operators in unitary space.** An operator $A$ is said to be *normal* if it commutes with its conjugate operator, that is, if $AA^* = A^*A$.

It is obvious that the matrix of a normal operator in any orthonormal basis is a normal matrix.

Self-conjugate operators belong to the class of normal operators. In the same class are contained unitary operators, that is, operators satisfying the condition $A^* = A^{-1}$.

The theory of eigenvalues and eigenvectors for normal operators bears a close resemblance to the corresponding theory for self-conjugate operators. Namely, the following theorem is true.

THEOREM 11.21. *In a space there exists an orthonormal basis consisting of the eigenvectors of a given normal operator.*

## 12. QUADRATIC FORMS

### 1. Transformation of a quadratic form to canonical form

THEOREM 12.1. *Every quadratic form with a real matrix of coefficients may be transformed to the following canonical form:*

$$F(x_1, x_2, \cdots, x_n) = \alpha_1 y_1^2 + \cdots + \alpha_n y_n^2 ,$$

*where* $y_1, y_2, \cdots, y_n$ *are variables connected with the initial variables* $x_1, x_2, \cdots, x_n$ *by a non-singular linear transformation.*

*Proof.* For $n = 1$ the theorem is trivial. Assume that for forms of $n - 1$ variables the theorem is proved. With this assumption we shall prove that the theorem is also true for forms of $n$ variables. Let

$$
\begin{aligned}
F(x_1, x_2, \cdots, x_n) &= a_{11}x_1^2 + a_{12}x_1x_2 + \cdots \\
&\quad + a_{1n}x_1x_n \\
&\quad + a_{21}x_2x_1 + a_{22}x_2^2 + \cdots \\
&\quad + a_{2n}x_2x_n \\
&\quad \cdot\ \cdot\ \cdot\ \cdot\ \cdot \\
&\quad + a_{n1}x_nx_1 + a_{n2}x_nx_2 + \cdots + a_{nn}x_n^2 \\
&= a_{11}x_1^2 + 2a_{12}x_1x_2 + \cdots + 2a_{1n}x_1x_n + \varphi(x_2, \cdots, x_n) .
\end{aligned}
$$

We assume first that $a_{11} \neq 0$. Then

$$
\begin{aligned}
F(x_1, x_2, \cdots, x_n) &= a_{11}\left(x_1 + \frac{a_{12}}{a_{11}}x_2 + \cdots + \frac{a_{1n}}{a_{11}}x_n\right)^2 \\
&\quad - a_{11}\left(\frac{a_{12}}{a_{11}}x_2 + \cdots + \frac{a_{1n}}{a_{11}}x_n\right)^2 + \varphi(x_2, \cdots, x_n) \\
&= a_{11}\left(x_1 + \frac{a_{12}}{a_{11}}x_2 + \cdots + \frac{a_{1n}}{a_{11}}x_n\right)^2 + F_1(x_2, \cdots, x_n) .
\end{aligned}
$$

On the basis of the induction assumption

$$F_1(x_2, \cdots, x_n) = \alpha_2 y_2^2 + \cdots + \alpha_n y_n^2 ,$$

where

$$
\begin{bmatrix} y_2 \\ \cdot \\ \cdot \\ \cdot \\ y_n \end{bmatrix} = B_1 \begin{bmatrix} x_2 \\ \cdot \\ \cdot \\ \cdot \\ x_n \end{bmatrix} ,
$$

and $B_1$ is a non-singular matrix.

We let

$$y_1 = x_1 + \frac{a_{12}}{a_{11}} x_2 + \cdots + \frac{a_{1n}}{a_{11}} x_n .$$

Then

$$\begin{bmatrix} y_1 \\ y_2 \\ \cdot \\ \cdot \\ \cdot \\ y_n \end{bmatrix} = \begin{bmatrix} 1 & \frac{a_{12}}{a_{11}} & \cdots & \frac{a_{1n}}{a_{11}} \\ 0 & & & \\ \cdot & & & \\ \cdot & & B_1 & \\ \cdot & & & \\ 0 & & & \end{bmatrix} \begin{bmatrix} x_1 \\ x_2 \\ \cdot \\ \cdot \\ \cdot \\ x_n \end{bmatrix} = B \begin{bmatrix} x_1 \\ x_2 \\ \cdot \\ \cdot \\ \cdot \\ x_n \end{bmatrix} .$$

It is clear that $B$ is a non-singular matrix. With the new variables the quadratic form has the desired form

$$F(x_1, x_2, \cdots, x_n) = \alpha_1 y_1^2 + \alpha_2 y_2^2 + \cdots + \alpha_n y_n^2 \text{ for } \alpha_1 = a_{11} .$$

If $a_{11} = 0$ then it is always possible to make a preliminary non-singular transformation of variables under which the coefficient for the square of the new first variable turns out to be different from zero (assuming that the quadratic form is not equal to zero identically). In fact, if $a_{11} = 0$ but $a_{kk} \neq 0$ for a certain $k$, then it suffices to change only the order of numbering for the variables. If all the coefficients $a_{11}, \cdots, a_{nn}$ are equal to zero, but $a_{ij} \neq 0$, then it is enough to let

$$x_1 = y_1 ,$$
$$\cdots \cdots ,$$
$$x_i = y_i + y_j ,$$
$$\cdots \cdots \cdots ,$$
$$x_j = y_j ,$$
$$\cdots \cdots \cdots ,$$
$$x_n = y_n .$$

This transformation is obviously non-singular. The transformed quadratic form will have a non-zero coefficient $2a_{ij}$ for $y_j^2$.

*Note.* This proof also gives a description of the computational process for transforming a quadratic form to a canonical form. It will often happen that the first case holds for all steps of the process and then the necessity for auxiliary transformations vanishes. In this case the matrix of the final transformation $B$ will be triangular.

**2. Positive definite quadratic forms.** We recall that the quadratic form $F(x_1, \cdots, x_n)$ is called positive definite if all its values are positive except the value for $x_1 = x_2 = \cdots = x_n = 0$.

THEOREM 12.2. *For a quadratic form to be positive definite it is necessary and sufficient that all its coefficients be positive after its transformation to canonical form.*

*Proof.* Let

$$F(x_1, x_2, \cdots, x_n) = \alpha_1 y_1^2 + \alpha_2 y_2^2 + \cdots + \alpha_n y_n^2 .$$

If $\alpha_1 > 0, \alpha_2 > 0, \cdots, \alpha_n > 0$, the $F(x_1, x_2, \cdots, x_n)$ is obviously positive definite. If $\alpha_i \leq 0$ for some $i$, then letting $y_1 = 0, \cdots, y_i = 1, \cdots, y_n = 0$ and finding corresponding values $x_1^0, \cdots, x_n^0$ of the variables $x_1, x_2, \cdots, x_n$ we obtain $F(x_1^0, \cdots, x_n^0) = \alpha_i \leq 0$. The theorem is proved.

The established criterion for positive definiteness is very convenient for practical control.

There exists another criterion for the positive definiteness of a quadratic form, the application of which does not demand a transformation to the canonical form.

THEOREM 12.3. (*Criterion of Sylvester*). *A quadratic form* $\sum a_{ij} x_i x_j$ *is positive definite if and only if all the determinants*

$$\Delta_1 = a_{11}, \Delta_2 = \begin{vmatrix} a_{11} & a_{12} \\ a_{21} & a_{22} \end{vmatrix}, \cdots, \Delta_n = \begin{vmatrix} a_{11} & \cdots & a_{1n} \\ \cdots & \cdots & \cdots \\ a_{n1} & \cdots & a_{nn} \end{vmatrix}$$

*are positive.*

*Proof.* Let $F(x_1, \cdots, x_n) = \sum_{i,j=1}^{n} a_{ij} x_i x_j$ be positive definite. Then $A = B'\Lambda B$, where $B$ is the matrix which transforms a form to the canonical form, $\Lambda = [\alpha_1, \cdots, \alpha_n]$ is a diagonal matrix which consists of the coefficients of the canonical representation. In view of the criterion proved above for positive definiteness all $\alpha_i > 0$. Therefore $\Delta_n = |A| = |B'| \cdot |\Lambda| \cdot |B| = \alpha_1 \cdot \alpha_2 \cdots \alpha_n |B|^2 > 0$. Moreover if the form $F(x_1, \cdots, x_n)$ is positive definite, then the forms

$$F_k(x_1, \cdots, x_k) = F(x_1, \cdots, x_k, 0, \cdots, 0) = \sum_{i,j=1}^{n} a_{ij} y_i y_j$$

will also be positive definite. Therefore all the determinants $\Delta_k (k = 1, 2, \cdots, n)$ are positive.

We now assume that $\Delta_1 > 0, \cdots, \Delta_n > 0$. Then the matrix $A$ may be expanded into the product

$$A = C\Lambda B$$

where $\Lambda = [\Delta_1, \Delta_2/\Delta_1, \cdots, \Delta_n/\Delta_{n-1}]$ and $C$ and $B$ are triangular matrices with diagonal elements one. Since the matrix $A$ is symmetric, it follows that $C = B'$. If one now makes a transformation, in the quadratic form, of the variables via the matrix $B^{-1}$ one arrives at a quadratic form with the matrix $\Lambda$, that is, at a canonical form with positive coefficients. Thus the form $F(x_1, \cdots, x_n)$ is positive definite.

## 3. Orthogonal transformation of a quadratic form to a canonical form

THEOREM 12.4. *Any quadratic form with a real coefficient matrix may* be transformed to canonical form by employment of a transformation of variables having an orthogonal matrix.

*Proof.* Let $A$ be the matrix of a quadratic form. Then there exists, in accordance with the algebraic formulation of theorem 11.3, an orthogonal matrix $P$ such that $P'AP = \Lambda$ is a diagonal matrix. Having effected, on the quadratic form, a transformation of variables with the matrix $P$, we arrive at a quadratic form in new variables and having the matrix $\Lambda$, that is, at the quadratic form $\lambda_1 y_1^2 + \lambda_2 y_2^2 + \cdots + \lambda_n y_n^2$. The theorem is proved.

We observe that the coefficients $\lambda_1, \cdots, \lambda_n$, obtained by transforming the quadratic form to the canonical form via the orthogonal transformation, are eigenvalues of the matrix of the quadratic form.

**4. Law of inertia for quadratic forms.** It is clear that a given quadratic form may be transformed to canonical form in an infinite number of ways. For example, one may first employ some kind of non-singular transformation of variables and then apply the method described in the first paragraph. It is then obvious that even the coefficient set $\alpha_1, \alpha_2, \cdots, \alpha_n$ may differ. However the following important theorem holds.

THEOREM 12.5. *The number of positive, negative, and zero coefficients in a canonical representation of a quadratic form does not depend on the method of transformation.*

*Proof.* Let

$$F(x_1, \cdots, x_n) = \alpha_1 y_1^2 + \cdots + \alpha_p y_p^2 - \alpha_{p+1} y_{p+1}^2 - \cdots - \alpha_{p+q} y_{p+q}^2$$
$$= \beta_1 z_1^2 + \cdots + \beta_r z_r^2 - \beta_{r+1} z_{r+1}^2 - \cdots - \beta_{r+s} z_{r+s}^2$$

be two canonical representations of a given quadratic form. In this notation we assume that all $\alpha_i > 0$ and $\beta_i > 0$, i.e. that we have thrown out all terms having zero coefficient. Thus the variables $y_{p+q+1}, \cdots, y_n$ and $z_{r+s+1}, \cdots, z_n$ do not enter into the transformed quadratic forms. We interpret the quadratic form as the functional of a vector $X$ whose coordinates in a certain given basis are the numbers $x_1, \cdots, x_n$. Then passing from the variables $x_1, \cdots, x_n$ to the variables $y_1, \cdots, y_n$, and then passing to the variables $z_1, \cdots, z_n$, may be interpreted as a coordinate transformation. We denote by $V_1, \cdots, V_n$ that basis in which the coordinates of the vector $X$ are $y_1, \cdots, y_n$ and by $W_1, \cdots, W_n$ that basis in which the coordinates are $z_1, \cdots, z_n$. We need to prove that $p = r$, $q = s$. We assume that $p < r$. We shall consider the subspace $Q$ spanned by the vectors $V_{p+1}, \cdots, V_{p+q}, \cdots, V_n$ and the subspace $P$ spanned by the vectors $W_1, \cdots, W_r$. The sum of the dimensions of these subspaces is equal to $n - p + r > n$.

From the theorem on the dimensions of a sum and an intersection we conclude that the dimension of $Q \cap P$ is greater than zero, since the dimension of the vector sum of $Q$ and $P$ may not be greater than $n$. Consequently, in $Q \cap P$ there exists at least one vector $X_0 \neq 0$. Let $y_1^0, \cdots, y_n^0$ and $z_1^0, \cdots,$ $z_n^0$ be its coordinates, respectively, in the bases $V_1, \cdots, V_n$ and $W_1, \cdots, W_n$, Since $X_0 \in Q$, then $y_1^0 = \cdots = y_p^0 = 0$; since $X_0 \in P$, then $z_{r+1}^0 = \cdots = z_n^0 = 0$. Consequently

$$F(X_0) = - \alpha_{p+1} y_{p+1}^{0\,2} - \cdots - \alpha_{p+q} y_{p+q}^{0\,2} ,$$
$$F(X_0) = \beta_1 z_1^{0\,2} + \cdots + \beta_r z_r^{0\,2} .$$

From the first equation we conclude that $F(X_0) \leq 0$; from the second it follows that $F(X_0) > 0$, since at least one of the coordinates $z_1^0, \cdots, z_r^0$ is different from zero (otherwise the vector $X_0$ would be null). The resultant contradiction shows that the assumption that $p < r$ was untenable. In exactly the same way the assumption that $p > r$ leads us to a contradiction. Consequently $p = r$ and the theorem's assertion as to the number of positive coefficients is proved.

To prove that $q = s$ it is sufficient to consider instead of the form $F$ the form $-F$.

### 5. Simultaneous transformation of two quadratic forms to a canonical form

THEOREM 12.6. *Let $F(x_1, \cdots, x_n)$ and $\Phi(x_1, \cdots, x_n)$ be two quadratic forms and let $\Phi(x_1, \cdots, x_n)$ be positive definite. Then both forms may be changed to canonical form by one and the same transformation of variables.*

*Proof.* We shall carry out some transformation of the form $\Phi(x_1, \cdots, x_n)$ to canonical form

$$\Phi(x_1, \cdots, x_n) = \alpha_1 y_1^2 + \cdots + \alpha_n y_n .$$

We shall make the same transformation on the form $F(x_1, \cdots, x_n)$. The form obtained will be denoted by $F_1(y_1, \cdots, y_n)$. Moreover in both quadratic forms we let $y_1 = \dfrac{z_1}{\sqrt{\alpha_1}}, \cdots, y_n = \dfrac{z_n}{\sqrt{\alpha_n}}$. The coefficients of this transformation are real, since $\alpha_1 > 0, \cdots, \alpha_n > 0$. After this transformation we obtain that

$$\Phi(x_1, \cdots, x_n) = z_1^2 + \cdots + z_n^2$$
$$F(x_1, \cdots, x_n) = F_2(z_1, \cdots, z_n) .$$

we now change the form $F_2(z_1, \cdots, z_n)$ to canonical form by the orthogonal transformation

$$(z_1, \cdots, z_n)' = P(t_1, \cdots, t_n)' .$$

After the transformation we get

$$F_2(z_1, \cdots, z_n) = \lambda_1 t_1^2 + \cdots + \lambda_n t_n^2 .$$

Under this transformation the form $z_1^2 + \cdots + z_n^2$ is transformed into $t_1^2 + \cdots + t_n^2$. In fact, the matrix of the form $z_1^2 + \cdots + z_n^2$ will be a unit matrix;

the matrix of the transformed form will also be a unit matrix, since $P'EP = E$ on the basis of the orthogonality of matrix $P$.

We observe that the numbers $\lambda_1, \cdots, \lambda_n$ may be determined without performing the indicated transformations. We shall denote the matrix of the form $F(x_1, \cdots, x_n)$ by $A$, the matrix of the form $\Phi(x_1, \cdots, x_n)$ by $B$, and the diagonal matrix $[\lambda_1, \cdots, \lambda_n]$ by $\Lambda$.

Let $C$ be the matrix of a linear transformation which produces a simultaneous change of both forms to the canonical form by the method indicated above. Then $C'AC = \Lambda$, $C'BC = E$, and consequently $\Lambda - tE = C'AC - tC'BC = C'(A - tB)C$. From this, $|\Lambda - tE| = |C'| |A - tB| \cdot |C|$ or $(\lambda_1 - t) \cdots (\lambda_n - t) = |C|^2 \cdot |A - tB|$.

Thus the numbers $\lambda_1, \cdots, \lambda_n$ turn out to be zeros of the polynomial $|A - tB|$. The equation $|A - tB| = 0$ is called the *generalized secular equation*.

This proof is essentially equivalent to the following argument.

Let $A$ be the matrix of the form $F(x_1, \cdots, x_n)$ and let $B$ be the matrix for the form $\Phi(x_1, \cdots, x_n)$. The matrix $B$ is positive definite by assumption. Let $Q$ be an orthogonal matrix which transforms the quadratic form with matrix $B^{-1/2}AB^{-1/2}$ into canonical form. Then $Q'B^{-1/2}AB^{-1/2}Q = \Lambda$, where $\Lambda$ is diagonal. We write $B^{-1/2}Q = C$. Then

$$C'AC = Q'B^{-1/2}AB^{-1/2}Q = \Lambda$$
$$C'BC = Q'B^{-1/2}BB^{-1/2}Q = Q'Q = E.$$

Thus the transformation of variables with the matrix $C$ carries the quadratic form $F$ to canonical form with coefficients $\lambda_1, \cdots, \lambda_n$ and the quadratic form $\Phi$ to canonical form with coefficients $1, \cdots, 1$.

**6. Hermitian forms.** The algebraic expression

$$\begin{aligned}F(z_1, \cdots, z_n) = &a_{11}z_1\bar{z}_1 + a_{12}z_1\bar{z}_2 + \cdots + a_{1n}z_1\bar{z}_n \\ &+ a_{21}z_2\bar{z}_1 + a_{22}z_2\bar{z}_2 + \cdots + a_{2n}z_2\bar{z}_n \\ &+ \cdots \cdots \cdots \cdots \\ &+ a_{n1}z_n\bar{z}_1 + a_{n2}z_n\bar{z}_2 + \cdots + a_{nn}z_n\bar{z}_n,\end{aligned}$$

where $z_1, \cdots, z_n$ are complex variables and the coefficients satisfy the condition $a_{ij} = \bar{a}_{ji}$ is called a *Hermitian form*. The matrix found of the coefficients of a Hermitian form is by definition a Hermitian matrix. All values of a Hermitian form are real, since

$$\overline{F(z_1, \cdots, z_n)} = \overline{\sum_{i,j} a_{ij}z_i\bar{z}_j} = \sum_{i,j} \bar{a}_{ij}\bar{z}_iz_j = F(z_1, \cdots, z_n).$$

Theorems analogous to theorems for quadratic forms hold for Hermitian forms. Namely,

THEOREM 12.7.  *Any Hermitian form may be changed by a non-singular transformation of variables to the canonical form*

$$\alpha_1 z_1\bar{z}_1 + \alpha_2 z_2\bar{z}_2 + \cdots + \alpha_n z_n\bar{z}_n.$$

A Hermitian form is called positive definite if all its values are positive except the value for

$$z_1 = z_2 = \cdots = z_n = 0 \,.$$

The coefficient matrix of a positive definite Hermitian form is called a *positive definite Hermitian matrix*

**THEOREM 12.8.** *For a Hermitian form to be positive definite it is necessary and sufficient that all the coefficients in its canonical expansion be positive.*

**THEOREM 12.9.** *A Hermitian form may be changed to canonical form with the aid of a unitary transformation of variables.*

Under such a transformation the coefficients in the canonical form will be the eigenvalues of the matrix of the Hermitian form.

**THEOREM 12.10.** (*Law of inertia*). *The number of positive, negative, and zero coefficients in a canonical representation of a Hermitian form does not depend on the mode of transformation.*

**THEOREM 12.11.** *Two Hermitian forms, one of which is positive definite, may be changed to canonical form by one and the same transformation of variables.*

Proofs of these theorems coincide almost exactly with proofs of the analogous theorems for quadratic forms.

## 13. THE CONCEPT OF A LIMIT IN LINEAR ALGEBRA

The concept of a limit for linear algebraic objects will be needed by us primarily to describe iterative methods. In view of the fact that numerical problems are formulated in terms of matrices we shall define the concept of a limit for columns, which we shall identify with the vectors of arithmetic space in their natural representation, and for square matrices. In the sequel on we shall use the term 'vector' primarily in this sense.

**1. Limits of vectors and matrices.** Let there be given a sequence of vectors $X^{(1)}, \cdots, X^{(k)}, \cdots$ with components $x_1^{(1)}, \cdots, x_n^{(1)}; \cdots, x_1^{(k)}, \cdots, x_n^{(k)};$ $\cdots$ If for each component there exists a limit, $\lim_{k\to\infty} x_i^{(k)} = x_i$, then the vector $X$ with components $x_1, \cdots, x_n$ is called the *limit* of the sequence $X^{(1)}, \cdots, X^{(k)}, \cdots$, and the sequence itself is said to *converge* to the vector $X$. This is denoted as $X^{(k)} \to X$ or $\lim_{k\to\infty} X^{(k)} = X$.

Thus if there exists a sequence of square matrices $A^{(1)}, \cdots, A^{(k)}, \cdots$ with elements $a_{ij}^{(1)}, \cdots, a_{ij}^{(k)}, \cdots$, then the limit of the sequence $A^{(1)}, \cdots, A^{(k)} \cdots$ is a matrix with elements $a_{ij} = \lim_{k\to\infty} a_{ij}^{(k)}$, assuming that all these limits exist.

Corresponding to this definition of a limit, an infinite series of vectors

$X^{(1)} + X^{(2)} + \cdots + X^{(k)} + \cdots$ is called *convergent* if $\lim\limits_{k \to \infty} (X^{(1)} + X^{(2)} + \cdots + X^{(k)})$ exists; this limit is called the *sum* of the given series. It is obvious that for the convergence of a series of vectors it is necessary and sufficient that all the series of corresponding components converge; the sums of these series are the components of the sum of the vector series.

The concept of convergence for a series of matrices is defined in an analogous manner.

The limit concept is readily extended to vectors and operators in any linear space by considering columns of coordinates (for the vectors) and by considering matrices (for operators) with respect to a certain basis. It is easy to prove that the choice of a basis does not influence convergence nor the result of extending the concept of limit.

In certain problems the concept of convergence for a sequence of vectors with respect to one direction turns out to be useful.

A sequence of vectors $X^{(1)}, \cdots, X^{(k)}, \cdots$ is said to be *directionally convergent* if the sequence of normalized vectors $\dfrac{1}{|X^{(1)}|}X^{(1)}, \cdots, \dfrac{1}{|X^{(k)}|}X^{(k)},$ $\cdots$ is convergent in the usual sense. It is clear that for directional convergence it is sufficient that the sequence $b_k X^{(k)}$ converge in the usual sense to a non-null vector. Here $b_k$ are any positive numbers. In fact, if $b_k X^{(k)} \to X \neq 0$, then $b_k |X^{(k)}| \to |X|$ and consequently

$$\frac{1}{|X^{(k)}|}X^{(k)} = \frac{1}{b_k |X^{(k)}|}b_k X^{(k)} \to \frac{1}{|X|}X .$$

It is easy to see that the basic operations (addition, subtraction, multiplication) on vectors and matrices are continuous. In particular, if $A^{(k)} \to A$, then $A^{(k)}X \to AX$ for any vector $X$. Conversely, if $A^{(k)}X \to AX$ for any vector $X$, then $A^{(k)} \to A$. In fact, letting $X = e_i = (0, \cdots, 1, \cdots, 0)'$ for $i = 1, 2, \cdots, n$, we find that every column of the matrix $A^{(k)}$ converges to the corresponding column of matrix $A$; consequently all elements of matrix $A^{(k)}$ converge to the corresponding elements of matrix $A$.

Moreover if a sequence of square matrices has a non-singular matrix $A$ as a limit, then for sufficiently large $k$ there will exist an inverse for $A^{(k)}$ and $\lim\limits_{k \to \infty} (A^{(k)})^{-1} = A^{-1}$.

If $A^{(k)} \to A$, then it is obvious that matrices $B^{(k)}$ adjoint to matrices $A^{(k)}$ converge to the matrix $B$ adjoint to $A$, since their elements are polynomials, respectively, of the elements of $A^{(k)}$ and the elements of $A$.

For the same reason $|A^{(k)}| \to |A| \neq 0$, and consequently, from some point on, $|A^{(k)}| \neq 0$.

Finally $(A^{(k)})^{-1} \to A^{-1}$, since

$$(A^{(k)})^{-1} = \frac{1}{|A^{(k)}|}B_k \to \frac{1}{|A|}B = A^{-1} ,$$

We note the following theorem.

**THEOREM 13.1.** *If a sequence of matrices $A^{(k)}$ has a non-singular matrix*

*A as a limit and if the vectors $F^{(k)}$ converge to $F$, then solutions of the system*

$$A^{(k)} X^{(k)} = F^{(k)}$$

*have a limit which is a solution of the system $AX = F$.*

In fact, $X^{(k)} = (A^{(k)})^{-1} F^{(k)} \to A^{-1} F$.

For matrices whose elements are differentiable functions of a certain parameter $t$ it is natural to define differentiation with respect to this parameter. Namely,

$$\frac{d}{dt} A(t) = \lim_{h \to 0} \frac{1}{h} (A(t + h) - A(t)) .$$

If

$$A(t) = \begin{vmatrix} a_{11}(t) & \cdots & a_{1n}(t) \\ \cdot & \cdots \cdots & \cdot \\ a_{n1}(t) & \cdots & a_{nn}(t) \end{vmatrix} ,$$

then obviously

$$A'(t) = \begin{vmatrix} a'_{11}(t) & \cdots & a'_{1n}(t) \\ \cdot & \cdots \cdots & \cdot \\ a'_{n1}(t) & \cdots & a'_{nn}(t) \end{vmatrix} .$$

It is easy to establish the following rules for differentiation:

$$\frac{d}{dt}(A_1 + A_2) = \frac{dA_1}{dt} + \frac{dA_2}{dt} ,$$

$$\frac{d}{dt}(cA) = c \frac{dA}{dt} ,$$

$$\frac{dt}{d}(A_1 A_2) = \frac{dA_1}{dt} A_2 + A_1 \frac{dA_2}{dt} .$$

**2. Norms of vectors.** In applied questions it will be important to discuss not only the fact of convergence of sequences or series itself but also the rate of convergence. With this in mind it is very useful to introduce so-called *norms* of vectors and matrices. One may introduce a norm by different definitions, and in individual cases one or another norm turns out to be more convenient.

In general, the *norm of a vector* $X$ is a non-negative number $||X||$ corresponding to this vector which satisfies the following conditions:

1. $||X|| > 0$ for $X \neq 0$ and $||0|| = 0$;
2. $||cX|| = |c| \cdot ||X||$ for any numerical multiplier $c$;
3. $||X + Y|| \leq ||X|| + ||Y||$ ("triangular inequality").

From conditions 2) and 3) it is easy to deduce that

$$||X - Y|| \geq |\, ||X|| - ||Y|| \,| .$$

Namely, $||X|| = ||X - Y + Y|| \leq ||X - Y|| + ||Y||$ and therefore

$$\| X - Y \| \geq \| X \| - \| Y \| .$$

Moreover

$$\| X - Y \| = \| Y - X \| \geq \| Y \| - \| X \| .$$

Consequently

$$\| Y - X \| \geq \| X \| - \| Y \| .$$

Every norm determines a "unit sphere"—a set of vectors whose norms do not exceed 1. The unit sphere is a centrally symmetric convex body, that is, a set such that together with each vector $X$ it contains a vector $- X$ (central symmetry) and together with any two vectors $X_1$ and $X_2$ it contains a vector $tX_1 + (1 - t)X_2$, $0 \leq t \leq 1$, resting on the segment which joins the end points of the vectors $X_1$ and $X_2$ (convexity). Conversely, any centrally symmetric convex body $V$ in real space (in complex space central symmetry must be replaced by the more general requirement—together with a vector $X$ the body contains the vector $\alpha X$, $| \alpha | = 1$) generates a norm $\| X \|_V$ which is defined as *inf t*, where $t > 0$, $(1/t)X \in V$.

It is not difficult to verify the axioms for such a general definition of norm

We shall henceforth make use of the following three norms of a vector

$$X = (x_1, x_2, \cdots, x_n)' ,$$

defined for complex as well as real arithmetic spaces.

1. *First norm (cubic).*

$$\| X \|_I = \max_i | x_i | .$$

The set of vectors of real space with a norm which does not exceed 1 fills up the unit cube

$$- 1 \leq x_1 \leq 1, \cdots, - 1 \leq x_n \leq 1 .$$

2. *Second norm (octahedral).*

$$\| X \|_{II} = | x_1 | + | x_2 | + \cdots + | x_n | .$$

The set of real vectors for which $\| X \|_{II} \leq 1$ fills up the $n$-dimensional analogue of the octahedron.

3. *Third norm (spherical).*

$$\| X \|_{III} = | X | = \sqrt{| x_1 |^2 + | x_2 |^2 + \cdots + | x_n |^2} .$$

This norm is none other than the length of the vector. The collection of vectors for which $| X | \leq 1$ fills up the sphere of unit radius.

Conditions 1–3 are satisfied for all three norms.

For the cubic and octahedral norms this is obvious. For the spherical norm, fulfillment of conditions 1 and 2 is obvious; condition 3 is satisfied on the basis of the Cauchy-Bunyakovskii (Schwarz) inequality. In fact

$$|X + Y|^2 = (X + Y, X + Y)$$
$$= (X, X) + (X, Y) + (Y, X) + (Y, Y)$$
$$\leq |X|^2 + 2|(X, Y)| + |Y|^2$$
$$\leq |X|^2 + 2|X| \cdot |Y| + |Y|^2$$
$$= (|X| + |Y|)^2 .$$

The norms introduced above are connected with the following inequalities:

$$||X||_{\mathrm{I}} \leq ||X||_{\mathrm{II}} \leq n\,||X||_{\mathrm{I}} , \tag{1}$$

$$||X||_{\mathrm{I}} \leq |X| \quad \leq \sqrt{n}\,||X||_{\mathrm{I}} , \tag{2}$$

$$\frac{1}{\sqrt{n}}||X||_{\mathrm{II}} \leq |X| \quad \leq ||X||_{\mathrm{II}} . \tag{3}$$

Inequalities (1) and (2) are obvious, as is the right side of inequality (3). The left inequality of (3) is easy to deduce from the Cauchy-Bunyakovskii inequality. Namely,

$$(|x_1| + |x_2| + \cdots + |x_n|)^2$$
$$= (|x_1| \cdot 1 + |x_2| \cdot 1 + \cdots + |x_n| \cdot 1)^2$$
$$\leq (|x_1|^2 + |x_2|^2 + \cdots + |x_n|^2)(1^2 + 1^2 + \cdots + 1^2)$$
$$= n(|x_1|^2 + \cdots + |x_n|^2) ,$$

from which we get the left inequality of (3) by taking the square root and dividing by $\sqrt{n}$.

It is easy to establish that a necessary and sufficient condition for the convergence of a sequence of vectors $X^{(k)}$ to a vector $X$ is that $||X^{(k)} - X|| \to 0$ for each of the three norms.

For the first norm this is obvious. For the remaining norms it follows from inequalities (1) and (2).

If $X^{(k)} \to X$, then $||X^{(k)}|| \to ||X||$. In fact, $||X^{(k)}|| - ||X|| \leq ||X - X^{(k)}|| \to 0$.

**3. Norms of matrices.** The *norm of a square matrix* $A$ is a non-negative number $||A||$ which satisfies the conditions

1. $||A|| > 0$, if $A \neq 0$ and $||0|| = 0$;
2. $||cA|| = |c| \cdot ||A||$;
3. $||A + B|| \leq ||A|| + ||B||$;
4. $||AB|| \leq ||A|| \cdot ||B||$.

Just as for the norm of vectors, the condition $||A^{(k)} - A|| \to 0$ is necessary and sufficient in order that $A^{(k)} \to A$; likewise from $A^{(k)} \to A$ it follows that $||A^{(k)}|| \to ||A||$.

Often the following two norms of matrices are required:

$$M(A) = n \max_{i,j} |a_{ij}|$$

and

$$N(A) = \sqrt{\sum_{i,j} |a_{ij}|^2} = \sqrt{\mathrm{Sp}A^*A} .$$

The fact that these two norms satisfy the first three conditions is obvious, since they [$M(A)$ to within a multiplier $n$] are cubic and spherical norms of the matrix considered as a vector in $n^2$-dimensional arithmetic space. It remains to verify that the fourth condition is satisfied. Let $A = (a_{ij})$, $B = (b_{ij})$. Then $AB = \left( \sum\limits_{s=1}^{n} a_{is}b_{sj} \right)$. We have

$$M(AB) = n \max \left| \sum_{s=1}^{n} a_{is}b_{sj} \right| \leq n \max \sum_{s=1}^{n} |a_{is}| \cdot |b_{sj}|$$

$$\leq n \sum_{s=1}^{n} \frac{1}{n} M(A) \cdot \frac{1}{n} M(B) = M(A) \cdot M(B) .$$

Furthermore

$$N^2(AB) = \sum_{i,j} \left| \sum_{s=1}^{n} a_{is}b_{sj} \right|^2 ,$$

By the Cauchy-Bunyakovskii inequality

$$\left| \sum_{s=1}^{n} a_{is}b_{sj} \right|^2 \leq \sum_{s=1}^{n} |a_{is}|^2 \sum_{t=1}^{n} |b_{tj}|^2 .$$

from which

$$N^2(AB) \leq \sum_{i,j,s,t} |a_{is}|^2 |b_{tj}|^2 = \sum_{i,s} |a_{is}|^2 \cdot \sum_{j,t} |b_{jt}|^2 = N^2(A)N^2(B) .$$

Thus

$$N(AB) \leq N(A)N(B) .$$

It is obvious that $N(A) \leq M(A)$ holds for any matrix $A$.

The norm of a matrix may be introduced in an infinite number of ways. Since in the majority of problems connected with approximation theory vectors, as well as matrices, take part simultaneously in the argument, it is advisable to introduce matrix norms so that a matrix norm may be intelligibly related to vector norms in the same discussion. We shall say that the norm of matrix is *compatible* with a given norm of a vector if for the matrix $A$ and the vector $X$, we obtain the inequality

$$|| AX || \leq || A || \cdot || X || .$$

Thus the norm $M(A)$ introduced above is compatible with the cubic, octahedral, and spherical norms of a vector and $N(A)$ is compatible with the spherical norm of a vector.

In fact, let $X = (x_1, \cdots, x_n)'$. Then

$$|| AX ||_{\mathrm{I}} = \max_i \left| \sum_j a_{ij}x_j \right| \leq \max_i \sum_j |a_{ij}| \cdot |x_j|$$

$$\leq \sum_j \frac{M(A)}{n} || X ||_{\mathrm{I}} = M(A) \cdot || X ||_{\mathrm{I}} .$$

Furthermore

$$|| AX ||_{\mathrm{II}} = \sum_i \left| \sum_j a_{ij}x_j \right| \leq \sum_{i,j} |a_{ij}| \cdot |x_j| \leq \sum_{i,j} \frac{M(A)}{n} \cdot |x_j|$$

$$= M(A) \sum_j |x_j| = M(A) \cdot || X ||_{\mathrm{II}} .$$

Thus the norm $M(A)$ is compatible with the cubic and octahedral norms of a vector.

Finally,

$$|AX|^2 = \sum_i \left|\sum_j a_{ij}x_j\right|^2 \le \sum_i \sum_j |a_{ij}|^2 \sum_s |x_s|^2$$

$$= N^2(A)\,|X|^2 \le M^2(A)\,|X|^2 \, .$$

It follows from these inequalities that both the norms $M(A)$ and $N(A)$ are compatible with the spherical norm of a vector.

We shall indicate a construction which makes it possible to set up a least norm of a matrix compatible with a given norm for vectors. Namely, we take for the norm of a matrix $A$ the maximum of the norms of the vectors $AX$, under the assumption that the vector $X$ runs over the set of all vectors whose norm is equal to one:

$$\|A\| = \max_{\|X\|=1} \|AX\| \, .$$

Because of the continuity of the norm, this maximum is attained for each matrix $A$, that is, a vector $X_0$ is found such that $\|X_0\| = 1$ and $\|AX_0\| = \|A\|$.

We shall prove that a norm constructed in this way satisfies all the conditions 1)–4), as well as the compatibility condition.

We shall begin by verifying the *first requirement*. Let $A \ne 0$. Then a vector $X$ can be found, $\|X\| = 1$, such that $AX \ne 0$ and consequently $\|AX\| \ne 0$. Therefore $\|A\| = \max_{\|X\|=1} \|AX\| > 0$. If $A = 0$, then $\|A\| = \max_{\|X\|=1} \|0X\| = 0$.

*Second requirement.* In view of the definition, $\|cA\| = \max_{\|X\|=1} \|cAX\| = |c| \max_{\|X\|=1} \|AX\|$ and consequently $\|cA\| = |c| \cdot \|A\|$.

We shall verify the compatibility condition. Let $Y \ne 0$ be any vector; then $X = 1/\|Y\| \cdot Y$ satisfies the condition $\|X\| = 1$. Consequently $\|AY\| = \|A(\|Y\|X)\| = \|Y\| \cdot \|AX\| \le \|Y\| \cdot \|A\|$.

*Third requirement.* For the matrix $A + B$ there exists a vector $X_0$ such that $\|A + B\| = \|(A + B)X_0\|$ and $\|X_0\| = 1$. Then $\|A + B\| = \|(A + B)X_0\| = \|AX_0 + BX_0\| \le \|AX_0\| + \|BX_0\| \le \|A\| \cdot \|X_0\| + \|B\| \cdot \|X_0\| = \|A\| + \|B\|$.

Finally, the *fourth requirement*. For the matrix $AB$ there exists a vector $X_0$ such that $\|X_0\| = 1$ and $\|ABX_0\| = \|AB\|$. Then $\|AB\| = \|ABX_0\| = \|A(BX_0)\| \le \|A\| \cdot \|BX_0\| \le \|A\| \cdot \|B\| \cdot \|X_0\| = \|A\| \cdot \|B\|$.

We have verified that all four requirements and the compatibility condition are satisfied. We shall call the norm of a matrix constructed in this way *subordinate* to the given norm of vectors. We shall prove that a subordinate norm is no greater than any norm compatible with the same norm. In fact, Let $L(A)$ be a norm compatible with some norm of a vector, let $\|A\|$ be a norm subordinate to the same norm of a vector. Then a vector $X_0$ is found with a norm equal to one such that

$$|| A || = || AX_0 || \, .$$

But

$$|| AX_0 || \leq L(A) \, || X_0 || = L(A) \, ,$$

from which it follows that

$$|| A || \leq L(A) \, .$$

It is obvious that for any norm of matrices which is subordinate to some norm of vectors, $|| E || = 1$. From here it follows that the norms $N(A)$ and $M(A)$ are not subordinate to any of the norms of vectors, since $M(E) = N^2(E) = n$.

We shall now construct the norms of matrices subordinate to all the norms introduced above for vectors.

I.  $|| X ||_{\mathrm{I}} = \max_i | x_i |$.

The norm of matrices subordinate to this norm of vectors is

$$|| A ||_{\mathrm{I}} = \max_i \sum_{k=1}^{n} | a_{ik} | \, .$$

In fact, let $|| X ||_{\mathrm{I}} = 1$. Then

$$|| AX ||_{\mathrm{I}} = \max_i \left| \sum_{k=1}^{n} a_{ik}x_k \right| \leq \max_i \sum_{k=1}^{n} | a_{ik} | \cdot | x_k | \leq \max_i \sum_{k=1}^{n} | a_{ik} | \, .$$

Consequently

$$\max_{|| X ||=1} || AX || \leq \max_i \sum_{k=1}^{n} | a_{ik} | \, .$$

We shall now prove that $\max_{|| X ||=1} || AX ||$ is actually equal to $\max_i \sum_{k=1}^{n} | a_{ik} |$. We construct for this a vector $X_0$ such that $|| X_0 ||_{\mathrm{I}} = 1$ and $|| AX_0 || = \max_i \sum_{k=1}^{n} | a_{ik} |$. Namely, let $\sum_{k=1}^{n} | a_{ik} |$ attain a greatest value for $i = j$; then as the component $x_k^{(0)}$ of the vector $X_0$ we take $x_k^{(0)} = \dfrac{| a_{jk} |}{a_{jk}}$, if $a_{jk} \neq 0$, and $x_k^{(0)} = 1$, if $a_{jk} = 0$. It is obvious that $|| X_0 || = 1$. Moreover $\left| \sum_{k=1}^{n} a_{ik}x_k^{(0)} \right| \leq \sum_{k=1}^{n} | a_{ik} | \leq \sum_{k=1}^{n} | a_{jk} |$ for $i \neq j$ and $\left| \sum_{k=1}^{n} a_{jk}x_k^{(0)} \right| = \sum_{k=1}^{n} | a_{jk} |$. Consequently

$$\max_i \left| \sum_{k=1}^{n} a_{ik}x_k^{(0)} \right| = \sum_{k=1}^{n} | a_{jk} | = \max_i \sum_{k=1}^{n} | a_{ik} | \, .$$

Thus $|| AX_0 ||_{\mathrm{I}} = \max_i \sum_{k=1}^{n} | a_{ik} |$, which was to be proved.

It is obvious that the norm $|| A ||_{\mathrm{I}}$ may be represented as

$$|| A ||_{\mathrm{I}} = \max_i || A^* e_i ||_{\mathrm{II}} \, .$$

II.  $|| X ||_{\mathrm{II}} = \sum_{i=1}^{n} | x_i |$.

The norm of matrices subordinate to this vector norm is

$$\| A \|_{\mathrm{II}} = \max_{k} \sum_{i=1}^{n} | a_{ik} | .$$

In fact, let $\| X \|_{\mathrm{II}} = 1$. Then

$$\| AX \| = \sum_{i=1}^{n} \left| \sum_{k=1}^{n} a_{ik} x_k \right| \leq \sum_{i=1}^{n} \sum_{k=1}^{n} | a_{ik} | \, | x_k | \leq \sum_{k=1}^{n} | x_k | \left( \sum_{i=1}^{n} | a_{ik} | \right)$$

$$\leq \left( \max_{k} \sum_{i=1}^{n} | a_{ik} | \right) \sum_{k=1}^{n} | x_k | = \max_{k} \sum_{i=1}^{n} | a_{ik} | .$$

We now form the vector $X_0$ in the following way: let $\sum_{i=1}^{n} | a_{ik} |$ attain its greatest value for the column with number $j$. We shall let $x_k^{(0)} = 0$ for $k \neq j$ and $x_j^{(0)} = 1$. It is clear that the vector constructed in this way has a norm equal to one. Moreover,

$$\| AX_0 \| = \sum_{i=1}^{n} \left| \sum_{k=1}^{n} a_{ik} x_k^{(0)} \right| = \sum_{i=1}^{n} | a_{ij} | = \max_{k} \sum_{i=1}^{n} | a_{ik} | .$$

Thus,

$$\max_{\| X \| = 1} \| AX \| = \max_{k} \sum_{i=1}^{n} | a_{ik} | ,$$

which was to be proved.

It is readily seen that the norm $\| A \|_{\mathrm{II}}$ may be represented as

$$\| A \|_{\mathrm{II}} = \max_{i} \| Ae_i \|_{\mathrm{II}} .$$

III.  $| X |^2 = \sum_{k=1}^{n} | x_k |^2 = (X, X)$.

The matrix norm subordinate to this vector norm is

$$\| A \| = \sqrt{\lambda_1} ,$$

where $\lambda_1$ is the greatest eigenvalue of the matrix $A^*A$. In fact,

$$\| A \| = \max_{|X|=1} | AX | .$$

But

$$| AX |^2 = (AX, AX) = (X, A^*AX) .$$

The matrix $A^*A$ is Hermitian. Let $\lambda_1$ be its greatest eigenvalue. Then for $| X | = 1$, $\max (X, A^*AX) = \lambda_1$. Consequently,

$$\| A \| = \sqrt{\lambda_1} .$$

Sometimes this norm is called the *upper bound* of the matrix. Correspondingly, the smallest eigenvalue of the matrix $A^*A$ is called the *lower bound* of the matrix $A$. It is obvious that a lower bound of the matrix $A$ is a number inverse to the upper bound of the inverse matrix.

Equating the different matrix norms leads to the following inequalities:[†]

---

† Some of these inequalities are contained in the work of Turing (1).

$$\frac{1}{n}M(A) \leqq \|A\|_{\mathrm{I}} \leqq M(A); \tag{1}$$

$$\frac{1}{n}M(A) \leqq \|A\|_{\mathrm{II}} \leqq M(A); \tag{2}$$

$$\frac{1}{n}M(A) \leqq \|A\| \leqq M(A); \tag{3}$$

$$\frac{1}{n}M(A) \leqq N(A) \leqq M(A); \tag{4}$$

$$\frac{1}{\sqrt{n}}N(A) \leqq \|A\| \leqq N(A); \tag{5}$$

$$\frac{1}{\sqrt{n}}N(A) \leqq \|A\|_{\mathrm{I}} \leqq \sqrt{n}\, N(A); \tag{6}$$

$$\frac{1}{\sqrt{n}}N(A) \leqq \|A\|_{\mathrm{II}} \leqq \sqrt{n}\, N(A); \tag{7}$$

$$\frac{1}{\sqrt{n}}\|A\| \leqq \|A\|_{\mathrm{I}} \leqq \sqrt{n}\,\|A\|; \tag{8}$$

$$\frac{1}{\sqrt{n}}\|A\| \leqq \|A\|_{\mathrm{II}} \leqq \sqrt{n}\,\|A\|; \tag{9}$$

$$\frac{1}{n}\|A\|_{\mathrm{I}} \leqq \|A\|_{\mathrm{II}} \leqq n\,\|A\|_{\mathrm{I}}. \tag{10}$$

The right-hand inequalities of (1), (2), (3), (4), and (5) have already been noted. They are all exact, since they are transformed into an equality for the matrix $A$ which has $a_{ij} = 1$ $(i, j = 1, \cdots, n)$.

We shall prove the left members of inequalities (1), (2), (3), (4), and (5). We have

$$\|A\|_{\mathrm{I}} = \max_i \sum_j |a_{ij}| \geq \max_{i,j} |a_{ij}| = \frac{1}{n}M(A),$$

$$\|A\|_{\mathrm{II}} = \max_j \sum_i |a_{ij}| \geq \max_{i,j} |a_{ik}| = \frac{1}{n}M(A),$$

$$\|A\| = \max_{|X|=1} |AX| \geq \max_i |Ae_i|$$

$$= \max_i \sqrt{|a_{i1}|^2 + \cdots + |a_{in}|^2} \geq \max_{i,j} |a_{ij}| = \frac{1}{n}M(A),$$

$$\|A\|^2 = \max \lambda_i \geq \frac{1}{n}(\lambda_1 + \cdots + \lambda_n) = \frac{1}{n}\mathrm{Sp}\, A^*A = \frac{1}{n}N^2(A).$$

Here $\lambda_1, \cdots, \lambda_n$ are eigenvalues of the matrix $A^*A$. Finally.

$$N(A) = \sqrt{\sum_{i,j} |a_{ij}|^2} \geq \max_{i,j} |a_{ij}| = \frac{1}{n}M(A).$$

These inequalities are also exact. The first four of them are transformed into an equality for $A = E$, the last one for

$$A = \begin{bmatrix} 1 & 0 & \cdots & 0 \\ 0 & 0 & \cdots & 0 \\ & & \cdot & \cdot & \cdot & \cdot \\ 0 & 0 & \cdots & 0 \end{bmatrix}.$$

We shall now prove inequality (8). We have

$$\| A \| = \max_{|X|=1} | A^*X | \geq \max_i | A^*e_i | \geq \max_i \frac{1}{\sqrt{n}} \| A^*e_i \|_{\text{II}} = \frac{1}{\sqrt{n}} \| A \|_{\text{I}}.$$

Furthermore,

$$\| A \| = \max_{|X|=1} | AX | \leq \max_{|X|=1} \sqrt{n} \, \| AX \|_{\text{I}}$$

$$\leq \max_{|X|=1} \sqrt{n} \, \| A \|_{\text{I}} \| X \|_{\text{I}} \leq \sqrt{n} \, \| A \|_{\text{I}},$$

from which $\dfrac{1}{\sqrt{n}} \| A \| \leq \| A \|_{\text{I}}$.

Replacing $A$ by $A^*$, one establishes inequality (9). From inequalities (8) and (9) inequality (10) follows directly.

From (8) and (9) and the right member of inequality (5) follow the right member inequalities (6) and (7).

We shall prove the left inequality (6). We have

$$\| A \|_{\text{I}} = \max_i \| A^*e_i \|_{\text{II}} \geq \max_i | A^*e_i |$$

$$\geq \frac{1}{\sqrt{n}} \sqrt{| A^*e_1 |^2 + \cdots + | A^*e_n |^2} = \frac{1}{\sqrt{n}} N(A).$$

Replacing $A$ by $A^*$ transforms inequality (6) into inequality (7).

Inequalities (6)—(10) are exact. The right inequality (6), left (7), right (8), left (9), and left (10) are attained for

$$A = \begin{bmatrix} 1 & 1 & \cdots & 1 \\ 0 & 0 & \cdots & 0 \\ & & \cdot & \cdot & \cdot & \cdot \\ 0 & 0 & \cdots & 0 \end{bmatrix},$$

the left (6), right (7), left (8), right (9), and right (10) for

$$A = \begin{bmatrix} 1 & 0 & \cdots & 0 \\ 1 & 0 & \cdots & 0 \\ & & \cdot & \cdot & \cdot & \cdot \\ 1 & 0 & \cdots & 0 \end{bmatrix}.$$

**4. Convergence of a geometric progression.** We shall now prove several theorems connected with the limit concept.

THEOREM 13.2. *In order that $A^m \to 0$ it is necessary and sufficient that all eigenvalues of the matrix $A$ be less than one in modulus.*

*Proof.* It is clear that for any fixed non-singular matrix $C$ the matrices

$A^m$ and $(C^{-1}AC)^m = C^{-1}A^mC$ do, or fail to, approach zero simultaneously. Therefore it suffices to prove the validity of the theorem for a Jordan canonical matrix. Moreover, in order that a sequence of quasi-diagonal matrices of the same structure approach zero it is necessary and sufficient that the sequence of separate boxes converge to zero. Thus we need to establish the condition of convergence to zero only for matrices $J^m$, where $J$ is a Jordan canonical box. Let

$$J = \begin{bmatrix} \lambda & & & & & \\ 1 & \lambda & & & & \\ & 1 & \lambda & & & \\ & & & \ddots & \ddots & \\ & & & & \ddots & \ddots \\ & & & & & 1 & \lambda \end{bmatrix}.$$

It is readily found that for $m \geq k - 1$

$$J^m = \begin{bmatrix} \lambda^m & & & \\ \binom{m}{1}\lambda^{m-1} & & \lambda^m & \\ \cdots\cdots\cdots\cdots\cdots\cdots\cdots\cdots & \\ \binom{m}{k-1}\lambda^{m-k+1} & \binom{m}{k-2}\lambda^{m-k+2} & \cdots & \lambda^m \end{bmatrix},$$

where $k$ is the order of the box and

$$\binom{m}{j} = \frac{m(m-1)\cdots(m-j+1)}{j!}.$$

For $J^m$ to converge to zero it is necessary that $|\lambda| < 1$, since $\lambda^m$ are the diagonal elements of $J^m$.

But this condition is also sufficient, since under this condition

$$\frac{m(m-1)\cdots(m-j+1)}{j!}\lambda^{m-j} \to 0$$

for $m \to \infty$, for all $j = 1, \cdots, k-2$.

The given conditions in theorem 13.2 are inconvenient for verifying this, since they require a knowledge of the eigenvalues of matrix $A$. Therefore we shall establish certain simpler sufficiency conditions in order that $A^m \to 0$.

THEOREM 13.3. *In order that $A^m \to 0$ it is sufficient that at least one of norms of $A$ be less that one.*

*Proof.* In view of the fourth requirement for a norm we have:

$$\| A^m \| \leq \| A^{m-1} \| \cdot \| A \| \leq \cdots \leq \| A \|^m.$$

Therefore if $\| A \| < 1$, then $\| A^m \| \to 0$ and, in view of what was said above, $A^m \to 0$.

Comparing theorems 13.2 and 13.3 we arrive at the following conclusion.

THEOREM 13.4. *The modulus of every eigenvalue of a matrix does not exceed any of its norms.*

*Proof.* Let $\|A\| = a$. We shall consider the matrix $B = \dfrac{1}{a + \varepsilon}A$, where $\varepsilon$ is any positive number. Then

$$\|B\| = \frac{a}{a + \varepsilon} < 1 .$$

Consequently, $B^m \to 0$ for $m \to \infty$. In view of theorem 13.2 its eigenvalues will be less than 1 in modulus. But it is obvious that the eigenvalues of matrix $B$ are equal to $1/(a + \varepsilon)\lambda_i$, where $\lambda_i$ are eigenvalues of matrix $A$. Thus $|\lambda_i|/(a + \varepsilon) < 1$, that is, $|\lambda_i| < a + \varepsilon$. Since it is possible to take $\varepsilon$ as small as we wish, $|\lambda_i| \leq a$.

THEOREM 13.5. *For the series*

$$E + A + A^2 + \cdots + A^m + \cdots \tag{1}$$

*to converge it is necessary and sufficient that $A^m \to 0$ for $m \to \infty$. In this case the sum of series (1) is equal to $(E - A)^{-1}$.*

*Proof.* The necessity of this condition is obvious. We shall show that it is also sufficient. Let $A^m \to 0$. On the basis of theorem 13.2 all the eigenvalues of matrix $A$ are less than 1 in modulus. Consequently $|E - A| \neq 0$ and therefore $(E - A)^{-1}$ exists.

We consider the identity

$$(E + A + \cdots + A^k)(E - A) = E - A^{k+1} .$$

Post-multiplication by $(E - A)^{-1}$ yields

$$E + A + \cdots + A^k = (E - A)^{-1} - A^{k+1}(E - A)^{-1} .$$

Hence it follows that for $k \to \infty$

$$E + A + \cdots + A^k \to (E - A)^{-1} ,$$

since $A^{k+1} \to 0$.

Consequently,

$$E + A + \cdots + A^m + \cdots = (E - A)^{-1} ,$$

which was to be proved.

On the basis of theorem 13.2 the necessary and sufficient condition for the convergence of series (1) is the inequality $|\lambda_i| < 1$ for all eigenvalues of matrix $A$. A sufficient indication of convergence, on the basis of theorem 13.3, is the inequality $\|A\| < 1$ for at least one of the norms. For satisfying this condition we may to utilize the following estimate for the rate of convergence of series (1).

THEOREM 13.6. *If $\|A\| < 1$, then*

$$\| (E - A)^{-1} - (E + A + \cdots + A^k) \| \le \frac{\| A \|^{k+1}}{1 - \| A \|} \cdot$$

*Proof.*  We have:

$$(E - A)^{-1} - (E + A + \cdots + A^k) = A^{k+1} + A^{k+2} + \cdots$$

This implies that

$$\| (E - A)^{-1} - (E + A + \cdots + A^k) \| \le \| A \|^{k+1} + \| A \|^{k+2} + \cdots = \frac{\| A \|^{k+1}}{1 - \| A \|} \cdot$$

The theorem is proved.

**5. Some estimates of eigenvalues.**  In the preceding section certain approximations for eigenvalues were obtained.  Namely, it was established that all eigenvalues were less in modulus than any norm of a matrix. Estimates using the first and second norm are especially convenient, since they are simply expressed by the matrix elements.

We shall now deduce some estimates which give more precise information on the distribution of the eigenvalues of a matrix.  These estimates are called Gershgorin estimates, since they appeared for the first time in his work (1).

THEOREM 13.7.  *Let $A = (a_{ij})$ be a matrix with arbitrary complex elements. All the eigenvalues of this matrix are located in a region $D$ which is the union of the circles*

$$| z - a_{ii} | \le R_i \quad (i = 1, \cdots, n),$$

*where*

$$R_i = \sum_{\substack{j=1 \\ j \ne i}}^{n} | a_{ij} |.$$

*Proof.*  Let $\lambda$ be any eigenvalue of the matrix $A$ and let

$$X = (x_1, \cdots, x_n)'$$

be its corresponding eigenvector.  Then

$$\sum_{j=1}^{n} a_{ij} x_j = \lambda x_i \quad (i = 1, \cdots, n)$$

or, what is the same,

$$\sum_{\substack{j=1 \\ j \ne i}}^{n} a_{ij} x_j = (\lambda - a_{ii}) x_i .$$

Let the index $i$ be chosen so that $x_i$ is the largest component in modulus of the vector $X$.  Then

$$\lambda - a_{ii} = \sum_{\substack{i=1 \\ j \ne i}}^{n} a_{ij} \frac{x_j}{x_i} ,$$

from which we obtain

$$|\lambda - a_{ii}| \le \sum_{\substack{j=1 \\ j\neq i}}^{n} |a_{ij}| \cdot \left|\frac{x_j}{x_i}\right| \le \sum_{\substack{j=1 \\ j\neq i}}^{n} |a_{ij}| = R_i .$$

Thus for each eigenvalue $\lambda$ we find a circle with center $a_{ii}$ and radius $R_i$ which contains this eigenvalue. Consequently all the eigenvalues are located in the union of all such circles.

*Note* 1. Gershgorin's region $D$ lies entirely in the closed circle $|z| \le \|A\|_{\mathrm{I}}$ and in fact touches its boundary, for if $|z - a_{ii}| \le R_i$, then $|z| \le |a_{ii}| + R_i$ $= \sum_{j=1}^{n} |a_{ij}|$ and there exists $z$ such that the inequality is changed to an equality. Consequently $|z| \le \max_i \sum_{j=1}^{n} |a_{ij}| = \|A\|_{\mathrm{I}}$, while a number $z$ is again found which realizes the equality.

Thus, although the Gershgorin region is, in general, part of the circle $|z| \le \|A\|_{\mathrm{I}}$, the estimate for the eigenvalue with greatest modulus is obtained the same way with the aid of the first norm.

*Note* 2. Taking, in place of matrix $A$, its transpose we obtain another region $D'$ which will be contained within the circle $|z| \le \|A\|_{\mathrm{II}}$.

*Note* 3. The Gershgorin region may be broken down into several connected parts. Each connected part will contain within itself as many eigenvalues as the circles which form it.

For a proof we consider the matrix

$$A_u = \begin{vmatrix} a_{11} & ua_{12} & \cdots & ua_{1n} \\ ua_{21} & a_{22} & \cdots & ua_{2n} \\ \cdot & \cdot & \cdots & \cdot \\ ua_{n1} & ua_{n2} & \cdots & a_{nn} \end{vmatrix} ,$$

where $u$ is a real parameter. It is clear that $A_1 = A$.

The Gershgorin region for the matrix $A_u$ is clearly the union of the circles $|z - a_{ii}| \le uR_i$, where $R_i$ are the radii of corresponding circles for matrix $A$. Thus the Gershgorin region $D_u$ for $A_u$ for $0 \le u \le 1$ is contained in the Gershgorin region $D$ for the matrix $A$. Therefore for a continuous variation of $u$ from 0 to 1 the eigenvalues of the matrices $A_u$ vary continuously without leaving the region $D$. Consequently inside each connected component of region $D$ is found a similar number of eigenvalues for any matrix $A_u, 0 \le u \le 1$.

But the eigenvalues of the matrix $A_0$ are diagonal elements $a_{ii}$ which are the centers of the circles forming $D$. Consequently the number of eigenvalues of matrix $A_0$ (and consequently also of matrix $A$) which lie inside a certain connected component of the region $D$ is precisely equal to the number of circles from whose union this component is obtained.

During recent years many works have been published which were devoted to redefining the region containing all the eigenvalues of a matrix.

## 14.  FUNCTIONAL GRADIENTS

**1.  Definition.**  Let $F(X)$ be a functional, in general non-linear, which is defined in real Euclidean space $R$ and which takes on real values.   Let a vector $X$ be given by its coordinates in a certain orthonormal basis.   Then the functional $F(X)$ is a function $F(x_1, \cdots, x_n)$ of the variable coordinates of the vector $X$.   We shall assume that we can differentiate the functional $F(X)$, for which it is sufficient to require that the function $F(x_1, \cdots, x_n)$ have continuous partial derivatives with respect to its arguments.

Let $Y$ be an arbitrary vector of unit length with coordinates $y_1, \cdots, y_n$.

The derivative of a functional $F$ at a point $X$ in the direction $Y$ is given by the expression

$$\frac{\partial F(X)}{\partial Y} = \lim_{t \to 0} \frac{F(X + tY) - F(X)}{t} = \frac{d}{dt} F(X + tY)|_{t=0} .$$

The derivative $\dfrac{\partial F(X)}{\partial Y}$ characterizes the rate of change of the functional $F$ for an increment of the argument in the direction of the vector $Y$.

Moreover we have

$$F(X + tY) = F(x_1 + ty_1, \cdots, x_n + ty_n)$$

and therefore

$$\frac{\partial F(X)}{\partial Y} = \frac{d}{dt} F(x_1 + ty_1, \cdots, x_n + ty_n)_{t=0}$$

$$= \frac{\partial F}{\partial x_1} y_1 + \cdots + \frac{\partial F}{\partial x_n} y_n = (Z, Y) ,$$

where $Z$ is the vector with coordinates $\dfrac{\partial F}{\partial x_1}, \cdots, \dfrac{\partial F}{\partial x_n}$.   The vector $Z$ is called the *gradient* of the functional $F(X)$.   From the last equality it follows that

$$\frac{\partial F(X)}{\partial Y} = |Z| \cos (Z, Y) ,$$

since $|Y| = 1$.   From here it follows that

$$- |Z| \leq \frac{\partial F(X)}{\partial Y} \leq |Z| ,$$

where $\partial F(X)/\partial Y = |Z|$ if the direction $Y$ coincides with the direction of the gradient and $\partial F(X)/\partial Y = - |Z|$ if the direction of $Y$ is opposite to the direction of the gradient.   Therefore the direction of the gradient is the direction of the greatest rate of increase of the functional $F$ at the given point, and the direction opposite to the gradient is the direction of the greatest rate of decrease.

**2.  Gradients of certain functionals.**  Let $A$ be a self-conjugate operator.

We shall determine the gradient of the functional

$$h(X) = (AX, X) .$$

We have

$$\frac{h(X + tY) - h(X)}{t} = \frac{(A(X + tY), X + tY) - (AX, X)}{t}$$

$$= 2(AX, Y) + t(AY, Y) ,$$

from which we obtain

$$\frac{\partial h(X)}{\partial Y} = 2(AX, Y) ,$$

and consequently the gradient of $(AX, Y)$ is equal to $2AX$. In particular the gradient of $(X, X)$ is equal to $2X$.

Since only the direction of the gradient will be important to us from now on, we shall neglect the positive multiplier 2 and we shall mean by the gradient of a functional $(AX, X)$ the vector $AX$.

Analogously, we find that the gradient of a functional $f(X) = (AX, X) - 2(F, X) + c$ is equal (with accuracy up to the multiplier 2) to the vector $AX - F$.

Finally we shall compute the gradient of the functional $\mu(X) = \dfrac{(AX, X)}{(X, X)}$

We have

$$\frac{\partial \mu(X)}{\partial Y} = \frac{(X, X)\frac{\partial}{\partial Y}(AX, X) - (AX, X)\frac{\partial}{\partial Y}(X, X)}{(X, X)^2}$$

$$- \frac{(X, X)2(AX, Y) - (AX, X) \cdot 2(X, Y)}{(X, X)^2}$$

$$= \frac{2}{(X, X)}(AX - \mu(X)X, Y) .$$

Consequently, with accuracy up to the positive multiplier $2/(X, X)$, the gradient of the functional $\mu(X)$ is equal to the vector $\xi = AX - \mu(X)X$. We observe two properties of this gradient:

1.  $(X, \xi) = 0$;
2.  $(\xi, \xi) = (\xi, AX)$.

In fact

$$(X, \xi) = (X, AX - \mu(X)X) = (X, AX) - \mu(X)(X, X) = 0 ,$$

from the definition of $\mu(X)$.

Furthermore,

$$(\xi, \xi) = (\xi, AX - \mu(X)X) = (\xi, AX) - \mu(X)(X, \xi) = (\xi, AX)$$

since $(X, \xi) = 0$.

The first of these properties has a simple geometric meaning, namely, the gradient $\xi$ is the projection of the vector $\eta = AX$ (the gradient of $(AX, X)$) on the subspace orthogonal to the vector $X$, since

$$\eta = \mu(X)X + \xi ,$$

while $\mu(X)X$ is proportional to $X$ and $\xi$ is orthogonal to $X$.

This result can readily be inferred from a purely geometric examination, without using the computations carried out above, but we shall not stop for this.

# Chapter II

# EXACT METHODS FOR SOLVING
# SYSTEMS OF LINEAR EQUATIONS

The present chapter is devoted to three closely related problems: the problem of solving a linear non-homogeneous algebraic system of equations, the problem of inverting a matrix, and the so-called elimination problem.

In theory these three problems are easily resolved. However, when high-order matrices are encountered in a problem, the actual solution of the problems may demand a great number of computational operations.

At the present time there exist a large number of methods for the numerical solution of systems of linear equations, and work on improving them is now going on intensively.

Numerical methods for solving these problems are divided into two categories: exact and iterative methods. By exact methods we mean methods which give a solution of the problem by using a finite number of elementary arithmetic operations. If the initial data defining the problem are given exactly (for example, if the data consist of whole or rational numbers represented as ordinary fractions) and if the computations are carried out exactly (for example, by the rules for computing with ordinary fractions), then the solution is also exact. In exact methods the number of computational operations necessary for solving a problem depends only on the type of computational scheme and on the order of the matrix which defines the given problem.

The first method of this type, historically, is a method based on the idea of eliminating the unknowns. In an applied solution of a linear non-homogeneous system, the algorithm of this method, associated with the name of Gauss, consists of a series of successive eliminations by means of which the given system is transformed into a system with a triangular matrix whose solution presents no difficulty.

At the present time many different computational schemes for applying Gauss's method are being used for all three given problems.

Among these schemes should be noted the so-called *compact schemes*, which demand a minimum of notation and recalling of intermediate results. In compact schemes the operation of accumulation is used, that is, computation of sums of form $\sum a_k b_k$. If, in applying this operation, one preserves as many of the significant digits of the individual elements as possible and rounds off only at the end of the summation process, the error due to rounding-off is usually significantly reduced.

Close to Gauss's method is a method based on the idea of bordering. Basic to this method is the concept of considering a given matrix as a set of matrices successively embedded in each other, starting from a matrix of order one.

Different modifications of the elimination method, as well as of the bordering method, are connected in essence with the expansion of a matrix into the product of two triangular matrices, described theoretically in pp. 17–20.

Related to exact methods for inverting matrices is a method based on the idea of gradually altering the matrix being inverted by successively correcting its inverse (reinforcement method).

In recent years there has been widespread use of exact methods based on the idea of setting up auxiliary systems of vectors orthogonal in one or another matrix. As will be pointed out below, Gauss's method may be included in this category of methods.

*Iterative methods* provide a means for determining an approximate solution of a system of linear equations. Solving a system by an iterative methods yields a limit for the successive approximations obtained by the employment of some uniform process. In applying iterative methods the convergence of these successive approximations and the *rate* of convergence are both essential. No iterative method is universally effective in this respect; giving a rapid convergence for some matrices, a certain process may provide slow convergence, or none at all, for other matrices. Therefore in applying iterative methods an important role is played by the preliminary preparation of the system, that is, the replacement of the given system by an equivalent one which is constructed so that the selected process converges for it as rapidly as possible; different methods of acceleration are also important.

This chapter of the book describes the simplest methods included in the category of exact methods. Other methods (exact and iterative) will be presented during the course of the rest of the book.

## 15. CONDITIONED MATRICES

In the numerical solution of systems of linear equations there arise, even using exact methods, several sources of inaccuracy in the solution. One of them is the need to round off numbers during the computational process. It may happen that it is necessary during the computation to cope with the disappearance of significant digits as a result of subtracting two quantities whose values are close to each other. The disappearance of significant figures may cause significantly lower accuracy in the result; because of this it may sometimes be necessary to change the computational scheme or to redivide the work so as to provide a greater number of significant digits in the intermediate computations.

A second source of error appears in the conditions governing a system of linear equations arising in the process of solving a practical problem, so that the elements of the coefficient matrix, as well as independent quantities, are known only *approximately*, say up to a certain degree of accuracy. Inaccuracy in the initial data may itself produce errors in the solution, since changing the coefficients of a system within the limits of the given accuracy involves changing the solution. Let us take a more detailed look at this cause of inaccuracy in the solution of a system.

The theoretical solution of the system $AX = F$ is given by the formula $X = A^{-1}F$, where $A^{-1}$ is the matrix inverse to $A$. As we know, $A^{-1}$ exists if and only if $|A| \neq 0$. However, if the elements of matrix $A$ are given approximately, it is possible that even the very question of whether the matrix $A$ has a non-zero determinant or not may be meaningless. That is, it may happen that in the exact computation of the determinant, proceeding from approximate values of the matrix elements, considered as if exact, the determinant turns out to be different from zero, but changing the elements within the limits of accuracy may lead to a matrix with determinant equal to zero. It is clear that a system with a matrix possessing such a property cannot be solved with any sort of confidence; in one case there exists a unique solution, in the other an infinite number of solutions, or none at all!

We shall call the inverse matrix *stable* if "small" changes in the initial matrix elements correspond to small changes in the inverse matrix elements.

It is obvious that for stability of an inverse matrix it is necessary in every case that the determinant of the matrix not be too small. (Or too large, which will make the determinant of the inverse small). It is difficult to determine the degree of "smallness" of the determinant which definitely rules out a stable inverse matrix; the concept "not too small" is in itself meaningless without further definition. Before considering a refinement of this concept let us look at the following example. Let

$$W = \begin{bmatrix} 5 & 7 & 6 & 5 \\ 7 & 10 & 8 & 7 \\ 6 & 8 & 10 & 9 \\ 5 & 7 & 9 & 10 \end{bmatrix}.$$

It is readily computed that $|W| = 1$. Also

$$W^{-1} = \begin{bmatrix} 68 & -41 & -17 & 10 \\ -41 & 25 & 10 & -6 \\ -17 & 10 & 5 & -3 \\ 10 & -6 & -3 & 2 \end{bmatrix}.$$

If we consider the elements of matrix $W$ to be given with absolute accuracy, then everything comes out fine. However, the determinant of matrix $W$ turns out to "sufficiently small"! In fact, the well-known estimate of Hadamard for the value of a determinant

$$\Delta \le \sqrt{\prod_{i=1}^{n} \sum_{j=1}^{n} |a_{ij}|^2}$$

gives

$$|W| \le \sqrt{2{,}534{,}437{,}350} \approx 50{,}000 ,$$

which means (in view of the attainability of Hadamard's estimate) that with just such sums of the squares of moduli of row elements, the modulus of the determinant may run as high as 50,000.

This provides a basis for assuming that the matrix $W^{-1}$ will have little stability, which is easily substantiated by the following argument. We shall observe how the determinant of matrix $W$ is changed by small changes in the matrix's first element. Let

$$W(\varepsilon) = \begin{bmatrix} 5 + \varepsilon & 7 & 6 & 5 \\ 7 & 10 & 8 & 7 \\ 6 & 8 & 10 & 9 \\ 5 & 7 & 9 & 10 \end{bmatrix}.$$

Then $|W(\varepsilon)| = 1 + 68\varepsilon$, from which it follows that for $\varepsilon = -\frac{1}{68} \approx -0.015$ the matrix $W(\varepsilon)$ will be singular. Thus if we consider an element of matrix $W$ known within 0.02, then in practice the matrix $W$ must be considered singular.

We shall show how all elements of the inverse matrix change with an insignificant change in the first element of matrix $W$. Let

$$W_1 = \begin{bmatrix} 5.0002 & 7 & 6 & 5 \\ 7 & 10 & 8 & 7 \\ 6 & 8 & 10 & 9 \\ 5 & 7 & 9 & 10 \end{bmatrix}.$$

Then

$$|W_1| = 1.0136 ,$$

$$W_1^{-1} = \begin{bmatrix} 67.088 & -40.450 & -16.772 & 9.866 \\ -40.450 & 24.664 & 9.862 & -5.916 \\ -16.772 & 9.862 & 4.943 & -2.966 \\ 9.866 & -5.916 & -2.966 & 1.980 \end{bmatrix}$$

and consequently,

$$W^{-1} - W_1^{-1} = \begin{bmatrix} 0.912 & -0.550 & -0.228 & 0.134 \\ -0.550 & 0.335 & 0.138 & -0.084 \\ -0.228 & 0.138 & 0.057 & -0.033 \\ 0.134 & -0.084 & -0.033 & 0.020 \end{bmatrix}.$$

We get an even greater divergence if we consider the matrix

$$W_2 = \begin{bmatrix} 4.99 & 7 & 6 & 5 \\ 7 & 10 & 8 & 7 \\ 6 & 8 & 10 & 9 \\ 5 & 7 & 9 & 10 \end{bmatrix},$$

for which $|W_2| = 0.320$ and

$$W_2^{-1} = \begin{bmatrix} 204.82 & -128.12 & -53.12 & 31.25 \\ -128.12 & 77.53 & 31.78 & -18.81 \\ -53.12 & 31.78 & 14.03 & -8.31 \\ 31.25 & -18.81 & -8.31 & 5.12 \end{bmatrix}.$$

It is interesting to note that the very definition of the elements of the matrices $W^{-1}$, $W_1^{-1}$, and $W_2^{-1}$ does not encounter the difficulties connected with, for example, the disappearance of significant figures.

We shall now consider for the general case how the elements of an inverse matrix can change for changes in the elements of the matrix itself. Let $A^{-1} = (\alpha_{ij})$.

From the equality

$$AA^{-1} = E$$

we have

$$\frac{\partial A}{\partial a_{ij}} A^{-1} + A \frac{\partial A^{-1}}{\partial a_{ij}} = 0,$$

from which

$$\frac{\partial (A^{-1})}{\partial a_{ij}} = -A^{-1} \frac{\partial A}{\partial a_{ij}} A^{-1}.$$

But

$$\frac{\partial A}{\partial a_{ij}} = e_{ij},$$

where $e_{ij}$ is a matrix, the element of whose $i$th row and $j$th column is equal to 1 and the rest of whose elements are equal to zero. Consequently,

$$\frac{\partial (A^{-1})}{\partial a_{ij}} = -A^{-1} e_{ij} A^{-1} = -A^{-1} e_{i1} e_{1j} A^{-1}$$

$$= -\begin{bmatrix} \alpha_{1i} & 0 & \cdots & 0 \\ \alpha_{2i} & 0 & \cdots & 0 \\ \cdot & \cdot & \cdot & \cdot \\ \alpha_{ni} & 0 & \cdots & 0 \end{bmatrix} \begin{bmatrix} \alpha_{j1} & \alpha_{j2} & \cdots & \alpha_{jn} \\ 0 & 0 & \cdots & 0 \\ \cdot & \cdot & \cdot & \cdot \\ 0 & 0 & \cdots & 0 \end{bmatrix}$$

$$= -\begin{bmatrix} \alpha_{1i}\alpha_{j1} & \alpha_{1i}\alpha_{j2} & \cdots & \alpha_{1i}\alpha_{jn} \\ \alpha_{2i}\alpha_{j1} & \alpha_{2i}\alpha_{j2} & \cdots & \alpha_{2i}\alpha_{jn} \\ \cdot & \cdot & \cdot & \cdot \\ \alpha_{ni}\alpha_{j1} & \alpha_{ni}\alpha_{j2} & \cdots & \alpha_{ni}\alpha_{jn} \end{bmatrix},$$

and therefore

$$\frac{\partial \alpha_{kl}}{\partial a_{ij}} = -\alpha_{ki}\alpha_{jl} \, ,$$

from which

$$d\alpha_{kl} = -\sum_{i,j} \alpha_{ki}\alpha_{jl}\, da_{ij} \, .$$

Thus the change in each element of the inverse matrix produced by a change in the element $a_{ij}$ is equal to this change multiplied by the product of a certain two elements of the inverse matrix. If the elements of the inverse matrix are large enough (which will be any time for a small determinant), then an insignificant error in the elements of the initial matrix involves significant changes in the elements of the inverse matrix. Of course, in certain cases errors due to changes in different elements of the matrix may compensate for each other.

We shall call a matrix *ill-conditioned* if its corresponding inverse matrix is unstable.

An ill-conditioned matrix may turn out to be singular in practice if its elements are given approximately.

It is natural that a system of linear equations with an ill-conditioned matrix has little stability, since its solution is greatly changed for small variations of the independent variables as well as the coefficients. We shall consider how significant this instability is. Let

$$AX = F$$

be a given system. Then $X = A^{-1}F$ and consequently,

$$\frac{\partial X}{\partial a_{ij}} = \frac{\partial A^{-1}}{\partial a_{ij}} F = - A^{-1}e_{ij}A^{-1}F = - A^{-1}e_{ij}X = - \begin{bmatrix} \alpha_{1i}x_j \\ \alpha_{2i}x_j \\ \cdots\cdots \\ \alpha_{ni}x_j \end{bmatrix},$$

from which

$$\frac{\partial x_k}{\partial a_{ij}} = - \alpha_{ki}x_j \, .$$

Analogously, from the formula

$$\frac{\partial X}{\partial f_i} = A^{-1}\frac{\partial F}{\partial f_i} = A^{-1}\begin{bmatrix} 0 \\ \cdot \\ \cdot \\ \cdot \\ 1 \\ \cdot \\ \cdot \\ 0 \end{bmatrix} = \begin{bmatrix} \alpha_{1i} \\ \cdot \\ \cdot \\ \alpha_{ii} \\ \cdot \\ \cdot \\ \alpha_{ni} \end{bmatrix}$$

it follows that $\partial x_k/\partial f_i = \alpha_{ki}$.

Thus if the elements of an inverse matrix are large, then a small change in either the coefficients of the system or the independent variables involves a significant variation in the solution. Thus, for example, for the matrix $W$ considered above the "close" constants

$$
\begin{aligned}
F_1 &= (23 \quad , \quad 32 \quad , \quad 33 \quad , \quad 31 \quad )' \\
F_2 &= (23.001, \quad 31.999, \quad 32.999, \quad 31.001)' \\
F_3 &= (23.01 \quad , \quad 31.99 \quad , \quad 32.99 \quad , \quad 31.01 \quad )' \\
F_4 &= (23.1 \quad , \quad 31.9 \quad , \quad 32.9 \quad , \quad 31.1 \quad )'
\end{aligned}
$$

will lead us far from a "close" solution

$$
\begin{aligned}
X_1 &= (1 \quad , \quad 1 \quad , \quad 1 \quad , \quad 1 \quad )' \\
X_2 &= (1.136, \quad 0.918, \quad 0.965, \quad 1.021)' \\
X_3 &= (2.36 \quad , \quad 0.18 \quad , \quad 0.65 \quad , \quad 1.21 \quad )' \\
X_4 &= (14.6 \quad , \quad -7.2, \quad -2.5, \quad 3.1 \quad )'
\end{aligned}
$$

The phenomenon of ill-conditioned matrices was apparently known a long time ago by Gauss. However studies of the quantitative characteristics of it have appeared only recently.

We have seen that the "smallness" of the determinant may indicate an ill-conditioned matrix. However, it is easy to see that the value of the determinant alone may not be the sole criterion. Thus, for example, it is obvious that matrices which differ by only a constant factor should be considered equally conditioned. Their corresponding determinants will differ by the $n$th degree of the multiplier. From this it follows that one must equate the value of the determinant with at least the $n$th degree of the largest element of the matrix or with the $n$th degree of some norm of the matrix. However even this is not sufficient. Thus of the matrices

$$
\begin{bmatrix} 20 & 0 & 0 \\ 0 & 1 & 0 \\ 0 & 0 & 0.05 \end{bmatrix} \quad \text{and} \quad \begin{bmatrix} 20 & 0 & 0 \\ 0 & 0.2 & 0 \\ 0 & 0 & 0.25 \end{bmatrix} ,
$$

the first is conditioned "worse" than the second, although they have identical largest elements as well as determinants.

The corresponding inverse matrices will be

$$
\begin{bmatrix} 0.05 & 0 & 0 \\ 0 & 1 & 0 \\ 0 & 0 & 20 \end{bmatrix} \quad \text{and} \quad \begin{bmatrix} 0.05 & 0 & 0 \\ 0 & 5 & 0 \\ 0 & 0 & 4 \end{bmatrix} ,
$$

and therefore identical changes in elements of the given matrices produce different changes in elements of the inverse matrices which are more significant in the first one.

In order to characterize matrices from the point of view of their condition several authors have defined different quantitative characteristics known as *condition numbers*. The two numbers of Turing[†] are

$$
\begin{aligned}
N\text{-number} &= -N(A)N(A^{-1}) = \nu(A) , \\
M\text{-number} &= -M(A)M(A^{-1}) = \mu(A) ,
\end{aligned}
$$

---

† Turing (1).

where $N(A) = \sqrt{\text{Sp } A'A}$ and $M(A) = n \cdot \max_{ij} |a_{ij}|$ are the matrix norms considered above; there are also the numbers due to Todd[†]

$$P\text{-number} = \frac{\max |\lambda_i|}{\min |\lambda_i|} = \rho(A),$$

where $\lambda_i$ are the eigenvalues of the matrix $A$, and the number

$$H\text{-number} = \|A\| \, \|A^{-1}\|,$$

where $\|A\|$ is the third matrix norm. It is easy to see that

$$H\text{-number} = \sqrt{\frac{\mu_1}{\mu_n}} = \eta(A),$$

where $\mu_1$ and $\mu_n$ are the largest and smallest eigenvalues of the matrix $A'A$, where $A'$ is the transpose of $A$. It is clear that

$$\eta(A) = \sqrt{\rho(A'A)}.$$

For symmetric matrices the $P$-number coincides with the $H$-number.

Condition numbers, of course, are not a comprehensive characterization of the condition of a matrix.

We note the inequalities relating these numbers:

$$\nu(A) \le \mu(A) \le n^2\nu(A),$$
$$\nu(A) \le \eta(A) \le n\nu(A),$$
$$\rho(A) \le \eta(A).$$

It is easy to figure out that for orthogonal matrices $\nu(A) = \eta(A) = \rho(A) = 1$. We shall show that all condition numbers will be not less than one. For $\rho(A)$ and $\eta(A)$ this is obvious. For $\nu(A)$ it follows from the identity

$$\nu(A) = \frac{1}{n}\sqrt{(\mu_1 + \cdots + \mu_n)\left(\frac{1}{\mu_1} + \cdots + \frac{1}{\mu_n}\right)}$$
$$= \frac{1}{n}\sqrt{n^2 + \sum_{i>j}\left(\frac{\sqrt{\mu_i}}{\sqrt{\mu_j}} - \frac{\sqrt{\mu_j}}{\sqrt{\mu_i}}\right)^2},$$

where $\mu_1, \cdots, \mu_n$ are eigenvalues of the matrix $A'A$.

The condition numbers $\nu(A)$ and $\eta(A)$ have the following meaning for probability theory. Consider a system of linear equations $AX = F$, where the vector $F$ is given exactly and the elements of the matrix $A$ are independent random values with a mean value $a_{ij}$ and with the same dispersion $\sigma^2$, assumed to be very small in comparison with the values of the coefficients. Then the $N$-number of condition shows *by how many times the ratio of the mean square of the errors for the unknowns to the mean square of the unknowns themselves exceeds the ratio of the mean square of the errors for the coefficients of the system to the mean square of the coefficients themselves. The H-number gives the ratio of the largest semi-axis to the smallest*

---

[†] Todd (1).

*for an ellipsoid of dispersion of a vector whose components are the errors of the unknowns.[†]*

It is easy to calculate that for the matrix $W$ considered above

$$P\text{-number} = H\text{-number} \approx \frac{30.28868}{0.01015005} \approx 2984 \text{ ,}$$

$$N\text{-number} = \tfrac{1}{4}\sqrt{933}\,\sqrt{9708} \approx 752 \text{ ,}$$

$$M\text{-number} = \tfrac{1}{4}\,40 \times 272 = 2720 \text{ .}$$

The condition numbers for the matrix $W$ are not yet very large. In actual calculations it is sometimes necessary to cope with systems whose matrices have condition numbers exceeding 20,000.

The following discussion answers to a certain degree the question of what kind of indeterminacy is involved in the ill condition of a system.

Let $A$ be a non-degenerate matrix which, for simplicity, we shall consider real, and let $U_1, U_2, \cdots, U_n$ be an orthonormal system of eigenvectors for a matrix $A'A$; let $\mu_1 . \mu_2, \cdots, \mu_n$ be their corresponding eigenvalues. Then the vectors $V_i = \dfrac{1}{\sqrt{\mu_i}} A U_i$ form an orthonormal system of eigenvectors for the matrix $AA'$. In fact

$$AA'V_i = \frac{1}{\sqrt{\mu_i}} AA'A U_i = \sqrt{\mu_i}\, A U_i = \mu_i V_i \text{ ;}$$

moreover

$$(V_i, V_j) = \frac{1}{\sqrt{\mu_i \mu_j}}(A U_i, A U_j) = \frac{1}{\sqrt{\mu_i \mu_j}}(A'A U_i, U_j)$$

$$= \frac{\mu_i}{\sqrt{\mu_i \mu_j}}(U_i, U_j) = \delta_{ij} \text{ .}$$

Let there now be given a system

$$AX = F$$

with the matrix $A$.

We decompose the vector $F$ into the vectors $V_1, V_2, \cdots, V_n$.

$$F = \sum_{i=1}^{n} c_i V_i \text{ .}$$

It is clear that $c_i = (F, V_i)$.

Furthermore we have a solution in the form

$$X = \sum_{i=1}^{n} d_i U_i \text{ .}$$

Substituting it into the system gives

$$\sum_{i=1}^{n} d_i \sqrt{\mu_i}\, V_i = \sum_{i=1}^{n} c_i V_i \text{ ,}$$

from which it follows that

[†] Turing (1) and D. K. Faddeev (4).

$$d_i = \frac{c_i}{\sqrt{\mu_i}} . \qquad (1)$$

It is possible to convince oneself that for small changes in matrix $A$, there will be in the vector systems $\{U_i\}$ and $\{V_i\}$, as well as in the eigenvalues $\mu_i$ of the matrix $A'A$, small changes independent of the condition of matrix $A$. However, for ill condition of matrix $A$, there are small numbers among the $\mu$ and even small changes in them may turn out to be large in the relative sense. The coefficients $d_i$ of formula (1) corresponding to the small $\mu_i$ turn out to be poorly defined, so that the solution "works loose" in the directions of the vectors $U_i$ corresponding to the small $\mu_i$, which we saw earlier on the basis of the probability discussion.

If the coefficients of the system are known exactly (which will occur, for example, when solving by the grid method boundary problems for equations of the elliptical type with constant coefficients), then the only source of indeterminacy will be inaccuracy in the coefficients $c_i$ corresponding to small $\mu_i$; this inaccuracy, according to formula (1), increases many times after passing to the coefficients $d_i$. However this may not hold if, as a condition of the problem, the vector of constants is orthogonal or almost orthogonal to the vectors $V_i$ corresponding to small $\mu_i$; therefore the solution remains stable for small allowable (without destroying the almost-orthogonality) changes in the constant.

## 16. GAUSS'S METHOD

In this paragraph we shall consider a simpler and more natural method for solving a system of equations based on the successive elimination of the unknowns. As was said above, it is associated with Gauss's name. The method has many different computational schemes. We shall begin by considering the so-called *single-division* scheme.

Let a system of equations be given,

$$a_{11}x_1 + a_{12}x_2 + \cdots + a_{1n}x_n = f_1 ,$$
$$a_{21}x_1 + a_{22}x_2 + \cdots + a_{2n}x_n = f_2 ,$$
$$\cdots \cdots \cdots \cdots \cdots , \qquad (1)$$
$$a_{n1}x_1 + a_{n2}x_2 + \cdots + a_{nn}x_n = f_n .$$

We assume that the coefficient $a_{11} \neq 0$. We shall divide the coefficients of the first of the equations of (1) (including the right member) by the coefficients $a_{11}$, which we shall call the *leading element* (on the first step), and we write

$$b_{1j} = \frac{a_{1j}}{a_{11}} \qquad (j > 1) ,$$
$$g_1 = \frac{f_1}{a_{11},} . \qquad (2)$$

We obtain the new equation

$$x_1 + b_{12}x_2 + \cdots + b_{1n}x_n = g_1 . \qquad (3)$$

We now eliminate the unknown $x_1$ from all equations (1), beginning with the second, by subtracting from each of them the product of equation (3) by the number $a_{21}, a_{31}, \cdots, a_{n1}$ respectively. The transformed equations will be

$$a_{22 \cdot 1}x_2 + \cdots + a_{2n \cdot 1}x_n = f_{2 \cdot 1} ,$$
$$\cdots \cdots \cdots \cdots \cdots , \qquad (4)$$
$$a_{n2 \cdot 1}x_2 + \cdots + a_{nn \cdot 1}x_n = f_{n \cdot 1} ,$$

where

$$a_{ij \cdot 1} = a_{ij} - a_{i1}b_{1j} \qquad (i, j \geq 2) ,$$
$$f_{i \cdot 1} = f_i - a_{i1}g_1 . \qquad (5)$$

We now divide the coefficients of the first of the transformed equations by the leading element of the second step $a_{22 \cdot 1}$, which we assume to be different from zero. We get the equation

$$x_2 + b_{23}x_3 + \cdots + b_{2n}x_n = g_2 ,$$

where

$$b_{2j} = \frac{a_{2j \cdot 1}}{a_{22 \cdot 1}} ,$$

$$g_2 = \frac{f_{2 \cdot 1}}{a_{22 \cdot 1}} .$$

Eliminating the unknown $x_2$ from equations (4), beginning with the second, we arrive at the equations

$$a_{33 \cdot 2}x_3 + \cdots + a_{3n \cdot 2}x_n - f_{3 \cdot 2} ,$$
$$\cdots \cdots \cdots \cdots \cdots \cdots ,$$
$$a_{n3 \cdot 2}x_3 + \cdots + a_{nn \cdot 2}x_n - f_{n \cdot 2} ,$$

where

$$a_{ij \cdot 2} = a_{ij \cdot 1} - a_{i2 \cdot 1}b_{2j} \qquad (i, j \geq 3) ,$$
$$f_{i \cdot 2} = f_{i \cdot 1} - a_{i2 \cdot 1}g_2 .$$

We continue the process according to this plan. At the $m$th step we get the equations

$$x_m + b_{m\,m+1}x_{m+1} + \cdots + b_{mn}x_n = g_m ,$$
$$a_{m+1\,m+1 \cdot m}x_{m+1} + \cdots + a_{m+1\,n \cdot m}x_n = f_{m+1 \cdot m} ,$$
$$\cdots \cdots \cdots \cdots \cdots \cdots \cdots , \qquad (6)$$
$$a_{n\,m+1 \cdot m}x_{m+1} + \cdots + a_{nn \cdot m}x_n = f_{n \cdot m} ,$$

where

$$b_{mj} = \frac{a_{mj \cdot m-1}}{a_{mm \cdot m-1}} \qquad (j \geq m+1) ,$$

$$g_m = \frac{f_{m \cdot m-1}}{a_{mm \cdot m-1}} ,$$

(7)

$$a_{ij \cdot m} = a_{ij \cdot m-1} - a_{im \cdot m-1} b_{mj} ,$$

$$f_{i \cdot m} = f_{i \cdot m-1} - a_{im \ m-1} g_m .$$

Combining all the first equations of each step we get the system

$$x_1 + b_{12}x_2 + b_{13}x_3 + \cdots + b_{1n}x_n = g_1 ,$$

$$x_2 + b_{23}x_3 + \cdots + b_{2n}x_n = g_2 ,$$

$$\cdot \quad \cdot \quad \cdot \quad \cdot \quad \cdot \quad \cdot \quad \cdot \quad \cdot \quad \cdot \quad \cdot \quad \cdot \quad \cdot \quad \cdot ,$$

(8)

$$x_n = g_n$$

with triangular matrix; this system is equivalent to the initial system. We observe that the process described is possible only under the condition that none of the leading elements $a_{mm \cdot m-1}$ are equal to zero, since we divide by them in the process.

From system (8) we find the values for the unknowns in sequence from $x_n$ to $x_1$ by the obvious formulas

$$x_m = g_m - b_{m \cdot m+1} x_{m+1} - \cdots - b_{mn}x_n .$$

Thus to solve the given system by the single-division scheme we first construct an auxiliary triangular system and then proceed to solve it. The

**TABLE II.1   The Single-division Scheme**

| $a_{11}$ | $a_{12}$ | $a_{13}$ | $a_{14}$ | $a_{15}$ | $a_{16}$ | 1 | 0.17 | $-0.25$ | 0.54 | 0.3 | 1.76 |
|---|---|---|---|---|---|---|---|---|---|---|---|
| $a_{21}$ | $a_{22}$ | $a_{23}$ | $a_{24}$ | $a_{25}$ | $a_{26}$ | 0.47 | 1 | 0.67 | $-0.32$ | 0.5 | 2.32 |
| $a_{31}$ | $a_{32}$ | $a_{33}$ | $a_{34}$ | $a_{35}$ | $a_{36}$ | $-0.11$ | 0.35 | 1 | $-0.74$ | 0.7 | 1.20 |
| $a_{41}$ | $a_{42}$ | $a_{43}$ | $a_{44}$ | $a_{45}$ | $a_{46}$ | 0.55 | 0.43 | 0.36 | 1 | 0.9 | 3.24 |
| 1 | $b_{12}$ | $b_{13}$ | $b_{14}$ | $b_{15}$ | $b_{16}$ | 1 | 0.12 | $-0.25$ | 0.54 | 0.3 | 1.76 |
| | $a_{22 \cdot 1}$ | $a_{23 \cdot 1}$ | $a_{24 \cdot 1}$ | $a_{25 \cdot 1}$ | $a_{26 \cdot 1}$ | | 0.9201 | 0.7875 | $-0.5738$ | 0.3590 | 1.4928 |
| | $a_{32 \cdot 1}$ | $a_{33 \cdot 1}$ | $a_{34 \cdot 1}$ | $a_{35 \cdot 1}$ | $a_{36 \cdot 1}$ | | 0.3687 | 0.9725 | $-0.6806$ | 0.7330 | 1.3936 |
| | $a_{42 \cdot 1}$ | $a_{43 \cdot 1}$ | $a_{44 \cdot 1}$ | $a_{45 \cdot 1}$ | $a_{46 \cdot 1}$ | | 0.3365 | 0.4975 | 0.7030 | 0.7350 | 2.2720 |
| | 1 | $b_{23}$ | $b_{24}$ | $b_{25}$ | $b_{26}$ | | 1 | 0.85589 | $-0.62363$ | 0.39017 | 1.62243 |
| | | $a_{33 \cdot 2}$ | $a_{34 \cdot 2}$ | $a_{35 \cdot 2}$ | $a_{36 \cdot 2}$ | | | 0.65693 | $-0.45067$ | 0.58914 | 0.79540 |
| | | $a_{43 \cdot 2}$ | $a_{44 \cdot 2}$ | $a_{45 \cdot 2}$ | $a_{46 \cdot 2}$ | | | 0.20949 | 0.91285 | 0.60371 | 1.72605 |
| | | 1 | $b_{34}$ | $b_{35}$ | $b_{36}$ | | | 1 | $-0.68602$ | 0.89681 | 1.21079 |
| | | | $a_{44 \ 3}$ | $a_{45 \cdot 3}$ | $a_{46 \cdot 3}$ | | | | 1.05656 | 0.41584 | 1.47240 |
| | | | 1 | $x_4$ | $\overline{x}_4$ | | | | 1 | 0.39358 | 1.39358 |
| | | 1 | | $x_3$ | $\overline{x}_3$ | | | 1 | | 1.16681 | 2.16682 |
| | 1 | | | $x_2$ | $\overline{x}_2$ | | 1 | | | $-0.36304$ | 0.63695 |
| 1 | | | | $x_1$ | $\overline{x}_1$ | 1 | | | | 0.44089 | 1.44089 |

process of finding the coefficients of the triangular system will be called the *forward course* and the process of obtaining its solution will be called the *return course*.

In Table II.1 the single-division scheme is carried out for a system of four equations by numerical example.

We shall explain Table II.1. In the scheme we denote the constant terms of the initial and transformed equations by using the same letters as the coefficients of the equations but adjoining a second index 5. This permits us to combine the formulas by which the coefficients and the constant terms of the given system are transformed. The sixth column is used for control. This control is based on the following circumstance. If in the given system one carries out the substitution $\bar{x}_i = x_i + 1$, then for determining $\bar{x}_i$ we obtain a system with the former coefficients and with constant terms equal to sums of elements in the rows of the matrix of coefficients (including the constant). Therefore, if we compile the sum of the elements of each row of the initial matrix (including constants) and perform on them all the operations which were performed on the remaining elements, then in the absence of computational errors we should obtain numbers equal to the sums of the elements of the newly constructed rows. At the end of the process, having finished the return course, the numbers $\bar{x}_i$ should be obtained equal to $x_i + 1$. Sometimes it is more advisable to use this idea for the control of each row constructed. With this in mind we write down in the control column the sums of the elements of the rows

**TABLE II.1a   A Single-division Scheme in the Symmetric Case**

| $a_{11}$ | $a_{12}$ | $a_{13}$ | $a_{14}$ | $a_{15}$ | $a_{16}$ | 1.00 | 0.42 | 0.54 | 0.66 | 0.3 | 2.92 |
|---|---|---|---|---|---|---|---|---|---|---|---|
| | $a_{22}$ | $a_{23}$ | $a_{24}$ | $a_{25}$ | $a_{26}$ | | 1.00 | 0.32 | 0.44 | 0.5 | 2.68 |
| | | $a_{33}$ | $a_{34}$ | $a_{35}$ | $a_{36}$ | | | 1.00 | 0.22 | 0.7 | 2.78 |
| | | | $a_{44}$ | $a_{45}$ | $a_{46}$ | | | | 1.00 | 0.9 | 3.22 |
| 1 | $b_{12}$ | $b_{13}$ | $b_{14}$ | $b_{15}$ | $b_{16}$ | 1 | 0.42 | 0.54 | 0.66 | 0.3 | 2.92 |
| | $a_{22\cdot1}$ | $a_{23\cdot1}$ | $a_{24\cdot1}$ | $a_{25\cdot1}$ | $a_{26\cdot1}$ | | 0.82360 | 0.09320 | 0.16280 | 0.37400 | 1.45360 |
| | | $a_{33\cdot1}$ | $a_{34\cdot1}$ | $a_{35\cdot1}$ | $a_{36\cdot1}$ | | | 0.70840 | -0.13640 | 0.53800 | 1.20320 |
| | | | $a_{44\cdot1}$ | $a_{45\cdot1}$ | $a_{46\cdot1}$ | | | | 0.56440 | 0.70200 | 1.29280 |
| | 1 | $b_{23}$ | $b_{24}$ | $b_{25}$ | $b_{26}$ | | 1 | 0.11316 | 0.19767 | 0.45410 | 1.76493 |
| | | $a_{33\cdot2}$ | $a_{34\cdot2}$ | $a_{35\cdot2}$ | $a_{36\cdot2}$ | | | 0.69785 | -0.15482 | 0.49568 | 1.03871 |
| | | | $a_{44\cdot2}$ | $a_{45\cdot2}$ | $a_{46\cdot2}$ | | | | 0.53222 | 0.62807 | 1.00547 |
| | | 1 | $b_{34}$ | $b_{35}$ | $b_{36}$ | | | 1 | -0.22185 | 0.71030 | 1.48844 |
| | | | $a_{44\cdot3}$ | $a_{45\cdot3}$ | $a_{46\cdot3}$ | | | | 0.49787 | 0.73804 | 1.23591 |
| | | | 1 | $x_4$ | $\overline{x}_4$ | | | | 1 | 1.48240 | 2.48240 |
| | | 1 | | $x_3$ | $\overline{x}_3$ | | | 1 | | 1.03917 | 2.03916 |
| | 1 | | | $x_2$ | $\overline{x}_2$ | | 1 | | | 0.04348 | 1.04348 |
| 1 | | | | $x_1$ | $\overline{x}_1$ | 1 | | | | -1.25780 | -0.25779 |

being constructed and the numbers obtained are compared only with the results of the control operations.

The number of multiplications and divisions necessary for finding a solution to a system of $n$ equation by the single-division plan is equal to $\frac{n}{3}(n^2 + 3n - 1)$.

When the matrix of coefficients of a system is symmetric, that is, $a_{ij} = a_{ji}$, we obviously have $a_{ij \cdot k} = a_{ji \cdot k}$. Therefore it is possible to omit the notation for elements arrayed below the diagonal. A single-division scheme suitable for the symmetric case is given in Table II.1a.

The first elements of a row have been omitted in noting the coefficients but are necessary to us for computing the elements of the auxiliary matrices. They are easily found as the upper element of the column which includes the diagonal element of the given row. The control column contains, as before, the sums of all the elements of each row, including those omitted in the notation.

The number of computational operations in a single-division scheme is significantly reduced if the matrix of coefficients for the system is almost triangular, and particularly if it is tridiagonal.

If it is necessary to solve several systems having one given matrix it is natural to look for the solutions all at the same time, writing out the constant terms in neighboring columns. The control sum is formed at the sum of the elements in the rows of the expanded matrix. The scheme for solving several such systems is given in Table II.2.

**TABLE II.2   Single-division Scheme for Several Systems of Equations**

| | | | | | | | |
|---|---|---|---|---|---|---|---|
| 1.00 | 0.42 | 0.54 | 0.66 | 0.25 | 0.3 | 0.15 | 3.32 |
| 0.42 | 1.00 | 0.32 | 0.44 | 0.45 | 0.5 | 0.30 | 3.43 |
| 0.54 | 0.32 | 1.00 | 0.22 | 0.65 | 0.7 | 0.45 | 3.88 |
| 0.66 | 0.44 | 0.22 | 1.00 | 0.85 | 0.9 | 0.60 | 4.67 |
| 1 | 0.42 | 0.54 | 0.66 | 0.25 | 0.3 | 0.15 | 3.32 |
| | 0.82360 | 0.09320 | 0.16280 | 0.34500 | 0.37400 | 0.23700 | 2.03560 |
| | 0.09320 | 0.70840 | −0.13640 | 0.51500 | 0.53800 | 0.36900 | 2.08720 |
| | 0.16280 | −0.13640 | 0.56440 | 0.68500 | 0.70200 | 0.50100 | 2.47880 |
| | 1 | 0.11316 | 0.19767 | 0.41889 | 0.45410 | 0.28776 | 2.47159 |
| | | 0.69785 | −0.15482 | 0.47596 | 0.49568 | 0.34218 | 1.85685 |
| | | −0.15482 | 0.53222 | 0.61680 | 0.62807 | 0.45415 | 2.07643 |
| | | 1 | −0.22185 | 0.68204 | 0.71030 | 0.49033 | 2.66082 |
| | | | 0.49787 | 0.72239 | 0.73804 | 0.53006 | 2.48838 |
| | | | 1 | 1.45096 | 1.48240 | 1.06466 | 4.99805 |
| | | 1 | | 1.00394 | 1.03917 | 0.72652 | 3.76964 |
| | 1 | | | 0.01847 | 0.04348 | −0.00490 | 1.05705 |
| 1 | | | | −1.25752 | −1.25780 | −0.94294 | −2.45828 |

The single-division scheme is very simple and convenient. However it is not universal, in the sense that to apply it is necessary that all the leading elements be non-zero. Sometimes this condition cannot be predicted without computations which, in some form or another, are equivalent to applying the scheme itself. Proximity of the leading elements to zero may be a cause of significant loss in accuracy.

Therefore it is advisable to modify the single-division scheme a little without predicting *a priori* the order of elimination of the unknowns.

The best variant is the *single-division scheme by principal elements*. In this scheme we choose as the unknown being eliminated, at the $m$th step, the one whose coefficient had the largest modulus in the previous step. When computing the principal elements by this scheme, disappearance of significant figures may arise only if the system is ill-conditioned, so that the resultant loss of accuracy is essentially inevitable.

We observe that in the example investigated the principal elements coincide with the coefficients $a_{mm \cdot m-1}$ so that both schemes coincide.

The successive elimination of unknowns which transforms a given system into a system with a triangular matrix may also lead to other computational schemes.

In the *division and subtraction scheme* all the equations are divided at each step by the coefficient of the unknown being eliminated, and then the elimination itself is carried out by subtracting one equation from all the rest.

In the *multiplication and subtraction scheme* one eliminates the unknown $x_1$ from the $i$th equation in the first step by multiplying this equation by $a_{11}$ and by subtracting the first equation multiplied by $a_{i1}$. The same method is applied in the following steps, so that the coefficient of the auxiliary equations $\tilde{a}_{ij \cdot m}$ are computed at the $m$th step by the formulas

$$\tilde{a}_{ij \cdot m} = \tilde{a}_{mm \cdot m-1}\tilde{a}_{ij \cdot m-1} - \tilde{a}_{im \cdot m-1}\tilde{a}_{mj \cdot m-1} \, ,$$

$$\tilde{f}_{im} = \tilde{a}_{mm \cdot m-1}\tilde{f}_{i \cdot m-1} - \tilde{a}_{im \cdot m-1}\tilde{f}_{m \cdot m-1} \, .$$

Other computational schemes also exist. In particular, it is possible to apply each of the schemes described with the order for eliminating unknowns prescribed earlier or else by choosing the order of elimination during the process, for example, according to the principal elements.

The computational schemes of Gauss's method are based on the transformation of a system to a system with a right triangular matrix by a linear combination of equations; this is equivalent to premultiplying the matrix of the system (multiplying at the same time the column of constants) by certain auxiliary matrices. It is worth noticing the modification of the elimination method by which elementary matrices of rotation are taken as the auxiliary matrices which determine the linear combination of equations. Computations for this are carried out in the following way. We take

$$c_{21} = \frac{a_{11}}{\sqrt{a_{11}^2 + a_{21}^2}} \, ;$$

$$s_{21} = -\frac{a_{21}}{\sqrt{a_{11}^2 + a_{21}^2}}$$

(if $a_{11} = a_{21} = 0$ we take $c_{21} = 1$, $s_{21} = 0$) and we replace the first two equations of the system by the equations

$$c_{21} y_1 - s_{21} y_2 = c_{21} f_1 - s_{21} f_2\,,$$

$$s_{21} y_1 + c_{21} y_2 = s_{21} f_1 + c_{21} f_2\,.$$

Here the left members of the first two equations of the initial system are denoted by $y_1$ and $y_2$. After the transformation in the second equation the coefficient for the unknown $x_1$ will be equal to zero.

Continuing in an analogous manner we connect the transformed first equation with the third initial one, the newly obtained first one with the fourth initial one, and so on. After the $n$-1th step of the process we get a system of the form

$$a_{11}^{(1)} x_1 + a_{12}^{(1)} x_2 + \cdots + a_{1n}^{(1)} x_n = f_1^{(1)}\,,$$

$$a_{22}^{(1)} x_2 + \cdots + a_{2n}^{(1)} x_n = f_2^{(1)}\,,$$

$$\cdot\ \cdot\ \cdot\ \cdot\ \cdot\ \cdot\ \cdot\ \cdot\ \cdot\ \cdot\ \cdot\ \cdot\ \cdot\,,$$

$$a_{n2}^{(1)} x_2 + \cdots + a_{nn}^{(1)} x_n = f_n^{(1)}\,.$$

We now apply the same process to the system with the first equation eliminated. After $n(n-1)/2$ steps we get a system with a triangular matrix which is solved by the usual "return course" procedure.

This elimination scheme requires approximately four times as many computational operations as the single-division scheme, but it is distinguished by its great stability and small sensitivity to "gaps" in the intermediary determinants of the system.

In Table II.3 is found a solution for the system which was solved earlier in Table II.1 by the single-division method. In the last column of the table are the numbers $c_{ij}$ and $s_{ij}$; in the unnumbered rows are the auxiliary values needed to compute them.

The resultant system with triangular matrix is placed in rows $1'''$, $2'''$, $3'''$, and $4'''$; the initial system is placed in rows 1, 2, 3, and 4. Rows $1'$, $2'$, $3'$, $4'$ and $1''$, $2''$, $3''$, $4''$ are occupied by intermediate linear combinations of the initial equations. It is convenient to obtain them in an order which coincides with the order of movement for the rows in the table. The solution of the system is placed in the last row of the table. The sixth column in all the numbered rows is the control column.

For transforming the system to an equivalent system with a triangular matrix it is possible to use other orthogonal matrices instead of elementary rotation matrices. Matrices of mappings of an $n$-1-dimensional surface turn out to be convenient.[†] Let $W$ be a vector of unit length orthogonal to the given $(n-1)$-dimensional surface $Q$. Then a matrix for mapping relative to $Q$ is $U = E - 2WW'$ (here $WW'$ is the product of a column by a row,

---

† Householder (12).

**TABLE II.3  Solutions of a Linear System by the Method of Rotation**

| | | | | | | | |
|---|---|---|---|---|---|---|---|
| 1 | 1.00 | 0.17 | −0.25 | 0.54 | 0.3 | 1.76 | 0.90503 |
| 2 | 0.47 | 1.00 | 0.67 | −0.32 | 0.5 | 2.32 | −0.42536 |
| | 1.2209 | 1.10494 | | | | | |
| 1′ | 1.10495 | 0.57922 | 0.05873 | 0.35260 | 0.48419 | 2.57969 | 0.99508 |
| 3 | −0.11 | 0.35 | 1.00 | −0.74 | 0.7 | 1.20 | 0.09906 |
| | 1.23301 | 1.11041 | | | | | |
| 1″ | 1.11041 | 0.54170 | −0.04062 | 0.42417 | 0.41247 | 2.44813 | 0.89610 |
| 4 | 0.55 | 0.43 | 0.36 | 1.00 | 0.9 | 3.24 | −0.44385 |
| | 1.53551 | 1.23916 | | | | | |
| 1‴ | 1.23916 | 0.67627 | 0.12339 | 0.82395 | 0.76908 | 3.63184 | |
| 2′ | | 0.83272 | 0.71271 | −0.51930 | 0.32491 | 1.35104 | 0.89900 |
| 3′ | | 0.40566 | 1.00090 | −0.70143 | 0.74452 | 1.44964 | −0.43795 |
| | | 0.85798 | 0.92627 | | | | |
| 2″ | | 0.92627 | 1.07907 | −0.77404 | 0.61816 | 1.84945 | 0.98799 |
| 4′ | | 0.14489 | 0.34063 | 0.70783 | 0.62342 | 1.81676 | −0.15455 |
| | | 0.87897 | 0.93753 | | | | |
| 2‴ | | 0.93754 | 1.11875 | −0.65536 | 0.70708 | 2.10800 | |
| 3″ | | | 0.58768 | −0.40316 | 0.52703 | 0.71154 | 0.96072 |
| 4″ | | | 0.16978 | 0.81895 | 0.52040 | 1.50913 | −0.27755 |
| | | | 0.37419 | 0.61171 | | | |
| 3‴ | | | 0.61172 | −0.16002 | 0.65077 | 1.10245 | |
| 4‴ | | | | 0.89868 | 0.35368 | 1.25236 | |
| | 0.44089 | −0.36302 | 1.16679 | 0.39356 | | | |

i.e., a matrix of rank 1).  In fact if the vector $Z$ is orthogonal to $W$,

$$UZ = Z - 2WW'Z = Z,$$

since $W'Z = (W, Z) = 0$ and $UW = W - 2WW'W = -W$, since $WW' = (W, W) = 1$.  The matrix $U$ is obviously orthogonal and $|U| = -1$.

The vector $W$ may always be chosen so that the matrix $U$ transforms any given vector $S$ into a vector of a given direction, for example, into a vector directed by one of the coordinate vectors $e$.  For this it is sufficient to take

$$W = \frac{1}{\rho}(S - \alpha e),$$

where $\alpha = \pm |S|$, $\rho = |S - \alpha e| = \sqrt{2\alpha^2 - 2\alpha(S, e)}$,

In fact for such a choice of $W$ we have

$$US = S - 2(W, S)W = S - \rho W = S - S + \alpha e = \alpha e ,$$

since $2(W, S) = \dfrac{1}{\rho}(2\alpha^2 - 2\alpha(S, e)) = \rho$ .

In general the choice of sign for $\alpha$ is immaterial. It is possible to order it so that the number $\rho$ is the larger of the two possible ones, for which one should take sign $\alpha = -$ sign $(S, e)$.

For any vector $Y$ we have

$$UY = Y - 2(W, Y)W . \qquad (9)$$

For transforming the system $AX = F$ to a triangular system we do the following. In the first step a vector $W$ is constructed according to the first column $S = A_1$ of a matrix $A$ and the vector $e = e_1$. After multiplying both parts of the system by the matrix $U$ (which is equivalent to applying formula (9) to the columns of matrix $A$ and to the vector $F$) we get a system with a matrix of the form

$$\begin{bmatrix} \alpha & v \\ 0 & B \end{bmatrix}$$

where $B$ is a square matrix of the $n - 1$th order. The process is repeated for the system with the matrix $B$, and so forth.

After the $n - 1$th step we come to a triangular system. It is easy to see that the matrix of this system, up to the signs of the rows, coincides with the matrix obtained by the method of rotation.

We shall carry out the result of the first step for the system of Table II.1. We have

$$\alpha = - 1.23915$$
$$\rho = 2.35569$$
$$W = (0.95034, 0.19952, -0.04670, 0.23348)'$$

$$\begin{bmatrix} \alpha & v \\ 0 & B \end{bmatrix} = \begin{bmatrix} -1.23917 & -0.67628 & -0.12339 & -0.82397 \\ -0.00001 & 0.82236 & 0.69658 & -0.60630 \\ 0.00001 & 0.39158 & 0.99378 & -0.67299 \\ -0.00001 & 0.22213 & 0.39110 & 0.66497 \end{bmatrix}$$

$$UF = (-0.76908, 0.27560, 0.75252, 0.63740)' .$$

The single-division method may be generalized in the following way for systems whose matrices are partitioned into cells which are square diagonal cells. Having partitioned both the solution and the column of constants into vectors (the dimension of the vectors being equal to the orders of the diagonal cells) we write the system in the form

$$A_{11}X_1 + A_{12}X_2 + \cdots + A_{1n}X_n = F_1 ,$$
$$A_{21}X_1 + A_{22}X_2 + \cdots + A_{2n}X_n = F_2 ,$$
$$\cdots \cdots \cdots \cdots \cdots \cdots \cdots ,$$
$$A_{n1}X_1 + A_{n2}X_2 + \cdots + A_{nn}X_n = F_n .$$

Let $|A_{11}| \neq 0$. We shall find the matrix $A_{11}^{-1}$ and premultiply the first equation of the system by it. We obtain the matrix equation

$$X_1 + B_{12}X_2 + \cdots + B_{1n}X_n = G_1 ,$$

where

$$B_{12} = A_{11}^{-1}A_{12} , \cdots, B_{1n} = A_{11}^{-1}A_{1n} , \quad G_1 = A_{11}^{-1}F_1 .$$

We premultiply this equation by $A_{21}, \cdots, A_{n1}$ and we subtract respectively from the 2nd, $\cdots$, $n$th equations of the initial system. We get the system

$$A_{22 \cdot 1}X_2 + \cdots + A_{2n \cdot 1}X_n = F_{2 \cdot 1} ,$$
$$\cdots \cdots \cdots \cdots \cdots \cdots \cdots ,$$
$$A_{n2 \cdot 1}X_2 + \cdots + A_{nn \cdot 1}X_n = F_{n \cdot 1} ,$$

where

$$A_{ij \cdot 1} = A_{ij} - A_{i1}B_{ij} \quad (i, j \geq 2) .$$
$$F_{i \cdot 1} = F_i - A_{i1}G_1 .$$

Continuing in an analogous way we finally arrive at a system with a quasi-triangular matrix having only diagonal cells

$$X_1 + B_{12}X_2 + \cdots + B_{1n}X_n = G_1 ,$$
$$X_2 + \cdots + B_{2n}X_n = G_2 ,$$
$$\cdots \cdots \cdots \cdots \cdots ,$$
$$X_n = G_n .$$

The unknown vectors $X_1, \cdots, X_n$ are found successively from $X_n$ to $X_1$ by the formulas

$$X_i = G_i - B_{i i+1}X_{i+1} - \cdots - B_{in}X_n .$$

Partitioning a matrix into cells generally does not change the number of elementary operations necessary for solving the system. However one may obtain a significant gain in the volume of work if there exists a partition under which simpler inversions of the leading cells arise.

## 17. COMPUTATION OF DETERMINANTS

Gauss's method for solving a linear system, developed in the preceding paragraph, may be also applied in computing determinants. We shall consider separately a description of the corresponding single-division scheme, since one often encounters the computation of determinants in practice. Let

$$\Delta = \begin{vmatrix} a_{11} & a_{12} & \cdots & a_{1n} \\ a_{21} & a_{22} & \cdots & a_{2n} \\ \cdot & \cdot & \cdots & \cdot \\ a_{n1} & a_{n2} & \cdots & a_{nn} \end{vmatrix}$$

and let $a_{11} \neq 0$. We normalize the element $a_{11}$. Then, using the notation of Section 16, we obtain

$$\varDelta = a_{11} \begin{vmatrix} 1 & b_{12} & \cdots & b_{1n} \\ a_{21} & a_{22} & \cdots & a_{2n} \\ \cdot & \cdot & \cdot & \cdot \\ a_{n1} & a_{n2} & \cdots & a_{nn} \end{vmatrix}$$

Then successively from each row we subtract the first row multiplied by the first element of this row. We obtain

$$\varDelta = a_{11} \begin{vmatrix} 1 & b_{12} & \cdots & b_{1n} \\ 0 & a_{22 \cdot 1} & \cdots & a_{2n \cdot 1} \\ \cdot & \cdot & \cdot & \cdot \\ 0 & a_{n2 \cdot 1} & \cdots & a_{nn \cdot 1} \end{vmatrix} = a_{11} \begin{vmatrix} a_{22 \cdot 1} & \cdots & a_{2n \cdot 1} \\ \cdot & \cdot & \cdot \\ \cdot & \cdot & \cdot \\ a_{n2 \cdot 1} & \cdots & a_{nn \cdot 1} \end{vmatrix}.$$

Then if $a_{22 \cdot 1} \neq 0$, we repeat the same procedure for the determinant of order $n - 1$ whose initial element is $a_{22 \cdot 1}$.

Continuing this process we see that the determinant sought for is equal to the product of the leading elements:

$$\varDelta = a_{11} a_{22 \cdot 1} \cdots a_{nn \cdot n-1} .$$

If at any step it turns out that $a_{ii \cdot i-1} = 0$ or $a_{ii \cdot i-1}$ is zero or very close to zero (which involves less favorable computational accuracy), then it is possible to change beforehand the order of the determinant's rows and columns so that a non-vanishing element appears in the left upper corner.

The best result in the sense of reliability is obtained if at each step of the process one transfers to the left upper corner the largest (in absolute value) available element of the subdeterminant under consideration (by the allowable transformations, of course!).

The number of multiplications and divisions necessary for computing a

**TABLE II.4**   **Computing a Determinant by the Single-division Scheme** (with elimination by rows)

| | | | | | | | | | |
|---|---|---|---|---|---|---|---|---|---|
| $a_{11}$ | $a_{12}$ | $a_{13}$ | $a_{14}$ | $a_{15}$ | **1.00** | 0.17 | $-0.25$ | 0.54 | 1.46 |
| $a_{21}$ | $a_{22}$ | $a_{23}$ | $a_{24}$ | $a_{25}$ | 0.47 | 1.00 | 0.67 | $-0.32$ | 1.82 |
| $a_{31}$ | $a_{32}$ | $a_{33}$ | $a_{34}$ | $a_{35}$ | $-0.11$ | 0.35 | 1.00 | $-0.74$ | 0.50 |
| $a_{41}$ | $a_{42}$ | $a_{43}$ | $a_{44}$ | $a_{45}$ | 0.55 | 0.43 | 0.36 | 1.00 | 2.34 |
| 1 | $b_{12}$ | $b_{13}$ | $b_{14}$ | $b_{15}$ | 1 | 0.17 | $-0.25$ | 0.54 | 1.46 |
| | $a_{22 \cdot 1}$ | $a_{23 \cdot 1}$ | $a_{24 \cdot 1}$ | $a_{25 \cdot 1}$ | | **0.9201** | 0.7875 | $-0.5738$ | 1.1338 |
| | $a_{32 \cdot 1}$ | $a_{33 \cdot 1}$ | $a_{34 \cdot 1}$ | $a_{35 \cdot 1}$ | | 0.3687 | 0.9725 | $-0.6806$ | 0.6606 |
| | $a_{42 \cdot 1}$ | $a_{43 \cdot 1}$ | $a_{44 \cdot 1}$ | $a_{45 \cdot 1}$ | | 0.3365 | 0.4975 | 0.7030 | 1.5370 |
| | 1 | $b_{23}$ | $b_{24}$ | $b_{25}$ | | 1 | 0.85589 | $-0.62363$ | 1.23226 |
| | | $a_{33 \cdot 2}$ | $a_{34 \cdot 2}$ | $a_{35 \cdot 2}$ | | | **0.65693** | $-0.45067$ | 0.20627 |
| | | $a_{43 \cdot 2}$ | $a_{44 \cdot 2}$ | $a_{45 \cdot 2}$ | | | 0.20949 | 0.91285 | 1.12234 |
| | | 1 | $b_{34}$ | $b_{35}$ | | | 1 | $-0.68602$ | 0.31399 |
| | | | $a_{44 \cdot 3}$ | $a_{45 \cdot 3}$ | | | | **1.05656** | 1.05656 |
| | | | | $\varDelta =$ | 1.00 $\times$ | 0.9201 $\times$ | 0.65693$\times$ | 1.05656 = | 0.63863 |

determinant of order $n$ is equal to $(n-1)(n^2 + n + 3)/3$.

In Table II.4 is given a scheme, and a numerical example, for computing a determinant.

The single-division scheme for computing a determinant may be applied not only via elimination by rows but also by columns (Table II.5). This is obviously equivalent to computing the determinant of the transposed matrix by rows.

After considering the process for computing a determinant we see that it coincides, with the exception of the last multiplication, with the forward course of the Gauss process, applied to a system having the matrix for which the determinant is being computed.

The well-known formulas of Cramer (Section 1) show that the solution for a linear system may be found in the form $x_i = \Delta_i/\Delta$; $i = 1, \cdots, n$, where $\Delta$ denotes the determinant determined by the system's coefficient, and where $\Delta_i$ is the determinant formed by replacing the elements of the $i$th column by the columns of constants. Thus it is necessary to compute $n + 1$ determinants of order $n$ in order to solve the system.

Comparing Gauss's process for solving a system with the process of computing the determinant, we see that the number of computations necessary to solve the systems exceeds only by a small amount the number needed to compute the determinant alone. Therefore it is not advisable to use Cramer's formulas for the numerical solution of the system. As a matter of fact, in Gauss's method the computations of all the determinants $\Delta$ and

**TABLE II.5  Computing a Determinant by the Single-division Scheme (with elimination by columns)**

| | | | | |
|---|---|---|---|---|
| 1.00 | 0.17 | −0.25 | 0.54 | 1.00 |
| 0.47 | 1.00 | 0.67 | −0.32 | 0.47 |
| −0.11 | 0.35 | 1.00 | −0.74 | −0.11 |
| 0.55 | 0.43 | 0.36 | 1.00 | 0.55 |
| 1.91 | 1.95 | 1.78 | 0.48 | 1.91 |
| | 0.9201 | 0.7875 | −0.5738 | 1 |
| | 0.3687 | 0.9725 | −0.6806 | 0.40072 |
| | 0.3365 | 0.4975 | 0.7030 | 0.36572 |
| | 1.6253 | 2.2575 | −0.5514 | 1.76644 |
| | | 0.65693 | −0.45067 | 1 |
| | | 0.20949 | 0.91285 | 0.31889 |
| | | 0.86643 | 0.46218 | 1.31891 |
| | | | 1.05656 | |
| | | | 1.05656 | |
| 1 × | 0.9201 × | 0.65693 × | 1.05656 = | 0.63863 |

$\Delta_i$ are carried out simultaneously, where as division by $\Delta = a_{11}.a_{22\cdot1}\cdots a_{nn\cdot n-1}$ is carried out gradually, by one factor at each step.

In computing a determinant one may also obtain zeros by means of a linear combination of the rows with rotation matrices. This requires a significantly larger number of operations than a single-division scheme but provides a much more stable computational process.

In computing a determinant it is also possible to partition its matrix into cells with square diagonal cells and then to "obtain zero" in a way similar to what is done in the corresponding transformational single-division scheme for solving systems of linear equations.

In the end the desired determinant turns out to be the product of the determinants of the "leading" cells

$$\Delta = |A_{11}| \cdot |A_{22\cdot1}| \cdots |A_{nn\cdot n-1}| .$$

## 18. COMPACT SCHEMES FOR SOLVING NON-HOMOGENEOUS LINEAR SYSTEMS

In Section 16 we saw that solving a system of linear equations by a single-division scheme is the same as determining the coefficients $a_{ij\cdot k}$ of the transformed equations (including constants) and the coefficients $b_{ij}$ of the equations in the final triangular system. To obtain a solution of the given system we need only the coefficients $b_{ij}$; the numbers $a_{ij\cdot k}$ play an auxiliary role and are necessary only for determining the numbers $b_{ij}$.

We shall show[†] that the number $b_{ij}$ may be obtained by an accumulation process which permits us to avoid computing and writing down all the coefficients $a_{ij\cdot k}$.

We choose elements of the first column of each auxiliary matrix $a_{ij\cdot j-1}$, $i \geq j$, denoting them by $c_{ij}$, $i \geq j$.

Analyzing the process of computing the coefficients of the auxiliary matrices, we see that

$$a_{ij\cdot k} = a_{ij\cdot k-1} - a_{ik\cdot k-1}b_{kj} = a_{ij\cdot k-1} - c_{ik}b_{kj}$$
$$= a_{ij\cdot k-2} - c_{ik-1}b_{k-1j} - c_{ik}b_{kj} = \cdots$$
$$= a_{ij} - c_{i1}b_{1j} - c_{i2}b_{2j} - \cdots - c_{ik}b_{kj} = a_{ij} - \sum_{l=1}^{k} c_{il}b_{lj} . \tag{1}$$

Thus any element $a_{ij\cdot k}$ is expressed by means of accumulation in terms of the elements $c_{ij}$ and numbers $b_{ij}$ to be determined.

In particular, for the elements $c_{ij}$, $i \geq j$, and $b_{ij}$, $i < j$, the recurrence formulas

$$c_{ij} = a_{ij\cdot j-1} = a_{ij} - \sum_{l=1}^{j-1} c_{il}b_{lj} \qquad (i \geq j) ,$$

$$b_{ij} = \frac{a_{ij\cdot i-1}}{a_{ii\cdot i-1}} = \frac{a_{ij} - \sum_{l=1}^{i-1} c_{il}b_{lj}}{c_{ii}} \qquad (i < j) . \tag{2}$$

---

[†] Dwyer (1).

hold. It is obvious that the constants of the transformed equations are also determined by these same formulas. The scheme for carrying out the forward course according to formulas (2) is called *compact*. The return course remains the same as in the detailed single-division scheme.

It is convenient to arrange computation by the compact scheme as is shown in Table II.6.

**TABLE II.6  Compact Scheme of the Single-division Method**

| | | | | | | | | | | | |
|---|---|---|---|---|---|---|---|---|---|---|---|
| $a_{11}$ | $a_{12}$ | $a_{13}$ | $a_{14}$ | $a_{15}$ $a_{16}$ | 1 | | 0.17 | −0.25 | 0.54 | 0.3 | 1.76 |
| $a_{21}$ | $a_{22}$ | $a_{23}$ | $a_{24}$ | $a_{25}$ $a_{26}$ | 0.47 | 1 | 0.67 | −0.32 | 0.5 | 2.32 |
| $a_{31}$ | $a_{32}$ | $a_{33}$ | $a_{34}$ | $a_{35}$ $a_{36}$ | −0.11 | 0.35 | 1 | −0.74 | 0.7 | 1.20 |
| $a_{41}$ | $a_{42}$ | $a_{43}$ | $a_{44}$ | $a_{15}$ $a_{46}$ | 0.55 | 0.43 | 0.36 | 1 | 0.9 | 3.24 |
| $c_{11}$ ⌊1 $b_{12}$ | $b_{13}$ | $b_{14}$ | $b_{15}$ $b_{16}$ | 1 ⌊1 0.17 | −0.25 | 0.54 | 0.3 | 1.76 | | | |
| $c_{21}$ | $c_{22}$ ⌊1 $b_{23}$ | $b_{24}$ | $b_{25}$ $b_{26}$ | 0.47 | 0.9201⌊1 | 0.85589 | −0.62363 | 0.39017 | 1.62243 | | |
| $c_{31}$ | $c_{32}$ | $c_{33}$ ⌊1 $b_{34}$ | $b_{35}$ $b_{36}$ | −0.11 | 0.3687 | 0.65693⌊1 | −0.68602 | 0.89681 | 1.21080 | | |
| $c_{41}$ | $c_{42}$ | $c_{43}$ | $c_{44}$ ⌊1 $b_{45}$ $b_{46}$ | 0.55 | 0.3365 | 0.20949 | 1.05657⌊1 | 0.39357 | 1.39357 | | |
| | | | 1 | $x_4$ $\overline{x}_4$ | | | | 1 | 0.39357 | 1.39357 | |
| | | 1 | | $x_3$ $\overline{x}_3$ | | | 1 | | 1.16681 | 2.16682 | |
| | 1 | | | $x_2$ $\overline{x}_2$ | | 1 | | | −0.36305 | 0.63694 | |
| 1 | | | | $x_1$ $\overline{x}_1$ | 1 | | | | 0.44089 | 1.44090 | |

Here we carry out the computation of elements $c$ and $b$ in succession by "corners" beginning with computation of the $c$ elements:

$$
\begin{array}{c|c|c|c}
c_{11} & b_{12} & b_{13} & b_{14} \quad \text{First Step} \\ \hline
c_{21} & c_{22} & b_{23} & b_{24} \quad \text{Second Step} \\ \hline
c_{31} & c_{32} & c_{33} & b_{34} \quad \text{Third Step} \\ \hline
c_{41} & c_{42} & c_{43} &
\end{array}
$$

Any element is obtained as the difference between a corresponding element $a$ and the sum of pairwise products of non-diagonal elements $c$ located in the given row (from the left) and elements $b$ located in the given column (above). Of course in computing the elements $b$ it is still necessary to carry out division by a corresponding diagonal element $c$.

Thus

$$c_{43} = a_{43} - c_{41}b_{13} - c_{42}b_{23}$$
$$= 0.36 + 0.55 \cdot 0.25 - 0.3365 \cdot 0.85589 = 0.20949 \, ,$$
$$b_{34} = \frac{a_{34} - c_{31}b_{14} - c_{32}b_{24}}{c_{33}}$$
$$= \frac{-0.74 + 0.11 \cdot 0.54 + 0.3687 \cdot 0.62363}{0.65693} = -0.68602 \, .$$

We shall say a few words about control as applied in computations by the compact scheme. Just as before, we form a column of the control sums and perform on them the same operations as on the column of constants.

When this is done every number of a transformed column should coincide with the sum of the elements of corresponding rows in matrix $B$, expanded by adding the transformed column of constants. In fact matrix $B$ is the coefficient matrix for the system obtained from the given one after finishing the forward course in the single-division scheme.

Computing by the compact scheme requires the fixing of the leading elements beforehand, so that it is impossible to give compact form to the scheme of principal elements.

The compact scheme may be extended with hardly any formal change to solving systems of linear equations whose matrices are partitioned into square cells along the main diagonal.

Denoting the matrix $A_{ij,j-1}$ by $C_{ij}$ we shall have, just as in the numerical case, that

$$C_{ij} = A_{ij} - \sum_{l=1}^{j-1} C_{il}B_{lj} \qquad (i \geq j) ,$$

$$B_{ij} = C_{ii}^{-1}\left(A_{ij} - \sum_{l=1}^{i-1} C_{il}B_{lj}\right) \qquad (i < j) .$$

These formulas permit us to compute in succession the matrices $C_{ij}$ and $B_{ij}$ in the same order as in the numerical case.

## 19. CONNECTION OF GAUSS'S METHOD WITH THE EXPANSION OF A MATRIX INTO FACTORS

In Section 1 it was pointed out that we may write a consistent system of $n$ linear equations in $n$ unknowns in the matrix form

$$AX = F \qquad\qquad (1)$$

where $A$ is the appropriate non-singular matrix and $X$ and $F$ are, respectively, the (ordered) columns containing the unknowns and the constant terms. We shall consider $X$ and $F$ to be vectors in an arithmetic space.

Gauss's method, carried out for a fixed order of the leading elements, consists in replacing a given system by an equivalent triangular system using a linear combination of the equation. This is the same as a linear combination of the rows of $A$. Besides dividing by the leading elements we must (using the single-division scheme) add to row elements numbers proportional to elements of the preceding rows, that is, to carry out on matrix $A$ *elementary transformations of type a) and b'*) of Chapter I, § 1.11 (page 14).

The result of several transformations of this type, as was shown there, is equivalent to premultiplying matrix $A$ by the left triangular matrix

$$\Gamma = \begin{bmatrix} \gamma_{11} & 0 & \cdots & 0 \\ \gamma_{21} & \gamma_{22} & \cdots & 0 \\ \cdot & \cdot & \cdot & \cdot \\ \gamma_{n1} & \gamma_{n2} & \cdots & \gamma_{nn} \end{bmatrix}. \qquad (2)$$

As a result of these transformations we arrive at a system with the right triangular matrix

$$B = \begin{bmatrix} 1 & b_{12} & \cdots & b_{1n} \\ 0 & 1 & \cdots & b_{2n} \\ \cdot & \cdot & \cdot & \cdot \\ 0 & 0 & \cdots & 1 \end{bmatrix}. \qquad (3)$$

Thus $\Gamma A = B$, that is, $A = \Gamma^{-1} B$, and consequently matrix $A$ has been expressed as the product of two triangular matrices.

The compact scheme realizes this expansion. In fact $\Gamma^{-1} = C$ where the elements of the matrix $C$ are determined by formulas (2) in Section 18, since from these formulas it follows that

$$a_{ij} = c_{ij} + \sum_{l=1}^{j-1} c_{il} b_{lj} = \sum_{l=1}^{j} c_{il} b_{lj} \qquad (i \geq j)$$

(from the formula for $c_{ij}$) and

$$a_{ij} = \sum_{l=1}^{i-1} c_{il} b_{lj} + c_{ii} b_{ij} = \sum_{l=1}^{i} c_{il} b_{lj} \qquad (i < j)$$

(from formulas for $b_{ij}$).

The last formulas mean that

$$A = CB .$$

Since the diagonal elements of matrix $B$ are equal to one, such an expansion is unique.

The compact scheme applied to a partitioned matrix obviously realizes its expansion as the product of two quasi-triangular matrices.

Compact notation for the schemes of Gauss's method, as distinct from single-division schemes, is similarly related to the expansion of a matrix into the product of two triangular matrices, but with another choice of the diagonal elements.

We observe that compact schemes fix the order of elimination and are therefore applied (this follows from theorem 1.1) only when none of the determinants

$$a_{11}, \begin{vmatrix} a_{11} & a_{12} \\ a_{21} & a_{22} \end{vmatrix}, \cdots, |A|$$

are equal to zero.

We shall show that if the matrix $A$ is symmetric, then

$$b_{ik} = \frac{c_{ki}}{c_{ii}} . \qquad (4)$$

In fact $A = CB$, $A' = B'C'$, and, since $A = A'$,

$$CB = B'C' = B' \begin{bmatrix} c_{11} & 0 & \cdots & 0 \\ 0 & c_{22} & \cdots & 0 \\ \multicolumn{4}{c}{\cdot \quad \cdot \quad \cdot \quad \cdot \quad \cdot \quad \cdot \quad \cdot} \\ 0 & 0 & \cdots & c_{nn} \end{bmatrix} \begin{bmatrix} 1 & \dfrac{c_{21}}{c_{11}} & \cdots & \dfrac{c_{n1}}{c_{11}} \\ 0 & 1 & \cdots & \dfrac{c_{n2}}{c_{22}} \\ \multicolumn{4}{c}{\cdot \quad \cdot \quad \cdot \quad \cdot \quad \cdot \quad \cdot \quad \cdot} \\ 0 & 0 & \cdots & 1 \end{bmatrix}.$$

From this it follows, in view of the uniqueness of the representation of matrix $A$ as the product of two triangular matrices, that

$$B = \begin{bmatrix} 1 & \dfrac{c_{21}}{c_{11}} & \cdots & \dfrac{c_{n1}}{c_{11}} \\ 0 & 1 & \cdots & \dfrac{c_{n2}}{c_{22}} \\ \multicolumn{4}{c}{\cdot \quad \cdot \quad \cdot \quad \cdot \quad \cdot \quad \cdot \quad \cdot} \\ 0 & 0 & \cdots & 1 \end{bmatrix}.$$

**TABLE II.7   The Compact Scheme for the Single-division Method in the Symmetric Case**

| | | | | | | | | | | | |
|---|---|---|---|---|---|---|---|---|---|---|---|
| $a_{11}$ | $a_{12}$ | $a_{13}$ | $a_{14}$ | $a_{15}$ | $a_{16}$ | 1.00 | 0.42 | 0.54 | 0.66 | 0.3 | 2.92 |
| $a_{21}$ | $a_{22}$ | $a_{23}$ | $a_{24}$ | $a_{25}$ | $a_{26}$ | 0.42 | 1.00 | 0.32 | 0.44 | 0.5 | 2.68 |
| $a_{31}$ | $a_{32}$ | $a_{33}$ | $a_{34}$ | $a_{35}$ | $a_{36}$ | 0.54 | 0.32 | 1.00 | 0.22 | 0.7 | 2.78 |
| $a_{41}$ | $a_{42}$ | $a_{43}$ | $a_{44}$ | $a_{45}$ | $a_{46}$ | 0.66 | 0.44 | 0.22 | 1.00 | 0.9 | 3.22 |
| $c_{11}$ 1 $b_{12}$ | $b_{13}$ | $b_{14}$ | $b_{15}$ | $b_{16}$ | 1.00 1 | 0.42 | 0.54 | 0.66 | 0.3 | 2.92 | |
| $c_{21}$ | $c_{22}$ 1 $b_{23}$ | $b_{24}$ | $b_{25}$ | $b_{26}$ | 0.42 | 0.82360 1 | 0.11316 | 0.19767 | 0.45410 | 1.76493 | |
| $c_{31}$ | $c_{32}$ | $c_{33}$ 1 $b_{34}$ | $b_{35}$ | $b_{36}$ | 0.54 | 0.09320 | 0.69785 1 | $-0.22185$ | 0.71030 | 1.48844 | |
| $c_{41}$ | $c_{42}$ | $c_{43}$ | $c_{44}$ 1 $b_{45}$ | $b_{46}$ | 0.66 | 0.16280 | $-0.15482$ | 0.49787 1 | 1.48240 | 2.48239 | |
| | | | 1 | $x_4$ $\bar{x}_4$ | | | | 1 | 1.48240 | 2.48239 | |
| | | 1 | | $x_3$ $\bar{x}_3$ | | | 1 | | 1.03917 | 2.03916 | |
| | 1 | | | $x_2$ $\bar{x}_2$ | | 1 | | | 0.04348 | 1.04348 | |
| 1 | | | | $x_1$ $\bar{x}_1$ | 1 | | | | $-1.25780$ | $-0.25779$ | |

Thus the elements of matrix $B$ are found by dividing the elements of matrix $C$ by the diagonal elements. Nevertheless the compact scheme requires $n^2$ notations, even in the case where matrix $A$ is symmetric (Table II.7), since in the recurrent relationships (2) of Section 18 both the numbers $c_{ki}$ and $b_{ik}$ are involved. It is possible to give recurrent relationships in which the elements $b_{ik}$ are expressed in terms of each other and in terms of the diagonal elements $c_{ii}$. However formulas so constructed turn out to be more complex.

## 20. THE SQUARE-ROOT METHOD

In this paragraph we shall show that if the matrix of a system is symmetric the finding of a solution may be further simplified,[†] since in this

---

† Banachiewicz, T., (4), (5).

case the matrix may be expressed as the product of a triangular and its transposed matrix.

Thus let

$$A = S'S,\qquad(1)$$

where

$$S = \begin{bmatrix} s_{11} & s_{12} & \cdots & s_{1n} \\ 0 & s_{22} & \cdots & s_{2n} \\ \cdot & \cdot & \cdot & \cdot \\ 0 & 0 & \cdots & s_{nn} \end{bmatrix}.\qquad(2)$$

We shall determine the elements of matrix $S$. In view of the rule for multiplying matrices we have;

$$a_{ij} = s_{1i}s_{1j} + s_{2i}s_{2j} + \cdots + s_{ii}s_{ij} \qquad (i < j),$$
$$a_{ii} = s_{1i}^2 + s_{2i}^2 + \cdots + s_{ii}^2 \qquad (i = j).$$

From here we obtain the formulas for determining $s_{ij}$:

$$s_{11} = \sqrt{a_{11}},$$

$$s_{1j} = \frac{a_{1j}}{s_{11}},$$

$$s_{ii} = \sqrt{a_{ii} - \sum_{l=1}^{i-1} s_{li}^2} \qquad (i > 1),$$

$$s_{ij} = \frac{a_{ij} - \sum_{l=1}^{i-1} s_{li}s_{lj}}{s_{ii}} \qquad (j > i),\qquad(3)$$

$$s_{ij} = 0 \qquad (i > j).$$

Moreover, solving the system is the same as solving two triangular systems. In fact the equality

$$AX = F$$

is equivalent to the two equalities

$$S'K = F \quad \text{and} \quad SX = K.$$

Elements of the vector $K$ are determined by recurrent formulas analogous to the formulas for $s_{ij}$. Namely,

$$k_1 = \frac{f_1}{s_{11}},$$

$$k_i = \frac{f_i - \sum_{l=1}^{i-1} s_{li}k_l}{s_{ii}} \qquad (i > 1).\qquad(4)$$

The final solution is found by the formulas

$$x_n = \frac{k_n}{s_{nn}},$$

$$x_i = \frac{k_i - \sum\limits_{l=i+1}^{n} s_{il} x_l}{s_{ii}} \qquad (i < n) . \tag{5}$$

The usual control is applied in the scheme and we confirm all matrix elements by forming the control elements. Analogous to the compact scheme, the control equation is

$$\bar{k}_i = \sum_{k=1}^{n} s_{ik} + k_i .$$

In the square-root method we have to write only $\dfrac{n(n+1)}{2}$ elements of matrix $S$ and $2n$ components of the vectors $K$ and $X$.

In Table II.8 a solution of the system is carried out by the square-root method.

**TABLE II.8  The Square-root Method**

| $a_{11}$ | $a_{12}$ | $a_{13}$ | $a_{14}$ | $f_1$ | $\bar{f}_1$ | 1.00 | 0.42 | 0.54 | 0.66 | 0.3 | 2.92 |
|---|---|---|---|---|---|---|---|---|---|---|---|
| | $a_{22}$ | $a_{23}$ | $a_{24}$ | $f_2$ | $\bar{f}_2$ | | 1.00 | 0.32 | 0.44 | 0.5 | 2.68 |
| | | $a_{33}$ | $a_{34}$ | $f_3$ | $\bar{f}_3$ | | | 1.00 | 0.22 | 0.7 | 2.78 |
| | | | $a_{44}$ | $f_4$ | $\bar{f}_4$ | | | | 1.00 | 0.9 | 3.22 |
| $s_{11}$ | $s_{12}$ | $s_{13}$ | $s_{14}$ | $k_1$ | $\bar{k}_1$ | 1.00 | 0.42 | 0.54 | 0.66 | 0.3 | 2.92 |
| | $s_{22}$ | $s_{23}$ | $s_{24}$ | $k_2$ | $\bar{k}_2$ | | 0.90752 | 0.10270 | 0.17939 | 0.41211 | 1.60173 |
| | | $s_{33}$ | $s_{34}$ | $k_3$ | $\bar{k}_3$ | | | 0.83537 | −0.18533 | 0.59336 | 1.24340 |
| | | | $s_{44}$ | $k_4$ | $\bar{k}_4$ | | | | 0.70560 | 1.04597 | 1.75157 |
| $x_1$ | $x_2$ | $x_3$ | $x_4$ | | | −1.25778 | 0.04348 | 1.03917 | 1.48238 | | |
| $\bar{x}_1$ | $\bar{x}_2$ | $\bar{x}_3$ | $\bar{x}_4$ | | | −0.25779 | 1.04349 | 2.03917 | 2.48238 | | |

Computation of the elements $s_{ij}$ (as well as the elements $k_i$ and $\bar{k}_i$) is carried out successively by rows. Any diagonal element is computed as the square root of the difference between the corresponding element $a$ and the sum of the squares of all computed elements $s$ located in the same column. Non-diagonal elements $s_{ij}$ are obtained by subtracting from the elements $a_{ij}$ the sum of the products of corresponding elements $s$ taken from the columns with numbers $i$ and $j$. The difference obtained is divided by the diagonal element of the row.

Thus,

$$s_{34} = \frac{a_{34} - s_{13}s_{14} - s_{23}s_{24}}{s_{33}} = \frac{0.22 - 0.54 \cdot 0.66 - 0.10270 \cdot 0.17939}{0.83537} = -0.18533 .$$

The return course is given by formulas (5).

The forward course of the square-root method is essentially equivalent to transforming a quadratic form with matrix $A$ to a sum of squares by transformation of variables via a triangular matrix.

If $A$ is a positive definite matrix, then the square-root method goes along without any complications. However if $A$ is not positive definite, then degeneracy of the process is possible since a certain coefficient $s_{ii}$ might turn out to be equal to zero or close to zero. It may also turn out that the root expressions for certain $s_{ii}$ are negative. However this will not produce essential difficulties.

Actually in this case purely imaginary numbers will appear in any row for which $s_{ii}^2 < 0$; operations with these numbers will not be any more complicated than with real numbers.

We shall illustrate what has been said with an example:

| 2 | −1 | 1 | 4 | 6 |
|---|---|---|---|---|
|   | −2 | 3 | 5 | 5 |
|   | 1 | 1 | 6 | 11 |
| 1.41421 | −0.70711 | 0.70711 | 2.82843 | 4.24265 |
|   | 1.58114 $i$ | −2.21359 $i$ | −4.42720 $i$ | −5.05965 $i$ |
|   |   | 2.32379 | 5.93857 | 8.26235 |
| 1.11111 | 0.77776 | 2.55555 |   |   |
| 2.11111 | 1.77776 | 3.55555 |   |   |

At the present time the square-root method is used extensively for solving symmetric systems and may be recommended to the reader as one of the most effective methods.

## 21. INVERSION OF MATRICES

As was said already in the introduction, the problem of solving a linear non-homogeneous system and the problem of inverting a matrix are closely connected with each other.

In fact if for a matrix $A$ its inverse matrix $A^{-1}$ is known, then premultiplying the equality

$$AX = F \qquad (1)$$

by $A^{-1}$ gives us

$$X = A^{-1}F . \qquad (2)$$

Conversely, determining the elements of the inverse matrix may lead to a solution of $n$ systems of the form

$$\sum_{k=1}^{n} a_{ik}\alpha_{kj} = \delta_{ij} \qquad \begin{array}{l} (j = 1, \cdots, n), \\ (i = 1, \cdots, n). \end{array} \qquad (3)$$

where $\delta_{ij}$ is the Kronecker symbol.

The last equation arises from the definition of the inverse matrix

$$AA^{-1} = E$$

and the rule for multiplying matrices.

*Exact Methods for Systems of Linear Equations* [*Chap. 2*]

A method being applied in structural mechanics for determining the solution to a system with the aid of so-called *influence numbers*[†] is none other than the solution of a system by means of constructing the inverse matrix. The influence numbers themselves are elements of the inverse matrix.

The numerical solution of $n$ systems of equations given by elements of the inverse matrix may be realized, for example, by the single-division method for several systems with a common coefficient matrix (see Table II.9).

**TABLE II.9   Inversion of a Matrix: Single-division Scheme**

| | | | | | | | | |
|---|---|---|---|---|---|---|---|---|
| 1.00 | 0.42 | 0.54 | 0.66 | 1 | 0 | 0 | 0 | 3.62 |
| 0.42 | 1.00 | 0.32 | 0.44 | 0 | 1 | 0 | 0 | 3.18 |
| 0.54 | 0.32 | 1.00 | 0.22 | 0 | 0 | 1 | 0 | 3.08 |
| 0.66 | 0.44 | 0.22 | 1.00 | 0 | 0 | 0 | 1 | 3.32 |
| 1 | 0.42 | 0.54 | 0.66 | 1 | 0 | 0 | 0 | 3.62 |
| | 0.82360 | 0.09320 | 0.16280 | −0.42 | 1 | 0 | 0 | 1.65960 |
| | 0.09320 | 0.70840 | −0.13640 | −0.54 | 0 | 1 | 0 | 1.12520 |
| | 0.16280 | −0.13640 | 0.56440 | −0.66 | 0 | 0 | 1 | 0.93080 |
| | 1 | 0.11316 | 0.19767 | −0.50996 | 1.21418 | 0 | 0 | 2.01506 |
| | | 0.69785 | −0.15482 | −0.49247 | −0.11316 | 1 | 0 | 0.93740 |
| | | −0.15482 | 0.53222 | −0.57198 | −0.19767 | 0 | 1 | 0.60275 |
| | | 1 | −0.22185 | −0.70570 | −0.16216 | 1.43297 | 0 | 1.34327 |
| | | | 0.49787 | −0.68624 | −0.22278 | 0.22185 | 1 | 0.81071 |
| | | | 1 | −1.37835 | −0.44746 | 0.44560 | 2.00856 | 1.62836 |
| | | 1 | | −1.01149 | −0.26143 | 1.53183 | 0.44560 | 1.70452 |
| | 1 | | | −0.12304 | −1.33221 | −0.26142 | −0.44746 | 1.50030 |
| 1 | | | | 2.50759 | −0.12303 | −1.01149 | −1.37834 | 0.99472 |

As a result we get a matrix consisting of the rows of the inverse matrix arrayed in the reverse order. For controlling the computation and estimating the accuracy of the result it is advisable to multiply $A$ by $A^{-1}$.

The theorem on expanding a matrix as the product of two triangular matrices makes it possible to set up a compact scheme for computing the elements of the inverse matrix.[‡] This scheme requires $2n^2$ entries in all, while $n^2$ of them will give the elements of the inverse matrix.

Let

$$A = CB,\qquad\qquad (4)$$

where elements of the triangular matrices $C$ and $B$ are defined by formulas (2), Section 18, which we rewrite here:

---

[†] A. A  Umanskii (1).

[‡] Waugh and Dwyer (1).

$$c_{ij} = a_{ij} - \sum_{l=1}^{j-1} c_{il}b_{lj} \qquad (i \geq j) ,$$

$$b_{ij} = \frac{a_{ij} - \sum_{l=1}^{i-1} c_{il}b_{lj}}{c_{ii}} \qquad (i < j; \; b_{ii} = 1) .$$

We shall denote elements of the inverse matrix $A^{-1} = D$ by $d_{ij}$.
We have, obviously, that

$$D = B^{-1}C^{-1} . \tag{5}$$

We shall show that elements $d_{ij}$ may be determined without inverting matrices $B$ and $C$.

Post-multiplying equation (5) by $C$ we obtain

$$DC = B^{-1} \tag{6}$$

The matrix $B^{-1}$ is also triangular with ones along the principal diagonal. Therefore we know $n(n + 1)/2$ of its elements (of them, $n(n - 1)/2$ will be zeros and the remaining $n$ will be ones).

Analogously, we pre-multiply equation (5) by $B$ and get

$$BD = C^{-1} . \tag{7}$$

Since the matrix $C^{-1}$ is triangular, then $n(n - 1)/2$ of its elements will be zeros.

It is easy to see that the system obtained by joining the $n(n + 1)/2$ equations of the system $DC = B^{-1}$ above and the $n(n - 1)/2$ equations of the system $BD = C^{-1}$ is a recurrent system which makes it possible to determine the $n^2$ elements of the inverse matrix.

We shall write it out for $n = 4$.

| | $i = 1$ | 2 | 3 | 4 |
|---|---|---|---|---|
| $c_{11}d_{i1} + c_{21}d_{i2} + c_{31}d_{i3} + c_{41}d_{i4} =$ | 1 | 0 | 0 | 0 |
| $c_{22}d_{i3} + c_{32}d_{i3} + c_{42}d_{i4} =$ | | 1 | 0 | 0 |
| $c_{33}d_{i3} + c_{43}d_{i4} =$ | | | 1 | 0 |
| $c_{44}d_{i4} =$ | | | | 1 |

| | $j =$ | 2 | 3 | 4 |
|---|---|---|---|---|
| $d_{1j} + b_{12}d_{2j} + b_{13}d_{3j} + b_{14}d_{4j} =$ | | 0 | 0 | 0 |
| $d_{2j} + b_{23}d_{3j} + b_{24}d_{4j} =$ | | | 0 | 0 |
| $d_{3j} + b_{34}d_{4j} =$ | | | | 0 |

From the equations of the first grouping, for $i = 4$, are determined successively $d_{44}$, $d_{43}$, $d_{42}$, and $d_{41}$. Then from equations of the second grouping, for $j = 4$, are determined $d_{34}$, $d_{24}$, and $d_{14}$. The process continues in an analogous way and we shall use the formulas of the first and second group in their turn. Namely, from the equations of the first group, for $i = 3$, are determined $d_{31}$, $d_{32}$, and $d_{33}$ and from the equations of the second group, for $j = 3$, are determined $d_{23}$ and $d_{13}$; from the equations of the first group, for $i = 2$, $d_{21}$ and $d_{22}$ and from the equations of the second group, for $j = 2$, $d_{12}$. Finally, from the equations of the first group, for $i = 1$, we determine $d_{11}$. Matrix inversion by a compact scheme is shown in Table II.10.

**TABLE II.10   A Compact Scheme for Inverting a Matrix**

| 1.00 | 0.42 | 0.54 | 0.66 | | | | |
|---|---|---|---|---|---|---|---|
| 0.42 | 1.00 | 0.32 | 0.44 | | | | |
| 0.54 | 1.32 | 1.00 | 0.22 | | | | |
| 0.66 | 0.44 | 0.22 | 1.00 | | | | |
| 1.00 ⌊1 | 0.42 | 0.54 | 0.66 | 2.50759 | −0.12303 | −1.01149 | −1.37834 |
| 0.42 | 0.82360⌋ 1 | 0.11316 | 0.19767 | −0.12303 | 1.33221 | −0.26143 | −0.44745 |
| 0.54 | 0.09320 | 0.69785 ⌊1 | −0.22185 | −1.01149 | −0.26143 | 1.53184 | 0.44560 |
| 0.66 | 0.16280 | −0.15482 | 0.49787 ⌊1 | −1.37834 | −0.44745 | 0.44560 | 2.00855 |

The compact scheme for inverting a matrix may be extended to partitioned matrices with square diagonal cells. Expanding a matrix $A$ into the product $CB$ of two quasi-triangular matrices, we look for the inverse matrix $D = A^{-1}$ also in partitioned form. Then, analogous to the numerical case, the cells $D_{ij}$ of the inverse matrix are found in succession from the relationships (we carry them out for $n = 4$):

$$
\begin{array}{lcccc}
 & i = 1 & 2 & 3 & 4 \\
D_{i1}C_{11} + D_{i2}C_{21} + D_{i3}C_{31} + D_{i4}C_{41} = E & 0 & 0 & 0 \\
D_{i2}C_{22} + D_{i3}C_{32} + D_{i4}C_{42} = & E & 0 & 0 \\
D_{i3}C_{33} + D_{i4}C_{43} = & & E & 0 \\
D_{i4}C_{44} = & & & E
\end{array}
$$

$$
\begin{array}{lccc}
 & j = & 2 & 3 & 4 \\
D_{1j} + B_{12}D_{2j} + B_{13}D_{3j} + B_{14}D_{4j} = & & 0 & 0 & 0 \\
D_{2j} + B_{23}D_{3j} + B_{24}D_{4j} = & & & 0 & 0 \\
D_{3j} + B_{34}D_{4j} = & & & & 0
\end{array}
$$

The order of determining matrices $D_{ij}$ is the same as in the case of the numerical matrix.

## 22.  THE PROBLEM OF ELIMINATION

In the simplest case this problem consists in computing the value of a linear form

$$c_1 x_1 + c_2 x_2 + \cdots + c_n x_n ,  \tag{1}$$

where $c_1, c_2, \cdots, c_n$ are given numbers and $x_1, x_2, \cdots, x_n$ is the solution of the system

$$
\begin{aligned}
a_{11}x_1 + a_{12}x_2 + \cdots + a_{1n}x_n &= f_1 , \\
a_{21}x_1 + a_{22}x_2 + \cdots + a_{2n}x_n &= f_2 , \\
\cdot \ \cdot \ \cdot \ \cdot \ \cdot \ \cdot \ & \cdot \ \cdot \ \cdot \ \cdot \ \cdot \ , \\
a_{n1}x_1 + a_{n2}x_2 + \cdots + a_{nn}x_n &= f_n ,
\end{aligned}
\tag{2}
$$

whose determinant is different from zero.

A natural solution course for this problem would consist of solving for the numbers $x_1, x_2, \cdots, x_n$ explicitly and substituting them into expression (1). However it is possible to avoid a complete solution process by using the following procedure.

We write out the coefficient matrix of system (2), adjoining on the right of it the column of constants; below this we write a row composed of the coefficients of the linear form being computed, taking these coefficients with signs reversed. In the right lower corner we place the number 0. We then get the scheme:

$$
\begin{array}{cccc||c}
a_{11} & a_{12} & \cdots & a_{1n} & f_1 \\
a_{21} & a_{22} & \cdots & a_{2n} & f_2 \\
\cdot & \cdot & \cdots & \cdot & \cdot \\
a_{n1} & a_{n2} & \cdots & a_{nn} & f_n \\
\hline\hline
-c_1 & -c_2 & \cdots & -c_n & 0
\end{array}
\tag{3}
$$

or in abbreviated notation

$$
\begin{array}{c||c}
A & F \\
\hline\hline
-C & 0
\end{array}
\tag{3'}
$$

Let $\gamma_1, \gamma_2, \cdots, \gamma_n$ denote the solution of the system of equations:

$$
\begin{aligned}
a_{11}\gamma_1 + a_{21}\gamma_2 + \cdots + a_{n1}\gamma_n &= c_1, \\
a_{12}\gamma_1 + a_{22}\gamma_2 + \cdots + a_{n2}\gamma_n &= c_2, \\
\cdot \quad \cdot \quad \cdot \quad \cdot \quad \cdot \quad \cdot \quad \cdot \quad \cdot \quad \cdot & \\
a_{1n}\gamma_1 + a_{2n}\gamma_2 + \cdots + a_{nn}\gamma_n &= c_n.
\end{aligned}
\tag{4}
$$

Then

$$
\begin{aligned}
f_1\gamma_1 + f_2\gamma_2 + \cdots f_n\gamma_n &= (a_{11}x_1 + \cdots + a_{1n}x_n)\gamma_1 \\
&+ (a_{21}x_1 + \cdots + a_{2n}x_n)\gamma_2 \\
&+ \cdot \quad \cdot \quad \cdot \quad \cdot \quad \cdot \quad \cdot \quad \cdot \\
&+ (a_{n1}x_1 + \cdots + a_{nn}x_n)\gamma_n \\
&= (a_{11}\gamma_1 + \cdots + a_{n1}\gamma_n)x_1 \\
&+ (a_{12}\gamma_1 + \cdots + a_{n2}\gamma_n)x_2 \\
&+ \cdot \quad \cdot \quad \cdot \quad \cdot \quad \cdot \quad \cdot \quad \cdot \\
&+ (a_{1n}\gamma_1 + \cdots + a_{nn}\gamma_n)x_n \\
&= c_1x_1 + c_2x_2 + \cdots + c_nx_n.
\end{aligned}
$$

Thus it is possible to replace the computation of the form $c_1x_1 + c_2x_2 + \cdots + c_nx_n$ by computation of the form $f_1\gamma_1 + f_2\gamma_2 + \cdots + f_n\gamma_n$. On the other hand it is obvious that if we multiply the first row of scheme (3) by $\gamma_1$, the second by $\gamma_2, \cdots$, the $n$th by $\gamma_n$ and add it to the last row, then we obtain a row whose elements placed beyond the double line are equal to zero; the element in the right lower corner is obviously equal to the number sought for. Conversely, if we choose a linear combination of $n$ rows by some means, so that adding it to the last row gives a row of zeros (up to the line), then the coefficients of this combination will be the solution of system (4) and consequently the element in the right lower corner will be equal to the number desired. This follows from the uniqueness of the solution for system (4). Thus there is no need to find the numbers $\gamma_1, \gamma_2, \cdots, \gamma_n$. We need only to annihilate the last row by adding a suitable linear combination of the first $n$ rows. It is possible to do this by the usual forward course of Gauss's process applied to scheme (3).

It is obvious that such a method may be applied to compute the linear nonhomogeneous form $c_1 x_1 + \cdots + c_n x_n + d$. The only difference will be that in the right lower corner we shall need to write the constant $d$ instead of 0 so that the initial scheme has the form:

$$
\begin{array}{cccc|c}
a_{11} & a_{12} & \cdots & a_{1n} & f_1 \\
\cdot & \cdot & \cdots & \cdot & \cdot \\
a_{n1} & a_{n2} & \cdots & a_{nn} & f_n \\
\hline
-c_1 & -c_2 & \cdots & -c_n & d
\end{array}
\tag{3''}
$$

By this method one may find the solution to system (2), as well, without applying the return course. In fact the expressions $x_1, x_2, \cdots, x_n$ are special cases of the linear form (1) with coefficients $(1, 0, \cdots, 0)$, $(0, 1, \cdots, 0)$, $\cdots$, $(0, 0, \cdots, 1)$. They may be determined simultaneously with an elimination scheme having simultaneously notation in the lower left corner of the rows $(-1, 0, \cdots, 0)$, $(0, -1, \cdots, 0)$, $\cdots$, $(0, 0, \cdots, -1)$ which together form the matrix $-E$. In the right lower corner we replace the number 0 by a null column.

Thus the initial scheme for solving the system has the form:

$$
\begin{array}{c|c}
A & F \\
\hline
-E & 0
\end{array}
\tag{5}
$$

Having obtained a null matrix in the lower left corner of the scheme by adding suitable linear combinations of the first $n$ rows, we obtain in the lower right corner a column consisting of the values of the unknown.

Inverting a matrix is equivalent, as we have seen, to solving $n$ systems of the special case where the constants form a unit matrix. They may be solved as an aggregate with the help of the scheme

$$
\begin{array}{c|c}
A & E \\
\hline
-E & 0
\end{array}
\tag{6}
$$

where $E$ as before denotes a unit matrix, and a null matrix or order $n$ is placed in the lower right corner. After annihilating all the rows in the lower left corner by adding suitable linear combinations of the first $n$ rows we obtain the matrix $A^{-1}$ in the lower right corner.

Scheme (6) may be varied in the following way. It is obvious that annihilating the matrix $-E$ located in the lower left corner of scheme (6) requires the same linear combination of the first $n$ rows of the scheme as is needed to obtain the unit matrix $E$ in the lower left corner in place of the matrix A. Therefore the computation may be carried out in the following way. We form a matrix $[A, E]$. By dividing the first row by $a_{11}$ we make the first element of the first row equal to one. We then obtain zero in the first column by adding the first row (multiplied by suitable factors) to each of the other rows. We then obtain 1 in place of the second element of the second row by dividing, and zero in place of all the remaining elements of the second column by adding, the second row to all the others (in particular, to the first). After $n$ steps we obtain the identity matrix in place of matrix $A$ and matrix $A^{-1}$ in place of $E$.

During the process it is possible to transpose rows as well, perhaps permitting us to avoid division by numbers close to zero, provided the matrix is not too ill-conditioned.

There is another interpretation for the above process. We recall that combining the rows linearly is equivalent to premultiplying by a certain matrix. We denote by $B_1, B_2, \cdots, B_n$ the sequence of matrices, multiplication by which corresponds to completing the 1st, 2nd, $\cdots$, $n$th steps of the process. It is clear that the matrices obtained during the process will be $[A, E], [B_1A, B_1], [B_2A, B_2], \cdots, [B_nA, B_n]$.

In matrix $B_1A$ the first column coincides with the first column of the identity matrix, in matrix $B_2A$ the first two columns coincide with columns of the identity matrix, etc.

We denote by $V_i^{(k)}$ the vector whose components are the elements of the $i$th row of matrix $B_k$, and by $A_j$ the $j$th column of matrix $A$. From what has been said above it is clear that

$$(A_j, V_i^{(k)}) = \delta_{ij} \qquad \text{for} \quad j < k .$$

The process of constructing vectors $V_i^{(k)}$ coincides exactly with that described in Section 9 for constructing a dual basis. A dual basis is set up here for a basis consisting of the columns of matrix $A$, proceeding from the system of coordinate vectors.

We shall write out the formulas corresponding to the process described.

$$V_1^{(0)} = (1, 0, \cdots, 0)', V_2^{(0)} = (0, 1, \cdots, 0)', \cdots, V_n^{(0)} = (0, 0, \cdots, 1)',$$

$$V_k^{(k)} = \frac{1}{(A_k, V_k^{(k-1)})} V_k^{(k-1)},$$

$$V_i^{(k)} = V_i^{(k-1)} - (A_k, V_i^{(k-1)})V_k^{(k)},$$

Instead of unit vectors it is possible to take any other system of linearly independent vectors.

However, an unsuccessful choice of the initial system of vectors may mean that one of the scalar products $(A_k, V_k^{(k-1)})$ turns out to be equal to zero or approaches zero. In particular, for a coordinate system of vectors this will hold if one of the principal minors of matrix $A$ is equal to zero or close to zero.

It should be noted that if one takes the column $A_k$ of matrix $A$ as the system of vectors $V_1^{(0)}$, $V_2^{(0)}$, $\cdots$, $V_n^{(0)}$, then it is easy to see that $(A_k, V_k^{(k-1)}) \neq 0$. This follows from the fact (Section 12) that the matrix

$$
\begin{bmatrix}
(A_1, A_1) & (A_1, A_2) & \cdots & (A_1, A_n) \\
(A_2, A_1) & (A_2, A_2) & \cdots & (A_2, A_n) \\
\cdot\,\cdot\,\cdot\,\cdot\,\cdot\,\cdot\,\cdot\,\cdot\,\cdot\,\cdot\,\cdot\,\cdot\,\cdot\,\cdot\,\cdot \\
(A_n, A_1) & (A_n, A_2) & \cdots & (A_n, A_n)
\end{bmatrix} = A'A
$$

is positive definite, and consequently all its principal minors are positive.

We also observe that the process works out successfully if for the vectors $V_1^{(0)}$, $\cdots$, $V_n^{(0)}$ we take the rows of a matrix which is close to the inverse matrix.

In the work of Hestenes (4) the process described is recommended for inverting ill-conditioned matrices, proceeding first from the columns of matrix $A$ and then, with the idea of increasing the accuracy (possibly several times), proceeding from the obtained approximation to the inverse matrix.

It is possible to give matrix form to the result of the elimination process. This in turn makes it possible to obtain certain generalizations.

Namely, we may write a solution of the system $x_1$, $\cdots$, $x_n$ in matrix form as the column

$$
\begin{bmatrix}
x_1 \\
x_2 \\
\vdots \\
x_n
\end{bmatrix} = A^{-1}F,
$$

and the value of the linear form $c_1 x_1 + \cdots + c_n x_n$ as the number $CA^{-1}F$.

Such a representation indicates a way to compute the more complex expression $CA^{-1}B$ where $C$ and $B$ are certain rectangular matrices, $C$ consisting of $n$ columns and $B$ of $n$ rows, while the number of columns of $B$ and the number of rows of $C$ is immaterial. In fact, the element of the $i$th row and $k$th column of the matrix $CA^{-1}B$ is the number $c_i A^{-1} b_k$ where $c_i$ is the $i$th row of matrix $C$ and $b_k$ is the $k$th column of matrix $B$.

We may therefore compute the elements of the matrix $CA^{-1}B$ by the elimination method applied to the scheme

$$
\begin{array}{c||c}
A & B \\
\hline
-C & 0
\end{array}. \tag{7}
$$

After annihilating the elements located in the lower left corner by adding a linear combination of the first $n$ rows we obtain the matrix $CA^{-1}B$ in the lower right corner.

It is obvious that with a simultaneous change of sign for all the elements of matrices $B$ and $C$ the result is not altered, so that scheme (7) is equivalent to the scheme.

$$\frac{A \,\|\, -B}{C \,\|\, 0}. \qquad (7')$$

The inverse matrix for the matrix $A$ may be computed with the aid of the scheme

$$\frac{BA \,\|\, B}{-E \,\|\, 0}, \qquad (8)$$

in which we may take an arbitrary non-degenerate matrix as the matrix $B$. Application of scheme (8) is essentially equivalent to application of the biorthogonalization process described above, proceeding from a system of vectors whose components form the rows of matrix $B$. In particular, the process for $B = A'$ is equivalent to the process of biorthogonalization for the columns of matrix $A$.

To compute the product $CA^{-1}B$ it is possible to use, besides scheme (7), the transposed scheme (accurate up to the signs of $B$ and $C$):

$$\frac{A' \,\|\, C'}{-B' \,\|\, 0}. \qquad (9)$$

When computing by this scheme we obtain in the lower right corner the matrix $(CA^{-1}B)'$, since $(CA^{-1}B)' = B'(A')^{-1}C'$.

In particular, for computing the value of the linear form $c_1 x_1 + c_2 x_2 + \cdots + c_n x_n$ it is also possible to use the scheme

$$\frac{A' \,\|\, C'}{-F' \,\|\, 0}, \qquad (9')$$

where $A'$ is the transpose of the matrix of coefficients of the system, $C'$ is a column containing the coefficients of the linear from being computed, and $F'$ is a row of constants. It is easy to establish the validity of this construction directly without reference to the results adduced. In fact if we multiply the first $n$ rows of scheme (9') by $x_1, x_2, \cdots, x_n$ and add them to the last one we get zero in the lower left corner and $c_1 x_1 + c_2 x_2 + \cdots + c_n x_n$ in the lower right corner.

Analogously, to solve the system $AX = F$ we may use the scheme

$$\frac{A' \,\|\, E}{-F' \,\|\, 0}. \qquad (10)$$

Schemes (7) and (9) may be applied in computing $A^{-1}B$. For this it is sufficient to let $C = E$ in these schemes.

Along with scheme (6) we may use the following scheme for inverting a matrix:

$$\frac{A' \,\|\, E}{-E \,\|\, 0}. \qquad (11)$$

After completing this process we obtain in the lower right corner a matrix

**TABLE II.11   Inverting a Matrix by the Method of Biorthogonalization of Columns**

| I | II | III | IV | V | VI | VII |
|---|---|---|---|---|---|---|
| 1.00 | 0.47 | −0.11 | 0.55 | | | |
| 0.17 | 1.00 | 0.35 | 0.43 | | | |
| −0.25 | 0.67 | 1.00 | 0.36 | | | |
| 0.54 | −0.32 | −0.74 | 1.00 | | | |
| 0.65125 | 0.30609 | −0.07164 | 0.35819 | 0.999997 | 1.5355 | 0.65125 |
| −0.37575 | 0.74350 | 0.41003 | 0.12984 | 0.000004 | 0.8380 | |
| −0.34958 | 0.62320 | 1.01095 | 0.30523 | −0.000004 | 0.1529 | |
| −0.12493 | −0.63252 | −0.66687 | 0.63429 | 0.000001 | 1.0210 | |
| 0.88455 | −0.15555 | −0.32623 | 0.27757 | −0.000002 | 0.54575 | |
| −0.42749 | 0.84589 | 0.46650 | 0.14772 | 1.000011 | 0.87896 | 1.13771 |
| 0.09879 | −0.26401 | 0.52166 | 0.15029 | −0.000010 | 1.04885 | |
| −0.38759 | −0.11279 | −0.38024 | 0.72505 | 0.000007 | −0.61442 | |
| 1.03020 | −0.54478 | 0.44286 | 0.49914 | −0.000002 | −0.55166 | |
| −0.74255 | 1.68785 | −1.19713 | −0.33157 | 0.000002 | 1.19330 | |
| 0.26402 | −0.70557 | 1.39414 | 0.40165 | 0.999997 | 0.37418 | 2.67251 |
| −0.36175 | −0.18186 | −0.24377 | 0.76437 | 0.000006 | −0.09798 | |
| 1.43425 | −0.34165 | 0.71514 | −0.35461 | 0.000009 | 0.90206 | |
| −0.91580 | 1.60075 | −1.31388 | 0.03450 | −0.000001 | −0.38678 | |
| 0.14682 | −0.76449 | 1.31516 | 0.64930 | 0.000001 | −0.26166 | |
| −0.44792 | −0.22518 | −0.30184 | 0.94645 | 0.999992 | 0.80762 | 1.23821 |

which is the transpose of $A^{-1}$.

Thus for each of the selected problems we have two schemes, determined either by the matrix $A$ or by its transpose. We observe that it is advisable to use a scheme which contains the least number of rows in the lower left corner. Thus in the problem of solving a linear system without the return course it is advisable to use the scheme with the transposed matrix; in computing the product $A^{-1}B$ it is useful to use the scheme with the transpose matrix when the number of columns in $B$ is less than $n$ and the scheme with the given matrix when the number of columns in $B$ is greater than $n$.

In Table II.11 the inversion of a matrix is given for the method of biorthogonalization of columns. In columns I, II, III, and IV are written the components of the columns in matrix $A$ and the components of the vectors $V_i^{(k)}$ which are being computed in sequence. In column V (for control) are written $(A_k, V_i^{(k)})$; in column VI, the numbers $c_{ik} = (A_k, V_i^{(k-1)})$; in column VII, the numbers $c_i = 1/c_{ii}$. In the last four rows of columns I-IV is given $A^{-1}$.

Control multiplication of the inverse matrix by matrix $A$ has shown that the maximum element $E - A^{-1}A$ is not greater than 0.00005 in modulus.

In Table II.12 we carry out the solution for a system according to scheme (10).

As usual the last column is for control.  Its elements are small sums.

We observe that in solving a system by scheme (10) the number of operations slightly exceeds the number of operations required by Gauss's method. However the uniformity of the process, namely, the absence of a return course, often makes this method more convenient.

In Table II.13 we compute the product $A^{-1}B$.  Here we take as $A$ the matrix

$$\begin{bmatrix} 1.00 & 0.42 & 0.54 & 0.66 \\ 0.42 & 1.00 & 0.32 & 0.44 \\ 0.54 & 0.32 & 1.00 & 0.22 \\ 0.66 & 0.44 & 0.22 & 1.00 \end{bmatrix},$$

and as $B$ the matrix

$$\begin{bmatrix} 0.25 & 0.30 & 0.15 & 0.20 \\ 0.45 & 0.50 & 0.30 & 0.40 \\ 0.65 & 0.70 & 0.45 & 0.60 \\ 0.85 & 0.90 & 0.60 & 0.80 \end{bmatrix}.$$

The computation of $A^{-1}B$ is carried out by scheme (9) for $C'=E$.  As a

**TABLE II.12  Solving a System of Linear Equations by the Elimination Method**

| 1.00 | 0.42 | 0.54 | 0.66 | 1 | 0 | 0 | 0 | 3.62 |
|---|---|---|---|---|---|---|---|---|
| 0.42 | 1.00 | 0.32 | 0.44 | 0 | 1 | 0 | 0 | 3.18 |
| 0.54 | 0.32 | 1.00 | 0.22 | 0 | 0 | 1 | 0 | 3.08 |
| 0.66 | 0.44 | 0.22 | 1.00 | 0 | 0 | 0 | 1 | 3.32 |
| −0.3 | −0.5 | −0.7 | −0.9 | 0 | 0 | 0 | 0 | −2.40 |
| 1 | 0.42 | 0.54 | 0.66 | 1 | 0 | 0 | 0 | 3.62 |
|  | 0.82360 | 0.09320 | 0.16280 | −0.42 | 1 | 0 | 0 | 1.65960 |
|  | 0.09320 | 0.70840 | −0.13640 | −0.54 | 0 | 1 | 0 | 1.12520 |
|  | 0.16280 | −0.13640 | 0.56440 | −0.66 | 0 | 0 | 1 | 0.93080 |
|  | −0.37400 | −0.53800 | −0.70200 | 0.30 | 0 | 0 | 0 | −1.31000 |
|  | 1 | 0.11316 | 0.19767 | −0.50996 | 1.21418 | 0 | 0 | 2.01506 |
|  |  | 0.69785 | −0.15482 | −0.49247 | −0.11316 | 1 | 0 | 0.93740 |
|  |  | −0.15482 | 0.53222 | −0.57698 | −0.19767 | 0 | 1 | 0.60275 |
|  |  | −0.49568 | −0.62807 | 0.10928 | 0.45410 | 0 | 0 | −0.56037 |
|  |  | 1 | −0.22185 | −0.70570 | −0.16216 | 1.43297 | 0 | 1.34327 |
|  |  |  | 0.49787 | −0.68624 | −0.22278 | 0.22185 | 1 | 0.81071 |
|  |  |  | −0.73804 | −0.24052 | 0.37372 | 0.71029 | 0 | 0.10546 |
|  |  |  | 1 | −1.37835 | −0.44746 | 0.44560 | 2.00856 | 1.62836 |
|  |  |  |  | −1.25780 | 0.04348 | 1.03916 | 1.48240 | 1.30725 |

result of the computations

$$A^{-1}B = \begin{bmatrix} -1.25753 & -1.25780 & -0.94295 & -1.25728 \\ 0.01847 & 0.04348 & -0.00491 & -0.00655 \\ 1.00394 & 1.03916 & 0.72652 & 0.96871 \\ 1.45096 & 1.48240 & 1.06466 & 1.41957 \end{bmatrix}.$$

**TABLE II.13  Computation of the Product $A^{-1}B$**

| | | | | | | | | |
|---|---|---|---|---|---|---|---|---|
| 1.00 | 0.42 | 0.54 | 0.66 | 1 | 0 | 0 | 0 | 3.62 |
| 0.42 | 1.00 | 0.32 | 0.44 | 0 | 1 | 0 | 0 | 3.18 |
| 0.54 | 0.32 | 1.00 | 0.22 | 0 | 0 | 1 | 0 | 3.08 |
| 0.66 | 0.44 | 0.22 | 1.00 | 0 | 0 | 0 | 1 | 3.32 |
| −0.25 | −0.45 | −0.65 | −0.85 | 0 | 0 | 0 | 0 | −2.20 |
| −0.30 | −0.50 | −0.70 | −0.90 | 0 | 0 | 0 | 0 | −2.40 |
| −0.15 | −0.30 | −0.45 | −0.60 | 0 | 0 | 0 | 0 | −1.50 |
| −0.20 | −0.40 | −0.60 | −0.80 | 0 | 0 | 0 | 0 | −2.00 |
| 1 | 0.42 | 0.54 | 0.66 | 1 | 0 | 0 | 0 | 3.62 |
| | 0.82360 | 0.09320 | 0.16280 | −0.42 | 1 | 0 | 0 | 1.65960 |
| | 0.09320 | 0.70840 | −0.13640 | −0.54 | 0 | 1 | 0 | 1.12520 |
| | 0.16280 | −0.13640 | 0.56440 | −0.66 | 0 | 0 | 1 | 0.93080 |
| | −0.34500 | −0.51500 | −0.68500 | 0.25 | 0 | 0 | 0 | −1.29500 |
| | −0.37400 | −0.53800 | −0.70200 | 0.30 | 0 | 0 | 0 | −1.31400 |
| | −0.23700 | −0.36900 | −0.50100 | 0.15 | 0 | 0 | 0 | −0.95700 |
| | −0.31600 | −0.49200 | −0.66800 | 0.20 | 0 | 0 | 0 | −1.27600 |
| | 1 | 0.11316 | 0.19767 | −0.50996 | 1.21418 | 0 | 0 | 2.01506 |
| | | 0.69785 | −0.15482 | −0.49247 | −0.11316 | 1 | 0 | 0.93740 |
| | | −0.15482 | 0.53222 | −0.57698 | −0.19767 | 0 | 1 | 0.60275 |
| | | −0.47596 | −0.61680 | 0.07406 | 0.41889 | 0 | 0 | −0.59980 |
| | | −0.49568 | −0.62807 | 0.10928 | 0.45410 | 0 | 0 | −0.56037 |
| | | −0.34218 | −0.45415 | 0.02914 | 0.28776 | 0 | 0 | −0.47943 |
| | | −0.45624 | −0.60554 | 0.03885 | 0.38368 | 0 | 0 | −0.63924 |
| | | 1 | −0.22185 | −0.70570 | −0.16216 | 1.43297 | 0 | 1.34327 |
| | | | 0.49787 | −0.68624 | −0.22278 | 0.22185 | 1 | 0.81071 |
| | | | −0.72239 | −0.26182 | 0.34171 | 0.68204 | 0 | 0.03954 |
| | | | −0.73804 | −0.24052 | 0.37372 | 0.71029 | 0 | 0.10546 |
| | | | −0.53006 | −0.21234 | 0.23227 | 0.49033 | 0 | −0.01979 |
| | | | −0.70676 | −0.28312 | 0.30970 | 0.65378 | 0 | −0.02639 |
| | | | 1 | −1.37835 | −0.44746 | 0.44560 | 2.00856 | 1.62836 |
| | | | | −1.25753 | 0.01847 | 1.00394 | 1.45096 | 1.21585 |
| | | | | −1.25780 | 0.04348 | 1.03916 | 1.48240 | 1.30725 |
| | | | | −0.94295 | −0.00491 | 0.72652 | 1.06466 | 0.84334 |
| | | | | −1.25728 | −0.00655 | 0.96871 | 1.41957 | 1.12447 |

## 23. CORRECTION OF THE ELEMENTS OF THE INVERSE MATRIX

Use of the schemes introduced above for inverting a matrix offers no certainty as to the accuracy of the results obtained, because of the inevitable rounding-off processes whose effect on the final result is difficult to evaluate. To estimate the accuracy of a matrix $D_0$ obtained from a given matrix $A$ by some sort of inversion process it is necessary to form the product $AD_0$. The discrepancy between this product and the unit matrix indicates the degree of inaccuracy in the results.

Let the control computation show us that the approximation of $D_0$ to $A^{-1}$ is such that $\| R_0 \| \leq k < 1$, where

$$R_0 = E - AD_0 . \tag{1}$$

As the norm of the matrices it is convenient to choose the first or second norm given in Section 13 as most easily computable.

Under this condition the elements of the inverse matrix $A^{-1}$ may be computed to whatever degree of accuracy is convenient with the help of the following iterative process.[†]

We form the sequence of matrices

$$\begin{aligned}
D_1 &= D_0(E + R_0) , & R_1 &= E - AD_1 , \\
D_2 &= D_1(E + R_1) , & R_2 &= E - AD_2 , \\
& \cdots \cdots \cdots \cdots \cdots \cdots , \\
D_m &= D_{m-1}(E + R_{m-1}) , & R_m &= E - AD_m .
\end{aligned} \tag{2}$$

We shall show that the matrix $R_m = E - AD_m$ is equal to $R_0^{2^m}$. In fact,

$$\begin{aligned}
R_m &= E - AD_m = E - AD_{m-1}(E + R_{m-1}) \\
&= E - (E - R_{m-1})(E + R_{m-1}) = R_{m-1}^2 = R_{m-2}^4 = \cdots = R_0^{2^m} .
\end{aligned} \tag{3}$$

From here it follows that

$$D_m = A^{-1}(E - R_0^{2^m}) . \tag{4}$$

This last formula shows that $D_m$ approaches $A^{-1}$ while the convergence of the process is very rapid. Let us make an estimate of the error, taking into account that $A^{-1} = D_0(AD_0)^{-1} = D_0(E - R_0)^{-1}$:

$$\| D_m - A^{-1} \| = \| -A^{-1}R_0^{2^m} \| = \| -D_0(E - R_0)^{-1}R_0^{2^m} \|$$

$$\leq \| D_0 \| \, \| (E - R_0)^{-1} \| \, \| R_0^{2^m} \| \leq \| D_0 \| \frac{k^{2^m}}{1 - k} . \tag{5}$$

It is clear from this estimate that as soon as the initial approximation is chosen so that $\| E - AD_0 \| \leq k < 1$, the number of correct decimal figures increases in geometric progression.

Successive approximations should be computed by removing the parentheses in formulas (2), that is,

$$D_m = D_{m-1}(E + R_{m-1}) = D_{m-1} + D_{m-1}(E - AD_{m-1}) . \tag{6}$$

---

[†] Hotelling (1). We note that Hotelling used the norm $N(A)$.

The second summand will here play the role of a small correction to the first.

Sometimes, starting with the matrix $R_0$, it is advisable to form the matrices $R_1 = R_0^2$, $R_2 = R_0^4 = (R_1)^2$ by successive squaring, and then to utilize formula (6).

*Note.* If $A$ is a symmetric matrix, as is the initial approximation $D_0$, all the following approximations will be symmetric matrices, although the matrix $R_0$ may turn out to be not a symmetric matrix. In fact from formula (6) it follows that $D_m = 2D_{m-1} - D_{m-1}AD_{m-1}$, from which, having assumed $A' = A$ and $D'_{m-1} = D_{m-1}$, we obtain

$$D'_m = 2D'_{m-1} - (D_{m-1}AD_{m-1})' = 2D'_{m-1} - D'_{m-1}A'D'_{m-1}$$
$$= 2D_{m-1} - D_{m-1}AD_{m-1} = D_m .$$

As an example we shall apply the process described for increasing the accuracy of the elements of the matrix $A^{-1}$, where

$$A = \begin{bmatrix} 1.00 & 0.42 & 0.54 & 0.66 \\ 0.42 & 1.00 & 0.32 & 0.44 \\ 0.54 & 0.32 & 1.00 & 0.22 \\ 0.66 & 0.44 & 0.22 & 1.00 \end{bmatrix}. \tag{7}$$

For the initial approximation we shall take the result of inverting matrix $A$ by the single-division method (retaining four figures after the decimal point) from Table II.9.

$$D_0 = \begin{bmatrix} 2.5076 & -0.1230 & -1.0115 & -1.3783 \\ -0.1230 & 1.3322 & -0.2614 & -0.4475 \\ -1.0115 & -0.2614 & 1.5318 & 0.4456 \\ -1.3783 & -0.4475 & 0.4456 & 2.0086 \end{bmatrix}.$$

Control computation gives the following value for $R_0$:

$$R_0 = 10^{-6} \begin{bmatrix} -52 & -18 & 20 & -50 \\ -60 & 8 & -10 & 10 \\ -18 & -34 & 26 & -10 \\ -66 & 20 & 10 & -54 \end{bmatrix}.$$

We see from this that $\|R_0\|_\mathrm{I} \le 0.000150$, $\|R_0\|_\mathrm{II} \le 0.000196$. For estimating the error we take $\|R_0\|_\mathrm{I}$. On the basis of formula (5) we have, remembering that $\|D_0\|_\mathrm{I} < 5$:

$$\|D_1 - A^{-1}\|_\mathrm{I} < 5\frac{(0.00015)^2}{1 - 0.00015} < 0.0000001 .$$

Thus $D_1$ gives a value for $A^{-1}$ which is correct at least up to one unit in the seventh decimal places for each of its elements.

Computing we get:

$$D_1 = \begin{bmatrix} 2.50758616 & -0.12303930 & -1.01148870 & -1.37834207 \\ -0.12303930 & 1.33221281 & -0.26142705 & -0.44745375 \\ -1.01148870 & -0.26142705 & 1.53182667 & 0.44560858 \\ -1.37834207 & -0.44745375 & 0.44560858 & 2.00855152 \end{bmatrix}. \tag{8}$$

Control computation of $E - AD_1$ gives:

$$E - AD_1 = 10^{-8} \begin{bmatrix} 2 & 0 & 0 & 1 \\ 1 & 0 & -1 & 1 \\ 1 & 0 & 0 & 0 \\ 1 & 0 & 0 & -1 \end{bmatrix}.$$

A more accurate value for the matrix $A^{-1}$ permits us to obtain a more accurate solution to the oft-considered system

$$
\begin{aligned}
x_1 + 0.42x_2 + 0.54x_3 + 0.66x_4 &= 0.3 \ , \\
0.42x_1 + \quad x_2 + 0.32x_3 + 0.44x_4 &= 0.5 \ , \\
0.54x_1 + 0.32x_2 + \quad x_3 + 0.22x_4 &= 0.7 \ , \\
0.66x_1 + 0.44x_2 + 0.22x_3 + \quad x_4 &= 0.9 \ .
\end{aligned}
\tag{9}
$$

Namely,

$$
\begin{aligned}
x_1 &= -\,1.2577938 \ , \\
x_2 &= 0.0434873 \ , \\
x_3 &= 1.0391663 \ , \\
x_4 &= 1.4823929 \ .
\end{aligned}
\tag{10}
$$

## 24. INVERTING A MATRIX BY PARTITIONING

When inverting a matrix it is sometimes advisable to partition it into cells beforehand. We shall consider formulas for inverting a matrix of order $n$, partitioned into four cells by the scheme

$$S = \left[ \begin{array}{c|c} A & B \\ \hline C & D \end{array} \right],$$

where $A$ and $D$ are square matrices of order $p$ and $q$; $p + q = n$.

We shall look for an inverse matrix which is also in the form of a partitioned matrix

$$S^{-1} = \left[ \begin{array}{c|c} K & L \\ \hline M & N \end{array} \right],$$

where $K$ and $N$ are again square matrices of orders $p$ and $q$.

In accordance with the rule for multiplying partitioned matrices the following matrix equalities must hold:

$$
\begin{aligned}
AK + BM &= E \ , \\
AL + BN &= 0 \ , \\
CK + DM &= 0 \ , \\
CL + DN &= E \ .
\end{aligned}
$$

Premultiplying the third equation by $BD^{-1}$ and subtracting from the first, we obtain

$$(A - BD^{-1}C)K = E ,$$

from which

$$K = (A - BD^{-1}C)^{-1} .$$

Moreover from the third equality we find that

$$M = -D^{-1}CK .$$

In a similar way we find from the second and fourth equalities that

$$N = (D - CA^{-1}B)^{-1}$$

and

$$L = -A^{-1}BN .$$

Of course these formulas have been derived on the assumption that all the indicated matrix inversion are realizable.

Thus inverting a matrix of order $n$ comes down to inverting four matrices, of which two have order $p$ and two have order $q$, and the several matrix multiplications.

It is possible to change the derived formulas so that in computing the matrices $K$, $L$, $M$ and $N$ we need to invert only two matrices of orders $p$ and $q$.

In fact it is easy to verify that

$$N = (D - CA^{-1}B)^{-1} ; \quad M = -NCA^{-1}$$
$$L = A^{-1}BN \qquad ; \quad K = A^{-1} - A^{-1}BM$$

and analogously

$$K = (A - BD^{-1}C)^{-1} ; \quad L = KBD^{-1}$$
$$M = -D^{-1}CK \qquad ; \quad N = D^{-1} - D^{-1}CL .$$

The last formulas show that the partitioning method is conveniently applied when any diagonal cells is easily inverted.

Let us take as an example the inversion of the matrix

$$\begin{vmatrix} 1.00 & 0.42 & 0.54 & 0.66 \\ 0.42 & 1.00 & 0.32 & 0.44 \\ \hline 0.54 & 0.32 & 1.00 & 0.22 \\ 0.66 & 0.44 & 0.22 & 1.00 \end{vmatrix} .$$

We carry out the computation in the following way.

1. We compute the matrix $A^{-1}$

$$A^{-1} = \begin{bmatrix} 1.21418 & -0.50996 \\ -0.50996 & 1.21418 \end{bmatrix}$$

and form the products

$$A^{-1}B = \begin{bmatrix} 0.49247 & 0.57698 \\ 0.11316 & 0.19767 \end{bmatrix} ;$$

$$CA^{-1} = \begin{bmatrix} 0.49247 & 0.11316 \\ 0.57698 & 0.19767 \end{bmatrix} ;$$

$$CA^{-1}B = \begin{bmatrix} 0.30215 & 0.37482 \\ 0.37482 & 0.46778 \end{bmatrix}.$$

By computing the last matrix twice, as $C(A^{-1}B)$ and $(CA^{-1})B$, we obtain a control for the previous computations.

2. We form the matrix

$$D - CA^{-1}B = \begin{bmatrix} 0.69785 & -0.15482 \\ -0.15482 & 0.53222 \end{bmatrix}$$

and find its inverse

$$N = (D - CA^{-1}B)^{-1} = \begin{bmatrix} 1.53183 & 0.44560 \\ 0.44560 & 2.00855 \end{bmatrix}.$$

3. We compute the matrices

$$M = -NCA^{-1} = \begin{bmatrix} -1.01148 & -0.26142 \\ -1.37834 & -0.44745 \end{bmatrix},$$

$$L = -A^{-1}BN = \begin{bmatrix} -1.01148 & -1.37834 \\ -0.26142 & -0.44745 \end{bmatrix},$$

$$K = A^{-1} - A^{-1}BM = \begin{bmatrix} 2.50758 & -0.12305 \\ -0.12305 & 1.33221 \end{bmatrix}.$$

Thus the inverse matrix we are seeking will be

$$\begin{bmatrix} 2.50758 & -0.12305 & -1.01148 & -1.37834 \\ -0.12305 & 1.33221 & -0.26142 & -0.44745 \\ -1.01148 & -0.26142 & 1.53183 & 0.44560 \\ -1.37834 & -0.44745 & 0.44560 & 2.00855 \end{bmatrix}.$$

The method presented coincides essentially with the compact scheme for inverting a partitioned matrix described above for the case where the matrix is partitioned into four cells.

## 25. THE BORDERING METHOD

In this paragraph we shall consider computational schemes for inverting a matrix and for solving a linear system; these schemes are based on the concept of bordering.

We shall consider the given matrix $A$ to be the result of bordering a matrix of order $n - 1$ whose inverse matrix is considered, already known. Let

$$A = A_n = \begin{bmatrix} a_{11} & a_{12} & \cdots & a_{1,n-1} & a_{1n} \\ a_{21} & a_{22} & \cdots & a_{2,n-1} & a_{2n} \\ \cdots & \cdots & \cdots & \cdots & \cdots \\ a_{n-1,1} & a_{n-1,2} & \cdots & a_{n-1,n-1} & a_{n-1,n} \\ a_{n1} & a_{n2} & \cdots & a_{n,n-1} & a_{nn} \end{bmatrix}$$

$$= \begin{bmatrix} A_{n-1} & u_n \\ v_n & a_{nn} \end{bmatrix}.$$

Here $A_{n-1}$ denotes the above mentioned matrix of order $n-1$,

$$v_n = (a_{n1}, \cdots, a_{n,n-1}),$$
$$u_n = (a_{1n}, \cdots, a_{n-1,n})'.$$

We are also looking for $A^{-1}$ in the form of a bordered matrix

$$D_n = A_n^{-1} = \begin{bmatrix} P_{n-1} & r_n \\ q_n & \dfrac{1}{\alpha_n} \end{bmatrix},$$

where $P_{n-1}$ is a matrix, $q_n$ a row, $r_n$ a column and $\dfrac{1}{\alpha_n}$ a number, all of which we need to determine.

By the rule for multiplying bordered matrices we have:

$$AA^{-1} = \begin{bmatrix} A_{n-1} & u_n \\ v_n & a_{nn} \end{bmatrix} \begin{bmatrix} P_{n-1} & r_n \\ q_n & \dfrac{1}{\alpha_n} \end{bmatrix}$$

$$= \begin{bmatrix} A_{n-1}P_{n-1} + u_nq_n, & A_{n-1}r_n + \dfrac{u_n}{\alpha_n} \\ v_nP_{n-1} + a_{nn}q_n, & v_nr_n + \dfrac{a_{nn}}{\alpha_n} \end{bmatrix} = \begin{bmatrix} E & 0 \\ 0 & 1 \end{bmatrix}.$$

Thus

$$A_{n-1}P_{n-1} + u_nq_n = E, \tag{1}$$

$$v_nP_{n-1} + a_{nn}q_n = 0, \tag{2}$$

$$A_{n-1}r_n + \dfrac{u_n}{\alpha_n} = 0, \tag{3}$$

$$v_nr_n + \dfrac{a_{nn}}{\alpha_n} = 1. \tag{4}$$

From equality (3) we have

$$r_n = -\dfrac{A_{n-1}^{-1}u_n}{\alpha_n}.$$

Substituting the value for $r_n$ in (4) we obtain

$$\alpha_n = a_{nn} - v_nA_{n-1}^{-1}u_n. \tag{5}$$

Moreover from (1) we have

$$P_{n-1} = A_{n-1}^{-1} - A_{n-1}^{-1}u_nq_n \tag{6}$$

and therefore, on the basis of (2) and (5):

$$v_nA_{n-1}^{-1} - v_nA_{n-1}^{-1}u_nq_n + a_{nn}q_n$$
$$= v_nA_{n-1}^{-1} - (a_{nn} - \alpha_n)q_n + a_{nn}q_n$$
$$= v_nA_{n-1}^{-1} + \alpha_nq_n = 0.$$

It follows that

$$q_n = -\dfrac{v_nA_{n-1}^{-1}}{\alpha_n}.$$

Finally

$$P_{n-1} = A_{n-1}^{-1} + \frac{A_{n-1}^{-1} u_n v_n A_{n-1}^{-1}}{\alpha_n} \, .$$

Thus we obtain finally:

$$A^{-1} = \begin{bmatrix} A_{n-1}^{-1} + \dfrac{A_{n-1}^{-1} u_n v_n A_{n-1}^{-1}}{\alpha_n} \, , & -\dfrac{A_{n-1}^{-1} u_n}{\alpha_n} \\[2ex] -\dfrac{v_n A_{n-1}^{-1}}{\alpha_n} \, , & \dfrac{1}{\alpha_n} \end{bmatrix}, \tag{7}$$

where $\alpha_n = a_{nn} - v_n A_{n-1}^{-1} u_n$.

It is clear that the formula constructed is a particular case of the formula for inverting a matrix by partitioning into cells for $p = n - 1$ and $q = 1$.

The derived formula lies at the basis of the successive bordering method for inverting a matrix. That is, inverse matrices are successively constructed for the matrices

$$(a_{11}) \, , \quad \begin{bmatrix} a_{11} & a_{12} \\ a_{21} & a_{22} \end{bmatrix}, $$

$$\begin{bmatrix} a_{11} & a_{12} & a_{13} \\ a_{21} & a_{22} & a_{23} \\ a_{31} & a_{32} & a_{33} \end{bmatrix}, $$

$$\cdots \cdots \cdots ,$$

each of which is obtained from the previous one by bordering. Each step in this process may be realized on the basis of formula (7). Namely, if $A_{n-1}^{-1}$ is already known it is necessary to perform the following operations in order to find $A_n^{-1}$:

1. Compute the column $A_n^{-1} {}_1 u_n$. Elements of the column $\beta_{1n}, \cdots, \beta_{n-1,n}$ are found using accumulation.
2. Compute the row $v_n A_{n-1}^{-1}$ with elements $\gamma_{n1}, \cdots, \gamma_{n,n-1}$.
3. Compute the number

$$\alpha_n = a_{nn} - \sum_{i=1}^{n-1} a_{ni} \beta_{in} = a_{nn} - \sum_{i=1}^{n-1} a_{in} \gamma_{ni} \, .$$

(A double computation of the number $\alpha_n$ is a good check of the previous computations.)

4. Finally, find the elements $d_{ik}$ of the inverse matrix by the formulas:

$$d_{ik} = d_{ik}' + \frac{\beta_{in} \gamma_{nk}}{\alpha_n} \qquad (i, k \leq n - 1)$$

$$d_{in} = \frac{\beta_{in}}{\alpha_n} \, ; \quad d_{nk} = \frac{\gamma_{nk}}{\alpha_n} \qquad \left( i, k \leq n - 1; \; d_{nn} = \frac{1}{\alpha_n} \right).$$

Here $d_{ik}'$ are elements of the matrix $A_{n-1}^{-1}$.

In the case of a symmetric matrix $A$ the scheme is obviously half as long.

In Table II.14 the inversion of matrix (7) Section 23 is given with the aid of successive bordering.

Each step of the process is formulated by the following scheme:

$$
\begin{array}{c||c||c}
A_{n-1}^{-1} & u_n & -A_{n-1}^{-1}u_n \\
\hline
v_n & a_{nn} & \\
\hline
-v_n A_{n-1}^{-1} & & a_n
\end{array}
$$

**TABLE II.14  Inverting a Matrix by the Bordering Method**

| | | | | |
|---|---|---|---|---|
| 1 | | | | |
| 1 | 0.42 | −0.42 | | |
| 0.42<br>−0.42 | 1 | 0.82360 | | |
| 1.21418<br>−0.50996 | −0.50996<br>1.21418 | 0.54<br>0.32 | −0.49247<br>−0.11316 | |
| 0.54<br>−0.49247 | 0.32<br>−0.11316 | 1 | 0.69785 | |
| 1.56172<br>−0.43010<br>−0.70570 | −0.43010<br>1.23253<br>−0.16216 | −0.70570<br>−0.16216<br>1.43297 | 0.66<br>0.44<br>0.22 | −0.68624<br>−0.22277<br>0.22186 |
| 0.66<br>−0.68624 | 0.44<br>−0.22277 | 0.22<br>0.22186 | 1 | 0.49787 |
| 2.50759<br>−0.12304<br>−1.01149<br>−1.37835 | −0.12304<br>1.33221<br>−0.26143<br>−0.44745 | −1.01149<br>−0.26143<br>1.53183<br>0.44562 | −1.37835<br>−0.44745<br>0.44562<br>2.00856 | |

The bordering method may be applied to solving a system of linear equations. It is particularly advisable to use this method in cases where it is necessary to solve a system for which the truncated system, obtained from the given one by striking out one equation and one unknown has already been solved. Such a situation is frequently encountered in applied work. For example, in solving problems in mathematical physics by the methods of B. G. Galerkin or Ritz it may turn out that a solution obtained as a result of using the $n-1$th coordinate function is sufficiently precise; if it is sufficient to add just one more coordinate function to set up a more accurate solution, then the new system for determining coefficients is obtained from the previous system by bordering.

The bordering method may be applied to solve a system of linear equations in the following way.

Let the system have the form

$$A_n X_n = F_n .$$

We write

$$A_n = \begin{bmatrix} A_{n-1} & u_n \\ v_n & a_{nn} \end{bmatrix} ;$$

$$F_n = \begin{bmatrix} F_{n-1} \\ f_n \end{bmatrix} .$$

Then

$$X_n = \begin{bmatrix} A_{n-1}^{-1} & 0 \\ 0 & 0 \end{bmatrix} \begin{bmatrix} F_{n-1} \\ f_n \end{bmatrix}$$

$$+ \frac{1}{\alpha_n} \begin{bmatrix} A_{n-1}^{-1} u_n v_n A_{n-1}^{-1} & - A_{n-1}^{-1} u_n \\ - v_n A_{n-1}^{-1} & 1 \end{bmatrix} \begin{bmatrix} F_{n-1} \\ f_n \end{bmatrix}$$

$$= \begin{bmatrix} A_{n-1}^{-1} F_{n-1} \\ 0 \end{bmatrix} + \frac{1}{\alpha_n} \begin{bmatrix} A_{n-1}^{-1} u_n v_n A_{n-1}^{-1} F_{n-1} - A_{n-1}^{-1} u_n f_n \\ - v_n A_{n-1}^{-1} F_{n-1} + f_n \end{bmatrix} .$$

But $A_{n-1}^{-1} F_{n-1}$ is a solution of the *truncated system*, that is,

$$a_{11} x_1 + \cdots + a_{1, n-1} x_{n-1} = f_1 ,$$
$$\cdots \cdots \cdots \cdots \cdots ,$$
$$a_{n-1,1} x_1 + \cdots + a_{n-1, n-1} x_{n-1} = f_{n-1} ,$$

which we denote by $X_{n-1}$.

By analogy

$$- A_{n-1}^{-1} u_n = - A_{n-1}^{-1} \begin{bmatrix} a_{1n} \\ \cdots \\ a_{n-1, n} \end{bmatrix} = Q_{n-1}$$

is a solution of a system with the same matrix of coefficients but having different constant terms.

Knowing $X_{n-1}$ and $Q_{n-1}$ we may easily compute $X_n$. In fact,

$$X_n = \begin{bmatrix} X_{n-1} \\ 0 \end{bmatrix} + \frac{1}{a_{nn} + v_n Q_{n-1}} \begin{bmatrix} - Q_{n-1} v_n X_{n-1} + Q_{n-1} f_n \\ - v_n X_{n-1} + f_n \end{bmatrix}$$

$$= \begin{bmatrix} X_{n-1} \\ 0 \end{bmatrix} + \frac{f_n - v_n X_{n-1}}{a_{nn} + v_n Q_{n-1}} \begin{bmatrix} Q_{n-1} \\ 1 \end{bmatrix} .$$

Thus to find $X_n$ we need to compute $Q_{n-1}$ as well as $X_{n-1}$.

If the truncated system had been solved by Gauss's method, then to find $Q_{n-1}$ it would have been necessary to add one more column to the scheme. Computing the forward course is realized only for this column using the left half of the scheme already known. Then $Q_{n-1}$ is determined by the usual return course.

## 26. THE ESCALATOR METHOD

In inverting a matrix by the bordering method the computation of the expressions $- A_{k-1}^{-1}u_k$ and $- v_k A_{k-1}^{-1}$ plays an essential role in passing from the inversion of matrix $A_{k-1}$ to the inversion of the bordered matrix $A_k$.

For this the components of the vector $-A_{k-1}^{-1}u_k$ are nothing less than a solution of the system of equations:

$$a_{11}z_1 + a_{12}z_2 + \cdots + a_{1,k-1}z_{k-1} + a_{1k} = 0 ,$$
$$\cdots \cdots \cdots \cdots \cdots \cdots \cdots \cdots ,$$
$$a_{k-1,1}z_1 + a_{k-1,2}z_2 + \cdots + a_{k-1,k-1}z_{k-1} + a_{k-1,k} = 0 ,$$

Analogously the components of $- v_k A_{k-1}^{-1}$ form a solution for the transposed system.

A sequential solution of such systems for $k = 2, \cdots, n, n + 1$ is based on the so-called escalator method for solving systems of linear equations.[†] This escalator method is closely connected with the bordering method.

An essential virtue of the method is the presence of a reliable control by which the accuracy of the computations may be regulated, attaining satisfaction of the control equalities by using a large number of significant figures. It should be observed however that the escalator method is not unique in this respect among exact methods for solving systems. Reliable control is also possessed by the method of biorthogonalization of columns considered above and by a whole series of other methods which will be examined later.

We shall describe the escalator method only as applied to systems with a symmetric matrix, giving a compact form to the computational scheme. The connection between the escalator method and Gauss's method will be thereby revealed.

Let there be given a system of equations

$$a_{11}x_1 + a_{12}x_2 + \cdots + a_{1n}x_n + a_{1,n+1} = 0 ,$$
$$\cdots \cdots \cdots \cdots \cdots \cdots \cdots \cdots \cdots , \tag{1}$$
$$a_{n1}x_1 + a_{n2}x_2 + \cdots + a_{nn}x_n + a_{n,n+1} = 0 ,$$

whose coefficient matrix is symmetric

We denote by $z_{1k}, \cdots, z_{k-1,k}$ the solution of the system:

$$a_{11}z_1 + \cdots + a_{1,k-1}z_{k-1} + a_{1k} = 0 ,$$
$$\cdots \cdots \cdots \cdots \cdots \cdots \cdots \cdots , \tag{2}$$
$$a_{k-1,1}z_1 + \cdots + a_{k-1,k-1}z_{k-1} + a_{k-1,k} = 0 .$$

The numbers $z_{i,n+1} = x_i$ obviously form a solution for system (1). Therefore if we establish a means for computing successively the numbers $z_{ik}$ for $i < k \leq n + 1$, then we shall thereby give a method for constructing the desired solution.

We now assume that all numbers $z_{ik}$ for $i < k \leq n$ have already been computed. We shall set up with their help the matrix

---

† Morris (4).

$$Z = \begin{bmatrix} 1 & z_{12} & z_{13} & \cdots & z_{1n} \\ 0 & 1 & z_{23} & \cdots & z_{2n} \\ 0 & 0 & 1 & \cdots & z_{3n} \\ \cdot & \cdot & \cdot & \cdot & \cdot & \cdot \\ 0 & 0 & 0 & \cdots & 1 \end{bmatrix}. \tag{3}$$

It is easy to see that the matrix $AZ$ has zeros above the principal diagonal. In fact,

$$C_1 = AZ$$

$$= \begin{bmatrix} a_{11}, & a_{11}z_{12} + a_{12}, & \cdots, a_{11}z_{1n} + a_{12}z_{2n} + \cdots + a_{1n} \\ a_{21}, & a_{21}z_{12} + a_{22}, & \cdots, a_{21}z_{1n} + a_{22}z_{2n} + \cdots + a_{2n} \\ \cdot & \cdot & \cdot \cdot \cdot \cdot \cdot \cdot \cdot \cdot \\ a_{n1}, & a_{n1}z_{12} + a_{n2}, & \cdots, a_{n1}z_{1n} + a_{n2}z_{2n} + \cdots + a_{nn} \end{bmatrix} \tag{4}$$

and from the definition of the numbers $z_{ij}$ all elements lying above the principal diagonal are equal to zero. The non-zero elements of the matrix $C_1$ are computed by the formulas

$$c_{ij} = a_{i1}z_{1j} + a_{i2}z_{2j} + \cdots + a_{i,j-1}z_{j-1,j} + a_{ij} \qquad i \geq j \,. \tag{5}$$

From here arises the connection between this method and Gauss's method, considered with respect to expanding a matrix into factors.

That is, from (4) we have

$$A = C_1 Z^{-1} \,. \tag{6}$$

But the matrix $Z^{-1}$ is a triangular matrix with unit diagonal elements and with zero elements below the diagonal and $C_1$ is a triangular matrix with zero elements above the diagonal. Comparing this expansion with the matrix expansion corresponding to the single-division of Gauss's method ($A = CB$) we obtain, from the uniqueness of such an expansion, that

$$Z^{-1} = B \,,$$
$$C_1 = C \,. \tag{7}$$

We shall represent the matrix $C$ in the form

$$C = \begin{bmatrix} 1 & 0 & 0 & \cdots & 0 \\ \gamma_{21} & 1 & 0 & \cdots & 0 \\ \gamma_{31} & \gamma_{32} & 1 & \cdots & 0 \\ \cdot & \cdot & \cdot & \cdot & \cdot & \cdot \\ \gamma_{n1} & \gamma_{n2} & \gamma_{n3} & \cdots & 1 \end{bmatrix} \begin{bmatrix} c_{11} & & & & \\ & c_{22} & & & \\ & & c_{33} & & \\ & & & \cdot & \\ & & & & c_{nn} \end{bmatrix} = \Gamma \Lambda \,. \tag{8}$$

Here $\gamma_{ij} = c_{ij}/c_{jj}$, $i > j$.

We shall show that the matrix $\Gamma = C\Lambda^{-1}$ is the transpose of the matrix $Z^{-1}$. In doing this we make essential use of the symmetry of matrix $A$. We have:

$$A = \Gamma \Lambda Z^{-1} = A' = (Z^{-1})' \Lambda \Gamma' \,.$$

But $(Z^{-1})'$ is a triangular matrix with zeros above the diagonal and $\Gamma'$ is a triangular matrix with zeros below and ones along the diagonal. From the uniqueness of the expansion, $\Gamma' = Z^{-1}$.

This circumstance allows us to give the recurrent formulas for sequential determination of the numbers $z_{ij}$. We assume that we have already computed the elements of the first $k$ columns of the matrix $Z$. Then by formulas (5) we may determine $k$ columns of the matrix $C$ and consequently $k$ columns of matrix $\Gamma$, that is, $k$ rows of matrix $Z^{-1}$. To continue this process we need to compute the elements of the $(k+1)$th column of the matrix $Z$. Since each diagonal element of this column is equal to one and the elements below the diagonal are equal to zero, we need only to find formulas for computing $z_{i,k+1}$ where $i \leq k$. From the equality $Z\Gamma' = E$ we obtain the following recurrent formulas from the rule for multiplying matrices:

$$\gamma_{k+1,1} + \gamma_{k+1,2}z_{12} + \cdots + \gamma_{k+1,k}z_{1k} + z_{1,k+1} = 0 ,$$
$$\gamma_{k+1,2} + \cdots + \gamma_{k+1,k}z_{2k} + z_{2,k+1} = 0 ,$$
$$\cdot \; \cdot \; \cdot \; \cdot \; \cdot \; \cdot \; \cdot \; \cdot \; \cdot \; \cdot \; \cdot \; \cdot \; \cdot \; ,$$
$$\gamma_{k+1,k} + z_{k,k+1} = 0 ,$$

$$(9)$$

These determine the elements of the $(k+1)$th column of matrix $Z$.

The reliable control referred to above is the actual reduction of the subdiagonal elements of the matrix $\Gamma'$ to zero.

We introduce the compact scheme of the escalator method for a system of four equations. (We recall that for uniformity of computation we shall write the constants as the left members of the equations).

| | | | | | | | | | |
|---|---|---|---|---|---|---|---|---|---|
| $a_{11}$ | $a_{12}$ | $a_{13}$ | $a_{14}$ | $a_{15}$ | 1 | $z_{12}$ | $z_{13}$ | $z_{14}$ | $x_1$ |
| $a_{21}$ | $a_{22}$ | $a_{23}$ | $a_{24}$ | $a_{25}$ | 0 | 1 | $z_{23}$ | $z_{24}$ | $x_2$ |
| $a_{31}$ | $a_{32}$ | $a_{33}$ | $a_{34}$ | $a_{35}$ | 0 | 0 | 1 | $z_{34}$ | $x_3$ |
| $a_{41}$ | $a_{42}$ | $a_{43}$ | $a_{44}$ | $a_{45}$ | 0 | 0 | 0 | 1 | $x_4$ |
| $c_{11}$   1 | $\gamma_{21}$ | $\gamma_{31}$ | $\gamma_{41}$ | $\gamma_{51}$ | | | | | |
| 0 | $c_{22}$   1 | $\gamma_{32}$ | $\gamma_{42}$ | $\gamma_{52}$ | | | | | |
| 0 | 0 | $c_{33}$   1 | $\gamma_{43}$ | $\gamma_{53}$ | | | | | |
| 0 | 0 | 0 | $c_{44}$   1 | $\gamma_{54}$ | | | | | |

The scheme is made up of three parts. In the first part is written the matrix of coefficients for the system (symmetric, under our condition). In the second part we gradually write the elements of the columns for the matrix $Z$. In the lower part we write the diagonal elements of matrix $C$ and the elements of matrix $\Gamma'$. We observe that elements of the $k$th row of matrix $\Gamma'$ are obtained by multiplying the columns of matrix $A$ by the $k$th column of matrix $Z$ and dividing the resultant sums by the element $c_{kk}$. A column of matrix $Z$ is filled in by formulas (9). In Table II.15 an illustrative example is given.

**TABLE II.15  Compact Scheme for the Escalator Method**

| 1.00 | 0.42 | 0.54 | 0.66 | -0.3 | 1 | -0.42 | -0.49247 | -0.68624 | -1.25780 |
|------|------|------|------|------|---|-------|----------|----------|----------|
| 0.42 | 1.00 | 0.32 | 0.44 | -0.5 | 0 | 1 | -0.11316 | -0.22278 | 0.04348 |
| 0.54 | 0.32 | 1.00 | 0.22 | -0.7 | 0 | 0 | 1 | 0.22185 | 1.03917 |
| 0.66 | 0.44 | 0.22 | 1.00 | -0.9 | 0 | 0 | 0 | 1 | 1.48240 |
| 1 | 1 0.42 | 0.54 | 0.66 | -0.3 | | | | | |
| 0 | 0.82360 1 | 0.11316 | 0.19767 | -0.45410 | | | | | |
| 0.000003 | 0.000003 | 0.69785 1 | -0.22185 | -0.71030 | | | | | |
| -0.000009 | -0.000009 | -0.000009 | 0.49787 1 | -1.48240 | | | | | |

## 27. PURCELL'S METHOD

With the solution $X = (x_1, \cdots, x_n)$ of a system of equations

$$a_{11}x_1 + \cdots + a_{1n}x_n - f_1 = 0 ,$$
$$\cdots \cdots \cdots \cdots \cdots \cdots \cdots ,$$
$$a_{n1}x_1 + \cdots + a_{nn}x_n - f_n = 0$$

(1)

there is associated[†] a vector $Z = (x_1, \cdots, x_n, 1)' = (X, 1)'$ of an arithmetic $(n + 1)$-dimensional space $R_{n+1}$ with a natural representation. The equations of the system are interpreted as orthogonality conditions for the vector $Z$ with respect to the vectors

$$A_i = (a_{i1}, \cdots, a_{in}, -f_i)' \qquad (i = 1, 2, \cdots, n) .$$

We shall look for a solution in the following way. We construct step by step bases for subspaces of decreasing dimensions

$$R_{n+1} = R^{(0)} \supset R^{(1)} \supset \cdots \supset R^{(n)} ,$$

where $R^{(k)}$ is a subspace consisting of vectors orthogonal to the vectors $A_1, \cdots, A_k$.

Each successive basis

$$V_{k+1}^{(k)}, \cdots, V_{n+1}^{(k)}$$

is constructed from the previous one

$$V_k^{(k-1)}, \cdots, V_{n+1}^{(k-1)}$$

in the form of binomial linear combinations

$$V_i^{(k)} = V_i^{(k-1)} - c_i^{(k)} V_k^{(k-1)} \qquad (i = k + 1, \cdots, n + 1) .$$

(2)

The coefficients $c_i^{(k)}$ are defined from the orthogonality condition for the vector $A_k$, which gives

$$c_i^{(k)} = \frac{(A_k, V_i^{(k-1)})}{(A_k, V_k^{(k-1)})} .$$

(3)

For the process to be feasible it is necessary that all scalar products

---

† Purcell (1).

**TABLE II.16   Solving a System of Equations by Purcell's Method**

| | $V_1^{(0)}$ | $V_2^{(0)}$ | $V_3^{(0)}$ | $V_4^{(0)}$ | $V_5^{(0)}$ | $V_2^{(1)}$ | $V_3^{(1)}$ | $V_4^{(1)}$ | $V_5^{(1)}$ | $V_3^{(2)}$ | $V_4^{(2)}$ | $V_5^{(2)}$ | $V_4^{(3)}$ | $V_5^{(3)}$ | $V_5^{(4)} = (X, 1)'$ |
|---|---|---|---|---|---|---|---|---|---|---|---|---|---|---|---|
| I | 1 | 0 | 0 | 0 | 0 | -0.17 | 0.25 | -0.54 | 0.3 | 0.39550 | -0.64602 | 0.23367 | -0.37470 | 0.58836 | 0.44089 |
| | 0 | 1 | 0 | 0 | 0 | 1 | 0 | 0 | 0 | -0.85589 | 0.62363 | 0.39017 | 0.03647 | -0.37740 | -0.36304 |
| | 0 | 0 | 1 | 0 | 0 | 0 | 1 | 0 | 0 | 1 | 0 | 0 | 0.68602 | 0.89681 | 1.16681 |
| | 0 | 0 | 0 | 1 | 0 | 0 | 0 | 1 | 0 | 0 | 1 | 0 | 1 | 0 | 0.39357 |
| | 0 | 0 | 0 | 0 | 1 | 0 | 0 | 0 | 1 | 0 | 0 | 1 | 0 | 1 | 1 |
| II | | | | | | | | | | -0.000001 | -0.000003 | -0.000001 | -0.000005 | -0.000005 | |
| | | | | | | | | | | -0.000005 | 0.000001 | -0.000005 | -0.000006 | -0.000006 | |
| | | | | | | | | | | | | | 0.000002 | 0.0000004 | |
| III | | | | | | 0.9201 | | | | 0.65693 | -0.68602 | -0.89681 | 1.05656 | -0.39357 | |
| 1 | 1 | 0.17 | -0.25 | 0.54 | -0.3 | | 0.85589 | -0.62363 | -0.39017 | | | | | | |

**TABLE II.17   Solving a System of Equations by Purcell's Method: Symmetric Coefficient Matrix**

| | $V_1^{(0)}$ | $V_2^{(0)}$ | $V_3^{(0)}$ | $V_4^{(0)}$ | $V_5^{(0)}$ | $V_2^{(1)}$ | $V_3^{(1)}$ | $V_4^{(1)}$ | $V_5^{(1)}$ | $V_3^{(2)}$ | $V_4^{(2)}$ | $V_5^{(2)}$ | $V_4^{(3)}$ | $V_5^{(3)}$ | $V_5^{(4)} = (X, 1)'$ |
|---|---|---|---|---|---|---|---|---|---|---|---|---|---|---|---|
| I | 1 | 0 | 0 | 0 | 0 | -0.42 | -0.54 | -0.66 | 0.3 | -0.49247 | -0.57698 | 0.10928 | -0.68623 | -0.24052 | -1.25779 |
| | 0 | 1 | 0 | 0 | 0 | 1 | 0 | 0 | 0 | -0.11316 | -0.19767 | 0.45410 | -0.22277 | 0.37372 | 0.04349 |
| | 0 | 0 | 1 | 0 | 0 | 0 | 1 | 0 | 0 | 1 | 0 | 0 | -0.22185 | 0.71029 | 1.03916 |
| | 0 | 0 | 0 | 1 | 0 | 0 | 0 | 1 | 0 | 0 | 1 | 0 | 1 | 0 | 1.48240 |
| | 0 | 0 | 0 | 0 | 1 | 0 | 0 | 0 | 1 | 0 | 0 | 1 | 0 | 1 | 1 |
| II | | | | | | | | | | 0.000003 | -0.000011 | 0.000002 | 0.000006 | -0.000001 | |
| | | | | | | | | | | 0.000003 | -0.000002 | -0.000002 | 0.000005 | -0.000008 | |
| | | | | | | | | | | | | | -0.000011 | -0.0000004 | |
| III | | | | | | 0.82360 | | | | 0.69786 | -0.22185 | -0.71029 | 0.49788 | -1.48240 | |
| 1 | 1 | 0.42 | 0.54 | 0.66 | -0.3 | | 0.11316 | 0.19767 | -0.45410 | | | | | | |

$(A_k, V_k^{(k-1)})$ be different from zero.

For the basis $R^{(0)} = R_{n+1}$ we take the natural basis.

$$V_1^{(0)} = (1, 0, \cdots, 0)', \cdots, V_{n+1}^{(0)} = (0, 0, \cdots, 1)' \; .$$

From the process it is clear that in all the steps the vector $V_{n+1}^{(k)}$ will have its $(n + 1)$th component equal to one. The unique basis vector $V_{n+1}^{(n)}$ of the subspace $R^{(n)}$ is orthogonal to all the vectors $A_1, \cdots, A_n$ and has its last component equal to one. Thus the vector $V_{n+1}^{(n)}$ gives a numerical solution for system (1).

Purcell's method is essentially very close to the methods connected with triangular factorization, and particularly with the escalator method when the matrix of the system is symmetric.

In fact if we form the matrix

$$Z = \begin{bmatrix} 1 & z_{12} & z_{13} & \cdot & \cdot & \cdot & z_{1n} \\ & 1 & z_{23} & \cdot & \cdot & \cdot & z_{2n} \\ & & \cdot & & \cdot & \cdot & \cdot \\ & & & \cdot & & \cdot & \cdot \\ & & & & \cdot & & \cdot \\ & & & & & \cdot & \vdots \\ & & & & & & 1 \end{bmatrix},$$

whose columns consist of the first $n$ components of the vectors $V_1^{(0)}, V_2^{(1)}, \cdots, V_n^{(n-1)}$, then this matrix will obviously be right triangular and matrix $AZ$, in view of the orthogonality condition, will be a left triangular matrix. If the matrix $A$ is symmetric, then the matrix $Z$ coincides with the corresponding matrix of the escalator method (Section 26, (3)).

In Tables II.16 and II.17 is carried out the solution of systems with asymmetric and symmetric matrices respectively, considered earlier in Tables II.1 and II.1a.

The scheme consists of three parts. In the first part are written the components of the basic vectors $V_i^{(k)}$ which are being computed in sequence; in the second part control of the computations is realized by verifying the fulfillment of the orthogonality conditions. In the first row of the third part are written $(A_k, V_k^{(k-1)})$; in the second row, the coefficients $c_i^{(k)}$.

## 28. THE REINFORCEMENT METHOD FOR INVERTING A MATRIX

The reinforcement method for inverting a matrix is based on the following idea. Let $B$ be a non-singular matrix whose inverse is known, let $u$ be a certain column, $v$ a certain row, $A = B + uv$. It is clear that

$$uv = \begin{bmatrix} u_1 \\ u_2 \\ \vdots \\ u_n \end{bmatrix} (v_1, v_2, \cdots, v_n) = \begin{bmatrix} u_1v_1 & u_1v_2 & \cdots & u_1v_n \\ u_2v_1 & u_2v_2 & \cdots & u_2v_n \\ \cdot & \cdot & \cdot & \cdot \\ u_nv_1 & u_nv_2 & \cdots & u_nv_n \end{bmatrix}$$

is a matrix of rank one.

We shall show that the inverse matrix for $A$ is found by the formula[†]

[†] Dwyer and Waugh (1).

$$A^{-1} = B^{-1} - \frac{1}{\gamma} B^{-1} uv B^{-1} , \tag{1}$$

where $\gamma = 1 + vB^{-1}u$. Of course it is assumed that $\gamma \neq 0$.
  In fact,

$$(B + uv)\left(B^{-1} - \frac{1}{\gamma} B^{-1} uv B^{-1}\right)$$

$$= E + uv B^{-1} - \frac{1}{\gamma} uv B^{-1} - \frac{1}{\gamma} u(vB^{-1}u)vB^{-1}$$

$$= E + uv B^{-1} - \frac{1}{\gamma} uv B^{-1} - \frac{1}{\gamma}(\gamma - 1)uv B^{-1} = E .$$

In this way the validity of formula (1) is proved. The connection established shows that the elements of the matrix $A^{-1}$ are found if the elements of the matrix $B^{-1}$ are known.
  In particular it is possible to apply the resultant formula in the case where matrix $A$ is obtained from matrix $B$ by changing one row, that is, if

$$A = B + V ,$$

where $V$ is a matrix whose elements are all equal to zero, except for the elements of the row being changed. Let this row have index $k$. Then

$$V = uv = e_k v ,$$

where $v$ is the non-zero row of matrix $V$ and $e_k$ is a column whose $k$th element is equal to one and whose remaining elements are all equal to zero. Consequently,

$$A^{-1} = B^{-1} - \frac{1}{1 + v(B^{-1}e_k)} (B^{-1}e_k)(vB^{-1})$$

$$= B^{-1} - \frac{1}{1 + v\beta_k} \beta_k(vB^{-1}) = B^{-1} - \frac{1}{1 + (v', \beta_k)} \beta_k(vB^{-1}) ,$$

where $\beta_k$ is the $k$th column of matrix $B^{-1}$. Denoting the $j$th column of matrix $A^{-1}$ by $\alpha_j$ we obtain

$$\alpha_j = \beta_j - \frac{(v', \beta_j)}{1 + (v', \beta_k)} \beta_k . \tag{2}$$

  The reinforcement method for inverting a matrix consists of the following. A given matrix $A = (a_{ij})$ is considered as the last term in the sequence $A_0 = E$, $A_1, \cdots, A_n = A$ while the passage from the matrix $A_{k-1}$ to the matrix $A_k$ is realized by replacing the $k$th row of matrix $A_{k-1}$ by the $k$th row of matrix $A$. Thus the matrix $A^{-1}$ is obtained as a result of an $n$-tuple application of the process described above. We shall write the formulas for passing to the $k$th step. Let $\alpha_j^{(k)}$ be the $j$th column of matrix $A_k^{-1}$. Then,

$$\alpha_j^{(k)} = \alpha_j^{(k-1)} - \frac{(v_k', \alpha_j^{(k-1)})}{1 + (v_k', \alpha_k^{(k-1)})} \alpha_k^{(k-1)} . \tag{3}$$

Here $v_k = (a_{k1}, \cdots, a_{k,k-1}, 0, a_{k,k+1}, \cdots, a_{kn})$.

In Table II.18 is given the inversion of the matrix of Table II.1a, but using the reinforcement method.

**TABLE II.18   Inverting a Matrix by the Reinforcement Method**

| | | | | | |
|---|---|---|---|---|---|
| | ‖1‖ | 0 | 0 | 0 | 0 |
| | 0 | 1 | 0 | 0 | 0.42 |
| | 0 | 0 | 1 | 0 | 0.54 |
| | 0 | 0 | 0 | 1 | 0.66 |
| 1 | 0 | 0.42 | 0.54 | 0.66 | |
| | 0 | 0.42 | 0.54 | 0.66 | |
| | 1 | ‖−0.42‖ | −0.54 | −0.66 | 0.42 |
| | 0 | 1 | 0 | 0 | 0 |
| | 0 | 0 | 1 | 0 | 0.32 |
| | 0 | 0 | 0 | 1 | 0.44 |
| 0.82360 | 0.42 | −0.17640 | 0.09320 | 0.16280 | |
| | 0.50996 | −0.21418 | 0.11316 | 0.19767 | |
| | 1.21418 | −0.50996 | ‖−0.49247‖ | −0.57698 | 0.54 |
| | −0.50996 | 1.21418 | ‖−0.11316‖ | −0.19767 | 0.32 |
| | 0 | 0 | 1 | 0 | 0 |
| | 0 | 0 | 0 | 1 | 0.22 |
| 0.69785 | 0.49247 | 0.11316 | −0.30215 | −0.15482 | |
| | 0.70570 | 0.16216 | −0.43297 | −0.22185 | |
| | 1.56172 | −0.43010 | −0.70569 | ‖−0.68623‖ | 0.66 |
| | −0.43010 | 1.23253 | −0.16215 | ‖−0.22277‖ | 0.44 |
| | −0.70570 | −0.16216 | 1.43297 | ‖0.22185‖ | 0.22 |
| | 0 | 0 | 0 | 1 | 0 |
| 0.49788 | 0.68624 | 0.22277 | −0.22185 | −0.50212 | |
| | 1.37832 | 0.44744 | −0.44559 | −1.00852 | |
| | 2.50756 | −0.12305 | −1.01147 | −1.37831 | |
| | −0.12305 | 1.33231 | −0.26141 | −0.44744 | |
| | −1.01148 | −0.26142 | 1.53182 | 0.44559 | |
| | −1.37832 | −0.44744 | 0.44559 | 2.00852 | |

Each step of the process arises from the following scheme:

| | $\alpha_1^{(k-1)}$ | $\cdots$ | $\|\alpha_k^{(k-1)}\|$ | $\cdots$ | $\alpha_n^{(k-1)}$ | $v_k'$ |
|---|---|---|---|---|---|---|
| $1 + \rho_k$ | $\rho_1$ | $\cdots$ | $\rho_k$ | $\cdots$ | $\rho_n$ | |
| | $w_1$ | $\cdots$ | $w_k$ | $\cdots$ | $w_n$ | |

Here

$$\rho_j = (v'_k, \alpha_j^{(k-1)}) \, ;$$

$$w_j = \frac{\rho_j}{1 + \rho_k} \, .$$

The matrix $A^{-1}$ is placed in the last part of the table.

The reinforcement method may be carried out according to another computational scheme which is slightly more extensive.[†] We write

$$c_{ij}^{(k-1)} = (v'_i, \alpha_j^{(k-1)}) \, .$$

In this notation

$$\alpha_j^{(k)} = \alpha_j^{(k-1)} - \frac{c_{kj}^{(k-1)}}{1 + c_{kk}^{(k-1)}} \alpha_k^{(k-1)} \, . \tag{4}$$

The numbers $c_{ij}^{(k)}$ are connected by the relationship

$$c_{ij}^{(k)} = c_{ij}^{(k-1)} - \frac{c_{ik}^{(k-1)} c_{kj}^{(k-1)}}{1 + c_{kk}^{(k-1)}} \, .$$

In fact,

$$c_{ij}^{(k)} = (v'_i, \alpha_j^{(k)}) = (v'_i, \alpha_j^{(k-1)}) - \frac{(v'_k, \alpha_j^{(k-1)})}{1 + (v'_k, \alpha_k^{(k-1)})} (v'_i, \alpha_k^{(k-1)})$$

$$= c_{ij}^{(k-1)} - \frac{c_{kj}^{(k-1)} c_{ik}^{(k-1)}}{1 + c_{kk}^{(k-1)}} \, .$$

Denoting by $\gamma_j^{(k)}$ the $j$th column of the matrix $C_k = (c_{ij}^{(k)})$ we shall have

$$\gamma_j^{(k)} = \gamma_j^{(k-1)} - \frac{c_{kj}^{(k-1)}}{1 + c_{kk}^{(k-1)}} \gamma_k^{(k-1)} \, . \tag{5}$$

Thus the matrix $C_k$ is obtained from the matrix $C_{k-1}$ in exactly the same way as the matrix $A_k^{-1}$ is obtained from $A_{k-1}^{-1}$.

For going from matrix $A_{k-1}^{-1}$ to matrix $A_k^{-1}$ (and from $C_{k-1}$ to $C_k$) it is necessary to know, besides matrix $A_{k-1}^{-1}$, only the elements of the $k$th row of matrix $C_{k-1}$. In the following step we need to know, correspondingly, the elements of the $k + 1$th row of matrix $C_k$; for the construction of this row by formula (5) we need to know in turn the elements of the $k + 1$th row of matrix $C_{k-1}$, and of forth. Elements of the first $k - 1$ rows of matrix $C_{k-1}$ are generally not necessary in later computations and there is no need to calculate them. For matrix $A_{k-1}^{-1}$, on the other hand, it is sufficient to calculate the elements of the first $k - 1$ rows, since the following rows coincide with the rows of the identity matrix. It is therefore useful to introduce for consideration the matrices $S_{k-1}$ and $\widetilde{S}_{k-1}$ "pasted together" from $A_{k-1}^{-1}$ and $C_{k-1}$. The first $S_{k-1}$ has its first $k - 1$ rows coinciding with the first $k - 1$ rows of matrix $A_{k-1}^{-1}$ and its last $n - k + 1$ rows are from $C_{k-1}$; the second $S_{k-1}$ has its first $k$ rows from $A_{k-1}^{-1}$ and its last $n - k$ rows from $C_{k-1}$. The matrix $\widetilde{S}_{k-1}$ is obtained from $S_{k-1}$ by replacing the $k$th

[†] A. P. Ershov (1).

row by the $k$th row of a unit matrix. It is obvious that elements of the $k$th row of matrix $S_{k-1}$ will be numbers $c_{kj}^{(k-1)}$.

On the basis of formulas (4) and (5) the matrix $S_k$ is obtained from matrix $\widetilde{S}_{k-1}$ by the formulas

$$\sigma_j^{(k)} = \widetilde{\sigma}_j^{(k-1)} - \frac{c_{kj}^{(k-1)}}{1 + c_{kk}^{(k-1)}} \sigma_k^{(k-1)} . \tag{6}$$

Here $\sigma_j^{(k)}$ denotes the $j$th column of the matrix $S_k$, $\widetilde{\sigma}_j^{(k-1)}$ denotes the $j$th column of matrix $\widetilde{S}_{k-1}$. It is obvious that for the initial matrix $S_0$ we need to take $S_0 = A - E$. The matrix $S_n = \widetilde{S}_n$ is equal to $A^{-1}$.

**TABLE II.19  Inversion of a Matrix Using a "Pasted" Scheme of the Reinforcement Method**

| | | | | | |
|---|---|---|---|---|---|
| | 1 | 0 | 0 | 0 | 1 |
| | 0.42 | 0 | 0.32 | 0.44 | 1.18 |
| | 0.54 | 0.32 | 0 | 0.22 | 1.08 |
| | 0.66 | 0.44 | 0.22 | 0 | 1.32 |
| 1 | 0 | 0.42 | 0.54 | 0.66 | |
| | 0 | 0.42 | 0.54 | 0.66 | 1.62 |
| | 1 | −0.42 | −0.54 | −0.66 | −0.62 |
| | 0 | 1 | 0 | 0 | 1 |
| | 0.54 | 0.09320 | −0.29160 | −0.13640 | 0.2052 |
| | 0.66 | 0.16280 | −0.13640 | −0.43560 | 0.2508 |
| 0.82360 | 0.42 | −0.17640 | 0.09320 | 0.16280 | |
| | 0.50996 | −0.21418 | 0.11316 | 0.19767 | 0.60661 |
| | 1.21418 | −0.50996 | −0.49247 | −0.57698 | −0.36523 |
| | −0.50996 | 1.21418 | −0.11316 | −0.19767 | 0.39339 |
| | 0 | 0 | 1 | 0 | 1 |
| | 0.57698 | 0.19767 | −0.15482 | −0.46778 | 0.15205 |
| 0.69785 | 0.49247 | 0.11316 | −0.30215 | −0.15482 | |
| | 0.70570 | 0.16216 | −0.43297 | −0.22185 | 0.21304 |
| | 1.56172 | −0.43010 | −0.70569 | −0.68623 | −0.26030 |
| | −0.43010 | 1.23253 | −0.16215 | −0.22277 | 0.41751 |
| | −0.70570 | −0.16216 | 1.43297 | 0.22185 | 0.78696 |
| | 0 | 0 | 0 | 1 | 1 |
| 0.49788 | 0.68624 | 0.22278 | −0.22185 | −0.50213 | |
| | 1.37832 | 0.44746 | −0.44559 | −1.00854 | 0.37165 |
| | 2.50756 | −0.12304 | −1.01147 | −1.37832 | −0.00527 |
| | −0.12305 | 1.33221 | −0.26141 | −0.44746 | 0.50029 |
| | −1.01148 | −0.26143 | 1.53182 | 0.44559 | 0.70450 |
| | −1.37832 | −0.44746 | 0.44559 | 2.00854 | 0.62835 |

In Table II.19 the inversion of a matrix is obtained by employing this variant of the reinforcement method. We shall explain the contents of Table II.19. In rows 1-4, 7-10, 13-16, 19-22 the matrices $S_0$, $\tilde{S}_1$, $\tilde{S}_2$, and $\tilde{S}_3$ are written in succession. It is advisable to enter rows which coincide with the identity matrix rows into the scheme beforehand. In rows 5, 11, 17, 23 we write $k$ rows of the matrices $S_{k-1}$; in rows 6, 12, 18, and 24 we write the corresponding factors $c_{kj}^{(k-1)}/[1 + c_{kk}^{(k-1)}]$. In the same rows, starting from the left of the scheme, we write the denominators $1 + c_{kk}^{(k-1)}$. Finally we write the matrix $S_n = A^{-1}$ in rows 25-28.

To each matrix $S_k$ is attached, for control purposes, a column $\tilde{\sigma}_{n+1}^{(k)}$ formed from partial sums. Elements of rows formed from the factors $c_{kj}^{(k-1)}/[1 + c_{kk}^{(k-1)}]$ are also summed. The control columns are connected (with the exception of one element always equal to one) by the relationship

$$\tilde{\sigma}_{n+1}^{(k)} = \sigma_{n+1}^{(k-1)} - \tilde{\sigma}_k^{(k-1)} \sum \frac{c_{kj}^{(k-1)}}{1 + c_{kk}^{(k-1)}} .$$

# Chapter III

# ITERATIVE METHODS FOR SOLVING SYSTEMS OF LINEAR EQUATIONS

We now pass on to a description of iterative methods for solving systems of linear equations. These methods provide a solution for a system in the form of a limit of a sequence of certain vectors which are constructed using a uniform process called an iterative process.

In contemporary literature a large number of iterative methods have been described based on different principles. As a rule the computational schemes for these methods are simple and convenient for use in applied computing. However, each iterative process has a limited area of application, since in the first place an iterative process may turn out to be inapplicable for a given system and in the second place the convergence of the process may be so slow that in practice it turns out to be impossible to achieve a satisfactory approximation to the solution.

We observe that although the problem of inverting a matrix is equivalent to solving $n$ partial systems with one and the same matrix, iterative methods are rarely used for this.

The present chapter is devoted to a description of the general principles for constructing iterative processes, a detailed examination of the simplest iterative process—the method of successive approximations in its different modifications, and to coordinate relaxation methods.

## 29. PRINCIPLES FOR SETTING UP ITERATIVE PROCESSES

The basic iterative processes for solving linear systems may be described by the following general scheme.

Let there be given a system of linear equations

$$AX = F \qquad (1)$$

with a non-singular matrix $A$. A sequence of vectors $X^{(1)}, X^{(2)}, \cdots, X^{(k)}$ is constructed according to the recurrence formulas

$$X^{(k)} = X^{(k-1)} + H^{(k)}(F - AX^{(k-1)}), \qquad (2)$$

where $H^{(1)}, H^{(2)}, \cdots$ is a certain sequence of matrices and $X^{(0)}$ is the initial approximation, generally arbitrary. A different choice of the sequence of matrices $H^{(k)}$ will lead to different iterative processes.

The iterative processes arising from formula (2) possess the property that for each of them an exact solution $X^*$ is a fixed point. This means that if $X^*$ is taken for initial approximation $X^{(0)}$, then all subsequent approximations will also be equal to $X^*$.

Conversely, every iterative process for which $X^*$ is a fixed point given by formulas

$$X^{(k)} = C^{(k)} X^{(k-1)} + Z^{(k)}, \qquad (3)$$

where $C^{(k)}$ is a sequence of matrices and $Z^{(k)}$ is a sequence of vectors, may be represented in the form (2). In fact for $X^*$ we have

$$X^* = C^{(k)} X^* + Z^{(k)},$$

from which

$$\begin{aligned}
X^{(k)} &= X^* + C^{(k)}(X^{(k-1)} - X^*) = X^{(k-1)} + (C^{(k)} - E)(X^{(k-1)} - X^*) \\
&= X^{(k-1)} + (E - C^{(k)})A^{-1}A(X^* - X^{(k-1)}) = X^{(k-1)} + H^{(k)}(F - AX^{(k-1)})
\end{aligned}$$

for

$$H^{(k)} = (E - C^{(k)})A^{-1}.$$

It is not hard to give necessary and sufficient conditions for iterative process (2) to converge to the solution for any initial vector.

In fact,

$$X^* - X^{(k)} = (X^* - X^{(k-1)}) - H^{(k)}(AX^* - AX^{(k-1)}) = (E - H^{(k)}A)(X^* - X^{(k-1)}).$$

From here

$$X^* - X^{(k)} = (E - H^{(k)}A)(E - H^{(k-1)}A)\cdots(E - H^{(1)}A)(X^* - X^{(0)}).$$

In order for $X^* - X^{(k)} \to 0$ for any initial vector $X^{(0)}$ it is necessary and sufficient (see Section 13, par. 1) that the matrix

$$T^{(k)} = (E - H^{(k)}A)(E - H^{(k-1)}A)\cdots(E - H^{(1)}A)$$

approach zero, for which it is sufficient in turn that any norm of the matrix $T^{(k)}$ approach zero. Of course this condition gives only a general point of view for setting up convergence conditions for concrete iterative processes.

The simplest iterative process is a *stationary iterative process* in which the matrices $H^{(k)}$ do not depend on the number of the step $k$. In particular, for $H^{(k)} = E$ one obtains the *classical process of successive approximations*. Any stationary process with $H \neq E$ may be considered as a process of successive approximations applied to the equivalent system

$$HAX = HF,$$

"*prepared*", so to say, for application of the method of successive approximations. Of course it is usually not necessary to realize such a preparation and consideration of such kinds of stationary processes gives only a convenient means for their theoretical investigation.

Close to the stationary iterative processes are *cyclic* ones in which the matrices $H^{(k)}$ are repeated periodically in a certain number of steps $p$. It is clear that it is possible to obtain from each cyclic process an equivalent stationary one, taking for one step of the stationary process the result of applying a complete cycle of $p$ steps from the initial cyclic process.

Non-stationary iterative processes may in turn be subdivided into two types. The first type consists of non-stationary iterative processes in the literal sense, in which changing the matrix $H^{(k)}$ occurs at each step. For the second type it is possible to include here stationary processes with an acceleration of convergence gained by replacing from time to time the stationary matrix $H$ which defines the process by certain specially selected matrices $H^{(\kappa)}$.

The choice of a matrix $H$ for a stationary process and of matrices $H^{(k)}$ for a non-stationary process may be effected in many different ways on the basis of different principles.

It is possible to construct matrices $H^{(k)}$ so that the iterative process converges to the solution for a broader class of systems of equations. The opposite point of view is also possible, on the basis of which, in constructing matrices $H^{(k)}$ the peculiarities of the given system are studied with a view towards obtaining the iterative process which yields the most rapid convergence. It is natural that to apply the iterative process constructed from the last principle it is necessary to obtain as much information as possible about the coefficient of the system, and particularly about the distribution of its eigenvalues.

An important principle for constructing iterative processes is the principle of *relaxation*. This is the principle of choosing the matrices $H^{(k)}$ from a certain class of matrices outlined beforehand so that at each step of the process any value which characterizes the exactness of the solution for the system is refined.

The most developed relaxation methods are the *coordinate* ones, in which the matrices $H^{(k)}$ are chosen so that at each step one or several components of the succesive approximations are changed, and the *gradient* methods in which the matrices $H^{(k)}$ are scalar matrices.

It is natural to judge the exactness of an approximate solution $X$ for a system $AX = F$ by the magnitude (in one or another sense) of the *error vector* $Y = X^* - X$. However, the error vector may not be computable without knowledge of the system's exact solution and may only be estimated. A vector characterizing the exactness of the approximate solution $X$ of the system $AX = F$ is the *residual vector* $r = F - AX$. It is clear that $r = AY$. Thus relaxation may be set up by lessening any norm of each of these vectors.

For positive definite matrices $A$ a convenient measure of accuracy is the so-called *error function*,

$$f(X) = (AY, Y) = (Y, r) = (A^{-1}r, r) \,.$$

In view of the positive definiteness of $A$ it always holds that $f(X) \geq 0$, and $f(X) = 0$ only for $X = X^*$. It is clear that

$$f(X) = (X^* - X, F - AX) = (X^*, F) - (X, F) - (X^*, AX) + (X, AX)$$
$$= (AX, X) - 2(X, F) + (X^*, F) \,.$$

The error function can not be computed if the exact solution is not known. However, its values differ only by a constant from values of the functional

$$f_0(X) = (AX, X) - 2(X, F) \,,$$

which may be computed without a knowledge of $X^*$. Therefore we may judge any decrease in the error function by comparing corresponding values of the functional $f_0(X)$.

Another important principle for constructing iterative processes is the principle of successive "*suppression of components*" for an error vector, expanding it in terms of the eigenvectors of the system's coefficient matrix.

Chapter VII will be devoted to gradient relaxation methods. In Chapter IX methods based on the idea of suppression of components will be examined.

## 30. METHOD OF SUCCESSIVE APPROXIMATIONS

The simplest iterative process is the process of successive approximations.

By the process of successive approximations is meant the following iterative process. The system of equations $AX = F$ is written in the form

$$X = BX + G \,, \tag{1}$$

where $B = E - A, G = F$ and the successive approximations are computed by the formula

$$X^{(k)} = BX^{(k-1)} + G \,, \tag{2}$$

beginning with a certain initial approximation $X^{(0)}$ which may in general be chosen arbitrarily. It is obvious that the process of successive approximations is a special case of the general iterative process (2) of Section 29; that is, this will be a stationary process in which $H = E$. In fact,

$$X^{(k)} = (E - A)X^{(k-1)} + G = X^{(k-1)} + (F - AX^{(k-1)}) \,.$$

It is easy to give a formula for the expression $X^{(k)}$ directly in terms of the initial approximation $X^{(0)}$. Namely

$$X^{(k)} = B^k X^{(0)} + (E + B + \cdots + B^{k-1})G \,. \tag{3}$$

In fact for $k = 1$ this is true and for the remaining $k$ formulas it is easy to verify by a proof employing mathematical induction.

We observe that if the process of successive approximations converges, then it converges to a solution of the system. In fact if $X^{(k)} \to X^*$, then passing to the limit in the equality

$$X^{(k)} = BX^{(k-1)} + G$$

gives $X^* = BX^* + G$, that is, $X^*$ satisfies the given system.

THEOREM 30.1. *For a successive approximation process to converge for any initial vector $X^{(0)}$ it is necessary and sufficient that all eigenvalues of the matrix B be less than one in modulus.*

*Proof.* Let $X^{(k)} \to X^*$. Then as we have seen before, $X^*$ is the solution to the system and consequently $X^* - X^{(k)} = B(X^* - X^{(k-1)}) = \cdots = B^k(X^* - X^{(0)})$, from which $B^k(X^* - X^{(0)}) \to 0$. Since this must hold for any vector $X^{(0)}$, it is necessary that $B^k \to 0$, for which in turn it is necessary that all eigenvalues of matrix $B$ be less than one in modulus (theorem 13.2).

The sufficiency of the condition arises directly from formula (3), since $B^k \to 0$ and $E + B + \cdots + B^{k-1} \to (E - B)^{-1} = A^{-1}$ if all the eigenvalues of matrix $B$ are less than one in modulus.

Since it is difficult to verify the condition in theorem 30.1, it is better to use sufficiency conditions, associated directly with the elements of the matrix, in order to estimate the convergence of a successive approximation process. Certain criteria of sufficiency may be gleaned from theorem 30.2.

THEOREM 30.2. *For a process of successive approximations to converge it is sufficient that any norm of the matrix B be less than one.*

*Proof.* If $||B|| < 1$, then all eigenvalues of matrix $B$ are less than one and therefore on the basis of theorem 30.1 the process of successive approximations converges.

We now give an estimate of the rate of convergence of the successive approximation process in terms of the norms. The choice of a matrix norm must agree with the selected vector norm.

THEOREM 30.3. *If $||B|| < 1$, then*

$$||X^* - X^{(k)}|| \le ||B||^k ||X^{(0)}|| + \frac{||G|| \, ||B||^k}{1 - ||B||} . \tag{4}$$

*Proof.* We have

$$
\begin{aligned}
||X^* - X^{(k)}|| &= ||(E - B)^{-1}G - (E + B + \cdots + B^{k-1})G - B^k X^{(0)}|| \\
&\le ||(E - B)^{-1}G - (E + B + \cdots + B^{k-1})G|| + ||B^k X^{(0)}|| \\
&\le ||(E - B)^{-1}G - (E + B + \cdots + B^{k-1})G|| + ||B^k|| \, ||X^{(0)}|| \\
&\le ||B||^k ||X^{(0)}|| + \frac{||G|| \, ||B||^k}{1 - ||B||} .
\end{aligned}
$$

It will often be important to compare the accuracy of two successive approximations, that is, to compare the values $X^* - X^{(k)}$ and $X^* - X^{(k-1)}$. Such a comparison may be carried out on the basis of following theorem.

THEOREM 30.4.   $|| X^* - X^{(k)} || \le || B || \, || X^* - X^{(k-1)} ||$.

*Proof.* From the equality

$$X^* = BX^* + G, \qquad X^{(k)} = BX^{(k-1)} + G$$

it follows that

$$X^* - X^{(k)} = B(X^* - X^{(k-1)}) .$$

From here

$$|| X^* - X^{(k)} || = || B(X^* - X^{(k-1)}) || \le || B || \, || X^* - X^{(k-1)} || . \tag{5}$$

The vector norms introduced by us in Section 13 (cubic, octahedral, and spherical) and the matrix norms consistent with them give the following easily verified sufficiency criteria for convergence of a process of successive approximations and an estimate of the rate of convergence.

I.   If $\sum_{j=1}^{n} | b_{ij} | \le \mu < 1$ for $i = 1, 2, \cdots n$, then the process of successive approximations converges, while

$$| x_i - x_i^{(k)} | \le \mu \max | x_j - x_j^{(k-1)} | , \tag{6}$$

where $X^* = (x_1, \cdots, x_n)'$ and $X^{(k)} = (x_1^{(k)}, \cdots, x_n^{(k)})'$.

II.   If $\sum_{i=1}^{n} | b_{ij} | \le \nu < 1$ for $j = 1, 2, \cdots, n$, then the process of successive approximations converges, while

$$\sum_{i=1}^{n} | x_i - x_i^{(k)} | \le \nu \sum_{i=1}^{n} | x_i - x_i^{(k-1)} | . \tag{7}$$

III.   If $\sum_{i=1}^{n} b_{ik}^2 \le \rho < 1$, then the process of successive approximations converges and

$$\sqrt{\sum_{i=1}^{n} (x_i - x_i^{(k)})^2} \le \sqrt{\rho} \, \sqrt{\sum_{i=1}^{n} (x_i - x_i^{(k-1)})^2} . \tag{8}$$

We shall point out one way of constructing sufficient conditions for convergence of a process of successive approximations.

In the equation

$$X = BX + G \tag{1}$$

with matrix

$$B = \begin{bmatrix} b_{11} & b_{12} \cdots b_{1n} \\ b_{21} & b_{22} \cdots b_{2n} \\ \cdot & \cdot \quad \cdot \quad \cdot \\ b_{n1} & b_{n2} \cdots b_{nn} \end{bmatrix}$$

we introduce new unknowns $x_i = p_i z_i$ where $p_i$ are certain positive numbers. Then system (1) is transformed into the system

$$p_i z_i = \sum_{j=1}^{n} b_{ij} p_j z_j + g_i$$

or

$$z_i = \sum_{j=1}^{n} b_{ij} \cdot \frac{p_j}{p_i} \, z_j \, \frac{1}{p_i} \, g_i \, . \qquad (9)$$

It is obvious that components of the successive approximations $X^{(k)}$ for system (1) and $Z^{(k)}$ for system (9) are also connected by the relationships $x_i^{(k)} = p_i z_i^{(k)}$, if these relationships hold for the initial approximations $X^{(0)}$ and $Z^{(0)}$. Therefore the processes of successive approximations for systems (1) and (9) converge or diverge simultaneously and consequently any sufficient condition for the convergence of the process of successive approximations for system (9) is also a sufficient condition for the convergence of system (1).

Thus if it is possible to indicate positive numbers $p_1, \cdots, p_n$ such that one of the conditions

1. $\displaystyle\sum_{j=1}^{n} |b_{ij}| \cdot \frac{p_j}{p_i} < 1 \qquad$ for $i = 1, \cdots, n$

2. $\displaystyle\sum_{i=1}^{n} |b_{ij}| \cdot \frac{p_j}{p_i} < 1 \qquad$ for $j = 1, \cdots, n \qquad (10)$

3. $\displaystyle\sum_{i,j=1}^{n} \frac{b_{ij}^2 p_j^2}{p_i^2} < 1 \, ,$

is satisfied, then the process of successive approximations for system (1) converges.

*Note.* For computing iterations in practice we may use two methods.

1. Let $X^{(0)} = G$. Then

$$X^{(k)} = G + BG + \cdots + B^k G \, .$$

To compute $X^{(k)}$ we compute successively the vectors $G, BG, \cdots, B^k G$ and find their sum. This is convenient because of the uniformity of the computation process and also because each subsequent entry is only a correction to the sum of the preceding ones. The inadequacy of this method arises from the possible accumulation of errors from rounding off with an increase in the number of items.

2. The computation is introduced directly by the formulas

$$X^{(k)} = BX^{(k-1)} + G \, .$$

Here each approximation is like the initial one and therefore there is no need in the first steps of the process to carry out the computation with great accuracy; resulting errors are smoothed out in the sequel.

As an example we shall find the solution for the system

$$0.78x_1 - 0.02x_2 - 0.12x_3 - 0.14x_4 = 0.76 \ ,$$
$$- 0.02x_1 + 0.86x_2 - 0.04x_3 + 0.06x_4 = 0.08 \ ,$$
$$- 0.12x_1 - 0.04x_2 + 0.72x_3 - 0.08x_4 = 1.12 \ ,$$
$$- 0.14x_1 + 0.06x_2 - 0.08x_3 + 0.74x_4 = 0.68 \ .$$

(11)

Solving this system by the single-division scheme we find

$$x_1 = 1.534965 \ ,$$
$$x_2 = 0.122010 \ ,$$
$$x_3 = 1.975156 \ ,$$
$$x_4 = 1.412955 \ .$$

For applying the successive approximation process we switch the system to the form $X = BX + G$, letting $B = E - A$. We obtain

$$x_1 = 0.22x_1 + 0.02x_2 + 0.12x_3 + 0.14x_4 + 0.76 \ ,$$
$$x_2 = 0.02x_1 + 0.14x_2 + 0.04x_3 - 0.06x_4 + 0.08 \ ,$$
$$x_3 = 0.12x_1 + 0.04x_2 + 0.28x_3 + 0.08x_4 + 1.12 \ ,$$
$$x_4 = 0.14x_1 - 0.06x_2 + 0.08x_3 + 0.26x_4 + 0.68 \ .$$

(12)

It is not hard to see that the sufficiency conditions for the convergence of the successive approximation process are satisfied.

**TABLE III.1  Computation Scheme for the Formula** $X^{(k)} = \sum\limits_{l=0}^{k} B^l G$

(1)  $B^l G, \ l = 0, \cdots, 14.$

| G | 0.76 | 0.08 | 1.12 | 0.68 | 2.64 |
|---|---|---|---|---|---|
| BG | 0.3984 | 0.0304 | 0.4624 | 0.3680 | 1.2592 |
| · | 0.195264 | 0.008640 | 0.207936 | 0.186624 | 0.598464 |
| · | 0.09421056 | 0.00223488 | 0.09692928 | 0.09197568 | 0.28535040 |
| · | 0.04527913 | 0.00055572 | 0.04589292 | 0.04472340 | 0.13645117 |
| · | 0.02174095 | 0.00013570 | 0.02188361 | 0.02160525 | 0.06536551 |
| · | 0.01043649 | 0.00003285 | 0.01047017 | 0.01040364 | 0.03134316 |
| · | 0.00500961 | 0.00000792 | 0.00501763 | 0.00500170 | 0.01503686 |
| · | 0.00240463 | 0.00000190 | 0.00240654 | 0.00240272 | 0.00721580 |
| · | 0.00115422 | 0.00000046 | 0.00115468 | 0.00115376 | 0.00346312 |
| · | 0.00055403 | 0.00000011 | 0.00055414 | 0.00055392 | 0.00166219 |
| · | 0.00026593 | 0.00000003 | 0.00026596 | 0.00026591 | 0.00079783 |
| · | 0.00012765 | 0.00000001 | 0.00012765 | 0.00012764 | 0.00038295 |
| · | 0.00006127 | | 0.00006127 | 0.00006127 | 0.00018381 |
| $B^{14}G$ | 0.00002941 | | 0.00002941 | 0.00002941 | 0.00008823 |

(2)  $X^{(k)} = \sum\limits_{l=0}^{k} B^l G$ for $k = 12, 13, 14.$

| | | | | |
|---|---|---|---|---|
| $X^{(12)}$ | 1.53484720 | 0.12200958 | 1.97503858 | 1.41283762 |
| $X^{(13)}$ | 1.53490847 | 0.12200958 | 1.97509985 | 1.41289889 |
| $X^{(14)}$ | 1.53493788 | 0.12200958 | 1.97512926 | 1.41292830 |

For comparing the course of an iterative process in different variants we carry it out by three methods.

1) Computation of successive approximations is carried out by the formula $X^{(k)} = \sum_{l=0}^{k} B^l G$ (see Table III.1).

2) Computation of successive approximations is carried out by the formula $X^{(k)} = BX^{(k-1)} + G$ for $X^{(0)} = G$ (see Table III.2).

3) We compute $X^{(k)} = BX^{(k-1)} + G$ again for $X^{(0)} = (1, 0, 0, 0)'$ (see Table III.3).

We shall explain Table III.1. The first part of the table contains the components of the vectors $B^l G$ being computed successively. The last column is for control. In it are written the numbers $\sum_{j=1}^{n} c_j x_j^{(k)}$, where $c_j = \sum_{i=1}^{n} b_{ij}$ (the numbers $c_j$ must be computed beforehand and written in as a supplementary row of matrix $B$). But it is obvious that $\sum_{j=1}^{n} c_j x_j^{(k)} = \sum_{i=1}^{n} x_j^{(k+1)}$ so that the

**TABLE III.2  Computation of Approximations by the Formula**
$X^{(k)} = BX^{(k-1)} + G; \; X^{(0)} = G.$

| $X^{(0)}$ | 0.76 | 0.08 | 1.12 | 0.68 | 2.64 |
|---|---|---|---|---|---|
| $X^{(1)}$ | 1.1584 | 0.1104 | 1.5824 | 1.0480 | 3.8992 |
| $X^{(2)}$ | 1.3537 | 0.1190 | 1.7903 | 1.2346 | 4.4977 |
| $X^{(3)}$ | 1.4479 | 0.1213 | 1.8873 | 1.3266 | 4.7830 |
| $X^{(4)}$ | 1.4932 | 0.1218 | 1.9332 | 1.3713 | 4.9195 |
| $X^{(6)}$ | 1.5149 | 0.1220 | 1.9551 | 1.3929 | 4.9849 |
| $X^{(6)}$ | 1.5253 | 0.1220 | 1.9655 | 1.4033 | 5.0162 |
| $X^{(7)}$ | 1.5303 | 0.1220 | 1.9705 | 1.4083 | 5.0312 |
| $X^{(8)}$ | 1.53273 | 0.12201 | 1.97292 | 1.41072 | 5.03838 |
| $X^{(9)}$ | 1.53380 | 0.12201 | 1.97408 | 1.41188 | 5.04187 |
| $X^{(10)}$ | 1.53445 | 0.12201 | 1.97464 | 1.41244 | 5.04354 |
| $X^{(11)}$ | 1.53472 | 0.12201 | 1.97491 | 1.41271 | 5.04434 |
| $X^{(12)}$ | 1.53485 | 0.12201 | 1.97504 | 1.41284 | 5.04473 |
| $X^{(13)}$ | 1.534910 | 0.122010 | 1.975101 | 1.412900 | 5.044920 |
| $X^{(14)}$ | 1.5349385 | 0.1220096 | 1.9751299 | 1.4129289 | 5.0450069 |

**TABLE III.3  Computation of Approximations by the Formula**
$X^{(k)} = BX^{(k-1)} + G; \; X^{(0)} = (1, 0, 0, 0)'$

| $X^{(0)}$ | 1 | 0 | 0 | 0 | 1 |
|---|---|---|---|---|---|
| $X^{(1)}$ | 0.98 | 0.10 | 1.24 | 0.82 | 3.14 |
| $X^{(2)}$ | 1.2412 | 0.1140 | 1.6544 | 1.1236 | 4.1332 |
| $X^{(12)}$ | 1.534769 | 0.122010 | 1.974961 | 1.412761 | 5.04451 |
| $X^{(13)}$ | 1.534871 | 0.122010 | 1.975063 | 1.412862 | 5.044805 |
| $X^{(14)}$ | 1.534920 | 0.122010 | 1.975111 | 1.412910 | 5.044951 |

elements of the control column are equal to the sums of the remaining elements lying in the same row. The second part of the table gives an approximate solution to the system which we obtain by summing corresponding components of the computed vectors.

By comparing Tables III.1, III.2, and III.3 with the result obtained by the single-division scheme, we see that convergence of the process in all three variants is approximately the same; the fourteenth approximation gives a result in the given example which is precise to within one digit in the fourth decimal place.

Convergence of a process of successive approximations may be sharply improved by applying different methods of acceleration. The process of successive approximations with acceleration methods applied is placed, in the majority of cases, into the general scheme of iterative processes whose stationary property has been disturbed. An expedient choice of acceleration methods requires preliminary information on the distribution of the eigenvalues of the matrix. We shall return to this question in Chapter IX.

### 31.  PREPARING A SYSTEM OF LINEAR EQUATIONS IN A FORM SUITABLE FOR APPLYING THE METHOD OF SUCCESSIVE APPROXIMATIONS.  THE SIMPLE ITERATION METHOD

The convergence conditions for the method of successive approximations require that the matrix of coefficients of the system $AX = F$ be, in one sense or another, close to the unit matrix. If this condition is not satisfied or is "poorly satisfied," then it is advisable to prepare the system beforehand for application of the successive approximation method. The preparation consists in passing from the given system $AX = F$ to the equivalent system

$$HAX = HF,$$

where $H$ is a certain non-singular matrix chosen so that the matrix $HA$ will be close to a unit matrix, that is, sc that the matrix $H$ will be close to $A^{-1}$.

Application of the method of successive approximations to the prepared system, as was pointed out, is equivalent to application of the stationary iterative process

$$X^{(k)} = X^{(k-1)} + H(F - AX^{(k-1)})$$

to the initial system.

The selection of the matrix $H$ may be realized using the special peculiarities of the given system. We shall consider certain very common preparation methods which require only superficial information about the coefficient matrix.

Let a matrix $A$ be positive definite. Then the system $AX = F$ may always be prepared for a form in which the method of successive approximations will be convergent. In fact, having computed, for example, the first norm $\mu$ of matrix $A$, we obtain that all the eigenvalues of matrix $A$ are included in the open interval $(0, \mu)$. We let

$$H = \frac{2}{\mu} E . \tag{1}$$

The system $AX = F$ is transformed to the form

$$X = \left( E - \frac{2}{\mu} A \right) X + \frac{2}{\mu} F = BX + G . \tag{2}$$

The eigenvalues of the matrix $B = E - \frac{2}{\mu} A$ will be included in the open interval $(-1, 1)$ and consequently the method of successive approximations will be convergent.

As an example we shall take system (9), Section 23. Here $\mu = 2.62$. Performing the computations we get

$$B = \begin{bmatrix} 0.23664122 & -0.32061069 & -0.41221374 & 0.50381679 \\ -0.32061069 & 0.23664122 & -0.24427481 & -0.33587786 \\ -0.41221374 & -0.24427481 & 0.23664122 & -0.16793893 \\ -0.50381679 & -0.33587786 & -0.16793893 & 0.23664122 \end{bmatrix} ,$$

$$\tag{3}$$

$$G = \begin{bmatrix} 0.22900763 \\ 0.38167939 \\ 0.53435115 \\ 0.68702290 \end{bmatrix} .$$

In Section 23 it was discovered that the solution of the system was $X = (-1.2577938, 0.0434873, 1.0391663, 1.4823929)'$.

In Table III.4 we make several successive approximations, letting $X^{(0)} = G$.

**TABLE III.4  Computation of Approximations by the Formula $X^{(k)} = BX^{(k-1)} + G$**

| $X^{(0)}$ | 0.22900763 | 0.38167939 | 0.53435115 | 0.68702290 | 1.83206107 |
|---|---|---|---|---|---|
| $X^{(1)}$ | −0.4055708 | 0.0372939 | 0.3577880 | 0.5162869 | 0.5057980 |
| $X^{(20)}$ | −1.2354227 | 0.0473327 | 1.0287985 | 1.4668834 | 1.3075919 |
| $X^{(25)}$ | −1.2505812 | 0.0442080 | 1.0348124 | 1.4760729 | 1.3045121 |
| $X^{(40)}$ | 1.2574335 | 0.0435403 | 1.0389819 | 1.4821204 | 1.3072091 |
| $X^{(50)}$ | −1.2577475 | 0.0434939 | 1.0391421 | 1.4823572 | 1.3072456 |
| $X^{(74)}$ | −1.2577935 | 0.0434874 | 1.0391661 | 1.4823926 | 1.3072526 |

From the results arrived at it is clear that in the given case the method of successive approximations converges quite slowly.

In applied problems one often encounters systems in which the diagonal elements of a matrix $A$ significantly dominate the remaining elements of the matrix. In this case the preparation of the system is realized in the following way.

We rewrite the system $AX = F$ in detailed form

$$\begin{aligned} a_{11}x_1 + a_{12}a_2 + \cdots + a_{1n}x_n &= f_1 , \\ a_{21}x_1 + a_{22}x_2 + \cdots + a_{2n}x_n &= f_2 , \\ \cdots \cdots \cdots \cdots \cdots \cdots \cdots & \\ a_{n1}x_1 + a_{n2}x_2 + \cdots + a_{nn}x_n &= f_n . \end{aligned} \tag{4}$$

We divide each equation of system (4) by its diagonal element. We obtain the system

$$x_1 + \frac{a_{12}}{a_{11}} x_2 + \cdots + \frac{a_{1n}}{a_{11}} x_n = \frac{f_1}{a_{11}} ,$$

$$\frac{a_{21}}{a_{22}} x_1 + x_2 + \cdots + \frac{a_{2n}}{a_{22}} x_n = \frac{f_2}{a_{22}} ,$$

$$\cdots \cdots \cdots \cdots \cdots \cdots ,$$

$$\frac{a_{n1}}{a_{nn}} x_1 + \frac{a_{n2}}{a_{nn}} x_2 + \cdots + x_n = \frac{f_n}{a_{nn}} ,$$

or in matrix notation

$$X = BX + G , \tag{5}$$

where

$$B = \begin{bmatrix} 0 & -\dfrac{a_{12}}{a_{11}} & \cdots & -\dfrac{a_{1n}}{a_{11}} \\ -\dfrac{a_{21}}{a_{22}} & 0 & \cdots & -\dfrac{a_{2n}}{a_{22}} \\ \cdots & \cdots & \cdots & \cdots \\ -\dfrac{a_{n1}}{a_{nn}} & -\dfrac{a_{n2}}{a_{nn}} & \cdots & 0 \end{bmatrix} ,$$

$$\tag{6}$$

$$G = \begin{bmatrix} \dfrac{f_1}{a_{11}} \\ \dfrac{f_2}{a_{22}} \\ \vdots \\ \dfrac{f_n}{a_{nn}} \end{bmatrix} .$$

To apply the iterative process, it is not really necessary to transform system (4) into system (5). Successive approximations may be computed by the formulas:

$$a_{11}x_1^{(k)} = f_1 - a_{12}x_2^{(k-1)} - \cdots - a_{1n}x_n^{(k-1)} ,$$

$$\cdots \cdots \cdots \cdots \cdots \cdots \cdots \cdots ,$$

$$a_{nn}x_n^{(k)} = f_n - a_{n1}x_1^{(k-1)} - \cdots - a_{n, n-1}x_{n-1}^{(k-1)} .$$

$$\tag{7}$$

The modification described for the process of successive approximations is called the method of *simple iteration* or Jacobi's method.

The above mentioned transformation of system (4) into system (5) is obviously equivalent to premultiplying system (4) by the matrix

$$H = \begin{bmatrix} \dfrac{1}{a_{11}} & & & 0 \\ & \dfrac{1}{a_{22}} & & \\ & & \ddots & \\ 0 & & & \dfrac{1}{a_{nn}} \end{bmatrix} .$$

Thus $H = D^{-1}$ where $D$ is the diagonal matrix $[a_{11}, \cdots, a_{nn}]$.

It follows that the necessary and sufficient condition for the convergence of a process of simple iteration is that all the eigenvalues of the matrix $B = E - D^{-1}A$ be less than one in modulus.

This condition may be restated in another form. Namely, $|B - tE| = |E - D^{-1}A - tE| = |D^{-1}| |D - A - tD| = (-1)^n |D^{-1}| |A - D + tD|$. Thus for the convergence of a simple iteration process it is necessary and sufficient that all roots of the equation

$$|A - D + tD| = \begin{vmatrix} a_{11}t & a_{12} \cdots a_{1n} \\ a_{21} & a_{22}t \cdots a_{2n} \\ \cdot \ \ \cdot \ \ \cdot \ \ \cdot \ \ \cdot \ \ \cdot \ \ \cdot \\ a_{n1} & a_{n2} \ \cdots a_{nn}t \end{vmatrix} = 0 \qquad (8)$$

be less than one in modulus.

If the coefficient matrix $A$ of a system is symmetric (or Hermitian) and has positive diagonal elements, it is possible to give the necessary and sufficient condition for the convergence of a simple iteration method in the following easily verified form (Yu. M. Gavrilov (2)).

*For the convergence of the simple iteration method on the system $AX = F$ having symmetric matrix $A$ and positive diagonal elements, it is necessary and sufficient that the matrices $A$ and $2D - A$ (differing from each other at most in the signs of non-diagonal elements) be positive definite.*

In this case, because of the positive definiteness of the matrix $D$, all the eigenvalues of the matrix $D^{-1}A$ are real (Theorem 11.14). Therefore for the process to converge it is necessary and sufficient that the eigenvalues of the matrix $D^{-1}A = E - (E - D^{-1}A)$ be included in the interval $(0,2)$, that is, that the eigenvalues of matrices $D^{-1}A$ and $2E - D^{-1}A$ be positive. But in view of theorem 11.16 this is equivalent to the positive definiteness of matrices $A$ and $2D - A$.

The sufficiency conditions introduced in Section 30 for the convergence of the successive iteration process, applied to system (5), give the following sufficiency conditions for the convergence of the simple iteration method:

$$\text{I.} \quad \sum_{j=1}^{n}{}' \left| \frac{a_{ij}}{a_{ii}} \right| < 1 \qquad (i = 1, 2, \cdots, n) ,$$

$$\text{II.} \quad \sum_{i=1}^{n}{}' \left| \frac{a_{ij}}{a_{ii}} \right| < 1 \qquad (j = 1, 2, \cdots, n) , \qquad (9)$$

$$\text{III.} \quad \sum_{i,j=1}^{n}{}' \left( \frac{a_{ij}}{a_{ii}} \right)^2 < 1 .$$

Here the stroke sign at the sum indicates that the values for $i = j$ are omitted in summing.

If we use the general indicating numbers of (10) Section 30, then letting

$$p_i = \frac{1}{|a_{ii}|} ,$$

we obtain the following sufficiency criteria for the convergence of simple iteration:

I. $\displaystyle\sum_{j=1}^{n}{}'\left|\frac{a_{ij}}{a_{jj}}\right| < 1 \qquad (i = 1, 2, \cdots, n)$,

II. $\displaystyle\sum_{i=1}^{n}{}'\left|\frac{a_{ij}}{a_{jj}}\right| < 1 \qquad (j = 1, 2, \cdots, n)$,  $\qquad\qquad(10)$

III. $\displaystyle\sum_{i,j=1}^{n}{}'\left(\frac{a_{ij}}{a_{jj}}\right)^2 < 1$.

We now assume that a system

$$AX = F$$

is given in which the elements of the principal diagonal of $A$ do not in general dominate the remaining elements.

The selection of the auxiliary matrix $H$ may be realized, for example, by a crude inversion of matrix $A$ according to Gauss's method.

It often turns out to be advisable to take as the matrix $H$ the matrix inverse to the matrix

$$\begin{bmatrix} a_{11} & a_{12} & & & & 0 \\ a_{21} & a_{22} & & & & \\ & & a_{33} & a_{34} & & \\ & & a_{43} & a_{44} & & \\ & & & & \ddots & \\ 0 & & & & & \ddots \end{bmatrix}.$$

Inverting such a matrix is not difficult, since it comes down to inverting matrices of the second order. Namely,

$$H = \begin{bmatrix} \dfrac{a_{22}}{\varDelta_1} & -\dfrac{a_{12}}{\varDelta_1} & & & 0 \\[2ex] -\dfrac{a_{12}}{\varDelta_1} & \dfrac{a_{11}}{\varDelta_1} & & & \\[2ex] & & \ddots & & \\ 0 & & & \ddots \end{bmatrix},$$

where $\varDelta_1$ is the determinant $\begin{vmatrix} a_{11} & a_{12} \\ a_{21} & a_{22} \end{vmatrix}$.

If the matrix $A$ is symmetric and $A = R + S$, where $R$ is a positive matrix whose inverse is known, the process of successive approximations applied to a system prepared in the form

$$X = -R^{-1}SX + R^{-1}F,$$

has the following simple convergence criterion. *For the process to converge it is necessary and sufficient that the matrices $R + S$ and $R - S$ be positive definite.*

In fact the eigenvalues of the matrix $R^{-1}S$ are real, since $R$ is positive definite (theorem 11.14). For the process to converge it is necessary and sufficient that the eigenvalues of the matrix $R^{-1}S$ be less than one in modulus, that is, included in the interval $(-1,1)$. For this it is necessary and sufficient in turn that all eigenvalues of the matrices $E + R^{-1}S$ and $E - R^{-1}S$ be positive. This holds if and only if (theorem 11.16) the matrices $R(E + R^{-1}S) = R + S$ and $R(E - R^{-1}S) = R - S$ are positive definite.

## 32. ONE-STEP CYCLIC PROCESSES

Let a system of linear equations $AX = F$ be represented as

$$X = BX + G , \qquad (1)$$

where $B = E - A, G = F$. We denote the components of the desired solution vectors by $x_1, \cdots, x_n$. A *one-step cyclic process* (often called Seidel's method) is a process of successive approximations with the difference that in computing the $k$th approximation for the $i$th component we consider the $k$ approximations for components $x_1^{(k)}, \cdots, x_{i-1}^{(k)}$ which have already been computed.

Computation of the successive approximations is carried out according to the formulas.

$$x_i^{(k)} = \sum_{j=1}^{i-1} b_{ij}x_j^{(k)} + \sum_{j=i}^{n} b_{ij}x_j^{(k-1)} + g_i \qquad (2)$$

instead of

$$x_i^{(k)} = \sum_{j=1}^{n} b_{ij}x_j^{(k-1)} + g_i$$

as in the method of successive approximations.

The one-step cyclic process may be interpreted in two ways, indicating the variety of the general iterative process. For one step of the process we can take the passage from the vector $(x_1^{(k)}, \cdots, x_{i-1}^{(k)}, x_i^{(k-1)}, \cdots, x_n^{(k-1)})'$ to the vector $(x_1^{(k)}, \cdots, x_i^{(k)}, x_{i+1}^{(k-1)}, \cdots, x_n^{(k-1)})'$ (or from the vector $(x_1^{(k)}, \cdots, x_n^{(k)})'$ to the vector $(x_1^{(k1)}, x_2^{(k)}, \cdots, x_n^{(k)})')$, or again, for one step, we may consider the result of applying a complete cycle, that is, passing from the vector $(x_1^{(k-1)}, \cdots, x_n^{(k-1)})'$ to the vector $(x_1^{(k)}, \cdots, x_n^{(k)})'$. In the first interpretation the method will not be stationary but will be cyclic. Namely, as is easy to see, in this case we take as the matrices determining the process $e_{11}, e_{22}, \cdots, e_{nn}$, respectively, where

$$e_{ii} = \begin{bmatrix} & & (i) & & \\ 0 & \cdots & 0 & \cdots & 0 \\ \cdot & \cdot & \cdot & \cdot & \cdot \\ 0 & \cdots & 1 & \cdots & 0 \\ \cdot & \cdot & \cdot & \cdot & \cdot \\ 0 & \cdots & 0 & \cdots & 0 \end{bmatrix} (i) . \qquad (3)$$

More precisely, $H_{n(k-1)+i} = e_{ii}$.

In the second interpretation the process will be stationary. We shall investigate this in greater detail.

The equation

$$X = BX + G$$

is written in the form

$$X = (M + N)X + G , \tag{4}$$

where

$$M = \begin{bmatrix} 0 & 0 & \cdots 0 & 0 \\ b_{21} & 0 & \cdots 0 & 0 \\ \cdot & \cdot & \cdot \cdot \cdot \cdot \cdot \cdot \cdot & \\ b_{n1} & b_{n2} & \cdots b_{n,n-1} & 0 \end{bmatrix}, \quad N = \begin{bmatrix} b_{11} & b_{12} & \cdots & b_{1n} \\ 0 & b_{22} & \cdots & b_{2n} \\ \cdot & \cdot & \cdot \cdot \cdot \cdot \cdot \cdot & \\ 0 & 0 & \cdots b_{nn} \end{bmatrix}. \tag{5}$$

In this notation the formulas

$$x_i^{(k)} = \sum_{j=1}^{i-1} b_{ij} x_j^{(k)} + \sum_{j=i}^{n} b_{ij} x_j^{(k-1)} + g_i$$

may be represented in matrix form as

$$X^{(k)} = MX^{(k)} + NX^{(k-1)} + G . \tag{6}$$

From here it follows that

$$X^{(k)} = (E - M)^{-1} N X^{(k-1)} + (E - M)^{-1} G .$$

Thus one complete cycle of a one-step cyclic process for the system (4) turns out to be equivalent to one step of the successive approximation process applied to the system

$$X = (E - M)^{-1} N X + (E - M)^{-1} G ,$$

which is equivalent to the initial system

$$X = (M + N)X + G$$

and may be obtained from it by premultiplying by the non-singular matrix $(E - M)^{-1}$.

From such a representation of the process it follows that for its convergence it is necessary and sufficient that all eigenvalues of the matrix $S = (E - M)^{-1} N$ be less than one in modulus. These eigenvalues are roots of the polynomial $|S - tE|$. Multiplying both parts of this equation by $|E - M|$ and making use of the theorem on the product of the determinants of two matrices we transform the equation to the form

$$|N - (E - M)t| = 0$$

or in detailed form to

$$\begin{vmatrix} b_{11} - t & b_{22} & \cdots & b_{1n} \\ b_{21}t & b_{12} - t & \cdots & b_{2n} \\ \cdot & \cdot & \cdot \cdot \cdot \cdot \cdot \cdot & \\ b_{n1}t & b_{n2}t & \cdots b_{nn} - t \end{vmatrix} = 0 . \tag{7}$$

Thus, for the convergence of a one-step cyclic process it is necessary and sufficient that all roots of equation (7) be less than one in modulus.

We now turn to a consideration of a sufficiency condition for the convergence of a one-step cyclic process.

Namely, let

$$\| B \|_I = \max_i \sum_{j=1}^{n} |b_{ij}| \leq \mu < 1 .$$

As we have seen in this case there holds for the successive approximation method the estimate

$$\| X^* - X^{(k)} \| \leq \mu \| X^* - X^{(k-1)} \| , \tag{8}$$

where as the vector norm we take the cubic norm, i.e. $\max_i |x_i|$.

We shall show that under this condition the one-step cyclic process converges and slightly better estimate holds for it.

In fact if $X^* = (x_1, \cdots, x_n)'$, then

$$x_i = \sum_{j=1}^{n} b_{ij} x_j + g_i \qquad (i = 1, \cdots, n) . \tag{9}$$

Subtracting equation (2) from (9) we obtain

$$x_i - x_i^{(k)} = \sum_{j=1}^{i-1} b_{ij} (x_j - x_j^{(k)}) + \sum_{j=i}^{n} b_{ij} (x_j - x_j^{(k-1)}) ,$$

$$|x_i - x_i^{(k)}| \leq \sum_{j=1}^{i-1} |b_{ij}| \, |x_j - x_j^{(k)}| + \sum_{j=i}^{n} |b_{ij}| \, |x_j - x_j^{(k-1)}| . \tag{10}$$

We write

$$\sum_{j=1}^{i-1} |b_{ij}| = \beta_i ,$$

$$\sum_{j=i}^{n} |b_{ij}| - \gamma_i .$$

Then from inequality (10) it follows that

$$|x_i - x_i^{(k)}| \leq \beta_i \| X^* \quad X^{(k)} \| + \gamma_i \| X^* \quad X^{(k-1)} \| .$$

Taking for $i$ that value $i_0$ for which $|x_i - x_i^{(k)}|$ attains a maximum, we get

$$\left\| X^* - X^{(k)} \right\| \leq \frac{\gamma_{i_0}}{1 - \beta_{i_0}} \left\| X^* - X^{(k-1)} \right\| , \tag{11}$$

since

$$\| X^* - X^{(k)} \| = \max_i |x_i - x_i^{(k)}| = |x_{i_0} - x_{i_0}^{(k)}| .$$

We write

$$\max \frac{\gamma_i}{1 - \beta_i} = \mu' .$$

Then

$$\| X^* - X^{(k)} \| \leq \mu' \| X^* - X^{(k-1)} \| . \tag{12}$$

We shall establish that

$$\mu' \leq \mu .$$

In fact,

$$\beta_i + \gamma_i = \sum_{j=1}^{n} |b_{ij}| \leq \mu$$

and

$$\beta_i + \gamma_i - \frac{\gamma_i}{1-\beta_i} = \frac{\beta_i(1-\beta_i-\gamma_i)}{1-\beta_i} \geq 0 .$$

It follows that

$$\mu \geq \max (\beta_i + \gamma_i) \geq \max \frac{\gamma_i}{1-\beta_i} = \mu' .$$

Here the equality sign is possible only if $\max_i \sum_{j=1}^{n} |b_{ij}|$ is achieved for $i = 1$ (or if $\beta_i = 0$) and a reduction of the estimate compared with estimate (8) will be best if we place the equations in order of increasing $\sum_{j=1}^{n} |b_{ij}|$, taking for the first one that equation in which this sum is least.

However, the one-step cyclic process does not always turn out to be more advantageous than the method of successive approximations. Sometimes the one-step cyclic process converges more slowly than the process of successive approximations. It is even possible that the one-step cyclic process diverges although the successive approximation method is convergent. Areas of convergence for these two processes are different and only partially overlap.

We shall carry out some example which show the difference in convergence areas for the one-step cyclic process and the process of successive approximations

*Example 1.* Let

$$B = \begin{bmatrix} 5 & -5 \\ 1 & 0.1 \end{bmatrix} .$$

Then the eigenvalues of the matrix $B$ are determined from the equation $(0.1 - t)(5 - t) + 5 = 0$ and therefore $\max |\lambda_i| > 1$. The process of successive approximations diverges.

We form the matrix

$$S = (E - M)^{-1}N = \begin{bmatrix} 5 & -5 \\ 5 & -4.9 \end{bmatrix} .$$

The eigenvalues of this matrix are determined from the equation $t^2 - 0.1t + 0.5 = 0$. It is obvious that $\max |\lambda_i| < 1$. The one-step cyclic process converges.

*Example 2.* Let

$$B = \begin{bmatrix} 2.3 & -5 \\ 1 & -2.3 \end{bmatrix} .$$

The eigenvalues of matrix $B$ are determined from the equation $-(2.3 - t) \cdot (2.3 + t) + 5 = t^2 - 0.29 = 0$; $\max |\lambda_i| < 1$. The process of successive approximations converges. It is not hard to show that in this case the one-step cyclic process diverges. In fact

$$S = (E - M)^{-1}N = \begin{bmatrix} 2.3 & -5 \\ 2.3 & -7.3 \end{bmatrix}$$

and the eigenvalues of this matrix are greater than one in modulus.

The one-step cyclic process is in theory identical with the process of successive approximations applied in an appropriate way to a given prepared system, and this fact was utilized by us to obtain convergence conditions for the processes. However, when one actually carries out the process the computational scheme does not coincide with the computational scheme for the equivalent successive approximation process; in particular, computation of the "preparing matrix" $H = (E - M)^{-1}$ must not be realized. This circumstance also compels us to separate the one-step cyclic process as an independent iterative method.

As an example we shall find the solution to system (12) Section 30, bringing it to the form

$$x_1 = 0.22x_1 + 0.02x_2 + 0.12x_3 + 0.14x_4 + 0.76 ,$$
$$x_2 = 0.02x_1 + 0.14x_2 + 0.04x_3 - 0.06x_4 + 0.08 ,$$
$$x_3 = 0.12x_1 + 0.04x_2 + 0.28x_3 + 0.08x_4 + 1.12 ,$$
$$x_4 = 0.14x_1 - 0.06x_2 + 0.08x_3 + 0.26x_4 + 0.68 .$$

The sufficiency conditions for the convergence of the one-step cyclic process are obviously satisfied.

As an initial approximation we take the vector composed of the constants. Successive approximations are given in Table III.5.

Comparing the successive approximations with the solution of the system found by the single-division method (cf. Section 30) we see that the fourteenth

**TABLE III.5  Computing the Solution of a System by the One-step Cyclic Process**

| $X^{(0)} = G$ | 0.76 | 0.08 | 1.12 | 0.68 |
|---|---|---|---|---|
| $X^{(1)}$ | 1.1584 | 0.1184 | 1.6317 | 1.1424 |
| $X^{(2)}$ | 1.3730 | 0.1208 | 1.8379 | 1.3090 |
| $X^{(3)}$ | 1.4683 | 0.1213 | 1.9204 | 1.3723 |
| $X^{(4)}$ | 1.5080 | 0.1216 | 1.9533 | 1.3969 |
| $X^{(5)}$ | 1.5242 | 0.1218 | 1.9665 | 1.4066 |
| $X^{(6)}$ | 1.5307 | 0.1219 | 1.9717 | 1.4104 |
| $X^{(7)}$ | 1.5333 | 0.1219 | 1.9738 | 1.4118 |
| $X^{(8)}$ | 1.5343 | 0.1220 | 1.9746 | 1.4125 |
| $X^{(9)}$ | 1.53469 | 0.12201 | 1.97493 | 1.41278 |
| $X^{(10)}$ | 1.53485 | 0.12201 | 1.97507 | 1.41289 |
| $X^{(11)}$ | 1.53492 | 0.12201 | 1.97512 | 1.41293 |
| $X^{(12)}$ | 1.534947 | 0.122009 | 1.975141 | 1.412945 |
| $X^{(13)}$ | 1.5349579 | 0.1220094 | 1.9751507 | 1.4129513 |
| $X^{(14)}$ | 1.5349622 | 0.1220095 | 1.9751541 | 1.4129538 |

approximation gives a solution accurate to three units in the sixth decimal place. The one-step cyclic process in the example considered converges more rapidly than the process of successive approximations (cf. Tables III.1, III.2 and III.3).

**TABLE III.6  Computing the Solution of a System by the One-step Cyclic Process**

| $X^{(0)} = G$ | 0.22900763 | 0.38167939 | 0.53435115 | 0.68702290 |
|---|---|---|---|---|
| $X^{(1)}$ | $-0.40557078$ | 0.24074649 | 0.65379631 | 0.86327494 |
| $X^{(20)}$ | $-1.2560487$ | 0.0439969 | 1.0383844 | 1.4810254 |
| $X^{(25)}$ | $-1.2574501$ | 0.0435866 | 1.0390124 | 1.4821242 |
| $X^{(40)}$ | $-1.2577909$ | 0.0434880 | 1.0391650 | 1.4823907 |
| $X^{(49)}$ | $-1.2577935$ | 0.0434873 | 1.0391661 | 1.4823927 |

The same conclusion will also be true when applied to system (3) Section 31; this is apparent by comparing Table III.4 and Table III.6 in which a solution to the system under consideration is given by the one-step cyclic process.

## 33.  NEKRASOV'S METHOD

Just as in the method of successive approximations it is possible to prepare a given system $AX = F$ in a form which is convenient for applying the one-step cyclic process using different means. The most widely used one is a modification of the one-step cyclic process which parallels the simple iteration method.

Necessary and sufficient conditions for the convergence of this modified one-step cyclic process were first announced by P. A. Nekrasov[†] and hence this is known as Nekrasov's method.

The system $AX = F$ is written in the form

$$a_{ii}x_i = -\sum_{j=1}^{i-1} a_{ij}x_j - \sum_{j=i+1}^{n} a_{ij}x_j + f_i$$

or equivalently in the form

$$x_i = -\sum_{j=1}^{i-1} \frac{a_{ij}}{a_{ii}} x_j - \sum_{j=i+1}^{n} \frac{a_{ij}}{a_{ii}} x_j + \frac{f_i}{a_{ii}}, \qquad (1)$$

and the successive approximations are determined by the formulas

$$x_i^{(k)} = -\sum_{j=1}^{i-1} \frac{a_{ij}}{a_{ii}} x_j^{(k)} - \sum_{j=i+1}^{n} \frac{a_{ij}}{a_{ii}} x_j^{(k-1)} + \frac{f_i}{a_{ii}}. \qquad (2)$$

Necessary and sufficient conditions for the convergence of this modification of the process are easily obtained from the necessary and sufficient conditions for convergence introduced above in the general case.

In fact the chosen preparation of the system $AX = F$ for the form

---

[†] P. A. Nekrasov (1).

$X = BX + G$ is based on a preliminary multiplication of the system by the diagonal matrix $D^{-1} = [a_{11}, \cdots, a_{nn}]^{-1}$. Consequemtly $B = E - D^{-1}A$. We write

$$L = \begin{bmatrix} 0 & 0 & \cdots & 0 \\ a_{21} & 0 & \cdots & 0 \\ \cdot & \cdot & \cdot & \cdot \\ a_{n1} & a_{n2} & \cdots & 0 \end{bmatrix},$$

$$R = \begin{bmatrix} 0 & a_{12} & \cdots & a_{1n} \\ 0 & 0 & \cdots & a_{2n} \\ \cdot & \cdot & \cdot & \cdot \\ 0 & 0 & \cdots & 0 \end{bmatrix}. \tag{3}$$

Then

$$A = L + D + R$$

so that in the previous notation

$$M = -D^{-1}L, \qquad N = -D^{-1}R,$$
$$S = (E - M)^{-1}N = -(E + D^{-1}L)^{-1}D^{-1}R = -(D + L)^{-1}R.$$

The characteristic polynomial

$$|-(D + L)^{-1}R - tE|$$

of the matrix

$$-(D + L)^{-1}R$$

after multiplication by

$$|-(D + L)|$$

takes the form

$$|R + (D + L)t|.$$

Consequently for Nekrasov's method to converge it is necessary and sufficient that all the roots of the equation (Nekrasov's equation)

$$\begin{vmatrix} a_{11}t & a_{12} & \cdots & a_{1n} \\ a_{21}t & a_{22}t & \cdots & a_{2n} \\ \cdot & \cdot & \cdot & \cdot \\ a_{n1}t & a_{n2}t & \cdots & a_{nn}t \end{vmatrix} = 0 \tag{4}$$

be less than one in modulus.

A great number of sufficiency indicators for the convergence of the method are given in the work of P. A. Nekrasov and Mehmke.[†]

In particular the indicators I, II, 1, and 2 in Section 30 will be sufficiency indicators.

If the coefficient matrix $A$ of the system $AX = F$ is symmetric or Hermitian there exists one other important condition for the convergence of Nekrasov's method. Namely, *if the matrix $A$ is positive definite, then Nekrasov's method*

---

† R. Mehmke and P. A. Nekrasov (1).

*for the system* $AX = F$ *converges. This condition with the assumption that the diagonal elements of matrix A are positive also turns out to be necessary.*[†]

For a proof we let

$$A = R^* + D + R ,\qquad (5)$$

where $D$ is the diagonal matrix containing the diagonal elements of the matrix $A$, $R$ is the triangular matrix formed by the elements of matrix $A$ lying above the principal diagonal, and $R^*$ is its conjugate. As we saw above, a necessary and sufficient condition for the convergence of Nekrasov's method is the requirement that all eigenvalues of the matrix $S = -(D + R^*)^{-1}R$ be less than one in modulus.

We shall give an estimate for the moduli of the eigenvalues for this matrix, from which it will follow directly that they are all less than one.

We denote by $\lambda_0$ the least eigenvalue of matrix $A$. Since matrix $A$ is positive definite, $\lambda_0 > 0$. Moreover we write $\Lambda_0 = \|R\| = \|R^*\|$. Let $U$ be an eigenvector of length one for the matrix $(D + R^*)^{-1}R$ belonging to a certain eigenvalue $\lambda$. Then

$$(D + R^*)^{-1}RU = \lambda U$$

or

$$RU = (D + R^*)\lambda U = \lambda DU + \lambda R^*U .$$

Furthermore,

$$(RU, U) = \lambda(DU, U) + \lambda(R^*U, U) = \lambda(AU, U) - \lambda(RU, U) .$$

We write

$$(AU, U) = p ,$$
$$(DU, U) = d ,$$
$$(RU, U) = a + ib ,$$
$$(R^*U, U) = (U, RU) = a - ib .$$

Then

$$\lambda = \frac{a + ib}{p - a - ib} ,$$

from which

$$|\lambda|^2 = \frac{a^2 + b^2}{(p - a)^2 + b^2} .$$

But

$$p = (AU, U) = (DU, U) + (RU, U) + (R^*U, U) = d + 2a ,$$

and therefore

$$(p - a)^2 = p^2 - 2ap + a^2 = p(p - 2a) + a^2 = pd + a^2 .$$

---

[†] Ostrowski (8).

Thus

$$|\lambda|^2 = \frac{a^2 + b^2}{pd + a^2 + b^2} \leq \frac{a^2 + b^2}{\lambda_0 d_0 + a^2 + b^2} \, ,$$

where $d_0$ is the least eigenvalue of matrix $D$, since $d \geq d_0$, $p \geq \lambda_0$. It is obvious that $d_0 = \min\limits_i a_{ii}$. Moreover,

$$a^2 + b^2 = |a + ib|^2 = |(RU, U)|^2 \leq |RU|^2 \, |U|^2 \leq \|R\|^2 = \Lambda_0^2 \, .$$

Consequently

$$|\lambda| \leq \frac{\Lambda_0}{\sqrt{\lambda_0 d_0 + \Lambda_0^2}} \tag{6}$$

We observe that $d_0 \geq \lambda_0$, since

$$d_0 = \min_i a_{ii} = \min \, (Ae_i \, , \, e_i) \geq \min_{|X|=1} \, (AX, \, X) = \lambda_0 \, .$$

Therefore the correct estimate[†]

$$|\lambda| \leq \frac{\Lambda_0}{\sqrt{\lambda_0^2 + \Lambda_0^2}} \tag{7}$$

is slightly more crude than estimate (6).

From these estimates it follows that all the eigenvalues of matrix $(D + R^*)^{-1}R$ are less than one in modulus. This proves the convergence of Nekrasov's process.

We shall prove the necessity of the stated conditions.

Let $A$ be a symmetric or Hermitian matrix with positive diagonal elements, and let

$$X = (x_1^{(k)}, \, \cdots, \, x_i^{(k)}, \, x_{i+1}^{(k-1)}, \, \cdots, \, x_n^{(k-1)})$$

and

$$X' = (x_1^{(k)}, \, \cdots, \, x_{i+1}^{(k)}, \, x_{i+2}^{(k-1)}, \, \cdots, \, x_n^{(k-1)})$$

be two adjacent successive approximations at the $k$th cycle. Then

$$X' = X + e_i \, D^{-1}(F - AX) \, .$$

Let $X^*$ be the solution to the system and let $Y = X^* - X$, $Y' = X^* - X'$ be the corresponding error vectors. Then

$$Y' = Y - e_i \, D^{-1}AY = Y - \frac{1}{a_{ii}} \, r_i e_i \, ,$$

where $r_i$ is the $i$th component of the vector $r = F - AX = AY$, $e_i$ is the vector whose $i$th component is equal to one with the remaining components zero.

We compute the value of the functional $f(X') = (AY', Y')$ (which coincides with the error function if $A$ is positive definite). We have

$$f(X') = (AY', \, Y') = (AY, \, Y) - 2\frac{r_i}{a_{ii}}(AY, \, e_i) + \frac{r_i^2}{a_{ii}^2} \, (Ae_i, \, e_i) \, .$$

---

[†] Reich (1), Schemeidler (1), Ostrowski (8).

But

$$(AY, e_i) = r_i , \qquad (Ae_i, e_i) = a_{ii} ,$$

and therefore

$$f(X') = (AY', Y') = (AY, Y) - \frac{r_i^2}{a_{ii}} \le (AY, Y) . \tag{8}$$

If the matrix $A$ is not positive definite and $|A| \ne 0$, then it is possible to find an initial vector $X^{(0)}$ so that $(AY^{(0)}, Y^{(0)}) < 0$. Then on the basis of (8) it will hold during the whole process that $(AY^{(k)}, Y^{(k)}) < (AY^{(0)}, Y^{(0)}) < 0$, and consequently the limit relationship $Y^{(k)} \to 0$ is impossible. Therefore the process will diverge.

The criterion established shows that if the matrix of a system is symmetric and has positive diagonal elements, then the convergence area for Nekrasov's method is broader than that of the simple iteration method.

For Nekrasov's method to converge it is necessary and sufficient in this case that matrix $A$ be positive definite; for convergence of the simple iteration method the necessary and sufficient condition is that the matrices $A$ and $2D - A$ be positive definite.

In Table III.7 system (9) Section 23 is solved by Nekrasov's method.

**TABLE III.7   Solution of a System by Nekrasov's Method**

| $X^{(0)}$ | 0 | 0 | 0 | 0 |
|---|---|---|---|---|
| $X^{(1)}$ | 0.3 | 0.374 | 0.41832 | 0.4454096 |
| $X^{(20)}$ | $-1.2577875$ | 0.0434903 | 1.0391641 | 1.4823879 |
| $X^{(22)}$ | $-1.2577922$ | 0.0434881 | 1.0391657 | 1.4823916 |
| $X^{(25)}$ | $-1.2577935$ | 0.0434874 | 1.0391662 | 1.4823927 |

## 34.  THE METHOD OF COMPLETE RELAXATION

Beginning with this paragraph and up through Section 38 we shall consider primarily systems with positive definite matrices, but in each instance describing also a case for which this demand is not satisfied.

Let $X^*$ be the exact solution of the system $AX = F$, where $A$ is positive definite, let $X$ be a certain vector, and let $f(X)$ be the error function. We pose the problem: how may we change the $i$th component of the vector $X$ so that for the perturbed vector $X'$ the value of the error function will be a minimum? Let

$$X' = X + \alpha e_i .$$

Then

$$Y' = Y - \alpha e_i ,$$

and

$$f(X') = (AY', Y') = (A(Y - \alpha e_i), Y - \alpha e_i)$$
$$= (AY, Y) - 2\alpha(AY, e_i) + \alpha^2(Ae_i, e_i) = (AY, Y) + \alpha^2 a_{ii} - 2\alpha r_i$$
$$= f(X) + \frac{1}{a_{ii}}(a_{ii}\alpha - r_i)^2 - \frac{r_i^2}{a_{ii}} , \tag{1}$$

where $r_i$ is the $i$th component of the residual vector for the approximation $X$.

It is clear that $f(X')$ will have a minimal value for

$$\alpha = \frac{r_i}{a_{ii}} ,$$

and this minimal value will be equal to

$$f(X) - \frac{r_i^2}{a_{ii}} .$$

We now compute the $i$th component of the residual for the approximation $X'$. It is equal to

$$(F - AX', e_i) = (F - AX - \alpha Ae_i, e_i) = (F - AX, e_i) - \alpha(Ae_i, e_i)$$
$$= r_i - \alpha a_{ii} = 0 ,$$

that is, approximation $X'$ satisfies the $i$th equation of the system $AX = F$. In other words the $i$th component of the approximation $X'$ may be computed from the $i$th equation of the system $AX = F$ in which the components of the vector $X$ have replaced the rest of the unknowns. One step in any cycle of Nekrasov's process proceeds in just this way. Similarly Nekrasov's method may be given the following interpretation: the error function is minimized at each step by changing one component of the preceding approximation; the numbers of these components pass from 1 to $n$ cyclically.

If, using the separate steps of Nekrasov's process, we reject the cyclic principal in chosing the unknowns to be changed, then we arrive at a more general group of methods known as methods of complete relaxation.

Under such a construction the choice of the sequential order in which the components are to be changed (leading indices) is quite arbitrary. Thus, for example, solving a system of ten equations with ten unknowns, enumerated by numbers from 0 to 9, we may take as the "direction" for the process the decimal notation of the number $e = 2.71828182845904523536 \cdots$, that is, we may change the second component at the first step, the seventh at the second step, the first at the third step, and so forth. It is clear that actually to carry out the relaxation process demands some plan for directing the process, e.g. a principle for choosing the sequence of number for the components being changed. More will be said below about certain principles for directing a relaxation process.

Of course not every process of complete relaxation converges to the solution. Thus, for example, if the chosen sequence of numbers for the varying components does not contain at least one number, then all corrections $X^{(k+1)} - X^{(k)}$ will be contained in an $(n - 1)$-dimensional subspace; if $X^* - X^{(0)}$ is not contained in this subspace, then the process may not converge to $X^*$.

A sufficient condition for the convergence of the process to the solution will be given in Section 37.

## 35. INCOMPLETE RELAXATION

Instead of minimizing the error function at each separate step of the process we may concern ourselves only with "shrinking" the function. Processes built on this principle are called *incomplete relaxation processes*.

We shall explain how to change one component of an approximation so that the error function decreases.

Let $X' = X + \alpha e_i$. Then

$$f(X') - f(X) = \frac{1}{a_{ii}} (a_{ii}\alpha - r_i)^2 - \frac{r_i^2}{a_{ii}} .$$

For the value of $f(X') - f(X)$ to be negative it is necessary and sufficient that

$$(a_{ii}\alpha - r_i)^2 < r_i^2 ,$$

or equivalently

$$|a_{ii}\alpha - r_i| < |r_i| ,$$

and consequently

$$\alpha = q \frac{r_i}{a_{ii}} , \qquad (1)$$

for $0 < q < 2$.

For $q = 1$ the complete minimization of the error function holds (or complete relaxation). Incomplete relaxation is called *under-relaxation* if $0 < q < 1$ and *over-relaxation* if $1 < q < 2$.

With incomplete relaxation the error function changes according to the formula

$$f(X') = f(X) - \frac{q(2-q)}{a_{ii}} r_i^2 . \qquad (2)$$

At each individual step of the process the method of complete relaxation is most advantageous of all, since it guarantees a maximum decrease in the error function for one step. However in carrying through a large number of steps it may turn out that incomplete relaxation provides a better result.

The number $q$ may change from step to step under incomplete relaxation. If the process of incomplete relaxation is taken as cyclic for constant or cyclically changing $q$, then the process may be considered as a special case of the general one-step cyclic process applied to a system prepared in the form

$$X = (E - QD^{-1}A)X + QD^{-1}F ,$$

where $D = [a_{11}, \cdots, a_{nn}]$, $Q = [q_1, \cdots, q_n]$ ($q_1, \cdots, q_n$ are values of the relaxation factors).

In fact formulas for computing components of the result of the $k$th cycle will be

$$x_i^{(k)} = x_i^{(k-1)} + q_i \frac{f_i - (a_{i1}x_1^{(k)} + \cdots + a_{ii-1}x_{i-1}^{(k)} + a_{ii}x_i^{(k-1)} + \cdots + a_{in}x_n^{(k-1)})}{a_{ii}}. \quad (3)$$

We observe that the last formulas determine the iterative process for systems without positive definite matrices as well. However, in this case it is of course impossible to talk about the relaxation properties of the process.

It is easy to find a necessary and sufficient condition for the convergence of the process. Using the notation of Sections 32 and 33, we have

$$M = -QD^{-1}L,$$
$$N = E - QD^{-1}(D + R)$$

and

$$S = (E - M)^{-1}N = (E + QD^{-1}L)^{-1}(E - Q - QD^{-1}R)$$
$$= (D + QL)^{-1}(D - DQ - QR),$$

so that a necessary and sufficient condition for the convergence of process (3) will be the requirement that all eigenvalues of the matrix $(D + QL)^{-1}$ $\cdot(D - DQ - QR)$ be less than one in modulus. The eigenvalues of this matrix are obviously roots of the equation

$$\begin{vmatrix} (t + q_1 - 1)a_{11} & q_1 a_{12} & \cdots & q_1 a_{1n} \\ t q_2 a_{21} & (t + q_2 - 1)a_{22} & \cdots & q_2 a_{2n} \\ \cdots \cdots \cdots \cdots \cdots \cdots \cdots \cdots \cdots \cdots \cdots \\ t q_n a_{n1} & t q_n a_{n2} & \cdots & (t + q_n - 1)a_{nn} \end{vmatrix} = 0 \quad (4)$$

The rate of convergence of the method will depend on the value of the greatest modulus of the roots of this equation.

The convergence criterion obtained for process (3) is mainly of interest in a case where the matrices of a system are not positive definite, since in the case of a positive definite matrix the cyclic process for incomplete relaxation always converges, as will be shown in Section 37. However, the criterion is interesting even in this case, since it provides a means for investigating the rate of convergence of the process.

We shall consider process (3) in more detail for $q_1 = q_2 = \cdots = q_n = q$.

We shall show that for a small deviation of $q$ from unity it is almost always possible to achieve a lowering of the greatest modulus among the roots of equation (4), and consequently to obtain a process with more rapid convergence than Nekrasov's process.

Let $q = 1 - \varepsilon$, where $\varepsilon$ is a small real number. Then equation (4) is equivalent to the equation

$$\begin{vmatrix} \dfrac{t - \varepsilon}{1 - \varepsilon} a_{11} & a_{12} & \cdots & a_{1n} \\ t a_{21} & \dfrac{t - \varepsilon}{1 - \varepsilon} a_{22} & \cdots & a_{2n} \\ \cdots \cdots \cdots \cdots \cdots \cdots \cdots \cdots \cdots \cdots \\ t a_{n1} & t a_{n2} & \cdots & \dfrac{t - \varepsilon}{1 - \varepsilon} a_{nn} \end{vmatrix} = 0. \quad (5)$$

We let $t = \lambda_0 + k\varepsilon$, where $\lambda_0$ is the root largest in modulus for Nekrasov's equation $|tL + tD + R| = 0$. Then with accuracy up to $\varepsilon^2$ we have

$$\frac{t - \varepsilon}{1 - \varepsilon} = \lambda_0 + (k - 1 + \lambda_0)\varepsilon ,$$

and equation (5) (within small second order variations) passes to the equation

$$| T_0 + \varepsilon(k - 1 + \lambda_0)D + k\varepsilon L | = 0 . \tag{6}$$

Here $T_0 = \lambda_0 L + \lambda_0 D + R$. It is easy to see that with accuracy up to small numbers of order 2 the following formula is valid:

$$| A + \varepsilon B | = | A | + \varepsilon Sp\, B\tilde{A} ,$$

where $\tilde{A}$ is the matrix adjoint to matrix $A$. Consequently equation (6) takes the form

$$| T_0 | + \varepsilon[(k - 1 + \lambda_0)Sp\, D\tilde{T}_0 + kSp\, L\tilde{T}_0] = 0(\varepsilon^2) .$$

Remembering that $| T_0 | = 0$ we obtain, with accuracy up to small first-order numbers, that

$$k = \frac{(1 - \lambda_0)Sp\, D\tilde{T}_0}{Sp\, D\tilde{T}_0 + Sp\, L\tilde{T}_0} . \tag{7}$$

Thus the number $\lambda' = \lambda_0 + k\varepsilon$, where $k$ is defined by formula (7), is an approximate value for the root of equation (5) which is greatest in modulus. We compare the moduli of $\lambda_0$ and $\lambda'$. It is clear that

$$| \lambda' |^2 = | \lambda_0 |^2 + 2\varepsilon\, Re(k\bar{\lambda}_0) + \varepsilon^2 | k |^2 .$$

Therefore for small enough $\varepsilon$ we can achieve that $| \lambda' |^2$ is less than $| \lambda_0 |^2$ only if $Re(k\bar{\lambda}_0) \neq 0$. If $Re(k\bar{\lambda}_0) < 0$, one should take $\varepsilon > 0$, that is, one should have recourse to under-relaxation: if $Re(k\bar{\lambda}_0) > 0$, then it is necessary to take $\varepsilon < 0$, i.e. to utilize over-relaxation.

Formula (7) is equivalent to the following formula

$$k = \frac{(1 - \lambda_0)(DX_0 , Y_0)}{((D + L)X_0 , Y_0)} , \tag{8}$$

where $X_0$ is a non-zero vector determined by the equation $T_0 X_0 = 0$ and $Y_0$ is a non-zero vector determined by the equation $T_0^* Y_0 = 0$. The vector $X_0$ and $Y_0$ are determined simply (accurate up to scalar multipliers) if we assume that $\lambda_0$ is a simple root of the equation $|tL + tD + R| = 0$. But, as is possible to show, only in this case does formula (7) make sense.

The question of choice of the factor $q$ which gives the most rapid convergence to the process is not solved in the general case.

For matrices of the second order, exhaustive research has been carried out by A. M. Ostrowski (7). Let

$$A = \begin{bmatrix} a_{11} & a_{12} \\ a_{21} & a_{22} \end{bmatrix} .$$

A non-zero root of the equation

$$\begin{vmatrix} a_{11}t & a_{12} \\ a_{21}t & a_{22}t \end{vmatrix} = 0$$

is obviously $\lambda_0 = u = a_{12}a_{21}/a_{11}a_{22}$, so that for Nekrasov's method to converge it is necessary and sufficient that $|u| < 1$.

For a positive definite matrix $A$ there holds the inequality

$$0 < u < 1 .$$

Applying formula (7) we get

$$k = 2(1 - u) ,$$
$$k\bar{\lambda}_0 = 2u(1 - u) .$$

Therefore for $0 < u < 1$ (particularly for positive definite matrices) over-relaxation (at least for small $\varepsilon$) gives more rapid convergence than complete relaxation but for $-1 < u < 0$ more rapid convergence results with under-relaxation.

In the same work of A. M. Ostrowski an optimal value for the factor $q$ is given. Namely,

$$q_0 = \frac{2}{1 + \sqrt{1 - u}} ,$$

while the corresponding value for the modulus of $\lambda'$ is

$$|\lambda'| = \frac{|u|}{1 + \sqrt{1 - u}} .$$

We observe that changing the speed of convergence for incomplete relaxation may be significant enough. Thus for

$$A = \begin{bmatrix} 1 & 0.6 \\ 0.6 & 1 \end{bmatrix}$$

we shall have $\lambda_0 = 9/25 = 0.36$ for complete relaxation. Optimal incomplete relaxation will occur for $q = 10/9$, while the corresponding value for $\lambda'$ is equal to $1/9 = 0.111\cdots$

For matrices which are not positive definite Ostrowski has shown in the same work that for $u > 1$ process (3) diverges for all values of $q$ in the interval (0,2). If $u < -1$, then process (3) will converge for $0 < q < 2/1 + \sqrt{|u|}$, so that the convergence area for process (3) with $q < 1$ will be broader than the convergence area for Nekrasov's process.

For positive definite matrices of the third order we may show by means of fairly unwieldy calculations[†] that

$$Re(k\bar{\lambda}_0) > 0$$

so that over-relaxation (at least for small $\varepsilon$) leads to more rapid convergence than does complete relaxation.

In Tables III.8, III.9 and III.10 the results of computing the solutions of system (9) Section 23 by Nekrasov's method are presented with incomplete

---

† D. K. Faddeev (3).

relaxation for $q = 0.8$, $q = 1.1$ and $q = 1.2$.

Comparing these tables with Table III.7 shows that in the given example over-relaxation gives a better result than complete relaxation. We observe that a further increase in the factor of relaxation leads to a worse result. Thus for $q = 1.3$ we have

$$X^{(19)} = (- 1.2577970, 0.0434871, 1.0391679, 1.4823915)'$$

and the result, analogous to $X^{(18)}$ Table III.9, is obtained only for $k = 26$.

**TABLE III.8  Solving a System by Nekrasov's Method with Under-relaxation** ($q = 0.8$)

| $X^{(0)}$ | 0 | 0 | 0 | 0 |
|-----------|---|---|---|---|
| $X^{(1)}$ | 0.24 | 0.31936 | 0.3745638 | 0.4149420 |
| $X^{(20)}$ | −1.2560832 | 0.0440084 | 1.0384154 | 1.4810565 |
| $X^{(25)}$ | −1.2575067 | 0.0435748 | 1.0390402 | 1.4821686 |
| $X^{(40)}$ | −1.2577924 | 0.0434877 | 1.0391657 | 1.4823918 |
| $X^{(45)}$ | −1.2577935 | 0.0434874 | 1.0391661 | 1.4823927 |

**TABLE III.9  Solving a System by Nekrasov's Method with Over-relaxation** ($q = 1.1$)

| $X^{(0)}$ | 0 | 0 | 0 | 0 |
|-----------|---|---|---|---|
| $X^{(1)}$ | 0.33 | 0.39754 | 0.4340459 | 0 4529715 |
| $X^{(16)}$ | −1.2577939 | 0.0434866 | 1.0391661 | 1.4823932 |
| $X^{(18)}$ | −1.2577935 | 0.0434873 | 1.0391661 | 1.4823928 |

**TABLE III.10  Solving a System by Nekrasov's Method with Over-relaxation** ($q = 1.2$)

| $X^{(0)}$ | 0 | 0 | 0 | 0 |
|-----------|---|---|---|---|
| $X^{(1)}$ | 0.36 | 0.41856 | 0.4459930 | 0.4561382 |
| $X^{(16)}$ | −1.2577988 | 0.0434852 | 1.0391678 | 1.4823963 |
| $X^{(18)}$ | −1.2577939 | 0.0434869 | 1.0391662 | 1.4823930 |
| $X^{(19)}$ | −1.2577936 | 0.0434873 | 1.0391661 | 1.4823928 |

## 36.  AN INVESTIGATION OF ITERATIVE METHODS FOR SYSTEMS WITH QUASI-TRIDIAGONAL MATRICES

In the present paragraph we shall investigate the rate of convergence for the simple iteration method (Jacobi's process) and for a cyclic relaxation method with a constant relaxation factor for systems with positive definite quasi-tridiagonal matrices of the form

$$A = \begin{bmatrix} D_1 & W_1 & & & & & \\ W_1' & D_2 & W_2 & & & & \\ & W_2' & D_3 & W_3 & & & \\ & & \cdot & \cdot & \cdot & & \\ & & & \cdot & \cdot & \cdot & \\ & & & & \cdot & & \\ & & & & \cdot & D_{m-1} & W_{m-1} \\ & & & & & W_{m-1}' & D_m \end{bmatrix}, \tag{1}$$

where $D_1, \cdots, D_m$ are diagonal matrices (possibly of different orders), and $W_1, W_2, \cdots, W_{m-1}$ are certain rectangular matrices. It is clear that all diagonal elements of matrix $A$ are positive.

The results which we shall expand here belong to Young (2).

Any matrix $A$ of the type indicated possesses "property $A$" (Young), meaning that numbers of the rows and columns may be broken down into two non-intersecting sets $P$ and $Q$ so that if $a_{ij} \neq 0$ and $i \neq j$, then $i \in P$ and $j \in Q$ or $i \in Q$ and $j \in P$. That is, such sets will be aggregates of row numbers corresponding to the odd cells $D_1, D_3, \cdots$ on the one hand, and to $D_2, D_4, \cdots$ on the other.

The converse is also true, that any symmetric matrix which possesses "property $A$" may be transformed to quasi-tridiagonal form by simultaneously changing the numbering of rows and columns; it is sufficient, for example, to give the first numbers to set $P$ and the following ones to set $Q$.

After such renumbering the matrix takes the form

$$\begin{bmatrix} \tilde{D}_1 & W \\ W' & \tilde{D}_2 \end{bmatrix}, \tag{2}$$

where $\tilde{D}_1$ and $\tilde{D}_2$ are diagonal matrices and $W$ is a certain rectangular matrix. In particular, a general quasi-tridiagonal matrix may be transformed to form (2), since

$$\tilde{D}_1 = \begin{bmatrix} D_1 & & & & \\ & D_3 & & & \\ & & \cdot & & \\ & & & \cdot & \\ & & & & D_{2k-1} \end{bmatrix}, \quad \tilde{D}_2 = \begin{bmatrix} D_2 & & & & \\ & D_4 & & & \\ & & \cdot & & \\ & & & \cdot & \\ & & & & D_{2k} \end{bmatrix}$$

$$W = \begin{bmatrix} W_1 & & & \\ W_2' & W_3 & & \\ & \cdot & \cdot & \\ & & \cdot & \cdot \\ & & W_{2k-2}' & W_{2k-1} \end{bmatrix} \quad (\text{if } m = 2k)$$

and

$$\tilde{D}_1 = \begin{bmatrix} D_1 & & & & \\ & D_3 & & & \\ & & \cdot & & \\ & & & \cdot & \\ & & & & D_{2k+1} \end{bmatrix}, \quad \tilde{D}_2 = \begin{bmatrix} D_2 & & & & \\ & D_4 & & & \\ & & \cdot & & \\ & & & \cdot & \\ & & & & D_{2k} \end{bmatrix}$$

$$W = \begin{bmatrix} W_1 & & & & \\ W_2' & W_3 & & & \\ & \cdot & \cdot & & \\ & & \cdot & \cdot & \\ & & & \cdot & \\ & & W_{2k-2}' & W_{2k-1} & \\ & & & W_{2k}' & \end{bmatrix} \qquad (\text{if } m = 2k+1).$$

Systems with positive definite matrices possessing "property $A$" arise, for example, in solving certain equations in partial derivatives of the elliptical type by difference methods. Setting up an enumeration in which a matrix becomes quasi-tridiagonal is connected with some natural choice for numbering the nodes.

We observe first of all that in investigating convergence for the simple iteration method, and for cyclic relaxation methods with a constant relaxation factor having matrices with positive diagonal elements, we may consider without loss of generality that all the diagonal elements are equal to one.

In fact if $D$ is a diagonal matrix containing the diagonal elements of matrix $A$ and if $\tilde{A} = D^{-\frac{1}{2}}AD^{-\frac{1}{2}}$, then $\tilde{A}$ has unit diagonal elements. The rate of convergence for the simple iteration method depends on the largest modulus of the eigenvalues for the matrix $B = E - D^{-1}A$. The rate of convergence for the cyclic relaxation method with constant relaxation factor $q$ is determined by the largest modulus of eigenvalues for the matrix

$$S_q = (D + qL)^{-1}(D - qD - qR)$$

where $D$, $L$ and $R$ from the diagonal, sub-diagonal, and above-diagonal parts of matrix $A$. For the matrix $\tilde{A} = D^{-\frac{1}{2}}AD^{-\frac{1}{2}}$ the diagonal, sub-diagonal, and above-diagonal parts will be correspondingly $E$, $D^{-\frac{1}{2}}LD^{-\frac{1}{2}}$, and $D^{-\frac{1}{2}}RD^{-\frac{1}{2}}$. Therefore

$$\tilde{B} = E - E^{-1}\tilde{A} = E - D^{-\frac{1}{2}}AD^{-\frac{1}{2}} = D^{\frac{1}{2}}(E - D^{-1}A)D^{-\frac{1}{2}} = D^{\frac{1}{2}}BD^{-\frac{1}{2}},$$
$$\tilde{S}_q = (E + qD^{-\frac{1}{2}}LD^{-\frac{1}{2}})^{-1}(E - qE - qD^{-\frac{1}{2}}RD^{-\frac{1}{2}})$$
$$= D^{\frac{1}{2}}(D + qL)^{-1}D^{\frac{1}{2}}D^{-\frac{1}{2}}(D - qD - qR)D^{-\frac{1}{2}} = D^{\frac{1}{2}}S_qD^{-\frac{1}{2}}.$$

Thus the matrices $\tilde{B}$ and $\tilde{S}_q$ constructed from matrix $\tilde{A}$ are similar to matrices $B$ and $S_q$ and therefore their eigenvalues coincide, respectively.

This observation allows us, investigating the indicated methods of positive quasi-tridiagonal matrices, to limit ourselves to considering matrices of the form

$$A = \begin{bmatrix} E_1 & W_1 & & & \\ W_1' & E_2 & \cdot & & \\ & \cdot & \cdot & \cdot & \\ & & \cdot & \cdot & \cdot \\ & & & \cdot & W_{m-1} \\ & & & W_{m-1}' & E_m \end{bmatrix},$$

where $E_1$, $E_2$, $\cdots$, $E_m$ are unit matrices.

If all the above-diagonal cells of a quasi-tridiagonal matrix are multiplied by some number $\alpha$ and all sub-diagonal ones by the inverse number $\alpha^{-1}$, than the determinant of the matrix does not change.

In fact it is clear that

$$\begin{vmatrix} D_1 & \alpha W_1 & & & \\ \alpha^{-1}W_1' & D_2 & \cdot & & \\ & \cdot & \cdot & \cdot & \\ & & \cdot & \cdot & \alpha W_{m-1} \\ & & & \alpha^{-1}W_{m-1}' & D_m \end{vmatrix} = \begin{vmatrix} E_1 & & & & \\ & \alpha^{-1}E_2 & & & \\ & & \cdot & & \\ & & & \cdot & \\ & & & & \alpha^{-m+1}E_m \end{vmatrix}$$

$$\times \begin{vmatrix} D_1 & W_1 & & & \\ W_1' & D_2 & \cdot & & \\ & \cdot & \cdot & \cdot & \\ & & \cdot & \cdot & W_{m-1} \\ & & & W_{m-1}' & D_m \end{vmatrix} \times \begin{vmatrix} E_1 & & & & \\ & \alpha E_2 & & & \\ & & \cdot & & \\ & & & \cdot & \\ & & & & \alpha^{m-1}E_m \end{vmatrix},$$

from which the validity of what was said follows directly.

This property of quasi-tridiagonal matrices permits us to associate the characteristic polynomial of matrix $S_q$ with the characteristic polynomial of matrix $B$. (We consider the diagonal elements of matrix $A$ as units). In fact,

$$|tE - S_q| = |t(E + qL) - (E - qE - qR)|$$

$$= \begin{vmatrix} (t+q-1)E_1 & qW_1 & & & \\ tqW_1' & (t+q-1)E_2 & \cdot & & \\ & \cdot & \cdot & \cdot & \\ & & \cdot & \cdot & qW_{m-1} \\ & & tqW_{m-1}' & (t+q-1)E_m \end{vmatrix}$$

$$= \begin{vmatrix} (t+q-1)E_1 & q\sqrt{t}\,W_1 & & & \\ q\sqrt{t}\,W_1' & (t+q-1)E_2 & \cdot & & \\ & \cdot & \cdot & \cdot & \\ & & \cdot & \cdot & q\sqrt{t}\,W_{m-1} \\ & & q\sqrt{t}\,W_{m-1}' & (t+q-1)E_m \end{vmatrix}$$

$$= q^n t^{n/2} \begin{vmatrix} \dfrac{t+q-1}{q\sqrt{t}}E_1 & W_1 & & & \\ W_1' & \dfrac{t+q-1}{q\sqrt{t}}E_2 & \cdot & & \\ & \cdot & \cdot & \cdot & \\ & & \cdot & \cdot & W_{m-1} \\ & & W_{m-1}' & \dfrac{t+q-1}{q\sqrt{t}}E_m \end{vmatrix} \qquad (3)$$

$$= q^n t^{n/2} F\left(\frac{t+q-1}{q\sqrt{t}}\right).$$

where $F(t)$ is the characteristic polynomial of the matrix

$$B = E - A = \begin{vmatrix} 0 & -W_1 & & & \\ -W_1' & 0 & \cdot & & \\ & \cdot & \cdot & \cdot & \\ & & \cdot & \cdot & -W_{m-1} \\ & & & -W_{m-1}' & 0 \end{vmatrix} .$$

The polynomial $F(t)$ possesses the property $F(-t) = (-1)^n F(t)$. In fact,

$$F(-t) = \begin{vmatrix} -tE_1 & W_1 & & & \\ W_1' & -tE_2 & \cdot & & \\ & \cdot & \cdot & \cdot & \\ & & \cdot & \cdot & W_{m-1} \\ & & & W_{m-1}' & -tE_m \end{vmatrix}$$

$$= (-1)^n \begin{vmatrix} tE_1 & -W_1 & & & \\ -W_1' & tE_2 & \cdot & & \\ & \cdot & \cdot & \cdot & \\ & & \cdot & \cdot & -W_{m-1} \\ & & & -W_{m-1}' & tE_m \end{vmatrix} = (-1)^n F(t) .$$

Consequently

$$F(t) = t^{n-2k}(t^2 - c_1)(t^2 - c_2) \cdots (t^2 - c_k) . \tag{4}$$

Here $k$ is the number of pairs of non-zero roots of the polynomial $F(t)$. Since matrix $B$ is symmetric, all roots of the polynomical $F(t)$ are real. Therefore $c_1 > 0$; $c_2 > 0$; $\cdots$; $c_k > 0$. We let $c_i = \mu_i^2$, $i = 1, 2, \cdots, k$; $\mu_j > 0$. Thus the eigenvalues of matrix $B$ are 0 (with multiplicity $n$-2k) and $\pm \mu_1$, $\pm \mu_2, \cdots, \pm \mu_k$. The null root may be absent for odd $n$.

Since the matrix $A = E - B$ is positive definite, all the numbers $1 \pm \mu_i > 0$, from which $\mu_i < 1$. Consequently the method of simple iteration converges under our conditions. The rate of convergence is determined by the largest of the number $\mu_i$, which we shall denote by $\widetilde{\mu}$.

The characteristic polynomial of matrix $S_q$, in view of (3) and (4), is

$$F_q(t) = q^n t^{n/2} \frac{(t + q - 1)^{n-2k}}{q^{n-2k} t^{(n/2)-k}} \prod_{i=1}^{k} \left[ \frac{(t + q - 1)^2}{tq^2} - \mu_i^2 \right]$$

$$= (t + q - 1)^{n-2k} \prod_{i=1}^{k} [(t + q - 1)^2 - q^2 \mu_i^2] .$$

In particular, for $q = 1$ (Nekrasov's method)

$$F_1(t) = t^{n-k} \prod_{i=1}^{k} (t - \mu_i^2) .$$

Its roots are 0 (with multiplicity $n - k$) and the numbers $\mu_1^2, \mu_2^2, \cdots, \mu_k^2$. From this we conclude that Nekrasov's method converges twice as rapidly as the method of simple iteration.

We shall now clarify the question as to the most advisable choice of the relaxation factor $q$.

To the null eigenvalues of matrix $B$ correspond the eigenvalues of matrix $S_q$ equal to $1 - q$. To the eigenvalues $\pm \mu_i$ correspond two eigenvalues of matrix $S_q$ determined from the quadratic equation

$$(t + q - 1)^2 - q^2 \mu_i^2 t = 0 .$$

The roots of this equation are

$$\left( \frac{\mu_i q \pm \sqrt{\mu_i^2 q^2 - 4q + 4}}{2} \right)^2 .$$

For $0 < q \leq \dfrac{2}{1 + \sqrt{1 - \mu_i^2}} = q_i$ these roots will be real and positive while the largest one will be

$$\left( \frac{\mu_i q + \sqrt{\mu_i^2 q^2 - 4q + 4}}{2} \right)^2 .$$

For $q_i < q < 2$ the roots become complex and their moduli equal to $q - 1$.

In the plane $q, t$ the third-order curve $(t + q - 1)^2 - q^2 \mu_i^2 t^2 = 0$ has a double point for $q = 0$, $t = 1$ and a convex loop in the band $0 \leq q \leq q_i$ which touches at $t = 0$ for $q = 1$ and at $q = q_i$ for $t = q_i - 1$.

Since every staight line parallel to axis $q$ intersects the curve at no more than two points, the straight line $t = 1$ which passes through the double point does not cut the curve there after. Therefore the entire loop of the curve is below the line $t = 1$ and the upper part MN of the loop is decreasing as $q$ goes from 0 to $q_i$. Thus a graph of the modulus for the roots corresponding to the given $\mu_i$ has the form portrayed by the broken curve in Fig. 1.

When $\mu_i$ increases, the point N is moved to the right and the piece MN of the curve is lowered. Thus the eigenvalue of matrix $S_q$ greatest in modulus corresponds to $\tilde{\mu}$.

The most advantageous value for the relaxation factor is obviously

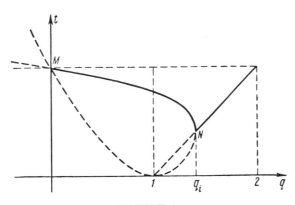

**FIGURE 1**

$$\tilde{q} = \frac{2}{1 + \sqrt{1 - \tilde{\mu}^2}} \, .$$

It is interesting to note that with such a choice of relaxation factor all eigenvalues of the matrix $\tilde{S}_q$ become equal in modulus to

$$\tilde{q} - 1 = \frac{1 - \sqrt{1 - \tilde{\mu}^2}}{1 + \sqrt{1 - \tilde{\mu}^2}} \, .$$

In fact to the null roots of matrix $B$ there correspond the roots $\tilde{q} - 1$; to the roots $\pm \mu_i$, for $\mu_i < \tilde{\mu}$, there correspond pairs of conjugate complex roots equal in modulus to $\tilde{q} - 1$; finally, to the roots $\pm \mu_i$ for $\mu_i = \tilde{\mu}$ both roots coincide and are equal to $\tilde{q} - 1$.

For $\tilde{q} < q < 2$, using the same considerations, all eigenvalues of matrix $S_q$ are equal modulus to $q = 1$.

The graph of the dependence on $q$ of the largest modulus of the eigenvalues for matrix $S_q$ has at the point $q = \tilde{q}$ a sharp minimum with a vertical tangent from the left and a tangent from the right which forms an angle $\pi/4$ with the abscissa. Therefore when $q$ diverges from $\tilde{q}$ we should obtain a sharp decrease in the rate of convergence, especially for a divergence in the direction of the decrease.

We observe in conclusion that the connection established between the eigenvalues of matrices $B$ and $S_q$ is preserved for any quasitridiagonal matrix with unit diagonal blocks without assuming symmetry and positive definiteness.

## 37. THE CONVERGENCE THEOREM

THEOREM 37.1. *If in a process of incomplete (or complete) relaxation for a system with a positive definite matrix we fulfill the conditions*:

1. *The sequence of leading indices $i_1, \cdots, i_k, \cdots$ has a recurrence interval, i.e. in each segment of length $l(i_{i+1}, \cdots, i_{k+l})$ of this sequence all the numbers $1, 2, \cdots, n$ are present at least once*:

2. *Relaxation factors satisfy the condition $\varepsilon < q_k < 1 - \varepsilon$ for $0 < \varepsilon < 1$,*

*then the process converges to the solution of the system. More than this there will exist a number $\theta$, $0 < \theta < 1$ such that $|X^* - X^{(k)}| \leq \theta^k$.*

*Proof.* Let the process for solving the system $AX = F$ proceed according to the formula

$$X^{(k)} = X^{(k-1)} + q_k \frac{r_{i_k}^{(k-1)}}{a_{i_k i_k}} e_{i_k} = X^{(k-1)} + m_k e_{i_k} \, , \qquad (1)$$

where

$$m_k = \frac{r_{i_k}^{(k-1)}}{a_{i_k i_k}} q_k \, . \qquad (2)$$

Here $i_k$ denotes the number of the component being changed at the $k$th step of the process. Let $f(X^{(k)})$ be the values of the error function at the $k$th step and $Y^{(k)}$ be the error vector at this step. Then

$$f(X^{(k-1)}) - f(X^{(k)}) = \frac{(r_{i_k}^{(k-1)})^2}{a_{i_k i_k}} q_k(2 - q_k) = \frac{2 - q_k}{q_k} a_{i_k i_k} m_k^2 , \tag{3}$$

$$Y^{(k)} = Y^{(k-1)} - m_k e_{i_k} . \tag{4}$$

The positive numbers $(2 - q_k)a_{i_k i_k}/q_k$ are bounded above and below so that there exist constants $\gamma_1$ and $\gamma_2$ such that

$$\gamma_1 m_k^2 < f(X^{(k-1)}) - f(X^{(k)}) < \gamma_2 m_k^2 ,$$

Since the series of positive numbers $\sum\limits_{k=1}^{\infty} f(X^{(k-1)}) - f(X^{(k)})$ obviously converges, then the series $\sum\limits_{k=0}^{\infty} \gamma_1 m_k^2$ will also be convergent, as will the series $\sum\limits_{k=0}^{\infty} m_k^2$, and consequently $m_k \to 0$. Since the component $r_{i_k}^{(k-1)}$ of the residual vector differs from $m_k$ by the factor $a_{i_k i_k}/q_k$ bounded above and below, we establish that $r_{i_k}^{(k-1)} \to 0$. Thus the component of the residual vector with a number equal to the component number for the approximation being changed in the following step approaches zero. For the process to converge it is sufficient to indicate that all the remaining components of the residual vector also approach zero for $k \to \infty$. Let $i$ be any natural number, $1 \le i \le n$, and let $k > 1$. We denote by $k_i$ the number of the last step preceding the $k$th for which the $i$th component of the approximation is changed. From the existence condition for the recurrence interval of lengh $l$ the inequality $0 \le k - k_i < l$ holds. In view of what has been proved $r_i^{(k_i-1)} \to 0$. More-over we have

$$|r_i^{(k)}| \le |r_i^{(k_i-1)}| + |r_i^{(k)} - r_i^{(k_i-1)}| \le c|m_{k_i}| + |r_i^{(k)} - r_i^{(k_i-1)}| .$$

We shall estimate the last entry. For this we first estimate $|r^{(k)} - r^{(m)}|$, where $m < k$. We have

$$|Y^{(k)} - Y^{(k-1)}| = |m_k e_{i_k}| = m_k .$$

Consequently,

$$|Y^{(k)} - Y^{(m)}| \le \sum\limits_{\nu=m+1}^{k} |m_\nu| ,$$

and therefore

$$|r^{(k)} - r^{(m)}| = |A(Y^{(k)} - Y^{(m)})| \le \|A\| \sum\limits_{\nu=m+1}^{k} |m_\nu| .$$

It follows that

$$|r_i^{(k)} - r_i^{(k_i-1)}| \le |r^{(k)} - r^{(k_i-1)}| \le \|A\| \sum\limits_{\nu=k_i}^{k} |m_\nu| \le \|A\| \sum\limits_{\nu=k-l}^{k} |m_\nu| ,$$

since $k_i \ge k - l$.

Consequently

$$|r_i^{(k)}| \le c|m_{k_i}| + \|A\| \sum\limits_{\nu=k-l}^{k} |m_\nu| \le c_1 \sum\limits_{\nu=k-l}^{k} |m_\nu|.$$

Thus $r_i^{(k)} \to 0$ for $k \to \infty$ and the convergence of the process is proved.

We estimate the rate of convergence. From the last inequality it follows that

$$|r_i^{(k)}|^2 \le c_1^2(l+1) \sum_{\nu=k-l}^{k} |m_\nu|^2 .$$

Consequently,

$$|r^{(k)}|^2 = \sum_{i=1}^{n} |r_i^{(k)}|^2 \le c_1^2 n(l+1) \sum_{\nu=k-l}^{k} |m_\nu|^2$$

and

$$f(X^{(k)}) = (A^{-1}r^{(k)}, r^{(k)}) \le \|A^{-1}\| \, |r^{(k)}|^2 \le c_2 M_k ,$$

where

$$c_2 = c_1^2 n(l+1) \|A^{-1}\| ,$$
$$M_k = \sum_{\nu=k-l}^{k} |m_\nu|^2 .$$

On the other side,

$$f(X^{(k)}) = \sum_{\nu=k}^{\infty} (f(X^{(\nu)}) - f(X^{(\nu+1)})) \ge \gamma_1 \sum_{\nu=k}^{\infty} |m_\nu|^2 \ge \frac{\gamma_1}{l+1} \sum_{\nu=k}^{\infty} M_\nu .$$

Thus

$$\sum_{\nu=k}^{\infty} M_\nu \le c_3 M_k \tag{5}$$

for all $k$ and for some $c_3 > 1$. We write

$$\sum_{\nu=k}^{\infty} M_\nu \le S_k .$$

Then inequality (5) can be written in the form

$$S_k \le c_3(S_k - S_{k+1})$$

from which

$$S_{k+1} \le \frac{c_3 - 1}{c_3} S_k = \theta_1 S_k \qquad \text{(where } 0 < \theta_1 < 1)$$

and consequently $S_k \le c_4 \theta_1^k$. Furthermore,

$$f(X^{(k)}) = \sum_{\nu=k+1}^{\infty} (f(X^{\nu-1}) - f(X^{(\nu)})) \le \gamma_2 \sum_{\nu=k+1}^{\infty} m_\nu^2 < \gamma_2 S_k < c_4 \gamma_2 \theta_1^k .$$

Thus,

$$f(X^{(k)}) = O(\theta_1^k) .$$

Besides this estimate, the estimates

$$|r^{(k)}| = O(\theta_1^{k/2}) ;$$
$$|Y^{(k)}| = O(\theta_1^{k/2}) = O(\theta^k) ,$$

are also correct where $\theta = \theta_1^{1/2}$.

Convergence of the process may be proved also under weaker assumptions regarding the relaxation factor (A. M. Ostrowski (8)).

From the proof of the theorem it follows, in particular, that the cyclic relaxation process with a constant relaxation factor $q$ satisfying the inequality $0 < q < 2$ converges.

We shall make one more observation concerning this case. The polynomial

$$F_q(t) = \begin{vmatrix} (t+q-1)a_{11} & qa_{12} & \cdots & qa_{1n} \\ tqa_{21} & (t+q-1)a_{22} \cdots & & qa_{2n} \\ \cdot \cdot \cdot \cdot \cdot \cdot \cdot \cdot \cdot \cdot \cdot \cdot \cdot \cdot \cdot \cdot \cdot \\ tqa_{n1} & tqa_{n2} & \cdots (t+q-1)a_{nn} \end{vmatrix},$$

whose maximum modulus among the roots determines the rate of convergence of the process, has leading coefficient $a_{11}a_{22}\cdots a_{nn}$ and constant term $(q-1)^n a_{11}a_{22}\cdots a_{nn}$. Therefore the product of its roots is equal to $(-1)^n(q-1)^n$. From this fact it follows that the greatest of its roots is not less than $|q-1|$ and may be equal to this number only if all its roots are equal in modulus to $|q-1|$.

If $0 < q < 2$, then the roots of the polynomial $F_q(t)$ are strictly less than one in modulus, since the relaxation process converges. In view of the continuous dependence of the roots on the polynomial's coefficients, (at the limits of the interval for $q$, i.e. $q = 0$ and $q - 2$, the roots of $F_q(t)$ will not exceed one. But for $q = 0$ and $q = 2$, $1 = |q-1|$ and consequently, in view of the observation made above, all roots of $F_q(t)$ are equal in modulus to one.

This circumstance for $q = 0$ is verified directly, since $F_0(t) = a_{11}\cdots a_{nn}(t-1)^n$. For $q = 2$ it is not trivial. It was established for quasi-tridiagonal matrices in Section 36 by direct calculation.

## 38. DIRECTED RELAXATION

Instead of assigning the sequence of leading indices *a priori* in carrying out a coordinate relaxation process (as was done, for example, in the cyclic process or in the process where direction is given by the decimal form of the number $e$) we may choose the index $i_k$ at each step from the results of the preceding step. Thus, for example, the most rapid decrease in the error function when passing from $X^{(k-1)}$ to $X^{(k)}$ is obtained if we choose the index $i_k$ so that the number $\dfrac{(r_{i_k}^{(k-1)})^2}{a_{i_k i_k}}$ will be the largest one among all numbers $\dfrac{(r_i^{(k-1)})^2}{a_{ii}}$, where $i = 1, \cdots, n$. In other words the index $i_k$ is chosen so that number $\dfrac{|r_{i_k}^{(k-1)}|}{\sqrt{a_{i_k i_k}}}$ will be a maximum. A relaxation process with such direction is called *Seidel's process*.

We associate Gauss's name with a process for which the leading index is chosen from the condition $\max \dfrac{|r_i^{(k-1)}|}{a_{ii}}$. Finally, the direction rule of Southwell is determined by computing $\max |r_i^{(k-1)}|$.

THEOREM 38.1   *If the leading index $i_k$ of a coordinate relaxation process for a system with a positive definite matrix $A$ is chosen so that*

$$|r_{i_k}^{(k-1)}| \geq \gamma |r_i^{(k-1)}|  \qquad (0 < \gamma \leq 1, i = 1, \cdots, n) \tag{1}$$

*and if $\varepsilon < q_k < 2 - \varepsilon$, then the incomplete relaxation process converges to the solution $X^*$ of the system and there exists a number $\theta$, $0 < \theta < 1$, such that $|X^* - X^{(k)}| < \theta^k$.*

Convergence of the incomplete relaxation precesses of Gauss, Seidel, and Southwell follows from this theorem, since these processes satisfy the condition of the theorem for $\gamma = \min a_{ii}/\max a_{ii}$ using Gauss's process, for $\gamma = \sqrt{\min a_{ii}/\max a_{ii}}$ using Seidel's process, and for $\gamma = 1$ using Southwell's method. We have

$$\frac{|r_{i_k}^{(k-1)}|}{a_{i_k i_k}} \geq \frac{|r_i^{(k-1)}|}{a_{ii}} \qquad \text{(for } i = 1, \cdots, n)$$

and therefore

$$|r_{i_k}^{(k-1)}| \geq \frac{a_{i_k i_k}}{a_{ii}} |r_i^{(k-1)}| \geq \frac{\min a_{ii}}{\max a_{ii}} |r_i^{(k-1)}| .$$

Analogous inequalities also hold for Seidel's method.

*Proof of the theorem.*   We shall use the same notation as in the proof of the theorem in the previous paragraph. Repeating the proof of this theorem we obtain that the series $\sum m_k^2$ converges, $m_k \to 0$ and $r_{i_k}^{(k-1)} \to 0$. Moreover from the inequality

$$|r_i^{(k-1)}| \leq \frac{1}{\gamma} |r_{i_k}^{(k-1)}| \tag{2}$$

we conclude that $|r_i^{(k-1)}| \to 0$ for all $i = 1, \cdots, n$. In this way the convergence of the process is proved.

To estimate the rate of convergence we use an inequality derived from (2).

$$|r^{(k-1)}| \leq \frac{n}{\gamma} |r_{i_k}^{(k-1)}| \leq c_1 m_k .$$

Furthermore,

$$f(X^{(k-1)}) = (A^{-1} r^{(k-1)}, r^{(k-1)}) \leq \| A^{-1} \| c_1^2 m_k^2 .$$

On the other hand,

$$f(X^{(k-1)}) - f(X^{(k)}) \geq \gamma_1 m_k^2 ,$$

and consequently

$$f(X^{(k-1)}) \geq \gamma_1 \sum_{\nu=k}^{\infty} m_\nu^2 .$$

Letting

$$s_k = \sum_{\nu=k}^{\infty} m_\nu^2$$

we obtain

$$\gamma_1 s_k \leq \| A^{-1} \| c_1^2 m_k^2$$

from which

$$s_k \leq c_2(s_k - s_{k+1}) \qquad (c_2 > 1)$$

and

$$s_{k+1} \leq \frac{c_2 - 1}{c_2} s_k = \theta_1 s_k \qquad (0 < \theta_1 < 1) .$$

Thus

$$s_k < c_3 \theta_1^k$$

and consequently

$$f(X^{(k)}) \leq \gamma_2 \sum_{\nu=k}^{\infty} m_\nu^2 = \gamma_2 s_k \leq c_4 \theta_1^k . \qquad (3)$$

Also valid are the estimates

$$| r^{(k)} | = O(\theta^k) ,$$
$$| Y^{(k)} | = O(\theta^k) ,$$
$$\theta = \theta_1^{1/2} .$$

In using relaxation methods with the directions of Southwell, Gauss, and Seidel it is necessary at each step to compute all components of the residual vector, but then to determine the leading index for the following step we calculate only one. Therefore computational schemes in which all these components are used are of some interest. For constructing such schemes one should use the relationship between two adjacent residual vectors.

That is, from the relationship

$$X^{(k)} = X^{(k-1)} + q_k \frac{r_{i_k}^{(k-1)}}{a_{i_k i_k}} e_{i_k}$$

it follows that

$$r^{(k)} = r^{(k-1)} - q_k \frac{r_{i_k}^{(k-1)}}{a_{i_k i_k}} A e_{i_k} = r^{(k-1)} - q_k \frac{r_{i_k}^{(k-1)}}{a_{i_k i_k}} A_{i_k}$$

where by $A_{i_k}$ is denoted a column with number $i_k$ of matrix $A$.

Passing to the components we get

$$r_i^{(k)} = r_i^{(k-1)} - r_{i_k}^{(k-1)} \frac{a_{i i_k}}{a_{i_k i_k}} q_k .$$

We let

$$r_i^{(k)} = r_i^{(k-1)} - r_{i_k}^{(k-1)} \frac{a_{ii_k}}{a_{i_k i_k}} q_k \, ,$$

$$r_i^{(k)} = \frac{r_i^{(k)}}{a_{ii}} \, , \tag{4}$$

$$\sigma_i^{(k)} = \frac{r_i^{(k)}}{\sqrt{a_{ii}}} \, .$$

Then

$$r_i^{(k)} = r_i^{(k-1)} - r_{i_k}^{(k-1)} \frac{a_{ii_k}}{a_{ii}} q_k \, ,$$

$$\sigma_i^{(k)} = \sigma_i^{(k-1)} - \sigma_i^{(k-1)} \frac{a_{ii_k}}{\sqrt{a_{ii}} \sqrt{a_{i_k i_k}}} q_k \, .$$

Letting

$$\frac{a_{ij}}{a_{jj}} = b_{ij} \, ;$$

$$\frac{a_{ij}}{a_{ii}} = c_{ij} \, ; \tag{5}$$

$$\frac{a_{ij}}{\sqrt{a_{ii}} \sqrt{a_{jj}}} = h_{ij} \, ,$$

we obtain

$$r_i^{(k)} = r_i^{(k-1)} - q_k r_{i_k}^{(k-1)} b_{ii_k} \, ,$$

$$r_i^{(k)} = r_i^{(k-1)} - q_k r_{i_k}^{(k-1)} c_{ii_k} \, , \tag{6}$$

$$\sigma_i^{(k)} = \sigma_i^{(k-1)} - q_k \sigma_{i_k}^{(k-1)} h_{ii_k} \, ,$$

or in vector form

$$r^{(k)} = r^{(k-1)} - q_k r_{i_k}^{(k-1)} B_{i_k} \, ,$$

$$r^{(k)} = r^{(k-1)} - q_k r_{i_k}^{(k-1)} C_{i_k} \, , \tag{6'}$$

$$\sigma^{(k)} = \sigma^{(k-1)} - q_k \sigma_{i_k}^{(k-1)} H_{i_k} \, .$$

Here $B_j$, $C_j$ and $H_j$ are the $j$th columns of matrices $(b_{ij})$, $(c_{ij})$ and $(h_{ij})$, respectively, which are obviously equal in turn to $D^{-1}A$, $AD^{-1}$ and $D^{-\frac{1}{2}}AD^{-\frac{1}{2}}$. In carrying out the selected process, the corresponding auxiliary matrix should be formed beforehand. The relaxation coefficients $q_1$, $q_2$, $\cdots$ should also be chosen a priori.

Moreover, the computations are arranged in the following way.

Using the initial approximation, we form $n$ numbers $r_i^{(0)}$ (or $r_i^{(0)}$ or $\sigma_i^{(0)}$). The largest of these in modulus is chosen, stressed ("earmarked"), and its number taken as $i_1$. By formulas (6) we form the numbers $r_i^{(1)}$ (or $r_i^{(1)}$ or $\sigma_i^{(1)}$); the largest of these in modulus is chosen, stressed, and its number taken as $i_2$ and so forth. After all this, when a sufficient number $N$ of steps has been completed, the approximation components are found by the formulas:

$$X_i^{(N)} = X_i^{(0)} + \frac{1}{a_{ii}} \sum' q_k r_i^{(k-1)}$$

or

$$X_i^{(N)} = X_i^{(0)} + \sum' q_k \gamma_i^{(k-1)}$$

or

$$X_i^{(N)} = X_i^{(0)} + \frac{1}{\sqrt{a_{ii}}} \sum' q_k \sigma_i^{(k-1)} \, .$$

The sum $\sum'$ is extended to those values of $k$ for which $r_i^{(k-1)}$ (or $\gamma_i^{(k-1)}$ or $\sigma_i^{(k-1)}$) will be stressed.

To lessen the influence of round-off errors the process should be interrupted from time to time, having computed an approximation, to find the residual directly and to begin the process anew.

The computational scheme becomes especially simple if all $a_{ii} = 1$. In this case all three direction methods coincide and there is no need to compute auxiliary matrices, since each of them coincides with the initial one.

If we transform the system $AX - F$, by substituting $X = D^{-1/2}Y$ and premultiplying by $D^{-1/2}$, into the form

$$D^{-1/2}AD^{-1/2}Y = D^{-1/2}F \, ,$$

the matrix of the system being symmetric and having diagonal elements equal to one, then application of the relaxation process with a direction is equivalent to applying Seidel's method to the initial system.

In Table III.11 are given the results of applying complete relaxation

TABLE III.11  Computing Successive Approximations by a Relaxation Process with Direction for $q = 1$

| $k$ | $X$ | 0 | 0 | 0 | 0 | $i_k$ |
|---|---|---|---|---|---|---|
| 1 | $F - AX$ | 0.3 | 0.5 | 0.7 | 0.9 | 4 |
| 2 | | −0.294 | 0.104 | 0.502 | 0 | 3 |
| 3 | | −0.56508 | −0.05664 | 0 | −0.11044 | 1 |
| 4 | | 0 | 0.186936 | 0.3051432 | 0.2625128 | 3 |
| 5 | | −0.16477733 | 0.08304778 | 0 | 0.19538130 | 4 |
| 6 | | −0.29372899 | −0.00291999 | −0.04298389 | 0 | 1 |
| 7 | | 0 | 0.12044619 | 0.11562976 | 0.19386113 | 4 |
| 8 | | −0.12794835 | 0.03514729 | 0.07298031 | 0 | 1 |
| $\sum r_{i_k}^{(k-1)}$ | | −0.98675734 | 0 | 0.80714320 | 1.28924243 | |
| $X$ | | −0.98675734 | 0 | 0.80714320 | 1.28924243 | |
| 9 | $F - AX$ | ............ | ............ | ............ | ............ | |
| ........ | | ............ | ............ | ............ | ............ | |
| 100 | | −0.00000007 | −0.00000004 | 0.00000003 | 0 | |
| | $X$ | −1.2577936 | 0.0434874 | 1.0391661 | 1.4823928 | |

($q_k = 1$) with direction to system (9) Section 23.

The successive residual vectors are computed by the formula

$$r^{(k)} = r^{(k-1)} - r_{i_k}^{(k-1)} A_{i_k} ,$$

where $A_j$ denotes the $j$th column of matrix $A$.

Thus a solution of the system, accurate to $2 \cdot 10^{-7}$ in each component, is obtained by 100 elementary steps, equivalent to about 25 simple iterations.

Applying the process with direction for $q_k = 1.2$ leads to the same result in 52 elementary steps; for $q_k = 0.8$ the result is obtained in 148 elementary steps.

## 39. RELAXATION IN TERMS OF THE RESIDUAL VECTOR'S LENGTH

We shall now consider processes based on the minimization or curtailing of the length of the residual vector by altering one component of the previous approximation at each step.

Let $X$ be a certain approximation and $i$ the identifying number for the component being changed. We let

$$X' = X + \delta e_i .$$

Then

$$r' = r - \delta A_i ,$$

and therefore

$$(r', r') = (r - \delta A_i , r - \delta A_i) = (r, r) - 2\delta(r, A_i) + \delta^2(A_i , A_i)$$

$$= (r, r) + \frac{1}{(A_i , A_i)} [(r, A_i) - \delta(A_i , A_i)]^2 - \frac{(r, A_i)^2}{(A_i , A_i)} .$$

Here $A_i$ is the $i$th column of matrix $A$. The value $(r', r')$ will be minimal for

$$\delta = \frac{(r, A_i)}{(A_i , A_i)}$$

so that in the case of complete relaxation we may take

$$X' = X + \frac{(r, A_i)}{(A_i , A_i)} e_i . \qquad (1)$$

If the given system is changed beforehand to some form such that all $(A_i , A_i) = 1$ (which it is always possible to do by substituting for the unknowns $X_i = \bar{x}_i / \sqrt{(A_i , A_i)}$ we shall have

$$X' = X + (r, A_i)e_i \qquad (2)$$

and

$$(r', r') = (r, r) - (r, A_i)^2 . \qquad (3)$$

Incomplete relaxation may be carried out in this case according to the formulas

$$X' = X + q(r, A_i)e_i \qquad (0 < q < 2) , \qquad (4)$$

while

$$(r', r') = (r, r) - q(2 - q)(r, A_i)^2 . \qquad (5)$$

It is not hard to see that the method of relaxation in terms of the residual vector's length is equivalent to the relaxation method in terms of the error function for the system $A'AX = A'F$ obtained from the given system $AX = F$ by a first Gauss transformation. In fact,

$$(r, r) = (AY, AY) = (A'AY, Y)$$

where $Y$ is the error vector, so that the square of the length of the residual vector for the system $AX = F$ is the error function for the system $A'AX = A'F$.

If the sequence of numbers for the components being altered is given *a priori* (for example, if we apply a cyclic process), then carrying out the process by formulas (2) or (4) does not require completing the Gauss transformation, that is, computing the matrix $A'A$. If we are to give a direction for the process, requiring at each step a maximum decrease in the residual vector's length by the choice of number of the component being altered, then as Pokorna (1) correctly noted it is advisable to compute beforehand the matrix $A'A$ and to introduce the process (for $(A_i, A_i) = 1$) by the formulas

$$r^{(k)} = r^{(k-1)} + r_{i_k}^{(k-1)} B_{i_k}$$

where $i_k$ is the number of the component of the vector $r^{(k-1)}$ largest in modulus and $B_i$ is the $i$th column of matrix $A'A$. This is equivalent to carrying out a relaxation process with direction for the system $A'AX = A'F$.

## 40. GROUP RELAXATION

Let $AX = F$ be a given system.

An individual step in the group relaxation method consists of the following. A group† $G$ of indices is separated out, and in passing from one approximation to the next only components with indices from the chosen set $G$ are altered. Such alteration is realized in such a way that equations whose indices are included in $G$ are satisfied exactly. In other words, the components being changed are solutions for the system of equations

$$\sum_{j \in G} a_{ij} x_j' = f_i - \sum_{m \in \bar{G}} a_{im} x_m \qquad (1)$$

where the index $i$ runs over the whole set $G$.

It is easy to see that if matrix $A$ is positive definite, then one step of the group relaxation minimizes the error function in the subspace spanned by unit vectors with indices which form the group $G$.

---

† The sets called "groups" are not to be identified with the concept of an "algebraic group".

The group relaxation method allows for many modifications, depending on the principle of selection for the groups $G$ at each step. The simplest modification is the cyclic grouping process in which all the indices are broken down once and for all into several non-intersecting groups; during the process these groups alternate cyclically. This process may be considered as a process of successive approximations, given a suitable preparation of the system. In order to indicate this we shall assume for simplicity that groups consist of successive indices and alternation is realized in their increasing order. We denote by $D$ a quasi-diagonal matrix "cut out" of matrix $A$ in accordance with the breaking-up of indices into groups, by $L$ a matrix consisting of the elements of matrix $A$ lying to the left of elements of matrix $D$, and by $R$ the matrix consisting of elements of $A$ lying to the right of elements of matrix $D$. Then

$$A = L + D + R \tag{2}$$

Let $X^{(k)}$ be the approximation obtained after completing $k$ cycles of the process. Then $X^{(k+1)}$ and $X^{(k)}$ are obviously related by

$$LX^{(k+1)} + DX^{(k+1)} + RX^{(k)} = F$$

and therefore

$$X^{(k+1)} = (L + D)^{-1}F - (L + D)^{-1}RX^{(k)} . \tag{3}$$

Thus the process being studied is equivalent to the process of successive approximations applied to a system prepared in the form

$$X = - (L + D)^{-1}RX + (L + D)^{-1}F .$$

From here arises a necessary and sufficient condition for the convergence of the process, namely, that the eigenvalues of the matrix $(L + D)^{-1}R$ be less than one in modulus. If matrix $A$ is a positive definite Hermitian matrix, then the process converges at all times. In fact, in this case the eigenvalues of the matrix $(L + D)^{-1}R = (R^* + D)^{-1}R$ do not exceed $\dfrac{\Lambda_0}{\sqrt{\lambda_0 d_0 + \Lambda_0^2}}$ in modulus. Here $\Lambda_0$ denotes $\| R \| = \| R^* \|$, $\lambda_0$ is the least eigenvalue of matrix $A$, and $d_0$ is the least eigenvalue of matrix $D$. The proof of this inequality is carried out analogously to the proof for Nekrasov's process. For this we use the fact that the cut-out matrix $D$ is always positive definite. The inequality $d_0 \geq \lambda_0$ is valid.

In free group relaxation we allow not only non-cyclic alternation of the sets chosen earlier but also change in their structure at each step. As we have seen, an individual step with selected group $G^{(k)}$ is carried out according to the formulas

$$\sum_{j \in G^{(k)}} a_{ij}x_j^{(k+1)} = f_i - \sum_{m \bar{\in} G^{(k)}} a_{im}x_m^{(k)} \qquad [i \in G^{(k)}]$$

$$x_j^{(k+1)} = x_j^{(k)} \qquad [\text{for } j \bar{\in} G^{(k)}]$$

These formulas are equivalent to the formulas

$$\sum_{j \in G^{(k)}} a_{ij}[x_j^{(k+1)} - x_j^{(k)}] = f_i - \sum_{m=1}^{n} a_{im}x_m^{(k)} = r_i^{(k)} \qquad [i \in G^{(k)}] \tag{4}$$

$$x_j^{(k+1)} = x_j^{(k)} \qquad [j \,\overline{\in}\, G^{(k)}] \,.$$

Writing

$$x_j^{(k+1)} - x_j^{(k)} = \delta_j^{(k)} \,,$$

we obtain

$$x_j^{(k+1)} = x_j^{(k)} + \delta_j^{(k)} \qquad [i \in G^{(k)}] \,,$$
$$x_j^{(k+1)} = x_j^{(k)} \qquad [j \,\overline{\in}\, G^{(k)}] \,,$$

where $\delta_j^{(k)}$ is the solution to the system

$$\sum_{j \in G^{(k)}} a_{ij}\delta_j^{(k)} = r_i^{(k)} \qquad [i \in G^{(k)}] \,. \tag{5}$$

For incomplete relaxation the computation is carried out according to the formulas

$$x_j^{(k+1)} = x_j^{(k)} + q_k\delta_j^{(k)} \qquad [j \in G^{(k)}] \,,$$
$$x_j^{(k+1)} = x_j^{(k)} \qquad [j \,\overline{\in}\, G^{(k)}] \,, \tag{6}$$

where $0 < q_k < 2$.

For positive definite matrices the following convergence theorem holds. *If there exists a number l such that any sequence of groups of length l contains each index at least once, and if $\varepsilon < q_k < 2 - \varepsilon$, then the process of free group (incomplete) relaxation converges to the solution of the system.* (Ostrowski (8)).

The proof of this theorem is analogous to the proof of the corresponding theorem for a one-step process.

As a rule convergence of a group process turns out to be more rapid in comparison with convergence for a corresponding one-step process.

The supplementary work involved in solving auxiliary systems is easily accomplished using sub-programs for solving a system of a fixed order.

Of course application of the group relaxation method is meaningful only for systems with a large number of equations.

# CHAPTER IV
## THE COMPLETE EIGENVALUE PROBLEM

By the *complete eigenvalue problem* is meant the problem of finding all the eigenvalues of a matrix $A$ as well as the eigenvectors (or vectors forming a canonical basis) belonging to them. We recall that the eigenvalues of a matrix $A$ are the roots of its characteristic polynomial, i.e. the roots of equation

$$| A - tE | = \begin{vmatrix} a_{11} - t & a_{12} & \cdots & a_{1n} \\ a_{21} & a_{22} - t & \cdots & a_{2n} \\ \cdot & \cdot & \cdots & \cdot \\ a_{n1} & a_{n2} & \cdots a_{nn} - t \end{vmatrix}$$

$$= (-1)^n [t^n - p_1 t^{n-1} - \cdots - p_n] = 0 .$$

Determining the components of an eigenvector involves solving a system of $n$ homogeneous equations in $n$ unknowns; to compute all the eigenvectors of a matrix we generally need to solve $n$ systems of the form

$$(A - \lambda_i E) X_i = 0 ,$$

where $X_i = (x_{1i}, \cdots, x_{ni})'$ is the eigenvector of matrix $A$ belonging to the eigenvalue $\lambda_i$.

As was already noted in Section 1, paragraph 8, the coefficients $p_i$ of the characteristic polynomial are, up to sign, sums of all principal minors of order $i$ of the matrix $A$. Direct computation of the coefficients $p_i$ is extremely unwieldy and requires a huge number of operations.

Therefore there have emerged, quite naturally, special computational methods which simplify the numerical solution of this kind of problem. The majority of methods giving a solution for the complete eigenvalue problem include a preliminary computation of the characteristic polynomial's

226

coefficients by some method which avoids the necessity of computing all the numerous minors. The eigenvalues are then computed by any convenient method which will yield the roots of the polynomial approximately. One of the best means for computing the approximate roots of the polynomial $\varphi(t) = a_0 t^n + a_1 t^{n-1} + \cdots + a_{n-1} t + a_n$ is *Newton's method*. That is, if $t_0$ is a certain initial approximation to a root, then the successive approximations $t_1, t_2, \cdots$ are computed by the formula

$$t_i = t_{i-1} - \frac{\varphi(t_{i-1})}{\varphi'(t_{i-1})} \qquad (i = 1, 2, \cdots) .$$

If $t_0$ has been taken sufficiently close to the desired root $\lambda$, then successive approximations $t_i$ converge to $\lambda$ with second-order convergence, i.e. the error in the last approximation will have the order of the square of the previous one

$$|\lambda - t_i| \leq c |\lambda - t_{i-1}|^2 ,$$

where $c$ is a certain constant.

Computation of the values for the polynomial $\varphi(t)$ and its derivative at a given point $t_0$ should be carried out by Horner's scheme:

$$
\begin{array}{llllll}
a_0 & a_1 & a_2 \cdots a_{n-1} & a_n \| t_0 \\
a_0 & b_1 & b_2 \cdots b_{n-1} & b_n \\
a_0 & c_1 & c_2 \cdots c_{n-1} ,
\end{array}
$$

which is filled in by the recurrent formulas

$$b_i = b_{i-1} t_0 + a_i \qquad (i = 1, 2, \cdots, n) ,$$
$$c_i = c_{i-1} t_0 + b_i \qquad (i = 1, 2, \cdots, n-1) .$$

Then

$$b_n = \varphi(t_0) ,$$
$$c_{n-1} = \varphi'(t_0) .$$

The numbers $a_0, b_1, \cdots, b_{n-1}$ will be coefficients of the quotient $\varphi_1(t)$ obtained by dividing the polynomial $\varphi(t)$ by $t - t_0$. Correspondingly, the numbers $a_0, c_1, \cdots, c_{n-1}$ will be coefficients of the quotient $\varphi_2(t)$ obtained by dividing $\varphi_1(t)$ by $t - t_0$.

In large measure the eigenvectors of a matrix may be determined from intermediate results of computations carried out to find the characteristic polynomial's coefficients. Of course, to determine the eigenvector belonging to a particular eigenvalue this eigenvalue must have already been computed. Methods of this type are called exact, i.e. if they are applied to matrices whose elements are given exactly (by rational numbers) and if the computations are *exact* (carried out by the rules of operation on ordinary fractions), then we shall obtain in our result exact values for the coefficients of the characteristic polynomial, and the components of the eigenvectors will turn out to be exact formulas expressed in terms of eigenvalues.

Along with exact methods for solving the eigenvalue problem there are *iterative* methods in which the eigenvalues are obtained as the limits of certain numerical sequences, as are the components of the eigenvectors belonging to them. In iterative methods, as a rule, the eigenvalues are computed directly without first computing the coefficients of the characteristic polynomial. This essentially simplifies the problem, since computing the roots of a polynomial is itself not always a minor task.

However, iterative methods are more applicable in solving the special eigenvalue problem. By the *special problem* we mean the problem of finding one or several eigenvalues and their corresponding eigenvectors.

The complete and special eigenvalue problems are completely different problems both in their methods of solution and in their area of application. Solving the complete problem for matrices for even a fairly low order inevitably turns out to be very unwieldy; the possibility of solving the complete one notwithstanding, is very valuable in practice.

The present chapter is devoted to a discussion of exact methods for solving the complete eigenvalue problem. Iterative methods of solution for the complete problem will be examined in Chapter VIII and the special problem will be studied beginning with the following chapter.

We observe that all the methods proposed below (both in this chapter and in the following ones), except for the methods of Leverrier (1840) and Jacobi (1846), appeared either in the nineteen-thirties or later.

In presenting these numerical methods we shall generally assume the elements of the matrices to be real.

## 41. STABILITY OF THE EIGENVALUE PROBLEM

In setting up the eigenvalue problem for matrices whose elements are given approximately the question naturally arises about the stability of the obtained solution, in other words, the question of how eigenvalues and eigenvectors change when the elements of a given matrix change within the limits of permissible error.

The fact that in particular cases the eigenvalue problem may not be stable is clear from the following illustration: We assume that a given matrix, if its numerical values are considered exact, has only simple eigenvalues. However for certain defined changes in its elements within given limits of accuracy we may get a matrix which has a multiple eigenvalue, or whose characteristic function has a non-linear elementary divisor. In this case the canonical form of the matrix undergoes a *qualitative* change, passing from purely diagonal form to the general canonical form. In particular, even the number of distinct eigenvectors may differ. Of course under these conditions the complete eigenvalue problem along with the definition of eigenvectors simply loses its meaning. Under conditions even closely resembling the given situation the problem of determining the eigenvectors then lacks stability.

Let $A$ be a given matrix and $A + dA$ a matrix close to it. We shall consider how the eigenvalues and eigenvectors of matrix $A$ change when it takes on the increment $dA$. We carry out the calculation under the assumption that all the eigenvalues of matrix $A$ are distinct, rejecting values of second-order smallness, i.e. we shall consider $dA$ (and correspondingly $dX$ and $d\lambda$) as differentials and not as finite increments.

Let

$$AX_i = \lambda_i X_i \qquad (i = 1, 2, \cdots, n) . \qquad (1)$$

Then

$$(dA)X_i + AdX_i = \lambda_i dX_i + d\lambda_i X_i . \qquad (2)$$

Let $V_1, \cdots, V_n$ be eigenvectors of the conjugate matrix $A^*$ corresponding to the eigenvalues $\bar{\lambda}_1, \cdots, \bar{\lambda}_n$. Then

$$((dA)X_i, V_j) + (A(dX_i), V_j) = \lambda_i(dX_i, V_j) + d\lambda_i(X_i, V_j) . \qquad (3)$$

Letting $i = j$ in equation (3) we obtain

$$((dA)X_i, V_i) + (A(dX_i), V_i) = (dX_i, \bar{\lambda}_i V_i) + d\lambda_i(X_i, V_i) ,$$

from which

$$d\lambda_i = \frac{((dA)X_i, V_i)}{(X_i, V_i)} , \qquad (4)$$

since $A^*V_i = \bar{\lambda}_i V_i$.

We now assume $i \neq j$. Then from the equalities $(X_i, V_j) = 0$ and $(A(dX_i), V_j) = \lambda_j(dX_i, V_j)$ we get

$$(\lambda_i - \lambda_j)(dX_i, V_j) = ((dA)X_i, V_j) ,$$

from which

$$(dX_i, V_j) = \frac{((dA)X_i, V_j)}{\lambda_i - \lambda_j} .$$

Let

$$dX_i = \sum_{j=1}^{n} \alpha_{ij} X_j . \qquad (5)$$

Then

$$(dX_i, V_j) = \alpha_{ij}(X_j, V_j)$$

and consequently

$$\alpha_{ij} = \frac{((dA)X_i, V_j)}{(X_j, V_j)(\lambda_i - \lambda_j)} \qquad (\text{for } i \neq j) . \qquad (6)$$

The coefficient $\alpha_{ii}$ naturally remains undefined in view of the non-uniqueness of the eigenvector, and without loss of generality we may assume that $\alpha_{ii} = 0$.

We now pass on to the estimates. From formula (4) we get

$$|d\lambda_i| \leq \frac{\|dA\| \cdot |X_i| \cdot |V_i|}{|(X_i, V_i)|} = c_i \|dA\| ,$$

where

$$c_i = \frac{|X_i| \cdot |V_i|}{|(X_i, V_i)|} . \tag{7}$$

It is clear that $c_i \geq 1$. If the eigenvectors are real, then

$$c_i = \frac{1}{|\cos \varphi_i|} ,$$

where $\varphi_i$ is the angle between the vectors $X_i$ and $V_i$.

We call the number $c_i$ the *distortion coefficient* for matrix $A$ corresponding to the eigenvalue $\lambda_i$. Thus the change in $\lambda_i$ for a given $\|dA\|$ may be proportionally greater than the distortion coefficient corresponding to this eigenvalue (i.e. $c_i$). For normal matrices, in particular for Hermitian and unitary matrices,

$$|d\lambda_i| \leq \|dA\| ,$$

since for normal matrices $X_i = V_i$. Therefore the problem of determining the eigenvalues is *always stable* for normal matrices. For arbitrary matrices the problem of determining eigenvalues will be unstable only for a large distortion coefficient.

Whatever applies in determining the eigenvectors, formulas (5) and (6) show that

$$|dX_i| \leq \|dA\| \cdot |X_i| \cdot \sum_{j=1}^{n}{}' \frac{c_j}{|\lambda_i - \lambda_j|} , \tag{8}$$

so that a problem may be unstable only if at least one of the distortion coefficients is large or if the eigenvalues are close to each other.

We now introduce the results concerning changes in the eigenvalues of a real matrix for random changes in the matrix elements. Let the elements of a matrix $A$ be independent random values with mean values $a_{ij}$ and with one and the same dispersion $\sigma^2$. Then any real eigenvalue will be a random value which has in the first approximation the dispersion $c(\lambda)\sigma^2$, where $c(\lambda)$ is the distortion coefficient corresponding to this eigenvalue.[†] In case $a_{ij} = a_{ji}$, i.e. the matrix of mean values is symmetric but non-symmetric variations are permitted, the dispersion of each eigenvalue is equal to $\sigma^2$. For symmetric matrices with permissible symmetric change in their elements such a result is obtained if we assume that the dispersion of the diagonal elements is equal to $\sigma^2$ and for the non-diagonal elements to $1/2\sigma^2$. In this case the eigenvalues are (in the first approximation) independent random quantities. As to the coefficients of the characteristic polynomial, their dispersions may be large enough for each separate coefficient, however the coefficients will thereby no longer be independent and their probable changes are related in such a way that they do not produce large changes in the eigenvalues. Nevertheless, careless rounding off of coefficients when computing them may destroy their mutual relationships and lead then to incorrect results for the eigenvalues.

---

[†] D. K. Faddeev (4).

We shall consider an example which illustrates what was said above. We take

$$A = \begin{bmatrix} 5 & 7 & 6 & 5 \\ 7 & 10 & 8 & 7 \\ 6 & 8 & 10 & 9 \\ 5 & 7 & 9 & 10 \end{bmatrix},$$

$$A + dA = \begin{bmatrix} 5.1 & 7 & 6 & 5 \\ 7 & 10 & 8 & 7 \\ 6 & 8 & 10 & 9 \\ 5 & 7 & 9 & 10 \end{bmatrix}.$$

As we have seen, matrix $A$ is ill conditioned. The characteristic polynomials for matrices $A$ and $A + dA$, respectively, will be

$$t^4 - 35t^3 + 146t^2 - 100t + 1$$

and

$$t^4 - 35.1t^3 + 149t^2 - 110.6t + 7.8 ,$$

and therefore the eigenvalues of the first matrix will be (up to three decimal places) the numbers 0.010, 0.843, 3.858, 30.289 and for the second matrix the numbers 0.079, 0.844, 3.874 and 30.303. We shall see that the results agree completely with what was said above. Changing one element of a matrix by 0.1 produces a change in the eigenvalues of at most 0.069 when the characteristic polynomial's coefficients have been significantly changed (absolutely as well as relatively) at most by 10.6. However, a significantly smaller change in the coefficients of the characteristic polynomial may lead to a larger change in all or some of the roots. Thus if we "round off" the coefficient 35.1, replacing it by 35.0, then, as is easily computed, the largest root of the polynomial $t^4 - 35t^3 + 149t^2 - 110.6t + 7.8$ will be 30.185. Thus the change in the largest root exceeds 0.1.

Errors arising from possible instability do not depend, of course, on the choice of method for the problem's numerical solution, as distinct from errors arising from inevitable rounding off of intermediate results. We shall not consider the question of estimating these errors.

We observe that ill condition of a matrix in the sense of solving the system is in no way connected with the stability of the eigenvalue problem. In fact ill condition means only that among the eigenvalues are small ones in comparison with other eigenvalues.

## 42. A. N. KRYLOV'S METHOD

The work of A. N. Krylov (1) appeared as the first in a large series of works devoted to transforming secular equations to polynomial form.

A. N. Krylov's idea consisted in a preliminary transformation of the equation

$$\varphi(t) = \begin{vmatrix} a_{11} - t & a_{12} & \cdots & a_{1n} \\ a_{21} & a_{22} - t & \cdots & a_{2n} \\ \multicolumn{4}{c}{\cdots} \\ a_{n1} & a_{n2} & \cdots & a_{nn} - t \end{vmatrix} = 0 \tag{1}$$

to a generally equivalent equation of the form

$$D(t) = \begin{vmatrix} b_{11} - t & b_{12} & \cdots & b_{1n} \\ b_{21} - t^2 & b_{22} & \cdots & b_{2n} \\ \multicolumn{4}{c}{\cdots} \\ b_{n1} - t^n & b_{n2} & \cdots & b_{nn} \end{vmatrix} = 0 , \tag{2}$$

whose expansion in powers of $t$ is obviously accomplished with more simplicity by expanding the determinant in terms of minors of the first column.

For effecting the indicated transformation A. N. Krylov introduced a differential equation associated with the given matrix; he simultaneously raised the question of finding a purely algebraic transformation which would change equation (1) into equation (2).

The works of N. N. Luzin (1), (2), I. N. Khlodovskii (1), F. P. Gantmacher (1), and D. K. Faddeev (1) have been devoted to clarifying the algebraic nature of Krylov's transformation. We shall give the Krylov method in its algebraic interpretation.

Setting the determinant

$$\varphi(t) = \begin{vmatrix} a_{11} - t & a_{12} & \cdots & a_{1n} \\ a_{21} & a_{22} - t & \cdots & a_{2n} \\ \multicolumn{4}{c}{\cdots} \\ a_{n1} & a_{n2} & \cdots & a_{nn} - t \end{vmatrix} \tag{3}$$

equal to zero is a necessary and sufficient condition for the system of homogeneous equations

$$\begin{aligned} tx_1 &= a_{11}x_1 + a_{12}x_2 + \cdots + a_{1n}x_n , \\ tx_2 &= a_{21}x_1 + a_{22}x_2 + \cdots + a_{2n}x_n , \\ &\cdots\cdots\cdots\cdots\cdots , \\ tx_n &= a_{n1}x_1 + a_{n2}x_2 + \cdots + a_{nn}x_n \end{aligned} \tag{4}$$

to have a non-trivial solution $x_1, x_2, \cdots, x_n$.

We shall transform system (4) in the following way. We multiply the first equation by $t$ and replace $tx_1, \cdots, tx_n$ by their expressions (4) in terms of $x_1, \cdots, x_n$.

This gives

$$t^2 x_1 = b_{21}x_1 + b_{22}x_2 + \cdots + b_{2n}x_n , \tag{5}$$

where

$$b_{2k} = \sum_{s=1}^{n} a_{1s} a_{sk} . \tag{6}$$

Moreover we multiply equation (5) by $t$ and again replace $tx_1, tx_2, \cdots, tx_n$ by their expressions in terms of $x_1, \cdots, x_n$. We obtain

$$t^3 x_1 = b_{31} x_1 + b_{32} x_2 + \cdots + b_{3n} x_n \,.$$

Repeating this process $n - 1$ times we pass from system (4) to the system

$$
\begin{aligned}
t x_1 &= b_{11} x_1 + b_{12} x_2 + \cdots + b_{1n} x_n \,, \\
t^2 x_1 &= b_{21} x_1 + b_{22} x_2 + \cdots + b_{2n} x_n \,, \\
&\cdot \cdot \cdot \cdot \cdot \cdot \cdot \cdot \cdot \cdot \cdot \cdot \cdot \,, \\
t^n x_1 &= b_{n1} x_1 + b_{n2} x_2 + \cdots + b_{nn} x_n \,,
\end{aligned}
\tag{7}
$$

whose coefficient $b_{ik}$ will be determined by the recurrent formulas

$$b_{1k} = a_{1k}$$

$$b_{ik} = \sum_{s=1}^{n} b_{i-1,s} a_{sk} \qquad (i = 2, \cdots, n;\ k = 1, \cdots, n) \tag{8}$$

It is obvious that the determinant of system (7) will have the form (2).

System (7) has a non-zero solution for all values $t$ which satisfy the equation $\varphi(t) = 0$. Thus $D(t)$ becomes zero for all $t$ which are roots of the equation $\varphi(t) = 0$.

We shall show that

$$\frac{D(t)}{\varphi(t)} = \begin{vmatrix} 1 & 0 & \cdots & 0 \\ b_{11} & b_{12} & \cdots & b_{1n} \\ \cdot & \cdot & \cdot \cdot \cdot & \cdot \\ b_{n-1,1} & b_{n-1,2} & \cdots & b_{n-1,n} \end{vmatrix} = N \,, \tag{9}$$

i.e. for $N \neq 0$, $D(t)$ differs from the desired characteristic polynomial by only a numerical factor.

Let all the roots of $\varphi(t)$ be distinct. Since all roots of $\varphi(t)$ are roots of $D(t)$, $D(t)$ is divisible by $\varphi(t)$. Since, moreover, the degrees of $D(t)$ and $\varphi(t)$ are the same, the quotient must be a constant (independent of $t$). Comparing the coefficients for $t^n$ we obtain

$$\frac{D(t)}{\varphi(t)} = N \,.$$

When $\varphi(t)$ has multiple roots the equation

$$D(t) = N\varphi(t) \tag{10}$$

is preserved, which follows moreover from considerations of continuity.

It is also possible to verify this equation directly by multiplying the determinants which enter into it, using relation (8) for this.

It is apparent from equation (10) that if $N = 0$, then $D(t)$ is identically equal to zero. In this case the indicated transformation gives nothing. However, even for $N = 0$, A. N. Krylov proposed a special method whose algebraic nature will be explained below.

We turn now to the coefficients $b_{ik}$ which determine $D(t)$. We introduce for consideration vectors $B_i$ with components $b_{i1}, b_{i2}, \cdots, b_{in}$. The equalities

$$b_{ik} = \sum_{s=1}^{n} b_{i-1,s} a_{sk} \qquad (i = 2, \cdots, n)$$

show that

$$B_i = A'B_{i-1} , \qquad (11)$$

where $A'$ is the transpose of the given matrix.

From equality (11) it follows that

$$B_i = A'^{i-1}B_1 \qquad (i = 2, \cdots, n) .$$

In turn, $B_1 = A'B_0$, where $B_0 = (1, 0, \cdots, 0)'$. Thus, finally,

$$B_i = A'^i B_0 \qquad (i = 1, 2, \cdots, n) . \qquad (12)$$

It is obvious that it is possible to transform system (4) using, for example, the second equation of this system. In this case $t$ goes into the second column of the determinant $D(t)$ and the coefficients $b_{ik}$ will be determined by formulas (12), where $B_0 = (0, 1, \cdots, 0)'$.

A. N. Krylov's method may be generalized in a natural way if we introduce for consideration the vector $B_0 = (b_{01}, b_{02}, \cdots, b_{0n})'$.

Let

$$u = b_{01}x_1 + b_{02}x_2 + \cdots + b_{0n}x_n , \qquad (13)$$

where $x_1, x_2, \cdots, x_n$ is a solution to system (4).

Then by repeating the previous arguments we obtain:

$$
\begin{aligned}
u &= b_{01}x_1 + b_{02}x_2 + \cdots + b_{0n}x_n , \\
tu &= b_{11}x_1 + b_{12}x_2 + \cdots + b_{1n}x_n , \\
t^2u &= b_{21}x_1 + b_{22}x_2 + \cdots + b_{2n}x_n , \\
&\quad \cdots \cdots \cdots \cdots \cdots \cdots \cdots , \\
t^nu &= b_{n1}x_1 + b_{n2}x_2 + \cdots + b_{nn}x_n ,
\end{aligned} \qquad (14)
$$

where $B_i = (b_{i1}, b_{i2}, \cdots, b_{in})' = A'^i B_0$.

Considering the $n + 1$ equations of (14) as a system of linear homogeneous equations with $n + 1$ unknowns $u, x_1, \cdots, x_n$ we obtained that a non-zero solution is possible if and only if the determinant

$$
D(t) = \begin{vmatrix}
1 & b_{01} & \cdots & b_{0n} \\
t & b_{11} & \cdots & b_{1n} \\
\cdot & \cdot & \cdot & \cdot \\
t^n & b_{n1} & \cdots & b_{nn}
\end{vmatrix} = 0 . \qquad (15)
$$

Repeating the former arguments we find that

$$D(t) = \varphi(t)N ,$$

where this time

$$
N = \begin{vmatrix}
b_{01} & b_{02} & \cdots & b_{0n} \\
b_{11} & b_{12} & \cdots & b_{1n} \\
\cdot & \cdot & \cdot & \cdot \\
b_{n-1,1} & b_{n-1,2} & \cdots & b_{n-1,n}
\end{vmatrix} . \qquad (16)
$$

Just as for the special case considered above the transformation gives nothing if $N = 0$.

We therefore assume from the beginning that $N \neq 0$. On the basis of the equality $D(t) = N\varphi(t)$ the coefficients $p_i$ of the characteristic polynomial are defined as the ratios $(-1)^{n-1}N_i/N$ where $N_i$ are the algebraic cofactors of the elements $t^{n-i}$ in the determinant $D(t)$. Determination of the coefficients for the characteristic polynomial by the ratios indicated also forms an essential part of A. N. Krylov's work. However, the investigation makes it possible to determine the desired coefficients without computing minors, essentially shortening thereby the number of necessary operations.

In view of the fact that row elements of determinant (16) are components of the vectors $B_0, B_1, \cdots, B_{n-1}$, the condition $N \neq 0$ is equivalent to the linear independence of these vectors. Therefore for $N \neq 0$ the vectors $B_0$, $B_1, \cdots, B_{n-1}$ form a basis of the space. Consequently the vector $B_n$ is their linear combination:

$$B_n = q_1 B_{n-1} + \cdots + q_n B_0 . \tag{17}$$

We shall show that the coefficients of this relationship are also the coefficients $p_i$ of the characteristic polynomial written in the form:

$$\varphi(t) = (-1)^n[t^n - p_1 t^{n-1} - \cdots - p_n] .$$

In fact by subtracting from the last row of the determinant $D(t)$ a linear combination of the preceding rows with corresponding coefficients $q_1, q_2, \cdots, q_n$ we get, on the basis of equations (17),

$$D(t) = \begin{vmatrix} 1 & b_{01} & \cdots & b_{0n} \\ \cdots & \cdots & \cdots & \cdots \\ t^{n-1} & b_{n-1,1} & \cdots & b_{n-1,n} \\ t^n - q_1 t^{n-1} - \cdots - q_n & 0 & \cdots & 0 \end{vmatrix}$$
$$= (-1)^n[t^n - q_1 t^{n-1} - \cdots - q_n]N .$$

From here

$$\varphi(t) = \frac{D(t)}{N} = (-1)^n[t^n - q_1 t^{n-1} - \cdots - q_n] ,$$

which we were also required to prove.

Equality (17) permits us to find the coefficients $q_1 = p_1, q_2 = p_2, \cdots, q_n = p_n$ as the solution to the system of linear equations equivalent to this vector equation.

Equations (17) relate A. N. Krylov's method with the Cayley-Hamilton relationship (applied to the matrix $A'$).

In fact from the equation

$$A'^n = p_1 A'^{n-1} + \cdots + p_n E$$

if follows that

$$A'^n B_0 = p_1 A'^{n-1} B_0 + \cdots + p_n B_0 ,$$

that is,

$$B_n = p_1 B_{n-1} + \cdots + p_n B_0 . \tag{17}$$

It is obvious that in place of system (17) for determining the coefficients $p_i$ we may use the system

$$C_n = p_1 C_{n-1} + \cdots + p_n C_0 , \qquad (17')$$

where the vectors $C_n$ are determined by the equalities $C_k = A^k C_0$.

To determine the coefficients $p_i$ by solving system (17) or (17') it is necessary to carry out $(3/2)n^2(n + 1)$ multiplications and divisions. In its initial form A. N. Krylov's method required $(1/3)(n^4 + 4n^3 + 2n^2 - n - 3)$ multiplications and divisions.

If $N = 0$ a system equivalent to equations (17) does not permit us to determine the coefficients of the characteristic polynomial, since the determinant of this system is just equal to $N$.

The algebraic aspect of A. N. Krylov's method consists in the fact that it is possible to determine the coefficients for the *polynomial of least degree* $\theta(\lambda)$ such that $\theta(A)C_0 = 0$, i.e. the coefficients of the minimum annihilating polynomial. In general this will be the minimum polynomial of the matrix and its roots will coincide with all the roots of the characteristic polynomial but will have lower multiplicity. However, for an unsuccessful choice of the vector $C_0$ we may obtain in place of the minimum polynomial some divisor of it and then some of the roots for the equation $|A - tE| = 0$ may be lost. As N. N. Luzin and I. N. Khlodovskii have shown[†], it is possible to obtain as the polynomial, for a special choice of the vector $C_0$, any divisor of the minimum polynomial. This result has already been noted in Section 8.

If the minimum polynomial of a matrix does not coincide with the characteristic polynomial, then $N = 0$ for any choice of the vector $C_0$. In fact in this case $\psi(A)C_0 = 0$ and, since the degree of the polynomial $\psi(t)$ is *less than n*, the vectors $C_0, AC_0, \cdots, A^{n-1}C_0$ are linearly dependent. Additional degeneracy may be avoided by changing the initial vector $C_0$.

Thus A. N. Krylov's method makes it possible to determine the coefficients of the characteristic polynomial if $N \neq 0$, or of one of the divisors of the minimum polynomial if $N = 0$.

In practice the circumstance $N = 0$ will appear during the forward course for the solution of system (17) by Gauss's method. Namely, all coefficients are included in part of the equations *simultaneously* so that these equations become identities $0 = 0$. The equations (let their number be equal to $n + m$) must be rejected; in the rest of the system it is necessary to reject the $n - m$ last columns beginning with the column of constants (that is, the column of components for the vector $C_n$). The last of the remaining columns, consisting of components of the vector $C_m$, must be taken as a constant in the new system. Solving the system gives coefficients for $C_m$ linearly dependent on $C_0, C_1, \cdots, C_{m-1}$, i.e. the coefficients of the minimum annihilating polynomial for the vector $C_0$.

We now consider two examples, both taken from the article of A. N. Krylov (1).

---

† N. N. Luzin (1), (2); I. N. Khlodovskii (1),

As the first example we shall determine the coefficients of the characteristic polynomial for the matrix

$$\begin{bmatrix} -5.509882 & 1.870086 & 0.422908 & 0.008814 \\ 0.287865 & -11.811654 & 5.711900 & 0.058717 \\ 0.049099 & 4.308033 & -12.970687 & 0.229326 \\ 0.006235 & 0.269851 & 1.397369 & -17.596207 \end{bmatrix},$$

taken by A. N. Krylov from the work of Leverrier (1). By Leverrier's computations

$$\varphi(t) = t^4 + 47.888430t^3 + 797.2789t^2 + 5349.457t + 12296.555;$$
$$\lambda_1 = -17.86303; \quad \lambda_2 = -17.15266; \quad \lambda_3 = -7.57404; \quad \lambda_4 = -5.29870.$$

In Table IV.1 we give the scheme for computing the coefficients $p_i$, considering them as solutions of system (17').

**TABLE IV.1 Computation of the Coefficients for the Characteristic Equation by A. N. Krylov's Method**

|    | $C_0$ | $C_1$ | $C_2$ | $C_3$ | $C_4$ | $\Sigma$ |
|----|----|----|----|----|----|----|
|    | 1 | -5.509882 | 30.917951 | -179.01251 | 1100.7201 | 948.11566 |
| I  | 0 | 0.287865 | -4.705449 | 66.38829 | -967.5973 | -905.62659 |
|    | 0 | 0.049099 | 0.334184 | -23.08728 | 576.5226 | 553.81860 |
|    | 0 | 0.006235 | 0.002224 | -0.649152 | -4.04004 | -4.68073 |
| II |   | -5.166683 | 26.548910 | -136.3606 | 705.6054 |  |
|    |   | 1 | -16.34603 | 230.62300 | -3361.2884 | -3146.0115 |
|    |   |   | 1.136758 | -34.41064 | 741.5585 | 708.18462 |
|    |   |   | 0.104141 | -2.087086 | 16.91759 | 14.93465 |
| III |  |   | 1 | -30.27086 | 652.3451 | 623.0742 |
|    |   |   |   | 1.065352 | -51.01828 | -49.95292 |
|    |   |   |   | 1 | -47.8887 | -46.8887 |
|    |   |   | 1 |   | -797.287 | -796.287 |
|    |   | 1 |   |   | -5349.53 | -5348.53 |
|    | 1 |   |   |   | -12296.8 | -12295.8 |

Here in part I are placed in succession the components being computed for the vectors $A^k C_0 (k = 0, 1, 2, 3, 4)$ and the usual control sums. In row II is given the control, analogous to that applied in Section 30 when computing successive iterations. In part III is contained the solution of the system obtained, which we find by the single-division scheme.

Final control of the computation for the coefficients $p_i$ is a comparison of the value $p_1$ with the spur of the matrix. Since Sp $A = -47.888430$, we see that the value $p_1$ found from the system's solution is accurate enough.

Comparison with Leverrier's data shows however that the coefficients are computed with a lower degree of accuracy. The known loss in accuracy is an inherent insufficiency in A. N. Krylov's method and is explained by the fact that the coefficients of the system which determines $p_i$ are quantities of different orders; this leads as a rule to a poor condition for the system.

A slightly better result may be obtained by applying the scheme of principal elements in solving the system.

As the second example we shall take another matrix from A. N. Krylov's article (1):

$$\begin{bmatrix} 5 & 30 & -48 \\ 3 & 14 & -24 \\ 3 & 15 & -25 \end{bmatrix}. \qquad (18)$$

In the table presented below the three first rows contain the coefficients of the system for determining $q_i$:

| 1 | 5 | $-29$ | 125 |
|---|---|-------|-----|
| 0 | 3 | $-15$ | 63 |
| 0 | 3 | $-15$ | 63 |
|   | 0 | 0 | 0 |

The first step of Gauss's process shows already that the case of degeneracy holds here. The reduced system

$$q_1 + 5q_2 = -29 ,$$
$$3q_2 = -15$$

gives $q_2 = -5$, $q_1 = -4$ as a solution. Thus in this case we have determined the coefficients of the *second degree* polynomial $t^2 + 5t + 4$, which is a divisor of the characteristic polynomial.

When using Krylov's method in practice the situation of exact degeneracy can occur only in special circumstances (for example, when in an essentially physical problem being solved to investigate the eigenvalues of a matrix, one must have, in the physical sense, a minimum polynomial distinct from the characteristic one). The case of approximate degeneracy is encountered more frequently. In this case the system of equations (17) or (17') turns out to be ill conditioned, but this does not destroy the process. In fact in carrying out the forward course process for solving the system by the elimination method, it is necessary to stop only when the coefficients of all the equations, beginning with a certain one, are close to zero within the limits of accuracy for the computation; it is then necessary to act as if they had become zero exactly.

In conclusion we observe that for a matrix of the form

$$\begin{bmatrix} a_{11} & a_{12} & \cdots & & a_{1n} \\ a_{21} & a_{22} & \cdots & & a_{2n} \\ 0 & a_{32} & \cdots & & a_{3n} \\ \cdot & \cdot & \cdot & \cdot & \cdot \\ 0 & 0 & \cdots & a_{n-1,n} & a_{nn} \end{bmatrix}.$$

under the condition that the elements $a_{21}$, $a_{32}$, $\cdots$, $a_{n-1,n}$ are different from zero, the system of equations (17) constructed from the initial vector

$$C_0 = \begin{bmatrix} 1 \\ 0 \\ \cdot \\ \cdot \\ \cdot \\ 0 \end{bmatrix}.$$

will be triangular, since the vector $AC_0$ has all its components except the first two equal to zero, the vector $A^2 C_0$ has all its components except the first three equal to zero, and so forth. In particular such conditions hold for a tridiagonal matrix.

## 43. DETERMINATION OF EIGENVECTORS BY A. N. KRYLOV'S METHOD

We shall assume that, for A. N. Krylov's method, the coefficients of the characteristic polynomial have been found and that all the eigenvalues have been computed and have turned out to be distinct. We shall show how to determine the eigenvectors of the matrix, using the computations which have been carried out. Let $C_0$ be the initial vector in A. N. Krylov's process, and let $X_1$, $X_2$, $\cdots$, $X_n$ be the eigenvectors of matrix $A$ belonging $\lambda_1$, $\lambda_2$, $\cdots$, $\lambda_n$. According to theorem 6.3 the vectors $X_1$, $\cdots$, $X_n$ are linearly independent.

We expand the vector $C_0$ in terms of the eigevectors:

$$C_0 = \alpha_1 X_1 + \alpha_2 X_2 + \cdots + \alpha_n X_n . \tag{1}$$

Then

$$C_1 = AC_0 = \alpha_1 \lambda_1 X_1 + \alpha_2 \lambda_2 X_2 + \cdots + \alpha_n \lambda_n X_n ,$$
$$\cdots \cdots \cdots \cdots \cdots \cdots , \tag{2}$$
$$C_{n-1} = A^{n-1} C_0 = \alpha_1 \lambda_1^{n-1} X_1 + \alpha_2 \lambda_2^{n-1} X_2 + \cdots + \alpha_n \lambda_n^{n-1} X_n .$$

The vectors $C_1$, $C_2$, $\cdots$, $C_{n-1}$ are computed in the process of finding the eigenvalues. We shall show that the eigenvectors may be obtained as linear combinations

$$\beta_{i0} C_{n-1} + \beta_{i1} C_{n-2} + \cdots + \beta_{in-1} C_0 \qquad (i = 1, 2, \cdots, n)$$

for a suitable choice of the coefficients $\beta_{ij}$. We consider the linear combination:

$$\begin{aligned}
&\beta_{10} C_{n-1} + \beta_{11} C_{n-2} + \cdots + \beta_{1,n-1} C_0 \\
&\quad = \alpha_1 (\beta_{10} \lambda_1^{n-1} + \beta_{11} \lambda_1^{n-2} + \cdots + \beta_{1,n-1}) X_1 \\
&\qquad + \alpha_2 (\beta_{10} \lambda_2^{n-1} + \beta_{11} \lambda_2^{n-2} + \cdots + \beta_{1,n-1}) X_2 \\
&\qquad + \cdots \cdots \cdots \cdots \cdots \cdots \\
&\qquad + \alpha_n (\beta_{10} \lambda_n^{n-1} + \beta_{11} \lambda_n^{n-2} + \cdots \beta_{1,n-1}) X_n \\
&\quad = \alpha_1 \varphi_1(\lambda_1) X_1 + \alpha_2 \varphi_1(\lambda_2) X_2 + \cdots + \alpha_n \varphi_1(\lambda_n) X_n ,
\end{aligned} \tag{3}$$

where

$$\varphi_1(t) = \beta_{10}t^{n-1} + \beta_{11}t^{n-2} + \cdots + \beta_{1,n-1} . \tag{4}$$

We select the coefficients $\beta_{10}, \cdots, \beta_{1,n-1}$ such that

$$\varphi_1(\lambda_1) \neq 0 ,$$
$$\varphi_1(\lambda_2) = \cdots = \varphi_1(\lambda_n) = 0 . \tag{5}$$

For this it is sufficient to take as $\varphi_1(t)$ the polynomial

$$\begin{aligned}
\varphi_1(t) &= (t - \lambda_2)\cdots(t - \lambda_n) \\
&= \frac{(t - \lambda_1)(t - \lambda_2)\cdots(t - \lambda_n)}{t - \lambda_1} = \frac{(-1)^n\varphi(t)}{t - \lambda_1} \\
&= \frac{t^n - p_1 t^{n-1} - \cdots - p_n}{t - \lambda_1} .
\end{aligned} \tag{6}$$

Here $\varphi(t)$ is the characteristic polynomial, whose coefficients and roots have already been computed.

The coefficients of quotient (6) are easily computed by Horner's scheme, i.e. by the recurrence formulas

$$\beta_{10} = 1 ,$$
$$\beta_{1j} = \lambda_1\beta_{1,j-1} - p_j \quad (j = 1, \cdots, n-1) . \tag{7}$$

Thus

$$\beta_{10}C_{n-1} + \beta_{11}C_{n-2} + \cdots + \beta_{1n-1}C_0 = \alpha_1\varphi_1(\lambda_1)X_1 ,$$

i.e. the linear combination formed by us is the eigenvector $X_1$ up to a numerical factor. Of course the coefficient $\alpha_1$ must be different from zero; this guarantees successful completion of A. N. Krylov's method. Since the eigenvector is determined up to a constant multiplier, we may accept the constructed linear combination as an eigenvector.

Analogously,

$$X_i = \sum_{j=0}^{n-1} \beta_{ij}C_{n-1-j} , \tag{8}$$

where

$$\beta_{i0} = 1, \beta_{ij} = \lambda_i\beta_{i,j-1} - p_j \quad (j = 1, \cdots, n - 1) . \tag{9}$$

As an example, we shall compute an eigenvector for Leverrier's matrix which belongs to the eigenvalue $\lambda_4 = -5.29870$.

We write out the characteristic equation with the coefficients taken from Table IV.1:

$$t^4 + 47.8887t^3 + 797.287t^2 + 5349.53t + 12296.8 = 0.$$

Computing the numbers $\beta_{4j}$ for $j = 0, 1, 2, 3$, we obtain

$$1; 42.5900; 571.615; 2320.71 .$$

We shall form linear combinations according to formula (8), putting the computations in the table:

**TABLE IV.2  Computation of an Eigenvector by A. N. Krylov's Method**

| $\beta_{43}C_0$ | $\beta_{42}C_1$ | $\beta_{41}C_2$ | $\beta_{40}C_3$ | $X_4$ | $\tilde{X}_4$ |
|---|---|---|---|---|---|
| 2320.71 | −3149.53 | 1316.80 | −179.01 | 308.97 | 1 |
| 0 | 164.548 | −200.405 | 66.388 | 30.531 | 0.098815 |
| 0 | 28.066 | 14.233 | −23.087 | 19.212 | 0.062181 |
| 0 | 3.5640 | 0.0947 | −0.6492 | 3.0095 | 0.009741 |

The last column contains the components of the eigenvector $X_4$, normalized so that its first component is equal to one. We shall see below that the component values for an eigenvector computed by A. N. Krylov's method agree well with values computed by other methods.

## 44.  HESSENBERG'S METHOD

In Hessenberg's method,[†] just as in Krylov's method, a null linear combination of vectors $C_0$, $AC_0$, $\cdots$, $A^{n-1}C_0$ is found. Whereas in Krylov's method the coefficients of such a linear combination are found by solving the system of linear equations (17), in Hessenberg's method the desired linear combination is obtained as the last vector in a recurrently constructed sequence of vectors $Z_1$, $\cdots$, $Z_n$, $Z_{n+1}$ for $Z_1 = C_0$ such that the vector $Z_{j+1}$ its first $j$ components equal to zero. Each successive vector is obtained by iteration of the preceding one, with a successive "correction" obtained by adding the suitable linear combination of all the previous vectors. In other words,

$$Z_{j+1} = \begin{bmatrix} 0 \\ \vdots \\ 0 \\ z_{j+1,j+1} \\ \vdots \\ z_{n,j+1} \end{bmatrix} = AZ_j + h_{1j}Z_1 + \cdots + h_{jj}Z_j. \qquad (1)$$

It is convenient to take as the vector $Z_1$ the vector $(1, 0, \cdots, 0)'$. The indicated process may not always be realizable. The natural course of the process is destroyed if $z_{ii} = 0$ is obtained at any step.

We shall consider the basic case in which $z_{11} \neq 0$, $\cdots$, $z_{nn} \neq 0$. In this case the vectors $Z_1$, $\cdots$, $Z_n$ turn out to be linearly independent, so that the matrix

$$Z = [Z_1, Z_2, \cdots, Z_n]$$

will be non-singular. It is clear that

$$Z_{j+1} = \varphi_j(A)Z_1,$$

---

† Zurmühl (3).

where $\varphi_j(t) = t^j + \cdots$ is a certain polynomial of degree $j$. Because of the linear independence of the vectors $Z_1, \cdots, Z_n$, the equality $f(A)Z_1 = 0$ is impossible only if the degree of polynomial $f$ is at least $n$. The polynomial $\varphi_n$ annihilates $Z_1$ and its degree is equal to $n$. Consequently the polynomial $\varphi_n$ coincides with the characteristic polynomial of matrix $A$.

The polynomials $\varphi_j$ are obviously related by the recurrent relationships

$$\varphi_j(t) = (t + h_{jj})\varphi_{j-1}(t) + h_{j-1,j}\varphi_{j-2}(t) + \cdots + h_{1j}\varphi_0(t) \tag{3}$$

where $\varphi_0(t) = 1$, $j = 1, \cdots, n$.

Thus the coefficients of the characteristic polynomial are determined as soon as all the coefficients $h_{ij}$ are known.

It is not hard to see that the system of vector equations

$$Z_2 = AZ_1 + h_{11}Z_1 \,,$$
$$Z_3 = AZ_2 + h_{12}Z_1 + h_{22}Z_2 \,,$$
$$\cdots \cdots \cdots \cdots \cdots \cdots \cdots \cdots \cdots \cdots ,$$
$$0 = Z_{n+1} = AZ_n + h_{1n}Z_1 + \cdots + h_{nn}Z_n$$

is equivalent to the matrix equation

$$AZ + ZH = 0 \tag{4}$$

where

$$H = \begin{bmatrix} h_{11} & h_{12} \cdots & h_{1n} \\ -1 & h_{22} \cdots & h_{2n} \\ 0 & -1 \cdots & h_{3n} \\ \cdot & \cdot \cdot \cdot \cdot \cdot \cdot & \cdot \cdot \\ 0 & 0 \cdots -1 & h_{nn} \end{bmatrix} . \tag{5}$$

The last equality allows us to determine successively all the elements of matrices $H$ and $Z$.

To clarify the computational scheme it is advisable to represent the equality $AZ + ZH = 0$ in the form

$$(A/Z)\left(\frac{Z}{H}\right) = 0 \,.$$

Here $(A/Z)$ and $(Z/H)$ are rectangular matrices consisting of matrices $A$, $Z$, and $H$, in the order indicated. We shall now form the following scheme:

$$
\begin{array}{ccccccc}
a_{11} & a_{12} \cdots a_{1n} & z_{11} & 0 & \cdots & & 0 \\
a_{21} & a_{22} \cdots a_{2n} & z_{21} & z_{22} & \cdots & & 0 \\
\cdot & \cdot \cdot \cdot \cdot \cdot \cdot \cdot \cdot & \cdot & \cdot & \cdot \cdot & \cdot & \cdot \\
a_{n1} & a_{n2} \cdots a_{nn} & z_{n1} & z_{n2} & \cdots & & z_{nn} \\
& & h_{11} & h_{12} & \cdots & & h_{1n} \\
& & -1 & h_{22} & \cdots & & h_{2n} \\
& & \cdot & \cdot & \cdot \cdot & \cdot & \cdot \\
& & 0 & 0 & \cdots & -1 & h_{nn}
\end{array}
$$

The first $n$ rows of this scheme are formed by the matrix $(A/Z)$, the last

**TABLE IV.3**   Determining the Coefficients of the Characteristic Polynomial by Hessenberg's Method

| $\left(\dfrac{A'}{Z'}\right)$ | | | | $\Sigma$ | $\left(\dfrac{Z}{H}\right)$ | | | |
|---|---|---|---|---|---|---|---|---|
| -5.509882 | 0.287865 | 0.049099 | 0.006235 | -5.166683 | 1 | 0 | 0 | 0 |
| 1.870086 | -11.811654 | 4.308033 | 0.269851 | -5.363684 | 0 | 0.287865 | 0 | 0 |
| 0.422908 | 5.711900 | -12.970687 | 1.397369 | -5.438510 | 0 | 0.049099 | 1.1367579 | 0 |
| 0.008814 | 0.058717 | 0.229326 | -17.556207 | -17.299350 | 0 | 0.006235 | 0.1041410 | 1.0653609 |
| 1 | 0 | 0 | 0 | 1 | -5.509882 | -0.55915162 | -0.48166191 | -0.00939009 |
| 0 | 0.287865 | 0.049099 | 0.006235 | 0.343199 | -1 | 10.836146 | -22.577119 | -0.21730602 |
| 0 | 0 | 1.1367579 | 0.1041410 | 1.2408990 | 0 | -1 | 13.924832 | -0.20553668 |
| 0 | 0 | 0 | 1.0653609 | 1.0653609 | 0 | 0 | -1 | 17.617570 |

| $k \diagdown i$ | 4 | 3 | 2 | 1 | 0 | $\varphi_i(1)$ | $i$ |
|---|---|---|---|---|---|---|---|
| | 0 | 0 | 0 | 0 | 1 | 1 | 0 |
| | 0 | 0 | 0 | 1 | 5.509882 | 6.509882 | 1 |
| | 0 | 0 | 1 | 16.346028 | 59.146734 | 76.492762 | 2 |
| | 0 | 1 | 30.270860 | 264.18531 | 698.72941 | 994.18558 | 3 |
| | 1 | 47.888430 | 797.27877 | 5349.4555 | 12296.551 | 18492.173 | 4 |

$n$ columns by the matrix $(Z/H)$. At the beginning of the process we know matrix $A$ and the first column of matrix $Z$. Multiplying the first row of matrix $(A/Z)$ by the first column of matrix $(Z/H)$ permits us to determine the element $h_{11}$; multiplying the remaining rows by the first column of matrix $(Z/H)$ gives the elements $z_{22}, \cdots, z_{n2}$ respectively. As soon as these elements have been determined, multiplying matrix $(A/Z)$ by the second column of matrix $(Z/H)$ gives successively the elements $h_{12}, h_{22}, z_{33}, \cdots, z_{n3}$. We then multiply the matrix $(A/Z)$ by the third, $\cdots$, $n$th columns of matrix $(Z/H)$. The computations permit the usual control using small sums.

In computing on desk calculators it is advisable to set up the computations in the form

$$\begin{bmatrix} A' & Z \\ Z' & H \end{bmatrix}$$

to replace row-column $n$ multiplication by multiplication of columns. The dual notation for matrix $Z$ is justified by the uniqueness property of the operations.

The coefficients of the characteristic polynomial are determined in the same way as the numbers $h_{ii}$ by the recurrent relationships arising from relationships (3). For control we compute recurrently the values of polynomials for $t = 1$, equal to the sums of their coefficients.

Table IV.3 displays the coefficients of the characteristic polynomial for Leverrier's matrix.

The left member of equation (4) was computed for control. The element greatest in modulus of the obtained matrix turned out to be equal to 0.0000004.

As soon as an eigenvalue is found it is easy to determine the eigenvector belonging to it. In fact matrix equation (4) shows that $A = -ZHZ^{-1}$, i.e. that matrix $A$ is similar to matrix $-H$. As a result the eigenvector $X_i$ of matrix $A$ belonging to the eigenvalue $\lambda_i$ is connected by the relationship

$$X_i = ZY_i$$

with the eigenvector $Y_i$ of matrix $H$ belonging the eigenvalue $-\lambda_i$. These eigenvectors for matrix $H$ are determined without difficulty since, given arbitrarily the last component of an eigenvector, we can determine the rest of its components from the system of linear equations:

$$-y_1 + (h_{22} + \lambda)y_2 + \cdots + h_{2n}y_n = 0 , \tag{6}$$
$$\cdots \cdots \cdots \cdots \cdots \cdots \cdots \cdots \cdots \cdots \cdots$$
$$-y_{n-1} + (h_{nn} + \lambda)y_n = 0$$

with a triangular matrix. The discarded first equation may be used for control.

As an example, we determine from Table IV.4 an eigenvector of Leverrier's matrix belonging to the lowest eigenvalue in modulus. The approximation to this eigenvalue computed as a zero of the polynomial

$$t^4 + 47.888430t^3 + 797.27877t^2 + 5349.4555t + 12296.551$$

**TABLE IV.4  Determining an Eigenvector by Hessenberg's Method**

| | $H + \lambda_4 E$ | | | $Y_4$ | $X_4$ | $\tilde{X}_4$ |
|---|---|---|---|---|---|---|
| 0.211182 | $-0.559152$ | $-0.481662$ | $-0.009390$ | 308.95220 | 308.9522 | 1.000000 |
| $-1$ | 5.537446 | $-22.577119$ | $-0.217306$ | 106.05866 | 30.5306 | 0.098820 |
| | $-1$ | 8.626132 | $-0.205537$ | 12.318870 | 19.2109 | 0.062181 |
| | | $-1$ | 12.318870 | 1.000000 | 3.0095 | 0.009741 |

is equal to $-5.298700$.

The result of substituting the components of the vector $Y_4$ into the first equation of system (6) yields $-0.0007$.

It is possible to approach the method from a slightly different point of view. Namely, the method may be interpreted as a method for changing a given matrix $A$ by a similarity transformation to the special form

$$
-\begin{bmatrix}
h_{11} & h_{12} \cdots & & h_{1n} \\
-1 & h_{22} \cdots & & h_{2n} \\
& \cdot \cdot \cdot \cdot \cdot \cdot \cdot \cdot \\
0 & 0 \cdots -1 & h_{nn}
\end{bmatrix},
$$

where the transformation matrix $Z$ is triangular. The polynomials $\varphi_1(t)$, $\varphi_2(t)$, $\cdots$, $\varphi_n(t)$, as is easy to see, will be characteristic polynomials for the matrices

$$
-[h_{11}], \quad -\begin{bmatrix} h_{11} & h_{12} \\ -1 & h_{22} \end{bmatrix}, \quad -\begin{bmatrix} h_{11} & h_{12} & h_{13} \\ -1 & h_{22} & h_{23} \\ 0 & -1 & h_{33} \end{bmatrix},
$$

and so forth. In fact by expanding the characteristic determinant of such a matrix in terms of elements of the last row we get the previous recurrent relationships (3).

We shall now consider some exceptional cases which may be encountered during the process.

It may happen that at the $i$th step of the process $z_{ii} = 0$ but at least one component of the vector $Z_i$ is different from zero. In this case we fill in the column for $Z_{i+1}$ and place zeros in it only where it is actually possible to attain the value zero by adding a linear combination of the preceding columns. This process is continued on in this way. If it then turns out that we construct $n$ non-zero vectors $Z_1, \cdots, Z_n$, than they will automatically be linearly independent and the matrix $Z$ (no longer triangular in the given case but obtained from a triangular matrix by rearranging columns) will bring about a similarity transformation of matrix $A$ into matrix $H$. But it may happen that at some step we encounter a null vector. This will obviously occur if and only if the vectors $Z_1, AZ_1, \cdots, A^{n-1}Z_1$ are linearly dependent, that is, if the minimum annihilating polynomial for $Z_1$ is only a divisor of the characteristic polynomial.

In this case the approach to the null vector shortens the process. The last polynomial $\varphi_i(t)$ will be the minimum annihilating polynomial for the

**TABLE IV.5  Hessenberg's Method for the Degenerate Case**

| 2 | 0 | 3 | −1 | 1 | 0 | 0 | 0 |
|---|---|---|----|---|---|---|---|
| 1 | 2 | 1 | 1 | 0 | 0 | 0 | 14 |
| −1 | 2 | 1 | 3 | 0 | 3 | 0 | 0 |
| −1 | 6 | 2 | 7 | 0 | −1 | 7/3 | 0 |
| 1 | 0 | 0 | 0 | −2 | 2 | 7/3 | −14 |
| 0 | 0 | 3 / 0 | −1 | −1 | −1/3 | −14/9 | −14/3 |
| 0 | 0 | 0 / 0 | 7/3 | 0 | −1 | −23/3 | −8 |
| 0 | 14 | 0 | 0 | 0 | 0 | −1 | −2 |

vector $Z_1$ and its roots will form only a part of the eigenvalue spectrum.

We shall consider two examples which illustrate both possible exceptional cases (Table IV.5 and Table IV.6).

From Table IV.5 we see that already in the first example $z_{22} = 0$. However the process can still be carried through to the end. We shall write a shorter course for the process. We shall determine successively: from $(1 \times 1)h_{11} = -2$, from $(2 \times 1)z_{22} = 0$, from $(3 \times 1)z_{32} = 3$. This allows us to consider $z_{33} = z_{34} = 0$. Moreover $(4 \times 1)$ we determine $z_{42} = -1$, from $(1 \times 2)$ $h_{12} = 2$, from $(2 \times 2)z_{23} = 0$, from $(3 \times 2)h_{22} = -1/3$, from $(4 \times 2)z_{43} = 7/3$. We let $z_{44} = 0$. From $(1 \times 3)$ we determine $h_{13} = 7/3$, from $(2 \times 3)z_{24} = 14$, from $(3 \times 3)h_{23} = -14/9$, from $(4 \times 3)h_{33} = -23/3$. Finally from $(1 \times 4)$ determine $h_{14} = -14$, from $(2 \times 4)h_{44} = -2$, from $(3 \times 4)h_{24} = -14/3$, from $(4 \times 4)h_{34} = -8$.

After determining matrix $H$ we determine the characteristic polynomial, as was shown above.

As the second example we shall consider matrix (18) of Section 42. In this case (Table IV.6) $Z_3 = (0, 0, 0)'$ and we compute only a divisor of the characteristic polynomial. Namely, we have $\varphi_0 = 1$, $\varphi_1 = t - 5$, $\varphi_2 = (t + 10)$ $(t - 5) + 54 = t^2 + 5t + 4$.

In concluding this paragraph we observe that when computing the coefficients of the characteristic polynomial it is possible to avoid using recurrent relationships for the polynomials $\varphi_i$. Instead of this one may find the characteristic polynomial $\varphi_n$ (or a divisor of it in the case of degeneracy)

**TABLE IV.6  Hessenberg's Method for the Degenerate Case**

| 5 | 30 | −48 | 1 | 0 | 0 |
|---|----|-----|---|---|---|
| 3 | 14 | −24 | 0 | 30 | 0 |
| 3 | 15 | −25 | 0 | −48 | 0 |
| 1 | 0 | 0 | −5 | 54 | |
| 0 | 30 | −48 | −1 | 10 | |
| 0 | 0 | 0 | 0 | −1 | |

directly by applying A. N. Krylov's method[†] to the matrix $-H$ which leads, as we saw above, to a solution for the triangular system. Thus in the last example it is necessary to apply A. N. Krylov's method to the matrix

$$B = \begin{bmatrix} 5 & -54 \\ 1 & -10 \end{bmatrix}.$$

We have

$$C_0 = \begin{bmatrix} 1 \\ 0 \end{bmatrix},$$

$$BC_0 = \begin{bmatrix} 5 \\ 1 \end{bmatrix},$$

$$B^2 C_0 = \begin{bmatrix} -29 \\ -5 \end{bmatrix}.$$

Consequently the coefficients of the polynomial $\varphi_2$ are determined from the system

$$p_2 - 5p_1 = -29 ,$$

$$p_1 = -5 ,$$

From which $p_1 = -5$, $p_2 = -4$, i.e. $\varphi_2 = t^2 + 5t + 4$.

## 45. SAMUELSON'S METHOD

A method close to A. N. Krylov's is the one proposed by Samuelson (1).

The computational scheme of this method goes as follows. We compute the rectangular matrix

$$\begin{bmatrix} R & : 0 & 0 & 0 & 0 & \cdots 0 & 0 & 1 & -a_{11} \\ RM & : 0 & 0 & 0 & 0 & \cdots 0 & 1 & -a_{11} & -RS \\ RM^2 & : 0 & 0 & 0 & 0 & \cdots 1 & -a_{11} & -RS & -RMS \\ \cdots & \vdots & \cdots & \cdots & \cdots & \cdots & \cdots & \cdots \\ RM^{n-1} & : 1 & -a_{11} & -RS & -RMS & \cdots & \cdots & \cdots & -RM^{n-2}S \end{bmatrix}, \quad (1)$$

where $R$, $S$, and $M$ are cells in the following partition of the given matrix:

$$A = \begin{bmatrix} a_{11} & a_{12} & \cdots & a_{1n} \\ a_{21} & a_{22} & \cdots & a_{2n} \\ \cdots & \cdots & \cdots & \cdots \\ a_{n1} & a_{n2} & \cdots & a_{nn} \end{bmatrix} = \begin{bmatrix} a_{11} & R \\ S & M \end{bmatrix}. \quad (2)$$

Moreover by elementary transformations (as was done in the elimination problem, Section 22) we need to get a null row in place of the row $RM^{n-1}$. Then the remaining elements of the last row generally give the coefficients of the characteristic polynomial. The elimination process, as we saw, was uniform and the result unique. This is basic merit of the scheme.

We will carry out this scheme via the transformation of the system of

---

[†] Saibel and Berger (1).

linear differential equations associated with the matrix to obtain a differential equation of order $n$ by a special elimination method. A brief algebraic foundation for the scheme follows.

Let

$$X_0 = \begin{bmatrix} x_{10} \\ x_{20} \\ \cdot\ \cdot \\ x_{n0} \end{bmatrix} = \begin{bmatrix} x_{10} \\ Y_0 \end{bmatrix} \tag{3}$$

be an arbitrary vector.

Furthermore let

$$AX_0 = \begin{bmatrix} x_{11} \\ x_{21} \\ \cdot \\ \cdot \\ x_{n1} \end{bmatrix} = \begin{bmatrix} x_{11} \\ Y_1 \end{bmatrix} ;$$

$$\cdots \cdots \cdots \cdots ;$$

$$A^{n-1}X_0 = \begin{bmatrix} x_{1,n-1} \\ x_{2,n-1} \\ \cdot \\ \cdot \\ x_{n,n-1} \end{bmatrix} = \begin{bmatrix} x_{1,n-1} \\ Y_{n-1} \end{bmatrix} ; \tag{4}$$

$$A^{n}X_0 = \begin{bmatrix} x_{1n} \\ x_{2n} \\ \cdot \\ \cdot \\ x_{nn} \end{bmatrix} = \begin{bmatrix} x_{1n} \\ Y_n \end{bmatrix} .$$

It follows from the construction that

$$x_{1k} = a_{11}x_{1,k-1} + RY_{k-1} , \tag{5}$$

$$Y_k = Sx_{1,k-1} + MY_{k-1} \qquad (k = 1, \cdots, n) . \tag{6}$$

Thus we have $n^2$ relationships ($n$ and $n(n-1)$) among the $n^2 + n$ values. These make it possible to eliminate from the system of equalities (5) and (6) the vectors $Y_1, \cdots, Y_n$, i.e. $n(n-1)$ values. As a result of this elimination there remain $n$ equalities which relate $2n$ numbers, namely, the components of the vector $Y_0$ and the numbers $x_{10}, \cdots, x_{1n}$.

We shall carry out this elimination. For $k = 1, 2, \cdots, n$ we have:

$$x_{1k} = a_{11}x_{1,k-1} + RY_{k-1}$$
$$= a_{11}x_{1,k-1} + RSx_{1,k-2} + RMY_{k-2}$$
$$= a_{11}x_{1,k-1} + RSx_{1,k-2} + RMSx_{1,k-3} + RM^2 Y_{k-3}$$
$$= \cdot\ \cdot\ \cdot\ \cdot\ \cdot\ \cdot\ \cdot\ \cdot\ \cdot\ \cdot\ \cdot\ \cdot\ \cdot\ \cdot$$
$$= a_{11}x_{1,k-1} + RSx_{1,k-2} + RMSx_{1,k-3}$$
$$+ \cdots + RM^{k-2}Sx_{10} + RM^{k-1}Y_0 ,$$

or

$$RM^{k-1}Y_0 = x_{1k} - a_{11}x_{1,k-1} - RS\,x_{1,k-2} - \cdots - RM^{k-2}Sx_{10}\,. \tag{7}$$

The coefficients of these $n$ equalities form the matrix (1).

Eliminating from these $n$ equalities the components of the vector $Y_0$ we obtain one linear relationship among the numbers $x_{10}, \cdots, x_{1n}$ with constant coefficients, independent of the choice of initial vector.

On the other hand, from the Cayley-Hamilton relationship we have

$$x_{1n} - p_1 x_{1,n-1} - \cdots - p_n x_{10} = 0\,.$$

This equality is also a linear dependence among the numbers $x_{10}, \cdots, x_{1n}$ with constant coefficients and is not dependent on the choice of the vector.

This dependence will coincide with the dependence obtained by the elimination method in the case where matrix $A$ is such that we may consider the numbers $x_{10}, \cdots, x_{1,n-1}$ to be independent variables, i.e. where we may assign arbitrary values to them independent of each other by a suitable choice of the remaining components for the initial vector $X_0$ or, in other words, by the vector $Y_0$.

It is possible to arrive at Samuelson's method more economically by the following relationship among coefficients of characteristic polynomials for the bordered matrices and matrices which are being bordered.

Let

$$\begin{aligned}
\varphi(t) &= (-1)^n(t^n + p_1 t^{n-1} + \cdots + p_n)\,, \\
f(t) &= (-1)^n(t^{n-1} + q_1 t^{n-2} \cdots + q_{n-1})
\end{aligned} \tag{8}$$

be the characteristic polynomials of matrices $A$ and $M$. (In place of the usual notation for polynomials we have changed signs for the coefficient $p_k$ and $q_k$).

Then the following relationships are valid:

$$\begin{aligned}
p_1 &= -a_{11} + q_1\,, \\
p_2 &= -RS - q_1 a_{11} + q_2\,, \\
p_3 &= -RMS - q_1 RS - q_2 a_{11} + q_3\,, \\
&\;\;\cdot\;\;\cdot\;\;\cdot\;\;\cdot\;\;\cdot\;\;\cdot\;\;\cdot\;\;\cdot\;\;\cdot\;\;\cdot\;\;\cdot\;\;\cdot\;\;\cdot\;\;\cdot\;\;, \\
p_{n-1} &= -RM^{n-3}S - q_1 RM^{n-4}S - \cdots + q_{n-1}\,, \\
p_n &= -RM^{n-2}S - q_1 RM^{n-3}S - \cdots - q_{n-2}RS - q_{n-1}a_{11}\,.
\end{aligned} \tag{9}$$

These relationships are obtained from the rule for expanding a bordered determinant.

Moreover if we apply A. N. Krylov's method to matrix $M$, taking $R'$ (with components $a_{12}, \cdots, a_{1n}$) for the initial vector, then the coefficients $q_1, q_2, \cdots, q_{n-1}$ will be determined from the system of equations

$$M'^{n-1}R' + q_1 M'^{n-1}R' + \cdots + q_{n-1}R' = 0\,. \tag{10}$$

The coefficients $p_1, p_2, \cdots, p_n$ are, because of relationship (9), linear non-homogeneous forms in $q_1, \cdots, q_{n-1}$ and consequently may be computed simultaneously by the elimination method (cf. Section 22). Of the two possible modifications of the elimination method, one should take the one in which

**TABLE IV.7  Determining Coefficients of the Characteristic Polynomial by Samuelson's Scheme**

| | | | | | | | |
|---|---|---|---|---|---|---|---|
| 1.870086 | 0.422908 | 0.008814 | 0 | 0 | 1 | 5.509882 | 8.811690 |
| −20.264529 | 5.208653 | 0.051697 | 0 | 1 | 5.509882 | −0.559152 | −9.053449 |
| 261.81060 | −183.23653 | −0.905064 | 1 | 5.509882 | −0.559152 | 5.577387 | 89.197123 |
| −3882.0495 | 3870.8749 | −10.72254 | 5.509882 | −0.559152 | 5.577387 | −66.36373 | −76.73275 |
| 1 | 0.226144 | 0.004713 | 0 | 0 | 0.534735 | 2.946325 | 4.711917 |
| | 9.791355 | 0.147204 | 0 | 1 | 16.346035 | 59.146736 | 86.43130 |
| | −242.44343 | −2.138977 | 1 | 5.509882 | −140.55844 | −765.80173 | −1144.4327 |
| | 4748.777 | 7.57356 | 5.509882 | −0.559152 | 2081.445 | 11371.416 | 18215.162 |
| | 1 | 0.015034 | 0 | 0.102131 | 1.669435 | 6.040710 | 8.827310 |
| | | 1.505942 | 1 | 30.270872 | 264.18511 | 698.72872 | 995.6906 |
| | | −63.82003 | 5.509882 | −485.5565 | −5846.330 | −17314.569 | −23703.765 |
| | | 1 | 0.664036 | 20.10095 | 175.42847 | 463.98116 | 661.17460 |
| | | | 47.88868 | 797.287 | 5349.52 | 12296.72 | 18492.42 |

the components of the vectors $R'$, $M'R'$, $\cdots$, $M'^{n-1}R'$ are arrayed in the scheme's rows. Then these rows, considered as matrices, are $R$, $RM$, $\cdots$, $RM^{n-1}$. The coefficients of relationships (9) thereupon turn out to be arrayed exactly in accordance with Samuelson's scheme. From the foundation of the method carried out above it is easy to explain its area of application. In fact it coincides with the area of application for A. N. Krylov's method for the matrix $M$, starting with vector $R'$.

As an example we again take Leverrier's matrix. We carry out the computations for the coefficients of the characteristic polynomial according to the scheme described (cf. Table IV.7). We first compute matrix (1), placing its elements in the first four rows. We then carry out the elimination as indicated in Section 22. The last row gives the desired values of the coefficients, which coincide almost exactly with the values computed by A. N. Krylov's method. The last column, as usual, is the control column.

The number of operations necessary to determine the coefficients of the characteristic polynomial is slightly less for Samuelson's method than for Krylov's. In fact forming matrix (1) requires $n(n-1)^2$ multiplications and the elimination process in Samuelson's scheme requires as many operations as solving the system by Krylov's method.

## 46. A. M. DANILEVSKII'S METHOD

An elegant and very effective method for computing the coefficients of the characteristic polynomial has been proposed by A. M. Danilevskii (1). The geometric meaning of this method is as follows. A given matrix $A$ is considered as the matrix of an operator in the basis

$$e_1 = (1, 0, \cdots, 0)',$$
$$e_2 = (0, 1, \cdots, 0)', \cdots,$$
$$e_n = (0, 0, \cdots, 1)'.$$

It is assumed that the vectors $e_1$, $Ae_1$, $\cdots$, $A^{n-1}e_1$ are linearly independent. Then

$$A^n e_1 = p_1 A^{n-1}e_1 + p_2 A^{n-2}e_1 + \cdots + p_n e_1.$$

It is clear that the coefficients $p_1$, $p_2$, $\cdots$, $p_n$ are the desired coefficients of the characteristic polynomial.

In the basis $e_1$, $Ae_1$, $\cdots$, $A^{n-1}e_1$ the operator considered will obviously have the so-called Frobenius matrix

$$P = \begin{bmatrix} 0 & 0 \cdots p_n \\ 1 & 0 \cdots p_{n-1} \\ 0 & 1 \cdots p_{n-2} \\ \cdot & \cdot \cdot \cdot \cdot \cdot \\ 0 & 0 \cdots p_1 \end{bmatrix},$$

clearly containing the desired coefficients of the characteristic polynomial.

Passing from the basis $e_1, e_2, \cdots, e_n$ to the basis $e_1, Ae_1, \cdots, A^{n-1}e_1$ is done gradually in $n-1$ steps. Each step consists in passing from the basis $e_1, Ae_1, \cdots, A^{k-1}e_1, e_{k+1}, \cdots, e_n$ to the basis $e_1, Ae_1, \cdots, A^{k-1}e_1, A^k e_1, e_{k+2}, \cdots, e_n$. To realize the whole process it is necessary to consider that all intermediate systems of vector are actually bases, i.e. they consist of linearly independent vectors. We shall examine below what happens in cases of degeneracy. Until then we shall consider only non-degenerate processes.

We denote by $A^{(k)}$ the matrix obtained at the $k$-1th step of the process so that $A = A^{(1)}$, $P = A^{(n)}$. The columns of matrix $A^{(k)}$ are the coordinates of the vectors $Ae_1, A^2e_1, \cdots, A^ke_1, Ae_{k+1}, \cdots, Ae_n$ in the basis $e_1, Ae_1, \cdots, A^{k-1}e_1, e_{k+1}, \cdots, e_n$. Therefore the first $k-1$ columns of matrix $A^{(k)}$ will coincide with the corresponding columns of the Frobenius matrix $P$. We have

$$A^{(k+1)} = S_k^{-1} A^{(k)} S_k$$

where $S_k$ is the corresponding matrix for transforming coordinates. It is clear that

$$S_k = \begin{bmatrix} 1 \cdots s_{1.k+1} \cdots 0 \\ 0 \cdots s_{2.k+1} \cdots 0 \\ \cdot \quad \cdot \quad \cdot \quad \cdot \quad \cdot \quad \cdot \\ 0 \cdots s_{n.k+1} \cdots 1 \end{bmatrix},$$

where $s_{1.k+1}, s_{2.k+1}, \cdots, s_{n.k+1}$ are the coordinates of the vector $A^k e_1$ in the basis $e_1, Ae_1, \cdots, A^{k-1}e_1, \cdots, e_n$. These coordinates, as we saw above, are none other than the elements $a_{ik}^{(k)}$ of the $k$th column of the matrix $A^{(k)}$. Moreover we have

$$S_k^{-1} = \begin{bmatrix} 1 \cdots - \dfrac{s_{1.k+1}}{s_{k+1.k+1}} \cdots 0 \\ 0 \cdots - \dfrac{s_{2.k+1}}{s_{k+1.k+1}} \cdots 0 \\ \cdot \quad \cdot \quad \cdot \quad \cdot \quad \cdot \quad \cdot \\ 0 \cdots - \dfrac{1}{s_{k+1.k+1}} \cdots 0 \\ \cdot \quad \cdot \quad \cdot \quad \cdot \quad \cdot \quad \cdot \\ 0 \cdots - \dfrac{s_{n.k+1}}{s_{k+1.k+1}} \cdots 1 \end{bmatrix}.$$

It is advisable to carry out the computation of matrix $A^{(k+1)}$ using two methods. At first the auxiliary matrix $B^{(k)} = S_k^{-1} A^{(k)}$ is computed. In view of the structure of matrix $S_k^{-1}$ this operation consists in multiplying the $(k+1)$th row of matrix $A^{(k)}$ by $\dfrac{1}{a_{k+1.k}^{(k)}}$ and subtracting from each of the remaining rows the obtained $(k+1)$th row of matrix $B^{(k)}$ multiplied by the corresponding element of the $k$th column in matrix $A^{(k)}$. It is obvious that as a result of these operations the first $(k-1)$ columns do not change; in the $k$th column, the $(k+1)$th place, has a one and all remaining elements become zero. The computations of the remaining of matrix $B^{(k)}$ will be binomial, recalling the computations for Gauss's method. The obtained matrix $B^{(k)}$ is post-multiplied by $S_k$. Only one column—the $k + 1$th—is thereby changed. Its elements will be, as is easy to see,

$$\sum_{j=1}^{n} b_{1j}^{(k)} a_{jk}^{(k)}, \sum_{j=1}^{n} b_{2j}^{(k)} a_{jk}^{(k)}, \cdots, \sum_{j=1}^{n} b_{nj}^{(k)} a_{jk}^{(k)}.$$

In other words the $(k + 1)$th column of matrix $A^{(k)}$ is the result of iterating the $k$th column of matrix $A^{(k)}$ with matrix $B^{(k)}$.

Thus the passage from matrix $A^{(k)}$ to matrix $A^{(k+1)}$ occurs according to the formulas

$$b_{k+1.j}^{(k)} = \frac{1}{a_{k+1,k}^{(k)}} \cdot a_{k+1.j}^{(k)},$$
$$b_{ij}^{(k)} = a_{ij}^{(k)} - a_{ik}^{(k)} b_{k+1.j}^{(k)} \qquad (i \neq k + 1),$$
$$a_{ij}^{(k+1)} = b_{ij}^{(k)} \qquad (j \neq k + 1),$$
$$a_{i,k+1}^{(k+1)} = \sum_{j=1}^{n} b_{ij}^{(k)} a_{jk}^{(k)}.$$

As an example we again take Leverrier's matrix.

We shall explain Table IV.8. In columns 2, 3, 4 and 5 are written successively matrices $A^{(1)}$, $B^{(1)}$, $B^{(2)}$ and $B^{(3)}$. Columns 6 and 7 are for control; column 6 contains the sums of the rows of the matrices $B^{(k)}$ and column 7 contains the sums of the rows of the matrices $A^{(k)}$. In column 1 are written the $k$ columns of matrix $A^{(k)}$ whose computation is accompanied by the usual control. Thus the coefficients of the characteristic polynomial are placed in the last four rows of column 1. Since they are computed simultaneously, the control coincidence of $p_1$ with the spur of the matrix is also an indicator of the computational accuracy for the remaining coefficients; the results obtained are closer to Leverrier's data than are the results found by Krylov's method or Samuelson's method.

The number of operations necessary in computing by A. M. Danilevskii's method is considerably less than for the two methods indicated. As we shall see below it is also less than that for other methods.

That is, the number of multiplications and divisions at the $k$th step will be

$$n - k + (n - k)(n - 1) + (n - k)n = 2n(n - k)$$

and therefore the more general number of multiplications and divisions will be $n^3 - n^2$.

A. M. Danilevskii's method permits us to compute the eigenvectors both for matrix $A$ itself and for its transpose. In fact since

$$S^{-1}AS = P,\tag{1}$$

where

$$S = S_1 S_2 \cdots S_{n-1},\tag{2}$$

then an eigenvector $U$ of matrix $A$ belonging to the eigenvalue $\lambda$ is expressed in terms of the eigenvector $Y$ of matrix $P$ by the formula

$$U = SY = S_1 S_2 \cdots S_{n-1} Y .$$

The components $y_1, \cdots, y_n$ of the vector $Y$ are found without difficutly as solutions of the system

$$p_n y_n = \lambda y_1 ,$$
$$y_1 + p_{n-1} y_n = \lambda y_2 ,$$
$$y_2 + p_{n-2} y_n = \lambda y_3 ,$$
$$\cdots \cdots \cdots ,$$
$$y_{n-1} + p_1 y_n = \lambda y_n .$$

Letting $y_n = 1$ we obtain successively

$$y_{n-1} = \lambda y_n - p_1 = \lambda - p_1 ,$$
$$y_{n-2} = \lambda y_{n-1} - p_2 = \lambda^2 - p_1 \lambda - p_2 ,$$
$$\cdots \cdots \cdots \cdots \cdots ,$$
$$y_1 = \lambda y_2 - p_{n-1} = \lambda^{n-1} - p_1 \lambda^{n-2} - \cdots - p_{n-1} .$$

The first equation of the system will be satisfied identically, since $\lambda^n - p_1 \lambda^{n-1} - \cdots - p_n = 0$.

To compute the components $y_i$ we must apply recurrent formulas. These formulas coincide with the formulas for Horner's scheme in dividing the characteristic polynomial by $t - \lambda$.

We write further

$$Y = Y^{(n)}, \; Y^{(n-1)} = S_{n-1} Y^{(n)}, \; Y^{(n-2)} = S_{n-2} Y^{(n-1)}, \; \cdots, \; Y^{(1)} = S_1 Y^{(2)} = U, .$$

The components of the vector $Y^{(k)}$ are computed according to components of the vector $Y^{(k+1)}$ by the binomial formulas

$$y_i^{(k)} = y_i^{(k+1)} + y_{k+1}^{(k+1)} a_{ik}^{(k)} \qquad (i \neq k + 1) ,$$
$$y_{k+1}^{(k)} = y_{k+1}^{(k+1)} a_{k+1,k}^{(k)} . \tag{3}$$

The formulas for computing the components of an eigenvector for the transposed matrix have a slightly simpler design. Passing to the transposed matrices in equality (1) we obtain

$$S' A' (S^{-1})' = P' ,$$

so that the eigenvector of matrix $A'$ belonging to the eigenvalue $\lambda$ is connected with the corresponding eigenvector $Z$ of matrix $P'$ by the relationship

$$V = (S^{-1})' Z = (S_1^{-1})' (S_2^{-1})' \cdots (S_{n-1}^{-1})' Z .$$

The components $z_1, \cdots, z_n$ of the vector $Z$ are found from the system

$$z_2 = \lambda z_1 ,$$
$$z_3 = \lambda z_2 ,$$
$$\cdots \cdots ,$$
$$z_n = \lambda z_{n-1} ,$$
$$p_n z_1 + p_{n-1} z_2 + \cdots p_1 z_n = \lambda z_n .$$

TABLE IV.8 Computing the Coefficients of the Characteristic Polynomial by A. M. Danilevskii's Method

| 1 | 2 | 3 | 4 | 5 | 6 | 7 |
|---|---|---|---|---|---|---|
| -5.509882 | -5.509882 | 1.870086 | 0.422908 | 0.008814 | -3.208074 | |
| 0.287865 | 0.287865 | -11.811654 | 5.711900 | 0.058717 | -5.753172 | |
| 0.049099 | 0.049099 | 4.308633 | -12.970687 | 0.229326 | -8.384229 | |
| 0.006235 | 0.006235 | 0.269851 | 1.397369 | -17.596207 | -15.922752 | |
| -59.146733 | 0 | -224.21096 | 109.75157 | 1.1326872 | -113.32670 | 51.737524 |
| -16.346028 | 1 | -41.031921 | 19.842287 | 0.20397409 | -19.985660 | 4.7002331 |
| 1.1367579 | 0 | 6.3225593 | -13.944923 | 0.21931108 | -7.402953 | -12.588854 |
| 0.10414109 | 0 | 0.52568503 | 1.2736523 | -17.597479 | -15.798142 | -16.219686 |
| -74.251862 | 1 | -258.39454 | 116.92259 | -16.041507 | | |
| -698.72936 | 0 | 0 | -615.81773 | 12.543678 | -603.27405 | -686.18568 |
| -264.18530 | 1 | 0 | -180.67895 | 3.3575611 | -176.32139 | -259.82774 |
| -30.270859 | 0 | 1 | -12.267276 | 0.19292681 | -11.074349 | -29.077932 |
| 1.0653607 | 0 | 0 | 2.5511798 | -17.617571 | -15.066391 | -16.552210 |
| -992.12016 | 1 | 1 | -806.21278 | -1.523405 | | |
| -12296.550 | 0 | 0 | 0 | -11541.147 | -11542.147 | -12296.550 |
| -5349.4555 | 1 | 0 | 0 | -4365.4005 | -4364.4005 | -5348.4555 |
| -797.27875 | 0 | 1 | 0 | -500.38776 | -499.38776 | -796.27875 |
| -47.888430 | 0 | 0 | 1 | -16.536719 | -15.536719 | -46.888430 |

Letting $z_1 = 1$, we obtain successively $z_2 = \lambda$, $z_3 = \lambda^2$, $\cdots$, $z_n = \lambda^{n-1}$. The last equation is satisfied identically.

Moreover

$$V = Z^{(1)}$$

where

$$Z = Z^{(n)},$$
$$Z^{(n-1)} = (S_{n-1}^{-1})' Z^{(n)}, \ Z^{(n-2)} = (S_{n-2}^{-1})' Z^{(n-1)}, \ \cdots, \ Z^{(1)} = (S_1^{-1})' Z^{(2)}.$$

At this point each transformation $(S_k^{-1})'$ will change only the $(k+1)$th component of the preceding vector so that the components of the vector $Z^{(k)}$ will be $z_1, z_2, \cdots, z_k, v_{k+1}, \cdots, v_n$. In this

$$v_{k+1} = \frac{1}{a_{k+1,k}^{(k)}} \left( -\sum_{i=1}^{k} a_{ik}^{(k)} z_i + z_{k+1} - \sum_{i=k+2}^{n} a_{ik}^{(k)} v_i \right) = \sum_{i=1}^{k+1} m_{ik} z_i + \sum_{i=k+2}^{n} m_{ik} v_i, \tag{4}$$

where

$$m_{ik} = -\frac{a_{ik}^{(k)}}{a_{k+1,k}^{(k)}} \qquad (i \neq k+1),$$

$$m_{k+1,k} = \frac{1}{a_{k+1,k}^{(k)}}.$$

Formulas (3) and (4) for computing the eigenvectors of matrix $A$ and its transpose are different in structure but require roughly the same number of computational operations. To find the eigenvectors of matrix $A$ we may use any of these formulas, of course, transposing matrix $A$ at will in the beginning of the process.

As an example we shall consider the computation of an eigenvector for Leverrier's matrix belonging to the eigenvalue $\lambda_4$, using the data of Table IV.8. An approximate value is obtained for $\lambda_4$ equal to $-5.298695$.

Using Horner's scheme,

| | | | | | |
|---|---|---|---|---|---|
| 1 | 47.888430 | 797.27875 | 5349.4555 | 12296.550 | $\|$   $-5.298695$ |
| 1 | 42.589735 | 571.60873 | 2320.6752 | $-0.0000789$, | |

we shall compute the components of the eigenvector $Y$ of matrix $P$. We obtain $Y = Y^{(4)} = (2320.6752; \ 571.60873; \ 42.589735; \ 1)'$. By formulas (3) we find successively

$$Y^{(3)} = (1621.9458; \quad 307.42343; \quad 12.318876; \quad 1.0653607)',$$
$$Y^{(2)} = (\ 893.32453; \quad 106.05874; \quad 14.003580; \quad 2.3482620)',$$
$$Y^{(1)} = U = (\ 308.95339; \quad 30.530599; \quad 19.210958; \quad 3.009538)'.$$

After normalizing to a unit first component we get

$$\tilde{U} = (1.000000; \quad 0.098819; \quad 0.062181; \quad 0.009741)'.$$

To find the eigenvector of the transposed matrix belonging to $\lambda_4$ we first compute the numbers $m_{ik}$ for $k = 1, 2, 3$, namely

|  $m_{i1}$ | $m_{i2}$ | $m_{i3}$ |
|---|---|---|
| 19.140507 | 52.031073 | 655.86178 |
| 3.4738506 | 14.379516 | 247.97733 |
| −0.170562 | 0.87969479 | 28.413718 |
| −0.021659 | −0.09161237 | 0.93864923 . |

We compute further

$$Z = (1, \quad -5.298695, \quad 28.076169, \quad -148.76706)'$$

and

$$V = (1, \quad 0.641980, \quad 0.535599, \quad 0.0138036)' .$$

We observe that in computing, particularly for the last component, there occurs a significant loss of significant digits, which gives an insufficiently reliable result.

It $\lambda$ is a multiple eigenvalue, then it is easy to compute for it the whole "tower" of root vectors. It will be the only one, since (in view of the assumption of linear independence for the vectors $e_1$, $Ae_1$, $\cdots$, $A^{n-1}e_1$) matrix $A$ has relatively prime elementary divisors.

The root vectors of matrix $A$ (or of its transpose $A'$) are computed by formulas (3) (or (4)) in which components of the eigenvectors $Y$ (or $Z$) must be replaced by components of the root vectors of matrix $P$ (or $P'$). The latter are easily found.

As was pointed out above, the components $y_1$, $y_2$, $\cdots$, $y_n$ of the eigenvector $Y$ of matrix $P$ are the coefficients for the quotient obtained by dividing the characteristic polynomial by $t - \lambda$. Correspondingly the components of the root vectors are the coefficients from dividing the characteristic polynomial by $(t - \lambda)^2$, $(t - \lambda)^3$, $\cdots$, $(t - \lambda)^m$ where $m$ is the multiplicity of the eigenvalue and the first component is always a constant. These components are found successively by Horner's scheme.

For the matrix $P'$ the root vectors are

$$Z = (1, \lambda, \lambda^2, \cdots, \lambda^{n-1}) ,$$
$$Z_1 = (0, 1, 2\lambda, \cdots, (n-1)\lambda^{n-2}) ,$$
$$Z_2 = \left( 0, 0, 1, \cdots, \frac{(n-1)(n-2)}{2} \lambda^{n-3} \right)$$
$$Z_{m-1} = (0, 0, 0, \cdots, 1, \cdots, C_{n-1}^{m-1}\lambda^{n-m}) .$$

As an example we consider the matrix

$$A = \begin{bmatrix} 13 & 16 & 16 \\ -5 & -7 & -6 \\ -6 & -8 & -7 \end{bmatrix} ,$$

whose eigenvalues are $\lambda_1 = \lambda_2 = 1$, $\lambda_3 = -3$.

The coefficients for the characteristic polynomial of the matrix are computed in Table IV.9. These computations show that the characteristic polynomial of the matrix is equal to $t^3 + t^2 - 5t + 3$.

**TABLE IV.9**   Determining the Coefficients of the Characteristic Polynomial by A. M. Danilevskii's Method

| 1 | 2 | 3 | 4 | 5 | 6 |
|---|---|---|---|---|---|
| 13 | 13 | 16 | 16 | 45 | |
| −5 | −5 | −7 | −6 | −18 | |
| −6 | −6 | −8 | −7 | −21 | |
| 8.6 | 0 | −2.2 | 0.4 | −1.8 | 9.0 |
| −1.2 | 1 | 1.4 | 1.2 | 3.6 | 1.0 |
| −3.2 | 0 | 0.4 | 0.2 | 0.6 | −3.0 |
| 4.2 | 1 | −0.4 | 1.8 | | |
| −3 | 0 | 0 | 0.9375 | 0.9375 | |
| 5 | 1 | 0 | 1.1250 | 2.1250 | |
| −1 | 0 | 1 | −0.0625 | 0.9375 | |

We shall find the eigenvector and root vector belonging to the eigenvalue $\lambda = 1$.

Applying Horner's scheme we have

$$
\begin{array}{l}
1, \quad 1, \quad -5, \quad 3 \;\|\; 1 \\
1, \quad 2, \quad -3 \\
1, \quad 3
\end{array}
$$

so that the eigenvector $Y$ of matrix $P$ is equal to $(-3, 2, 1)'$ and the root vector $\tilde{Y} = (0, 3, 1)'$.

Therefore the eigenvector for matrix $A$ will be

$$U = S_1 S_2 Y = (16, -4, -8,)'$$

and the root vector

$$\tilde{U} = S_1 S_2 \tilde{Y} = (32, -9, -14)' \ .$$

The algebraic presentation of A. M. Danilevskii's method given in the book of V. N. Faddeeva (1) is equivalent to passing from the basis $e_1, e_2,$ $\cdots, e_n$ to the basis $A'^{n-1}e_n, A'^{n-2}e_n, \cdots, A'e_n, e_n,$ as a result of which Frobenius' canonical matrix is obtained in slightly different form.

We observe one detail connected with Danilevskii's method and Hessenberg's method. The elements of the $k$th column of matrix $A^{(k)}$ are by their construction coefficients in the equation

$$A^k e_1 = a_{1k}^{(k)} e_1 + a_{2k}^{(k)} A e_1 + \cdots + a_{kk}^{(k)} A^{k-1}e_1 + a_{k+1k}^{(k)} e_{k+1} + \cdots + a_{nk}^{(k)} e_n \ ,$$

from which it follows that the polynomial

$$\varphi_k(t) = t^k - a_{kk}^{(k)} t^{k-1} - \cdots - a_{1k}^{(k)}$$

possesses the property that the vector $\varphi_k(A)e_1 = Z_{k+1}$ has its first $k$ components in the initial basis equal to zero. Therefore the vector $Z_{k+1}$ and the polynomial $\varphi_k(t)$ coincide with the corresponding vector and polynomial for Hessenberg's method. In the same way all numbers forming the $k$th column of the matrix $A^{(k)}$ are encountered in computations carried out by Hessenberg's method—the upper $k$ as coefficients (with reverse signs) of the polynomials $\varphi_k(t)$ and the lower $n - k$ as components of the $Z_{k+1}$.

We turn now to a consideration of possible degeneracies in the process. It may turn out that at some step $a^{(k)}_{k+1,k} = 0$. This shows that the system of vectors $e_1, Ae_1, \cdots, A^k e_1, e_{k+2}, \cdots, e_n$ is linearly dependent. If it then turns out that at least one of the numbers $a^{(k)}_{jk} \neq 0$ for $j > k + 1$, then in the matrix $A^{(k)}$ it is necessary to rearrange the $k + 1$th and $j$th rows and simultaneously the $k + 1$th and $j$th columns. Such a rearrangement is equivalent to passing from the basis $e_1, Ae_1, \cdots, A^{k-1} e_1, e_{k+1}, \cdots, e_j, \cdots, e_n$ to the basis $e_1, Ae_1, \cdots, A^{k-1} e_1, e_j, \cdots, e_{k+1}, \cdots, e_n$. It is easy to see that if $e_1, Ae_1, \cdots, A^{n-1} e_1$ is linearly independent, then such a $j$ must necessarily be found. After completing the transformation the process is continued. In computing the eigenvectors the transformation which has been made must of course be considered. (At a suitable moment it is necessary to rearrange the $k + 1$th and $j$th components of the corresponding vectors.)

The indicated transformation may be used even if $a^{(k)}_{k+1k} \neq 0$ but among the numbers $a^{(k)}_{jk}, j > k + 1$, is a number significantly larger in modulus than $a^{(k)}_{k+1k}$, since the transformation increases the accuracy of the computation. If such a transformation is made at each step, we obtain a scheme similar to the principal elements scheme for Gauss's method.

If indeed all $a^{(k)}_{jk} = 0$ for $j \geq k + 1$, meaning that the vectors $e_1, Ae_1, \cdots, A^k e_1, \cdots, A^k e_1$ are already linearly dependent, then the matrix $A^{(k)}$ has the form

$$
\left|
\begin{array}{cccc|c}
0 & 0 \cdots 0 & a^{(k)}_{k1} & & \\
1 & 0 \cdots 0 & a^{(k)}_{k2} & & C \\
\cdot & \cdot \cdot \cdot \cdot \cdot & \cdot & & \\
0 & 0 \cdots 1 & a^{(k)}_{kk} & & \\
\hline
 & 0 & & & A^{(k)}_2
\end{array}
\right|
=
\begin{bmatrix} A^{(k)}_1 & C \\ 0 & A^{(k)}_2 \end{bmatrix},
$$

and consequently the characteristic polynomial of matrix $A$ is equal to the product of the characteristic polynomials of matrices $A^{(k)}_1$ and $A^{(k)}_2$.

The matrix $A^{(k)}_1$ is already a Frobenius canonical matrix. It corresponds to the induced operator in the invariant subspace spanned by the vectors, $e_1, \cdots, A^{k-1} e_1$.

It is again necessary to apply to matrix $A^{(k)}_2$ the general transforming process.

Thus in the case considered the process of finding the characteristic polynomial is only simplified. However, computation of the canonical basis (and in particular of the eigenvectors) is made slightly more complex. We shall not stop to solve this question.

A. M. Danilevskii's method permits the generalization: One may pass to the canonical form by going from the basis $e_1, e_2, \cdots, e_n$ to the basis $f, Af, \cdots, A^{n-1} f$ where $f$ is a certain vector generally chosen arbitrarily. This demands only that the vectors $f, Af, \cdots, A^{n-1}f$ be linearly independent. If the elementary divisors of the matrix are relatively prime, such a vector may always be found.

This variant requires $n$ steps already, since one more "null" step is added which consists in passing from the basis $(e_1, e_2, \cdots, e_n)$ to the basis $(f_1, e_2, \cdots, e_n)$; this is accomplished by the similarity transformation

$$A_1 = S_0^{-1} A S_0$$

where $S_0$ is a matrix whose first column consists of the components of the vector $f$ and whose remaining ones coincide with columns of a unit matrix. Carrying the process out further does not differ from what has been described above. Bauer (6) recommends that the initial vector be chosen so that its coordinates in eigenvectors corresponding to the largest eigenvalues will be small. Methods for constructing such vectors will be described in Chapter IX.

## 47. LEVERRIER'S METHOD AND D. K. FADDEEV'S MODIFICATION

In this paragraph we shall present a method known as Leverrier's method (1) which requires a larger number of operations than all the methods considered above but which is completely insensitive to the special peculiarities of the matrix, particularly to "gaps" in intermediate determinants.

Let

$$\varphi(t) = (-1)^n [t^n - p_1 t^{n-1} - p_2 t^{n-1} - \cdots - p_n] \qquad (1)$$

be the characteristic polynomial of a matrix and let $\lambda_1, \lambda_2, \cdots, \lambda_n$ be its roots, some of which may be equal. We write

$$\sum_{i=1}^{n} \lambda_i^k = s_k . \qquad (2)$$

Then the relationships known as Newton's formulas are valid:

$$k p_k = s_k - p_1 s_{k-1} - \cdots - p_{k-1} s_1 \qquad (k = 1, \cdots, n) . \qquad (3)$$

If the numbers $s_k$ are known, then by solving the recurrent system (3) we may find the necessary coefficients $p_k$.

We shall show how the numbers $s_k$ are determined. We have

$$s_1 = \lambda_1 + \lambda_2 + \cdots + \lambda_n = \mathrm{Sp}\, A .$$

Moreover in view of Section 1, paragraph 10, the characteristic numbers of the matrix $A^k$ will be $\lambda_1^k, \lambda_2^k, \cdots, \lambda_n^k$. Consequently,

$$s_k = \lambda_1^k + \lambda_2^k + \cdots + \lambda_n^k = \mathrm{Sp}\, A^k . \qquad (4)$$

Thus the computational process comes down to computing successively the powers of matrix $A$, then to computing their spurs and finally to solving the recurrent system (3). To compute $n$ powers of matrix $A$ (for the last matrix $A^n$ only the diagonal elements need to be computed) requires a large number of operations (uniform, it is true) and therefore Leverrier's method is much more laborious than the methods explained above. Its value, as was already mentioned, lies in its universality. The number of necessary multiplications for Leverrier's method is equal to

$$\tfrac{1}{2}(n-1)(2n^3 - 2n^2 + n + 2) .$$

We observe that in computing the powers of a matrix it is useful to establish control using the column formed from the sums of the elements of each row of matrix $A$. The result of multiplying matrix $A$ by this column should coincide with the analogous column for the matrix $A^2$.

In fact let $\Sigma_1$ be the column of sums for matrix $A$ and let $\Sigma_2$ be the column of sums for the matrix $A^2$. Let $U = (1, 1, \cdots, 1)'$. Then

$$\Sigma_1 = AU ,$$
$$\Sigma_2 = A^2 U ,$$

that is,

$$\Sigma_2 = A\Sigma_1 . \tag{5}$$

It is obvious that what was said above is also true for other powers.

We now present a variation of the method proposed by D. K. Faddeev[†] which, besides simplifying the computation of the coefficients of the characteristic polynomial, permits us to determine the inverse matrix and the eigenvectors of the matrix.

In place of the sequence $A, A^2, \cdots, A^n$ we shall compute the sequence $A_1, A_2, \cdots, A_n$ which is constructed in the following way:

$$
\begin{array}{lll}
A_1 = A, & \mathrm{Sp}\, A_1 = q_1, & B_1 = A_1 - q_1 E \\[4pt]
A_2 = AB_1, & \dfrac{\mathrm{Sp}\, A_2}{2} = q_2, & B_2 = A_2 - q_2 E \\[10pt]
\cdots\cdots\cdots\cdots\cdots\cdots\cdots\cdots\cdots, & & \\[6pt]
A_{n-1} = AB_{n-2}, & \dfrac{\mathrm{Sp}\, A_{n-1}}{n-1} = q_{n-1}, & B_{n-1} = A_{n-1} - q_{n-1}E \\[10pt]
A_n = AB_{n-1}, & \dfrac{\mathrm{Sp}\, A_n}{n} = q_n, & B_n = A_n - q_n E .
\end{array}
\tag{6}
$$

We shall prove that

1. $q_1 = p_1, q_2 = p_2, \cdots, q_n = p_n;$
2. $B_n$ is a null matrix;
3. if $A$ is a non-singular matrix, then

$$A^{-1} = \frac{B_{n-1}}{p_n} .$$

(If the matrix $A$ is singular, then $(-1)^{n-1}B_{n-1}$ will be the matrix adjoint to matrix $A$).

---

† D. K. Faddeev and I. S. Sominskii, *Collected Problems in Higher Algebra*, 1949, problem No. 979. Cf. also Souriau (1) and Frame (1).

**TABLE IV.10**   Determining the Coefficients of the Characteristic Polynomial and Elements of the Inverse Matrix by D. K. Faddeev's Method

| A′ | | | | B₁ | | | | Σ |
|---|---|---|---|---|---|---|---|---|
| −5.509882 | 0.287865 | 0.049099 | 0.006235 | 42.378548 | 1.870086 | 0.422908 | 0.008814 | 44.68036 |
| 1.870086 | −11.811654 | 4.308033 | 0.269851 | 0.287865 | 36.076776 | 5.711900 | 0.058717 | 42.13526 |
| 0.422908 | 5.711900 | −12.970687 | 1.397369 | 0.049099 | 4.308033 | 34.917743 | 0.229326 | 39.50420 |
| 0.008814 | 0.058717 | 0.229326 | −17.596207 | 0.006235 | 0.269851 | 1.397369 | 30.292223 | 31.96568 |

$P_1 = -47.888430$    $A_2$

| | | | |
|---|---|---|---|
| −232.94165 | −400.96516 | −427.95884 | −532.69187 |

| B₂ | | | | Σ |
|---|---|---|---|---|
| 564.33711 | 58.98700 | 23.13088 | 0.42522 | 646.8802 |
| 9.07995 | 396.31360 | 132.18346 | 2.39755 | 539.9746 |
| 2.68546 | 99.69550 | 369.31992 | 4.22567 | 475.9266 |
| 0.30081 | 11.01857 | 25.74858 | 264.58689 | 301.6549 |

$P_2 = -792.27876$    $A_3$

| | | | |
|---|---|---|---|
| −3091.3122 | −4094.0411 | −4213.8419 | −4649.1712 |

| B₃ | | | | Σ |
|---|---|---|---|---|
| 2258.1433 | 458.3882 | 276.1613 | 6.2599 | 2998.953 |
| 70.5604 | 1255.4144 | 556.3836 | 11.4757 | 1893.834 |
| 32.0618 | 419.6360 | 1135.6136 | 16.2164 | 1603.527 |
| 4.4283 | 52.7398 | 98.8129 | 700.2843 | 856.265 |

$P_3 = -5349.4555$    $A_4$

| A⁻¹ | | | |
|---|---|---|---|
| −0.183640 | −0.037278 | −0.022458 | −0.000509 |
| −0.005738 | −0.102095 | −0.045247 | −0.000933 |
| −0.002607 | −0.034126 | −0.092352 | −0.001319 |
| −0.000360 | −0.004289 | −0.008036 | −0.056950 |

| | | | |
|---|---|---|---|
| −12296.551 | 0.000 | 0.000 | 0.000 |
| 0.001 | −12296.551 | 0.000 | 0.000 |
| 0.001 | 0.000 | −12296.550 | 0.001 |
| 0.001 | 0.000 | 0.000 | −12296.551 |

$P_4 = -12296.551$

We shall first prove a) by the method of mathematical induction. It is obvious that $p_1 = \mathrm{Sp}\, A = q_1$. We assume that $q_1 = p_1$, $q_2 = p_2$, $\cdots$, $q_{k-1} = p_{k-1}$ and we shall prove that $q_k = p_k$. According to our construction:

$$A_k = A^k - q_1 A^{k-1} - \cdots - q_{k-1} A = A^k - p_1 A^{k-1} - \cdots - p_{k-1} A .$$

Consequently,

$$\mathrm{Sp}\, A_k = kq_k = \mathrm{Sp}\, A^k - p_1 \mathrm{Sp}\, A^{k-1} - \cdots - p_{k-1} \mathrm{Sp}\, A = s_k - p_1 s_{k-1} - \cdots - p_{k-1} s_1 .$$

From here $kq_k = kp_k$ on the basis of Newton's formulas and consequently $q_k = p_k$.

Moreover in view of the Cayley-Hamilton relationship

$$B_n = A^n - p_1 A^{k-1} - \cdots - p_n E = 0 . \tag{7}$$

Finally from equation (7) it follows that

$$A B_{n-1} = A_n = B_n + p_n E = p_n E ,$$

so that

$$A^{-1} = \frac{1}{p_n} B_{n-1} . \tag{8}$$

The equality $A_n = p_n E$ may be used for control of the computation; it is obvious that the divergence of $A_n$ from the scalar matrix is a measure of the computational accuracy. Besides this final control it is advisable to use a special control, forming columns of sums for the matrices $B_i$. In this the relationship

$$\Sigma_{i+1} = A \Sigma_i - p_{i+1} \begin{bmatrix} 1 \\ 1 \\ \cdot \\ \cdot \\ \cdot \\ 1 \end{bmatrix} ,$$

$$\Sigma_1 = A \Sigma_0 - p_1 \begin{bmatrix} 1 \\ 1 \\ \cdot \\ \cdot \\ \cdot \\ 1 \end{bmatrix} , \tag{9}$$

is valid, where $\Sigma_i$ is the column of sums for matrix $B_i$ and $\Sigma_0$ is the analogous column for matrix $A$.

Formula (8) gives the algorithm for inverting the matrices. For matrices which are not of a very high order the indicated method is very convenient when it is necessary to solve the inversion problem for a matrix as well as the problem of finding eigenvalues.

The number of operations needed to obtain the coefficients $p_i$ (also including the computation of the matrix $B_n$) is equal to $(n-1)n^3$ multiplications.

In Table IV.10 the computational scheme by D. K. Faddeev's method for coefficients of the characteristic polynomial and elements of the inverse matrix is given for the previous example.

We now pass on to determining the eigenvectors of matrix $A$.

Let the eigenvalues be already computed and distinct. We construct the matrix

$$Q_k = \lambda_k^{n-1}E + \lambda_k^{n-2}B_1 + \cdots + B_{n-1} \qquad (10)$$

where $B_i$ are matrices computed while finding the coefficients of the characteristic polynomial and $\lambda_k$ is the $k$th eigenvalue of the matrix $A$.

Assuming that all $\lambda_1, \cdots, \lambda_n$ are distinct, one may prove that the matrix $Q_k$ is non-zero.

We shall show that *every* column of $Q_k$ consists of components of the eigenvector belonging to the eigenvalue $\lambda_k$.

In fact,

$$(\lambda_k E - A)Q_k = (\lambda_k E - A)(\lambda_k^{n-1}E + \lambda_k^{n-2}B_1 + \cdots + B_{n-1})$$
$$= (\lambda_k^n E + \lambda_k^{n-1}(B_1 - A) + \lambda_k^{n-2}(B_2 - AB_1) + \cdots - AB_{n-1}$$
$$= \lambda_k^n E - p_1\lambda_k^{n-1}E - p_2\lambda_k^{n-2}E - \cdots - p_n E = 0 .$$

From here it follows that $(\lambda_k E - A)u = 0$, where $u$ is any column of the matrix $Q_k$, i.e. it follows that

$$\lambda_k u = A u . \qquad (11)$$

This equality shows that $u$ is an eigenvector.

*Note 1.* When computing the eigenvectors in the manner described it is not necessary, of course, to find all the columns of matrix $Q_k$. It should be limited to the computation of one column; its elements are obtained as a linear combination of the analogous columns of the matrices $B_i$ with the previous coefficients.

*Note 2.* To compute column $u$ of matrix $Q_k$ it is convenient to use the recurrence formula:

$$u_0 = e ;$$
$$u_i = \lambda_k u_{i-1} + b_i \qquad (12)$$

where $b_i$ is the column of matrix $B_i$ taken by us and $e$ is the analogous column of a unit matrix.

Then

$$u = u_{n-1} .$$

As an example we shall compute the eigenvector $X_4$ of Leverrier's matrix belonging to the eigenvalue $\lambda_4 = -5.29870$.

**TABLE IV.11  Determining an Eigenvector by D. K. Faddeev's Method**

| I | II | III | IV | V | VI |
|---|---|---|---|---|---|
| 2258.1433 | −2990.2530 | 1189.8294 | −148.7675 | 308.9522 | 1 |
| 70.5604 | −48.1119 | 8.0822 | 0 | 30.5307 | 0.098820 |
| 32.0618 | −14.2294 | 1.3785 | 0 | 19.2109 | 0.062181 |
| 4.4283 | −1.5939 | 0.1751 | 0 | 3.0095 | 0.009741 |

In columns I, II and III are placed the components of the first column of the matrices $B_i$ multiplied by the corresponding powers of $\lambda_4$; in column IV are the components of the vector $\lambda_4^3(1, 0, 0, 0)'$. Column V contains the components of the vector $X_4$; column VI contains its components after normalization.

## 48. THE ESCALATOR METHOD[†]

An original method for determining the eigenvalues and eigenvectors of a matrix is known as the escalator method. This method gives an inductive construction whereby, knowing the eigenvalues and eigenvectors of the matrix $A_{k-1}$ and its transpose, it is possible to set up an equation to determine the eigenvalues of the matrix $A_k$ obtained from $A_{k-1}$ by bordering; it is then possible to compute by simple formulas the components of the eigenvectors for the matrix $A_k$ and its transpose. Application of the escalator method is begun by looking for the eigenvectors of a second-order matrix. This problem is very easily solved.

The great value of the method is existence of a powerful control which makes it possible for the computation to be verified at each step in terms of its own calculations and without loss of significant figures. Besides this, the very form of the equation for determining the eigenvalues turns out to be very convenient in applying Newton's method.

The method is based on the use of orthogonality properties for the eigenvectors of the matrix and of its transpose.

We shall not carry out the general induction from the $k$th step to the $k + 1$th step but shall limit ourselves to considering the passage from a third-order matrix to a fourth-order one. As a convenience the components of the vectors will be denoted by different letters, contrary to ordinary usage. We assume that all eigenvalues of matrix $A_3$ are real and distinct.

Let $\lambda_r (r = 1, 2, 3)$ be the eigenvalues of matrices $A_3$ and $A_3'$ where

$$A_3 = \begin{bmatrix} a_{11} & a_{12} & a_{13} \\ a_{21} & a_{22} & a_{23} \\ a_{31} & a_{32} & a_{33} \end{bmatrix}. \qquad (1)$$

Furthermore let $X_r = x_r, y_r, z_r)'$ and $X_r' = (x_r', y_r', z_r')'$ $(r = 1, 2, 3)$ be the set of eigenvectors for these matrices.

These eigenvectors may be normalized so that

$$\begin{bmatrix} x_1' & x_2' & x_3' \\ y_1' & y_2' & y_3' \\ z_1' & z_2' & z_3' \end{bmatrix} \begin{bmatrix} x_1 & y_1 & z_1 \\ x_2 & y_2 & z_2 \\ x_3 & y_3 & z_3 \end{bmatrix} = \begin{bmatrix} x_1 & y_1 & z_1 \\ x_2 & y_2 & y_2 \\ x_3 & y_3 & z_3 \end{bmatrix} \begin{bmatrix} x_1' & x_2' & x_3' \\ y_1' & y_2' & y_3' \\ z_1' & z_2' & z_3' \end{bmatrix} = E. \qquad (2)$$

This follows from the orthogonality properties for eigenvectors of a matrix and its transpose established in Section 10, paragraph 3.

Let $A_4$ be a fourth-order matrix obtained from $A_3$ by bordering and let

---

† Morris and Head (1), (2); Morris (5).

$X = (x, y, z, u)'$ be its eigenvector belonging to the eigenvalue $\lambda$.
We have

$$
\begin{aligned}
\lambda x &= a_{11}x + a_{12}y + a_{13}z + a_{14}u , \\
\lambda y &= a_{21}x + a_{22}y + a_{23}z + a_{24}u , \\
\lambda z &= a_{31}x + a_{32}y + a_{33}z + a_{34}u , \\
\lambda u &= a_{41}x + a_{42}y + a_{43}z + a_{44}u .
\end{aligned}
\tag{3}
$$

We multiply the first three equations of system (3) by $x'_r$, $y'_r$ and $z'_r$ respectively and add. We obtain

$$
\begin{aligned}
\lambda(xx'_r + yy'_r + zz'_r) &= (a_{11}x'_r + a_{21}y'_r + a_{31}z'_r)x \\
&+ (a_{12}x'_r + a_{22}y'_r + a_{32}z'_r)y + (a_{13}x'_r + a_{23}y'_r + a_{33}z'_r)z \\
&+ (a_{14}x'_r + a_{24}y'_r + a_{34}z'_r)u,
\end{aligned}
$$

from which, since $(x'_r, y'_r, z'_r)'$ is an eigenvector for matrix $A'_3$,

$$
\lambda(xx'_r + yy'_r + zz'_r) = \lambda_r(xx'_r + yy'_r + zz'_r) + (a_{14}x'_r + a_{24}y'_r + a_{34}z'_r)u
$$

and consequently,

$$
xx'_r + yy'_r + zz'_r = -\frac{P'_r u}{\lambda_r - \lambda} ,
\tag{4}
$$

where

$$
P'_r = a_{14}x'_r + a_{24}y'_r + a_{34}z'_r .
\tag{5}
$$

Let

$$
P_r = a_{41}x_r + a_{42}y_r + a_{43}z_r .
\tag{6}
$$

Then in view of orthogonality properties (2) the following relationship is valid:

$$
\sum_{r=1}^{3} P_r(x'_r x + y'_r y + z'_r z) = P ,
\tag{7}
$$

where

$$
P = a_{41}x + a_{42}y + a_{43}z = -(a_{44} - \lambda)u .
\tag{8}
$$

In fact,

$$
\begin{aligned}
\sum_{r=1}^{3} (a_{41}x_r + a_{42}y_r &+ a_{43}z_r)(x'_r x + y'_r y + z'_r z) \\
&= a_{41}(x_1 x'_1 + x_2 x'_2 + x_3 x'_3)x + a_{41}(x_1 y'_1 + x_2 y'_2 + x_3 y'_3)y \\
&+ a_{41}(x_1 z'_1 + x_2 z'_2 + x_3 z'_3)z + \cdots = a_{41}x + a_{42}y + a_{43}z .
\end{aligned}
$$

Replacing the expression $x'_r x + y'_r y + z'_r z$ in equation (7) by $-P'_r u/(\lambda_r - \lambda)$ according to (4) we obtain the following equality for determining the eigenvalues of matrix $A_4$:

$$
a_{44} - \lambda = \sum_{r=1}^{3} \frac{P_r P'_r}{\lambda_r - \lambda} .
\tag{9}
$$

We shall call equation (9) the *escalator form of the characteristic equation* or the *escalator equation*. Moreover by multiplying (4) successively by $x_r$, $y_r$, and $z_r(r = 1, 2, 3)$ and adding we obtain, again considering orthogonality properties (2):

$$\frac{x}{u} = -\sum_{r=1}^{3}\frac{P'_r x_r}{\lambda_r - \lambda} \; ;$$

$$\frac{y}{u} = -\sum_{r=1}^{3}\frac{P'_r y_r}{\lambda_r - \lambda} \; ; \qquad (10)$$

$$\frac{z}{u} = -\sum_{r=1}^{3}\frac{P'_r z_r}{\lambda_r - \lambda} \; .$$

Analogously,

$$\frac{x'}{u'} = -\sum_{r=1}^{3}\frac{P_r x'_r}{\lambda_r - \lambda} \; ,$$

$$\frac{y'}{u'} = -\sum_{r=1}^{3}\frac{P_r x'_r}{\lambda_r - \lambda} \; , \qquad (11)$$

$$\frac{z'}{u'} = -\sum_{r=1}^{3}\frac{P_r z'_r}{\lambda_r - \lambda} \; .$$

Thus by finding the eigenvalue $\lambda$ from equation (9) we find by formulas (10) and (11) the eigenvectors $A_4$ and $A'_4$ which belong to this value, accurate within a numerical factor. To continue the process we must still normalize them in the sense of formula (2).

It is not hard to verify (again using properties (2)) that

$$\frac{xx' + yy' + zz'}{uu'} = \sum_{r=1}^{3}\frac{P_r P'_r}{(\lambda_r - \lambda)^2} \; .$$

Consequently,

$$\frac{xx' + yy' + zz' + uu'}{uu'} = 1 + \sum_{r=1}^{3}\frac{P_r P'_r}{(\lambda_r - \lambda)^2} \; .$$

We thus satisfy the normalization condition for

$$\frac{1}{uu'} = 1 + \sum_{r=1}^{3}\frac{P_r P'_r}{(\lambda_r - \lambda)^2} \; .$$

We observe that if the right side minus the left side of the escalator equation is denoted by $f(\lambda)$,

$$f(\lambda) = -a_{44} + \lambda + \sum_{r=1}^{3}\frac{P_r P'_r}{\lambda_r - \lambda} \; , \qquad (12)$$

then

$$f'(\lambda) = 1 + \sum_{r=1}^{3}\frac{P_r P'_r}{(\lambda_r - \lambda)^2} \; .$$

Thus

$$\frac{1}{uu'} = f'(\lambda) \ .$$

Without loss of generality we may consider that $u = \pm u'$, choosing the sign so that $1/u^2 = \pm f'(\lambda)$ is positive. This gives

$$u = u' = \frac{1}{\sqrt{f'(\lambda)}} \qquad \text{[if } f'(\lambda) > 0];$$

$$u = -u' = \frac{1}{\sqrt{-f'(\lambda)}} \qquad \text{[if } f'(\lambda) < 0] \ . \qquad (13)$$

In conclusion we observe the control equalities:

$$\sum_{r=1}^{3} P_r x_r' = a_{41} \ ; \quad \sum_{r=1}^{3} P_r y_r' = a_{42} \ ; \quad \sum_{r=1}^{3} P_r z_r' = a_{43} \ ;$$

$$\sum_{r=1}^{3} P_r' x_r = a_{14} \ ; \quad \sum_{r=1}^{3} P_r' y_r = a_{24} \ ; \quad \sum_{r=1}^{3} P_r' z_r = a_{34} \ ; \qquad (14)$$

$$\lambda_1 + \lambda_2 + \lambda_3 + \lambda_4 = \operatorname{Sp} A_4 \ .$$

These last equalities show that all the "new" elements of matrix $A_4$ are required by us for control. A close coincidence of the control formulas guarantees correct computations at each step. At the end of the process it is useful to verify the fulfillment of the orthogonality conditions for the resulting vectors.

We have described the process only for matrices of the fourth order; the passage to the general case is obvious. In observance of the control equations the method guarantees a high accuracy for both the eigenvalues and the components of the eigenvectors belonging to them.

In the case of a symmetric matrix the escalator process is naturally simplified, since all values marked by a stroke (i.e. related to the transposed matrix) will coincide with the corresponding values without strokes. It is possible to conclude from the form of the escalator equation in this case that the characteristic roots of successively bordered matrices separate each other. This circumstance greatly facilitates the determination of the roots which are usually found by Newton's method.

We observe that the escalator form of the characteristic equation turns out to be more convenient in applying Newton's method than the detailed one, since the computation of $f(\lambda)$ and $f'(\lambda)$ is accomplished very easily.

Without going into details we observe that in the case where successive escalator equations have the same or complex roots the process described must be changed in a certain way.[†]

We shall find the eigenvalues and eigenvectors for Leverrier's matrix using the escalator method.

The solution will consist of three stages.

*Stage I.* For the matrix

$$\begin{bmatrix} -5.509882 & 1.870086 \\ 0.287865 & -11.811654 \end{bmatrix}$$

the equation for determining the eigenvalues will be

$$\lambda^2 + 17.321536\lambda + 64.542487 = 0 \ ;$$

---

[†] Morris and Head (2).

its roots will be

$$\lambda_1 = -11.895952\ , \quad \lambda_2 = -5.425584\ .$$

For control we form $\lambda_1 + \lambda_2 = -17.321536$ and compute the spur of matrix $A_2$:

$$\text{Sp } A_2 = -17.321536\ .$$

Moreover we compute the eigenvectors of matrices $A_2$ and $A_2'$, solving the corresponding systems, and normalize the vectors obtained:

| $X_1$ | $X_2$ | $X_1'$ | $X_2'$ | |
|---|---|---|---|---|
| −0.061767 | 4.679234 | −0.210926 | 0.210926 | I |
| 0.210926 | 0.210926 | 4.679234 | 0.061767 | |
| 0.905643 | 1.138422 | 26.638114 | 0.442009 | $P_i$ & $P_i'$ |

The first stage is completed.

*Stage II.* We form the matrix

$$\begin{bmatrix} -5.509882 & 1.870086 & 0.422908 \\ 0.287865 & -11.811654 & 5.711900 \\ 0.049099 & 4.308033 & -12.970687 \end{bmatrix}.$$

We write out on a separate sheet the coefficients $a_{13}$, $a_{23}$ and $a_{31}$, $a_{32}$ introduced again as columns and, adding them to the columns of eigenvectors of matrix $A_2$, find the values $P_i'$ and $P_i$ (by accumulation). As a convenience in further computations we write them out along with the eigenvectors in scheme (I), placing them in a row.

We may now write the escalator equation for matrix $A_3$:

$$f(\lambda) = 12.970687 + \lambda + \frac{24.124621}{-11.895952 - \lambda} + \frac{0.503193}{-5.425584 - \lambda} = 0\ .$$

We determine its roots by Newton's method, writing the computation down according to the scheme:

| $\lambda$ | −15 | −16.651 | −17.3458 | −17.3975 | −17.397655 | |
|---|---|---|---|---|---|---|
| $-11.895952 - \lambda$ | 3.104 | 4.755 | 5.4498 | 5.501548 | 5.501703 | |
| $-5.425584 - \lambda$ | 9.574 | 11.225 | 11.9202 | 11.971916 | 11.972071 | |
| $12.970687 + \lambda$ | −2.029 | −3.680 | −4.3751 | −4.426813 | −4.426968 | |
| $\dfrac{P_1 P_1'}{-11.895958 - \lambda}$ | 7.772 | 5.074 | 4.4267 | 4.385061 | 4.384936 | |
| $\dfrac{P_2 P_2'}{-5.425584 - \lambda}$ | 0.053 | 0.045 | 0.0422 | 0.042031 | 0.042031 | II |
| $f(\lambda)$ | 5.796 | 1.439 | 0.0938 | 0.000279 | −0.000001 | |
| $\dfrac{P_1 P_1'}{(-11.985952 - \lambda)^2}$ | 2.504 | 1.067 | 0.8123 | 0.797059 | 0.797014 | |
| $\dfrac{P_2 P_2'}{(-5.425584 - \lambda)^2}$ | 0.006 | 0.004 | 0.0035 | 0.003511 | 0.003511 | |
| $f'(\lambda)$ | 3.510 | 2.071 | 1.8158 | 1.800570 | 1.800525 | |
| $\Delta\lambda$ | −1.651 | −0.6948 | −0.0517 | −0.000155 | 0.000000 | |

Thus $\lambda_1 = -17.397655$.

Analogously we find $\lambda_2 = -7.594378$ and $\lambda_3 = -5.300190$. For control:

$$\lambda_1 + \lambda_2 + \lambda_3 = -30.292223, \quad \operatorname{Sp} A_3 = -30.292223 .$$

We now pass on to determining the components of the eigenvectors for matrices $A_3$ and $A_3'$ which are found, accurate up to a multiplier, by formulas analogous to formulas (10) and (11). For this it is convenient to form the auxiliary schemes (III) and (IV).

| $P_i' x_i$ | $P_i' y_i$ | $P_i x_i'$ | $P_i y_i'$ | |
|---|---|---|---|---|
| $-1.645356$ | $5.618671$ | $-0.191024$ | $4.237716$ | III |
| $2.068264$ | $0.093231$ | $0.240123$ | $0.070318$ | |
| $0.422908$ | $5.711902$ | $0.049099$ | $4.308034$ | |

| $\dfrac{1}{\lambda_i + 17.397655}$ | $\dfrac{1}{\lambda_i + 7.594378}$ | $\dfrac{1}{\lambda_i + 5.300190}$ | |
|---|---|---|---|
| $0.181762$ | $-0.232473$ | $-0.151613$ | IV |
| $0.083528$ | $0.461086$ | $-7.974863$ | |

Scheme (III) contains the products of the numbers $P_i$ and $P_i'$ by the components of the corresponding vectors and is obtained from scheme (I); the last row sets up the control (for example, $\sum\limits_{i=1}^{2} P_i' x_i = a_{12}$). Scheme (IV) contains the factors $\dfrac{1}{\lambda_i - \lambda}$, where in place of $\lambda$ we take in succession the three roots which have been computed.

Moreover the normalizing factors are determined from scheme (II) and from two analogous schemes which serve to compute two other roots. Since $f'(\lambda_i) > 0$,

$$D_i = \frac{1}{\sqrt{f'(\lambda_i)}} \quad \text{and} \quad z_i = z_i' = D_i \quad (i = 1, 2, 3) .$$

Computing, we get $D_1 = 0.745248$, $D_2 = 0.644055$, $D_3 = 0.172627$.

Utilizing the previous schemes we may find without difficulty the components of the eigenvectors for matrices $A_3$ and $A_3'$ (in the final scheme we shall write them after multiplication by the corresponding normalizing factor).

| $X_1$ | $X_2$ | $X_3$ | $X_1'$ | $X_2'$ | $X_3'$ | |
|---|---|---|---|---|---|---|
| $0.094129$ | $-0.860553$ | $2.804268$ | $0.010928$ | $-0.099909$ | $0.325572$ | |
| $-0.766896$ | $0.813572$ | $0.275404$ | $-0.578409$ | $0.613612$ | $0.207717$ | |
| $0.745248$ | $0.644059$ | $0.172627$ | $0.745248$ | $0.644055$ | $0.172627$ | |
| $0.835026$ | $1.114160$ | $0.333026$ | $0.137039$ | $0.182847$ | $5.054654$ | $P_i$ & $P_i'$ |

The second stage is completed.

*Stage III.* Having computed the values $P_i$ and $P_i'$ we write the escalator equation for the matrix $A_4$:

$$17.596207 + \lambda + \frac{0.114431}{-17.397655 - \lambda} + \frac{0.203721}{-7.594378 - \lambda} + \frac{0.018201}{-5.300190 - \lambda} = 0$$

and compute its roots:

$$\lambda_1 = -17.863262 , \qquad \lambda_2 = -17.152427 ,$$
$$\lambda_3 = -7.574044 , \qquad \lambda_4 = -5.298698 ,$$
$$\sum_{i=1}^{4} \lambda_i = -47.888431 ; \quad \mathrm{Sp}\, A_4 = -47.888430 .$$

We then compute the eigenvectors of the matrix $A_4$, normalizing them in the usual way:

| $X_1$ | $X_2$ | $X_3$ | $X_4$ |
|---|---|---|---|
| -0.019873 | 0.032932 | -0.351235 | 1.135218 |
| 0.169807 | -0.261310 | 0.328467 | 0.112183 |
| -0.187215 | 0.236640 | 0.260927 | 0.070591 |
| 0.808482 | 0.586694 | 0.045005 | 0.011058 |

and the eigenvectors of the matrix $A_4'$:

| $X_1'$ | $X_2'$ | $X_3'$ | $X_4'$ |
|---|---|---|---|
| -0.014058 | 0.023297 | -0.248476 | 0.803091 |
| 0.780383 | -1.200908 | 1.509559 | 0.515564 |
| -1.140762 | 1.441927 | 1.589924 | 0.430123 |
| 0.808482 | 0.586694 | 0.045005 | 0.011058 |

The final control for the computation is the computation of the product of two corresponding matrices which consist of the components of the eigenvectors. We obtain in place of the identity matrix the matrix

$$\begin{bmatrix} 1.000005 & -0.000004 & 0.000000 & 0.000002 \\ -0.000003 & 1.000004 & -0.000002 & -0.000003 \\ -0.000002 & 0.000000 & 0.999994 & 0.000000 \\ -0.000001 & 0.000000 & 0.000004 & 1.000006 \end{bmatrix} .$$

Finally, as a comparison, we represent the escalator equation as an ordinary polynomial:

$$t^4 + 47.888430t^3 + 797.27877t^2 + 5349.4556t + 12296.550 = 0 ,$$

and we normalize the eigenvector belonging to $\lambda_4$ so that its first component will be equal to one. This gives $X_4 = (1;\ 0.098820;\ 0.062183;\ 0.009741)'$.

We shall see that the coefficients of the equation described coincide with Leverrier's data with greater accuracy than for computation by other methods. The orthogonality relationships are also satisfied with significant accuracy.

## 49. THE INTERPOLATION METHOD

The method introduced by us in the previous paragraphs solved the problem of bringing the secular equation to polynomial form. We shall apply the interpolation method developed in this paragraph to the more general problem of expanding a determinant of the form

$$F(t) = \begin{vmatrix} f_{11}(t) & \cdots & f_{1n}(t) \\ \vdots & & \vdots \\ f_{n1}(t) & \cdots & f_{nn}(t) \end{vmatrix} \tag{1}$$

($f_{ik}(t)$ is a given polynomial in $t$), in particular, to expanding the characteristic determinant $D(t) = |A - tE|$ and the determinant $|A - Bt|$, where $A$ and $B$ are given matrices.

The method essentially consists of the following. Let it be known that $F(t)$ is a polynomial whose degree does not exceed the number $k$. As is well known, such a polynomial is determined completely by its values at $k + 1$ points and may be reconstructed from such values using one or another interpolation formula.

Therefore for a clear representation of $F(t)$ it is necessary to compute the value of $k + 1$ numerical determinants

$$F(\lambda_i) = \begin{vmatrix} f_{11}(\lambda_i) & \cdots & f_{1n}(\lambda_i) \\ \vdots & & \vdots \\ f_{n1}(\lambda_i) & \cdots & f_{nn}(\lambda_i) \end{vmatrix} \qquad (i = 0, 1, \cdots, k), \tag{2}$$

where $\lambda_0, \lambda_1, \cdots, \lambda_k$ are certain numbers generally chosen arbitrarily.

Computation of the necessary determinants may be accomplished, for example, according to the scheme presented in Section 17.

To construct the polynomial $F(t)$ from its values it is most advantageous to use Newton's interpolation formula applied for equidistant abscissas $\lambda_i$. We shall carry out Newton's formula for $\lambda_i = i, i = 0, \cdots, k$:

$$F(t) = \sum_{i=0}^{k} \frac{\Delta^i F(0)}{i!} t(t-1) \cdots (t-i+1), \tag{3}$$

where $\Delta^i F(l)$ denotes the $i$th difference of computed values for the polynomial $F(t)$; this difference is determined from the recurrent formula

$$\Delta^i F(l) = \Delta^{i-1} F(l+1) - \Delta^{i-1} F(l).$$

We let

$$\frac{t(t-1) \cdots (t-i+1)}{i!} = \sum_{m=1}^{i} c_{mi} t^m.$$

Then formula (3) is transformed to:

$$F(t) = \sum_{i=0}^{k} \Delta^i F(0) \left( \sum_{m=1}^{i} c_{mi} t^m \right) = F(0) + \sum_{m=1}^{k} \left( \sum_{i=m}^{k} c_{mi} \Delta^i F(0) \right) t^m . \tag{4}$$

This formula is called the interpolation formula of A. A. Markov.

In the work of Sh. E. Mikeladze (1), where an interpolation method for finding polynomial determinants was described, formula (4) was chosen as an interpolation formula.

The table of coefficients for Markov's interpolation formula for $m \leq i \leq 20$ is presented in V. N. Faddeeva's book (1).

In using an interpolation formula it is advisable for control of the supporting values of determinant (1) to compute at least one value of $F(t)$, namely, for our case, $F(k + 1)$, since, $\Delta^{k+1} F(0)$ must be equal to zero and $\Delta^k F(0)$ and $\Delta^k F(1)$ must be equal to each other.

The interpolation method requires a great number of operations. Thus to compute the coefficients of the characteristic polynomial using interpolation formula (4) requires first of all computing $(n + 1)$ $n$th-order determinants. This requires $(n + 1)(n - 1)(n^2 + n + 3)/3$ multiplications and divisions. If we take the coefficients for the interpolation formula from the table, then to obtain the coefficients of the characteristic polynomial it is still necessary to carry out $[n(n + 1)]/2$ multiplications.

Thus the total number of necessary multiplication and division operations:

$$\frac{n + 1}{3}(n - 1)(n^2 + n + 3) + \frac{n(n + 1)}{2}$$

greatly increases the number of operations necessary to compute the same coefficients by the methods of A. M. Danilevskii and A. N. Krylov.

Besides this, the indicated method does not allow us to simplify in any way the problem of finding the eigenvectors of the matrix, whereas in computing by the Danilevskii and Krylov methods the problem of determining the eigenvectors of a matrix is greatly simplified. Nevertheless the interpolation method is interesting as a method which makes it possible to solve more general problems.

As an example we again carry out the computation for Leverrier's matrix. Omitting the tiresome computation of determinants we find that

$$\varphi(0) = 12296.55 , \quad \varphi(1) = 18492.17 , \quad \varphi(2) = 26583.68 ,$$
$$\varphi(3) = 36894.41 , \quad \varphi(4) = 49771.69 .$$

We then form the table of differences:

| $t$ | $\varphi(t)$ | $\Delta$ | $\Delta^2$ | $\Delta^3$ | $\Delta^4$ |
|---|---|---|---|---|---|
| 0 | 12296.55 |          |         |        |    |
|   |          | 6195.62  |         |        |    |
| 1 | 18492.17 |          | 1895.89 |        |    |
|   |          | 8091.51  |         | 323.33 |    |
| 2 | 26583.68 |          | 2219.82 |        | 24 |
|   |          | 10310.73 |         | 347.33 |    |
| 3 | 36894.41 |          | 2566.55 |        |    |
|   |          | 12877.28 |         |        |    |
| 4 | 49771.69 |          |         |        |    |

(We observe that when computing the coefficients of the characteristic polynomial it is possible to avoid computing extra values of $\varphi(t)$ for control, since in this case the equality $\Delta^k F(0) = (-1)^n n!$ is a reliable control).

Finally we compute the coefficients of the characteristic polynomial, setting up the computations by the scheme:

| $i$ | $\Delta^i \varphi(0)$ | $c_{4i}$ | $c_{3i}$ | $c_{2i}$ | $c_{1i}$ |
|---|---|---|---|---|---|
| 1 | 6195.62 | | | | 1.00000000 |
| 2 | 1895.89 | | | 0.50000000 | -0.50000000 |
| 3 | 323.33 | | 0.16666667 | -0.50000000 | 0.33333333 |
| 4 | 24.00 | 0.04166667 | -0.25000000 | 0.45833333 | -0.25000000 |
| | | | 47.8883 | 797.280 | 5349.45 |

The coefficient $p_4 = \varphi(0) = 12296.55$.

We can always apply the interpolation method; in particular, the case where the characteristic polynomial has multiple roots does not differ in any way from the other cases.

If in place of the numbers $0, \cdots, k$ we takes as the interpolation nodes the numbers $t_i = a + hi$, then formula (4) is modified in the following way:

$$F(t) = F(a) + \sum_{m=1}^{k} \left( \sum_{i=m}^{k} c_{mi} h^i \Delta^i F(a) \right)(t - a)^m . \tag{5}$$

Sometimes it may be advisable to take as interpolation abscissas non-equidistant numbers. In this case it is possible to use Newton's general interpolation formula. However, in the case of non-equidistant abscissas it is more convenient to construct the necessary polynomial by the method of undetermined coefficients. That is,

$$F(t) = a_0 t^k + a_1 t^{k-1} + \cdots + a_k .$$

Then to determine the numbers $a_j$, $j = 0, \cdots, k$, we obtained the system of algebraic equations

$$F(\lambda_i) = a_0 \lambda_i^k + a_1 \lambda_i^{k-1} + \cdots + a_k ,$$

which may be solved by any of the methods presented.

In particular, it is convenient to apply the interpolation method in expanding the determinant $|A - Bt|$ when matrix $B$ has a small determinant. If, however, the determinant of $B$ is not a small number, then it is better to determine the coefficients of the desired polynomial by the transformation

$$|A - Bt| = |B| |AB^{-1} - tE| .$$

The matrix $AB^{-1}$ may be found by the elimination method (Section 22).

The interpolation method turns out to be useful in computing the eigenvalues without computing the characteristic polynomial.

The following method exists for determining the roots of a polynomial[†] $f(t)$ (or of any other analytic function[‡]) and is a simple generalization of the well-known method of false position. Three arbitrary values $t_0, t_1, t_2$ are taken as an independent variable and the corresponding values of the function $y_0 = f(t_0)$, $y_1 = f(t_1)$ and $y_2 = f(t_2)$ are computed. An interpolation polynomial of the second order is constructed from the resultant values and its roots are found. The root which turns out to be closer to $t_2$ than any other is taken as the next approximation $t_3$. By the same means we then construct $t_4$, starting from the triple of numbers $t_1, t_2, t_3$, and so forth.

There is no known proof of the convergence of this process for arbitrary initial approximations. A proof is known only under the assumption that the initial approximations are close enough to the root which is being computed.[‡] However, in practice the process turns out to be always convergent.

Choosing an interpolation polynomial of the second order (and not of the first or higher) is convenient in that it is possible, without important complications, to pass to the complex plane from the real axis even if $f(t)$ has real coefficients and the initial approximation is taken to be real.

The computational formulas are as follows. We let $t_1 - t_0 = \Delta_1$, $t_2 - t_1 = \Delta_2$, and $t_3 - t_2 = \Delta_3$. Then $\Delta_3$ is a root of the quadratic equation

$$az^2 + bz + c = 0 , \tag{6}$$

where

$$a = \Delta_1(y_2 - y_1) + \Delta_2(y_0 - y_1) ,$$
$$b = \Delta_1(\Delta_1 + 2\Delta_2)(y_2 - y_1) + \Delta_2^2(y_0 - y_1) , \tag{7}$$
$$c = \Delta_1\Delta_2(\Delta_1 + \Delta_2)y_2 .$$

The formulas become even more useful if we introduce the relationships for error correction letting

$$\tilde{\delta}_2 = \frac{\Delta_2}{\Delta_1} ;$$
$$\tilde{\delta}_3 = \frac{\Delta_3}{\Delta_2} \tag{8}$$

For $\tilde{\delta}_3$ we obtain the quadratic equation

$$\alpha\tilde{\delta}^2 + \beta\tilde{\delta} + \gamma = 0 \tag{9}$$

with coefficients

$$\alpha = \delta_2(y_2 - y_1) + \delta_2^2(y_0 - y_1) ,$$
$$\beta = (1 + 2\delta_2)(y_2 - y_1) + \delta_2^2(y_0 - y_1) , \tag{10}$$
$$\gamma = (1 + \delta_2) y_2 .$$

In solving the quadratic equation the formula

† Muller, D., Math. Tables Other Aids Comput., 1956, 10, 208-215.
‡ Frank, W. L., J. Assoc. Comput. Machinery, 1958, 5, No. 2, 154-160.

$$\delta = \frac{2\gamma}{-\beta \pm \sqrt{\beta^2 - 4\alpha\gamma}}, \qquad (11)$$

is convenient while the sign for the square root should be chosen so that of the two possible values for the denominator we obtain the greater one in modulus.

A new correction is obtained by the formula $\Delta_3 = \Delta_2 \delta_3$ and a new approximation by

$$t_3 = t_2 + \Delta_3 . \qquad (12)$$

In applying this method to compute eigenvalues[†] there is no need to compute the coefficients of the characteristic polynomial. Computation of the characteristic polynomial's values for fixed values of $t$ comes down to computation of the numerical determinants $|A - tE|$; this may be carried out without particular difficulty, for example, by the elimination method.

The method described is especially applicable if the matrix $A$ has a form convenient for computing the determinants $|A - tE|$, for example, if the matrix is almost triangular or tridiagonal. These considerations provide a basis for recommending application of quadratic interpolation after transforming the initial matrix to almost triangular form (for example, by Hessenberg's method or by Givens' method of rotations, cf. Section 51) or to tridiagonal form (for example, by the biorthogonal algorithm as in Section 63 or by the method of rotations in the symmetric case).

As an example we compute two eigenvalues of matrix (4), Section 51. In Table IV.13 this matrix is changed to tridiagonal form by a similarity transformation.

We take $t_0 = 0$, $t_1 = 0.5$, $t_2 = 1$.

Then $y_0 = 0.28615247$, $y_1 = -0.01927552$, $y_2 = -0.07370353$. Carrying out the computations from formulas (10), (11) and (12) we obtain $t_3 = 1.26691227$. Computing the determinant $|A - t_3E|$ we obtain $y_3 = -0.31978253$. We again apply formulas (10), (11) and (12), increasing all their indices by one. We obtain $t_4 = 0.84456137$.

After six steps the process is stabilized by $t_8 = 0.79670667$. This value coincides with $\lambda_2$ within the limits of accuracy.

We now compute another root, starting from the same initial approximations $t_0 = 0$, $t_1 = 0.5$, $t_2 = 1$. We find the values of $y_i$ by the formula

$$y_i = \frac{1}{t_i - \lambda_2} \cdot |A - t_iE| .$$

After three steps of the process we obtain

$$\lambda_3 = 0.63828382$$

with satisfactory accuracy.

---

† Frank (1).

## 50. THE METHOD OF ORTHOGONALIZATION OF SUCCESSIVE ITERATIONS

This method, like the methods of A. N. Krylov and Hessenberg, is aimed toward finding a linear combination for a sequence of iterations of an arbitrary vector for a matrix $A$ which is equal to zero. Whereas in Krylov's method this is done by solving a linear system, and in Hessenberg's method by a gradual accumulation of null components in the "corrected" iterations, in this paragraph we shall apply the orthogonalization process toward the same goal.

That is, starting with a vector $X_1$ we construct its iteration $AX_1$ and orthogonalize it with the vector $X_1$, i.e. we construct a vector $X_2 = AX_1 + g_{11}X_1$ such that $(X_1, X_2) = 0$. This will be satisfied if

$$g_{11} = -\frac{(AX_1, X_1)}{(X_1, X_1)}.$$

Furthermore, we construct the vector $AX_2$ and orthogonalize it with the vectors $X_1$ and $X_2$. As a result we get the vector

$$X_3 = AX_2 + g_{21}X_1 + g_{22}X_2$$

where

$$g_{21} = -\frac{(AX_2, X_1)}{(X_1, X_1)},$$

$$g_{22} = -\frac{(AX_2, X_2)}{(X_2, X_2)}.$$

The process may be continued in a natural way by the formulas

$$X_{i+1} = AX_i + g_{i1}X_1 + g_{i2}X_2 \cdots + g_{ii}X_i,$$

$$g_{ik} = -\frac{(AX_i, X_k)}{(X_k, X_k)} \qquad (k = 1, 2, \cdots, i)$$

until we get the null vector. In every case this occurs by the $n$th step of the process but may happen even earlier if the minimum annihilating polynomial for $X_1$ is not the characteristic polynomial. It is clear that

$$X_{i+1} = \varphi_i(A)X_1,$$

where the polynomials $\varphi_i(t)$ are connected with each other by the recurrent relationships

$$\varphi_i(t) = (t + g_{ii})\varphi_{i-1}(t) + \cdots + g_{i1}\varphi_0(t) \quad (1 < i \le n).$$

Thus, having computed all the coefficients $g_{ij}$, we may compute successively all the polynomials $\varphi_0 = 1, \varphi_1(t), \cdots, \varphi_n(t) = \varphi(t)$. When the process ends at an earlier time we obtain the minimum annihilating polynomial for the vector $X_1$.

We observe that in the normal case $AX + XG = 0$, where $X$ is a non-singular matrix with columns $X_1, \cdots, X_n$ and

**TABLE IV.12  Determining the Coefficients of the Characteristic Polynomial by the Method of Orthogonalization of Iterations**

| | $X_1$ | $AX_1$ | $X_2$ | $AX_2$ | $X_3$ | $AX_3$ | $X_4$ | $AX_4$ | $X_5$ |
|---|---|---|---|---|---|---|---|---|---|
| I | 0<br>1<br>0<br>0 | 1.870086<br>−11.811654<br>4.308033<br>0.269851 | 1.870086<br>0<br>4.308033<br>0.269851 | −8.4796731<br>25.161231<br>−55.724444<br>1.2832177 | 13.118302<br>0<br>−5.9701610<br>4.3997776 | −74.766345<br>−30.066321<br>79.090169<br>−85.680123 | 15.920690<br>0<br>−3.6292098<br>−52.393399 | −89.717741<br>−19.223057<br>35.839866<br>916.95301 | −0.000054<br>0<br>−0.000349<br>0.000028 |
| II | 1 | −5.363684 | 0<br>22.129190 | −37.759668 | 0<br>−0.000013<br>227.09071 | −111.422620 | 0<br>−0.000107<br>0.000083<br>3011.7078 | 843.85208 | |
| III | −11.811654<br>25.161231<br>−30.066321<br>−19.223056 | | −255.57418<br>177.78270<br>234.06012 | | −1829.9620<br>2643.4751 | | −49600.723 | | |
| IV | 11.811654<br>−25.161231<br>30.066321<br>19.223057 | | 11.549188<br>−8.0338548<br>−10.576985 | | 8.0582864<br>−11.640613 | | 16.469301 | | |
| V | 1 | | 1<br>11.811654 | | 1<br>23.360842<br>111.25378 | | 1<br>31.419128<br>291.46828<br>831.68803 | | 1<br>47.888429<br>797.27874<br>5349.4554<br>12296.550 |

$$G = \begin{bmatrix} g_{11} \cdots & & g_{n1} \\ -1 \cdots & & g_{n2} \\ \cdot \cdot \cdot \cdot \cdot \cdot \cdot \cdot \\ 0 \cdots -1 & & g_{nn} \end{bmatrix}.$$

Thus the matrix $A$ is similar to the matrix $-G$. This permits us to determine the eigenvectors of matrix $A$ accurately by the same process as in Hessenberg's method.

In Table IV.12 is given a numerical illustration of the method for the example of Leverrier's matrix.

The table consists of five parts. In the first part are placed the mutually orthogonal vectors $X_1, \cdots, X_4$ and their iterations. Parts II, III and IV contain $(X_i, X_k)$, $(AX_i, X_k)$ and $g_{ik}$ respectively. In part V are computed the coefficients of the characteristic polynomial by recurrence for formulas. The usual control by sums is carried out in computing the iterations. From Table IV.12 it is apparent that the coefficients of the characteristic polynomial are computed with a sufficient degree of accuracy. It is also possible to compute theoretically the null matrix $AX + XG$ as a control.

The method of orthogonalization of iterations is laborious enough in the general case. The volume of computational work in application is greater than, for example, in the application of Hessenberg's method.

However, when matrix $A$ is symmetric the picture is greatly simplified; in this case the matrix $G$ will be tridiagonal.

In fact matrix $X$, whose columns are pairwise orthogonal may be represented as

$$X = UD$$

where $U$ is an orthogonal matrix and $D = [d_1, \cdots, d_n]$, $d = \sqrt{(X_i, X_i)}$. From the equality $AX + XG = 0$ it follows that $DGD^{-1} = -U^{-1}AU$. If $A$ is symmetric, then the matrix $-U^{-1}AU$ is also symmetric. Consequently $d_i g_{ij} d_j^{-1} = d_j g_{ji} d_i^{-1}$. Since $g_{ij} = 0$ for $j - i > 1$, then $g_{ji} = 0$ for $j - i > 1$, i.e. $G$ is actually a tridiagonal matrix.

The method of orthogonalization of iterations as applied to a symmetric matrix is called the method of minimal iterations. It was described for the first time by Lanczos in his well-known works (2)-(5). Among computational methods of linear algebra the method of minimal iterations occupies an exceptionally important place in view of its many connections with other contemporary methods, exact and iterative.

Chapter VI and part of chapter VII will be devoted to the method of minimal iterations and its generalizations.

## 51. TRANSFORMATION OF SYMMETRIC MATRICES TO TRIDIAGONAL FORM BY MEANS OF ROTATIONS

By a rotation we shall mean a transformation of coordinates with the elementary matrix of rotation

$$
T_{ij} = \begin{bmatrix}
1 & & & & & & & & \\
& \ddots & & & & & & & \\
& & \ddots & & & & & & \\
& & & c & \cdots & -s & & & \\
& & & & \ddots & & \ddots & & \\
& & & & 1 & & & & \\
& & & & & \ddots & & \ddots & \\
& & & s & \cdots & & c & & \\
& & & & & & & \ddots & \\
& & & & & & & & 1
\end{bmatrix}
\begin{matrix} \\ \\ \\ \cdots i \\ \\ \\ \\ \cdots j \\ \\ \\ \end{matrix}
\qquad (1)
$$

where $c^2 + s^2 = 1$.

A rotation may be interpreted geometrically as a change in the basis vectors $e_i$ and $e_j$ by a certain angle, carried out in the plane spanned by the vectors $e_i$ and $e_j$. The matrix $T_{ij}$ is orthogonal.

We shall show[†] that it is possible to bring any symmetric matrix to tridiagonal form by a sequence of rotations, i.e. by a sequence of similarity transformations with matrices of the form $T_{ij}$.

We shall carry out the necessary calculations. Let $A$ be a symmetric matrix, $B = A T_{ij}$, $C = T'_{ij} B = T'_{ij} A T_{ij}$. It is easy to see that all the columns of matrix $B$ coincide with the columns of matrix $A$ with the exception of the $i$th and $j$th columns; these are obtained from the corresponding columns of matrix $A$ by the formulas:

$$
\begin{aligned}
B_i &= cA_i + sA_j, \\
B_j &= -sA_i + cA_j.
\end{aligned}
\qquad (2)
$$

The rows of matrix $C$ coincide in turn with the rows of matrix $B$ with the exception of the $i$th and $j$th rows: these are obtained from the corresponding rows of matrix $B$ by the same formulas:

$$
\begin{aligned}
C^i &= cB^i + sB^j, \\
C^j &= -sB^i + cB^j.
\end{aligned}
\qquad (3)
$$

To construct the rows $C^j$ and $C^i$ it is necessary to compute only four elements $c_{ii}, c_{ij}, c_{ji}$ and $c_{jj}$ (where $c_{ji}$ is only for control, since $c_{ji} = c_{ij}$, but they are obtained by dissimilar computations). The remaining elements of the rows $C^i$ and $C^j$ are only theoretically equal to the corresponding elements of the columns $B_i$ and $B_j$ but the same operations are performed in computing them.

Let $1 < i < j$. We shall show that $c$ and $s$ may be chosen so that $c_{i-1,j} = 0$. In fact $c_{i-1j} = b_{i-1j} = -sa_{i-1i} + ca_{i-1j}$ so that it is sufficient to take

$$
\frac{s}{c} = \frac{a_{i-1j}}{a_{i-1i}}
$$

---

† Givens (1), (2).

and consequently the choice of sign for the denominator is immaterial.

The whole process of bringing a symmetric matrix to tridiagonal form goes along in this way. By these transformations the elements of the first row, beginning with the third one, are annihilated in turn by $T_{23}$, $\cdots$, $T_{2n}$. Then the elements of the second row, beginning with the fourth, are annihilated by $T_{34}$, $\cdots$, $T_{3n}$. It is clear that the elements of the first row will not be changed any more by this. In fact the first two elements of the first row will not be changed under transformations by $T_{34}$, $\cdots$, $T_{3n}$. The remaining elements, equal to zero, will undergo a linear homogeneous transformation and will therefore remain equal to zero. Furthermore, by using $T_{45}$, $\cdots$, $T_{4n}$ the elements of the third row, beginning with the fifth, are annihilated etc.

From what has been said above it is clear that each successive transformation will not change the elements annihilated earlier. Thus we pass from the given symmetric matrix $A$ to a tridiagonal matrix $S$ in at most $[(n-1)(n-2)]/2$ transformations.

We shall illustrate the method's computational scheme by the example of a matrix

$$A = \begin{bmatrix} 1.00 & 0.42 & 0.54 & 0.66 \\ 0.42 & 1.00 & 0.32 & 0.44 \\ 0.54 & 0.32 & 1.00 & 0.22 \\ 0.66 & 0.44 & 0.22 & 1.00 \end{bmatrix} \tag{4}$$

which has been encountered earlier in examples of solving a system of linear equations.

The characteristic polynomial of this matrix is equal to

$$t^4 - 4t^3 + 4.752t^2 - 2.111856t + 0.28615248 .$$

Here all the coefficients are computed exactly.

The eigenvalues of matrix (4), computed accurately within $5 \cdot 10^{-9}$, are

$$\lambda_1 = 2.32274880 , \quad \lambda_2 = 0.79670669 ,$$
$$\lambda_3 = 0.63828380 , \quad \lambda_4 = 0.24226071 .$$

The tridiagonalization process is carried out in Table IV.13.

The table consists of four parts. In part II are placed, along with the given matrix $A$, the results of successive similarity transformations using $T_{23}$, $T_{24}$ and $T_{34}$. The last matrix is the desired one $S$. Parts I and IV are auxiliary, part III is for control.

We shall describe one step for filling in the table (consisting in accomplishing a transformation by the matrix $T_{ij}$). In part II all the elements of the previous matrix are rewritten, except for elements lying in the two rows and two columns with numbers $i$ and $j$. The $i$th and $j$th columns are then transformed according to formulas (2) and the elements of the transformed columns, except for elements with indices $ii$, $ij$, $ji$ and $jj$, are written down in suitable places in part II. The four separate elements are

TABLE IV.13  Transforming a Symmetric Matrix to Tridiagonal Form using a Sequence of Rotations

**Panel A**

I
| 1.00 | 0.42 | 0 54 | 0.66 |
|------|------|------|------|
| 0.42 | 1.00 | 0.32 | 0.44 |
| 0.54 | 0.32 | 1.00 | 0.22 |
| 0.66 | 0.44 | 0.22 | 1.00 |

4.00000000

II
| 1 | 0 54 | 0.66 | 0.86653332 | −0.59289121 |
|---|------|------|------------|-------------|
| 0.68410525 | 0.32 | 0.44 | 0.98581321 | 0.36134790 |
| 0 | 1.00 | 0.22 |  |  |
| 0.66 | 0.22 | 1.00 |  |  |

| 0 68410525 | 0 66 |
|------------|------|
| 1.31015383 | 0.44379135 |
| −0.07876923 | −0.21224804 |
| 0.44379135 | 1 |

0.68410525 / 0 / −0.07876923 / 0.68984614

III
2.18    2.08
| 2.3592812 | 1.8915433 |
| 2.98024313 | |
3.99999997

IV
| 0.4680 | 0.68410526 | 0.90359999 | 0.95057876 |
|--------|------------|------------|------------|
| 0.61394061 | 0.78935221 | 0.71967235 | 0.69431385 |

**Panel B**

I
| 1.25101197 | −0.59027359 | 0.51483019 | −0.46952426 |
|------------|-------------|------------|-------------|
| 1.01369821 | 0.41154187 | 0.31656653 | 0.63863277 |

II
| 1.00 | 0.95057877 | 0.00000000 | 1 | 0.95057877 | 0 |
|------|------------|------------|---|------------|---|
| 0.95057877 | 1.60414343 | −0.13906436 | 0.68410525 | 1.60414343 | −0.24693573 |
| 0 | −0.20405479 | −0.09805848 | 0 | −0.20405479 | 0.60370643 |
| 0.00000000 | −0.13906436 | 0.70601043 | 0 | −0.13906436 | −0.02833780 |

−0.13906436 / −0.09805848 / 0.70601043

0 / −0.24693573 / 0.60370643 / −0.02833780

0 / 0.00000000 / −0.02833781 / 0.79215011

−0.46952426 / 0.63863277

III
3.01123416
| 0.38773287 | 0.46888759 |
| −0.27679020 | −0.27679020 |
3.99999997

IV
| 0.060977254 | 0.24693573 | 0.58446099 | 0.16910852 |
|-------------|------------|------------|------------|
| 0.82634777 | 0.56316014 | | |

introduced into part I. The resultant matrix is filled in further by symmetry. The remaining four elements are then constructed from formulas (2) upon the numbers placed in part I. In the fourth part are written the coefficient $c$ and $s$ which determine the rotation matrices (in the last row) and the numbers necessary for their computation. Control (part III) is accomplished by computing corresponding column sums (for control of the operations on columns) and by computing the spurs of the constructed matrices; these should be equal to each other.

After the tridiagonal matrix similar to the initial matrix $A$ has been constructed, the search for eigenvalues may be carried out in various ways.

The most direct method is the construction of the characteristic polynomial $\varphi(t)$ for the matrix $S$ (and consequently also for the matrix $A$ similar to it) from the recurrent formulas

$$\varphi_0 = 1, \quad \varphi_i(t) = (t - s_{ii})\varphi_{i-1}(t) - s_{i-1i}^2 \varphi_{i-2}(t), \quad \varphi_n(t) = (-1)^n \varphi(t)$$

and then finding of its roots.

It is convenient to compute the coefficients of the polynomials $\varphi_i(t)$ in the following way. The coefficients are arrayed according to the scheme

| $\varphi_0$ | $\varphi_1$ | $\cdots$ | $\varphi_{n-1}$ | $\varphi_n$ | |
|---|---|---|---|---|---|
| | | | | 1 | $t^n$ |
| | | | 1 | $\pi_{n-1}^{(n)}$ | $t^{n-1}$ |
| | | | . | . | . |
| | | | . | . | . |
| | | | . | . | . |
| | | 1 | $\cdots \pi_1^{(n-1)}$ | $\pi_1^{(n)}$ | $t$ |
| | 1 | $\pi_0^{(1)}$ | $\cdots \pi_0^{(n-1)}$ | $\pi_0^{(n)}$ | $t^0$ |

The scheme is then filled in from left to right according to the recurrent formula

$$\pi_j^{(i+1)} = -s_{i-1i}^2 \pi_j^{(i-1)} - s_{ii}\pi_j^{(i)} + \pi_{j-1}^{(i)} .$$

The coefficients involved in the formula obviously enter the scheme with the following positioning:

$$\begin{array}{ccc} \pi_j^{(i-1)} & \pi_j^{(i)} & \pi_j^{(i+1)} \\ & \pi_{j-1}^{(i)} & \end{array}$$

Givens[†] recommends another method which avoids computation of the characteristic polynomial's coefficients. This method utilizes the fact that the polynomials $\varphi_0, \varphi_1, \cdots, \varphi_n$ form a Sturm sequence.

Finally, as we saw in Section 49, it is convenient to carry out the computation of roots by the method of quadratic interpolation.

The eigenvectors for matrix $S$ may be computed just as in Hessenberg's method, i.e. by solving the corresponding triangular system

$$\begin{aligned} (s_{11} - \lambda_i)v_1 + s_{12}v_2 &= 0 , \\ s_{12}v_1 + (s_{22} - \lambda_i)v_2 + s_{23}v_3 &= 0 , \\ \cdot\;\cdot\;\cdot\;\cdot\;\cdot\;\cdot\;\cdot\;\cdot\;\cdot\;\cdot\;\cdot\;\cdot\;\cdot\;&, \\ s_{n-1n}v_{n-1} + (s_{nn} - \lambda_i)v_n &= 0 \end{aligned} \qquad (5)$$

---

† Givens (2).

for components $v_1, \cdots, v_n$ of the eigenvector $V_i$ belonging to $\lambda_i$.

However, it is convenient here to have the first component given (but not the last, as in Hessenberg's method) and then to compute successively the second, third, etc.

It also turns out to be possible to give concise formulas for the eigenvector's components, corresponding to an eigenvalue $\lambda_i$. Namely,

$$v_k = \frac{1}{s_{12}s_{23} \cdots s_{k-1k}} \varphi_{k-1}(\lambda_i) . \tag{6}$$

In fact substituting these values for the components in the left part of the $k$th equation of system (5) gives

$$s_{k-1k} \frac{1}{s_{12} \cdots s_{k-2k-1}} \varphi_{k-2}(\lambda_i) + (s_{kk} - \lambda_i) \frac{1}{s_{12} \cdots s_{k-1k}} \varphi_{k-1}(\lambda_i)$$

$$+ s_{kk+1} \frac{1}{s_{12} \cdots s_{kk+1}} \varphi_k(\lambda_i) = -\frac{1}{s_{12} \cdots s_{k-1k}} [(\lambda_i - s_{kk})\varphi_{k-1}(\lambda_i)$$

$$- \varphi_k(\lambda_i) - s_{k-1k}^2 \varphi_{k-2}(\lambda_i)] = 0 \qquad\qquad \text{(for } 2 \le k \le n-1\text{)} .$$

Thus the second to the $n$-1th equations are satisfied. It is obvious that the first equation is also satisfied. The last is a consequence of the remaining ones.

However, it is less advisable to use formulas (6) than to solve numerically system (5), both with respect to the volume of computations and the reliability of the results.

To pass from the eigenvectors of matrix $S$ to the eigenvectors of matrix $A$ it is necessary to use the relationship

$$S = [T_{23} \cdots T_{n-1,n}]' A T_{23} \cdots T_{n-1,n} ,$$

from which it follows that each eigenvector $U$ of matrix $A$ is expressed in terms of the corresponding eigenvector $V$ of matrix $S$ according to the formula

$$U = T_{23} T_{24} \cdots T_{n-1,n} V ; \tag{7}$$

i.e. $U$ is obtained from $V$ by a sequence of multiplications by the rotation matrices $T_{ij}$. For each separate multiplication only two components of the preceding vector—the $i$th and the $j$th—will be changed, according to the formulas

$$v_i' = cv_i - sv_j , \qquad\qquad v_j' = sv_i + cv_j . \tag{8}$$

Here we denote $v_i$ and $v_j$ the components of the preceding vector and by $v_i'$ and $v_j'$ the components of the following one.

Although the number of multiplications in this process is very significant, the rounding-off errors accumulate very slowly, since they are multiplied by the coefficients $c$ and $s$ which are less than one in modulus.

Finally we observe that if it turns out that one or several elements $s_{k-1,k}$ are equal to zero, then the matrix $S$ is broken up into two or several

Jacobian boxes and the problem of computing the eigenvalues and eigenvectors is facilitated. This occurrence will probably hold if the initial matrix has multiple eigenvalues.

We complete this paragraph by computing the characteristic polynomial for the matrix carried to tridiagonal form in Table IV.13; we shall determine the eigenvalues of this matrix and shall compute the two eigenvectors belonging to the greatest and least eigenvalues.

**TABLE IV.14** **Computation of the Characteristic Polynomial's Coefficients using Recurrent Formulas**

| $k$ \ $i$ | 0 | 1 | 2 | 3 | 4 |
|---|---|---|---|---|---|
| 4 | | | | | 1 |
| 3 | | | | 1 | −3.99999997 |
| 2 | | | 1 | −3.20784986 | 4.75199990 |
| 1 | | 1 | −2.60414343 | 2.21170431 | −2.11185592 |
| 0 | 1 | −1 | 0.70054343 | −0.36194532 | 0.28615247 |
| | 1 | | | | |
| | 0.90360000 | 1.60414343 | | | |
| | | 0.060977255 | 0.60370643 | | |
| | | | 0.0008030309 | 0.79215011 | |

Here the coefficients of the successive polynomials $\varphi_i$ (by columns) are placed in the first part of the table. In the second part are written the coefficients of the recurrence formulas for $s_{ii}$ and $s^2_{i-1i}$ (computed from data in Table IV.13). Thus we find that the desired characteristic polynomial will be

$$t^4 - 3,99999997t^3 + 4.75199990t^2 - 2.11185592t + 0.28615247 .$$

Its greatest and least roots will be

$$\lambda_1 = 2.32274880 \quad \text{and} \quad \lambda_2 = 0.24226070 .$$

To determine the eigenvectors belonging to them we first find the corresponding eigenvectors of matrix $S$, solving system (5). We obtain

$$V_1 = (1, \quad 1.3915194, \quad 0.19994896, \quad 0.00370089)' ,$$
$$V_4 = (1, \quad -0.79713467, \quad -0.54680289, \quad -0.02817925)' .$$

Moreover we compute successively

$$T_{34}V_1 = (1, \quad 1.3915194, \quad -0.16731157, \quad -0.10954506)' ,$$
$$T_{24}T_{34}V_1 = (1, \quad 1.0774967, \quad -0.16731157, \quad 0.88731461)' ,$$
$$T_{23}T_{24}T_{34}V_1 = U_1 = (1, \quad 0.793587, \quad 0.747805, \quad 0.887315)' .$$

and

$$T_{34} V_4 = (1, \quad -0.79713467, \quad -0.43597992, \quad 0.33122345)' \,,$$

$$T_{24} T_{34} V_4 = (1, \quad -0.34370275, \quad -0.43597992, \quad -0.79183400)' \,,$$

$$T_{23} T_{24} T_{34} V_4 = U_4 = (1, \quad 0.133129, \quad -0.538968, \quad -0.791834)' \,.$$

We observe that the process indicated may be applied to a non-symmetric matrix only when, instead of a tridiagonal matrix, we obtain the almost triangular matrix

$$\begin{bmatrix} S_{11} & S_{12} & 0 & \cdots 0 & 0 \\ S_{21} & S_{22} & S_{23} & \cdots 0 & 0 \\ \cdot & \cdot & \cdot & \cdots & \cdot \\ S_{n-1,1} & S_{n-1,2} & S_{n-1,3} & \cdots S_{n-1,n-1} & S_{n-1,n} \\ S_{n1} & S_{n2} & S_{n3} & \cdots S_{n,n-1} & S_{nn} \end{bmatrix}.$$

Thus in the case of a non-symmetric matrix the sequence of rotations brings it to almost the same form as in Hessenberg's method and in the method of orthogonalization of iterations. The problem of eigenvalues is solved analogously to the indicated methods.

## 52. IMPROVING THE ACCURACY OF THE COMPLETE EIGENVALUE PROBLEM

Let $A$ be a given matrix whose eigenvalues are pairwise distinct. Let us assume approximate values $\lambda_1, \cdots, \lambda_n$ for the eigenvalues of matrix $A$ approximate eigenvectors $U_1, \cdots, U_n$ of matrix $A$, as well as approximate eigenvectors $V_1, \cdots, V_n$ of the conjugate matrix $A^*$. We pose the problem of increasing the accuracy for all the sets of values enumerated.

We shall look for more accurate values in the form

$$\begin{aligned} \tilde{\lambda}_i &= \lambda_i + \Delta\lambda_i \,, \\ \tilde{U}_i &= U_i + \Delta U_i \,, \\ \tilde{V}_i &= V_i + \Delta V_i \,, \end{aligned} \tag{1}$$

assuming both the numbers $\Delta\lambda_i$ and the components of the vectors $\Delta U_i$ and $\Delta V_i$ to be small.

Without loss of generality we may consider that

$$\Delta U_i = \sum_{\substack{j=1 \\ j\neq i}}^{n} h_{ij} U_j \,. \tag{2}$$

In fact the eigenvectors are determined accurately within a scalar multiple and we therefore have a right to consider that the $i$th coordinate of the more accurate eigenvector with respect to the basis $U_1, \cdots, U_n$ is equal to one. On the same basis we may consider that

$$\Delta V_i = \sum_{\substack{j=1 \\ j\neq i}}^{n} k_{ij} V_j \,. \tag{3}$$

It is obvious that the coefficients $h_{ij}$ and $k_{ij}$ will be small numbers.

We shall express the corrections $\Delta\lambda_i$ and the coefficients $h_{ij}$ and $k_{ij}$ by residuals of the solution known to us for the complete eigenvalue problem, i.e. by

$$r_i = AU_i - \lambda_i U_i ,$$
$$r_i^* = A^*V_i - \bar{\lambda}_i V_i . \tag{4}$$

The equation

$$A\tilde{U}_i = \tilde{\lambda}_i \tilde{U}_i$$

is written as

$$AU_i + A\Delta U_i = \lambda_i U_i + \lambda_i \Delta U_i + \Delta\lambda_i U_i + \Delta\lambda_i \Delta U_i .$$

Introducing the residual $r_i$ into this equality and throwing out the last element of the right side of the equation we obtain

$$A\Delta U_i - \lambda_i \Delta U_i \approx - r_i + \Delta\lambda_i U_i . \tag{5}$$

We now form scalar products with the vectors $V_j (j = 1, \cdots, n)$. Then

$$(A\Delta U_i, V_j) - \lambda_i(\Delta U_i, V_j) \approx - (r_i, V_j) + \Delta\lambda_i(U_i, V_j) . \tag{6}$$

But to within small second-order numbers

$$(A\Delta U_i, V_j) = (\Delta U_i, A^*V_j) \approx (\Delta U_i, \bar{\lambda}_j V_j) = \lambda_j(U_i, V_j) .$$

Therefore

$$(\lambda_j - \lambda_i)(\Delta U_i, V_j) \approx - (r_i, V_j) + \Delta\lambda_i(U_i, V_j) . \tag{7}$$

Letting $i = j$ we get

$$\Delta\lambda_i \approx \frac{(r_i, V_i)}{(U_i, V_i)} \tag{8}$$

We now consider $i \neq j$. Since to within small second-order numbers

$$(\Delta U_i, V_j) \approx h_{ij}(U_j, V_j) ,$$
$$\Delta\lambda_i(U_i, V_j) \approx 0 ,$$

then from equation (7) we obtain

$$h_{ij} = \frac{(r_i, V_j)}{(\lambda_i - \lambda_j)(U_j, V_j)} . \tag{9}$$

Analogously

$$k_{ij} = \frac{(r_i^*, U_j)}{(\bar{\lambda}_i - \bar{\lambda}_j)(V_j, U_j)} . \tag{10}$$

We conclude from the form of these formulas that in increasing the accuracy of eigenvalues and eigenvectors we use essentially the results of the control computations carried out after obtaining the initial approximation to the solution of the complete eigenvalue problem.

We observe that the coefficients $h_{ij}$ and $k_{ij}$ are connected here by the easily verified relationship

$$(U_j, V_j)h_{ij} + (U_i, V_i)\bar{k}_{ji} = - (U_i, V_j)$$

so that to compute the coefficients $k_{ij}$ it is not even necessary to compute the residual $r_i^*$.

The more accurate eigenvalues may also be computed by the formula

$$\tilde{\lambda}_i = \lambda_i + \Delta\lambda_i = \lambda_i + \frac{(r_i, V_i)}{(U_i, V_i)} = \frac{(AU_i, V_i)}{(U_i, V_i)}, \tag{11}$$

from which it is apparent that to obtain a more accurate eigenvalue it is sufficient to know only the approximate values for the eigenvectors $U_i$ and $V_i$.

We observe that the process indicated is none other than an application of Newton's one-step method to the non-linear system

$$A\tilde{U}_i = \tilde{\lambda}_i\tilde{U}_i,$$
$$A^*\tilde{V}_i = \tilde{\lambda}_i\tilde{V}_i.$$

For a symmetric matrix we may consider that $U_i = V_i$ and therefore

$$\Delta\lambda_i = \frac{(r_i, U_i)}{(U_i, U_i)},$$

$$h_{ij} = \frac{(r_i, U_j)}{(\lambda_i - \lambda_j)(U_j, U_j)}.$$

We shall give the results of increasing the accuracy of the complete eigenvalue problem for Leverrier's matrix. As initial approximations we take the escalator method's data, rounded off to three decimal places. We have

$$\lambda_1 = -17.863, \quad \lambda_2 = -17.152,$$
$$\lambda_3 = -7.574, \quad \lambda_4 = -5.299.$$

We have likewise

| $U_1$ | $U_2$ | $U_3$ | $U_4$ | $V_1$ | $V_2$ | $V_3$ | $V_4$ |
|---|---|---|---|---|---|---|---|
| −0.020 | 0.033 | −0.351 | 1.135 | −0.014 | 0.023 | −0.248 | 0.803 |
| 0.170 | −0.261 | 0.328 | 0.112 | 0.780 | −1.201 | 1.510 | 0.516 |
| −0.187 | 0.237 | 0.261 | 0.071 | −1.141 | 1.442 | 1.590 | 0.430 |
| 0.808 | 0.587 | 0.045 | 0.011 | 0.808 | 0.587 | 0.045 | 0.011 |
| 0.999111 | 1.000543 | 0.999344 | 0.999848 | | | | |

In the last partial row are placed the corresponding scalar products $(U_j, V_j)$ for $j = 1, 2, 3, 4$. We carry out the computation for $i = 1$. We have

| $r_1$ | $(r_1, V_j)$ | $X\dfrac{(\lambda_1 - \lambda_j)}{(U_j, V_j)}$ | $h_{1j}$ | $\Delta U_1$ | $\tilde{U}_1$ | $c\tilde{U}_1$ |
|---|---|---|---|---|---|---|
| −0.00110982 | −0.00026259 | 0 | | 0.00013616 | −0.01986384 | −0.019873 |
| 0.00228956 | −0.00014959 | −0.711386 | 0.00021028 | −0.00027565 | 0.16972435 | 0.169806 |
| 0.00181651 | 0.00662120 | −10.282240 | −0.00064394 | −0.00012429 | −0.18712429 | −0.187214 |
| 0.00001071 | 0.00107144 | −12.562090 | −0.00008529 | 0.00009352 | 0.80809352 | 0.808482 |

In the last column is placed the eigenvector belonging to the eigenvalue $\lambda_1$, normalized so that its last component coincides with the last component of the eigenvector computed by the escalator method.

From formula (8) we find

$$\varDelta\lambda_1 = -0.000263 ;$$
$$\tilde{\lambda}_1 = -17.863263 .$$

Likewise we carry out the results of computations for $i = 2, 3, 4$.

| $\tilde{U}_2$ | $\tilde{U}_3$ | $\tilde{U}_4$ |
|---|---|---|
| 0.032933 | −0.351235 | 1.135218 |
| −0.261309 | 0.328466 | 0.112182 |
| 0.236640 | 0.260925 | 0.070589 |
| 0.586694 | 0.045006 | 0.011058 |

$$\tilde{\lambda}_2 = -17.152428 ,$$
$$\tilde{\lambda}_3 - -7.574043 ,$$
$$\tilde{\lambda}_4 = -5.298696 .$$

The values obtained for the eigenvalues and the components of the eigenvectors are already correct within $2 \cdot 10^{-6}$.

# Chapter V

# THE SPECIAL EIGENVALUE PROBLEM

The present chapter is devoted to the special eigenvalue problem. This consists, as was said above, in determining one or several (not many as a rule) eigenvalues of a matrix and the eigenvectors belonging to them. The originality of the special problem consists in the fact that methods for its solution must be based on indirect considerations which use certain properties of eigenvalues and eigenvectors. All methods for solving the special problem are iterative methods.

To set up these methods two essentially different basic ideas are used.

We shall explain the first idea under the assumption that there exists a basis of eigenvectors in the space. Starting with a certain vector, generally arbitrary, an infinite sequence of vectors is constructed such that in this sequence one which is formed as an expansion in eigenvectors predominates more and more. Then the constructed sequence will converge directionally to the individual eigenvector. The eigenvalue is also determined along the way. Processes based on this idea may be applied even in the absence of a basis of eigenvectors. In this case it is possible to use as their basis an expansion in terms of a canonical basis. A certain modification thereby permits us to compute several vectors from a canonical basis.

The second idea is based on the extremal properties of eigenvalues and is applicable only to symmetric matrices. This idea is close to the idea of relaxation for solving a linear system of equations. Methods based on this idea give a sequence of vectors which best realize the maximum (or minimum) of the quotient $(AX, X)/(X, X)$.

The choice of corrections for passing from the previous vector to the following one may be accomplished in different ways. The most important group of methods using this idea, the corrections being taken in the direction of the functional gradient $(AX, X)/(X, X)$, will be considered in Chapter VII. In the present chapter we shall consider, in brief, methods analogous to methods of coordinate relaxation, simple and group.

## 53. DETERMINING THE EIGENVALUE OF MAXIMAL MODULUS OF A MATRIX, USING SUCCESSIVE ITERATIONS

In the present paragraph we shall present a method which allows us to compute the eigenvalue which is greatest in modulus and the eigenvector belonging to it by computing a sequence of iterations starting from an arbitrary vector. The method to be presented is called the *power method* and it is the simplest iterative process for solving the special eigenvalue problem. We apply it for an arbitrary matrix although the course of the iterative process essentially depends on how the largest eigenvalue of the matrix enters into its Jordan canonical form. In connection with this we must differentiate several possible cases. However we shall not investigate the problem in general but limit ourselves to considering only the most important special cases.

To simplify the presentation we shall assume that all the eigenvalues of the matrix except possibly the greatest in modulus have linear elementary divisors, although the conclusions we draw will hold even without this assumption. Likewise we shall consider that the elements of the matrices under consideration are real.

1. **The eigenvalue of maximal modulus is real and simple.** In this case there corresponds to the eigenvalue of maximal modulus one linear elementary divisor so that on the basis of the agreement just formulated all elementary divisors of the matrix are linear. Therefore there exists a basis of eigenvectors $U_1, \cdots, U_n$ belonging to the eigenvalues $\lambda_1, \lambda_2, \cdots, \lambda_n$, arrayed in order of decreasing moduli, where $|\lambda_1| > |\lambda_2|$ but some of the rest may be equal to each other. We take an arbitrary vector $Y_0$ and form the sequence of its iterations with the matrix $A$

$$AY_0, A^2Y_0, \cdots, A^kY_0, \cdots$$

We write the expansion of the vector $Y_0$ in terms of eigenvectors:

$$Y_0 = a_1U_1 + a_2U_2 + \cdots + a_nU_n. \tag{1}$$

Among the numbers $a_i$ certain ones may be equal to zero. We assume however that $a_1 \neq 0$.[†]

It is obvious that

$$AY_0 = a_1\lambda_1U_1 + a_2\lambda_2U_2 + \cdots + a_n\lambda_nU_n,$$
$$\cdots \cdots \cdots \cdots \cdots \cdots \cdots \cdots \cdots, \tag{2}$$
$$A^kY_0 = a_1\lambda_1^kU_1 + a_2\lambda_2^kU_2 + \cdots + a_n\lambda_n^kU_n.$$

---

[†] Since $a_1 = (V_1, Y_0)$ where $V_1$ is the first eigenvector of the transposed matrix, then the requirement $a_1 \neq 0$ will be satisfied when the vector $Y_0$ is not orthogonal to the vector $V_1$.

We write $A^k Y_0 = Y_k = (y_{1k}, y_{2k}, \cdots, y_{nk})'$ and we clarify the structure of the components for the vector $Y_k$. Let

$$U_1 = (u_{11}, u_{21}, \cdots, u_{n1})',$$
$$U_2 = (u_{12}, u_{22}, \cdots, u_{n2})', \cdots,$$
$$U_n = (u_{1n}, u_{2n}, \cdots, u_{nn})'.$$

Then from (2) we obtain

$$y_{ik} = a_1 u_{i1} \lambda_1^k + a_2 u_{i2} \lambda_2^k + \cdots + a_n u_{in} \lambda_n^k.$$

The coefficient for $\lambda_1^k$ in at least one of the components is not equal to zero, since $a_1 \neq 0$ by assumption and the vector $U_1$ is non-zero. Let $y_k$ (we omit the first index) be any of the components for the vector $Y_k$ for which the coefficient for $\lambda_1^k$ is different from zero. Then

$$y_k = c_1 \lambda_1^k + c_2 \lambda_2^k + \cdots + c_n \lambda_n^k, \tag{3}$$

while the coefficient $c_i$ does not depend on the index $k$ and $c_1 \neq 0$.

We shall consider the ratio of the components of two adjacent iterations.

$$\frac{y_{k+1}}{y_k} = \frac{c_1 \lambda_1^{k+1} + c_2 \lambda_2^{k+1} + \cdots + c_n \lambda_n^{k+1}}{c_1 \lambda_1^k + c_2 \lambda_2^k + \cdots + c_n \lambda_n^k}$$
$$= \lambda_1 \frac{1 + b_2 \alpha_2^{k+1} + b_3 \alpha_3^{k+1} + \cdots + b_n \alpha_n^{k+1}}{1 + b_2 \alpha_2^k + b_3 \alpha_3^k + \cdots + b_n \alpha_n^k} \tag{4}$$

where

$$b_i = \frac{c_i}{c_1},$$
$$\alpha_i = \frac{\lambda_i}{\lambda_1}. \tag{5}$$

Carrying out the division and retaining elements up to order $\alpha_2^{2k}$ and $\alpha_3^k$ inclusive we obtain

$$\frac{y_{k+1}}{y_k} = \lambda_1 [1 - b_2' \alpha_2^k - b_3' \alpha_3^k + b_2 b_2' \alpha_2^{2k}] + O(\alpha_3^k + \alpha_2^{2k}), \tag{6}$$

where

$$b_2' = b_2(1 - \alpha_2),$$
$$b_3' = b_3(1 - \alpha_3). \tag{7}$$

From here we see that if $k$ is sufficiently large, then

$$\lambda_1 \approx \frac{y_{k+1}}{y_k}. \tag{8}$$

Since all components of the vector $U_1$ are usually different from zero, then as a rule we may take as $y_k$ any component of the vector $Y_k$. Thus the first eigenvalue is approximately equal to the ratio of any corresponding components of two adjacent (sufficiently high) iterations for an arbitrary vector with matrix $A$.

To carry out the iterations in practice we must compute the ratios $y_{k+1}/y_k$ for several components. Good coincidence of these ratios will show that in expression (6) a difference in the values of coefficients $b_2'$ and $b_3'$ has already ceased to play a noticeable role.

The rate of convergence of the process in the case being considered is determined by the magnitude of the ratio $\lambda_2/\lambda_1$ and may be slow if this ratio is close to one.

To avoid the growth of components it is sometimes advisable in computing iterations by one method or another to normalize the vectors obtained at each step. Convenient in normalizing are: the division of a vector by its first component; or by the largest component; or finally normalizing to unit length. For this we obtain in place of the sequence $Y_k$ the sequence $\tilde{Y}_k = \mu_k Y_k$, where $\mu_k$ are normalizing factors, and to obtain $\lambda_1$ we need to take the ratios of components of the vectors $A\tilde{Y}_k$ and $\tilde{Y}_k$.

Although it is improbable, it may happen that the initial vector $Y_0$ is chosen unsuccessfully, that is, so that the coefficient $a_1$ is equal to zero or very close to zero. In this case there will not be a clear picture of the directional convergence of the iterations. In fact in the first steps of the iteration the predominant element will be the one depending on $\lambda_2$ (if $a_2 \neq 0$). However later on even if $a_1$ is exactly equal to zero, then after several steps of the iteration the element depending on $\lambda_1$ shows up, thanks to rounding-off errors, first with a very small coefficient; in later iterations this element will grow fairly rapidly in comparison with the rest. The "battle for predominance" of elements depending on $\lambda_1$ and $\lambda_2$ produces confusion during the process. For an unsuccessful choice of the initial vector, in the sense indicated, it is necessary to change it.

The process described likewise lets us determine all the components of the eigenvector belonging to the largest eigenvalue. Namely, the ratios of components of the vector $Y_k$ approach the ratios of components of this eigenvector.

In fact for $a_1 \neq 0$

$$Y_k = A^k Y_0 = \lambda_1^k \left[ a_1 U_1 + a_2 \left(\frac{\lambda_2}{\lambda_1}\right)^k U_2 + \cdots + a_n \left(\frac{\lambda_n}{\lambda_1}\right)^k U_n \right]$$

$$= a_1 \lambda_1^k \left[ U_1 + O\left(\frac{\lambda_2}{\lambda_1}\right)^k \right]. \tag{9}$$

*Example 1.* We shall try to determine the first eigenvalue for Leverrier's matrix.

We take the vector $(1, 0, 0, 0)'$ as our initial one and form twenty iterations, normalizing them at each step by dividing by the first component. We present only the last two iterations:

| $\tilde{Y}_{19}$ | $A\tilde{Y}_{19}$ | component ratios |
|---|---|---|
| 1.00000 | $-$ 17.4655 | $-17.466$ |
| $-8.20321$ | 143.3809 | $-17.479$ |
| 8.17013 | $-143.0881$ | $-17.514$ |
| $-7.95957$ | 149.2676 | $-18.753$ |
| $-6.99265$ | 132.0949 | |

From the data presented we see that the ratios of the different components are still further from each other; this shows that the process has not yet stopped. In fact to three places $\lambda_1 = -17.863$. The iteration process converges slowly because the second eigenvalue $\lambda_2 = -17.152$ differs little in modulus from the first.

*Example 2.* We shall determine the first eigenvalue and the eigenvector belonging to it for the matrix

$$\begin{bmatrix} -5.509882 & 1.870086 & 0.422908 \\ 0.287865 & -11.811654 & 5.711900 \\ 0.049099 & 4.308033 & -12.970687 \end{bmatrix}.$$

We take as the initial vector $(1, 0, 0)'$. We introduce the table of iterations beginning with the twelfth iteration:

| $\tilde{Y}_{12}$ | $A\tilde{Y}_{12}$ | $\tilde{Y}_{13}$ | $A\tilde{Y}_{13}$ | $\tilde{Y}_{14}$ | $A\tilde{Y}_{14}$ |
|---|---|---|---|---|---|
| 1.0000000 | −17.351783 | 1.0000000 | −17.378482 | 1.0000000 | −17.389552 |
| −8.1139091 | 141.126754 | −8.1332710 | 141.483894 | −8.1413264 | 141.632991 |
| 7.8783245 | −137.093170 | 7.9008117 | −137.468256 | 7.9102568 | −137.625469 |
| 0.7644154 | −13.318199 | 0.7675407 | −13.362844 | 0.7689304 | −13.382030 |

We find the ratios of corresponding components for the 12th, 13th, 14th, and 15th iterations:

$$\begin{matrix} -17.351783 & -17.378482 & -17.389552 \\ -17.393189 & -17.395694 & -17.396795 \\ -17.401310 & -17.399257 & -17.398356 \end{matrix}$$

The last three ratios allow us to consider that $\lambda_1 = -17.39$ or that $\lambda_1 = -17.40$. As we saw in Section 48, $\lambda_1 = -17.3977$ to four places.

Moreover we find the following values for components of the eigenvector:

$$\begin{matrix} 1.00000 & 1.00000 & 1.00000 \\ -8.13327 & -8.14133 & -8.14472 \\ 7.90081 & 7.91026 & 7.91426. \end{matrix}$$

We see that the last result is already close enough to the exact value, since in Section 48 it was found that $U_1 = (0.094129, -0.766896, 0.745248)'$ or after corresponding normalization $U_1 = (1.00000, -8.14729, 7.91730)'$.

*Example 3.* We shall find the first eigenvalue and the vector belonging to it for the matrix

$$\begin{bmatrix} 0.22 & 0.02 & 0.12 & 0.14 \\ 0.02 & 0.14 & 0.04 & -0.06 \\ 0.12 & 0.04 & 0.28 & 0.08 \\ 0.14 & -0.06 & 0.08 & 0.26 \end{bmatrix}.$$

In Table III. 1 were computed 14 iterations, starting from the vector $(0.76, 0.08, 1.12, 0.68)'$.

Computing the ratios of the components of the 14th and 13th iterations we find the value 0.4800 for $\lambda_1$.

(We ignore the second component of iterations because of its smallness in comparison with the rest of the components).

The ratios of the components of the 7th and 6th iterations give for $\lambda_1$ the values

$$0.4800, \quad 0.4792, \quad 0.4808.$$

For the components of the eigenvector we find from the 14th iteration the following values:

$$1.0000, \quad 0.0000, \quad 1.0000, \quad 1.0000.$$

It is not hard to verify that the exact value is $\lambda_1 = 0.48$ and the eigenvector belonging to it has the components 1, 0, 1, 1.

**2. The largest eigenvalue is real, multiple, but the elementary divisors corresponding to it are linear.** In this case formula (3) remains true but several of the first elements may be joined together so that

$$y_k = c_1 \lambda_1^k + c_{r+1} \lambda_{r+1}^k + \cdots + c_n \lambda_u^k ,$$

where $r$ is the multiplicity of $\lambda_1$.

All later arguments remain in force and therefore

$$\frac{y_{k+1}}{y_k} = \lambda_1 + O\left(\frac{\lambda_{r+1}}{\lambda_1}\right)^k . \tag{10}$$

Thus in this case, under the condition $a_1 \neq 0$, the ratio $y_{k+1}/y_k$ gives an approximate value for the largest eigenvalue. The question of the multiplicity of a root may not be solved without a more detailed investigation. We shall return again to this question below.

The vectors $Y_k = A^k Y_0$, just as in the previous case, converge directionally to one of the eigenvectors belonging to $\lambda_1$, namely, to the eigenvector which lies in the cyclic subspace generated by the vector $Y_0$. Starting with different initial vectors we generally arrive at different eigenvectors.

*Example 4.* We shall determine the first eigenvalue of the matrix

$$\begin{bmatrix} 1.022551 & 0.116069 & -0.287028 & -0.429969 \\ 0.228401 & 0.742521 & -0.176368 & -0.283720 \\ 0.326141 & 0.097221 & 0.197209 & -0.216487 \\ 0.433864 & 0.148965 & -0.193686 & 0.006472 \end{bmatrix} .$$

Solving the characteristic equation

$$\lambda^4 - 1.968753\lambda^3 + 1.391184\lambda^2 - 0.415291\lambda + 0.044360 = 0 ,$$

we obtain for eigenvalues the values

$$\lambda_1 = \lambda_2 = 0.667483 ,$$
$$\lambda_3 = 0.346148 ,$$
$$\lambda_4 = 0.287639 .$$

We determine $\lambda_1$ using the power method, taking for the initial vector $(1, 1, 1, 1)'$.

We write the table of iterations beginning with the 9th iteration:

| $\tilde{Y}_9$ | $A\tilde{Y}_9$ | $\tilde{Y}_{10}$ | $A\tilde{Y}_{10}$ | $\tilde{Y}_{11}$ | $A\tilde{Y}_{11}$ |
|---|---|---|---|---|---|
| 1.000000 | 0.666160 | 1.000000 | 0.666822 | 1.000000 | 0.667151 |
| 1.844723 | 1.230507 | 1.847165 | 1.232545 | 1.848387 | 1.233563 |
| 0.676506 | 0.449420 | 0.674643 | 0.449211 | 0.673660 | 0.449088 |
| 0.875250 | 0.583298 | 0.875613 | 0.584025 | 0.875834 | 0.584399 |
| 4.396479 | 2.929385 | 4.397421 | 2.932603 | 4.397881 | 2.934201 |

We compute the component ratios for these iterations:

| | | |
|---|---|---|
| 0.666160 | 0.666822 | 0.667151 |
| 0.667042 | 0.667263 | 0.667373 |
| 0.664325 | 0.665850 | 0.666639 |
| 0.666466 | 0.666990 | 0.667249 |

The last four ratios give the value 0.667 for $\lambda_1$, accurate to three places.

**3. The two eigenvalues of maximal modulus are real and have opposite signs.** From equality (3) we see that in this case the odd and even iterations have different coefficients for corresponding powers of $\lambda_1$, since

$$y_{2k} = (c_1 + c_2)\lambda_1^{2k} + c_3\lambda_3^{2k} + \cdots + c_n\lambda_n^{2k} ,$$
$$y_{2k+1} = (c_1 - c_2)\lambda_1^{2k+1} + c_3\lambda_3^{2k+1} + \cdots + c_n\lambda_n^{2k+1} ,$$

and therefore two adjacent iterations may not be used to determine $\lambda_1$. However we may determine $\lambda_1^2$ by one of the following formulas:

$$\lambda_1^2 \approx \frac{y_{2k+2}}{y_{2k}} \quad \text{or} \quad \lambda_1^2 \approx \frac{y_{2k+1}}{y_{2k-1}} . \tag{11}$$

In finding the eigenvectors belonging to $\lambda_1$ and $\lambda_2 = -\lambda_1$ it is advisable to construct the vectors $Y_k + \lambda_1 Y_{k-1}$ and $Y_k - \lambda_1 Y_{k-1}$. The ratios of the components of these vectors will approach respectively the ratio of the components of vectors $U_1$ and $U_2$ belonging to the eigenvalues $\lambda_1$ and $\lambda_2$.

In fact, on the strength of the equality

$$Y_k = a_1\lambda_1^k U_1 + a_2(-\lambda_1)^k U_2 + a_3\lambda_3^k U_3 \cdots \tag{12}$$

we have

$$Y_k + \lambda_1 Y_{k-1} = 2a_1\lambda_1^k U_1 + a_3(\lambda_3 + \lambda_1)\lambda_3^{k-1}U_3 + \cdots$$
$$= \lambda_1^k\left[2a_1 U_1 + O\left(\frac{\lambda_3}{\lambda_1}\right)^k\right]$$

$$Y_k - \lambda_1 Y_{k-1} = 2a_2(-\lambda_1)^k U_2 + a_3(\lambda_3 - \lambda_1)\lambda_3^{k-1}U_3 + \cdots$$
$$= (-\lambda_1)^k\left[2a_2 U_2 + O\left(\frac{\lambda_3}{\lambda_1}\right)^k\right].$$

$$\tag{13}$$

*Example 5.* It is not hard to discover that the eigenvalues of the matrix

$$A = \begin{bmatrix} 4.2 & -3.4 & 0.3 \\ 4.7 & -3.9 & 0.3 \\ -5.6 & 5.2 & 0.1 \end{bmatrix}$$

are $\lambda_1 = -\lambda_2 = 0.5$, $\lambda_3 = 0.4$, while to the eigenvalue 0.5 belongs the eigenvector $(1, 1, -1)'$ and to the eigenvalue $-0.5$ belongs the eigenvector $(-2/3, -5/6, 1)' \approx (-0.667, -0.833, 1)'$.

Carrying out the computations we obtain

| $Y_0$ | $Y_{23}$ | $Y_{24}$ | $Y_{25}$ |
|---|---|---|---|
| 0.2 | $0.25972708 \cdot 10^{-6}$ | $0.22548439 \cdot 10^{-6}$ | $0.65159766 \cdot 10^{-7}$ |
| 0.4 | $0.23588520 \cdot 10^{-6}$ | $0.23740533 \cdot 10^{-6}$ | $0.59199296 \cdot 10^{-7}$ |
| 0.6 | $-0.21119890 \cdot 10^{-6}$ | $-0.24898850 \cdot 10^{-6}$ | $-0.53103718 \cdot 10^{-7}$ |
| | $0.28441339 \cdot 10^{-6}$ | $0.21390121 \cdot 10^{-6}$ | $0.71255344 \cdot 10^{-7}$. |

For the ratios of corresponding components of the vectors $Y_{25}$ and $Y_{23}$ we obtain

$$0.25087782, \quad 0.25096655, \quad 0.25143936,$$

from which we find for $\lambda_1$ the three approximate values

$$0.500877, \quad 0.500966, \quad 0.501437,$$

so that $\lambda_1 \approx 0.501$ within three decimal places.

We find furthermore

| $Y_{25} + 0.501 Y_{23}$ | $\tilde{U}_1$ | $Y_{25} - 0.501 Y_{24}$ | $\tilde{U}_2$ |
|---|---|---|---|
| 0.1780 | 1.000 | $-0.0479$ | $-0.669$ |
| 0.1781 | 1.000 | $-0.0597$ | $-0.834$ |
| $-0.1778$ | $-0.998$ | 0.0716 | 1.000. |

Thus the components of the eigenvectors are determined accurately within $2 \cdot 10^{-3}$.

**4. The eigenvalues of maximal modulus form simple complex-conjugate pairs.** Let $\lambda_1$ and $\lambda_2$ be conjugate complex eigenvalues which are the largest in modulus and let $|\lambda_3| < |\lambda_1|$. According to formula (3)

$$y_k = c_1 \lambda_1^k + c_2 \lambda_2^k + \cdots + c_n \lambda_n^k,$$

while in this case $c_1$ and $c_2$ are conjugate complex. Let

$$c_1 = R e^{i\alpha}, \quad c_2 = R e^{-i\alpha},$$
$$\lambda_1 = r e^{i\theta}, \quad \lambda_2 = r e^{-i\theta}; \tag{14}$$

then

$$y_k = 2 R r^k \cos(k\theta + \alpha) + c_3 \lambda_3^k \cdots \tag{15}$$

The presence of the factor $\cos(k\theta + \alpha)$ will cause strong oscillation in the values $y_k$ both in magnitude and sign. Thus the presence of complex roots greatest in modulus is revealed immediately when forming the iterations. We let

$$p = -(\lambda_1 + \lambda_2) = -2r\cos\theta \,,$$
$$q = \lambda_1\lambda_2 = r^2 \,. \tag{16}$$

Then $\lambda_1$ and $\lambda_2$ will be roots of the quadratic equation

$$t^2 + pt + q = 0 \,.$$

The coefficients $p$ and $q$ may be determined from the following considerations. Let $k$ be so large that $y_k \approx c_1\lambda_1^k + c_2\lambda_2^k$. Then

$$y_{k+1} + py_k + qy_{k-1} \approx c_1\lambda_1^{k-1}[\lambda_1^2 + p\lambda_1 + q] + c_2\lambda_2^{k-1}[\lambda_2^2 + p\lambda_2 + q] = 0 \,. \tag{17}$$

Here the approximate equality is valid within $O(|\lambda_3^k|)$.

The analogous approximate equation

$$z_{k+1} + pz_k + qz_{k-1} = 0 \tag{18}$$

will be valid for any other component $z_k$ of the vector $Y_k = A^k Y_0$. It is also possible to take $z_k = y_{k+1}$ for $z_k$ or, more generally, to take any component of the vector $Z_k = A^k Z_0$, where $Z_0$ is the arbitrary initial vector.

From equalities (17) and (18) we obtain in general

$$p \approx -\frac{y_{k-1}z_{k+1} - z_{k-1}y_{k+1}}{y_{k-1}z_k - z_{k-1}y_k} \,,$$
$$q \approx \frac{y_kz_{k+1} - z_k y_{k+1}}{y_{k-1}z_k - z_{k-1}y_k} \,. \tag{19}$$

In particular, if we take $z_k = y_{k+1}$, we obtain

$$p \approx -\frac{y_{k-1}y_{k+2} - y_k y_{k+1}}{y_{k-1}y_{k+1} - y_k^2} \,,$$
$$q \approx \frac{y_k y_{k+2} - y_{k+1}^2}{y_{k-1}y_{k+1} - y_k^2} \,. \tag{20}$$

We shall give a more detailed justification for formulas (19) and (20). This will allow us to clarify their conditions of applicability and to estimate the error.

Let

$$y_k = c_1\lambda_1^k + c_2\lambda_2^k + O(|\lambda_3|^k) \,,$$
$$z_k = d_1\lambda_1^k + d_2\lambda_2^k + O(|\lambda_3|^k) \,.$$

Then

$$y_{k-1}z_{k+1} - z_{k-1}y_{k+1}$$
$$= (c_1\lambda_1^{k-1} + c_2\lambda_2^{k-1} + O(|\lambda_3|^k))(d_1\lambda_1^{k+1} + d_2\lambda_2^{k+1} + O(|\lambda_3|^k))$$
$$- (c_1\lambda_1^{k+1} + c_2\lambda_2^{k+1} + O(|\lambda_3|^k))(d_1\lambda_1^{k-1} + d_2\lambda_2^{k-1} + O(|\lambda_3|^k))$$
$$= (c_1d_2 - c_2d_1)(\lambda_2^2 - \lambda_1^2)\lambda_1^{k-1}\lambda_2^{k-1} + O(|\lambda_1^k\lambda_3^k|) \,.$$

Analogously

$$y_{k-1}z_k - z_{k-1}y_k = (c_1d_2 - c_2d_1)(\lambda_2 - \lambda_1)\lambda_1^{k-1}\lambda_2^{k-1} + O(|\lambda_1^k\lambda_3^k|) \,,$$
$$y_kz_{k+1} - z_k y_{k+1} = (c_1d_2 - c_2d_1)(\lambda_2 - \lambda_1)\lambda_1^k\lambda_2^k + O(|\lambda_1^k\lambda_3^k|) \,.$$

Therefore

$$
-\frac{y_{k-1}z_{k+1} - z_{k-1}y_{k+1}}{y_{k-1}z_k - z_{k-1}y_k} = -(\lambda_1 + \lambda_2) + O\left(\left|\frac{\lambda_3}{\lambda_2}\right|^k\right) = p + O\left(\left|\frac{\lambda_3}{\lambda_2}\right|^k\right),
$$

$$
\frac{y_k z_{k+1} - z_k y_{k+1}}{y_{k-1}z_k - z_{k-1}y_k} = \lambda_1\lambda_2 + O\left(\left|\frac{\lambda_3}{\lambda_2}\right|^k\right) = q + O\left(\left|\frac{\lambda_3}{\lambda_2}\right|^k\right),
$$
(19')

provided $c_1 d_2 - c_2 d_1 \neq 0$. Analogously

$$
-\frac{y_{k-1}y_{k+2} - y_k y_{k+1}}{y_{k-1}y_{k+1} - y_k^2} = p + O\left(\left|\frac{\lambda_3}{\lambda_2}\right|^k\right),
$$

$$
\frac{y_k y_{k+2} - y_{k+1}^2}{y_{k-1}y_{k+1} - y_k^2} = q + O\left(\left|\frac{\lambda_3}{\lambda_2}\right|^k\right).
$$
(20')

We observe that for formulas (20') the condition $c_1 d_2 - c_2 d_1 \neq 0$ is always satisfied, since $d_1 = \lambda_1 c_1$, $d_2 = \lambda_2 c_2$, so that $c_1 d_2 - c_2 d_1 = c_1 c_2(\lambda_2 - \lambda_1) \neq 0$.

The eigenvalues $\lambda_1$ and $\lambda_2$ may be computed without computing $p$ and solving the quadratic equation. Namely, having determined $q = r^2$ by one of the formulas (19) or (20) we find $r$ and compute the expression[†]

$$
\mu_k = \frac{1}{2}[ry_{k-1} + r^{-1}y_{k+1}]
$$

$$
\approx \frac{1}{2}[r(c_1\lambda_1^{k-1} + c_2\lambda_2^{k-1}) + r^{-1}(c_1\lambda_1^{k+1} + c_2\lambda_2^{k+1})]
$$
(21)

$$
= \frac{1}{2}[c_1 r^k e^{ik\theta} + c_2 r^k e^{-ik\theta}][e^{-i\theta} + e^{i\theta}] \approx y_k \cos\theta .
$$

From here we find

$$
\cos\theta \approx \frac{\mu_k}{y_k}
$$
(22)

with accuracy to magnitudes of order $(\lambda_3/\lambda_1)^k$.

After the eigenvalues $\lambda_1$ and $\lambda_2$ have been determined the eigenvectors corresponding to them are easily determined as well. That is, from the approximate equalities

$$
Y_k \approx a_1\lambda_1^k U_1 + a_2\lambda_2^k U_2 ,
$$

$$
Y_{k+1} \approx a_1\lambda_1^{k+1}U_1 + a_2\lambda_2^{k+1}U_2
$$

we find

$$
Y_{k+1} - \lambda_2 Y_k \approx a_1\lambda_1^k(\lambda_1 - \lambda_2)U_1 ,
$$

$$
Y_{k+1} - \lambda_1 Y_k \approx a_2\lambda_2^k(\lambda_2 - \lambda_1)U_2 ,
$$
(23)

from which it follows that $Y_{k+1} - \lambda_2 Y_k$ and $Y_{k+1} - \lambda_1 Y_k$ are the eigenvectors corresponding to the eigenvalues $\lambda_1$ and $\lambda_2$, accurate to within small values.

*Example 6.* The eigenvalues of the matrix

$$
A = \begin{vmatrix} 26 & -54 & 4 \\ 13 & -28 & 3 \\ 26 & -56 & 5 \end{vmatrix}
$$

---

† Aitken (5).

are $\lambda_1 = 1 + 5i$, $\lambda_2 = 1 - 5i$, and $\lambda_3 = 1$. The eigenvector belonging to $\lambda_1$ (normalized suitably) is $U_1 = (1, \ 0.53974564 - 0.09141494i, \ 1.03656599 + 0.01589827i)'$.

We compute the eigenvalues $\lambda_1$ and $\lambda_2$ and the eigenvectors belonging to them by the power method. We have

| $Y_0$ | $Y_8$ | $Y_9$ | $Y_{10}$ |
|---|---|---|---|
| 0.2 | 1293880.4 | 3669538.0 | −26301835 |
| 0.4 | 654932.2 | 2528583.4 | −11971081 |
| 0.6 | 1348746.6 | 3708420.2 | −27650581 |
| | 3297559.2 | 9906541.6 | −65923497 |

Taking the first component for $y_k$ and the second component for $z_k$ we obtain

$$p \approx -\frac{17367716 \cdot 10^5}{86838591 \cdot 10^4} = -1.9999997 \ ,$$

$$q \approx \frac{22578047 \cdot 10^6}{86838591 \cdot 10^4} = 26.000015 \ ,$$

from which

$$\lambda_1 \approx 0.9999999 + 5.0000015i \ ,$$
$$\lambda_2 \approx 0.9999999 - 5.0000015i \ .$$

Furthermore,

$$(Y_{10} - \lambda_2 Y_9) = (-22632297 + 18347696i,$$
$$-9442498 + 12642921i, \ -23942161 + 18542107i)' \ ,$$

so that after the corresponding normalizing

$$\tilde{U}_1 = (1, 0.5250 - 0.1330i, 1.0391 + 0.0231i)'.$$

Using $Y_{18}$, $Y_{19}$, and $Y_{20}$ we obtain

$$\lambda_1 = 1.00000000 + 5.00000000i \ ,$$

$$U_1 = (1, 0.5397452 - 0.0914155i, 1.0365660 + 0.0158983i)'.$$

**5. The eigenvalue of maximal modulus is real and is found in a second-order Jordan box.** We have already observed that the course of the iterative power process essentially depends on the structure of the Jordan canonical form associated with the given matrix. On this point we shall show by a very simple example the character of those changes which arise if there corresponds to the eigenvalue greatest in modulus a non-linear elementary divisor.

Namely, we consider the case where $\lambda_1$ is real and belongs, in Jordan canonical form, to the box

$$\begin{bmatrix} \lambda_1 & 0 \\ 1 & \lambda_1 \end{bmatrix}$$

and the next eigenvalue $\lambda_2$ is less than $\lambda_1$ in modulus. For simplicity of computation we shall assume as before that to all remaining eigenvalues there correspond linear elementary divisors.

In the case under consideration we shall take, instead of a basis of eigenvectors, a canonical basis $U_1, U_2, \cdots, U_n$. The influence of the matrix $A$ on the vectors of this basis is expressed by the formulas

$$A U_1 = \lambda_1 U_1 + U_2 ,$$
$$A U_2 = \lambda_1 U_2 ,$$
$$A U_3 = \lambda_3 U_3 ,$$
$$\cdots \cdots ,$$
$$A U_n = \lambda_n U_n ,$$

and consequently

$$A^k U_1 = \lambda_1^k U_1 + k\lambda_1^{k-1} U_2 ,$$
$$A^k U_2 = \lambda_1^k U_2 ,$$
$$A^k U_3 = \lambda_3^k U_3 , \qquad (24)$$
$$\cdots \cdots \cdots \cdots ,$$
$$A^k U_n = \lambda_n^k U_n .$$

Let $Y_0$ be the initial vector. We shall assume that the projection of the vector $Y_0$ into the root subspace corresponding to the eigenvalue $\lambda_1$ is different from zero and is not an eigenvector. We take this projection for the first vector $U_1$ of the canonical basis. Then

$$Y_0 = U_1 + a_3 U_3 + \cdots + a_n U_n$$

and, from (24),

$$Y_k = A^k Y_0 = \lambda_1^k U_1 + k\lambda_1^{k-1} U_2 + \lambda_3^k a_3 U_3 + \cdots + \lambda_n^k a_n U_n .$$

Any component of the vector $Y_k$ will have the form (omitting the first index, as hitherto):

$$y_k = c_1 \lambda_1^k + c_2 k \lambda_1^{k-1} + c_3 \lambda_3^k + \cdots + c_n \lambda_n^k . \qquad (25)$$

The ratio $y_{k+1}/y_k$ approaches $\lambda_1$, as before, but more slowly than any geometric progression because of the presence of the factor $k$ in the second item. Namely:

$$\frac{y_{k+1}}{y_k} = \lambda_1 \left( 1 + O\left( \frac{1}{k} \right) \right) .$$

It becomes almost impossible to determine $\lambda_1$ from the ratio $y_{k+1}/y_k$ in practice.[†]

---

† We observe that if the box to which $\lambda_1$ belongs in canonical Jordan form has a more involved structure, then in expression (6) there also appear other powers of $\lambda_1$ multiplied by corresponding binomial coefficients:

$$y_k = c_1 \lambda_1^k + c_2 k \lambda_1^{k-1} + c_3 \frac{k(k-1)}{2} \lambda_1^{k-2} + \cdots + c_n \lambda_n^k .$$

The ratio $y_{k+1}/y_k$ approaches $\lambda_1$ still more slowly.

To determine the eigenvalue $\lambda_1$ it is necessary to do the same as in determining a complex pair of eigenvalues, i.e. to look for the coefficients $p = -2\lambda_1$ and $q = \lambda_1^2$ of the quadratic equation whose double root is $\lambda_1$. Thus let

$$y_k = c_1\lambda_1^k + kc_2\lambda_1^{k-1} + c_3\lambda_3^k + \cdots .$$

Then

$$y_{k+1} + py_k + qy_{k-1} = c_1\lambda_1^{k-1}(\lambda_1^2 + p\lambda_1 + q)$$
$$+ c_2\lambda_1^{k-2}[(k+1)\lambda_1^2 + pk\lambda_1 + q(k-1)] + O(\lambda_3^k) = O(\lambda_3^k) .$$

Analogously

$$z_{k+1} + pz_k + qz_{k-1} = O(\lambda_3^k) ,$$

where $z_k$ is determined just as in the previous point.

From the approximate equations obtained we find

$$p \approx -\frac{y_{k-1}z_{k+1} - z_{k-1}y_{k+1}}{y_{k-1}z_k - z_{k-1}y_k} ,$$

$$q \approx \frac{y_kz_{k+1} - z_ky_{k+1}}{y_{k-1}z_k - z_{k-1}y_k} .$$

(19'')

It is easy to verify that these equalities will be correct within magnitudes of order $(\lambda_3/\lambda_1)^k$. This is done exactly as in the previous case.

In determining the eigenvalue $\lambda_1$ it is obviously sufficient to determine one of the coefficients $p$ or $q$. However the coincidence of the numbers $-p/2$ and $\sqrt{q}$ serves as a check on the validity of the hypothesis that the eigenvalue $\lambda_1$ is contained in a canonical box.

The hypothesis we have made may be supported by other means as well. That is, finding $\lambda_1 \approx \sqrt{q}$ we may construct the so-called $\lambda$-differences

$$\Delta y_k = y_{k+1} - \lambda_1 y_k ,$$
$$\Delta^2 y_k = \Delta y_{k+1} - \lambda_1\Delta y_k .$$

(26)

It is easy to discover that

$$\Delta y_k = c_1\lambda_1^{k+1} + c_2(k+1)\lambda_1^k + c_3\lambda_3^{k+1} + \cdots + c_n\lambda_n^{k+1} - c_1\lambda_1^{k+1}$$
$$- c_2k\lambda_1^k - c_3\lambda_3^k\lambda_1 - \cdots - c_n\lambda_n^k\lambda_1 = c_2\lambda_1^k + O(\lambda_3^k) ,$$

(27)

$$\frac{\Delta y_{k+1}}{\Delta y_k} = \lambda_1 + O\left(\frac{\lambda_3}{\lambda_1}\right)^k ,$$

i.e. $\Delta y_{k+1}/\Delta y_k$ approaches $\lambda_1$ quite rapidly. Coincidence of the limit of $\Delta y_{k+1}/\Delta y_k$ with the value for $\lambda_1$ computed earlier and the rapid convergence of $\Delta y_{k+1}/\Delta y_k$ to $\lambda_1$ serves to support the assumption that $\lambda_1$ is contained in a first-order box.

Moreover it is clear that

$$\Delta^2 y_k = O(\lambda_3^k) ,$$

i.e. the second $\lambda$-difference is small in comparison with the component $y_k$ itself.

The eigenvector $U_2$ corresponding to the eigenvalue $\lambda_1$ is easily determined. From the equation

$$Y_k = \lambda_1^k U_1 + k\lambda_1^{k-1} U_2 + O(\lambda_3^k)$$

it follows that

$$Y_{k+1} - \lambda_1 Y_k = \lambda_1^k U_2 + O(\lambda_3^k) , \tag{28}$$

i.e. the vector $Y_{k+1} - \lambda_1 Y_k = \tilde{U}_2$ is approximately equal to the eigenvector corresponding to $\lambda_1$. After normalization the accuracy of the approximate equality will be of order $(\lambda_3/\lambda_1)^k$. The root vector corresponding to the eigenvalue $\lambda_1$ is determined accurately within an element proportional to the eigenvector $U_2$. It is possible to take as one of the possible approximate values for the root vector the very vector

$$Y_k \approx c_1\lambda_1^k U_1 + c_1 k\lambda_1^{k-1} U_2 .$$

However for large values of $k$ this root vector, thanks to the factor $k$ in the second item, is greatly "stretched" in the direction of the eigenvector $U_2$. As an approximate root vector it is more advisable to take

$$k\lambda_1 Y_{k-1} - (k-1)Y_k = Y_k - k(Y_k - \lambda_1 Y_{k-1}) = Y_k - k\tilde{U}_2 = \lambda_1^k \tilde{U}_1 . \tag{29}$$

The vector obtained differs only by a scalar multiple from the projection $U_1$ of the initial vector $Y_0$ on the root subspace corresponding to the eigenvalue $\lambda_1$.

*Example 7.* The eigenvalues of the matrix

$$A = \begin{bmatrix} -9 & -2 & -9 \\ -13 & -2 & -12 \\ 16 & 4 & 16 \end{bmatrix}$$

are $\lambda_1 = \lambda_2 = 2$; $\lambda_3 = 1$. The eigenvector corresponding to $\lambda_1 = 2$ is $U_2 = (-2/3, -5/6, 1)' = (-0.666667, -0.833333, 1)'$.

We carry out the computations by the power method. For the initial vector $Y_0$ we take the vector $(1, 0, 1)'$. Its projection on the root subspace is (after normalizing):

$$\left(-\frac{5}{7}, -\frac{6}{7}, 1\right)' \approx (-0.714286, -0.857143, 1)' .$$

We have

| $Y_0^{(0)}$ | $Y^{(18)}$ | $Y^{(19)}$ | $Y^{(20)}$ |
|---|---|---|---|
| 1 | 8126470 | 17301510 | 36700166 |
| 0 | 10223622 | 21757958 | 46137350 |
| −1 | −12320776 | −26214408 | −55574536 |
| | 6029316 | 12845060 | 27262980 |

The convergence of the process turns out to be slow, giving us a reason to suspect the presence of multiple eigenvalues.

To compute $p$ and $q$ we take the first component for $y_k$ and the second for $z_k$. Then

$$p = -\frac{-27483387 \cdot 10^4}{-68705321 \cdot 10^3} = -4.0001831 , \qquad -\frac{p}{2} = 2.000092 ,$$

$$q = \frac{-27484802 \cdot 10^4}{-68705321 \cdot 10^3} = 4.0003891 , \qquad \sqrt{q} = 2.000097 .$$

If we take the first component for $y_k$ but the third for $z_k$, then the coincidence of $-p/2$ and $\sqrt{q}$ will be even better. Namely,

$$p = -4.0000648, \qquad -\frac{p}{2} = 2.000032,$$

$$q = 4.0001373, \qquad \sqrt{q} = 2.000034.$$

Thus we may consider $\lambda_1 = 2.00003$.

Furthermore,

| $\tilde{U}_2 = Y_{20} - 2.00003\,Y_{19}$ | $\tilde{U}_1 = Y_{20} - 20\tilde{U}_2$ | $U_2$ | $U_1$ |
|---|---|---|---|
| 2096627 | −5232374 | −0.666668 | −0.714401 |
| 2620781 | −6278270 | −0.833334 | −0.857202 |
| −3144934 | 7324144 | 1.000000 | 1.000000 |

Comparing the values found for $U_2$ and $U_1$ with the exact ones we find good coincidence for $U_2$ and slightly worse for $U_1$. The latter may be explained by the not very successful choice of the initial vector whose projection on the corresponding root subspace turns out to be quite close to the eigenvector.

**6. There are two eigenvalues of maximal modulus whose moduli are close.** To determine the eigenvalues in points 4 and 5 we in fact applied the same method and the basis for applying this method served to disturb the convergence of the sequence $y_{k+1}/y_k$. From the results of point 1 it follows that the reason for the poor convergence of this sequence may be not only the equality of the largest eigenvalue moduli but also their proximity to each other. In this case the method consists of using formulas (19) or (20) may also be applied; however the roots of the formed quadratic equation will then already be real and inevitably close in modulus. Solving the quadratic equation we then determine them with accuracy of order $(\lambda_3/\lambda_2)^k$

We observe that if the sequences of ratios of components converge rapidly, the indicated method will not allow us to determine *simultaneously* $\lambda_1$ and $\lambda_2$ with satisfactory accuracy, since the formulas for determining the coefficients $p$ and $q$ will contain numbers close to zero in the numerator and denominator.

The process will likewise have poor convergence if $|\lambda_2|$ does not exceed $|\lambda_3|$ sufficiently.

*Example 8.* As an illustration of the method indicated we shall compute the eigenvalue greatest in modulus for Leverrier's matrix.

We shall use for this the normalized iterations of the vector $(1, 0, 0, 0)'$ which we computed for example 1.

We have

| $\tilde{Y}_{18}$ | $\tilde{Y}_{19}$ | $\tilde{Y}_{20}$ |
|---|---|---|
| 1.0000000 | 1.0000000 | 1.0000000 |
| −8.1970024 | −8.2032008 | −8.2093658 |
| 8.1476812 | 8.1701265 | 8.1926032 |
| −7.3754981 | −7.9595730 | −8.5494157 |
| −6.4248193 | −6.9926473 | −7.5631783 |

We write similarly the normalizing factors

$$\rho_{18} = -17.450861,$$
$$\rho_{19} = -17.458270,$$
$$\rho_{20} = -17.465517.$$

It is obvious that

$$p = \rho_{20}\tilde{p} ,$$
$$q = \rho_{19}\rho_{20}\tilde{q},$$

where $\tilde{p}$ and $\tilde{q}$ are formed from components of the normalized iterations according to formulas (19').

If we take the first component for $y_k$ and the second for $z_k$, then

$$\tilde{p} = -\frac{0.0123634}{0.0061984} = -1.994611 , \quad p = 34.836912 ,$$

$$\tilde{q} = \frac{0.0061650}{0.0061984} = 0.994611 , \quad q = 303.27451 .$$

Thus for determining $\lambda_1$ and $\lambda_2$ we have the equation

$$t^2 + 34.836912t + 303.27451 = 0 ,$$

from which

$$\lambda_1 = -17.418456 - \sqrt{0.12810} \approx -17.4185 - 0.3579 = -17.7764 ,$$
$$\lambda_2 = -17.0606 .$$

Taking the first component for $y_k$ and taking for $z_k$ first the third and then the fourth component we obtain

$$\lambda_1 = -17.831 , \quad \lambda_2 = -17.125 ;$$
$$\lambda_1 = -17.866 , \quad \lambda_2 = -17.148 .$$

Comparing the values obtained for $\lambda_1$ we see that all three values are closer to each other than the values obtained in example 1. The last value $\lambda_1 = -17.866$, accurate to $3 \cdot 10^{-3}$, coincides with the exact value. $\lambda_2$ is thereby determined with roughly the same accuracy as $\lambda_1$.

## 54. IMPROVING THE CONVERGENCE OF THE POWER METHOD

In this paragraph two methods for improving the convergence of the power method will be presented for the case where the largest eigenvalue is real and simple.

**1. The scalar product.** This method is especially convenient for application to symmetric matrices; however we present it without this assumption.

Let there be computed along with the iterative sequence of a vector with a matrix $A$

$$Y_0, \ Y_1 = A Y_0, \ Y_2 = A^2 Y_0, \ \cdots, \ Y_k = A^k Y_0, \ \cdots \tag{1}$$

the iterative sequence with the matrix $A'$

$$Z_0, \ Z_1 = A' Z_0, \ Z_2 = A'^2 Z_0, \ \cdots, \ Z_k = A'^k Z_0, \ \cdots \tag{1'}$$

We introduce the bases $U_1, \cdots, U_n$ and $V_1, \cdots, V_n$ formed from the eigenvectors of the matrices $A$ and $A'$. We shall assume that these bases satisfy the orthogonalization and normalization conditions in the sense of Section 10, paragraph 3.

Let

$$\begin{aligned}
Y_0 &= a_1 U_1 + a_2 U_2 + \cdots + a_n U_n, \\
Z_0 &= b_1 V_1 + b_2 V_2 + \cdots + b_n V_n.
\end{aligned} \tag{2}$$

Then

$$\begin{aligned}
(Y_k, Z_k) &= (A^k Y_0, A'^k Z_0) = (A^{2k} Y_0, Z_0) \\
&= (a_1 \lambda_1^{2k} U_1 + a_2 \lambda_2^{2k} U_2 + \cdots + a_n \lambda_n^{2k} U_n, b_1 V_1 + b_2 V_2 + \cdots + b_n V_n).
\end{aligned}$$

Moreover on the basis of the orthogonalization and normalization properties for the system of vectors $U_1, \cdots, U_n$ and $V_1, \cdots, V_n$ we have

$$(Y_k, Z_k) = a_1 b_1 \lambda_1^{2k} + a_2 b_2 \lambda_2^{2k} + \cdots + a_n b_n \lambda_n^{2k}. \tag{3}$$

Analogously,

$$(Y_{k-1}, Z_k) = a_1 b_1 \lambda_1^{2k-1} + a_2 b_2 \lambda_2^{2k-1} + \cdots + a_n b_n \lambda_n^{2k-1}. \tag{4}$$

From equalities (3) and (4) we obtain:

$$\begin{aligned}
\frac{(Y_k, Z_k)}{(Y_{k-1}, Z_k)} &= \frac{a_1 b_1 \lambda_1^{2k} + a_2 b_2 \lambda_2^{2k} + \cdots + a_n b_n \lambda_n^{2k}}{a_1 b_1 \lambda_1^{2k-1} + a_2 b_2 \lambda_2^{2k-1} + \cdots + a_n b_n \lambda_n^{2k-1}} \\
&= \lambda_1 + O\left(\frac{\lambda_2}{\lambda_1}\right)^{2k}.
\end{aligned} \tag{5}$$

It is apparent from this estimate that forming the scalar product lessens the number of iterative steps necessary to determine $\lambda_1$ to the given accuracy by almost half. However an additional computation of sequence (1') is then required.

In the case of a symmetric matrix for $Z_0 = Y_0$ the sequences (1) and (1') coincide and therefore in this case application of the scalar product method is particularly expedient. Starting with a certain step of the process we must compute the corresponding scalar products and determine $\lambda_1$ in terms of their ratios. Namely,

$$\lambda_1 \approx \frac{(A^k Y_0, A^k Y_0)}{(A^{k-1} Y_0, A^k Y_0)} . \tag{6}$$

Thus in example 3, Section 53, we easily compute

$$(A^7 Y_0, A^7 Y_0) = 0.00007528987 ,$$
$$(A^7 Y_0, A^6 Y_0) = 0.00015685433 ,$$

which gives the value 0.479999 for $\lambda_1$ (instead of the values 0.4800, 0.4792, 0.4808 found from the ratios of components $A^7 Y_0$ and $A^6 Y_0$). As a second example we consider the matrix

$$\begin{bmatrix} 1.0000000 & 0 & 1.0000000 & 0 \\ 1.0000000 & 0.7777778 & 0.3333333 & 0.3333333 \\ 0 & -0.0252525 & 0.5555556 & -0.0252525 \\ 0 & -0.8888889 & -8.6444444 & 0.1111111 \end{bmatrix} \tag{7}$$

whose eigenvalues are 1, 2/3, 4/9, and 1/3.

To determine $\lambda_1$ we form the iterations $A^k Y_0$, taking as $Y_0$ the vector $(1, 1, 1, 1)'$.

We carry out the 17th, 18th, 19th and 20th iterations:

| $A^{17} Y_0$ | $A^{18} Y_0$ | $A^{19} Y_0$ | $A^{20} Y_0$ |
|---|---|---|---|
| 4.6731097 | 4.6760089 | 4.6779433 | 4.6792336 |
| 8.3733415 | 8.3912886 | 8.4032694 | 8.4112637 |
| 0.0028992 | 0.0019344 | 0.0012903 | 0.0008605 |
| −8.3861607 | −8.3998278 | −8.4089592 | −8.4150555 |
| 4.6631897 | 4.6694041 | 4.6735438 | 4.6763023. |

The ratios of components of these iterations will be

| | | |
|---|---|---|
| 1.000620 | 1.000414 | 1.000276 |
| 1.002143 | 1.001428 | 1.000951 |
| 0.667219 | 0.667029 | 0.666899 |
| 1.001630 | 1.001087 | 1.000725. |

Here the ratios of the third components differ greatly from the rest in view of the disappearance of significant figures. The last column gives $\lambda_1 \approx 1.001$; the value found coincides with the exact one to within one unit in the third place.

We shall now show how it is possible to make this value more precise by applying the scalar product method.

For this we form iterations of the vector $(1, 1, 1, 1)'$ with the transposed matrix $A'$.

Computing we obtain

$$A'^{20} Y_0 = (0.7961118, -0.0002189, 3.9939022, -0.1134904)' .$$

Furthermore

$$(A^{20}Y_0, A'^{20}Y_0) = 4.681817 \text{ and } (A^{19}Y_0, A'^{20}Y_0) = 4.681816 .$$

The ratio

$$\frac{(A^{20}Y_0, A'^{20}Y_0)}{(A^{19}Y_0, A'^{20}Y_0)} = 1.000000$$

gives the value 1.000000 for $\lambda_1$, accurate to within six decimal places.

*Note.* If in finding the iterations we normalize them, then to determine $\lambda_1$ it is necessary to use one of the formulas

$$\lambda_1 = \frac{(A\tilde{Y}_{k-1}, \hat{Z}_k)}{(\tilde{Y}_{k-1}, \hat{Z}_k)} \tag{8}$$

or

$$\lambda_1 = \frac{(\tilde{Y}_k, A'\tilde{Z}_{k-1})}{(\tilde{Y}_k, \tilde{Z}_{k-1})} . \tag{9}$$

Here we denote by $\tilde{Y}_k$ and $\tilde{Z}_k$ the normalized iterations $A^k Y_0$ and $A'^k Z_0$. The normalizing method used for these formulas is immaterial.

2. *The $\delta^2$-process*[†]. We only apply this method in the case where $|\lambda_1| > |\lambda_2| > |\lambda_3|$ and $\lambda_1$ and $\lambda_2$ are real.

We assume that we have determined the sequence of values

$$y_k, y_{k+1}, y_{k+2}, \cdots, \tag{10}$$

regarding which it is known that

$$y_k = c_1\lambda_1^k + c_2\lambda_2^k + \cdots + c_n\lambda_n^k . \tag{11}$$

We may take for $y_k$, for example, any component of the vector $Y_k = A^k Y_0$, the scalar product of corresponding iterations, etc. Then, as was pointed out in Section 53 and in paragraph 1, Section 54, it is possible to determine approximately the first eigenvalue $\lambda_1$ as the ratio $u_k = y_{k+1}/y_k$.

Moreover in Section 53 it was shown that

$$u_k = \lambda_1[1 - b_2'\alpha_2^k - b_3'\alpha_3^k + b_2 b_2'\alpha_2^{2k}] + O(\alpha_3^k + \alpha_2^{2k}) , \tag{12}$$

where $b_2' = (c_2/c_1)(1-\alpha_2)$, $b_3' = (c_3/c_1)(1 - \alpha_3)$ and $\alpha_i = \lambda_i/\lambda_1$.

If the convergence of the sequence $u_k, u_{k+1}, u_{k+2}, \cdots$ is not rapid enough, then it may be greatly improved by the following method which is known as the $\delta^2$-process. We write

$$P(u_k) = \frac{\begin{vmatrix} u_k & u_{k+1} \\ u_{k+1} & u_{k+2} \end{vmatrix}}{u_k - 2u_{k+1} + u_{k+2}} . \tag{13}$$

We shall show that

$$P(u_k) = \lambda_1 + O\left(\frac{\lambda_2}{\lambda_1}\right)^{2k} + O\left(\frac{\lambda_3}{\lambda_1}\right)^k . \tag{14}$$

---

[†] Aitken (5).

With this in mind we let

$$u_k = \lambda_1(1 + \varepsilon_k) ,$$

so that

$$\begin{vmatrix} u_k & u_{k+1} \\ u_{k+1} & u_{k+2} \end{vmatrix} = \lambda_1^2 \begin{vmatrix} 1 + \varepsilon_k & 1 + \varepsilon_{k+1} \\ 1 + \varepsilon_{k+1} & 1 + \varepsilon_{k+2} \end{vmatrix} .$$

Breaking up the last determinant into the sum of four others we obtain

$$\begin{vmatrix} u_k & u_{k+1} \\ u_{k+1} & u_{k+2} \end{vmatrix} = \lambda_1^2 \left[ \varepsilon_k - 2\varepsilon_{k+1} + \varepsilon_{k+2} + \begin{vmatrix} \varepsilon_k & \varepsilon_{k+1} \\ \varepsilon_{k+1} & \varepsilon_{k+2} \end{vmatrix} \right] .$$

But

$$u_k - 2u_{k+1} + u_{k+2} = \lambda_1[\varepsilon_k - 2\varepsilon_{k+1} + \varepsilon_{k+2}] .$$

Thus

$$P(u_k) = \lambda_1 \left[ 1 + \frac{\begin{vmatrix} \varepsilon_k & \varepsilon_{k+1} \\ \varepsilon_{k+1} & \varepsilon_{k+2} \end{vmatrix}}{\varepsilon_k - 2\varepsilon_{k+1} + \varepsilon_{k \mid 2}} \right] .$$

Computing, we find that

$$\frac{\begin{vmatrix} \varepsilon_k & \varepsilon_{k+1} \\ \varepsilon_{k+1} & \varepsilon_{k+2} \end{vmatrix}}{\varepsilon_k - 2\varepsilon_{k+1} + \varepsilon_{k+2}} \approx A\alpha_3^k + B\alpha_2^{2k} ,$$

where

$$A = \frac{-b_3'(\alpha_2 - \alpha_3)^2}{(1 - \alpha_2)^2} ,$$

$$B = b_2 b_2' \alpha_2^2 .$$

Thus,

$$P(u_k) = \lambda_1 + O\left(\frac{\lambda_3}{\lambda_1}\right)^k + O\left(\frac{\lambda_2}{\lambda_1}\right)^{2k} .$$

From here it follows that the error in determining $\lambda_1$ using the $\delta^2$-process may be far less than in finding it directly from the sequence $u_k$, $u_{k+1}$, $\cdots$

We observe that in actually carrying out the $\delta^2$-process by formula (13) a loss of significant figures occurs in both the numerator and denominator. This phenomenon may be avoided in the following way. Let $u_k = c + l_k$ where $c$ is a number formed from the dots in the sequence $u_k$. Then it is not hard to verify that

$$P(u_k) = c + P(l_k) ,$$

where the second item plays the role of a small correction with respect to the first.

*Note.* In finding the first eigenvalue of a symmetric matrix one must apply the $\delta^2$-process not to the iterative components but to the corresponding scalar products.

We shall show an application of the $\delta^2$-process for the examples of Section 53.

In example 2, Section 53, applying the $\delta^2$-process for the given ratios gives the following values for $\lambda_1$:

$$-17.3974$$
$$-17.3977$$
$$-17.3977.$$

Thus to four decimal places $\lambda_1 = -17.3977$.

Analogously, applying the $\delta^2$-process to the given ratios in example 4, Section 53, gives the following values for $\lambda_1$:

$$0.66748, \qquad 0.66748,$$
$$0.66748, \qquad 0.66748.$$

Thus $\lambda_1$ is determined already to five places. (The exact value is $\lambda_1 = 0.667483$).

The $\delta^2$-process may likewise be applied in determining the components of the first eigenvector. We shall show two different variations of this process which depend on whether it is possible to consider $\lambda_1$ as known with sufficient accuracy or not.

1. Assume that we know only the iterative sequence $A^k Y_0$ where for computational convenience every iteration is normalized by dividing by a component $z_k$ of fixed subscript.

We denote any other component of the vector $Y_k$ by $y_k$.

If

$$y_k = c_1 \lambda_1^k + c_2 \lambda_2^k + \cdots + c_n \lambda_n^k \,,$$
$$z_k = b_1 \lambda_1^k + b_2 \lambda_2^k + \cdots + b_n \lambda_n^k \,,$$

then the division indicated brings $z_k$ to one and $y_k$ to $v_k$ where

$$v_k = \frac{y_k}{z_k} = \frac{c_1 \lambda_1^k + c_2 \lambda_2^k + \cdots + c_n \lambda_n^k}{b_1 \lambda_1^k + b_2 \lambda_2^k + \cdots + b_n \lambda_n^k}$$

$$= \frac{c_1}{b_1} + \frac{c_2 b_1 - b_2 c_1}{b_1^2} \left( \frac{\lambda_2}{\lambda_1} \right)^k + O \left( \frac{\lambda_3}{\lambda_1} \right)^k \,.$$

Simple computation shows that

$$P(v_k) = \frac{c_1}{b_1} + O \left( \frac{\lambda_3}{\lambda_1} \right)^k + O \left( \frac{\lambda_2}{\lambda_1} \right)^{2k} \,.$$

Thus if the $\delta^2$-process is applied to all components of a normalized vector $A^k Y_0$, then we find the ratios of coefficients associated with powers of $\lambda_1$ in expressions for $y_k$. These coefficients are proportional to the components of the eigenvector.

Thus application of the $\delta^2$-process for the eigenvector of example 2 in the preceding paragraph gives for the vector components

$$1.00000, \qquad -8.14718, \qquad 7.91721,$$

which are significantly closer to the exact values than the values computed directly from the same iterations.

2. If $\lambda_1$ is known accurately enough, it is possible to set up a process of improvement in the following way. We multiply all components of the vectors $A^{k-1}Y_0$, $A^k Y_0$, $A^{k+1}Y_0$ by $\lambda_1$, $1$, $\lambda_1^{-1}$ respectively and then apply the $\delta^2$-process to them. Since

$$y_{k-1}\lambda_1 = \lambda_1^k c_1 + \lambda_2^{k-1}\lambda_1 c_2 + \cdots + \lambda_n^{k-1}\lambda_1 c_n \, ,$$

$$y_k = \lambda_1^k c_1 + \lambda_2^k c_2 + \cdots + \lambda_n^k c_n \, ,$$

$$y_{k+1}\lambda_1^{-1} = \lambda_1^k c_1 + \frac{\lambda_2^{k+1}}{\lambda_1} c_2 + \cdots + \frac{\lambda_n^{k+1}}{\lambda_1} c_n \, ,$$

then

$$\begin{vmatrix} \lambda_1 y_{k-1} & y_k \\ y_k & \lambda_1^{-1} y_{k+1} \end{vmatrix} = [c_1 c_2 \lambda_1^{k-1}\lambda_2^{k-1}(\lambda_1 - \lambda_2)^2 + \cdots$$

$$+ c_1 c_n \lambda_1^{k-1}\lambda_n^{k-1}(\lambda_1 - \lambda_n)^2] \times \left[1 + O\left(\frac{\lambda_3}{\lambda_1}\right)^k\right].$$

Furthermore,

$$\lambda_1 y_{k-1} - 2y_k + \lambda_1^{-1} y_{k+1} = c_2 \frac{\lambda_2^{k-1}}{\lambda_1}(\lambda_1 - \lambda_2)^2 + \cdots$$

$$+ c_n \frac{\lambda_n^{k-1}}{\lambda_1}(\lambda_1 - \lambda_n)^2 \, .$$

From here

$$\frac{\begin{vmatrix} \lambda_1 y_{k-1} & y_k \\ y_k & \lambda_1^{-1} y_{k+1} \end{vmatrix}}{\lambda_1 y_{k-1} - 2y_k + \lambda_1^{-1} y_{k+1}} = c_1 \lambda_1^k \left[1 + O\left(\frac{\lambda_3}{\lambda_1}\right)^k\right].$$

The ratios of the resultant numbers which are computed for different components give the ratios of the eigenvector components.

## 55. MODIFICATIONS OF THE POWER METHOD

1. **The power method with translation.** It is known that the eigenvalues $\lambda_i$ of a matrix $A$ are connected with the eigenvalues $\mu_i$ of a matrix $B = A - cE$ by the simple relationships

$$\lambda_i = \mu_i + c \, ,$$

and the eigenvectors of both matrices coincide. This fact makes it possible to determine the eigenvalues of matrix $A$ by applying the power method to the matrix $A - cE$. Such a variation of the power method is called the *power method with translation.* The translation alters the inter-relationship among the eigenvalue moduli while in the presence of complex roots, even for real $c$, changing this inter-relationship may be quite involved. If all eigenvalues are real, then by translation it is possible to make the largest one in modulus the smallest algebraically as well as the largest algebraically. That is, (cf. fig. 2) for $c < c_0 = (\lambda_1 + \lambda_n)/2$ the largest one in modulus will be $\mu_1$, for $c > c_0$ the largest in modulus will be $\mu_n$.

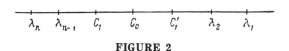

**FIGURE 2**

The optimal value of $c$ for computing $\lambda_1$ is, as is easy to see, $c_1 = (\lambda_n + \lambda_2)/2$, since for such a choice of $c$ the convergence of the power iterations for determining $\mu_1$ will be the most rapid. Accordingly the optimal value for computing $\lambda_n$ will be $c_1' = (\lambda_{n-1} + \lambda_1)/2$. Of course these considerations have only a theoretical value, since as a rule we do not know even roughly $\lambda_1$ or $\lambda_n$. However they do make it possible at any rate to make certain recommendations for an advisable choice of $c$. Thus if the matrix $A$ is positive definite, then it is sensible to make a translation for some positive $c$ and, having tried to compute the iterations, to judge the advisability of this translation by their behavior. Nevertheless it is sometimes possible to use crude estimates for $\lambda_2$ and $\lambda_n$.

We shall compare the courses of the power process with and without translation; such translation will be close to the optimal one for matrix (4), Section 51.

Here

$$\frac{\lambda_2 + \lambda_4}{2} = \frac{0.7967 + 0.2423}{2} = 0.5195 \ .$$

We carry out the process with translation, letting $c = 0.5$.

Iterating the vector $(0.3, 0.5, 0.7, 0.9)'$ with the matrix $A$ we obtain

| $Y_0$ | $Y_{10}$ | $Y_{11}$ | $Y_{12}$ |
|---|---|---|---|
| 0.3 | $0.31005932 \cdot 10^4$ | $0.72018993 \cdot 10^4$ | $0.16728203 \cdot 10^5$ |
| 0.5 | $0.24605904 \cdot 10^4$ | $0.57153328 \cdot 10^4$ | $0.13275282 \cdot 10^5$ |
| 0.7 | $0.23186358 \cdot 10^4$ | $0.53856100 \cdot 10^4$ | $0.12509420 \cdot 10^5$ |
| 0.9 | $0.27512042 \cdot 10^4$ | $0.63903554 \cdot 10^4$ | $0.14843190 \cdot 10^5$ |
| | $1.06310236 \cdot 10^4$ | $2.46931975 \cdot 10^4$ | $0.57356095 \cdot 10^5.$ |

For the component ratios we obtain respectively

| | |
|---|---|
| 2.3227488 | 2.3227488 |
| 2.3227485 | 2.3227487 |
| 2.3227494 | 2.3227489 |
| 2.3227484 | 2.3227889. |

In the last column the seventh place is already stabilized. We obtain the same accuracy in the component ratios from the eighth and ninth iterations with the matrix $A_1 = A - 0.5E$. In fact,

| $Y_0$ | $Y_8$ | $Y_9$ |
|---|---|---|
| 0.3 | $0.82648864 \cdot 10^2$ | $0.15064813 \cdot 10^3$ |
| 0.5 | $0.65589056 \cdot 10^2$ | $0.11955238 \cdot 10^3$ |
| 0.7 | $0.61805179 \cdot 10^2$ | $0.11265531 \cdot 10^3$ |
| 0.9 | $0.73335608 \cdot 10^2$ | $0.13367238 \cdot 10^3$ |
| | $2.83378707 \cdot 10^2$ | $0.51652820 \cdot 10^3$, |

so that for $0.5 + (y_i)_9/(y_i)_8$ we obtain the values

$$2.3227489, \qquad 2.3227487,$$
$$2.3227489, \qquad 2.3227486.$$

To determine the smallest eigenvalue the translation will be optimal for

$$\frac{2.32 + 0.64}{2} = 1.48 .$$

We carry out the result of the computations for $c = 1.5$. Namely,

| $1.5 + (y_i)_{20}/(y_i)_{19}$ | $1.5 + (y_i)_{45}/(y_i)_{44}$ |
|---|---|
| 0.2434512 | 0.2422607 |
| 0.2492658 | 0.2422606 |
| 0.2406113 | 0.2422607 |
| 0.2409250 | 0.2422607. |

**2. Raising a matrix to a power.** In constructing higher iterations of a vector it is sometimes advisable to raise the given matrix to a power beforehand. It is most simple to compute the successive powers of a matrix $A$, $A^2$, $A^4$, $A^8$, $A^{16}$, $\cdots$. However squaring a matrix is obviously equivalent in volume of work to forming $n$ iterations of the vector, so that computing the matrix $A^{2^k}$ is equivalent to constructing $kn$ iterations. Correspondingly, computing $A^{2^k} Y_0$ is equivalent to computing $kn + 1$ iterations and therefore an advantage in work volume is obtained if $kn + 1 < 2^k$, i.e. if the number of iterations necessary to obtain the required accuracy exceeds $n \log_2 n$.

We may limit ourselves to computing a certain fixed power for matrix $A$ and may then form iterations using the power computed. For example, having computed $A^8$ we may find $A^8 Y_0$, then $A^8(A^8 Y_0) = A^{16} Y_0$, and finally $A^{17} Y_0 = A(A^{16} Y_0)$.

The powers of a matrix may be used indirectly as well for determining the eigenvalue largest in modulus in the case where it is simple and real. Namely,

$$|\lambda_1| \approx \sqrt[2^k]{\operatorname{Sp} A^{2^k}} . \tag{1}$$

This follows from the fact that

$$\operatorname{Sp} A^m = \lambda_1^m + \lambda_2^m + \cdots + \lambda_n^m ,$$

and consequently

$$\sqrt[m]{\operatorname{Sp} A^m} = \lambda_1 \sqrt[m]{1 + \left(\frac{\lambda_2}{\lambda_1}\right)^m + \cdots + \left(\frac{\lambda_n}{\lambda_1}\right)^m} = \lambda_1 + O\left(\frac{1}{m}\left(\frac{\lambda_2}{\lambda_1}\right)^m\right) .$$

This method is essentially equivalent to applying Lobachevskii's method in finding the largest root of the characteristic equation for a matrix.

Slightly more convenient than formula (1) is the formula

$$\lambda_1 \approx \frac{\text{Sp } A^{2k+1}}{\text{Sp } A^{2k}} \, ,$$

since the additional computation of $\text{Sp } A^{2k+1}$ is equivalent (in volume of work) to one iteration of the vector with the matrix $A^{2k}$. The method using spurs may be extended to the case of multiple and complex roots.[†]

## 56. APPLICATION OF THE POWER METHOD TO FINDING SEVERAL EIGENVALUES

In Section 53 we considered several cases where the eigenvalue greatest in modulus was not isolated, i.e. where there existed another eigenvalue equal or close in modulus to the first. The method applied there consisted in computing the coefficients of a quadratic equation whose roots were the two eigenvalues largest in modulus. This method may be generalized in a natural way. We assume that the elementary divisors of the matrix are relatively prime and that $|\lambda_1| \geq |\lambda_2| \geq \cdots \geq |\lambda_r| > |\lambda_{r+1}| \geq \cdots \geq |\lambda_n|$. We denote by $U_1, \cdots, U_n$ a canonical basis of the space. We assume that the initial vector $Y_0$ is taken such that all components in its expansion in terms of vectors of the canonical basis will be different from zero, i.e.

$$Y_0 = a_1 U_1 + a_2 U_2 + \cdots + a_r U_r + a_{r+1} U_{r+1} + \cdots + a_n U_n$$
$$a_i \neq 0 \ (i = 1, 2, \cdots, n) \, . \tag{1}$$

Let $t^r + b_1 t^{r-1} + \cdots + b_r = (t - \lambda_1) \cdots (t - \lambda_r)$ be a polynomial whose roots are $\lambda_1, \cdots, \lambda_r$. Then, as is easy to see,

$$Y_{k+r} + b_1 Y_{k+r-1} + \cdots + b_r Y_k \approx 0 \tag{2}$$

accurate (for each component) to magnitudes of order $|\lambda_{r+1} + \varepsilon|^k$.

In fact in the linear combination described all $U_1, \cdots, U_r$ formed by the vectors disappear; the coefficients for $U_{r+1}, \cdots, U_n$ are multiplied by no more than $\lambda_{r+1}^{k+2}$ and possibly by a quantity of an order with power $k$ (if the matrix has non-linear elementary divisors). The vector equation (2) is equivalent to a system of $n$ equations for corresponding components. Taking any $r$ of them we obtain a system of $r$ linear equations relative to $b_1, \cdots, b_r$. For simplicity we shall consider that the components have been numbered so that the chosen ones will be the $r$ first components of the vectors $Y_{k+r}, \cdots, Y_k$. Solving the system

$$y_{1k+r} + b_1 y_{1k+r-1} + \cdots + b_r y_{1k} = 0 \, ,$$
$$\cdots \cdots \cdots \cdots \cdots \cdots \cdots \cdots \cdots , \tag{3}$$
$$y_{rk+r} + b_1 y_{rk+r-1} + \cdots + b_r y_{rk} = 0 \, ,$$

[†] Frazer, Duncan and Collar (1).

by Cramer's formulas we obtain for the coefficients $b_1, \cdots, b_r$ the approximate values

$$b_1 = \frac{\begin{vmatrix} -y_{1k+r} & y_{1k+r-2} \cdots y_{1k} \\ \cdots \cdots \cdots \cdots \\ -y_{rk+r} & y_{rk+r-2} \cdots y_{rk} \end{vmatrix}}{\begin{vmatrix} y_{1k+r-1} y_{1k+r-2} \cdots y_{1k} \\ \cdots \cdots \cdots \cdots \\ y_{rk+r-1} y_{rk+r-2} \cdots y_{rk} \end{vmatrix}}, \cdots, b_r = \frac{\begin{vmatrix} y_{1k+r-1} \cdots -y_{1k+r} \\ \cdots \cdots \cdots \cdots \\ y_{rk+r-1} \cdots -y_{rk+r} \end{vmatrix}}{\begin{vmatrix} y_{1k+r-1} \cdots y_{1k} \\ \cdots \cdots \cdots \\ y_{rk+r-1} \cdots y_{rk} \end{vmatrix}}. \qquad (4)$$

It is possible to show that these equalities are correct to within quantities of order $((|\lambda_{r+1}| + \varepsilon)/|\lambda_r|)^k$. For $k = 2$ these estimates were carried out above.

Having determined the coefficients $b_1, \cdots, b_r$ we find the eigenvalues as roots of the polynomial $t^r + b_1 t^{r-1} + \cdots + b_r$. If equal ones turn up among them, this will verify the presence of non-linear elementary divisors for matrix $A$. We observe that to determine the coefficients $b_1, \cdots, b_r$ we may take instead of $r$ different components of the vectors $Y_{k+r}, \cdots, Y_k$ any one component of the vectors $Y_{k+r}, \cdots, Y_k; Y_{k+r+1}, \cdots, Y_{k+1}; \cdots; Y_{k+2r-1}, \cdots, Y_{k+r-1}$.

In actual computations it is not necessary to compute the determinants. The system described may be solved numerically by one of the methods presented above.

We observe that some kind of satisfactory result is obtained if the $r$ eigenvalues being determined are close in modulus and the next $(r + 1)$ are far away from them. If this does not hold, then the system obtained in (3) will be very ill conditioned and the determinants in formulas (4) will be very close to zero.

Speaking theoretically, the indicated process makes it possible to construct the whole characteristic polynomial (more correctly, the minimum annihilating polynomial for the vector $Y_0$), taking $r = n$. In this case we pass to Krylov's method, satisfied starting with the initial vector $A^k Y_0$. Of course here it is advisable to assume $k = 0$ so as to avoid computing extra iterations. For increasing $k$ the condition of the system degenerates to the same thing as in Krylov's method.

In the case where $|\lambda_1| > |\lambda_2| > \cdots > |\lambda_r| > |\lambda_{r+1}| \geq \cdots \geq |\lambda_n|$, it is possible to change this process slightly. That is, in this case it is sufficient to compute only the constants of the successive polynomials $(t - \lambda_1)(t - \lambda_2)$, $\cdots, (t - \lambda_1)(t - \lambda_2) \cdots (t - \lambda_r)$, since these numbers are equal to within a sign to the product of successive eigenvalues. We write out the corresponding formulas[†]:

$$\lambda_1 \approx \frac{y_{1k+1}}{y_{1k}},$$

$$\lambda_1 \lambda_2 \approx \frac{\begin{vmatrix} y_{1k+2} & y_{1k+1} \\ y_{2k+2} & y_{2k+1} \end{vmatrix}}{\begin{vmatrix} y_{1k+1} & y_{1k} \\ y_{2k+1} & y_{2k} \end{vmatrix}},$$

---

† Aitken (5).

$$\cdots \cdots \cdots,$$

$$\lambda_1\lambda_2 \cdots \lambda_r \approx \frac{\begin{vmatrix} y_{1k+r} \cdots y_{1k+1} \\ \cdots \cdots \\ y_{rk+r} \cdots y_{rk+1} \end{vmatrix}}{\begin{vmatrix} y_{1k+r-1} \cdots y_{1k} \\ \cdots \cdots \\ y_{rk+r-1} \cdots y_{rk} \end{vmatrix}}.$$

We observe that even determining the product of two eigenvalues runs up against a barrier in the form of disappearing significant figures, since for a sufficiently large number of iterations the rows of the determinants become almost proportional (in the case where the first eigenvalue is far removed from the second). Therefore as a rule even the second eigenvalue is determined using the power method with a much lower degree of accuracy than the first.

In Section 53 we considered already some examples for computing the coefficients of equations whose roots are the eigenvalues of maximal modulus for $r = 2$. We shall limit ourselves to these examples and shall illustrate only the second method described here. That is, we determine the second eigenvalue for matrix (7), Section 54. Taking $k = 18$ we find, starting from the first two components of corresponding vectors,

$$\lambda_1\lambda_2 \approx \frac{\begin{vmatrix} 4.6792336 & 4.6779433 \\ 8.4112637 & 8.4032694 \end{vmatrix}}{\begin{vmatrix} 4.6779433 & 4.6760089 \\ 8.4032694 & 8.3912886 \end{vmatrix}} = \frac{-0.0265541}{-0.0397902} = 0.667353.$$

Starting from the second and fourth components for the same iterations we obtain

$$\lambda_1\lambda_2 \approx 0.666110 .$$

In Section 54 we computed, using these same iterations, that $\lambda_1 = 1.001$. This gives for $\lambda_2$ the values 0.6667 or 0.6654. The exact value is $\lambda_2 = 0.66666\cdots$

Finally we determine $\lambda_2$ by the ratio of determinants formed from corresponding components of adjacent iterations. Using the first components of the 17th, 18th, 19th, and 20th iterations we obtain

$$\lambda_1\lambda_2 \approx \frac{\begin{vmatrix} 4.6792336 & 4.6779433 \\ 4.6779433 & 4.6760089 \end{vmatrix}}{\begin{vmatrix} 4.6779433 & 4.6760089 \\ 4.6760089 & 4.6731097 \end{vmatrix}} = \frac{-0.0030156}{-0.0045170} = 0.667611 ,$$

so that $\lambda_2 \approx 0.6669$.

## 57. THE STEPPED POWER METHOD

Determining two or more eigenvalues using the power method runs into two difficulties, as we just saw. These are: first, the possible loss of significant figures in forming the needed linear combinations and, second, the absence of a criterion by which the realization of satisfactory accuracy may be judged. The slightly more laborious *stepped power method* is free from both these insufficiencies; we shall now examine this method. We shall investigate it in detail as applied to the problem of determining the two eigenvalues largest in modulus and the eigenvectors belonging to them (or the eigenvector and root vector if the eigenvalue largest in modulus lies in a second-order canonical box) and shall only touch on its generalization to the problem of determining the first $r$ eigenvalues. We shall consider three modifications of the method. For this we shall consider that $|\lambda_1| \geq |\lambda_2| > |\lambda_3| \geq \cdots \geq |\lambda_n|$. For simplicity of presentation we shall likewise assume that the eigenvalues, starting with $\lambda_3$, have linear elementary divisors.

1. **The completely stabilized stepped method.** Let $X_0$ and $Y_0$ be two arbitrary vectors. We form the vectors $AX_0$ and $AY_0$ and construct their linear combinations $X_1$ and $Y_1$ such that the first two components of the vectors $X_1$ and $Y_1$ form the matrix

$$\begin{bmatrix} 1 & 0 \\ 0 & 1 \end{bmatrix}.$$

For this it is necessary to multiply the rectangular matrix formed from the components of vectors $AX_0$ and $AY_0$ by the second-order matrix inverse to the matrix formed from the first two components of the vectors $AX_0$ and $AY_0$.

This may be done, for example, as follows. We construct the vector $\tilde{X}_1$, dividing all components of the vector $AX_0$ by the first component. The vector $Y_1$ is constructed by subtracting from the vector $AY_0$ the vector $\tilde{X}_1$ multiplied by the first component of the vector $AY_0$ and by dividing all components of the resultant vector by its second component. Finally the vector $X_1$ is constructed by subtracting from the vector $\tilde{X}_1$ the vector $Y_1$ multiplied by the second component of the vector $\tilde{X}_1$.

Furthermore, the process is repeated. That is, the vectors $X_k$ and $Y_k$ are constructed as linear combinations of the vectors $AX_{k-1}$ and $AY_{k-1}$ such that their first two components form a second-order unit matrix.

THEOREM 57.1. *Let the eigenvalues of a matrix $A$ satisfy the inequalities*

$$|\lambda_1| \geq |\lambda_2| > |\lambda_3| \geq \cdots \geq |\lambda_n|.$$

*Then if 1) the determinant of the first two components of the eigenvectors $U_1$ and $U_2$ belonging to the eigenvalues $\lambda_1$ and $\lambda_2$ or of the eigenvector $U_1$ and the root vector $U_2$ belonging to the eigenvalue $\lambda_1 = \lambda_2$ is non-zero; 2) the determinant $c_1 d_2 - c_2 d_1$ of coefficients of the expansion*

$$X_0 = c_1 U_1 + c_2 U_2 + \cdots + c_n U_n,$$
$$Y_0 = d_1 U_1 + d_2 U_2 + \cdots + d_n U_n$$

*of the vectors $X_0$ and $Y_0$ in terms of eigenvectors (root vectors) is non-zero; 3) all the determinants formed from the first two components of the vectors $A^k X_0$ and $A^k Y_0$ differ from zero, then the vector sequences $X_k$ and $Y_k$ have limits $X$ and $Y$ and these limit vectors lie in the invariant subspace spanned by the vectors $U_1$ and $U_2$.*

*Proof.* From the process of constructing the vectors $X_k$ and $Y_k$ it is clear that the vectors $X_k$ and $Y_k$ are linear combinations of the vectors $A^k X_0$ and $A^k Y_0$. Let

$$F_k = \begin{bmatrix} x_{1k} & y_{1k} \\ x_{2k} & y_{2k} \\ \cdot & \cdot \\ \cdot & \cdot \\ \cdot & \cdot \\ x_{nk} & y_{nk} \end{bmatrix}$$

be a two-column matrix formed from the components of the vectors $A^k X_0$ and $A^k Y_0$. Let

$$\Phi_k = \begin{bmatrix} 1 & 0 \\ 0 & 1 \\ \xi_{3k} & \eta_{3k} \\ \cdot & \cdot \\ \cdot & \cdot \\ \xi_{nk} & \eta_{nk} \end{bmatrix}$$

be a matrix formed from the components of the vectors $X_k$ and $Y_k$. It follows from the construction that the matrix $\Phi_k$ is obtained from the matrix $F_k$ by a linear combining of columns which is equivalent to multiplying on the right by a certain second-order matrix. We obviously need to take for this matrix the matrix

$$\begin{bmatrix} x_{1k} & y_{1k} \\ x_{2k} & y_{2k} \end{bmatrix}^{-1}$$

so that

$$\Phi_k = F_k \begin{bmatrix} x_{1k} & y_{1k} \\ x_{2k} & y_{2k} \end{bmatrix}^{-1}.$$

The third condition of the theorem guarantees the existence of the matrices

$$\begin{bmatrix} x_{1k} & y_{1k} \\ x_{2k} & y_{2k} \end{bmatrix}^{-1},$$

i.e. guarantees infinite continuation of the process.

We now pass on to estimates for the components of the vectors $X_k$ and $Y_k$ or, what is the same thing, to estimates for elements of the matrix $\Phi_k$. For this we compute the elements of $\Phi_k$ in terms of $x_{ik}$ and $y_{ik}$. We have

$$\begin{bmatrix} x_{1k} & y_{1k} \\ x_{2k} & y_{2k} \end{bmatrix}^{-1} = \frac{1}{\begin{vmatrix} x_{1k} & y_{1k} \\ x_{2k} & y_{2k} \end{vmatrix}} \begin{bmatrix} y_{2k} & -y_{1k} \\ -x_{2k} & x_{1k} \end{bmatrix},$$

and consequently,

$$\xi_{ik} = \frac{x_{ik}y_{2k} - y_{ik}x_{2k}}{x_{1k}y_{2k} - x_{2k}y_{1k}} \qquad \text{(for } i \geq 3\text{)}. \tag{1}$$

$$\eta_{ik} = \frac{y_{ik}x_{1k} - x_{ik}y_{1k}}{x_{1k}y_{2k} - x_{2k}y_{1k}} \qquad \text{(for } i \geq 3\text{)}. \tag{2}$$

We assume first that $U_1$ and $U_2$ are eigenvectors possibly corresponding to equal eigenvalues $\lambda_1$ and $\lambda_2$. From the expansions

$$\begin{aligned} X_0 &= c_1 U_1 + c_2 U_2 + \cdots + c_n U_n, \\ Y_0 &= d_1 U_1 + d_2 U_2 + \cdots + d_n U_n \end{aligned} \tag{3}$$

it follows that

$$\begin{aligned} A^k X_0 &= c_1 \lambda_1^k U_1 + c_2 \lambda_2^k U_2 + \cdots + c_n \lambda_n^k U_n, \\ A^k Y_0 &= d_1 \lambda_1^k U_1 + d_2 \lambda_2^k U_2 + \cdots + d_n \lambda_n^k U_n. \end{aligned}$$

Consequently,

$$\begin{aligned} x_{ik} &= c_1 \lambda_1^k u_{1i} + c_2 \lambda_2^k u_{2i} + \cdots + c_n \lambda_n^k u_{ni}, \\ y_{ik} &= d_1 \lambda_1^k u_{1i} + d_2 \lambda_2^k u_{2i} + \cdots + d_n \lambda_n^k u_{ni} \end{aligned} \tag{4}$$

and therefore

$$\begin{aligned} x_{ik}y_{2k} - x_{2k}y_{ik} &= (c_1 \lambda_1^k u_{1i} + c_2 \lambda_2^k u_{2i} + \cdots + c_n \lambda_n^k u_{ni})(d_1 \lambda_1^k u_{12} + d_2 \lambda_2^k u_{22} \\ &\quad + \cdots + d_n \lambda_n^k u_{n2}) - (c_1 \lambda_1^k u_{12} + c_2 \lambda_2^k u_{22} + \cdots + c_n \lambda_n^k u_{n2})(d_1 \lambda_1^k u_{1i} + \cdots \\ &\quad + d_n \lambda_n^k u_{ni}) = (c_1 d_2 - c_2 d_1)(u_{22} u_{1i} - u_{12} u_{2i}) \lambda_1^k \lambda_2^k \left[ 1 + O\left( \left| \frac{\lambda_3}{\lambda_2} \right|^k \right) \right]. \end{aligned}$$

On the basis of the theorem's second condition $c_1 d_2 - c_2 d_1 \neq 0$. Therefore

$$\xi_{ik} = \frac{u_{1i}u_{22} - u_{2i}u_{12}}{u_{11}u_{22} - u_{12}u_{21}} + O\left( \left| \frac{\lambda_3}{\lambda_2} \right|^k \right). \tag{5}$$

On the basis of the first condition of the theorem $u_{11}u_{22} - u_{12}u_{21} \neq 0$. Analogously

$$\eta_{ik} = \frac{u_{11}u_{2i} - u_{21}u_{1i}}{u_{11}u_{22} - u_{12}u_{21}} + O\left( \left| \frac{\lambda_3}{\lambda_2} \right|^k \right). \tag{6}$$

In the case where $\lambda_1 = \lambda_2$ and this eigenvalue lies in a second-order canonical box we have that

$$\xi_{ik} = \frac{u_{1i}u_{22} - u_{2i}u_{12}}{u_{11}u_{22} - u_{12}u_{21}} + O\left( k \left| \frac{\lambda_3}{\lambda_2} \right|^k \right), \tag{5'}$$

$$\eta_{ik} = \frac{u_{11}u_{2i} - u_{21}u_{1i}}{u_{11}u_{22} - u_{12}u_{21}} + O\left( k \left| \frac{\lambda_3}{\lambda_2} \right|^k \right). \tag{6'}$$

Passing to the limit in equations (5) and (6) (or (5′) and (6′)) for $k \to \infty$ we obtain

$$\xi_{ik} \to x_i = \frac{u_{1i} u_{22} - u_{2i} u_{12}}{u_{11} u_{22} - u_{12} u_{21}} ,$$

$$\eta_{ik} \to y_i = \frac{u_{11} u_{2i} - u_{2i} u_{1i}}{u_{11} u_{22} - u_{12} u_{21}} .$$

Thus the limit vectors (with components $x_i$ and $y_i$) are equal to

$$X = \frac{u_{22}}{u_{11} u_{22} - u_{12} u_{21}} U_1 - \frac{u_{12}}{u_{11} u_{22} - u_{12} u_{21}} U_2 ,$$

$$Y = -\frac{u_{21}}{u_{11} u_{22} - u_{12} u_{21}} U_1 + \frac{u_{11}}{u_{11} u_{22} - u_{12} u_{21}} U_2 .$$

$$(7)$$

The theorem is proved. Estimates for the rate of convergence have also been given in the course of the proof.

*Note.* We may always satisfy conditions 2 and 3 of the theorem by a suitable choice of initial vectors $X_0$ and $Y_0$. Condition 1 may be fulfilled by changing when necessary the numbering of components for vectors of the space; this is equivalent to changing simultaneously the rows and columns of matrix $A$. It is not hard to prove further that if conditions 1 and 2 are satisfied, then condition 3 will be satisfied for large enough $k$.

The first modification of the stepped power method to be considered is also based on the theorem just proved.

The sequence of vectors $X_k$ and $Y_k$ is constructed until the equalities

$$X_{k+1} \approx X_k$$
$$Y_{k+1} \approx Y_k$$

turn out to be fulfilled with sufficient accuracy.

Such a stabilization occurs if the conditions of theorem 57.1 are satisfied. On the basis of the theorem the subspace spanned by the limit vectors $X$ and $Y$ coincides with the invariant subspace spanned by the vectors $U_1$ and $U_2$. Therefore our problem now is to solve the complete eigenvalue problem in this two-dimensional subspace. We take for a basis of this subspace the limit vectors $X$ and $Y$. Then

$$AX = \alpha X + \beta Y$$
$$AY = \gamma X + \delta Y ,$$

so that the matrix

$$L = \begin{bmatrix} \alpha & \gamma \\ \beta & \delta \end{bmatrix} \qquad (8)$$

is the matrix of the induced operator in the subspace considered.

The numbers $\alpha$, $\beta$, $\gamma$, and $\delta$ are easily determined. In fact, $\alpha$ and $\beta$ are the first two components of the vector $AX$ and $\gamma$ and $\delta$ are first two components of the vector $AY$. Approximate values for them were already obtained by us as the corresponding components of the vectors $AX_k$ and $AY_k$. The desired eigenvalues coincide with the eigenvalues of the matrix $L$. The coordinates of the vectors $U_1$ and $U_2$ relative to the basis of $X$ and $Y$ are equal to the components of the eigenvectors (or eigenvector and root vector) of the matrix $L$.

Thus if $\lambda_1 \neq \lambda_2$, we easily compute that

$$U_1 = X + \frac{\lambda_1 - \alpha}{\gamma} Y,$$

$$U_2 = X + \frac{\lambda_2 - \alpha}{\gamma} Y. \tag{9}$$

We shall not carry out an example illustrating this modification of the stepped power method, since we are only considering it as a theoretical base for subsequent modifications which are more suitable for numerical realization.

**2. Bauer's stepped iterations.**[†] Starting from the two vectors $Z_0$ and $Y_0$ we form two sequences of vectors $Z_k$ and $Y_k$ in the following manner. The vector $Z_{k+1}$ is obtained from the vector $AZ_k$ by dividing by its first component. The vector $Y_{k+1}$ is obtained as a linear combination of the vectors $AY_k$ and $AZ_k$ such that the first component of the vector $Y_{k+1}$ is equal to zero and the second is equal to one. In other words, the vectors $Y_k$ coincide with the corresponding vectors of the previous modification for $X_0 = Z_0$ and the vectors $Z_k$ coincide with the vectors, normalized to a unit first component, of the usual power method.

Therefore under the conditions of the previous theorem the vectors $Y_k$ are stabilized. The behavior of the vectors $Z_k$ depends on the mutual placing of the first two eigenvalues which may be *a priori* unknown. That is, the sequence $Z_k$ will be stabilized if $|\lambda_1| > |\lambda_2|$ but possibly very slowly if $|\lambda_1|$ is close to $|\lambda_2|$ or if $\lambda_1 = \lambda_2$ and will not be stabilized if $|\lambda_1| = |\lambda_2|$, but $\lambda_1 \neq \lambda_2$, i.e. if $\lambda_1 = -\lambda_2$ and if $\lambda_1, \lambda_2$ form a complex conjugate pair.

Nevertheless, using the vectors $Z_k$ and $Y_k$ for large enough $k$ we may determine $\lambda_1$ and $\lambda_2$ as well as the eigenvectors (or eigenvector and root vector) belonging to them; this is no more complex than using the completely stabilized iterations $X_k$ and $Y_k$ of the previous modification.

It is clear that for $Z_0 = X_0$ the vectors $Z_k$ and the vectors $X_k$ are connected by the relationship

$$X_k = Z_k - \varepsilon_k Y_k,$$

where $\varepsilon_k$ is the second component of the vector $Z_k$. Therefore for large enough $k$ we may consider that

$$Z_k - \varepsilon_k Y \approx X,$$

where $X$ and $Y$ are the limit vectors of the stabilizing process, i.e. we

---

† Bauer (7).

may consider the vector $Z_k$ to fall in the subspace spanned by the vectors $X$ and $Y$.

We take the vectors $Z_k$ and $Y$ for a basis of this subspace. We shall consider how matrix $A$ operates on this basis. For all $k$ we have the relationships

$$AZ_k = \rho_k Z_{k+1},$$
$$AY_k = \sigma_k Z_{k+1} + \tau_k Y_{k+1}.$$

The coefficients $\rho_k$, $\sigma_k$, and $\tau_k$ are easily determined by comparing the first two components in these relationships.

For a sufficiently large choice of $k$ we may replace the vectors $Y_k$ and $Y_{k+1}$ by $Y$, the vector $Z_{k+1}$ by $Z_k + (\varepsilon_{k+1} - \varepsilon_k)Y$ so that

$$AZ_k = \rho_k Z_k + \rho_k(\varepsilon_{k+1} - \varepsilon_k)Y,$$
$$AY = \sigma_k Z_k + [\tau_k + \sigma_k(\varepsilon_{k+1} - \varepsilon_k)]Y. \tag{10}$$

Consequently the eigenvalues of matrix $A$ subject to determination are equal to the eigenvalues of the matrix

$$M_k = \begin{bmatrix} \rho_k & \sigma_k \\ \rho_k(\varepsilon_{k+1} - \varepsilon_k) & \tau_k + \sigma_k(\varepsilon_{k+1} - \varepsilon_k) \end{bmatrix}, \tag{11}$$

and the eigenvectors (or eigenvector and root vector) have coordinates in the basis $Z_k$ and $Y$ equal to the components of the eigenvectors (or eigenvector and root vector) of the matrix $M_k$.

It is easy to compute that the characteristic polynomial of matrix $M_k$ is

$$t^2 - [\rho_k + \tau_k + \sigma_k(\varepsilon_{k+1} - \varepsilon_k)]t + \rho_k \tau_k. \tag{12}$$

We observe that if not only the second column but also the first is stabilized in the process, then the matrix $M_k$ is stabilized and the limit matrix $M$ has the triangular form

$$\begin{bmatrix} \rho & \sigma \\ 0 & \tau \end{bmatrix}$$

**3. Non-stabilizing iterations.** Here $V_0$ and $W_0$ are certain initial vectors and the vector sequences are constructed from the formulas

$$V_{k+1} = AV_k,$$
$$W_{k+1} = AW_k - \eta_k V_{k+1}, \tag{13}$$

where $\eta_k$ is the ratio of the first two components of the vectors $AW_k$ and $V_{k+1}$, so that the first component of the vector $W_{k+1}$ is equal to zero for $k \geq 0$.

It is clear that the vectors $V_k$ and $W_k$ differ only in normalization from the vectors $Z_k$ and $Y_k$ of the previous modification. That is,

$$V_k = a_k Z_k,$$
$$W_k = c_k Y_k, \tag{14}$$

where $a_k$ is the first component of the vector $V_k$ and $c_k$ is the second component of the vector $W_k$. Therefore the vectors $V_k$ and $W_k$ may also be taken for a basis of the invariant subspace spanned by the vectors $U_1$ and $U_2$ for sufficiently large $k$.

It is easy to deduce the connection between the numbers $\rho_k$, $\sigma_k$, and $\tau_k$ of the previous modification and the first components of the vectors $V_k$, $V_{k+1}$, $W_k$, and $W_{k+1}$. Namely,

$$\rho_k = \frac{a_{k+1}}{a_k},$$

$$\tau_k = \frac{c_{k+1}}{c_k}, \tag{15}$$

$$\sigma_k = \eta_k \frac{a_{k+1}}{c_k}.$$

Finally, $\varepsilon_k = b_k/a_k$ where $b_k$ is the second component of the vector $V_k$.

Relationships (10) for vectors $Z_k$ and $Y_k$ which hold for large enough $k$ are transformed into the relationships

$$A V_k = \frac{a_{k+1}}{a_k} V_k + \frac{1}{c_k}\left(b_{k+1} - b_k \frac{a_{k+1}}{a_k}\right) W_k,$$

$$A W_k = \eta_k \frac{a_{k+1}}{a_k} V_k + \left[\frac{c_{k+1}}{c_k} + \frac{\eta_k}{c_k}\left(b_{k+1} - b_k \frac{a_{k+1}}{a_k}\right)\right] W_k,$$

Therefore the desired eigenvalues of matrix $A$ are equal to the corresponding eigenvalues of the matrix

$$N_k = \begin{bmatrix} \dfrac{a_{k+1}}{a_k} & \eta_k \dfrac{a_{k+1}}{a_k} \\ \dfrac{1}{c_k}\left(b_{k+1} - b_k \dfrac{a_{k+1}}{a_k}\right) & \dfrac{c_{k+1}}{c_k} + \dfrac{\eta_k}{c_k}\left(b_{k+1} - b_k \dfrac{a_{k+1}}{a_k}\right) \end{bmatrix}, \tag{16}$$

i.e. to the roots of the quadratic equation

$$t^2 - \left[\frac{a_{k+1}}{a_k} + \frac{c_{k+1}}{c_k} + \frac{\eta_k}{c_k}\left(b_{k+1} - b_k \frac{a_{k+1}}{a_k}\right)\right]t + \frac{a_{k+1}c_{k+1}}{a_k c_k} = 0. \tag{17}$$

The eigenvectors are determined analogously to the previous modification.

In the modification considered, on satisfying the conditions of theorem 57.1, the vectors $W_k$ converge only directionally and the vectors $V_k$ may converge directionally at a rapid or slow rate or they may not converge at all.

Stabilization of the constant in equation (17) may serve as the criterion which permits us to complete the process.

In the case where $|\lambda_1| > |\lambda_2|$ the vectors $V_k$ converge directionally so that $b_{k+1} \approx b_k a_{k+1}/a_k$ and therefore we may take $\lambda_1 = a_{k+1}/a_k$, $\lambda_2 = c_{k+1}/c_k$ for the eigenvalues.

The modification indicated is a bit more simple than the previous one in forming the vector sequences but is less convenient in the concluding operations. It is therefore advisable, having started the process in this modification, to pass then to the former one by a corresponding normalization.

As an example we determine the first two eigenvalues for Leverrier's matrix and the eigenvectors belonging to them using the third modification of the stepped power method.

We take $V_0 = (1, 1, 1, 1)'$, $W_0 = (0, 1, 1, 1)'$.  Then

| $V_1 = A V_0$ | $A W_0$ | $W_1$ | $V_{17}$ | $W_{17}$ |
|---|---|---|---|---|
| −3.208074 | 2.301808 | −0.00000001 | −0.10419891·10⁹ | 0 |
| −5.753172 | −6.041037 | −10.168965 | 0.72583945·10⁹ | −0.15619598·10⁸ |
| −8.384229 | −8.433328 | −14.449051 | −0.37864184·10⁹ | 0.57277646·10⁸ |
| −15.922752 | −15.928987 | −27.353636 | −0.11545164·10¹¹ | −0.14996613·10¹⁰ |
| −33.268227 | −28.101544 |  |  |  |
| −0.71750465 |  |  | −0.91870763·10 |  |

| $V_{18}$ | $W_{18}$ | $V_{19}$ | $W_{19}$ |
|---|---|---|---|
| 0.16696157·10¹⁰ | 0 | −0.26535495·10¹¹ | 0 |
| −0.11444017·10¹¹ | 0.29882041·10⁹ | 0.17832344·10¹² | −0.57619870·10¹⁰ |
| 0.53854629·10¹⁰ | −0.10954099·10¹⁰ | −0.72561121·10¹¹ | 0.21119162·10¹¹ |
| 0.20281721·10¹² | 0.28675616·10¹¹ | −0.35643659·10¹³ | −0.55281857·10¹² |
| −0.10903620·10 |  | −0.13126160·10 |  |

The eigenvalues of the matrix $N_{17}$ are determined from the equation

$$t^2 + 35.015145t + 306.38882 = 0 ,$$

the eigenvalues of the matrix $N_{18}$ from the equation

$$t^2 + 35.015455t + 306.39423 = 0.$$

We see that the process has been sufficiently stabilized.

For eigenvalues we obtain

$$\lambda_1 = -17.8629 , \quad \lambda_2 = -17.1522 \quad \text{(for } k = 17) ,$$

or

$$\lambda_1 = -17.8631 , \quad \lambda_2 = -17.1523 \quad \text{(for } k = 18) .$$

To determine the eigenvectors belonging to these eigenvalues we first find (for $k = 18$) the eigenvectors of the matrix $N_k$. It is not hard to figure out that $X_1 = (1, -9.4429)'$, $X_2 = (1, -6.0357)'$. We now have

$$U_1 = V_{18} - 9.4429\,W_{18} \qquad \tilde{U}_1 \qquad U_2 = V_{18} - 6.0357\,W_{18} \qquad \tilde{U}_2$$

| | | | |
|---|---|---|---|
| $0.166962 \cdot 10^{10}$ | $-0.019861$ | $0.166962 \cdot 10^{10}$ | $0.032937$ |
| $-0.142657 \cdot 10^{11}$ | $0.169702$ | $-0.132477 \cdot 10^{11}$ | $-0.261341$ |
| $0.157293 \cdot 10^{11}$ | $-0.187112$ | $0.119970 \cdot 10^{11}$ | $0.236668$ |
| $-0.679638 \cdot 10^{11}$ | $0.808482$ | $0.297402 \cdot 10^{11}$ | $0.586694.$ |

All three modifications of the two-step power method may be generalized in the form of an $r$-step power method, permitting us to determine, generally speaking, $r$ eigenvalues and their eigenvectors for a canonical basis. The first modification gives a stabilizing process if $|\lambda_r| > |\lambda_{r+1}|$. Application of the first modification lets us construct an invariant subspace spanned by the vectors $U_1, \cdots, U_r$ of the canonical basis corresponding to the eigenvalues $\lambda_1, \cdots, \lambda_r$. Thus applying the first modification brings the special problem for an $n$th-order matrix to that of solving the complete problem for a matrix of order $r$. For $r = n$ the first modification obviously loses its interest. Therefore it may be applied only for $r$ significantly less than $n$.

The case is somewhat different with the second modification and with the third, which differs little from it. Here for $r = n$ we pass to the so-called triangular power method which solves the complete eigenvalue problem for a matrix and which will be presented by us in Section 78, Chapter VIII. For large $r$, in particular for $r = n$, the method gives good results only in the case where all eigenvalues to be determined are real and distinct. The presence of complex roots and eigenvalues belonging to higher-order Jordan boxes makes the process much more difficult, a fact of which we were convinced even when considering the case $r = 2$.

## 58. THE λ-DIFFERENCE METHOD

The $\lambda$-difference method makes it possible, knowing the eigenvalue $\lambda_1$ largest in modulus, to find the next eigenvalue $\lambda_2$ and the eigenvector belonging to it under the condition that

$$|\lambda_1| > |\lambda_2| > |\lambda_3|.$$

The method goes as follows.

Let the sequence

$$y_1, y_2, \cdots, y_m, \cdots, y_k, \cdots \qquad (1)$$

be computed and let $\lambda_1 \approx y_{k+1}/y_k$ be determined from it. Here $y_k$ is any component of the vector $Y_k = A^k Y_0$.

We form the difference

$$\Delta y_k = y_{k+1} - \lambda_1 y_k = c_2(\lambda_2 - \lambda_1)\lambda_2^k + \cdots + c_n(\lambda_n - \lambda_1)\lambda_n^k. \qquad (2)$$

Since $\lambda_2$ is greater in modulus than all the remaining eigenvalues, and $c_2 \neq 0$, then the first element of this difference will predominate and we may determine $\lambda_2$ in the same way as we determined $\lambda_1$. Namely,

$$\lambda_2 \approx \frac{y_{k+1} - \lambda_1 y_k}{y_k - \lambda_1 y_{k-1}} \tag{3}$$

However with such a determination of $\lambda_2$ we run up against disappearing significant figures, since in the numerator and denominator of ratio (3) we have to compute values close to each other. In practice it is advisable, finding $\lambda_1$ from the ratio of $y_{k+1}$ and $y_k$, to *go back* and determine $\lambda_2$ from the ratio

$$\lambda_2 \approx \frac{y_{m+1} - \lambda_1 y_m}{y_m - \lambda_1 y_{m-1}} \tag{4}$$

taking as $m$ the smallest of the numbers for which the predominance of $\lambda_2$ over the following eigenvalues already begins to have an effect. The method indicated gives fairly crude values for $\lambda_2$ although these are often sufficient in practice. It is theoretically possible, using an analogous process, to determine the following eigenvalues as well.

It is obvious that to determine the second eigenvector it is necessary to carry out the process of forming the $\lambda$-difference in the sequence $AY_0$, $A^2Y_0, \cdots, A^kY_0, \cdots$. In fact the difference

$$A^{k+1}Y_0 - \lambda_1 A^k Y_0 = a_2(\lambda_2 - \lambda_1)\lambda_2^k X_2 + \cdots + a_n(\lambda_n - \lambda_1)\lambda_n^k X_n$$

shows that the components of the vector $X_2$ may be found in the same way as we determined the components of the vector $X_1$ in Section 53.

For an example we shall determine the second eigenvalue of matrix (7), Section 54.

As $\lambda_1$ we shall take both the value obtained directly from the ratios of the components for the 20th and 19th iterations ($\lambda_1 \approx 1.001$) and the more accurate value using the scalar product ($\lambda_1 \approx 1.000000$).

Taking the first component of the vector $A^k Y_0$ for $y_k$ we obtain (for $\lambda_1 \approx 1.000000$), considering the 17th, 18th and 19th iterations ($m = 18$):

$$\lambda_2 \approx \frac{y_{m+1} - \lambda_1 y_m}{y_m - \lambda_1 y_{m-1}} \approx \frac{4.677943 - 4.676009}{4.676009 - 4.673110} = \frac{0.001934}{0.002899} = 0.6671 \ .$$

Analogously, taking the fourth component of the vector $A^k Y_0$ for $y_k$ we obtain

$$\lambda_2 \approx \frac{-0.009131}{-0.013667} \approx 0.6681 \ .$$

Thus knowing a sufficiently accurate value for $\lambda_1$ made it possible to determine $\lambda_2$ with comparative accuracy as well (to three decimal places). The exact value is $\lambda_2 = 0.666 \cdots$

If we take the cruder value $\lambda_1 \approx 1.001$ for $\lambda_1$, then by computing the previous ratio we run into the described phenomenon of disappearing significant figures. In this case we need to take a number significantly less than 20 for $m$.

Thus, considering the 9th, 10th and 11th iterations of the vector $A^k Y_0$,

$$
\begin{array}{ccc}
A^9 V_0 & A^{10} Y_0 & A^{11} Y_0 \\
4.4665336 & 4.5365193 & 4.5841480 \\
7.1243407 & 7.5407651 & 7.8281626 \\
0.0699857 & 0.0476287 & 0.0321941 \\
-7.4707539 & -7.7678185 & -7.9777169 \\
\hline
4.1901061 & 4.3570946 & 4.4667878
\end{array}
$$

we obtain, computing the values $y_{m+1} - \lambda_1 y_m$:

$$
\begin{array}{cc}
m = 9 & m = 10 \\
0.06552 & 0.04309 \\
0.40930 & 0.27986 \\
-0.02242 & -0.01548 \\
-0.28959 & -0.20213.
\end{array}
$$

The ratios of these values give for $\lambda_2$ the values

$$0.658, \qquad 0.684, \qquad 0.690, \qquad 0.698.$$

Thus knowing a very crude value for $\lambda_1$ allows us at any rate, using the earlier iterations, to obtain for $\lambda_2$ a value accurate to three units of the second place.

We may obtain approximate values for the components of the second eigenvector as corresponding ratios of components of the vector $A^{m+1} Y_0 - \lambda_1 A^m Y_0$. Taking $m = 9$ we obtain, using the components of the vector $A^{10} Y_0 - \lambda_1 A^9 Y_0$ computed earlier, the following values for the components of the eigenvector (after normalizing):

$$1.00; \qquad 6.49; \qquad -0.36; \qquad -4.69.$$

For the matrix being considered the second eigenvector has components

$$1, \frac{31}{5} = 6.2, \; -\frac{1}{3} = -0.333 \cdots, \; -\frac{71}{15} = \; 4.733 \cdots,$$

so that the computational results agree well enough with the exact values.

The $\lambda$-difference method may be generalized in the following manner. Let $\lambda_1$ and $\lambda_2$ be known; it is required to determine $\lambda_3$, where it is likewise known that $|\lambda_3| > |\lambda_4|$. In this case we form the "second $\lambda_1\lambda_2$-difference", i.e.

$$
\begin{aligned}
\Delta^2 y_k &= (y_{k+2} - \lambda_1 y_{k+1}) - \lambda_2(y_{k+1} - \lambda_1 y_k) \\
&= y_{k+2} - (\lambda_1 + \lambda_2)y_{k+1} + \lambda_1\lambda_2 y_k \, .
\end{aligned}
$$

It is easy to see that

$$\Delta^2 y_k = c_3(\lambda_3 - \lambda_1)(\lambda_3 - \lambda_2)\lambda_3^k + \cdots + c_n(\lambda_n - \lambda_1)(\lambda_n - \lambda_2)\lambda_n^k$$

and for large enough $k$

$$\lambda_3 \approx \frac{\Delta^2 y_{k+1}}{\Delta^2 y_k}$$

The loss of significant digits in forming the second $\lambda_1\lambda_2$-difference will be even more marked than in using the first $\lambda_1$-difference so that in the case of real $\lambda_1$ and $\lambda_2$ the method is hardly applicable. Relatively good results are obtained only in the case where $\lambda_1$ and $\lambda_2$ form a complex pair. In this case it is advisable to look for the second $\lambda_1\lambda_2$-difference in the form

$$\Delta^2 y_k = y_{k+2} + p y_{k+1} + q y_k ,$$

where $p$ and $q$ are the coefficients of the quadratic trinomial $(t - \lambda_1)(t - \lambda_2)$ and may be computed from formulas (19), Section 53.

Moreover if a complex pair follows the simple real root $\lambda_1$ or the complex pair $\lambda_1$ and $\lambda_2$, then by using $\lambda$-differences instead of iterations in formulas (19), Section 53, we may obtain the coefficients of the quadratic trinomial $(t - \lambda_2)(t - \lambda_3)$ or $(t - \lambda_3)(t - \lambda_4)$ respectively.


## 59. THE METHOD OF EXHAUSTION

The methods of exhaustion and reduction (Section 60) make it possible to determine the next eigenvalue and its associated eigenvector after preceding eigenvalues and their eigenvectors have been found with a sufficient degree of accuracy. Unlike the methods of Sections 56 and 58, applied toward the same goal, applying the methods of exhaustion and reduction does not involve a loss of accuracy. Therefore these methods make it possible, in particular, to solve the complete eigenvalue problem using a sequence of solutions to special problems.

For simplicity of presentation we shall assume that all eigenvalues of a matrix $A$ are real.

To carry out one step of the exhaustion method for a matrix $A$ it is necessary to know beforehand not only some eigenvalue $\lambda_1$ (not necessarily largest in modulus) and its associated eigenvector $U_1 = (u_1, u_2, \cdots, u_n)'$ but also the eigenvector $V_1 = (v_1, v_2, \cdots, v_n)'$ of the matrix $A'$ belonging to the eigenvalue $\lambda_1$. We shall assume, besides this, that all eigenvalues of matrix $A$ are pairwise distinct so that there exist bases $U_1, \cdots, U_n$ and $V_1, \cdots, V_n$ consisting of eigenvectors of the matrices $A$ and $A'$ and satisfying the normalization conditions $(U_i, V_i) = 1$ for $i = 1, 2, \cdots, n$.

We form the matrix product $U_1 V_1'$ where $V_1'$ is a row formed from components of the vector $V_1$.

This will be the square matrix

$$U_1 V_1' = \begin{bmatrix} u_1v_1 & u_1v_2 & \cdots & u_1v_n \\ u_2v_1 & u_2v_2 & \cdots & u_2v_n \\ \cdot & \cdot & \cdots & \cdot \\ u_nv_1 & u_nv_2 & \cdots & u_nv_n \end{bmatrix}. \tag{1}$$

We observe that the matrix product $V_1' U_1$ is equal to the number one, since it is equal to the scalar product $(U_1, V_1)$.

Furthermore we form the matrix

$$A_1 = A - \lambda_1 U_1 V_1' .$$

We shall prove that the matrix $A_1$ possesses the same eigenvalues and eigenvectors as matrix $A$, excluding the first eigenvalue, in place of which there is an eigenvalue equal to zero. In fact,

$$A_1 U_1 = A U_1 - \lambda_1 (U_1 V_1')U_1 = A U_1 - \lambda_1 U_1 (V_1' U_1) = A U_1 - \lambda_1 U_1 = 0$$
$$A U_i = A U_i - \lambda_1 (U_1 V_1')U_i = A U_i - \lambda_1 U_1 (V_1' U_i) = \lambda_i U_i ,$$

since $V_1' U_1 = 1$, $V_1' U_i = (V_1, U_i) = 0$ on the basis of the orthogonal properties of the vectors $U_1, U_2, \cdots, U_n$ and $V_1, V_2, \cdots, V_n$.

The indicated property of matrix $A$ makes it possible, starting from the vector sequence $A_1 Y_0, \cdots, A_1^m Y_0, \cdots$, to determine $\lambda_2$ and $U_2$ in the same way as we determined $\lambda_1$ and $U_1$ from the sequence $A Y_0, \cdots, A^k Y_0, \cdots$, since the eigenvalue $\lambda_2$ will be the first eigenvalue for matrix $A_1$. We call this the *exhaustion process*.

We shall show that

$$A_1^m Y_0 = A^m Y_0 - \lambda_1^m U_1 V_1' Y_0 , \tag{2}$$

i.e. that in practical applications of the indicated process there is no need to compute the matrix $A_1$ explicitly and to form the vector sequence $A_1 Y_0, \cdots, A_1^m Y_0, \cdots$ but it is sufficient to compute only two adjacent vectors $A_1^{m+1} Y_0$ and $A_1^m Y_0$ by formula (2).

To establish equalities (2) we shall introduce the so-called bilinear expansion of the matrix $A$.

On the basis of the orthogonal properties of the system of eigenvectors for matrix $A$ and its transpose we may write the matrix equality

$$E = U_1 V_1' + U_2 V_2' + \cdots + U_n V_n' .$$

Multiplying the members of this equality on the left by $A$ and replacing $A U_i$ by $\lambda_i U_i (i = 1, 2, \cdots, n)$ we obtain that

$$A - \lambda_1 U_1 V_1' + \lambda_2 U_2 V_2' + \cdots + \lambda_n U_n V_n' .$$

The exhaustion process eliminates the first element in this expansion so that

$$A_1 = \lambda_2 U_2 V_2' + \cdots + \lambda_n U_n V_n' .$$

Moreover

$$A^m = \lambda_1^m U_1 V_1' + \lambda_2^m U_2 V_2' + \cdots + \lambda_n^m U_n V_n' .$$

Analogously,

$$A_1^m = \lambda_2^m U_2 V_2' + \cdots + \lambda_n^m U_n V_n' .$$

Therefore

$$A_1^m = A^m - \lambda_1^m U_1 V_1' ,$$

from which follows equality (2).

Thus it is possible to apply the exhaustion method in two variations. In one of them it is necessary to compute the vector $U_1 V_1' Y_0$, to form the vectors $A_1^{m+1} Y_0$ and $A_1^m Y_0$ from formula (2) and then to determine $\lambda_2$ and $U_2$ in the usual power method way.

In this variation a decisive loss of significant figures occurs and we must therefore take for $m$ a number which is significantly less than the number of iterations applied to determine the number $\lambda_1$ and components of the eigenvectors $U_1$ and $V_1$. In using this variation the accuracy obtained is not high.

The second variation consists of actually constructing the matrix $A_1$ and computing iterations using $A_1$. This variation requires a large volume of computations but guarantees significantly better results as far as accuracy is concerned.

In both variations it is possible to apply methods for improving convergence.

*Example.* We again consider matrix (7) of Section 54.

The method of exhaustion for determining the second eigenvalue requires a knowledge of both the first eigenvalue and the eigenvectors belonging to it for both matrix $A$ and matrix $A'$. Therefore in using this method it is necessary to compute, along with the sequence of iterations $A^k Y_0$, the iterative sequence $A'^k Y_0$. Thus in determining $\lambda_1$ we may always improve the accuracy of the value for $\lambda_1$ using scalar product method.

In our example, using twenty iterations of the vector $Y_0 = (1, 1, 1, 1)'$ with matrix $A$ and matrix $A'$, we obtained for $\lambda_1$ the value $\lambda_1 = 1.000000$ (cf. paragraph 1, Section 54).

For the components of the eigenvectors for matrices $A$ and $A'$ we obtain, normalizing the components of the vectors $A^{20} Y_0$ and $A'^{20} Y_0$, the values:

$$
\begin{array}{ll}
1.00000, & 1.00000, \\
1.79757, & -0.00027, \\
0.00018, & 5.01676, \\
-1.79838, & -0.14256.
\end{array}
$$

The exact values for the components of the first eigenvector of matrix $A$ are 1, 1.8, 0, and $-1.8$. Following the theory we must first of all normalize the vectors $U_1$ and $V_1$ so that $(U_1, V_1) = 1$. Computing the normalization factor we obtain $c = 0.795678$. Thus for the components of the first eigenvectors of matrices $A$ and $A'$ we obtain the values:

$$
\begin{array}{ll}
1.00000, & 0.79568, \\
1.79757, & -0.00021, \\
0.00018, & 3.99173, \\
-1.79838, & -0.11343.
\end{array}
$$

We now form the matrix product $U_1 V_1'$. Namely:

$$
U_1 V_1' = \begin{bmatrix}
0.79568 & -0.00021 & 3.99173 & -0.11343 \\
1.43029 & -0.00038 & 7.17541 & -0.20390 \\
0.00014 & 0 & 0.00072 & -0.00002 \\
-1.43093 & 0.00038 & -7.17865 & 0.20399
\end{bmatrix}.
$$

Furthermore we form the matrix $A_1$,

$$A_1 = A - \lambda_1 U_1 V_1' = \begin{bmatrix} 0.20432 & 0.00021 & -2.99173 & 0.11343 \\ -0.43029 & 0.77816 & -6.84208 & 0.53723 \\ -0.00014 & -0.02525 & 0.55484 & -0.02523 \\ 1.43093 & -0.88927 & -1.46579 & -0.09288 \end{bmatrix},$$

and form iterations of the vector $Y_0 = (1, 1, 1, 1)'$ with this matrix. We carry out the 17th and 18th iterations with matrix $A_1$.

| $A_1^{17} Y_0$ | $A_1^{18} Y_0$ |
|---|---|
| −0.00869 | −0.00580 |
| −0.05388 | −0.03597 |
| 0.00290 | 0.00193 |
| 0.04107 | 0.02741 |
| −0.01861 | −0.01242. |

The ratios of components of the 17th and 18th iterations give for $\lambda_2$ the values:

$$0.667, \quad 0.668, \quad 0.666, \quad 0.667.$$

From the 18th iteration we obtain $U_2 = (1.00, 6.20, -0.333, -4.73)'$ by normalizing to a unit first component.

We see that both the second eigenvalue $\lambda_2$ itself and the components of the second eigenvector are determined by the method of exhaustion more accurately than by the $\lambda$-difference method. However this method requires much additional work in the case of a non-symmetric matrix. In the case of a symmetric matrix the method of exhaustion may be recommended.

In computing the second eigenvalue and the components of the second eigenvector we may use the modification described for the exhaustion method for which the vector $A_1^k Y_0$ is computed without iteration with matrix $A_1$ by formula (2). Computing we have

$$U_1 V_1' Y_0 - (1.67377, 8.40142, 0.00084, -8.40521)'.$$

We now compute $A_1^k Y_0$ from formula (2) for $k = 9$ and $k = 10$. This gives:

| $A_1^9 Y_0$ | $A_1^{10} Y_0$ |
|---|---|
| −0.20724 | −0.13725 |
| −1.27708 | −0.86065 |
| 0.06915 | 0.04679 |
| 0.93446 | 0.63739. |

The component ratios are equal to:

$$0.662, \quad 0.674, \quad 0.677, \quad 0.682$$

so that $\lambda_2 \approx 0.67$ to within $10^{-2}$. Moreover, normalizing $A_1^{10} Y_0$, we obtain $U_2 = (1.00, 6.27, -0.34, -4.64)'$.

The method of exhaustion may be applied with hardly any change to matrices with complex elements and to real matrices with complex eigen-values. To substantiate this we need to consider in place of the rows $V_i'$ the rows $V_i^*$ formed from complex numbers conjugate to components of the eigenvectors $V_i$ of the matrix $A^*$ belonging to the eigenvalues $\overline{\lambda}_i$.

To apply the exhaustion method in the case of a real matrix for which a pair of conjugate complex values with eigenvectors corresponding to them is defined we must carry out the exhaustion process according to the formula

$$A_1 = A - \lambda_1 U_1 V_1^* - \overline{\lambda_1 U_1 V_1^*} = A - 2\mathrm{Re}(\lambda_1 U_1 V_1^*) .$$

This permits us to remain with the class of real matrices.


## 60.   THE METHOD OF REDUCTION

Let the first eigenvalue $\lambda_1$ and the eigenvector $U_1 = (u_1, \cdots, u_n)'$ belonging to it be computed for a matrix $A$. We consider the matrix

$$P = \begin{bmatrix} u_1 & 0 \cdots 0 \\ u_2 & 1 \cdots 0 \\ \cdot & \cdot \cdot \cdot \cdot \cdot \\ u_n & 0 \cdots 1 \end{bmatrix} .$$

It is not hard to verify that

$$P^{-1} = \begin{bmatrix} \dfrac{1}{u_1} & 0 \cdots 0 \\ -\dfrac{u_2}{u_1} & 1 \cdots 0 \\ \cdot & \cdot \cdot \cdot \cdot \cdot \\ -\dfrac{u_n}{u_1} & 0 \cdots 1 \end{bmatrix} .$$

The matrix $P^{-1}AP$ is similar to matrix $A$ and therefore the eigenvalues of both matrices are the same.

But

$$P^{-1}AP = \begin{bmatrix} \lambda_1 & \dfrac{a_{12}}{u_1} & \cdots & \dfrac{a_{1n}}{u_1} \\ 0 & a_{22} - \dfrac{u_2}{u_1}a_{12} & \cdots & a_{2n} - \dfrac{u_2}{u_1}a_{1n} \\ \cdot & \cdot \cdot \cdot \cdot \cdot \cdot \cdot \cdot \cdot \cdot \cdot \cdot \\ 0 & a_{n2} - \dfrac{u_n}{u_1}a_{12} & \cdots & a_{nn} - \dfrac{u_n}{u_1}a_{1n} \end{bmatrix} = \begin{bmatrix} \lambda_1 & b_{12} \cdots b_{1n} \\ 0 & \\ \cdot & B \\ \cdot & \\ 0 & \end{bmatrix} .$$

Thus

$$| P^{-1}AP - tE | = (\lambda_1 - t) | B - tE | .$$

Consequently the eigenvalue for matrix $A$, excepting $\lambda_1$, will be the eigen-values of the $(n-1)$th-order matrix $B$. If we normalize the vector $U_1$ so that $u_1 = 1$, then we shall have

$$B = \begin{bmatrix} a_{22} - u_2 a_{12} & \cdots & a_{2n} - u_2 a_{1n} \\ \cdots & \cdots & \cdots \\ a_{n2} - u_n a_{12} & \cdots & a_{nn} - u_n a_{1n} \end{bmatrix}.$$

To find $\lambda_2$ we obviously need to construct the sequence $BY_0, \cdots, B^k Y_0$ and find $\lambda_2$ as ratio of any components of the vectors constructed.

Furthermore, let $Z$ be a certain eigenvector of matrix $B$ corresponding to the eigenvalue $\lambda$. Then the matrix $P^{-1}AP$ will have the eigenvector $\begin{bmatrix} z_1 \\ Z \end{bmatrix}$. We shall find $z_1$.

We have

$$\begin{bmatrix} \lambda_1 & a_{12} & \cdots & a_{1n} \\ 0 & & & \\ \vdots & & B & \\ 0 & & & \end{bmatrix} \begin{bmatrix} z_1 \\ Z \end{bmatrix} = \begin{bmatrix} \lambda_1 z_1 + a_{12} z_2 + \cdots + a_{1n} z_n \\ BZ \end{bmatrix} = \lambda \begin{bmatrix} z_1 \\ Z \end{bmatrix},$$

Equating the first components in this vector equation we have

$$\lambda_1 z_1 + a_{12} z_2 + \cdots + a_{1n} z_n = \lambda z_1$$

from which

$$z_1 = \frac{a_{12} z_2 + \cdots + a_{1n} z_n}{\lambda - \lambda_1}.$$

Finally, the eigenvector $U$ of matrix $A$ is determined by the formula

$$U = P \begin{bmatrix} z_1 \\ Z \end{bmatrix} = \begin{bmatrix} z_1 \\ u_2 z_1 + z_2 \\ \cdots \\ u_n z_1 + z_n \end{bmatrix}$$

*Example.* We shall determine the second eigenvalue and the components of its eigenvector for matrix (7) of Section 54. As we have seen, the first eigenvalue of this matrix was determined as $\lambda_1 \approx 1.001$ (paragraph 1, Section 54) and the components of its eigenvector as 1, 1.79757, 0.00018, $-1.79838$ (Section 59).

To determine $\lambda_2$ we compute the matrix

$$B = \begin{bmatrix} 0.77778 & -1.46424 & 0.33333 \\ -0.02525 & 0.55538 & -0.02525 \\ -0.88889 & -6.84606 & 0.11111 \end{bmatrix}.$$

We then form iterations of the vector $Y_0 = (1, 1, 1)'$ with matrix $B$. We present the results for the 15th and 16th iterations:

| $B^{15} Y_0$ | $B^{16} Y_0$ |
|---|---|
| $-0.09565$ | $-0.06388$ |
| $0.00725$ | $0.00484$ |
| $0.06339$ | $0.04243$ |
| $-0.02502$ | $-0.01661.$ |

We shall determine $\lambda_2$ from the ratio of components of the 16th and 15th iterations. This gives:

$$0.668, \qquad 0.668, \qquad 0.669.$$

Moreover we determine the components of the second eigenvector, assuming $Z = B^{16} Y_0$. For this we first determine

$$z_1 = \frac{a_{12}z_2 + a_{13}z_3 + a_{14}z_4}{\lambda_2 - \lambda_1} = \frac{0.00484}{-0.333} \approx -0.0145 \ .$$

We then compute the components $u_2$, $u_3$ and $u_4$ of the vector $U_2$: $u_2 = -0.0900$, $u_3 = 0.00484$, $u_4 = 0.0685$. Thus the components of the vector $U_2$ are $-0.0145$; $-0.0900$; $0.00484$; $0.0685$, so that after normalization $U_2 = (1, 6.21, -0.333, -4.72)'$.

## 61. COORDINATE RELAXATION

The method described in this paragraph is applied only in the case where a matrix $A$ is symmetric. To a certain degree it is close to iterative methods for solving linear systems based on the relaxation of one functional or another. In the case given the role of such a functional is played by the Rayleigh quotient.

We compute first of all the change in the quotient $\mu(X) = (AX, X)/(X,X)$ for a change in $X$ in a determined direction. Let

$$X' = X + \alpha Y , \tag{1}$$

where $Y$ is a certain fixed vector determining the direction of the change in $X$. Then

$$(AX', X') = (AX + \alpha A Y, X + \alpha Y) = (AX, X) + \alpha(AX, Y)$$
$$+ \alpha(A Y, X) + \alpha^2(A Y, Y) = (AX, X) + 2\alpha(AX, Y) + \alpha^2(A Y, Y)$$

and consequently

$$\mu(X') = \frac{(AX', X')}{(X', X')} = \frac{(AX, X) + 2\alpha(AX, Y) + \alpha^2(A Y, Y)}{(X, X) + 2\alpha(X, Y) + \alpha^2(Y, Y)} \ . \tag{2}$$

We now take the factor $\alpha$ such that the quotient $\mu(X')$ attains a largest value. Computing the derivative of $\mu(X')$ with respect to $\alpha$ we obtain

$$\frac{d\mu(X')}{d\alpha} = \frac{[2(AX, Y) + 2\alpha(A Y, Y)] [(X, X) + 2\alpha(X, Y) + \alpha^2(Y, Y)]}{[(X, X) + 2\alpha(X, Y) + \alpha^2(Y, Y)]^2}$$
$$- \frac{[2(X, Y) + 2\alpha(Y, Y)][(AX, X) + 2\alpha(AX, Y) + \alpha^2(A Y, Y)]}{[(X, X) + 2\alpha(X, Y) + \alpha^2(Y, Y)]^2}$$

and therefore for determining $\alpha$ we shall have the equation

$$[(AX, Y)(Y, Y) - (A Y, Y)(X, Y)]\alpha^2 + [(AX, X)(Y, Y) - (A Y, Y)(X, X)]\alpha$$
$$+ (AX, X)(X, Y) - (AX, Y)(X, X) = a\alpha^2 + b\alpha + c = 0 \ . \tag{3}$$

Investigation of equation (3) indicates that its roots are always real and distinct. Of course it may happen that the coefficient for $\alpha^2$ becomes zero. Then we must consider as a condition that one of the roots of the equation is equal to infinity. This will occur if one of the extremums is attained for the vector $Y$. A second exception is the case where $\mu(X')$ turns out to be independent of $\alpha$. In this case equation (3) becomes the identity $0 = 0$. The geometric meaning of what has been said is very simple. For clarity we assume that the matrix $A$ is positive definite; this does not destroy the generality, since any symmetric matrix may be made positive definite by adding a scalar matrix. Such a transformation does not bring about a change in the eigenvectors but all values for Rayleigh's quotient (in particular, all eigenvalues) are changed by a constant. Under this assumption the Rayleigh quotient $\mu(X) = (AX, X)/(X, X)$ is obviously $1/\rho^2$ where $\rho$ is the length of a vector coming from the coordinate origin in the direction of the vector $X$ until intersecting the ellipsoid $(AX, X) = 1$. The vectors $X' = X + \alpha Y$ lie in the plane (in a two-dimensional subspace) spanned by the vectors $X$ and $Y$. This plane intersects the ellipsoid $(AX, X) = 1$ along the ellipse while the Rayleigh quotient attains a maximum for the vector directed along the minor axis of this ellipse and a minimum for a vector directed along the major axis (cf. Fig. 3). It may happen that the vector

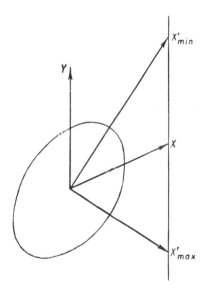

**FIGURE 3**

$Y$ is directed along one of the axes of this ellipse. In this case one of the roots of equation (3) becomes infinite. Finally it may happen that the ellipse degenerates into a circle. Then equation (3) becomes an identity and $\alpha$ may be taken arbitrarily. We now observe that if for certain $X$ and $Y$ a circular segment is not obtained and we make small enough deformations in the vectors $X$ and $Y$, then the direction of the elliptic axes changes as little as we wish.

Simple computation indicates that the extremal value $\mu'$ is equal to

$$\frac{\delta + a\alpha}{d} , \tag{4}$$

where $d = (X, X)(Y, Y) - (X, Y)^2$, $\delta = (AX, X)(Y, Y) - (AX, Y)(X, Y)$ and $a = (AX, Y)(Y, Y) - (AY, Y)(X, Y)$ is the coefficient for $\alpha^2$ in equation (3). Since $d > 0$, then it follows from (4) that to obtain the maximum $\mu'$ it is necessary to take the larger root of equation (3) if $a > 0$ and the smaller root of the equation if $a < 0$.

The *coordinate relaxation method* consists in taking one of the coordinate vectors $e_i$ for the vector $Y$ at each step of the process.

Let $X = (x_1, \cdots, x_n)'$. We introduce the notation

$$F = AX = (f_1, \cdots, f_n)' ,$$
$$p = (AX, X) , \tag{5}$$
$$q = (X, X) .$$

Then

$$(X, AY) = (X, Ae_i) = (AX, e_i) = f_i ,$$
$$(AY, Y) = (Ae_i, e_i) = a_{ii} ,$$
$$(Y, Y) = (e_i, e_i) = 1 , \tag{6}$$
$$(X, Y) = (X, e_i) = x_i .$$

The quadratic equation for determining $\alpha$ turns into the equation

$$\alpha^2[f_i - a_{ii}x_i] + \alpha[p - a_{ii}q] + px_i - f_iq = 0 . \tag{7}$$

Choosing the corresponding root we determine the following approximation

$$X' = X + \alpha e_i . \tag{8}$$

We have for this approximation

$$F' = AX' = AX + \alpha Ae_i = F + \alpha A_i , \tag{9}$$

where $A_i$ is the $i$th column of matrix $A$. Furthermore,

$$p' = (AX', X') = (F', X') = p + 2\alpha f_i + \alpha^2 a_{ii} ,$$
$$q' = (X', X') = q + 2\alpha x_i + \alpha^2 . \tag{10}$$

Thus the values $p'$ and $q'$ and the components of the vector $F'$ needed to carry out the following step are easily determined.

If we change the $j$th component, $j \neq i$, at the next step, then in the quadratic equation it is necessary to replace $a_{ii}$ by $a_{jj}$, $x_i$ by $x'_j$, $f_i$ by $f'_j$, $q$ by $q'$ and $p$ by $p'$.

The choice of numeration for the components being changed is immaterial. The simplest method is a cyclic alternation of indices. In view of the possible accumulation of rounding-off errors it is necessary from time to time to compute the values $q$, $p$ and $f_i$ $(i = 1, 2, \cdots, n)$ directly from the formulas which define them.

We also make note of the relationship

$$\mu' = \frac{p - f_i x_i + a\alpha}{q - x_i^2} = \frac{p'}{q'} , \qquad (11)$$

which may be used as a check on the computations.

Maximization of the Rayleigh quotient by changing the vector in a given direction may be accomplished slightly differently than was described above.

Essentially we need to find the maximum of the Rayleigh quotient $(AX', X')/(X', X')$ in the plane $X' = c_1 X + c_2 Y$ spanned by the given vectors $X$ and $Y$.

Assuming $X'$ to be a normalized vector $((X', X') = 1)$ we arrive at the problem of finding a maximum for the quadratic form

$$(AX, X)c_1^2 + 2(AX, Y)c_1 c_2 + (AY, Y)c_2^2 \qquad (12)$$

under the condition

$$(X, X)c_1^2 + 2(X, Y)c_1 c_2 + (Y, Y)c_2^2 = 1 . \qquad (13)$$

Solving this problem by means of Lagrange multipliers we arrive at the system of equations for determining the coefficients $c_1$ and $c_2$

$$[(AX, X) - \mu(X, X)]c_1 + [(AX, Y) - (X, Y)\mu]c_2 = 0 ,$$
$$[(AX, Y) - (X, Y)\mu]c_1 + [(AY, Y) - (Y, Y)\mu]c_2 = 0 , \qquad (14)$$

from which it follows that the Lagrange multiplier $\mu$ is a root of the quadratic equation

$$\begin{vmatrix} (AX, X) - (X, X)\mu & (AX, Y) - (X, Y)\mu \\ (AX, Y) - (X, Y)\mu & (AY, Y) - (Y, Y)\mu \end{vmatrix} = 0 . \qquad (15)$$

The larger root of this quadratic equation will give the value for the desired maximum $\mu'$; the coefficients $c_1$ and $c_2$ are determined from system (14) after substituting the resultant value $\mu'$ in it in place of $\mu$.

For the ratio of $c_1$ and $c_2$ we obtain

$$\alpha = \frac{c_2}{c_1} = \frac{(AX, X) - (X, X)\mu'}{(X, Y)\mu' - (AX, Y)} = \frac{(X, Y)\mu' - (AX, Y)}{(AY, Y) - (Y, Y)\mu'} . \qquad (16)$$

Since in the end we do not need to normalize the vector $X'$, we may let

$$X' = X + \alpha Y .$$

In the case of coordinate relaxation, when $Y = e_i$, we find using the previous notation that $\mu'$ is the larger root of the quadratic equation

$$(q - x_i^2)\mu^2 - (a_{ii}q + p - 2x_i f_i)\mu + a_{ii}p - f_i^2 = 0 \qquad (17)$$

and

$$X' = X + \alpha e_i , \qquad (18)$$

where

$$\alpha \frac{q\mu' - p}{f_i - x_i\mu'} = \frac{f_i - x_i\mu'}{\mu' - a_{ii}} \, . \tag{19}$$

Passing to the next step is accomplished just as in the previous scheme. A good control check is the satisfying of the equality $\mu' = p'/q'$.

The cyclic coordinate process turns out to be almost always convergent but we shall give the proof of convergence under certain limitations.

THEOREM 61.1. *Let a matrix* $A$ *be symmetric and let its largest eigenvalue* $\lambda_1$ *be simple. Let the initial vector* $X_0$ *be chosen such that* 1) $(AX_0, X_0)/$ $(X_0, X_0) > \lambda_2$, *where* $\lambda_2$ *is the second (in decreasing order) eigenvalue of matrix* $A$, *and that* 2) $(AX_0, X_0)/(X_0, X_0) > (Ae_i, e_i) = a_{ii}$ *for all* $i = 1, 2, \cdots, n$. *Under these conditions the successive approximations* $X_k$ *in the cyclic coordinate relaxation process converge directionally to the eigenvector belonging to the eigenvalue* $\lambda_1$.

*Proof.* We shall assume that at each step of the process the successive approximations are normalized to unit length and of two mutually opposite normalized vectors the one in a certain fixed hemisphere is chosen. We denote by $D$ the set of vectors of length one which satisfy conditions 1) and 2) of the theorem and which lie in the chosen hemisphere. From the theorem conditions it follows that the initial approximation belongs in the region $D$. All subsequent approximations will also belong in the region $D$, since the Rayleigh quotient does not increase during the process.

We consider in more detail an individual step in the cyclic coordinate process. Let $X$ be a certain vector of the region $D$. We denote by $\varphi_i(X)$ the normalized vector, taken in the chosen hemisphere, for which $(AX, X)$ attains a maximum in the subspace spanned by the vectors $X$ and $e_i$. On the basis of the second condition of the theorem $(AX, X) > (Ae_i, e_i)$ and therefore the vector $\varphi_i(X)$ is determined simply. From the geometric considerations introduced above it is obvious that the non-linear operator $\varphi_i$ is continuous in the region $D$.

In the new notation the process may be described in the following way:

$$X_1 = \varphi_1(X_0), \ X_2 = \varphi_2(X_1), \ \cdots, \ X_n = \varphi_n(X_{n-1})$$
$$X_{n+1} = \varphi_1(X_n), \ \cdots, \ X_{nk+j} = \varphi_j(X_{nk+j-1}), \ \cdots \qquad (j = 1, 2, \cdots, n).$$

We denote by $\mu_0, \mu_1, \cdots$ the values $(AX_0, X_0), (AX_1, X_1), \cdots$ of the functional $(AX, X)$. These values satisfy the inequalities $\mu_0 \leq \mu_1 \leq \mu_2 \leq \cdots$ and are bounded above by $\lambda_1$. Consequently there exists $\lim_{k \to \infty} \mu_k = \mu$.

We consider the sequence of vectors $X_1, X_{n+1}, X_{2n+1}, \cdots$. All the vectors of this sequence are found in the region $D$. On the basis of the boundedness and closure of the unit sphere it is possible to extract from this sequence the convergent sequence $X_{nk_j+1}$. Let $Y$ be the limit of this sequence. The vector $Y$ is found inside the region $D$, since $(AY, Y) = \mu > \mu_0 > \max (\lambda_2, (Ae_i, e_i))$. From the continuity of the operator $\varphi_2$ in the region $D$ we have

$$\tilde{Y} = \varphi_2(Y) = \lim_{j \to \infty} \varphi_2(X_{nk_j+1}) = \lim_{j \to \infty} X_{nk_j+2} \,.$$

Consequently $(A\tilde{Y}, \tilde{Y}) = \lim (AX_{nk_j+2}, X_{nk_j+2}) = \mu$. But by the definition of the operator $\varphi_2$ there is obtained with the vector $Y$ a maximum $(AZ, Z)$ in the plane spanned by $Y$ and $e_2$ and this maximum is equal to $\mu = (AY, Y) > (Ae_2, e_2)$. Consequently $\tilde{Y} = Y$. We prove in just the same way that $\varphi_3(Y) = Y, \cdots, \varphi_n(Y) = Y, \varphi_1(Y) = Y$ so that the vector $Y$ turns out to be stationary for all operators $\varphi_i(i = 1, 2, \cdots, n)$.

This means that by changing each separate coordinate of the vector $Y$ Rayleigh's quotient is not increased. Consequently all the partial derivatives of the Rayleigh quotient at the point $Y$ are equal to zero and thus $Y$ is an eigenvector. But on the basis of conditions 1) and 2) of the theorem this can be the only eigenvector corresponding to the largest eigenvalue.

Thus the natural limit point of the sequence $X_1, X_{n+1}, X_{2n+1}, \cdots$ is the eigenvector belonging to the largest eigenvalue.

The remaining sequences $X_i, X_{n+i}, X_{2n+i}, \cdots$ also converge to the same vector and therefore the whole sequence $X_1, X_2, \cdots$ does too. The theorem is proved.

*Note.* Condition 2) of the theorem is not essential, since it will generally be satisfied automatically after carrying out the first cycle of the process for any choice of the initial vector. But if we remove the first limitation, then in repeating the proof we obtain that every limit point of the sequence $X_0, X_1, X_2, \cdots$ is an eigenvector of matrix $A$. If we assume that the eigenvalues of matrix $A$ are distinct, then since the Rayleigh quotient is equal at all limit points we may conclude that the limit point is unique, i.e. the sequence $X_0, X_1, \cdots$ converges to a certain eigenvector. However this eigenvector does not have to belong to the largest eigenvalue; this may be illustrated by the following example. Let

$$A = \begin{bmatrix} 35 & -22 & 8 \\ -22 & 38 & 14 \\ 8 & 14 & 53 \end{bmatrix}.$$

If we take as the initial vector $X_0 = (2/3, -1/3, 2/3)'$, then it turns out that all subsequent approximations will be equal to $X_0$. It is easy to verify that $X_0$ is the eigenvector belonging to the eigenvalue $\lambda_2 = 54$. The largest eigenvalue is $\lambda_1 = 63$.

Closer analysis shows that the sequence $X_0, X_1, \cdots$ can converge to an eigenvector which does not correspond to the largest eigenvalue only if the sequence is stabilized, starting with a certain vector.

Using coordinate one-step relaxation to find the largest eigenvalue is generally unprofitable, since it is necessary to solve quadratic equation (7) at each little step during the process.

However we shall present an example (matrix (4), Section 51) which shows the course of the process, omitting the intermediate computations.

| | | | | |
|---|---|---|---|---|
| $X_0$ | 0.34 | 0.20 | 0.90 | 0.75 | 1.8300504 |
| $X_1$ | 0.59995498 | 0.41211685 | 0.76621642 | 0.70021179 | 2.2138031 |
| $X_2$ | 0.69781651 | 0.51354408 | 0.67966297 | 0.67841258 | 2.2977743 |
| $X_3$ | 0.73541175 | 0.56051428 | 0.62985366 | 0.67161815 | 2.3165637 |
| $X_4$ | 0.75062330 | 0.58274350 | 0.60231592 | 0.67072572 | 2.3211276 |
| $X_5$ | 0.75694325 | 0.59351423 | 0.58729670 | 0.67163815 | 2.3223053 |
| $X_6$ | 0.75958696 | 0.59882656 | 0.57913506 | 0.67281146 | 2.3226235 |
| $X_7$ | 0.76068465 | 0.60148118 | 0.57469926 | 0.67377537 | 2.3227126 |
| $X_8$ | 0.76113138 | 0.60282111 | 0.57228443 | 0.67445635 | 2.3227381 |
| $X_9$ | 0.76130659 | 0.60350310 | 0.57096708 | 0.67490251 | 2.3227456 |
| $X_{10}$ | 0.76137102 | 0.60386377 | 0.57024369 | 0.67517902 | 2.3227477 |
| $X_{11}$ | 0.76139227 | 0.60403891 | 0.56985092 | 0.67535101 | 2.3227485 |
| $\tilde{X}_{11}$ | 1.0000 | 0.7933 | 0.7484 | 0.8870 | |

In the last column are given the values $\mu(X_i)$.

*Group coordinate relaxation* consists of the following. A maximum (or minimum) of the quotient $(AX, X)/(X, X)$ is sought by *simultaneously* changing several coordinates of the preceding approximation. The chosen groups of coordinates are changed from step to step.

We describe one step of the process.[†] Let $X$ be a certain approximation to the eigenvector belonging to the largest eigenvalue $\lambda_1$ found in the previous $k$ steps. For definiteness we assume that in the step considered the coordinates being changed have lower indices $1, \cdots, r$. The subspace $Q$ is spanned by the vectors $X, e_1, \cdots, e_r$ and we shall look for a maximum $\mu(X')$ in this subspace. Let

$$X' = c_0 X + c_1 e_1 + \cdots + c_r e_r . \tag{20}$$

Then

$$(AX', X') = \sum_{i,j=0}^{r} \gamma_{ij} c_i c_j , \tag{21}$$

where

$$\begin{aligned} \gamma_{00} &= (AX, X) , \\ \gamma_{0i} &= (AX, e_i) = (X, Ae_i) = (X, A_i) , \\ \gamma_{ij} &= (Ae_i, e_j) = a_{ij} \quad (i, j = 1, \cdots, r) . \end{aligned} \tag{22}$$

Here we denote by $A_i$ the $i$th column of matrix $A$. In turn

$$(X', X') = \sum_{i,j=0}^{r} \beta_{ij} c_i c_j , \tag{23}$$

where

$$\begin{aligned} \beta_{00} &= (X, X) = x_1^2 + \cdots + x_n^2 , \\ \beta_{0i} &= (X, e_i) = x_i , \\ \beta_{ij} &= (e_i, e_j) = \delta_{ij} \quad (i, j = 1, \cdots, r) \end{aligned} \tag{24}$$

---

[†] Hestenes (1).

and $\delta_{ij}$ is the Kronecker symbol. We write

$$\Gamma = (\gamma_{ij}),$$
$$B = (\beta_{ij}). \qquad (25)$$

It is easy to see that the maximum $\mu(X')$ is equal to the largest root $\mu'$ of the equation

$$|\Gamma - tB| = 0, \qquad (26)$$

and the vector which realizes this maximum is a solution of the linear homogeneous system with matrix $\Gamma - \mu'B$. It is advisable to normalize the solution obtained so that $c_0 = 1$.

Thus one step of group coordinate relaxation is equivalent to solving the generalized special eigenvalue problem for a matrix of order $r + 1$. However in the given case the generalized problem reduces easily to the ordinary one.

To do this in the subspace spanned by $X, e_1, \cdots, e_r$ we choose the orthonormal basis $Z = \tilde{X}/|\tilde{X}|, e_1, \cdots, e_r$ where $\tilde{X} = (0, 0, \cdots, 0, x_{r+1}, \cdots, x_n)'$.

We shall consider the vector which realizes the maximum for the Rayleigh quotient in the form

$$X' = s_0 Z + s_1 e_1 + \cdots + s_r e_r. \qquad (27)$$

Then

$$\frac{(AX', X')}{(X', X')} = \frac{\sum\limits_{i,j=0}^{r} \sigma_{ij} s_i s_j}{s_0^2 + \cdots + s_r^2}, \qquad (28)$$

where

$$\sigma_{00} = (AZ, Z),$$
$$\sigma_{0i} = \sigma_{i0} - (Z, A_i), \qquad (29)$$
$$\sigma_{ij} = a_{ij} \quad \text{(for } i, j = 1, \cdots, r\text{)}.$$

Therefore the numbers $s_0, \cdots, s_r$ are the components of the eigenvector for the matrix

$$S = \begin{bmatrix} \sigma_{00} & \sigma_{01} & \cdots & \sigma_{0r} \\ \sigma_{10} & \sigma_{11} & \cdots & \sigma_{1r} \\ \cdots & \cdots & \cdots & \cdots \\ \sigma_{r0} & \sigma_{r1} & \cdots & \sigma_{rr} \end{bmatrix}, \qquad (30)$$

corresponding to the largest eigenvalue. The latter is equal to the desired maximum.

Thus in the basis chosen each separate step of the group relaxation method requires solving the special eigenvalue problem for a matrix of order $r + 1$.

The method of group coordinate relaxation is applicable only for matrices of high order.

## 62. IMPROVING THE ACCURACY OF AN INDIVIDUAL EIGENVALUE AND ITS ASSOCIATED EIGENVECTOR

**1. Derwidue's method.**[†] Let $\lambda$ and $U$ be approximate values for a certain eigenvalue of a matrix $A$ and its associated eigenvector. We denote by $\lambda^* = \lambda + \varDelta\lambda$ and $U^* = U + \varDelta U$ the exact values for them. Let

$$AU - \lambda U = r \,. \tag{1}$$

For determining $\varDelta\lambda$ and $\varDelta U$ we have the non-linear equation

$$AU^* = \lambda^* U^*$$

or

$$A(U + \varDelta U) = (\lambda + \varDelta\lambda)(U + \varDelta U) \,. \tag{2}$$

Eliminating elements of second-order smallness we have

$$AU + A\varDelta U = \lambda U + \varDelta\lambda U + \lambda \varDelta U$$

or, recalling (1),

$$r + A\varDelta U - \varDelta\lambda U - \lambda\varDelta U = 0 \,. \tag{3}$$

Without loss of generality we may consider that the first component of the vector $\varDelta U$ is equal to zero, since the eigenvector $U^*$ is determined accurately to within a constant multiplier. Writing equation (3) in components we obtain a system of $n$ linear equations with $n$ unknowns $\varDelta\lambda$, $\varDelta u_2$, $\cdots$, $\varDelta u_n$. This system has the form

$$
\begin{aligned}
- u_1\varDelta\lambda + a_{12}\varDelta u_2 + \cdots + a_{1n}\varDelta u_n &= -r_1 \,, \\
- u_2\varDelta\lambda + (a_{22} - \lambda)\varDelta u_2 + \cdots + a_{2n}\varDelta u_n &= -r_2 \,, \\
\cdot\ \cdot\ \cdot\ \cdot\ \cdot\ \cdot\ \cdot\ \cdot\ \cdot\ \cdot\ \cdot\ \cdot\ \cdot\ \cdot\ \cdot\ \cdot\ \cdot\ \cdot\ &, \\
- u_n\varDelta\lambda + a_{n2}\varDelta u_2 + \cdots + (a_{nn} - \lambda)\varDelta u_n &= -r_n \,.
\end{aligned}
\tag{4}
$$

Finding $\varDelta\lambda$, $\varDelta u_2$, $\cdots$, $\varDelta u_n$ we may then repeat the process.

As an example we shall improve the accuracy of an eigenvalue and its associated eigenvector for the matrix of example 2, Section 53. There it was found that $\lambda_1 = -17.39$ and $U_1 = (1.00000, -8.14472, 7.91426)'$. Computing the residual we obtain $r = (-0.00420498, 0.0592605, -0.0630314)'$. Below is presented the solution of the system for determining $\varDelta\lambda$, $\varDelta u_2$ and $\varDelta u_3$ by the single-division method:

| | | | | |
|---|---|---|---|---|
| $-1$ | 1.870086 | 0.422908 | 0.00420498 | 1.297199 |
| 8.14472 | 5.578346 | 5.711900 | $-0.05926046$ | 19.375706 |
| $-7.91426$ | 4.308033 | 4.419313 | 0.06303143 | 0.8761174 |
| 1 | $-1.870086$ | $-0.422908$ | $-0.00420498$ | $-1.297199$ |
| | 20.809673 | 9.156367 | $-0.02501208$ | 29.941029 |
| | $-10.492314$ | 1.072309 | 0.02975214 | $-9.390253$ |
| | 1 | 0.4400053 | $-0.00120195$ | 1.438804 |
| | | 5.6889803 | 0.01714090 | 5.706124 |
| | | 1 | 0.00301300 | 1.0030130 |
| | 1 | | $-0.0025277$ | 0.9974724 |
| 1 | | | $-0.0076578$ | 0.9923423 |

† Derwidue (2).

Thus

$$\Delta\lambda = -0.007658 ,$$

$$\Delta u_2 = -0.00253 ,$$

$$\Delta u_3 = 0.00301 ,$$

and therefore we obtain the improved values

$$\lambda_1 = -17.397658 ,$$

$$U_1 = (1.00000, -8.14725, 7.91727)' .$$

The values obtained are already significantly close to the exact ones (cf. Section 53).

**2. Wielandt's method.** To improve the accuracy of an individual eigenvalue and its associated eigenvector we may apply the following modification of the power method first proposed by Wielandt.[†]

Let $\mu$ be an approximation to the eigenvalue $\lambda$ of a matrix $A$ to which there corresponds an eigenvector $U$. Then the number $\nu = 1/(\lambda - \mu)$ will be an eigenvalue for the matrix $B = (A - \mu E)^{-1}$, to which will correspond, as is easy to verify, the eigenvector $U$. The remaining eigenvalues for matrix $B$ will be $\nu_i = 1/(\lambda_i - \mu)$ and if $\mu$ is sufficiently close to $\lambda$, then $\nu$ will significantly predominate over the rest of matrix $B'$s eigenvalues. Therefore applying the power method with matrix $B$ for a large number of iterations gives a good approximation for the number $\nu$. The eigenvalue $\lambda$ is found from the formula

$$\lambda = \mu + \frac{1}{\nu} \tag{5}$$

The iterations with matrix $B = (A - \mu E)^{-1}$ require either an actual inversion of the matrix $A - \mu E$ or the less laborious solution of the system of linear equations

$$(A - \mu E)U_i = U_{i-1} , \tag{6}$$

which immediately gives the iteration of the vector $U_{i-1}$ with the matrix $B$. For the vector $U_0$ we should take the known approximation to the eigenvector.

The determinant of $A - \mu E$ will be close to zero and system (6) turns out to be ill conditioned. However the character of the ill condition here will be such that the absolute values of the components for the vector $U_i$ are determined with little accuracy; the ratios of components for this vector will as a rule be determined with good accuracy. For computing the eigenvector components only this is necessary, since the latter are determined accurately only to within a scalar multiplier.

One more modification of the process is possible in which the iterations are obtained as a result of solving the system

$$(A - \mu_i E)U_i = U_{i-1}, \tag{7}$$

where $\mu_i$ is an approximation to the eigenvalue $\lambda$, made more accurate in the preceding step. This accelerates the convergence of the process. Of

---

† Bodewig (6), pp. 293–294.

course this is essentially small, since the process already converges very rapidly even without it.

We observe further that for $\mu$ very close to $\lambda$ a solution of the system

$$(A - \mu E)U_i = U_{i-1} \tag{8}$$

hardly differs from the solution of the homogeneous system

$$(A - \mu E)U = 0 \tag{9}$$

for the eigenvector coordinates.

The eigenvalue $\nu$ is determined as a ratio of corresponding components of the vectors $U_i$ and $U_{i-1}$. From what has been said above concerning the ill condition of the system these ratios will not be determined with very high accuracy but this does not spoil things since the number $1/\nu$ plays the role of a small correction in the formula $\lambda = \mu + 1/\nu$ for determining the eigenvalue of the matrix.

Wielandt's method will be illustrated by the example of the previous point.

Let $\mu = -17.39$, $U_0 = (1.00, -8.14, 7.91)'$. Using the single-division method to solve the corresponding system

| 11.880118 | 1.870086 | 0.422908 | 1.00 | 15.173112 |
|---|---|---|---|---|
| 0.287865 | 5.578346 | 5.711900 | −8.14 | 3.438111 |
| 0.049099 | 4.308033 | 4.419313 | 7.91 | 16.686445 |
| 1 | 0.15741308 | 0.03559796 | 0.08417425 | 1.2771853 |
|  | 5.5330323 | 5.7016526 | −8.1642308 | 3.0704541 |
|  | 4.3003042 | 4.4175652 | 7.9058671 | 16.6237365 |
|  | 1 | 1.0304751 | −1.4755437 | 0.5549315 |
|  |  | −0.01379120 | 14.251154 | 14.237362 |
|  |  | 1 | −1033.3512 | −1032.3512 |
|  | 1 |  | 1063.3671 | 1064.3671 |
| 1 |  |  | −130.5185 | −129.5185 |

we obtain $U_1 = (1.0000, -8.14725, 7.91728)'$ and $1/\nu = -0.007655$ (or $-0.007655$, $-0.007662$). This gives $\lambda = -17.397655$. The vector $U_1$ is also much closer to the eigenvector than is $U_0$.

**3. The perturbation method.** The perturbation method was worked out for bounded operators in a Hilbert space.[†] As applied to matrices it consists basically of the following.

Let a symmetric matrix $A_0$ have a simple eigenvalue $\lambda_0$ and an eigenvector $X_0$ corresponding to it. We denote by $R$ the matrix determined by the conditions $R(A_0 - \lambda_0 E)Y = Y$ for any $Y$ orthogonal to $X_0$ and $RX_0 = 0$. The matrix $R$ exists, since the operator with matrix $A_0 - \lambda_0 E$ will be non-

---

† F. Rellich, Math. Ann., 1936, 113 (4), pp. 600–619. M. K. Gavurin, Vestn. Leningr. un-ta, 1952, 9, pp. 77–95.

degenerate in a subspace orthogonal to the vector $X_0$. Let $A$ be some other symmetric matrix with an eigenvalue $\lambda$ and a corresponding eigenvector $X$ while $(X, X_0) \neq 0$, permitting us to consider without loss of generality that $X = X_0 + Z$ for $(Z, X_0) = 0$.

We assume $B = A - A_0$ and $\mu = \lambda - \lambda_0$. Then the following relationships hold:

$$((\mu E - B)X, X_0) = 0 , \tag{10}$$

$$(E - R(\mu E - B))X = X_0 . \tag{11}$$

In fact $(\mu E - B)X = (\lambda E - A)X - (\lambda_0 E - A_0)X = (A_0 - \lambda_0 E)X$, so that $((\mu E - B)X, X_0) = (X, (A_0 - \lambda_0 E)X_0) = 0$. Furthermore,

$$(E - R(\mu E - B))X = X - R(A_0 - \lambda_0 E)X = X - Z = X_0 .$$

If the matrix $E - R(\mu E - B)$ is non-degenerate (which will hold, for example, if $|\mu|$ and $\|B\|$ are small enough), then formula (11) is equivalent to the formula

$$X = (E - R(\mu E - B))^{-1}X_0 . \tag{12}$$

For small $|\mu|$ and $\|B\|$ we may expand the right part of equation (12) into a series. Limiting ourselves to at most second-order elements and remembering that $RX_0 = 0$, we obtain the approximate formula

$$X \approx X_0 - RBX_0 + RBRBX_0 - \mu R^2 BX_0 , \tag{13}$$

which together with (10) gives

$$\mu \approx \frac{(BX_0, X_0) - (RBX_0, BX_0) + (BRBX_0, RBX_0)}{\|X_0\|^2 + \|BRX_0\|^2} . \tag{14}$$

In applying the expanded form to the problem of increasing the accuracy of an individual eigenvalue for a symmetric matrix we use the method of "false perturbation," also proposed by M. K. Gavurin.[†]

Let the approximate normalized eigenvector $X_0$ and the approximate eigenvalue $\lambda_0 - (AX_0, X_0)$ be known for a symmetric matrix $A$. We construct the matrix

$$A_0 = A - X_0 r' - r X_0' ,$$

where $r = AX_0 - \lambda_0 X_0$. It is clear that $A_0$ is symmetric and it is easily verified that the vector $X_0$ is an eigenvector for it belonging to the eigenvalue $\lambda_0$.

The approximate formulas (13) and (14) are transformed into

$$X \approx X_0 - Rr - \mu R^2 r , \tag{15}$$

$$\mu \approx -\frac{(Rr, r)}{1 + \|Rr\|^2} . \tag{16}$$

Having made the eigenvalue and eigenvector more accurate by formulas

---

[†] M. K. Gavurin, *Uspekhi matem. nauk*, 1957, 12, No. 1, pp. 173-175.

(15) and (16) we may repeat the process until we achieve the accuracy required.

The vectors $Rr$ and $R^2r$ are found from the systems of linear equations

$$(A_0 - \lambda_0 E)Z = r , \tag{17}$$
$$(Z, X_0) = 0$$

and

$$(A_0 - \lambda_0 E)Z = Rr , \tag{18}$$
$$(Z, X_0) = 0$$

respectively.

Since $|A_0 - \lambda_0 E| = 0$, one of the first $n$ equations in systems (17) and (18) may be eliminated and used only for control.

If in formula (15) we eliminate the element $\mu R^2 r$ of second-order smallness, then the need to solve the second system disappears.

We observe that the method may be applied starting with fairly crude approximations for $X_0$.

L. A. Rukhovets directed our attention to the means described here for realizing the perturbation method.

For matrix (4) in Section 51 we have $\lambda = 0.24226071$, $X = (0.718846, 0.095699, -0.387435, -0.569207)'$.

Starting from the approximation $X_0 = (0.704361, 0.176090, -0.440225, -0.528271)'$ obtained by normalizing the vector $(0.8, 0.2, -0.5, -0.6)'$ and $\lambda_0 = (AX_0, X_0) = 0.248992$ we obtain, after a single application of formulas (15) and (16) and normalizing, $X_1 = (0.718839, 0.095717, -0.387445, -0.569206)'$, $\lambda_1 = 0.24226072$.

In carrying out the iterative process by $n$ repeated applications of formulas (15) and (16) the rate of convergence has order $q^{5^n}$; if we use the simplified formula (15) with the discarded element $\mu R^2 r$, then the rate of convergence will have order $q^{3^n}$. Here $q = \|r\|/\tau - 2\|r\|$ where $r$ is the residual of the initial approximation and $\tau$ is the distance from the eigenvalue being improved to its nearest neighbor. These estimates were given to us by M. K. Gavurin.

# THE METHOD OF MINIMAL ITERATIONS AND OTHER METHODS BASED ON THE IDEA OF ORTHOGONALIZATION

The methods to be described in this and following chapters are applicable to solving a system of linear equations and to solving the eigenvalue problem, complete (Chapter VI) and special (Chapter VII). Although certain of the methods described in Chapter VII (particularly the method of steepest descent) appeared earlier historically than did the methods of Chapter VI we shall present these latter ones first, since their appearance allowed the further construction and development of the former.

Despite the fact that the methods in Chapter VI are exact and the methods in Chapter VII iterative, their computational schemes have essentially the same design.

## 63. THE METHOD OF MINIMAL ITERATIONS

**1. Constructing basis vectors in the non-degenerate case.** Let a matrix $A$ be symmetric. The method of minimal iterations[†] for solving the complete eigenvalue problem is none other than the method of orthogonalizing successive iterations of a certain initial vector (Section 50). However we shall present it independently of the results in Section 50 in view of the many important peculiarities of the method which arise and whose presence permits us to generalize the method and broaden its area of application.

We shall assume at this point that the symmetric matrix $A$ has pairwise distinct eigenvalues $\lambda_1, \lambda_2, \cdots, \lambda_n$. Let $U_1, U_2, \cdots, U_n$ be their corresponding eigenvectors, which we shall consider to be normalized.

We select a certain initial vector $X$. It may be easily represented in the

---

† Lanczos (2).

form

$$X = a_1 U_1 + a_2 U_2 + \cdots + a_n U_n .\tag{1}$$

Moreover we shall assume for the present that all coefficients $a_i \neq 0$. Then the vector system $X, AX, \cdots, A^{n-1}X$ will be linearly independent.

THEOREM 63.1. *If the vectors* $X, AX, \cdots, A^{n-1}X$ *are linearly independent and the vectors* $p_0, p_1, \cdots, p_{n-1}$ *are obtained from them by an orthogonalization process, then these vectors are determined by the trinomial recurrence formulas*

$$p_{i+1} = Ap_i - \alpha_i p_i - \beta_i p_{i-1} , \qquad (i = 1, 2, \cdots, n-2)$$
$$p_0 = X , \quad p_1 = Ap_0 - \alpha_0 p_0 ,\tag{2}$$

*where*

$$\alpha_i = \frac{(Ap_i, p_i)}{(p_i, p_i)} \qquad (i = 0, 1, \cdots, n-2) ,$$

$$\beta_i = \frac{(Ap_i, p_{i-1})}{(p_{i-1}, p_{i-1})} = \frac{(p_i, Ap_{i-1})}{(p_{i-1}, p_{i-1})} = \frac{(p_i, p_i)}{(p_{i-1}, p_{i-1})} \qquad (i = 1, 2, \cdots, n-2) .\tag{3}$$

*Proof.* Let $X, AX, \cdots, A^{n-1}X$ be linearly independent and let the vectors $p_0, p_1, \cdots, p_{n-1}$ be obtained from them by orthogonalization. Then $p_i = A^i X + c_1^{(i)} A^{i-1} X + \cdots c_i^{(i)} X$.

From here it follows that the vector $p_{i+1} - Ap_i$ belongs to the subspace spanned by the vectors $X, AX, \cdots, A^i X$ or that, equivalently, it belongs to the subspace spanned by the vectors $p_0, p_1, \cdots, p_i$. Thus the vector $p_{i+1}$ is expressed in terms of the preceding ones by the formulas

$$p_{i+1} = Ap_i - \gamma_i^{(i)} p_i - \cdots - \gamma_0^{(i)} p_0 ,$$

where $\gamma_i^{(i)}, \cdots, \gamma_0^{(i)}$ are certain numbers. From the orthogonality relationships of the vector $p_{i+1}$ to the preceding vectors we have

$$\gamma_j^{(i)} = \frac{(Ap_i, p_j)}{(p_j, p_j)} .$$

But for $j = 0, 1, \cdots, i-2$ the equalities $(Ap_i, p_j) = 0$ hold, since $(Ap_i, p_j) = (p_i, Ap_j)$ and the vector $Ap_j$ is a linear combination of the vectors $p_{j+1}, p_j, \cdots, p_0$, each of which is orthogonal to the vector $p_i$ for $j \leq i-2$. Consequently only the two coefficients

$$\alpha_i = \gamma_i^{(i)} = \frac{(Ap_i, p_i)}{(p_i, p_i)}$$

and

$$\beta_i = \gamma_{i-1}^{(i)} = \frac{(Ap_i, p_{i-1})}{(p_{i-1}, p_{i-1})} = \frac{(p_i, Ap_{i-1})}{(p_{i-1}, p_{i-1})}$$

remain non-zero.

Moreover,

$$(p_i, Ap_{i-1}) = (p_i, p_i + \alpha_{i-1}p_{i-1} + \beta_{i-1}p_{i-2}) = (p_i, p_i) \qquad (4)$$

and consequently

$$\beta_i = \frac{(p_i, p_i)}{(p_{i-1}, p_{i-1})} .$$

From the last formula for $\beta_i$ it follows that $\beta_i > 0$. For a positive definite matrix the coefficients $\alpha_i$ will also be positive.

It is obvious that the vectors $p_i$ are represented as

$$p_i = p_i(A)X = p_i(A)p_0 \qquad (i = 1, \cdots, n-1), \qquad (5)$$

where $p_i(t) = t^i + \cdots$ is a certain polynomial of degree $i$.

The polynomials $p_i(t)$ may be computed along with the computation of the vectors $p_i$ from the formulas

$$p_{i+1}(t) = (t - \alpha_i)p_i(t) - \beta_i p_{i-1}(t) \qquad (6)$$

in which the coefficients $\alpha_i$ and $\beta_i$ have the previous values.

We observe that the vector $p_n = Ap_{n-1} - \alpha_{n-1}p_{n-1} - \beta_{n-1}p_{n-2}$ where

$$\alpha_{n-1} = \frac{(Ap_{n-1}, p_{n-1})}{(p_{n-1}, p_{n-1})} \quad \text{and} \quad \beta_{n-1} = \frac{(p_{n-1}, p_{n-1})}{(p_{n-2}, p_{n-2})}$$

will be orthogonal to $n$ vectors $p_0, p_1, \cdots, p_{n-1}$ and consequently the vector $p_n$ is equal to zero. Correspondingly, the polynomial $p_n(t)=(t - \alpha_{n-1})p_{n-1}(t) - \beta_{n-1}p_{n-2}(t)$ coincides with the characteristic polynomial. In this way we obtain a simple and convenient algorithm for computing the coefficients of the characteristic polynomial.

We have already established by two methods (Section 50) that the recurrence relationships connecting the vectors $p_i$ and the polynomials $p_i(t)$ are trinomials. We consider this fact from still another point of view. From equation (5) and the expansion

$$p_0 = a_1 U_1 + a_2 U_2 + \cdots + a_n U_n$$

it follows that

$$p_i = a_1 p_i(\lambda_1)U_1 + a_2 p_i(\lambda_2)U_2 + \cdots + a_n p_i(\lambda_n)U_n .$$

From here, remembering the condition $|U_i| = 1$, we obtain

$$(p_j, p_i) = a_1^2 p_j(\lambda_1) p_i(\lambda_1) + \cdots + a_n^2 p_j(\lambda_n)p_i(\lambda_n) = 0 .$$

This last equation is equivalent to the equation

$$\int_m^M p_j(t)p_i(t)dF(t) = 0 . \qquad (7)$$

Here the weight function of the Stieltjes integral is determined as follows:

$$F(t) = 0 \qquad (m \le t \le \lambda_1) ,$$
$$F(t) = a_1^2 \qquad (\lambda_1 < t \le \lambda_2) ,$$
$$F(t) = a_1^2 + a_2^2 \qquad (\lambda_2 < t \le \lambda_3) ,$$
$$\cdots \cdots \cdots \cdots ,$$
$$F(t) = a_1^2 + a_2^2 + \cdots + a_n^2 \qquad (\lambda_n < t \le M) , \qquad (8)$$

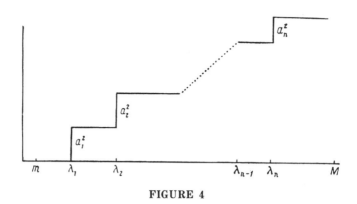

**FIGURE 4**

where $m$ and $M$ are the spectral limits for matrix $A$ (Fig. 4).

Condition (7) determines the system of polynomials $p_i(t)$ which is orthogonal to the integral weight $F(t)$. But from the theory of orthogonal polynomials it is known[†] that these polynomials are related by the trinomial recurrence relationships.

We shall consider a number of properties of the vectors $p_0, \cdots, p_{n-1}$ and the polynomials $p_0(t), p_1(t), \cdots, p_{n-1}(t), p_n(t)$.

THEOREM 63.2.   *The roots of the polynomials* $p_n(t), \cdots, p_0(t)$ *are real and separate each other.*

In fact from relationships (6) and the positivity of the coefficients $\beta_i$ the sequence of polynomials $p_n(t), p_{n-1}(t), \cdots, p_0(t)$ is a Sturm sequence with positive leading coefficients

THEOREM 63.3.   *The vectors* $p_0, \cdots, p_{i-1}$ *form an orthogonal basis of the subspace* $P_i$ *spanned by the vectors* $X, AX, \cdots, A^{i-1}X$.

*Proof.*   The vectors $p_0, \cdots, p_{i-1}$ belong to the subspace $P_i$, are linearly independent and are orthogonal.

The theorem that follows describes the properties of the vectors $p_i$ which give us the right to call them minimal iterations. That is, the property we shall consider was used as a basis for constructing the system of vectors $p_0, \cdots, p_{n-1}$ by Lanczos (2), the author of the first publication devoted to the method of minimal iterations.

THEOREM 63.4   *Among all vectors of the form* $Z = A^iX + Y$, *where* $Y \in P_i$, *the vector* $p_i$ *has the shortest length.*

[†] I. P. Natanson, *Konstruktivnaya teoriya funktsii* (Constructive Function Theory), 1949, Part II, Chapter IV.

In fact $p_i = A^i X + Y_0$ where $Y_0$ is a certain defined vector in $P_i$ and consequently $Z = p_i + Y - Y_0$.

But

$$|Z|^2 = (Z, Z) = (p_i + Y - Y_0, p_i + Y - Y_0) = |p_i|^2 + |Y - Y_0|^2,$$

since $(p_i, Y - Y_0) = 0$. Thus $|Z|^2 \geq |p_i|^2$ and the equality sign is possible only if $Y - Y_0 = 0$, i.e. $Z = p_i$.

THEOREM 63.5. *An operator with matrix $A$ has in the basis $p_0, p_1, \cdots, p_{n-1}$ the tridiagonal Jacobian matrix*

$$J = \begin{bmatrix} \alpha_0 & \beta_1 & & & & \\ 1 & \alpha_1 & \beta_2 & & & \\ & 1 & \alpha_2 & \cdot & & \\ & & \cdot & \cdot & \cdot & \\ & & & \cdot & \cdot & \cdot \\ & & & \cdot & \alpha_{n-2} & \beta_{n-1} \\ & & & & 1 & \alpha_{n-1} \end{bmatrix}. \tag{9}$$

In fact,

$$\begin{aligned} Ap_0 &= \alpha_0 p_0 + p_1, \\ Ap_1 &= \beta_1 p_0 + \alpha_1 p_1 + p_2, \\ &\cdots\cdots\cdots\cdots\cdots\cdots\cdots, \\ Ap_{n-2} &= \beta_{n-2} p_{n-3} + \alpha_{n-2} p_{n-2} + p_{n-1}, \\ Ap_{n-1} &= \beta_{n-1} p_{n-2} + \alpha_{n-1} p_{n-1}, \end{aligned}$$

i.e. the coordinates of the vectors $Ap_0, Ap_1, \cdots, Ap_{n-1}$ in the basis $p_0, p_1, \cdots, p_{n-1}$ coincide with the columns of matrix $J$. The positivity of the numbers $\beta_i$ has already been noted. The theorem is thus proved.

COROLLARY

$$\alpha_0 + \alpha_1 + \cdots + \alpha_{n-1} = SpA. \tag{10}$$

Actually the matrices $A$ and $J$ are similar and therefore their spurs are identical.

Equation (10) gives a good conclusive control when actually using the minimal iteration method.

THEOREM 63.6

$$p_i(t) = \begin{vmatrix} t - \alpha_0 & \beta_1 & & & \\ 1 & t - \alpha_1 & \beta_2 & & \\ & \cdot & \cdot & \cdot & \\ & & \cdot & \cdot & \cdot \\ & & & \cdot & \beta_{i-1} \\ & & & 1 & t - \alpha_{i-1} \end{vmatrix}.$$

*Proof.* The polynomials

$$\tilde{p}_i(t) = \begin{vmatrix} t - \alpha_0 & \beta_1 & & & & \\ 1 & t - \alpha_1 & \beta_2 & & & \\ & \cdot & \cdot & \cdot & & \\ & & \cdot & \cdot & \cdot & \\ & & & \cdot & \cdot & \beta_{i-1} \\ & & & & 1 & t - \alpha_{i-1} \end{vmatrix}$$

are connected by the same recurrence relationships as the polynomials $p_i(t)$; it is easy to convince ourselves of this by expanding the determinant in terms of elements from the last column. It follows directly that $\tilde{p}_1(t) = p_1(t)$ and $\tilde{p}_2(t) = p_2(t)$. Therefore $\tilde{p}_i(t) = p_i(t)$ for all $i = 1, 2, \cdots, n$

THEOREM 63.7. *The vectors* $p_0, \cdots, p_{n-1}$ *satisfy the relationships*

$$(Ap_i, p_j) = 0 \qquad (\text{for } |i - j| > 1). \tag{11}$$

For $j = 0, \cdots, i - 2$ the validity of the theorem was already established in proving theorem 63.1. For $i < j$ affirmation of the theorem follows from the equalities

$$(Ap_i, p_j) = (p_i, Ap_j).$$

In conclusion we observe that the vector $p_1$ coincides with the gradient $\xi$ of the functional $\mu(X) = (AX, X)/(X, X)$ at the point $p_0$. The basic properties of the gradient

$$(X, \xi) = 0,$$
$$(\xi, AX) = (\xi, \xi)$$

are special case of the properties

$$(p_i, p_j) = 0,$$
$$(Ap_{i-1}, p_i) = (p_i, p_i) \tag{12}$$

for the system of vectors $p_0, p_1, \cdots, p_{n-1}$.

**2. Completing a system of basis vectors in case of degeneracy.** In the previous paragraph we assumed first that all eigenvalues of matrix $A$ were distinct and second that all the coefficients $a_i$ in the expansion of the chosen initial vector in terms of eigenvectors were different from zero. These conditions guaranteed the linear independence of the vectors $X, AX, \cdots, A^{n-1}X$.

If at least one of these conditions is not satisfied, then the vectors $X, AX, \cdots, A^{n-1}X$ will no longer be linearly independent. If we apply an orthogonalization process to them, then the process breaks off at a certain step, i.e. we find that $p_r = 0$ for a certain $r < n$. In this case the polynomial $p_r(t)$ will obviously be the minimum annihilating polynomial for the vector $p_0$.

We shall show how in a degenerate case we can build up the system of vectors $p_0, \cdots, p_{r-1}$ to an orthogonal basis of the whole space.

We again take an arbitrary vector $Y$ and form the vector

$$p_0^{(1)} = Y - \sum_{i=0}^{r-1} c_i p_i , \tag{13}$$

determining the coefficients $c_i$ from the orthogonality condition for the vector $p_0^{(1)}$ with the vectors $p_0, \cdots, p_{r-1}$ constructed earlier. This gives

$$c_i = \frac{(Y, p_i)}{(p_i, p_i)} \qquad (i = 0, \cdots, r-1) . \tag{14}$$

The vector $p_0^{(1)}$ is orthogonal by construction to all the vectors $p_0, \cdots, p_{r-1}$, i.e. it belongs to the subspace which is orthogonally complementary to $P_r$. Thus all its iterations will also be orthogonal to $P_r$, since $P_r$ is invariant and by theorem 11.1 its orthogonal complement is also invariant. Therefore by applying the method of minimal iterations to the vector $p_0^{(1)}$ we construct the system of vectors $p_0^{(1)}, \cdots, p_{r-1}^{(1)}$ pairwise orthogonal not only to each other but also to all the vectors $p_0, \cdots, p_{r-1}$. If $r+1 < n$, we continue the completion process until we arrive at a basis of the whole space. The whole space is thereupon broken down naturally into the direct sum of several pairwise orthogonal invariant subspaces $P_r, P_l \cdots$. The characteristic polynomial of the matrix will be equal to the product of the minimal polynomials which annihilate successive initial vectors.

**3. Determination of eigenvectors.** We shall first consider the non-degenerate case. Let the orthogonal basis of the space $p_0, \cdots, p_{n-1}$, the system of orthogonal polynomials $p_0(t), \cdots, p_{n-1}(t), p_n(t)$, be constructed and let us assume we have found the roots of the characteristic polynomial $p_n(t)$. Let $U_i$ be the eigenvector belonging to the eigenvalue $\lambda_i$. Then

$$U_i = c_{i0} p_0 + \cdots + c_{in-1} p_{n-1} . \tag{15}$$

From the orthogonality of the vectors $p_0, \cdots, p_{n-1}$,

$$c_{ij} = \frac{(U_i, p_j)}{(p_j, p_j)} . \tag{16}$$

Thus to determine the eigenvector $U_i$ we need to compute only the constants $(U_i, p_j)$. This is done without difficulty from the following considerations. From the expansion of the initial vector $p_0$ in eigenvectors of the matrix

$$p_0 = a_1 U_1 + \cdots + a_n U_n \qquad (a_i \neq 0, |U_i| = 1)$$

we obtain

$$p_j = a_1 p_j(\lambda_1) U_1 + \cdots + a_n p_j(\lambda_n) U_n$$

and consequently

$$(U_i, p_j) = a_i p_j(\lambda_i) .$$

Thus, within a constant multiplier,

**TABLE VI.1**   **Computation of the Basis Vectors and Coefficients of the Characteristic Polynomial by the Method of Minimal Iterations**

| | $p_0$ | $Ap_0$ | $p_1$ | $Ap_1$ | $p_2$ | $Ap_2$ | $p_3$ | $Ap_3$ | $p_4$ |
|---|---|---|---|---|---|---|---|---|---|
| I | 1 | 2.62 | 0.32 | 0.1640 | -0.02929952 | -0.0001850 | -0.001685676 | -0.001540834 | $-0.52\cdot10^{-9}$ |
| | 1 | 2.18 | -0.12 | -0.0472 | -0.03163768 | -0.01495082 | 0.007179741 | 0.004809803 | $0.39\cdot10^{-9}$ |
| | 1 | 2.08 | -0.22 | -0.0812 | -0.01816908 | -0.02671150 | -0.006295284 | -0.004731764 | $0.81\cdot10^{-9}$ |
| | 1 | 2.32 | 0.02 | 0.1300 | 0.07910628 | 0.0185082 | 0.000801219 | 0.00162796 | $0.31\cdot10^{-9}$ |
| II | | 9.20 | 0 | 0.1656 | 0 | $-0.16\cdot10^{-8}$ | 0 | $0.2\cdot10^{-8}$ | |
| | | | | | $-0.16\cdot10^{-8}$ | | $0.16\cdot10^{-8}$ | | |
| | | | | | | | $0.12\cdot10^{-9}$ | | |
| III | 4.0 | 9.20 | 0.1656 | 0.078608 | 0.0084473237 | 0.0042745183 | 0.00009462737 | 0.00006809030305 | |
| | | | 0.1656 | | 0.0084473229 | | 0.00009462885 | | |
| IV | 2.3 | | 0.47468599 | | 0.50602042 | | 0.71929364 | | 4.00000005 |
| | 0.0414 | | 0.05101041 | | 0.01120624 | | | | |
| V | 1 | | | | 1 | | 1 | | 1 |
| | -2.3 | | | | -2.77468599 | | -3.28070641 | | -4.00000005 |
| | | | | | 1.05037778 | | 2.40341514 | | 4.75200016 |
| | | | | | | | -0.41418866 | | -2.11185609 |
| | | | | | | | | | 0.28615248 |

$$U_i = \frac{p_0(\lambda_i)}{(p_0, p_0)} p_0 + \cdots + \frac{p_{n-1}(\lambda_i)}{(p_{n-1}, p_{n-1})} p_{n-1} . \tag{17}$$

We observe that the values $p_j(\lambda_i)$ may be computed directly by the recurrent relationships for the polynomials $p_j(t)$.

In the degenerate case the vectors computed by the formula

$$U_i = \frac{p_0(\lambda_i)}{(p_0, p_0)} p_0 + \cdots + \frac{p_{r-1}(\lambda_i)}{(p_{r-1}, p_{r-1})} p_{r-1} , \tag{17'}$$

where $\lambda_i$ are the roots of the polynomial $p_r(t)$ will be eigenvectors lying in the invariant subspace $P_r$. The complete system of eigenvectors is determined if we apply the method described to all the subspaces $P_r, P_l, \cdots$, into whose direct sum the space $R$ is broken up in the process of completing the orthogonal basis.

As an example we shall determine all the eigenvalues and their associated eigenvectors for matrix (4) in Section 51:

$$A = \begin{bmatrix} 1.00 & 0.42 & 0.54 & 0.66 \\ 0.42 & 1.00 & 0.32 & 0.44 \\ 0.54 & 0.32 & 1.00 & 0.22 \\ 0.66 & 0.44 & 0.22 & 1.00 \end{bmatrix}$$

The computation of an orthogonal basis and the coefficients of the characteristic polynomial is given in Table VI.1. As the initial vector we take $p_0 = (1, 1, 1, 1)'$.

In the first part of the table are placed the components of the vectors $p_0, Ap_0, p_1, Ap_1, \cdots, p_4$ being successively computed. In the second part are written the results of the control check on the fulfillment of orthogo-

**TABLE VI.2  Computation of the Expansion Coefficients**

| $i$ | 1 | 2 | 3 | 4 |
|---|---|---|---|---|
| $\lambda_i$ | 2.32274880 | 0.79670672 | 0.63828385 | 0.24226068 |
| $p_0(\lambda_i)$ | 1 | 1 | 1 | 1 |
| $p_1(\lambda_i) = \lambda_i - \alpha_0$ | 0.02274880 | −1.5032933 | −1.6617162 | −2.0577393 |
| $\lambda_i - \alpha_1$ | 1.84806281 | 0.32202073 | 0.16359786 | −0.23242531 |
| $p_2(\lambda_i)$ | 0.00064121125 | 0.52549161 | −0.31325321 | 0.43687069 |
| $\lambda_i - \alpha_2$ | 1.81672838 | 0.29068630 | 0.13226343 | −0.26375974 |
| $p_3(\lambda_i)$ | 0.0000044810604 | −0.076069604 | 0.043332881 | −0.010262774 |
| $\lambda_i - \alpha_3$ | 1.60345516 | 0.07741308 | −0.08100979 | −0.47703296 |
| $p_4(\lambda_i)$ | $-0.3877 \cdot 10^{-9}$ | $0.2760 \cdot 10^{-8}$ | $0.3062 \cdot 10^{-8}$ | $0.3658 \cdot 10^{-8}$ |
| $p_4(\lambda_i)$ | 5.348 | −0.134 | 0.106 | −4.457 |
| $p_0(\lambda_i)/(p_0, p_0)$ | 0.25 | 0.25 | 0.25 | 0.25 |
| $p_1(\lambda_i)/(p_1, p_1)$ | 0.13737198 | 9.0778581 | 10.034518 | −12.425962 |
| $p_2(\lambda_i)/(p_2, p_2)$ | 0.075907030 | −62.208059 | −37.083131 | 51.717053 |
| $p_3(\lambda_i)/(p_3, p_3)$ | 0.047337111 | −803.58551 | 457.76070 | −108.41408 |

**TABLE VI.3   Eigenvectors of Matrix** $A$

| $U_1$ | $U_2$ | $U_3$ | $U_4$ |
|---|---|---|---|
| 1.000000 | 0.061916 | −0.447443 | 1.000000 |
| 0.793587 | −0.291847 | 1.000000 | 0.133129 |
| 0.747804 | 1.000000 | 0.042210 | −0.538970 |
| 0.887315 | −0.651533 | −0.425676 | −0.791834 |

nality conditions for the vectors $p_i$ and the usual control sums for the vectors $Ap_i$. In the third part of the table are placed the scalar products $(p_0, p_0)$, $(p_0, Ap_0)$, $(p_1, p_1)$, $\cdots$, while $(p_i, p_i) = (Ap_{i-1}, p_i)$ are computed in two ways. In the fourth part are found the coefficients $\alpha_i$ and $\beta_i$. In the last part of the table are written the coefficients for the polynomials $p_i(t)$ which are computed recurrently, just as in Section 51. As a result of the computations we obtain

$$\varphi(t) = p_4(t) = t^4 - 4.00000005t^3 + 4.75200016t^2$$
$$- 2.11185609t + 0.28615248 .$$

The values obtained for the coefficients compare well with the exact ones (cf. Section 51). Computing the roots of the last polynomial we obtain for the eigenvalues

$$\lambda_1 = 2.32274880; \quad \lambda_2 = 0.79670672; \quad \lambda_3 = 0.63828385; \quad \lambda_4 = 0.24226068 .$$

To find the associated eigenvectors beforehand we find the coefficients of the expansion in Table VI.2.

Finally we compute from formulas (17) all the eigenvectors of matrix $A$ and normalize them to a unit first norm.

**TABLE VI.4   The Method of Minimal Iterations for the Degenerate Case**

| | $p_0$ | $Ap_0$ | $p_1$ | $Ap_1$ | $p_2$ |
|---|---|---|---|---|---|
| | 1 | 19 | 6 | 42 | 0 |
| I | 1 | 13 | 0 | 24 | 0 |
| | 1 | 7 | −6 | 6 | 0 |
| II | | 39 | 0 | 72 | |
| III | 3 | 39 | 72 | 216 | |
| | | | 72 | | |
| IV | 13 | | 3 | | |
| | 24 | | | | |
| V | | | | | 1 |
| | | | 1 | | −16 |
| | 1 | | −13 | | 15 |

As a second example we consider the matrix

$$A = \begin{bmatrix} 10 & 6 & 3 \\ 6 & 5 & 2 \\ 3 & 2 & 2 \end{bmatrix}.$$

For the initial vector we take $p_0 = (1, 1, 1)'$.

We shall carry out the computation according to the scheme described above (we do not abandon the scheme although there is no need here for the control computations, since the computations are carried out exactly).

From the above table it is apparent that in the given case the degeneracy holds, since already $p_2 = 0$.

We observe that the sum $\alpha_1 + \alpha_2$ is equal to the spur of the operator induced in the subspace spanned by the vectors $p_0$ and $p_1$ and does not coincide with the spur of matrix $A$.

For completing the basis of the space we take the new initial vector

$$Y = (1, 0, 0)'.$$

Then for $p_0^{(1)}$ it is necessary to take the vector

$$p_0^{(1)} = Y - \frac{(Y, p_0)}{(p_0, p_0)} p_0 - \frac{(Y, p_1)}{(p_1, p_1)} p_1 = \left( \frac{1}{6}, -\frac{1}{3}, \frac{1}{6} \right)'$$

and to carry out the computation according to the previous scheme. We immediately obtain that $Ap_0^{(1)} = p_0^{(1)}$ so that $p_0^{(1)}$ is the eigenvector belonging to the eigenvalue $\lambda_3 = 1$.

It is not hard to compute that $\lambda_1 = 15$, $\lambda_2 = 1$. The eigenvectors belonging to them (in the subspace spanned by $p_0$ and $p_1$) are computed from formulas (17′)

$$U_1 = \frac{1}{3} p_0 + \frac{1}{36} p_1 = \left( \frac{1}{2}, \frac{1}{3}, \frac{1}{6} \right)',$$

$$U_2 = \frac{1}{3} p_0 - \frac{1}{6} p_1 = \left( -\frac{2}{3}, \frac{1}{3}, \frac{4}{3} \right)'.$$

**4. The solution of a linear system.** Knowing the basis $p_0, \cdots, p_{n-1}$ allows us also to find a solution of the linear system

$$AX = F. \tag{18}$$

In fact let

$$X = \sum_{i=0}^{n-1} a_i p_i \tag{19}$$

be the expansion of the desired solution in terms of basis vectors. Our problem is to find the coefficients $a_i$. Let

$$F = b_0 p_0 + \cdots + b_{n-1} p_{n-1}.$$

Then

$$b_i = \frac{(F, p_i)}{(p_i, p_i)} .$$

From equation (18) we have

$$\sum_{i=0}^{n-1} a_i A p_i = \sum_{i=0}^{n-1} \frac{(F, p_i)}{(p_i, p_i)} p_i . \qquad (20)$$

But from the recurrent relationship (2) we have

$$A p_i = p_{i+1} + \alpha_i p_i + \beta_i p_{i-1} .$$

Substituting this expression in (20) and equating coefficients for the vectors $p_i$ $(i = 0, 1, \cdots, n - 1)$ we obtain for determining the coefficients $a_i$ a system with tridiagonal matrix

$$
\begin{aligned}
\alpha_0 a_0 + \beta_1 a_1 \qquad\qquad &= \frac{(F, p_0)}{(p_0, p_0)} , \\
\alpha_0 + \alpha_1 a_1 + \beta_2 a_2 \qquad &= \frac{(F, p_1)}{(p_1, p_1)} , \\
\cdot \qquad\qquad &\qquad \cdot \\
\cdot \qquad\qquad &\qquad \cdot \\
\cdot \qquad\qquad &\qquad \cdot \\
a_{n-2} + \alpha_{n-1} a_{n-1} &= \frac{(F, p_{n-1})}{(p_{n-1}, p_{n-1})} .
\end{aligned}
\qquad (21)
$$

Solving it with respect to $a_0, \cdots, a_{n-1}$ we obtain the desired solution by formula (19).

However, as we shall see below, a minor change in the method permits us to avoid solving the tridiagonal system and to obtain concise formulas for the coefficients of the solution's expansion according to some other system of basis vectors.

In the degenerate case the matrix of an auxiliary system is partitioned into several tridiagonal boxes to facilitate solving the system.

## 64. THE BIORTHOGONAL ALGORITHM

The biorthogonal algorithm may be considered as a generalization of the minimal iteration method for the case of a non-symmetric matrix. The diversity in the possible reasons for degeneracy is considerably increased here in comparison with the symmetric case; this slightly complicates the theory of the method.

**1. The normal course of the process.** Let $A$ be a non-symmetric matrix and let $p_0$ and $\tilde{p}_0$ be two arbitrary initial vectors. We set up the two vector sequences

$$
\begin{aligned}
p_1 &= A p_0 - \alpha_0 p_0 , & \tilde{p}_1 &= A' \tilde{p}_0 - \alpha_0 \tilde{p}_0 , \\
p_{i+1} &= A p_i - \alpha_i p_i - \beta_i p_{i-1} , & \tilde{p}_{i+1} &= A' \tilde{p}_i - \alpha_i \tilde{p}_i - \beta_i \tilde{p}_{i-1} ,
\end{aligned}
\qquad (1)
$$

where

$$\alpha_i = \frac{(Ap_i, \tilde{p}_i)}{(p_i, \tilde{p}_i)} = \frac{(p_i, A'\tilde{p}_i)}{(p_i, \tilde{p}_i)},$$

$$\beta_i = \frac{(Ap_i, \tilde{p}_{i-1})}{(p_{i-1}, \tilde{p}_{i-1})} = \frac{(p_i, A'\tilde{p}_{i-1})}{(p_{i-1}, \tilde{p}_{i-1})} = \frac{(\tilde{p}_i, Ap_{i-1})}{(p_{i-1}, \tilde{p}_{i-1})} = \frac{(A'\tilde{p}_i, p_{i-1})}{(p_{i-1}, \tilde{p}_{i-1})} = \frac{(p_i, \tilde{p}_i)}{(p_{i-1}, \tilde{p}_{i-1})} \tag{2}$$

It is clear that $p_i = p_i(A)p_0$, $\tilde{p}_i = p_i(A')\tilde{p}_0$ where $p_i(t)$ are polynomials connected by the recurrent relationships

$$p_{i+1}(t) = (t - \alpha_i)p_i(t) - \beta_i p_{i-1}(t) . \tag{3}$$

From formulas (2) it is apparent that the process is broken off as soon as the equality $(p_r, \tilde{p}_r) = 0$ holds for the first time so that until the process breaks off $(p_i, \tilde{p}_i) \neq 0$. By direct computation it is possible to verify that for the constructed systems of vectors the biorthogonality conditions are satisfied, i.e. that $(p_i, \tilde{p}_j) = 0$ for $i \neq j$. Therefore the vectors $p_0, p_1, \cdots, p_{r-1}$ as well as the vectors $\tilde{p}_0, \tilde{p}_1, \cdots, \tilde{p}_{r-1}$ are linearly independent. In fact if

$$\gamma_0 p_0 + \cdots + \gamma_{r-1} p_{r-1} = 0 ,$$

then forming the scalar products with $\tilde{p}_i$ we obtain

$$\gamma_i (p_i, \tilde{p}_i) = 0$$

and since $(p_i, \tilde{p}_i) \neq 0$, then $\gamma_i = 0$, $i = 0, 1, \cdots, r - 1$. In the same way we may convince ourselves of the linear independence of the vectors $\tilde{p}_0, \tilde{p}_1, \cdots, \tilde{p}_{r-1}$. From here it follows that $r \leq n$, i.e. the process of constructing the vectors $p_0, p_1, \cdots$ and $\tilde{p}_0, \tilde{p}_1, \cdots$ must break off no later than at the $n$th step.

We shall consider that the biorthogonal algorithm proceeds normally if it breaks off at the $n$th step and that it is degenerate if it breaks off earlier.

In the normal course of the process both the vector $p_n$ and the vector $\tilde{p}_n$ are equal to zero. In fact the vector $p_n$ is orthogonal to the $n$ linearly independent vectors $\tilde{p}_0, \tilde{p}_1, \cdots, \tilde{p}_{n-1}$ and the vector $\tilde{p}_n$ is orthogonal to the $n$ linearly independent vectors $p_0, p_1, \cdots, p_{n-1}$.

We shall show how the complete eigenvalue problem is solved for the normal course of the biorthogonal algorithm.

We first consider the simplest case where the eigenvalues of matrix $A$ are distinct.

Let $U_1, \cdots, U_n$ be the eigenvectors of matrix $A$ and let $V_1, \cdots, V_n$ be the eigenvectors of matrix $A'$ where

$$p_0 = a_1 U_1 + \cdots + a_n U_n ,$$
$$\tilde{p}_0 = \tilde{a}_1 V_1 + \cdots + \tilde{a}_n V_n .$$

It is easy to see that both the coefficients $a_1, \cdots, a_n$ and the coefficients $\tilde{a}_1, \cdots, \tilde{a}_n$ are all different from zero, since if among the coefficients $a_1, \cdots, a_n$ (or $\tilde{a}_1, \cdots, \tilde{a}_n$) were found zero values, then the vectors $p_0, \cdots, p_{n-1}$ (or $\tilde{p}_0, \cdots, \tilde{p}_{n-1}$) would be linearly dependent, since they are all contained in a

**TABLE VI.5 Computation of Dual Bases and Coefficients of the Characteristic Polynomial**

| | $p_0$ | $\tilde{p}_0$ | $Ap_0$ | $A'\tilde{p}_0$ | $p_1$ | $\tilde{p}_1$ | $Ap_1$ | $A'\tilde{p}_1$ | $p_2$ |
|---|---|---|---|---|---|---|---|---|---|
| I | 0<br>1<br>0<br>0 | 0<br>1<br>1<br>0 | 1.870086<br>-11.811654<br>4.308033<br>0.269851 | 0.336964<br>-7.503621<br>-7.258787<br>0.288043 | 1.870086<br>-4.308033<br>4.308033<br>0.269851 | 0.336964<br>0<br>0.244834<br>0.288043 | -16.536065<br>76.046226<br>-74.283593<br>0.12069068 | -1.8428148<br>1.7626333<br>-2.6306581<br>-5.0093475 | 8.6346549<br>16.298938<br>-16.298938<br>3.7527931 |
| II | | | -5.363684 | -14.137401 | 0 | 0 | -14.652741 | -7.7201870 | 0<br>-0.55·10⁻⁶ |
| III | | | -7.503621 | -7.503621 | 1.7626333 | 1.7626333 | -23.724444 | -23.724444 | 36.894861 |
| IV | -7.503621<br>1.7626333 | | | | -13.459659<br>20.931671 | | | | -7.577879<br>-7.926152 |
| V | | | | | 1<br>7.503621 | | | | 1<br>20.963280<br>99.233547 |

| | $\tilde{p}_2$ | $Ap_2$ | $A'\tilde{p}_2$ | $p_3$ | $\tilde{p}_3$ | $Ap_3$ | $A'\tilde{p}_3$ | $p_4$ |
|---|---|---|---|---|---|---|---|---|
| I | 2.6926057<br>0<br>-1.0979092<br>-1.1323869 | -23.955388<br>-282.90935<br>282.90935<br>-84.358433 | -14.896906<br>-0.00000058<br>13.796999<br>19.697668 | 2.332957<br>-69.223641<br>69.223641<br>-61.568653 | -1.5458854<br>-0.00000058<br>0.3523911<br>5.0873557 | -113.57591<br>1210.1007<br>-1210.1007<br>1161.4402 | 8.5666662<br>0.0000087<br>1.8843879<br>-89.450974 | 0.0000285<br>0.0000180<br>-0.0000180<br>-0.0000070 |
| II | 0<br>-0.58·10⁻⁶ | -108.31382 | 18.597760 | 0<br>3.93·10⁻⁶<br>-2.89·10⁻⁶ | -0.58·10⁻⁶<br>4.45·10⁻⁶<br>-4.04·10⁻⁶ | 1047.8644 | -78.999913 | |
| III | 36.894869 | -279.58481 | -279.58481 | -292.43429 | -292.43418 | 5657.8054 | 5657.8053 | |
| IV | | | | -19.347271 | | | | |
| V | | | | 1<br>28.541159<br>237.15908<br>594.91649 | | | | 1<br>47.888430<br>797.27877<br>5349.4556<br>12296.551 |

subspace of dimension less than $n$.

From the assumption that the biorthogonal algorithm has a normal course we construct two systems of basis vectors $p_0, p_1, \cdots, p_{n-1}$ and $\tilde{p}_0, \tilde{p}_1, \cdots, \tilde{p}_{n-1}$ (dual, to within normalization of the basis). At the same time we construct the system of polynomials $p_0(t), p_1(t), \cdots, p_{n-1}(t), p_n(t)$ where $p_n(t)$ is the characteristic polynomial. Let its roots be known. We shall look for the eigenvectors of matrices $A$ and $A'$ in the form

$$U_i = c_{i0} p_0 + \cdots + c_{in-1} p_{n-1} ,$$
$$V_i = \tilde{c}_{i0} \tilde{p}_0 + \cdots + \tilde{c}_{in-1} \tilde{p}_{n-1} . \tag{4}$$

From the orthogonal properties of the basis vectors

$$c_{ik} = \frac{(U_i, \tilde{p}_k)}{(p_k, \tilde{p}_k)} = \tilde{a}_i \frac{p_k(\lambda_i)}{(p_k, \tilde{p}_k)} ,$$

$$\tilde{c}_{ik} = \frac{(V_i, p_k)}{(p_k, \tilde{p}_k)} = a_i \frac{p_k(\lambda_i)}{(p_k, \tilde{p}_k)} . \tag{5}$$

Eliminating the non-zero multipliers we obtain

$$U_i = \frac{p_0(\lambda_i)}{(p_0, \tilde{p}_0)} p_0 + \cdots + \frac{p_{n-1}(\lambda_i)}{(p_{n-1}, \tilde{p}_{n-1})} p_{n-1} ,$$

and $\qquad\qquad\qquad\qquad\qquad\qquad\qquad\qquad\qquad\qquad\qquad (6)$

$$V_i = \frac{p_0(\lambda_i)}{(p_0, \tilde{p}_0)} \tilde{p}_0 + \cdots + \frac{p_{n-1}(\lambda_i)}{(p_{n-1}, \tilde{p}_{n-1})} \tilde{p}_{n-1} .$$

As an illustration we shall solve the complete eigenvalue problem for Leverrier's matrix. The computation of the dual bases and coefficients of the characteristic polynomial are given in Table VI.5 whose structure coincides with the structure of Table VI.1 for the minimal iteration method. Of course as a check other than satisfying the biorthogonality conditions it is advisable to compute the theoretically equal values $(Ap_i, \tilde{p}_i)$ and $(p_i, A'\tilde{p}_i)$ as well as $(p_i, \tilde{p}_i)$ by different formulas. Final control is obtained by computing

$$\sum_{i=0}^{n-1} \alpha_i = SpA . \tag{7}$$

From Table VI.5 it is apparent that the characteristic polynomial is equal to

$$\varphi(t) = p_4(t) = t^4 + 47.888430t^3 + 797.27877t^2 + 5349.4556t + 12296.551 .$$

Thus the values for coefficients of the characteristic polynomial found by the biorthogonal algorithm almost coincide with the corresponding values found by the escalator method (cf. Section 48). Furthermore we have

$$\lambda_1 = - 17.863262;$$
$$\lambda_2 = - 17.152427;$$
$$\lambda_3 = - 7.574044;$$
$$\lambda_4 = - 5.298698 .$$

**TABLE VI.6   Computation of the Expansion Factors**

| $i$ | 1 | 2 | 3 | 4 |
|---|---|---|---|---|
| $\lambda_i$ | $-17.863262$ | $-17.152427$ | $-7.574044$ | $-5.298698$ |
| $p_0(\lambda_i)$ | 1 | 1 | 1 | 1 |
| $\lambda_i - \alpha_0 = p_1(\lambda_i)$ | $-10.359641$ | $-9.648806$ | $-0.070423$ | $2.204923$ |
| $\lambda_i - \alpha_1$ | $-4.403603$ | $-3.692768$ | $5.885615$ | $8.160961$ |
| $p_2(\lambda_i)$ | $43.857113$ | $33.868169$ | $-2.177116$ | $16.231657$ |
| $\lambda_i - \alpha_2$ | $-10.285383$ | $-9.574548$ | $0.003835$ | $2.274181$ |
| $p_3(\lambda_i)$ | $-234.24261$ | $-122.30678$ | $1.465722$ | $-9.157839$ |
| $\lambda_i - \alpha_3$ | $1.484009$ | $2.194844$ | $11.773227$ | $14.048573$ |
| $p_4(\lambda_i)$ | $0.00000250$ | $-0.0000468$ | $0.000125$ | $0.0000109$ |
| $p_4'(\lambda_i)$ | $-91.8$ | $80.7$ | $-224$ | $338$ |
| $p_0(\lambda_i)/(p_0, \tilde{p}_0)$ | 1 | 1 | 1 | 1 |
| $p_1(\lambda_i)/(p_1, \tilde{p}_1)$ | $-5.8773660$ | $-5.4740858$ | $-0.03995329$ | $1.2509255$ |
| $p_2(\lambda_i)/(p_2, \tilde{p}_2)$ | $1.1887052$ | $0.9179644$ | $-0.05900865$ | $0.4399436$ |
| $p_3(\lambda_i)/(p_3, \tilde{p}_3)$ | $0.8010094$ | $0.4182368$ | $-0.00501214$ | $0.0313159$ |

When computing the eigenvectors we first compute the expansion's factors in Table VI.6; this requires no explanation.

We then compute the eigenvectors of matrices $A$ and $A'$, normalizing them according to the normalization of the escalator method (Tables VI.7 and VI.8). Comparison with the results of the escalator method indicates good coincidence.

The biorthogonal algorithm is not complicated if the eigenvalues of the matrix are complex.

**TABLE VI.7   Eigenvectors of Matrix $A$**

| $U_1$ | $U_2$ | $U_3$ | $U_4$ |
|---|---|---|---|
| $-0.019873$ | $0.032933$ | $-0.351235$ | $1.135218$ |
| $0.169806$ | $-0.261308$ | $0.328468$ | $0.112182$ |
| $-0.187214$ | $0.236639$ | $0.260923$ | $0.070589$ |
| $0.808482$ | $0.586694$ | $0.045007$ | $0.011058$ |

**TABLE VI.8   Eigenvectors of Matrix $A'$**

| $V_1$ | $V_2$ | $V_3$ | $V_4$ |
|---|---|---|---|
| $-0.014059$ | $0.023297$ | $-0.248476$ | $0.803091$ |
| $0.780381$ | $-1.200904$ | $1.509558$ | $0.515562$ |
| $-1.140762$ | $1.441927$ | $1.589924$ | $0.430126$ |
| $0.808484$ | $0.586694$ | $0.045006$ | $0.011058$ |

In Table VI.9 we consider as an example the matrix

$$A = \begin{bmatrix} 4 & 3 \\ -3 & 4 \end{bmatrix}$$

whose eigenvalues are $\lambda_1 = 4 + 3i$, $\lambda_2 = 4 - 3i$.

We conclude from Table VI.9 that $p_2(t) = t^2 - 8t + 25$ from which $\lambda_1 = 4 + 3i$, $\lambda_2 = 4 - 3i$. The expansion coefficients for $U_1$ and $U_2$ will be 1, $(-3 + 3i)/-18$ and 1, $(-3 - 3i)/-18$. From here

$$U_1 = (1, i)',$$
$$U_2 = (1, -i)'.$$

**TABLE VI.9  Computation of a Dual Pair of Bases and Coefficients of the Characteristic Polynomial**

| | $p_0$ | $\tilde{p}_0$ | $Ap_0$ | $A'\tilde{p}_0$ | $p_1$ | $\tilde{p}_1$ | $Ap_1$ | $A'\tilde{p}_1$ | $p_2$ | $\tilde{p}_2$ |
|---|---|---|---|---|---|---|---|---|---|---|
| I | 1 | 1 | 7 | 4 | 0 | -3 | -18 | -21 | 0 | 0 |
|   | 1 | 0 | 1 | 3 | -6 | 3 | -24 | 3 | 0 | 0 |
| II |  |  | 8 | 7 | 0 | 0 | -42 | -18 |  |  |
| III | 1 |  | 7 | 7 | -18 | -18 -18 | -18 | -18 |  |  |
| IV | 1 -18 |  |  |  | 1 |  |  |  |  |  |
| V | 1 |  |  |  | 1 -7 |  |  |  | 1 -8 25 |  |

We shall now assume that among the eigenvalues of matrix $A$ (and consequently of $A'$) are some which are repeated. The corresponding elementary divisors must be relatively prime, since in the opposite case the minimum polynomial of matrix $A$ (and $A'$) would not coincide with the characteristic one and this would lead to a degeneracy for the biorthogonal algorithm no later than at the $m$th step (where $m$ is the degree of the minimum polynomial); this contradicts the assumption that the algorithm's course was non-degenerate. The biorthogonal algorithm permits us to determine the whole canonical basis of a space. Namely, let $\lambda_1, \cdots, \lambda_s$ be distinct eigenvalues of a matrix with multiplicities $n_1, n_2, \cdots, n_s$ where $n_1 + \cdots + n_s = n$. Then there will be $s$ boxes with orders $n_1, \cdots, n_s$ in the Jordan canonical form. There corresponds to each box a canonical basis $U_i^{(0)}, \cdots, U_i^{(n_i-1)}$ for $i = 1, 2, \cdots, s$ where

$$A U_i^{(0)} - \lambda_i U_i^{(0)} = 0,$$
$$A U_i^{(1)} - \lambda_i U_i^{(1)} = U_i^{(0)},$$
$$\cdots \cdots \cdots \cdots \cdots \cdots,$$
$$A U_i^{(n_i-1)} - \lambda_i U_i^{(n_i-1)} = U_i^{(n_i-2)}.$$

The vector $U_i = U_i^{(0)}$ is the eigenvector belonging to the eigenvalue $\lambda_i$.

We carry out the biorthogonal algorithm. It is clear that the last polynomial $p_n(t)$ obtained during the course of the biorthogonal algorithm will be the characteristic polynomial of matrix $A$ (and $A'$). The eigenvectors of matrix $A$ (and $A'$) will also be determined by the previous formulas (6) which however will require substantiation. It is easiest to prove their validity by direct verification. Let

$$U_i = \frac{p_0(\lambda_i)}{(p_0, \tilde{p}_0)} p_0 + \cdots + \frac{p_{n-1}(\lambda_i)}{(p_{n-1}, \tilde{p}_{n-1})} p_{n-1}.$$

On the strength of the relationship

$$A p_i = p_{i+1} + \alpha_i p_i + \beta_i p_{i-1}$$

we have

$$AU_i = \frac{p_0(\lambda_i)}{(p_0, \tilde{p}_0)} (p_1 + \alpha_0 p_0) + \frac{p_1(\lambda_i)}{(p_1, \tilde{p}_1)} (p_2 + \alpha_1 p_1 + \beta_1 p_0) + \cdots$$

$$+ \frac{p_{n-1}(\lambda_i)}{(p_{n-1}, \tilde{p}_{n-1})} (\alpha_{n-1} p_{n-1} + \beta_{n-1} p_{n-2}) = \left[ \frac{\alpha_0 p_0(\lambda_i)}{(p_0, \tilde{p}_0)} + \frac{\beta_1 p_1(\lambda_i)}{(p_1, \tilde{p}_1)} \right] p_0$$

$$+ \left[ \frac{p_0(\lambda_i)}{(p_0, \tilde{p}_0)} + \frac{\alpha_1 p_1(\lambda_i)}{(p_1, \tilde{p}_1)} + \frac{\beta_2 p_2(\lambda_i)}{(p_2, \tilde{p}_2)} \right] p_1 + \cdots$$

$$+ \left[ \frac{p_{n-2}(\lambda_i)}{(p_{n-2}, \tilde{p}_{n-2})} + \frac{\alpha_{n-1} p_{n-1}(\lambda_i)}{(p_{n-1}, \tilde{p}_{n-1})} \right] p_{n-1}.$$

But

$$\frac{\alpha_0 p_0(\lambda_i)}{(p_0, \tilde{p}_0)} + \frac{\beta_1 p_1(\lambda_i)}{(p_1, \tilde{p}_1)} = \frac{\alpha_0 + p_1(\lambda_i)}{(p_0, \tilde{p}_0)} = \frac{\alpha_0 + \lambda_i - \alpha_0}{(p_0, \tilde{p}_0)} = \lambda_i \frac{1}{(p_0, \tilde{p}_0)} = \lambda_i \frac{p_0(\lambda_i)}{(p_0, \tilde{p}_0)}$$

$$\frac{p_0(\lambda_i)}{(p_0, \tilde{p}_0)} + \frac{\alpha_1 p_1(\lambda_i)}{(p_1, \tilde{p}_1)} + \frac{\beta_2 p_2(\lambda_i)}{(p_2, \tilde{p}_2)} = \frac{\beta_1 p_0(\lambda_i) + \alpha_1 p_1(\lambda_i) + p_2(\lambda_i)}{(p_1, \tilde{p}_1)} = \lambda_i \frac{p_1(\lambda_i)}{(p_1, \tilde{p}_1)}$$

$$\cdots \cdots \cdots \cdots \cdots \cdots \cdots \cdots$$

$$\frac{p_{n-2}(\lambda_i)}{(p_{n-2}, \tilde{p}_{n-2})} + \frac{\alpha_{n-1} p_{n-1}(\lambda_i)}{(p_{n-1}, \tilde{p}_{n-1})} = \frac{\beta_{n-1} p_{n-2}(\lambda_i) + \alpha_{n-1} p_{n-1}(\lambda_i)}{(p_{n-1}, \tilde{p}_{n-1})} = \lambda_i \frac{p_{n-1}(\lambda_i)}{(p_{n-1}, \tilde{p}_{n-1})}.$$

Consequently $AU_i = \lambda_i U_i$, i.e. $U_i$ is actually an eigenvector belonging to the eigenvalue $\lambda_i$.

Likewise true are the exact formulas

$$V_i = \frac{p_0(\lambda_i)}{(p_0, \tilde{p}_0)} \tilde{p}_0 + \cdots + \frac{p_{n-1}(\lambda_i)}{(p_{n-1}, \tilde{p}_{n-1})} \tilde{p}_{n-1} \qquad (i = 1, 2, \cdots, s)$$

for the eigenvectors of matrix $A'$.

We now show that the vectors

$$U_i^{(1)} = \frac{p_1'(\lambda_i)}{(p_1, \tilde{p}_1)} p_1 + \cdots + \frac{p_{n-1}'(\lambda_i)}{(p_{n-1}, \tilde{p}_{n-1})} p_{n-1},$$

$$U_i^{(2)} = \frac{p_2''(\lambda_i)}{2!(p_2, \tilde{p}_2)} p_2 + \cdots + \frac{p_{n-1}''(\lambda_i)}{2!(p_{n-1}, \tilde{p}_{n-1})} p_{n-1}, \qquad (8)$$

$$\cdots \cdots \cdots \cdots \cdots \cdots \cdots \cdots$$

$$U_i^{(n_i-1)} = \frac{p_{n_i-1}^{(n_i-1)}(\lambda_i)}{(n_i - 1)!(p_{n_i-1}, \tilde{p}_{n_i-1})} p_{n_i-1} + \cdots + \frac{p_{n-1}^{(n_i-1)}(\lambda_i)}{(n_i-1)!(p_{n-1}, \tilde{p}_{n-1})} p_{n-1}$$

are root vectors belonging to the eigenvalue $\lambda_i$ and forming a canonical basis of the corresponding subspace.

We prove the first of these formulas. To simplify the computation we introduce the notation

$$d_1 = \frac{p_1'(\lambda_i)}{(p_1, \tilde{p}_1)}, \quad \cdots, \quad d_{n-1} = \frac{p_{n-1}'(\lambda_i)}{(p_{n-1}, \tilde{p}_{n-1})}.$$

We have

$$
\begin{aligned}
AU_i^{(1)} - \lambda_i U_i^{(1)} &= d_1 A p_1 + \cdots + d_{n-1} A p_{n-1} - \lambda_i d_1 p_1 - \cdots - \lambda_i d_{n-1} p_{n-1} \\
&= d_1 \beta_1 p_0 + (d_1 \alpha_1 + d_2 \beta_2 - \lambda_i d_1) p_1 + (d_1 + d_2 \alpha_2 + d_3 \beta_3 - \lambda_i d_2) p_2 \\
&\quad + \cdots + (d_{n-2} + d_{n-1} \alpha_{n-1} - \lambda_i d_{n-1}) p_{n-1} \\
&= b_0 p_0 + b_1 p_1 + \cdots + b_{n-1} p_{n-1}.
\end{aligned}
$$

But

$$b_0 = d_1 \beta_1 = \frac{p_1'(\lambda_i)}{(p_1, \tilde{p}_1)} \frac{(p_1, \tilde{p}_1)}{(p_0, \tilde{p}_0)} = \frac{1}{(p_0, \tilde{p}_0)} = \frac{p_0(\lambda_i)}{(p_0, \tilde{p}_0)},$$

$$
\begin{aligned}
b_1 &= d_1 \alpha_1 + d_2 \beta_2 - \lambda_i d_1 = \frac{p_2'(\lambda_i)(p_2, \tilde{p}_2)}{(p_2, \tilde{p}_2)(p_1, \tilde{p}_1)} + \frac{(\alpha_1 - \lambda_i)p_1'(\lambda_i)}{(p_1, \tilde{p}_1)} \\
&= \frac{1}{(p_1, \tilde{p}_1)} [p_2'(\lambda_i) + (\alpha_1 - \lambda_i)p_1'(\lambda_i)] \\
&= \frac{1}{(p_1, \tilde{p}_1)} [p_2'(\lambda_i) + (\alpha_1 - \lambda_i)] = \frac{p_1(\lambda_i)}{(p_1, \tilde{p}_1)},
\end{aligned}
$$

since $p_2'(t) = t - \alpha_1 + p_1(t)$. Furthermore,

$$
\begin{aligned}
b_2 &= d_1 + d_2 \alpha_2 + d_3 \beta_3 - \lambda_i d_2 \\
&= \frac{1}{(p_2, \tilde{p}_2)} [\beta_2 + (\alpha_2 - \lambda_i)p_2'(\lambda_i) + p_3'(\lambda_i)] = \frac{p_2(\lambda_i)}{(p_2, \tilde{p}_2)},
\end{aligned}
$$

since $p_3'(t) - (t - \alpha_2)p_2'(t) - \beta_2 + p_2(t)$.

Analogously we obtain

$$b_k = \frac{p_k(\lambda_i)}{(p_k, \tilde{p}_k)} \qquad (k = 3, \cdots, n - 2).$$

Finally

$$
\begin{aligned}
b_{n-1} &= d_{n-2} + d_{n-1} \alpha_{n-1} - \lambda_i d_{n-1} = \frac{(\alpha_{n-1} - \lambda_i)p_{n-1}'(\lambda_i)}{(p_{n-1}, \tilde{p}_{n-1})} + \frac{p_{n-2}'(\lambda_i)}{(p_{n-2}, \tilde{p}_{n-2})} \\
&= \frac{1}{(p_{n-1}, \tilde{p}_{n-1})} [(\alpha_{n-1} - \lambda_i)p_{n-1}'(\lambda_i) + \beta_{n-1}p_{n-2}'(\lambda_i)] = \frac{p_{n-1}(\lambda_i)}{(p_{n-1}, \tilde{p}_{n-1})}
\end{aligned}
$$

since

$$p_n'(t) = p_{n-1}(t) + (t - \alpha_{n-1})p_{n-1}'(t) - \beta_{n-1}p_{n-2}'(t) \quad \text{and} \quad p_n'(\lambda_i) = 0,$$

since $\lambda_i$ is an $n_i$-tuple root of the polynomial $p_n(t)$, $p_i \geq 2$.

Thus

$$AU_i^{(1)} - \lambda_i U_i^{(1)} = U_i^{(0)}$$

which is what we were to prove

The remaining formulas are verified in an analogous way.

As an example we consider the matrix

$$A = \begin{bmatrix} 4 & 5 & -2 \\ -2 & -2 & 1 \\ -1 & -1 & 1 \end{bmatrix}$$

whose normal form consists of one canonical box belonging to the triple eigenvalue $\lambda_1 = 1$.

**TABLE VI.10**    **Computation of a Dual Pair of Bases and the Coefficients of the Characteristic Polynomial**

| | $p_0$ | $\tilde p_0$ | $Ap_0$ | $A'\tilde p_0$ | $p_1$ | $\tilde p_1$ | $Ap_1$ | $A'\tilde p_1$ | $p_2$ | $\tilde p_2$ | $Ap_2$ | $A'\tilde p_2$ | $p_3$ | $\tilde p_3$ |
|---|---|---|---|---|---|---|---|---|---|---|---|---|---|---|
|     | 1 | 1 | 4 | 4 | 0 | 0 | -8 | -8 | 0 | 0 | 0 | 0 | 0 | 0 |
| I   | 0 | 0 | -2 | 5 | -2 | 5 | 3 | -8 | -1/4 | 1/8 | -1/8 | 0 | 0 | 0 |
|     | 0 | 0 | -1 | -2 | -1 | -2 | 1 | 3 | -5/8 | -1/4 | -3/8 | -1/8 | 0 | 0 |
| II  |   |   | 1 | 7 | 0 |   | -4 | -13 | 0 | 0 | -4/8 | -1/8 |   |   |
|     |   |   |   |   |   |   |   |   | 0 | 0 |   |   |   |   |
| III | 1 |   | 4 | 4 | -8 | -8 | 13 | 13 | 1/8 | 1/8 | 5/64 | 5/64 |   |   |
| IV  | 4 |   |   | -13/8 |   |   |   |   | 5/8 |   |   |   |   |   |
|     | -8 |   |   | -1/64 |   |   |   |   |   |   |   |   |   |   |
|     |   |   |   |   |   |   |   |   |   |   |   |   | 1 |   |
| V   |   |   |   |   |   |   |   |   | 1 |   |   |   | -3 |   |
|     |   |   |   |   | 1 |   |   |   | -19/8 |   |   |   | 3 |   |
|     | 1 |   |   |   | -4 |   |   |   | 3/2 |   |   |   | -1 |   |

According to Table VI.10, $p_3(t) = t^3 - 3t^2 + 3t - 1$ and $\lambda_1 = \lambda_2 = \lambda_3 = 1$.

We compute further

$$p_0(1) = 1 , \qquad p_0'(1) = 0 , \qquad p_0''(1) = 0 ,$$
$$p_1(1) = -3 , \quad p_1'(1) = 1 , \qquad p_1''(1) = 0 ,$$
$$p_2(1) = \frac{1}{8} , \quad p_2'(1) = -\frac{3}{8} , \quad p_2''(1) = 2$$

and

$$U_1^{(0)} = \frac{1}{1}p_0 + \frac{-3}{-8}p_1 + \frac{1/8}{1/8}p_2 = (1, -1, -1)' ,$$

$$U_1^{(1)} = \frac{1}{-8}p_1 + \frac{-3/8}{1/8}p_2 = (0, 1, 2)' ,$$

$$U_1^{(2)} = \frac{2}{2!\,1/8}p_2 = (0, -2, -5)' .$$

Analogously,

$$V_1^{(0)} = \tilde{p}_0 + \frac{3}{8}\tilde{p}_1 + \tilde{p}_2 = (1, \quad 2, \; -1)',$$

$$V_1^{(1)} = \quad -\frac{1}{8}\tilde{p}_1 - 3\tilde{p}_2 = (0, \; -1, \quad 1)',$$

$$V_1^{(2)} = \qquad\qquad 8\tilde{p}_2 = (0, \quad 1, \; -2)'.$$

As a second example we consider a matrix whose canonical form consists of two boxes (one of which is of the first order), namely

$$A = \begin{bmatrix} 13 & 16 & 16 \\ -5 & -7 & -6 \\ -6 & -8 & -7 \end{bmatrix}.$$

Its eigenvalues will be $\lambda_1 = \lambda_2 = 1$, $\lambda_3 = -3$.

**TABLE VI.11  Computation of a Dual Pair of Bases and Coefficients of the Characteristic Polynomial**

| | $p_0$ | $\tilde{p}_0$ | $Ap_0$ | $A'\tilde{p}_0$ | $p_1$ | $\tilde{p}_1$ | $Ap_1$ | $A'\tilde{p}_1$ | $p_2$ | $\tilde{p}_2$ | $Ap_2$ | $A'\tilde{p}_2$ | $p_3$ | $\tilde{p}_3$ |
|---|---|---|---|---|---|---|---|---|---|---|---|---|---|---|
| | 1 | 1 | 13 | 13 | 0 | 0 | -176 | -176 | 0 | 0 | 0 | 0 | 0 | 0 |
| I | 0 | 0 | -5 | 16 | -5 | 16 | 71 | -240 | 16/11 | -192/11 | -16/11 | 64/11 | 0 | 0 |
| | 0 | 0 | -6 | 16 | -6 | 16 | 82 | -208 | -16/11 | 160/11 | -16/11 | 32/11 | 0 | 0 |
| II | | | 2 | 45 | 0 | 0 | -23 | -624 | 0 | 0 | -32/11 | 96/11 | | |
| | | | | | | | | | 0 | 0 | | | | |
| III | | | | | -176 | | | | | -512/11 | | | | |
| | 1 | | 13 | 13 | -176 | -176 | 2448 | 2448 | -512/11 | -512/11 | 512/121 | 512/121 | | |
| IV | 13 | | | | -153/11 | | | | -1/11 | | | | | |
| | -176 | | | | 32/121 | | | | | | | | | |
| V | | | | | 1 | | | | 1 | | | | 1 | |
| | 1 | | | | -13 | | | | 10/11 | | | | 1 | |
| | | | | | | | | | -53/11 | | | | 5 | |
| | | | | | | | | | | | | | 3 | |

According to Table VI.11, $p_3(t) = t^3 + t^2 - 5t + 3$, $\lambda_1 = \lambda_2 = 1$, and $\lambda_3 = -3$. We compute further

$$p_0(\lambda_1) = 1, \qquad p_0'(\lambda_1) = 0, \qquad p_0(\lambda_2) = 1,$$
$$p_1(\lambda_1) = -12, \qquad p_1'(\lambda_1) = 1, \qquad p_1(\lambda_2) = -16,$$
$$p_2(\lambda_1) = -\frac{32}{11}, \qquad p_2'(\lambda_1) = \frac{32}{11}, \qquad p_2(\lambda_2) = \frac{16}{11}$$

and

$$U_1^{(0)} = (1, \; -\tfrac{1}{4}, \; -\tfrac{1}{2})';$$
$$U_1^{(1)} = (0, \; 1, \; -2)';$$
$$U_2^{(0)} = (1, \; -\tfrac{1}{2}, \; -\tfrac{1}{2})'.$$

**2. The degenerate course of the algorithm.** We shall study the possible causes of breakdown for the algorithm. We assume that the algorithm breaks down at the $r$th step so that $(p_i, \tilde{p}_i) \neq 0$, $i = 0, \cdots, r - 1$, and $(p_r, \tilde{p}_r) = 0$. As we have seen, the systems of vectors $p_0, \cdots, p_{r-1}$ and $\tilde{p}_0, \cdots, \tilde{p}_{r-1}$ are linearly independent. In view of the fact that the vectors $p_0, Ap_0, \cdots, A^{r-1}p_0$ are expressed linearly in terms of $p_0, p_1, \cdots, p_{r-1}$, and conversely, we conclude that the vectors $p_0, Ap_0, \cdots, A^{r-1}p_0$ are linearly independent. In exactly the same way the vectors $\tilde{p}_0, A'\tilde{p}_0, \cdots, A'^{r-1}\tilde{p}_0$ will also be linearly independent.

We denote by $P_r$ the subspace spanned by the vectors $p_0, p_1, \cdots, p_{r-1}$ (or, what is the same thing, by the vectors $p_0, Ap_0, \cdots, A^{r-1}p_0$), by $\tilde{P}_r$ the subspace spanned by the vectors $\tilde{p}_0, \tilde{p}_1, \cdots, \tilde{p}_{r-1}$, and by $Q_r$ and $\tilde{Q}_r$ their orthogonal complements.

Then $P_r \cap \tilde{Q}_r = 0$ and $\tilde{P}_r \cap Q_r = 0$. In fact from the conditions $(p_i, \tilde{p}_j) = 0$ for $i \neq j$ and $(p_i, \tilde{p}_i) \neq 0$ it follows that in $P_r$ there does not exist a single vector (except the null one) orthogonal to all the vectors of $\tilde{P}_r$ and in $\tilde{P}_r$ there does not exist a single vector (except the null one) orthogonal to all the vectors of $P_r$.

Four cases are possible for the breakdown of the algorithm at the $r$th step.

1. $p_r = \tilde{p}_r = 0$ (*bilateral breakdown of the algorithm*). In this case $p_r = p_r(A)p_0 = 0$ but because of the linear independence of the vectors $p_0, Ap_0, \cdots, A^{r-1}p_0$ it is impossible for any polynomial $\omega(t)$ of degree less than $r$ to be such that $\omega(A)p_0 = 0$. Consequently the polynomial $p_r(t)$ is the minimum annihilating polynomial for the vector $p_0$. It is likewise established that $p_r(t)$ is the minimum annihilating polynomial for the vector $\tilde{p}_0$ (with respect to matrix $A'$).

The subspace $P_r$ will be invariant with respect to $A$, that is, it will be a cyclic subspace generated by the vector $p_0$. The subspace $\tilde{P}_r$ will be invariant with respect to $A'$. Correspondingly $\tilde{Q}_r$ will be invariant for $A$ and $Q_r$ invariant for $A'$. The dimensions of the subspaces $P_r$ and $\tilde{P}_r$ are the same; consequently the sum of the dimensions of $P_r$ and $\tilde{Q}_r$ is equal to the dimension of the whole space and, since $P_r \cap \tilde{Q}_r = 0$, we find that the whole space is a *direct sum of the invariant* (w. r. t. $A$) *subspaces $P_r$ and $\tilde{Q}_r$*. Thus the whole space is a direct sum of the invariant (w. r. t. $A'$) subspaces $\tilde{P}_r$ and $Q_r$.

Thus in the case of bilateral breakdown the algorithm gives the minimum annilating polynomial for the vector $p_0$. Thereupon it turns out to be possible to expand the space into the direct sum of two invariant subspaces, one of which is cyclic.

2. $p_r = 0$, $\tilde{p}_r \neq 0$ (*unilateral breakdown*). In this case $p_r(t)$ will be the minimum annihilating polynomial for $\tilde{p}_0$. The subspace $\tilde{P}_r$ will not be invariant for $A'$ and $\tilde{Q}_r$ will not be invariant for $A$. As before $R = P_r + \tilde{Q}_r$ but this expansion is not an expansion into a direct sum of the invariant subspaces.

An analogous result holds in the case

3.   $p_r \neq 0$, $\tilde{p}_r = 0$.

In the last case

4.   $p_r \neq 0$, $\tilde{p}_r \neq 0$ but $(p_r, \tilde{p}_r) = 0$ (*dead-end breakdown*). The algorithm "runs into a dead-end", since as a result of carrying it out we do not get either a divisor of the characteristic polynomial of an invariant subspace.

It is possible to show that unilateral and dead-end breakdowns of the algorithm are exceptions and they may be avoided by a suitable choice of the initial vectors $p_0$ and $\tilde{p}_0$. Namely, the vectors $p_0$ and $\tilde{p}_0$ may always be chosen so that as a result of carrying out the algorithm the minimum polynomial of matrix $A$ and a canonical basis of the cyclic subspace generated by the initial vector will be determined.

In the case of bilateral breakdown it is always possible to complete the obtained systems of vectors $p_0, \cdots, p_{r-1}$ and $\tilde{p}_0, \cdots, \tilde{p}_{r-1}$ to biorthogonal bases of the space. The completing is done analogously to the method of minimal iterations.

## 65.   THE METHOD OF *A*-MINIMAL ITERATIONS

We assume that a matrix $A$ is positive definite. Then as we have seen (theorem 11.19) any system of linearly independent vectors may undergo an $A$-orthogonalizations process. If the $A$-orthogonalization process is carried out for the system of vectors $q_0, Aq_0, \cdots, A^{n-1}q_0$, then we shall arrive at the method of $A$-minimal iterations.

This method may be applied to solving the complete eigenvalue problem analogous to the method of minimal iterations described above. As applied to solving a system of equations with matrix $A$ the method of $A$-minimal iterations turns out to be more convenient.

In fact let $q_0, \cdots, q_{n-1}$ be the basis obtained in the $A$-orthogonalization process. If we look for a solution of the system

$$AX = F \tag{1}$$

in the form

$$X = \sum_{i=0}^{n-1} a_i q_i , \tag{2}$$

then from equation

$$\sum_{i=0}^{n-1} a_i A q_i = F$$

we obtain for the coefficients $a_i$ the evident formulas

$$a_i = \frac{(F, q_i)}{(Aq_i, q_i)} . \tag{3}$$

We give the formulas for computing the inverse matrix. The $A$-orthogonality conditions for the vectors $q_0, \cdots, q_{n-1}$ may be written in the matrix form

$$Q'AQ = \Lambda ,\tag{4}$$

where $Q$ is a matrix with columns $q_0, \cdots, q_{n-1}$ and $\Lambda = [(q_0, Aq_0), \cdots, (q_{n-1}, Aq_{n-1})]$. From formula (4) it follows directly that

$$A^{-1} = Q\Lambda^{-1}Q' .\tag{5}$$

It is easy to see that a system of $A$-orthogonal (conjugate) vectors is constructed from the trinomial recurrence relationships

$$q_{i+1} = Aq_i - \gamma_i q_i - \delta_i q_{i-1} , \quad q_1 = Aq_0 - \gamma_0 q_0 ,\tag{6}$$

where

$$\begin{aligned}
\gamma_i &= \frac{(Aq_i, Aq_i)}{(Aq_i, q_i)} = \frac{(Aq_i, Aq_i)}{(Aq_i, Aq_{i-1})} , \\
\delta_i &= \frac{(Aq_i, Aq_{i-1})}{(Aq_{i-1}, q_{i-1})} = \frac{(Aq_i, q_i)}{(Aq_{i-1}, q_{i-1})} .
\end{aligned}\tag{7}$$

This construction process for the vectors $q_i$ will break down only if a certain vector $q_r$ turns out to be zero. As a rule this happens for $r = n$. Premature breakdown implies the linear dependence of successive iterations $q_0, Aq_0, \cdots, A^r q_0$, i.e. the minimum annihilating polynomial for the vector $q_0$ does not coincide with the characteristic polynomial. From the construction,

$$q_i = q_i(A)q_0\tag{8}$$

where $q_i(t)$ are polynomials satisfying the recurrence relationships

$$q_0 = 1 , \quad q_1(t) = t - \gamma_0 , \quad q_{i+1}(t) = (t - \gamma_i)q_i(t) - \delta_i q_{i-1}(t) .\tag{9}$$

The polynomials $q_i(t)$ will obviously satisfy the orthogonality relationships

$$\int_m^M q_i(t)q_j(t)t\,dF(t) = 0 \qquad (i \neq j) ,$$

where $F(t)$ is the weight function for the polynomials $p_i(t)$ in the method of minimal iterations.

The vectors $q_0, \cdots, q_{n-1}$ possess a series of properties analogous to the properties of the vectors $p_0, \cdots, p_{n-1}$.

In particular, the roots of the polynomials $q_0(t), \cdots, q_{n-1}(t), q_n(t)$ are real and separate each other; the vectors $q_0, \cdots, q_{i-1}$ form an $A$-orthogonal basis of the subspace $Q_i$ spanned by the vectors $q_0, Aq_0, \cdots, A^{i-1}q_0$; among all vectors of the form $A^i q_0 + Z$, where $Z \in Q_i$, the vector $q_i$ has the shortest $A$-length.

Moreover if the process continues without degeneracy, then the operator with matrix $A$ has in the basis $q_0, \cdots, q_{n-1}$ the tridiagonal Jacobian matrix

$$J = \begin{bmatrix}
\gamma_0 & \delta_1 & & & & \\
1 & \gamma_1 & \delta_2 & & & \\
& \cdot & \cdot & \cdot & & \\
& & \cdot & \cdot & \cdot & \\
& & & \cdot & \cdot & \cdot \\
& & & \cdot & \gamma_{n-2} & \delta_{n-1} \\
& & & & 1 & \gamma_{n-1}
\end{bmatrix}\tag{10}$$

From this it follows that

$$\sum_{i=0}^{n-1} r_i = SpA .$$ (11)

Finally $(Aq_i, Aq_j) = 0$ for $|i - j| > 1$.

In the case of premature breakdown of the process we may always complete the system of $A$-orthogonal vectors to a basis of the whole space, as was done in the method of minimal iterations.

It is necessary to make such a completion if the problem arises of inverting a matrix or of solving a large system of equations with the same coefficient matrix $A$ but with different constant terms.

In solving such a system, completion turns out to be unnecessary if we take the constants of the system for the initial vector $q_0$, since for such a choice of the initial vector the premature breakdown of the process only decreases the volume of computations.

In fact let the process of $A$-minimal iterations breakdown at the $r$th step. Then $q_r(t) = t^r + d_1 t^{r-1} + \cdots + d_{r-1}t + d_r$ is the minimum annihilating polynomial for the vector $F$ so that

$$A^r F + d_1 A^{r-1} F + \cdots + d_{r-1} AF + d_r F = 0 ,$$

from which

$$X = A^{-1}F + \frac{1}{d_r}[- A^{r-1}F - \cdots - d_{r-1}F] .$$

Therefore $X$ belongs to the subspace $Q_r$.

The vectors $q_0, \cdots, q_{r-1}$ form a basis of the subspace $Q_r$ so that the solution $X$ is written as

$$X = \sum_{i=0}^{r-1} a_i q_i .$$

For the coefficients $a_i$ the previous formulas

$$a_i = \frac{(F, q_i)}{(Aq_i, q_i)} \qquad (i = 0, 1, \cdots, r - 1) .$$

are preserved.

With similar success we may take, instead of the constant $F$ as the initial vector $q_0$, any residual $r_0 = F - AX_0$ where $X_0$ is an arbitrary vector. In fact the system $AX = F$ is equivalent to the system $A(X - X_0) = r_0$ whose constant is $r_0$. In this case the system's solution is obtained in the form

$$X = X_0 + \sum_{i=0}^{r-1} a_i q_i$$

where

$$a_i = \frac{(r_0, q_i)}{(Aq_i, q_i)} .$$

If matrix $A$ is symmetric but not positive definite, then construction of the vectors $q_i$ is likewise possible. However, in this case a "dead-end"

TABLE VI.12 Solving a Linear System by the Method of A-Minimal Iterations

| | $q_0$ | $Aq_0$ | $q_1$ | $Aq_1$ | $q_2$ | $Aq_2$ | $q_3$ | $Aq_3$ | $X$ |
|---|---|---|---|---|---|---|---|---|---|
| I | 1 | 2.62 | 0.302 | 0.11684 | -0.06303198 | -0.01323917 | 0.000105619 | -0.001164594 | -1.2577936 |
| | 1 | 2.18 | -0.138 | -0.08644 | -0.01622351 | -0.00529574 | 0.007640798 | 0.004960303 | 0.0434874 |
| | 1 | 2.08 | -0.238 | -0.11864 | 0.00841478 | -0.01345978 | -0.006534421 | -0.004349251 | 1.0391662 |
| | 1 | 2.32 | 0.002 | 0.08824 | 0.07888289 | 0.03199469 | -0.001440546 | 0.000553541 | 1.4823928 |
| II | | 9.20 | 0 | 0 | $0.78 \cdot 10^{-8}$ | $1 \cdot 10^{-9}$ | $-0.98 \cdot 10^{-10}$ | $-1 \cdot 10^{-9}$ | |
| | | | | | $0.38 \cdot 10^{-9}$ | | $-0.13 \cdot 10^{-9}$ | | |
| | | | | | | | $-0.59 \cdot 10^{-10}$ | | |
| III | 9.2 | 21.3256 | 0.075627200 | 0.042985206 | 0.0033309791 | 0.0014081463 | 0.00006540106 | | |
| | | | | | 0.0033309789 | | 0.00006540117 | | |
| IV | 2.318 | | 0.56838283 | | 0.42274246 | | | | |
| | 0.008220348 | | 0.04404472 | | | | | | |
| V | 0.24 | | -0.1432 | | 0.049863598 | | -0.0020185014 | | |
| | 0.26086957 | | -1.8934986 | | 14.969652 | | -30.863885 | | |

termination of the process may occur in which the scalar product $(Aq_i, q_i)$ may breakdown to zero at some step although $q_i \neq 0$. Just as in the bi-orthogonal algorithm the dead-end termination is connected with an unsuccessful choice of the initial vector and may be eliminated by re-selecting it.

We shall consider examples for solving a system and for inverting a matrix. We observe that the computational scheme for constructing the vectors $q_0, \cdots, q_{n-1}$ hardly differs from the computational scheme for the minimal iteration method. A good final control for the non-degenerate course of the process is satisfaction of equality (11):

$$\sum_{i=0}^{n-1} r_i = SpA$$

In Table VI.12 is presented the solution of system (9), Section 23.

Parts I-IV of the table are analogous to the corresponding parts of Table VI.1. In part V are written $(F, q_i)$ and the coefficients $a_i$.

The solution obtained (cf. Section 23) is accurate to within $2 \cdot 10^{-7}$ in each component.

We shall also compute the inverse matrix for matrix $A$. It is advisable to compute beforehand the matrix $\tilde{Q} = Q\Lambda^{-1}$ with elements $\tilde{q}_{ij} = q_{ij}/(Aq_i, q_i)$ where $q_{ij}$ is the $j$th component of the vector $q_i$. We have

$$\tilde{Q} = \begin{bmatrix} 0.10869565 & 3.9932723 & -18.922959 & 1.6149668 \\ 0.10869565 & -1.8247403 & -4.8704929 & 116.83158 \\ 0.10869565 & -3.1470159 & 2.5262182 & -99.914532 \\ 0.10869565 & 0.0264455 & 23.681593 & -22.026662 \end{bmatrix}.$$

Then $A^{-1} = \tilde{Q}Q'$, according to formula (5). By computing we obtain

$$A^{-1} = \begin{bmatrix} 2.507586 & -0.123039 & -1.011489 & -1.378342 \\ -0.123039 & 1.332213 & -0.261427 & -0.447454 \\ -1.011489 & -0.261427 & 1.531827 & 0.445608 \\ -1.378342 & -0.447454 & 0.445608 & 2.008551 \end{bmatrix}.$$

Discrepancies with the values of Section 23 are observed only in three elements and do not exceeding $1 \cdot 10^{-6}$ in any element.

In the case of the normal course for the algorithm all eigenvalues of the matrix are found as roots of the polynomial $q_n(t)$. The following formula is valid for the eigenvectors:

$$U_i = \frac{q_0(\lambda_i)}{(Aq_0, q_0)} q_0 + \cdots + \frac{q_{n-1}(\lambda_i)}{(Aq_{n-1}, q_{n-1})} q_{n-1}. \tag{12}$$

In the case of breakdown in the algorithm (not dead-end) the last polynomial gives the minimum annihilating polynomial for $q_0$ and the eigenvectors found according to the formula

$$U_i = \frac{q_0(\lambda_i)}{(Aq_0, q_0)} q_0 + \cdots + \frac{q_{r-1}(\lambda_i)}{(Aq_{r-1}, q_{r-1})} q_{r-1}, \tag{12'}$$

form a basis of the subspace $Q_r$. The completion process may be used to determine all eigenvalues and the complete system of eigenvectors.

We shall determine the coefficients of the characteristic polynomial and one of the eigenvectors for the matrix considered above. We use computations carried out on the basis of Table VI.12. First of all,

$$(Aq_3, Aq_3) = 0.000045183277 ,$$

$$\gamma_3 = 0.69087467 ,$$

$$\delta_3 = 0.01963390 .$$

We then compute recurrently the coefficients of the characteristic polynomial:

|  |  |  |  | 1 |
|---|---|---|---|---|
|  |  |  | 1 | $-4.00000001$ |
|  |  | 1 | $-3.30912534$ | $4.75200003$ |
|  | 1 | $-2.88638288$ | $2.48544305$ | $-2.11185600$ |
| 1 | $-2.318$ | $1.30929117$ | $-0.45139731$ | $0.28615248$ |

Thus

$$\varphi(t) = q_4(t) = t^4 - 4.00000001t^3 + 4.75200003t^2$$
$$- 2.11185600t + 0.28615248 .$$

One of the roots of this polynomial is $\lambda_3 = 0.63828371$. We shall determine the eigenvector belonging to this eigenvalue. We first compute the expansion coefficients by the scheme:

$$
\begin{array}{rl}
\lambda_3 & 0.63828371 \\
q_0(\lambda_3) & 1 \\
q_1(\lambda_3) = \lambda_3 - \gamma_0 & -1.6797163 \\
\lambda_3 - \gamma_1 & 0.06990083 \\
q_2(\lambda_3) & -0.12563391 \\
\lambda_3 - \gamma_2 & 0.21554125 \\
q_3(\lambda_3) & 0.046903344 \\
\lambda_3 - \gamma_3 & -0.05259096 \\
q_4(\lambda_3) & -0.8263\cdot 10^{-8} \\
q_4'(\lambda_3) & 0.106 \\
q_0(\lambda_3)/(q_0, q_0) & 0.10869565 \\
q_1(\lambda_3)/(q_1, q_1) & -22.210478 \\
q_2(\lambda_3)/(q_2, q_2) & -37.716811 \\
q_3(\lambda_3)/(q_3, q_3) & 717.17535 .
\end{array}
$$

The eigenvector is now found from formula (12). Namely,

$$U_3 = (-4.145756, 9.265433, 0.391085, -3.944060)'$$
$$= 9.265433(-0.447443, 1.000000, 0.042209, -0.425675)' .$$

The largest discrepancy from the values of Section 63 does not exceed $1\cdot 10^{-6}$.

If the matrix of the system $AX = F$ is not symmetric, then we may still find a solution of the system of equations by the method of $A$-minimal iterations; we must apply this to an equivalent system obtained by a first or second Gauss transformation.

In both cases the computational formulas may be transformed so as to avoid actually computing the corresponding matrices $A'A$ or $AA'$.

For the first transformation we have

$$
\begin{aligned}
& q_{i+1} = A'Aq_i - \gamma_i q_i - \delta_i q_{i-1} = A'v_i - \gamma_i q_i - \delta_i q_{i-1} , \\
& v_i = Aq_i , \\
& \gamma_i = \frac{(A'Aq_i, A'Aq_i)}{(A'Aq_i, q_i)} = \frac{(A'v_i, A'v_i)}{(v_i, v_i)} , \\
& \delta_i = \frac{(A'Aq_i, q_i)}{(A'Aq_{i-1}, q_{i-1})} = \frac{(v_i, v_i)}{(v_{i-1}, v_{i-1})} , \\
& X = \sum_{i=0}^{n-1} \frac{(A'F, q_i)}{(A'Aq_i, q_i)} q_i = \sum_{i=0}^{n-1} \frac{(F, v_i)}{(v_i, v_i)} q_i .
\end{aligned}
\tag{13}
$$

The method described will be called the *method of $A'A$-minimal iterations*.

As a check we may use the pairwise orthogonality of the vectors $v_i$. In fact

$$ (v_i, v_j) = (Aq_i, Aq_j) = (A'Aq_i, q_j) = 0 . $$

We observe that

$$ A^{-1} = V\Lambda^{-1}Q' \tag{14} $$

where the matrices $V$ and $Q$ are formed from the columns $v_0, v_1, \cdots, v_{n-1}$, and $q_0, q_1, \cdots, q_{n-1}$ respectively and $\Lambda = [(v_0, v_0), \cdots, (v_{n-1}, v_{n-1})]$.

The second Gauss transformation leads analogously to the *method of $AA'$-minimal iterations*.

The computational formulas for the method are

$$
\begin{aligned}
& w_i = A'q_i , \\
& q_{i+1} = Aw_i - \gamma_i q_i - \delta_i q_{i-1} , \\
& \gamma_i = \frac{(Aw_i, Aw_i)}{(w_i, w_i)} , \\
& \delta_i = \frac{(w_i, w_i)}{(w_{i-1}, w_{i-1})} , \\
& X = A' \sum_{i=0}^{n-1} a_i q_i = \sum_{i=0}^{n-1} a_i w_i , \\
& a_i = \frac{(F, q_i)}{(w_i, w_i)} .
\end{aligned}
\tag{15}
$$

It is easy to see that $(w_i, w_j) = 0$.

In Tables VI.13 and VI.14 is given the solution of the system (according to the data of Table II.1) by the methods of $A'A$- and $AA'$-minimal iterations. In the second part of each table the control relationships are satisfied for orthogonality.

**TABLE VI.13**   **The Solution of a System of Equations by the Method of $A'A$-Minimal**

| | $q_0$ | $v_0$ | $A'v_0$ | $q_1$ | $v_1$ | $A'v_1$ |
|---|---|---|---|---|---|---|
| I | 1 | 1.46 | 3.5474 | 0.6194927 | 0.45089334 | 0.34115826 |
| | 1 | 1.82 | 3.2494 | 0.3214927 | 0.36342271 | 0.16392188 |
| | 1 | 0.50 | 2.1963 | −0.7311073 | −0.13031765 | −0.19255928 |
| | 1 | 2.34 | 2.1760 | −0.7519073 | −0.53614308 | −0.31252088 |
| II | | 6.12 | 11.1696 | | $-2 \cdot 10^{-8}$ | $-2 \cdot 10^{-8}$ |
| III | | 11.1696 | 32.703553 | | 0.63981296 | 0.27800772 |
| IV | 2.9279073 | | | 0.43451405 | | |
| | 0.057281636 | | | 0.015893004 | | |
| V | 0.34056725 | 3.804 | | −0.40132318 | −0.25677177 | |

**TABLE VI.14**   **The Solution of a System of Equations by the Method of $AA'$-Minimal**

| | $q_0$ | $w_0$ | $Aw_0$ | $q_1$ | $w_1$ | $Aw_1$ |
|---|---|---|---|---|---|---|
| I | 1 | 0.45 | 0.3099 | −0.45860226 | −0.43566002 | −0.19902223 |
| | 0 | −0.26 | −0.3100 | −0.31000000 | −0.43491841 | −1.04409661 |
| | 0 | −0.61 | −0.4101 | −0.41010000 | −0.42229262 | −0.80751843 |
| | −1 | −0.46 | −0.5439 | 0.22460226 | 0.37963104 | −0.19902223 |
| II | | −0.88 | −0.9541 | | $-0.26 \cdot 10^{-8}$ | −2.24965950 |
| III | | 0.8538 | 0.65614723 | | 0.70140446 | 1.8214434 |
| IV | 0.76850226 | | | 2.5968518 | | |
| | 0.82150909 | | | 0.026084155 | | |
| V | −0.6 | −0.70274069 | | −0.37750864 | −0.53821819 | |

**Iterations**

| $q_2$ | $v_2$ | $A'v_2$ | $q_3$ | $v_3$ | $X$ |
|---|---|---|---|---|---|
| 0.01469834 | −0.03114716 | −0.03875003 | −0.09027375 | −0.03919682 | 0.4408884 |
| −0.03305285 | 0.03309338 | 0.04742345 | 0.13603752 | 0.03886464 | −0.3630309 |
| 0.06783548 | 0.08653546 | 0.10756822 | −0.07316431 | −0.02885813 | 1.1667982 |
| −0.04308823 | −0.02479610 | −0.11624169 | 0.01788792 | 0.00039434 | 0.3935674 |
| | −0.2·10$^{-8}$ | | | 0.2·10$^{-8}$ | |
| | −5·10$^{-8}$ | | | 1·10$^{-8}$ | |
| | | | | −2·10$^{-8}$ | |
| | 0.010168550 | 0.028833601 | | 0.0038797981 | |
| 2.8355666 | | | | | |
| 4.4707328 | 0.04546087 | | −3.1374083 | −0.0012172511 | |

**Iterations**

| $q_2$ | $w_2$ | $Aw_2$ | $q_3$ | $w_3$ | $X$ |
|---|---|---|---|---|---|
| 0.17039078 | 0.05128257 | 0.025854965 | 0.009691839 | 0.023329767 | 0.4408884 |
| −0.23907255 | −0.10313037 | −0.038447305 | 0.009101047 | 0.010441822 | −0.3630310 |
| 0.25745049 | 0.06879630 | 0.014310165 | −0.017488523 | −0.008946124 | 1.1667983 |
| 0.03922808 | 0.01722895 | 0.025854972 | 0.013521269 | 0.028784034 | 0.3935672 |
| | −0.73·10$^{-8}$ | 0.027572797 | | 0.143·10$^{-8}$ | |
| | 0.55·10$^{-8}$ | | | 0.753·10$^{-8}$ | |
| | | | | −0.105·10$^{-9}$ | |
| | 0.018295543 | 0.0030199349 | | 0.0015618634 | |
| 0.16506397 | | | | | |
| 0.14710157 | 8.0402954 | | 0.0073852512 | 4.7284873 | |

## 66. THE $A$-BIORTHOGONAL ALGORITHM

Along with the symmetrization methods described above for solving a system with a non-symmetric matrix we may use a modification of the biorthogonal algorithm, the $A$-biorthogonal algorithm. Essentially it consists of constructing two systems of vectors $q_0, \cdots, q_{n-1}$ and $\tilde{q}_0, \cdots, \tilde{q}_{n-1}$ such that $(Aq_i, \tilde{q}_j) = 0$ for $i \neq j$ and $(Aq_i, \tilde{q}_i) \neq 0$; these form a dual pair of conjugate bases.

The vector systems are constructed from the recurrent relationships

$$q_1 = Aq_0 - \gamma_0 q_0,$$
$$q_{i+1} = Aq_i - \gamma_i q_i - \delta_i q_{i-1},$$
$$\tilde{q}_1 = A'\tilde{q}_0 - \gamma_0 \tilde{q}_0,$$
$$\tilde{q}_{i+1} = A'\tilde{q}_i - \gamma_i \tilde{q}_i - \delta_i \tilde{q}_{i-1}$$

(1)

where

$$\gamma_i = \frac{(Aq_i, A'\tilde{q}_i)}{(Aq_i, \tilde{q}_i)},$$
$$\delta_i = \frac{(Aq_i, \tilde{q}_i)}{(Aq_{i-1}, \tilde{q}_{i-1})}.$$

(2)

After constructing the dual pair of conjugate bases the solution of the linear system is computed by the formula

$$X = \sum_{i=0}^{n-1} \frac{(F, \tilde{q}_i)}{(Aq_i, \tilde{q}_i)} q_i.$$

(3)

We shall not investigate the possible degeneracies in the process, since this would be almost a literal repetition of paragraph 2, Section 64. We observe only that unilateral and dead-end terminations of the process may be avoided by changing to another system of initial vectors.

The $A$-biorthogonal algorithm makes it possible to solve the complete eigenvalue problem. First, in the normal course of the algorithm the polynomial $q_n(t)$ in the sequence of polynomials constructed by the formulas

$$q_{i+1}(t) = tq_i(t) - \gamma_i q_i(t) - \delta_i q_{i-1}(t), \quad q_1(t) = t - \gamma_0$$

(4)

is the characteristic polynomial. Finding its roots enables us to construct the corresponding eigenvectors by the formulas

$$U_i = \frac{q_0(\lambda_i)}{(Aq_0, \tilde{q}_0)} q_0 + \cdots + \frac{q_{n-1}(\lambda_i)}{(Aq_{n-1}, \tilde{q}_{n-1})} q_{n-1}.$$

(5)

The canonical basis in the case where matrix $A$ is not brought to diagonal form is found from formulas analogous to the formulas of the biorthogonal algorithm.

In Table VI.15 is given the solution of the system with the data of Table II.1 using the $A$-biorthogonal algorithm.

## 67. BINOMIAL FORMULAS FOR THE METHOD OF MINIMAL ITERATIONS AND THE BIORTHOGONAL ALGORITHM

The systems of vectors $p_0, \cdots, p_{n-1}$ and $q_0, \cdots, q_{n-1}$ are closely related to each other and they may be computed simultaneously from simpler formulas than the trinomial ones which determine each system separately. The following theorems are valid.

THEOREM 67.1. *If $A$ is a positive definite matrix, and the vectors $X$, $AX$, $\cdots, A^{n-1}X$ are linearly independent, and the vectors $p_0, \cdots, p_{n-1}$ are obtained from them by the orthogonalization process, and the vectors $q_0, \cdots, q_{n-1}$ by the A-orthogonalization process, then among the vectors of the constructed systems there exist the following binomial relationships*:

$$p_0 = q_0 ,$$
$$p_{i+1} = Aq_i - \rho_i p_i , \qquad (1)$$
$$q_{i+1} = p_{i+1} - \sigma_{i+1} q_i ,$$

*where*

$$\rho_i = \frac{(p_i, Aq_i)}{(p_i, p_i)} = \frac{(q_i, Aq_i)}{(p_i, p_i)} ,$$

$$\sigma_{i+1} = \frac{(p_{i+1}, p_{i+1})}{(p_i, Aq_i)} = \frac{(p_{i+1}, p_{i+1})}{(q_i, Aq_i)} . \qquad (2)$$

*Proof.* The equality $p_0 = q_0$ follows directly from the construction. We denote by $P_i$ the subspace spanned by the vectors $X$, $AX$, $\cdots$, $A^i X$ and by $\tilde{P}_i$ the set of vectors $A^i X + Y$ where $Y \in P_{i-1}$. The vectors $p_0, \cdots, p_i$ as well as the vectors $q_0, \cdots, q_i$ form bases of the subspace $P_i$ where $p_i \in \tilde{P}_i$, $q_i \in \tilde{P}_i$. Furthermore if $Z \in \tilde{P}_i$, then $AZ \in \tilde{P}_{i+1}$ and in particular $Ap_i \in \tilde{P}_{i+1}$ and $Aq_i \in \tilde{P}_{i+1}$. We consider the vector $p_{i+1} - Aq_i$. The vector $Aq_i$ is orthogonal to the vectors $q_0, \cdots, q_{i-1}$ and consequently to all vectors of the subspace $P_{i-1}$, in particular, to the vectors $p_0, \cdots, p_{i-1}$. The vector $p_{i+1}$ is also orthogonal to $p_0, \cdots, p_{i-1}$. Consequently the vector $p_{i+1} - Aq_i$ is also orthogonal to $p_0, \cdots, p_{i-1}$. On the other hand, the vector $p_{i+1} - Aq_i$, as the difference of two vectors from the set $\tilde{P}_{i+1}$, belongs to the subspace $P_i$. Therefore $p_{i+1} - Aq_i = -\rho_i p_i$ where $\rho_i$ is some number. It is clear that

$$\rho_i = -\frac{(p_{i+1} - Aq_i, p_i)}{(p_i, p_i)} = \frac{(Aq_i, p_i)}{(p_i, p_i)} ,$$

since $(p_{i+1}, p_i) = 0$.

We now consider the vector $q_{i+1} - p_{i+1}$. This vector belongs to the subspace $P_i$. Moreover the vector $p_{i+1}$ is orthogonal to all the vectors of the subspace $P_i$ and in particular to the vectors $Aq_0, \cdots, Aq_{i-1}$ so that the vector $p_{i+1}$ is A-orthogonal to the vectors $q_0, \cdots, q_{i-1}$. The vector $q_{i+1}$ is A-orthogonal to the vectors $q_0, \cdots, q_{i-1}$ by construction. Consequently the

**TABLE VI.15  Solving a System of Linear Equations using the A-Biorthogonal Algorithm**

| 0 | $q_0$ | $\tilde{q}_0$ | $Aq_0$ | $A'\tilde{q}_0$ | $q_1$ | $\tilde{q}_1$ | $Aq_1$ | $A'\tilde{q}_1$ |
|---|---|---|---|---|---|---|---|---|
| I | 0<br>1<br>0<br>0 | 0<br>1<br>1<br>0 | 0.17<br>1.00<br>0.35<br>0.43 | 0.36<br>1.35<br>1.67<br>-1.06 | 0.17<br>-0.1406667<br>0.35<br>0.43 | 0.36<br>0.2093333<br>0.5293333<br>-1.06 | 0.29078666<br>0.03613330<br>-0.03613334<br>0.58901332 | -0.18284001<br>-0.00000004<br>0.19798661<br>-1.3242933 |
| II |  | 1.35 | 1.95 | 2.32 | $-0.45\cdot10^{-7}$ | $-0.45\cdot10^{-7}$ | 0.87979994 | -1.3091467 |
| III |  |  | 1.5399 |  | -0.53123360 | -0.53123360 | -0.84034775 |  |
| IV | 1.1406667<br>-0.39350637 |  |  |  | 1.5818799<br>0.1623868 |  |  |  |
| V | 1.2<br>0.88888889 |  |  |  | -0.37080004<br>0.69799809 |  |  |  |

| | $q_2$ | $\tilde{q}_2$ | $Aq_2$ | $A'\tilde{q}_2$ | $q_3$ | $\tilde{q}_3$ | $Aq_3$ | $X$ |
|---|---|---|---|---|---|---|---|---|
| I | 0.02186708<br>0.65215750<br>-0.58979130<br>-0.09119504 | -0.75231677<br>0.06236619<br>-0.24584873<br>0.35249939 | 0.23093636<br>0.29645727<br>-0.29645722<br>-0.01106529 | -0.50208664<br>0.00000002<br>0.11091559<br>0.10821921 | 0.16530023<br>-0.81490787<br>0.67245006<br>0.07771139 | 0.77785489<br>-0.14245783<br>0.45252968<br>-0.33270445 | -0.09938247<br>-0.31154287<br>0.31154285<br>0.06029815 | 0.4408885<br>-0.3630309<br>1.1667983<br>0.3935673 |
| II | $0.452\cdot10^{-7}$<br>$-0.284\cdot10^{-8}$ | $0.213\cdot10^{-7}$<br>$-0.128\cdot10^{-7}$ | 0.21987112 | -0.28295182 | $-0.149\cdot10^{-7}$<br>$-0.284\cdot10^{-7}$<br>$0.701\cdot10^{-8}$ | $-0.242\cdot10^{-7}$<br>$-0.734\cdot10^{-8}$<br>$0.706\cdot10^{-8}$ | -0.03908434 |  |
| III | -0.08626526 | -0.08626526 | -0.15002936 |  | 0.09097897 | 0.09097898 |  |  |
| IV | 1.7391620 |  |  |  |  |  |  |  |
| V | -0.04935660<br>0.57214921 |  |  |  | 0.17046432<br>1.8736671 |  |  |  |

vector $q_{i+1} - p_{i+1}$ is also $A$-orthogonal to the vectors $q_0, \cdots, q_{i-1}$. Therefore $q_{i+1} - p_{i+1} = -\sigma_{i+1}q_i$ where $\sigma_{i+1}$ is some number. It is obvious that

$$\sigma_{i+1} = -\frac{(q_{i+1} - p_{i+1}, Aq_i)}{(Aq_i, q_i)} = \frac{(p_{i+1}, Aq_i)}{(Aq_i, q_i)} .$$

But $Aq_i - p_{i+1} \in P_i$ and $q_i - p_i \in P_{i-1}$ so that $(p_{i+1}, Aq_i) = (p_{i+1}, q_{i+1})$ and $(Aq_i, q_i) = (Aq_i, p_i)$. Consequently

$$\sigma_{i+1} = \frac{(p_{i+1}, p_{i+1})}{(q_i, Aq_i)} = \frac{(p_{i+1}, p_{i+1})}{(p_i, Aq_i)}; \quad \rho_i = \frac{(q_i, Aq_i)}{(p_i, p_i)} .$$

The theorem is proved.

*Note 1.* If $A$ is symmetric but not positive definite, then the theorem remains true under the condition that the $A$-orthogonalization process does not have a dead-end termination.

*Note 2.* The polynomials $p_i(t)$ and $q_i(t)$ are also related by the binomial relationships

$$p_{i+1}(t) = tq_i(t) - \rho_i p_i(t) ,$$
$$q_{i+1}(t) = p_{i+1}(t) - \sigma_{i+1} q_i(t) . \tag{3}$$

It is easy to establish the connections between the coefficients of the binomial formulas $\rho_i$ and $\sigma_i$ and the coefficients of the trinomial formulas $\alpha_i$ and $\beta_i$ or $\gamma_i$ and $\delta_i$. Namely,

$$\alpha_0 = \rho_0 ,$$
$$\alpha_i = \rho_i + \sigma_i \quad (i = 1, 2, \cdots, n-1) , \tag{4}$$
$$\beta_i = \rho_{i-1}\sigma_i \quad (i = 1, 2, \cdots, n-1) ,$$

and

$$\gamma_i = \rho_i + \sigma_{i+1} \quad (i = 0, 1, \cdots, n-1) ,$$
$$\delta_i = \rho_i \sigma_i \quad (i = 1, 2, \cdots, n-1) . \tag{5}$$

In fact, because of the binomial relationships (1), we have

$$p_{i+1} = Aq_i - \rho_i p_i ,$$
$$Aq_i = Ap_i - \sigma_i Aq_{i-1} , \tag{6}$$
$$p_i = Aq_{i-1} - \rho_{i-1} p_{i-1} .$$

Eliminating $Aq_i$ and $Aq_{i-1}$ from these relationships, we obtain

$$p_{i+1} = Ap_i - \sigma_i(p_i + \rho_{i-1}p_{i-1}) - \rho_i p_i$$
$$= Ap_i - (\sigma_i + \rho_i)p_i - \sigma_i \rho_{i-1} p_{i-1}$$
$$= Ap_i - \alpha_i p_i - \beta_i p_{i-1} .$$

From this we obtain, because of the linear independence of the vectors $p_i$ and $p_{i-1}$,

$$\alpha_i = \sigma_i + \rho_i ,$$
$$\beta_i = \rho_{i-1}\sigma_i .$$

**TABLE VI.16**    **Simultaneous Computation of Biorthogonal and A-Biorthogonal**

| | $p_0$ | $\tilde{p}_0$ | $q_0$ | $\tilde{q}_0$ | $Aq_0$ | $A'\tilde{q}_0$ |
|---|---|---|---|---|---|---|
| I | 0 | 0 | 0 | 0 | 1.870086 | 0.336964 |
| | 1 | 1 | 1 | 1 | −11.811654 | −7.503621 |
| | 0 | 1 | 0 | 1 | 4.308033 | −7.258787 |
| | 0 | 0 | 0 | 0 | 0.269851 | 0.288043 |
| II | | | | | −5.363684 | −14.137401 |
| III | 1 | | | | −7.503621 | −7.503621 |
| | | | | | −7.503621 | −7.503621 |
| IV | 7.503621 | | −0.23490436 | | | |
| V | 1 | | 1 | | | |

| | $p_2$ | $\tilde{p}_2$ | $q_2$ | $\tilde{q}_2$ | $Aq_2$ | $A'\tilde{q}_2$ |
|---|---|---|---|---|---|---|
| I | 8.6346552 | 2.6926058 | 11.594561 | 3.2259404 | −49.43280 | −17.688366 |
| | 16.298936 | 0 | 9.852133 | 0.3717982 | −166.93761 | 0 |
| | −16.298936 | −1.0979091 | −9.480335 | −0.3385964 | 166.93761 | 1.934482 |
| | 3.7527932 | −1.1323868 | 4.179904 | −0.6764826 | −84.06708 | 11.876143 |
| II | 0 | 0 | $2.12 \cdot 10^{-6}$ | $0.31 \cdot 10^{-6}$ | −133.49987 | 1.122259 |
| | $0.07 \cdot 10^{-6}$ | $0.07 \cdot 10^{-6}$ | $-1.24 \cdot 10^{-6}$ | $0.66 \cdot 10^{-6}$ | | |
| III | 36.894859 | 36.894860 | | | −221.18891 | |
| | | 36.894860 | | | −221.18892 | |
| IV | −5.9951147 | | 1.3221017 | | | |
| V | 1 | | 1 | | | |
| | 20.963280 | | 22.546044 | | | |
| | 99.233549 | | 111.48181 | | | |

**Systems**

| $p_1$ | $\tilde{p}_1$ | $q_1$ | $\tilde{q}_1$ | $Aq_1$ | $A'\tilde{q}_1$ |
|---|---|---|---|---|---|
| 1.870086 | 0.336964 | 1.8700860 | 0.33696400 | −16.096774 | −1.7636605 |
| −4.308033 | 0 | −4.0731286 | 0.23490436 | 73.271617 | 0.00000001 |
| 4.308033 | 0.244834 | 4.3080330 | 0.47973836 | −73.271617 | −4.3357788 |
| 0.269851 | 0.288043 | 0.2698510 | 0.28804300 | 0.18407987 | −4.9416849 |
| 0 | 0 | $-0.29\cdot10^{-6}$ | $0.01\cdot10^{-6}$ | −15.9126943 | −11.0411242 |
| 1.7626333 | 1.7626333 | | | −23.310394 | −23.310394 |
| | 1.7626333 | | | −23.310394 | −23.310394 |
| −13.224755 | | −1.5827643 | | | |
| 1 | | 1 | | | |
| 7.503621 | | 7.738525 | | | |

| $p_3$ | $\tilde{p}_3$ | $q_3$ | $\tilde{q}_3$ | $Aq_3$ | $p_4$ |
|---|---|---|---|---|---|
| 2.332948 | −1.5458854 | −12.996241 | −5.8109067 | −48.220650 | −0.00008 |
| −69.223619 | 0 | −82.249141 | −0.4915550 | 1430.8088 | 0.00006 |
| 69.223619 | 0.3523910 | 81.757586 | 0.8000499 | −1430.8088 | −0.00006 |
| −61.568654 | 5.0873542 | −67.094912 | 5.9817370 | 1272.5854 | −0.00001 |
| 0 | 0 | $-5.86\cdot10^{-6}$ | $0.32\cdot10^{-6}$ | | |
| $-4.78\cdot10^{-6}$ | $1.03\cdot10^{-6}$ | $61.46\cdot10^{-6}$ | $11.03\cdot10^{-6}$ | | |
| $-0.84\cdot10^{-6}$ | $2\cdot48\cdot10^{-6}$ | $6.43\cdot10^{-6}$ | $2.86\cdot10^{-6}$ | | |
| −292.43424 | −292.43426 | | | 6044.4321 | |
| | −292.43424 | | | 6044.4321 | |
| −20.669372 | | | | | |
| | | | | | 1 |
| 1 | | 1 | | | 47.888429 |
| 28.541159 | | 27.219057 | | | 797.27875 |
| 237.15908 | | 207.35092 | | | 5349.4555 |
| 594.91651 | | 447.52622 | | | 12296.551 |

**TABLE VI.17   Simultaneous Computation of Orthogonal and $A$-Orthogonal Bases**

| | $p_0$ | $q_0$ | $Aq_0$ | $p_1$ | $q_1$ | $Aq_1$ | $p_2$ |
|---|---|---|---|---|---|---|---|
| I | 1 | 1 | 2.62 | 0.32 | 0.302 | 0.11684 | −0.02929952 |
| | 1 | 1 | 2.18 | −0.12 | −0.138 | −0.08644 | −0.03163768 |
| | 1 | 1 | 2.08 | −0.22 | −0.238 | −0.11864 | −0.01816908 |
| | 1 | 1 | 2.32 | 0.02 | 0.002 | 0.08824 | 0.07910628 |
| II | | | 9.20 | 0 | 0 | 0 | 0 |
| | | | | | | | −0.16·10$^{-8}$ |
| III | 4 | 9.2 | | 0.1656 | 0.0756272 | | 0.0084473237 |
| | | | | | 0.0756272 | | |
| IV | 2.3 | 0.018 | | 0.45668599 | 0.11169690 | | 0.39432359 |
| V | | | | | | | 1 |
| | | | | 1 | 1 | | −2.77468599 |
| | 1 | 1 | | −2.3 | −2.318 | | 1.05037778 |

Analogously, eliminating $p_{i+1}$ and $p_i$ from the relationships

$$q_{i+1} = p_{i+1} - \sigma_{i+1}q_i \, ,$$
$$p_{i+1} = Aq_i - \rho_i p_i \, ,$$
$$q_i = p_i - \sigma_i q_{i-1} \, ,$$

we obtain

$$q_{i+1} = Aq_i - \rho_i q_i - \rho_i \sigma_i q_{i-1} - \sigma_{i+1} q_i$$
$$= Aq_i - (\rho_i + \sigma_{i+1})q_i - \rho_i \sigma_i q_{i-1}$$
$$= Aq_i - \gamma_i q_i - \delta_i q_{i-1}$$

from which

$$\gamma_i = \rho_i + \sigma_{i+1} \, ,$$
$$\delta_i = \rho_i \sigma_i \, .$$

The derived relationships among the coefficients show, in particular, that

$$\sum_{i=0}^{n-1} \rho_i + \sum_{i=1}^{n} \sigma_i = \mathrm{Sp}\, A \, .$$

Binomial formulas also exist for simultaneously constructing biorthogonal and $A$-biorthogonal systems in the case where both processes have not degenerated.   Namely if

$$p_0 = q_0 \quad \text{and} \quad \tilde{p}_0 = \tilde{q}_0 \, ,$$

then

| $q_2$ | $Aq_2$ | $p_3$ | $q_3$ | $Aq_3$ | $p_4$ |
|---|---|---|---|---|---|
| $-0.06303198$ | $-0.01323917$ | $-0.001685678$ | $0.000105622$ | $-0.001164592$ | $0.28\cdot10^{-9}$ |
| $-0.01622351$ | $-0.00529574$ | $0.007179744$ | $0.007640798$ | $0.004960304$ | $0.52\cdot10^{-9}$ |
| $0.00841478$ | $-0.01345978$ | $-0.006295283$ | $-0.006534422$ | $-0.004349251$ | $0.75\cdot10^{-9}$ |
| $0.07888289$ | $0.03199469$ | $0.000801218$ | $-0.001440547$ | $0.000553542$ | $0.75\cdot10^{-9}$ |
| $0.78\cdot10^{-8}$ | $0.1\cdot10^{-8}$ | $0.1\cdot10^{-8}$ | $0.25\cdot10^{-8}$ | $0.2\cdot10^{-8}$ | |
| $0.38\cdot10^{-9}$ | | $0.38\cdot10^{-9}$ | $0.25\cdot10^{-9}$ | | |
| | | $-0.11\cdot10^{-10}$ | $0.64\cdot10^{-12}$ | | |
| $0.0033309790$ | | $0.0000094662773$ | $0.000065400114$ | | |
| $0.0033309791$ | | | $0.000065400112$ | | |
| $0.02841890$ | | $0.69087470$ | | | |
| | | $1$ | $1$ | | $1$ |
| $1$ | | $-3.28070648$ | $-3.90912538$ | | $-4.00000008$ |
| $-2.88638289$ | | $2.40341533$ | $2.48544316$ | | $4.75200027$ |
| $1.30929119$ | | $-0.41418874$ | $-0.45139736$ | | $-2.11185621$ |
| | | | | | $0.28615252$ |

$$p_{i+1} = Aq_i - \rho_i p_i ,$$
$$q_{i+1} = p_{i+1} - \sigma_{i+1}q_i ,$$
$$\tilde{p}_{i+1} = A'\tilde{q}_i - \rho_i \tilde{p}_i ,$$
$$\tilde{q}_{i+1} = \tilde{p}_{i+1} - \sigma_{i+1}\tilde{q}_i$$

where

$$\rho_i = \frac{(p_i, A'\tilde{q}_i)}{(p_i, \tilde{p}_i)} = \frac{(\tilde{p}_i, Aq_i)}{(p_i, \tilde{p}_i)} ,$$

$$\sigma_{i+1} = \frac{(p_{i+1}, \tilde{p}_{i+1})}{(p_i, A'\tilde{q}_i)} = \frac{(\tilde{p}_{i+1}, \tilde{p}_{i+1})}{(\tilde{p}_i, Aq_i)} . \tag{7}$$

Relationships (4) and (5) are preserved as before. The proof does not differ essentially from the proof carried out above.

In Table VI.16 is given the simultaneous computation of biorthogonal and $A$-biorthogonal bases for Leverrier's matrix; in Table VI.17 is given the same computation for matrix (4), Section 51.

## 68. THE METHOD OF CONJUGATE DIRECTIONS AND ITS GENERAL PROPERTIES

Methods of conjugate directions for solving systems of linear equation

$$AX = F \tag{1}$$

are methods in which the solution $X$ is represented as a linear combination of vectors orthogonal in a certain metric which is connected in some way with the matrix of the system. The terminology "conjugate directions" originates with the fact that the directions of vectors orthogonal in a certain metric $R$ are conjugate with respect to the second-order surface

$$(RX, X) = \text{const.} \tag{2}$$

The method of $A$-minimal iterations as well as the methods of $A'A$- and $AA'$-minimal iterations are methods involving conjugate directions.

Each individual method is characterized by the choice of metric and of the initial system of vectors undergoing orthogonalization. In constructing a system of $R$-orthogonal vectors the formulas of theorem 11.20 are often used.

Conjugate direction methods are contained in the following general scheme.

Let $R = CAB$ be a positive definite matrix. Furthermore let a system of $R$-orthogonal vectors $s_1, \cdots, s_n$ (vectors of $R$-conjugate directions) be constructed. We shall look for a solution of the system in the form

$$X = X_0 + B \sum_{i=1}^{n} a_i s_i . \tag{3}$$

Here $X_0$ is the initial approximation, generally chosen arbitrarily. We usually take $X_0 = 0$.

Substituting (3) in the system gives

$$AB \sum_{i=1}^{n} a_i s_i = F - AX_0 = r_0 .$$

Multiplying the last equation by $C$, we obtain

$$R \sum_{i=1}^{n} a_i s_i = Cr_0 ,$$

from which the coefficients $a_i$ are easily determined. Namely,

$$a_i = \frac{(Cr_0, s_i)}{(Rs_i, s_i)} . \tag{4}$$

Solution (3) may also be represented as

$$X = X_0 + \sum_{i=1}^{n} a_i Bs_i . \tag{5}$$

The vectors $Bs_1, \cdots, Bs_n$ in turn are orthogonal in the metric defined by the matrix $R_1 = B'^{-1}RB^{-1}$ which is obviously positive definite. In fact,

$$(R_1 Bs_1, Bs_j) = (B'R_1 Bs_i, s_j) = (Rs_i, s_j) = 0 \qquad (i \neq j) .$$

In the well-known special methods of conjugate directions either $R = A$ ($A$ is positive definite) or $R = A'A$ or $R = AA'$. In the first two cases $B = E$ and the vector systems $s_1, \cdots, s_n$ and $Bs_1, \cdots, Bs_n$ coincide. In the last case $R_1 = E$ and the system $Bs_1, \cdots, Bs_n$ is constructed more simply than is the system $s_1, \cdots, s_n$. It thus turns out to be possible to eliminate the vectors $s_1, \cdots, s_n$ from the formulas for solving the system.

The solution in form (3) may be represented as the last element in a sequence of vectors $X_0, X_1, \cdots, X_n$ where

$$X_i = X_0 + \sum_{j=1}^{i} a_j Bs_j = X_{i-1} + a_i Bs_i$$

and the vectors $X_0, X_1, \cdots, X_n$ are considered to be *successive approximations* to the solution. The corresponding successive residuals $r_0, r_1, \cdots$ will be related to each other by the formulas

$$r_i = r_{i-1} - a_i ABs_i . \tag{6}$$

In fact

$$r_i = F - AX_i = F - AX_{i-1} - a_i ABs_i = r_{i-1} - a_i ABs_i .$$

In constructing the $i$th approximation $X_i$ there is no need to know the whole system of conjugate directions, since it is determined by only the first $i$ vectors of the system. Therefore the properties of the successive approximations $X_0, \cdots, X_i$ will not depend on how the system of vectors $Bs_1, \cdots, Bs_i$ (or $s_1, \cdots, s_i$) is continued up to the complete one. The constructed system of approximations will possess the following extremal properties.

THEOREM 68.1. *Among all the vectors of the form $Z = X_0 + V$, where $V$ belongs to the subspace spanned by the vectors $Bs_1, \cdots, Bs_i$, the vector $X_i$ differs least in $R_1$-length from the exact solution. In other words, the generalized error function $f_{R_1}(Z) = (R_1(X - Z), X - Z)$ will take on its least value for $Z = X_i$.*

*Proof.* Let $Z = X_0 + \sum_{j=1}^{i} \gamma_j Bs_j$. Then

$$f_{R_1}(Z) = (R_1(X - Z), X - Z)$$
$$= \left( R_1 \left( \sum_{j=1}^{n} a_j Bs_j - \sum_{j=1}^{i} \gamma_j Bs_j \right), \sum_{j=1}^{n} a_j Bs_j - \sum_{j=1}^{i} \gamma_j Bs_j \right)$$
$$= \sum_{j=1}^{i} (a_j - \gamma_j)^2 (Rs_j, s_j) + \sum_{j=i+1}^{n} a_j^2 (Rs_j, s_j) \geq \sum_{j=i+1}^{n} a_j^2 (Rs_j, s_j) ,$$

where the equality is attained for $\gamma_j = a_j$. Thus the minimum for $f_{R_1}(Z)$ is achieved for $Z = X_i = X_0 + \sum_{j=1}^{i} a_j Bs_j$.

THEOREM 68.2. *The generalized error function decreases with increasing index $i$.*

The truth of the theorem follows directly from theorem 68.1. It is also easy to verify it directly by computation. Namely,

$$f_{R_1}(X_{i-1}) - f_{R_1}(X_i) = \frac{(Cr_0, s_i)^2}{(s_i, Rs_i)} .$$

We note further the properties of successive residuals $r_0, r_1, \cdots, r_i$ and of the vectors $s_1, \cdots, s_i$.

THEOREM 68.3.   *The following equalities hold*:

$$(Cr_i, s_j) = 0 \qquad (j = 1, \cdots, i)$$
$$(Cr_i, s_j) = (Cr_0, s_j) \qquad (i = 1, \cdots, j - 1) . \tag{7}$$

*Proof.*  We have

$$r_i = F - AX_i = A(X - X_i) = \sum_{k=i+1}^{n} \frac{(Cr_0, s_k)}{(s_k, Rs_k)} ABs_k .$$

Consequently

$$Cr_i = \sum_{k=i+1}^{n} \frac{(Cr_0, s_k)}{(s_k, Rs_k)} Rs_k .$$

From here $(Cr_i, s_j) = 0$ if $j < i + 1$.  Furthermore,

$$(Cr_i, s_j) = \frac{(Cr_0, s_j)}{(Rs_j, s_j)} (Rs_j, s_j) = (Cr_0, s_j)$$

if $j \geq i + 1$.  Thus the theorem is proved.

The methods of conjugate directions possess the following unsatisfactory property.  In the case where a system of vectors undergoing the generalized orthogonalization process is far from being orthogonal with respect to the chosen metric, a significant loss in accuracy may occur in carrying out the orthogonalization process.  The situation may be corrected by "pre-orthogonalization," i.e. by actually constructing a new orthogonal system starting from that which has already been constructed.

This pre-orthogonalization must be carried out according to the usual formulas for generalized orthogonalization; this requires computing certain small correction factors from a triangular system with a strongly predominant principal diagonal.

The desired solution of the system must then be found from formulas (3) and (4) (or (5) and (6)) as applied to the newly constructed vector system.

In the case where $A$ is a positive definite matrix we may take $R = A$, $B = C = E$.  In this case the process of solving the system $AX = F$ has a simple geometric meaning.

We consider an $n$-dimensional point space in which a Cartesian coordinate system has been chosen.  To each point of the space we associate a vector running from the coordinate origin to the point.  We relate the point and its corresponding vector, in the sense that operations on points will be visualized as the comparable operations on the corresponding vectors.

We consider the surface given by the vector equation

$$f(X) = (A(X - X^*), X - X^*) = c .$$

(Here $X^*$ is the system's exact solution).  It is clear that this surface is, for $c > 0$, an ellipsoid $W_n$ with center at the point $X^*$.  For different values of $c$ the equation defines similar ellipsoids with a common center.

The straight lines $S_1, S_2, \cdots, S_n$ passing through the center of the ellipsoid in the directions of the $A$-orthogonal vectors $s_1, s_2, \cdots, s_n$ form a system of conjugate diameters for the ellipsoid $f(X) = c$.

Let $X_0$ be an initial approximation. We take $c > f(X_0)$. Under this assumption the point $X_0$ will be inside the ellipsoid $f(X) = c$. We draw through the point $X_0$ the straight line $\widetilde{P}_1$ parallel to $S_1$, and then the plane $\widetilde{P}_2$ parallel to $S_1$ and $S_2$, and then the three-dimensional plane $\widetilde{P}_3$ parallel to $S_1, S_2$, and $S_3$, etc. The straight line $\widetilde{P}_1$ is cut by the figure $f(X) \leq c$ of the ellipsoids $f(X) = c$ into the segment $W_1$, the plane $\widetilde{P}_2$ into the plane ellipse $W_2$, the three-dimensional plane $\widetilde{P}_3$ into the three-dimensional ellipsoid $W_3$, and so on. We denote by $X_1$ the mid-point of the segment $W_1$, by $X_2$ the center of the ellipse $W_2$, by $X_3$ the center of the ellipsoid $W_3$, etc. The last point $X_n$ of this series will be the center of the $n$-dimensional ellipsoid $W_n$, i.e. the solution $X^*$ of the system $AX = F$.

It is clear geometrically that of all the vectors issuing from the center of the ellipsoid $W_n$ and resting on the $k$-dimensional plane $\widetilde{P}_k$, the vector $X_n - X^*$ directed toward the center $X_k$ of the ellipsoid $W_k$ has the shortest $A$-length. But the $k$th approximation in the method of conjugate directions possesses just this extremal property. Therefore the centers $X_1, X_2, \cdots, X_n$ represent nothing less than the successive approximations of this method. The actual process of successive construction of the points $X_0, X_1, \cdots, X_n$ may be described in the following way. The chord $\overline{W}_1$ of the ellipsoid $W_n$ parallel to the diameter $S_1$ is made to pass through $X_0$. The mid-point of this chord is $X_1$. The chord $\overline{W}_2$ parallel to $S_2$ is made to pass through $X_1$. This chord will be the diameter conjugate to chord $\overline{W}_1$ in the ellipse $W_2$. Its mid-point will be the center of the ellipse $W_2$, i.e. the point $X_2$. Further, the chord $\overline{W}_3$ parallel to $S_3$ is passed through $X_2$. This chord is the diameter of the ellipsoid $W_3$ and is conjugate to its plane intersection $W_2$; its mid-point $X_3$ is the center of the ellipsoid, and so on (cf. Fig. 5).

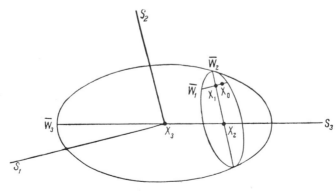

**FIGURE 5**

In concluding this paragraph we introduce formulas for finding the inverse matrix. The $R$-orthogonality condition for the vectors $s_1, \cdots, s_n$ is written in matrix form as

$$S'RS = \Lambda ,$$

where $S$ is a matrix with columns $s_1, \cdots, s_n$ and $\Lambda = [(s_1, Rs_1), \cdots, (s_n, Rs_n)]$. From this it follows that

$$S'CABS = \Lambda$$

and  (8)

$$A^{-1} = BS\Lambda^{-1}S'C = (BS)\Lambda^{-1}(C'S)' .$$

If $R = A$, $B = C = E$, then

$$A^{-1} = S\Lambda^{-1}S' .  (9)$$

## 69. SOME CONJUGATE DIRECTIONS METHODS

Several methods using conjugate directions have been described and studied in contemporary mathematical literature. In these methods either the system of unit vectors $e_1, \cdots, e_n$ or the system of successive iteration for a certain arbitrary vector, and with matrix $R$, has been taken as the system of vectors $Z_1, \cdots, Z_n$ undergoing the $R$-orthogonalization process.

**1. The method of orthogonal vectors.** In this method[†] $A$ is positive definite, $R = A$, and $Z_1 = e_1, \cdots, Z_n = e_n$. The corresponding formulas are obtained from the formulas of Section 68 for $C = B = E$ and $R_1 = A$. The directional vectors will be the $A$-orthogonal vectors

$$s_i = e_i - \gamma_{i1}s_1 - \cdots - \gamma_{ii-1}s_{i-1}$$

for

$$\gamma_{ij} = \frac{(Ae_i, s_j)}{(Ae_j, s_j)} \quad \binom{i = 2, \cdots, n}{j = 1, \cdots, i-1} .  (1)$$

If $X_0 = 0$, then

$$X = \sum_{i=1}^{n} a_i s_i ,  (2)$$

where

$$a_i = \frac{(F, s_i)}{(s_i, As_i)} = \frac{(F, s_i)}{(s_i, Ae_i)} .$$

We observe that the vectors $Ae_i$ are none other than the columns of matrix $A$. Thus

$$\gamma_{ij} = \frac{(A_i, s_j)}{(A_j, s_j)} ,$$

$$a_i = \frac{(F, s_i)}{(s_i, A_i)} .  (3)$$

---

[†] Fox, Huskey, Wilkinson (1).

Control of the computations is accomplished by computing the values $(A_i, s_j)$ for $j > i$; these should theoretically be equal to zero.

The computation of the solution by the method of orthogonal vectors is contained in the scheme

| | $A_1$ | $A_2$ | $A_3$ | $A_4$ | $F$ | $s_1$ | $s_2$ | $s_3$ | $s_4$ | $X$ |
|---|---|---|---|---|---|---|---|---|---|---|
| I | $A_1$ | $A_2$ | $A_3$ | $A_4$ | $F$ | $s_1$ | $s_2$ | $s_3$ | $s_4$ | $X$ |
| II | control | | | | | | $(s_i, As_j)$ | | | |
| III | | | | | | | $(A_i, s_i)$ | | | |
| IV | | | | | | | $\gamma_{ij}$ | | | |
| V | | | | | | | $a_i$ | | | |

**TABLE VI.18  Solving a System of Equations by the Method of Orthogonal Vectors**

| | $A_1$ | $A_2$ | $A_3$ | $A_4$ | $F$ | $s_1$ | $s_2$ | $s_3$ | $s_4$ | $X$ |
|---|---|---|---|---|---|---|---|---|---|---|
| I | 1.00 | 0.42 | 0.54 | 0.66 | 0.3 | 1 | $-0.42$ | $-0.4924721$ | $-0.6862368$ | $-1.2577937$ |
| | 0.42 | 1.00 | 0.32 | 0.44 | 0.5 | 0 | 1 | $-0.1131617$ | $-0.2227744$ | 0.0434872 |
| | 0.54 | 0.32 | 1.00 | 0.22 | 0.7 | 0 | 0 | 1 | 0.2218557 | 1.0391663 |
| | 0.66 | 0.44 | 0.22 | 1.00 | 0.9 | 0 | 0 | 0 | 1 | 1.4823929 |
| II | | | | | | | 0 | $-0.14 \cdot 10^{-7}$ | $0.30 \cdot 10^{-7}$ | |
| | | | | | | | | $0.18 \cdot 10^{-7}$ | $-0.32 \cdot 10^{-7}$ | |
| | | | | | | | | | $0.20 \cdot 10^{-7}$ | |
| III | | | | | | 1.00 | 0.82360 | 0.6978533 | 0.4978712 | |
| IV | | | | | | 0.42 | | | | |
| | | | | | | 0.54 | 0.1131617 | | | |
| | | | | | | 0.66 | 0.1976688 | $-0.2218557$ | | |
| V | | | | | | 0.3 | 0.4541039 | 0.7102890 | 1.4823929 | |

In Table VI.18 is given the solution for the system of equations (9), Section 23, by the method of orthogonal vectors.

We now establish the connection between the method of orthogonal vectors and the escalator method (Section 26).

The relationships

$$s_1 = e_1 ,$$
$$s_2 = e_2 - \gamma_{21}s_1 ,$$
$$\cdots\cdots\cdots ,$$
$$s_n = e_n - \gamma_{n1}s_1 - \cdots - \gamma_{nn-1}s_{n-1}$$

may be written in the form

$$e_1 = s_1 ,$$
$$e_2 = \gamma_{21}s_1 + s_2 ,$$
$$\cdots\cdots\cdots ,$$
$$e_n = \gamma_{n1}s_1 + \gamma_{n2}s_2 + \cdots + \gamma_{nn-1}s_{n-1} + s_n$$

or abbreviated as

$$E = \Gamma S' , \qquad (4)$$

where

$$\Gamma = \begin{bmatrix} 1 & 0 & \cdots & 0 & 0 \\ \gamma_{21} & 1 & \cdots & 0 & 0 \\ \cdot & \cdot & \cdot \cdot \cdot \cdot & \cdot & \cdot \\ \gamma_{n1} & \gamma_{n2} & \cdots \gamma_{nn-1} & 1 \end{bmatrix} ,$$

and $S$ is a matrix whose columns are the directional vectors. Thus

$$S' = \Gamma^{-1}$$

and consequently $S$ is a right triangular matrix with units along the principal diagonal. Moreover, the orthogonality of the vectors $s_i$ to the vectors $As_j$ for $i \neq j$, may be abbreviated to the matrix equation

$$S'AS = \Lambda ,$$

where $\Lambda$ is a diagonal matrix. In fact the element of the $i$th row and $j$th column of the matrix $S'AS$ is obviously equal to $(s_i, As_j)$ and therefore equal to zero for $i \neq j$. Thus

$$A = S'^{-1}\Lambda S^{-1} = \Gamma \Lambda S^{-1} .$$

Just such an expansion forms the basis of the escalator method for $S = Z$ in the notation of Section 26. In computing by both methods the elements of matrices $\Gamma$, $\Lambda$ and $S$ are determined. Basically the computational schemes for the methods coincide.

**2. The method of $A$-minimal iterations.** This method is contained in the general scheme of conjugate directions methods for $R = A$, $A$ positive definite, and $Z_1 = q_0, \cdots, Z_n = A^{n-1}q_0$ (under the assumption of linear independence for the vectors $q_0, \cdots, A^{n-1}q_0$). As in the previous method $C = B = E$ and $R_1 = A$. The vector system for the conjugate directions $q_0, \cdots, q_{n-1}$ is constructed directly from the trinomial recurrence formulas (Section 65) without computing the vectors $Aq_0, \cdots, A^{n-1}q_0$ and without $A$-orthogonalization of the system $q_0, Aq_0, \cdots, A^{n-1}q_0$.

**3. The method of conjugate gradients.** In the method of conjugate gradients[†] $R = A$, $A$ is positive definite, $B = C = E$, and $R_1 = A$. The $A$-orthogonal vectors $s_1, \cdots, s_n$ are theoretically constructed by $A$-orthogonalizing the residuals $r_0, r_1, \cdots r_{n-1}$ of the successive approximations $X_0, X_1, \cdots, X_{n-1}$; these are determined by the formulas for the conjugate directions method, i.e. by the formulas

$$X_i = X_0 + \sum_{j=1}^{i} a_j s_j \qquad (5)$$

where

$$a_j = \frac{(r_0, s_j)}{(s_j, As_j)} .$$

---

[†] Stiefel (1), Hestenes and Stiefel (1).

Thus here the system of vectors undergoing the orthogonalization process is not given beforehand but is constructed along with the construction of conjugate directions vectors and their corresponding successive approximations. We observe that the name of the method is connected with the fact that the residual $r_i$ is the gradient of the error function computed at the point $X_i$. It is obvious that the method of conjugate gradients is naturally bound up with solving only one system $AX = F$, although knowledge of the basis $s_1, \cdots, s_n$ also allows us to solve systems whose constants differ from $F$.

The process of constructing successive approximations breaks down as soon as a certain residual $r_k$ turns out to be equal to zero, i.e. as soon as the approximation coincides with the exact solution of the system. Another reason is conceivable *a priori* for the process to stop without reaching the exact solution. Namely, the process may stop if a certain residual $r_k$ turns out to be a linear combination of the preceding ones. Then we obtain $s_{k+1} = 0$ and $r_{k+1} = r_k$. However, this is impossible in view of the following theorem.

THEOREM 69.1. *The non-zero residuals* $r_0, r_1, \cdots, r_{k-1}$ *are mutually orthogonal.*

*Proof.* We consider $(r_i, r_j)$ for $i > j$. Since the vectors $s_1, \cdots, s_{j+1}$ are obtained by $A$-orthogonalization of the vector system $r_0, \cdots, r_j$ we conclude that the vector $r_j$ is a linear combination of the vectors $s_1, \cdots, s_{j+1}$ and therefore $(r_i, r_j)$ is a linear combination of the numbers $(r_i, s_k)$ for $k = 1, \cdots, j + 1$. But, by theorem 68.3, $(r_i, s_k) = 0$ for $k \leq i$, and in particular for $k = 1, \cdots, j + 1$.

It follows from the proven theorem that if $r_k \neq 0$, then $r_k$ can not be a linear combination of the residuals $r_0, \cdots, r_{k-1}$, so that reducing the residual to zero is the only cause for the process to stop.

We now derive the computational formulas for the process. We observe first of all that from the formulas

$$X_i = X_{i-1} + a_i s_i \, ,$$

$$a_i = \frac{(r_0, s_i)}{(As_i, s_i)} \tag{6}$$

it follows that

$$r_i = r_{i-1} - a_i As_i \tag{7}$$

where $a_i \neq 0$, since $r_{i-1} \neq r_i$.

Assume that we have already constructed the vectors $s_1, \cdots, s_i$ and $r_0, \cdots, r_i$; we shall show that the next directional vector, $s_{i+1}$, is constructed from the formula

$$s_{i+1} = r_i + b_i s_i \tag{8}$$

where

$$b_i = -\frac{(r_i, As_i)}{(s_i, As_i)} .$$

To demonstrate this it suffices to prove that the vector $s_{i+1}$ so constructed will be $A$-orthogonal to the vectors $s_1, s_2, \cdots, s_i$. It is clear that

$$(s_{i+1}, As_i) = (r_i, As_i) - \frac{(r_i, As_i)}{(s_i, As_i)}(s_i, As_i) = 0 .$$

We now consider $(s_{i+1}, As_j)$ for $j = 1, 2, \cdots, i - 1$. We observe first of all that $(s_{i+1}, As_j) = (r_i, As_j)$. Furthermore, from formula (7), we have $As_j = (1/a_j)(r_{j-1} - r_j)$, so that

$$(s_{i+1}, As_j) = \frac{1}{a_j}(r_i, r_{j-1}) - \frac{1}{a_j}(r_i, r_j) = 0$$

since $j < i$, $j - 1 < i$ and the residuals are orthogonal. Thus formula (8) holds.

We observe that the coefficients $a_i$ and $b_i$ may also be computed from the formulas

$$a_i = \frac{(s_i, r_{i-1})}{(s_i, As_i)} = \frac{(r_{i-1}, r_{i-1})}{(s_i, As_i)} ,$$

$$b_i = \frac{(r_i, r_i)}{(r_{i-1}, r_{i-1})} ,$$

$$(9)$$

whose validity is easily established.

In actually applying the method of coefficients, one should compute from the formula

$$a_i = \frac{(r_{i-1}, r_{i-1})}{(s_i, As_i)} ,$$

using the formula

$$a_i = \frac{(s_i, r_{i-1})}{(s_i, As_i)}$$

for control. The formula

$$a_i = \frac{(r_0, s_i)}{(s_i, As_i)}$$

is sensitive to rounding-off errors and it is not advisable to apply it.

Thus the computational formulas for the method of conjugate gradients are as follows. An arbitrary approximation $X_0$ to the solution of the system $AX = F$ is chosen and the residual $r_0 = F - AX_0$ is computed. Moreover, from the recurrent relationships

$$s_1 = r_0 ,$$

$$r_i = r_{i-1} - a_i As_i ,$$

$$a_i = \frac{(r_{i-1}, r_{i-1})}{(s_i, As_i)} = \frac{(s_i, r_{i-1})}{(s_i, As_i)} ,$$

$$(10)$$

$$s_{i+1} = r_i + b_i s_i ,$$

$$b_i = \frac{(r_i, r_i)}{(r_{i-1}, r_{i-1})} = -\frac{(r_i, As_i)}{(s_i, As_i)}$$

**TABLE VI.19  The Solution of a Linear System by the Method of Conjugate Gradients**

| | $r_0$ | $s_1$ | $As_1$ | $r_1$ | $s_2$ | $As_2$ |
|---|---|---|---|---|---|---|
| I | 0.3 | 0.3 | 1.482 | $-0.44866929$ | $-0.40302542$ | $-.012915356$ |
| | 0.5 | 0.5 | 1.246 | $-0.12944800$ | $-0.05337488$ | $-0.03272691$ |
| | 0.7 | 0.7 | 1.220 | $0.08368655$ | $0.19018892$ | $0.02000433$ |
| | 0.9 | 0.9 | 1.472 | $0.15638246$ | $0.29331408$ | $0.04567392$ |
| II | | | 5.420 | $1\cdot10^{-8}$ | $4\cdot10^{-8}$ | $-0.09620222$ |
| III | 1.64 | | 3.2464 | $0.24951983$ | $0.24951983$ | $-0.07100037$ |
| IV | | 0.15214624 | 0.5051749 6 | | $0.001730603$ | $3.5143454$ |

| | $r_2$ | $s_3$ | $As_3$ | $r_3$ | $s_4$ | $As_4$ | $X$ |
|---|---|---|---|---|---|---|---|
| I | 0.00522093 | 0.00452345 | 0.00343581 | 0.00058168 | 0.000628773 | 0.00039923 | $-1.2577936$ |
| | $-0.01443433$ | $-0.01452670$ | $-0.00983301$ | $-0.00115718$ | $-0.001308415$ | $-0.00079361$ | 0.0434875 |
| | 0.01338442 | 0.01371356 | 0.01071043 | $-0.00107748$ | $-0.000934710$ | $-0.00073895$ | 1.0391663 |
| | $-0.00413147$ | $-0.00362386$ | $-0.00401315$ | 0.00128733 | 0.00124960 | 0.00088325 | 1.4823930 |
| II | $-11\cdot10^{-8}$ | $-29\cdot10^{-8}$ | 0.00030008 | $28\cdot10^{-8}$ | $62\cdot10^{-8}$ | $-0.00025008$ | |
| | $3\cdot10^{-8}$ | $1\cdot10^{-7}$ | | $-4\cdot10^{-8}$ | $1\cdot10^{-8}$ | | |
| | | | | $2\cdot10^{-11}$ | $2\cdot10^{-11}$ | | |
| III | 0.0004318197 | 0.0004318197 | 0.0003198041 | 0.0000749560 | 0.0000449560 | 0.0000308381 | |
| IV | | 0.01041083 | 1.3502631 | | | 1.4578070 | |

are constructed the directional vectors $s_1$, $s_2$, $\cdots$ and the residuals $r_1$, $r_2$, $\cdots$. The process ends when for a certain $m \leq n$ it turns out that $r_m = 0$. The system's solution is obtained by the formula

$$X = X_0 + \sum_{i=1}^{m} a_i s_i . \tag{11}$$

As a rule the process is carried through without degeneracy, so that $m = n$.

Computational control is achieved by computing the scalar products $(s_i, As_j)$ or $(r_i, r_j)$, which should equal zero.

From formulas (10) it is apparent that the computational scheme for the method of conjugate gradients is close to the binomial form for the method of minimal iterations.

In Table VI.19 there is given the solution of system (9) of Section 23 by the method of conjugate gradients.

We now establish the connection between the method of conjugate gradients and the method of $A$-minimal iterations.

From the recurrent relationships for constructing the vectors $r_0$, $r_1$, $\cdots$, $r_i$ and $s_1$, $s_2$, $\cdots$, $s_{i+1}$ it is clear that both vector systems consist of linear combinations of the vectors $r_0$, $Ar_0$, $\cdots$, $A^i r_0$.

The vectors $r_0$, $r_1$, $\cdots$, $r_i$ are orthogonal. Consequently they differ only by scalar multipliers from the vectors $p_0$, $p_1$, $\cdots$, $p_i$ of the method of minimal iterations, constructed starting with $p_0 = r_0$. Thus $r_i = k_i p_i$. The vectors $s_1$, $\cdots$, $s_{i+1}$ are $A$-orthogonal so that they differ only by scalar multipliers from the vectors $q_0 = r_0$, $q_1$, $\cdots$, $q_i$ of the $A$-minimal iterations method, i.e. $s_{i+1} = l_i q_i$. According to the computational formulas for the method of conjugate gradients, we obtain

$$l_i q_i = k_i p_i + b_i l_{i-1} q_{i-1} ,$$
$$k_i p_i = k_{i-1} p_{i-1} - a_i l_{i-1} A q_{i-1} .$$

Setting these formulas equal to the binomial formulas for the method of minimal iterations, we conclude that

$$k_i = l_i \quad \text{and} \quad k_i = -a_i l_{i-1} ,$$

from which

$$k_i = l_i = (-1)^i a_1 a_2 \cdots a_i .$$

Moreover,

$$\rho_{i-1} = \frac{k_{i-1}}{k_i} = -\frac{1}{a_i} ,$$
$$\sigma_i = \frac{b_i l_{i-1}}{l_i} = -\frac{b_i}{a_i} . \tag{12}$$

The coefficients $\gamma_i$ and $\delta_i$ of the method of $A$-minimal iterations are expressed in terms of the coefficients $a_i$ and $b_i$ via the formulas

$$\gamma_i = -\frac{1 + b_{i+1}}{a_{i+1}} ,$$
$$\delta_i = \frac{b_i}{a_i a_{i+1}} . \tag{13}$$

We observe that the successive approximations $X_i$ computed by the method of conjugate gradients will coincide with the successive approximations computed by the method of $A$-minimal iterations starting from an initial approximation $X_0$ and $q_0 = r_0$. Actually the directional vectors corresponding to these methods differ only by normalization.

This identification also follows from theorem 68.1. In fact the approximations $X_i$ of both methods minimize the error function among the vectors $X_0 + V$, where $V$ belongs to the subspace spanned by $r_0, Ar_0, \cdots, A^{i-1}r_0$.

The method of conjugate gradients may also be applied in solving the complete eigenvalue problem. It is clear that for the vectors $r_i$ and $s_i$ the representation

$$r_i = r_i(A)r_0 \,,$$
$$s_i = s_i(A)r_0 \tag{14}$$

holds where $r_i(t)$ and $s_i(t)$ are polynomials defined by the recurrent formulas

$$r_0(t) = s_1(t) = 1 \,,$$
$$r_i(t) = r_{i-1}(t) - a_i t s_i(t) \,, \tag{15}$$
$$s_{i+1}(t) = r_i(t) + b_i s_i(t) \,.$$

For a non-degenerate course of the process, the polynomial $r_n(t)$ will differ only by a scalar multiplier from the characteristic polynomial of matrix $A$. This is clear even if the polynomials $r_i(t)$ differ only by constant factors from the polynomials $p_i(t)$ of the method of minimal iterations.

It is easy to verify that if $\lambda_i$ is an eigenvalue for matrix $A$, then

$$U_i = \frac{r_0(\lambda_i)}{(r_0, r_0)} r_0 + \cdots + \frac{r_{n-1}(\lambda_i)}{(r_{n-1}, r_{n-1})} r_{n-1}$$

is the eigenvector belonging to $\lambda_i$. This follows from the analogous formula for the method of minimal iterations.

Moreover it is clear from the recurrent relationships that $r_i(1) = 1$ for $i = 0, 1, \cdots, n$. Therefore $r_i(t) = p_i(t)/p_i(1)$. We observe further that the polynomials $s_i(t)$ for $i = 1, \cdots, n$ differ only by normalization from the polynomials $q_{i-1}(t)$ of the $A$-minimal iterations method.

**4. The orthogonalization of columns method.** Here $A$ is any non-singular matrix, $R = A'A$, $Z_1 = e_1, \cdots, Z_n = e_n$. In this case $C = A'$, $B = E$, and $R_1 = R$.

The formulas for the method will be (we consider $X_0 = 0$):

$$X = \sum_{i=1}^{n} a_i s_i \,,$$

$$s_i = e_i - \gamma_{i1}s_1 - \cdots - \gamma_{i,i-1}s_{i-1} \,,$$

$$\gamma_{ij} = \frac{(s_j, A'Ae_i)}{(s_j, A'Ae_j)} = \frac{(As_j, A_i)}{(As_j, A_j)} \,, \tag{16}$$

$$a_i = \frac{(A'F, s_i)}{(s_i, A'Ae_i)} = \frac{(F, As_i)}{(As_i, A_i)} \,.$$

Here $A_i = Ae_i$ is the $i$th column of matrix $A$.

The vectors $As_1, \cdots, As_n$ form an orthogonal system (since $(As_j, As_i) = (s_j, A'As_i) = (s_j, Rs_j) = 0$ for $j \neq i$) and

$$As_i = A_i - \gamma_{i1}As_1 - \cdots - \gamma_{i,i-1}As_{i-1} .$$

Therefore the vectors $As_1, \cdots, As_n$ are actually found by orthogonalizing the columns of matrix $A$.

The directional vectors $s_1, \cdots, s_n$ may be constructed, then, from the formula

$$s_i = e_i - \gamma_{i1}s_1 - \cdots - \gamma_{i,i-1}s_{i-1} ,$$

using the already computed orthogonalization coefficients $\gamma_{ij}$. However, the construction of the vectors $s_1, \cdots, s_n$ may be avoided. That is, it is not hard to verify that the desired unknowns are determined by the recurrent formulas

$$\begin{aligned}
x_n &= a_n , \\
x_{n-1} &= a_{n-1} - \gamma_{n,n-1}x_n , \\
&\cdot \cdot \cdot \cdot \cdot \cdot \cdot \cdot \cdot \cdot , \\
x_1 &= a_1 - \gamma_{21}x_2 - \cdots - \gamma_{n1}x_n .
\end{aligned} \qquad (17)$$

In essence the method of orthogonalizing the columns is equivalent to applying the method of orthogonal vectors to the system obtained from the given system by a first Gauss transformation.

In Table VI.20 is given an example of solving a system of equations by the orthogonalization of columns method using the data of Table II.1.

Table VI.20 is filled in according to the scheme.

| | $A_1$ | $A_2$ | $A_3$ | $A_4$ | $F$ | $As_1$ | $As_2$ | $As_3$ | $As_4$ |
|---|---|---|---|---|---|---|---|---|---|
| I | | | | | | | | | |
| II | control | | | | | $(As_i, As_j)$ | | | |
| III | | | | | | $(As_i, As_i)$ | | | |
| | | | | | | $(A_i, As_i)$ | | | |
| IV | | | | | | $\gamma_{ij}$ | | | |
| V | | | | | | $a_1$ | $a_2$ | $a_3$ | $a_4$ |
| VI | | | | | | $x_1$ | $x_2$ | $x_3$ | $x_4$ |

**5. The orthogonalization of rows method.**[†] Here $A$ is any non-singular matrix, $R = AA'$, and $Z_1 = e_1, \cdots, Z = e_n$. In this case $C = E$, $B = A'$, and $R_1 = E$.

The formulas for the method are

$$X = \sum_{i=1}^{n} a_i A's_i ,$$

$$a_i = \frac{(F, s_i)}{(AA'r_i, e_i)} = \frac{(F, s_i)}{(A's_i, A'e_i)} \qquad (18)$$

Here the vectors $s_i$ are obtained by the formula

---
† J. Schröder (1).

TABLE VI.20  The Solution of a System of Equations by the Orthogonalization of Column Method

|   | $A_1$ | $A_2$ | $A_3$ | $A_4$ | $F$ | $As_1$ | $As_2$ | $As_3$ | $As_4$ |
|---|---|---|---|---|---|---|---|---|---|
| I | 1 | 0.17 | -0.25 | 0.54 | 0.3 | 1 | -0.3757506 | 0.0988025 | -0.3617379 |
|   | 0.47 | 1 | 0.67 | -0.32 | 0.5 | 0.47 | 0.7434972 | -0.2640083 | -0.1818692 |
|   | -0.11 | 0.35 | 1 | -0.74 | 0.7 | -0.11 | 0.4100326 | 0.5216659 | -0.2437552 |
|   | 0.55 | 0.43 | 0.36 | 1 | 0.9 | 0.55 | 0.1298372 | 0.1502995 | 0.764696 |
| II |   |   |   |   |   |   | $-0.420 \cdot 10^{-7}$ | $0.750 \cdot 10^{-7}$ | $-0.720 \cdot 10^{-7}$ |
|   |   |   |   |   |   |   |   | $-0.389 \cdot 10^{-7}$ | $0.222 \cdot 10^{-7}$ |
|   |   |   |   |   |   |   |   |   | $-0.376 \cdot 10^{-7}$ |
| III |   |   |   |   |   | 1.5355 | 0.8789610 | 0.3741876 | 0.8076082 |
|   |   |   |   |   |   | 1.5355 | 0.8789610 | 0.3741875 | 0.8076081 |
| IV |   |   |   |   |   | 0.5457506 | 1.1932893 |   |   |
|   |   |   |   |   |   | 0.0995767 | -0.6990200 | -0.2616262 |   |
|   |   |   |   |   |   | 0.6649300 |   |   |   |
| V |   |   |   |   |   | 0.6206447 | 0.7541856 | 1.0638310 | 0.3935672 |
| VI |   |   |   |   |   | 0.4408886 | -0.3630312 | 1.1667985 | 0.3935672 |

**TABLE VI.21**   The Solution of a System of Equations by the Orthogonalization of Rows Method

| | $A^1$ | $A^2$ | $A^3$ | $A^4$ | $A's_1$ | $A's_2$ | $A's_3$ | $A's_4$ | $X$ |
|---|---|---|---|---|---|---|---|---|---|
| I | 1 | 0.47 | -0.11 | 0.55 | 1 | 0.2532972 | 0.1949144 | -0.2455381 | 0.4408885 |
| | 0.17 | 1 | 0.35 | 0.43 | 0.17 | 0.9631605 | -0.3293997 | 0.0239030 | -0.3030308 |
| | -0.25 | 0.67 | 1 | 0.36 | -0.25 | 0.7241757 | 0.2979181 | 0.4495452 | 1.1667984 |
| | 0.54 | -0.32 | 0.74 | 1 | 0.54 | -0.4370195 | -0.1193276 | 0.6552979 | 0.3935672 |
| II | 0.3 | 0.5 | 0.7 | 0.9 | 0.3 | 0.4349892 | 0.5061646 | 0.6654999 | |
| III | | | | | | $1.05 \cdot 10^{-7}$ | $2.20 \cdot 10^{-7}$ | $0.240 \cdot 10^{-7}$ | |
| | | | | | | | $2.87 \cdot 10^{-7}$ | $0.614 \cdot 10^{-7}$ | |
| | | | | | | | | $-0.263 \cdot 10^{-7}$ | |
| IV | | | | | 1.3830 | 1.7072541 | 0.2494901 | 0.6923665 | |
| | | | | | 1.3830 | 1.7072541 | 0.2494900 | 0.6923665 | |
| V | | | | | 0.2167028 | | | | |
| | | | | | -0.5062184 | 0.7947344 | | | |
| | | | | | 0.7759219 | 0.2209139 | -0.1864445 | | |
| VI | | | | | 0.2169197 | 0.2547888 | 2.0287971 | 0.9611960 | |

$$s_i = e_i - \gamma_{i1}s_1 - \cdots - \gamma_{i,i-1}s_{i-1}$$

where

$$\gamma_{ij} = \frac{(s_j, AA'e_i)}{(s_j, AA'e_j)} = \frac{(A's_j, A'e_i)}{(A's_j, A'e_j)} = \frac{(A's_j, A^i)}{(A's_j, A^j)} \qquad (19)$$

and $A'e_i = A^i$ is the $i$th row of matrix $A$.

Thus a basic role in the method's formulas is played by the auxiliary vectors $A's_i$ which obviously can be obtained by orthogonalizing the rows $A^i$ of matrix $A$ via the formulas

$$A's_i = A^i - \gamma_{i1}A's_1 - \cdots - \gamma_{i,i-1}A's_{i-1} \qquad (20)$$

with the former values of the coefficients.

The vectors $s_1, \cdots, s_n$ themselves are needed only to compute the numerators $F_i = (F, s_i)$ of the coefficients $a_i$. Just as in the preceding method, they can be eliminated. In fact we have

$$F_i = (F, s_i) = (F, e_i - \gamma_{i1}s_1 - \cdots - \gamma_{i,i-1}s_{i-1})$$
$$= f_i - \gamma_{i1}F_1 - \cdots - \gamma_{i,i-1}F_{i-1}. \qquad (21)$$

Here $f_i = (F, e_i)$ is the $i$th component of the vector $F$. Thus the numbers $F_i$ are found while computing the orthogonalized rows $A's_i$ by the same recurrent formulas.

The method is equivalent to applying the method of orthogonal vectors to the system obtained by a second Gauss transformation.

In Table VI.21 is given the solution for the system by the orthogonalization of rows method. The table is filled in according to the scheme

| | $A^1$ | $A^2$ | $A^3$ | $A^4$ | $A's_1$ | $A's_2$ | $A's_3$ | $A's_4$ | $X$ |
|---|---|---|---|---|---|---|---|---|---|
| I | | | | | | | | | |
| II | $f_1$ | $f_2$ | $f_3$ | $f_4$ | $F_1$ | $F_2$ | $F_3$ | $F_4$ | |
| III | control | | | | | $(A's_i, A's_j)$ | | | |
| VI | | | | | | $(A's_i, A's_i)$ | | | |
| | | | | | | $(A's_i, A^i)$ | | | |
| V | | | | | | $\gamma_{ij}$ | | | |
| VI | | | | | $a_1$ | $a_2$ | $a_3$ | $a_4$ | |

**6. The method of $A'A$-minimal iterations.** Here $A$ is a non-singular matrix, $R = A'A$, $Z_1 = q_0$, $Z_2 = Rq_0, \cdots, Z_n = R^{n-1}q_0$; also $C = A'$, $B = E$, and $R_1 = R$.

Formulas of the method and an example are given in Section 65.

**7. The method of $AA'$-minimal iterations.** Here $A$ is a non-singular matrix, $R = AA'$, $Z_1 = q_0$, $Z_2 = Rq_0, \cdots, Z_n = R^{n-1}q_0$; also $C = E$, $B = A'$, and $R_1 = E$.

Formulas of the method and an example are given in Section 65.

**8. The method of conjugate gradients after a first Gauss transformation.** Let $A$ be a non-singular matrix. Applying the method of conjugate gradients to the system

*Minimal Iterations Based on Orthogonalization* [*Chap.* 6]

TABLE VI.22 The Solution of a System of Linear Equations by the Method of Conjugate Gradients After a First Gauss Transformation

| | $r_0$ | $s_1$ | $As_1$ | $r_1$ | $A'r_1$ | $s_2$ | $As_2$ | $r_2$ | $A'r_2$ |
|---|---|---|---|---|---|---|---|---|---|
| I | 0.3 | 0.953 | 1.04047 | -0.05773067 | -0.09325850 | -0.07812450 | -0.05997035 | 0.10480066 | 0.01012190 |
| | 0.5 | 1.183 | 2.36831 | -0.31426387 | -0.12740998 | -0.10862350 | -0.08061954 | -0.09576921 | -0.04239773 |
| | 0.7 | 1.284 | 1.30906 | 0.24992368 | 0.14521847 | 0.16560887 | 0.02926481 | 0.17061034 | 0.06001929 |
| | 0.9 | 0.384 | 1.87908 | 0.25394144 | 0.13838779 | 0.14448585 | 0.11442846 | -0.05618198 | -0.09519513 |
| II | | 3.804 | 6.59692 | | $0.7 \cdot 10^{-7}$ | | 0.00310338 | | $-0.51 \cdot 10^{-6}$ |
| | | | | | | | | | $0.65 \cdot 10^{-8}$ |
| III | | 4.103810 | 11.936050 | | 0.065170035 | | 0.024046255 | | 0.014564448 |
| IV | | 0.015880373 | 0.34381642 | | | 0.22348381 | 2.7101948 | | |

| | $s_3$ | $As_3$ | $r_3$ | $A'r_3$ | $s_4$ | $As_4$ | X |
|---|---|---|---|---|---|---|---|
| I | -0.00733766 | -0.07689831 | 0.15056225 | 0.07946753 | 0.07448233 | 0.03405157 | 0.4408886 |
| | -0.06667332 | 0.01501777 | -0.10470617 | -0.05324086 | -0.09853868 | -0.02368062 | -0.3630310 |
| | 0.09703019 | 0.12105128 | 0.09857365 | -0.01644605 | 0.04947622 | 0.02229369 | 1.1667984 |
| | -0.06290488 | -0.06067925 | -0.02007222 | 0.02179287 | -0.02094468 | -0.00453959 | 0.3935672 |
| II | -0.00150851 | | | $0.35 \cdot 10^{-6}$ | | | |
| | | | | $-0.88 \cdot 10^{-8}$ | | | |
| | | | | $-0.13 \cdot 10^{-8}$ | | | |
| III | 0.024474267 | 0.024474267 | | 0.0099850792 | | 0.0022378977 | |
| IV | 0.67939954 | 0.59509231 | | | | 4.4215959 | |

$$A'AX = A'F$$

obtained from the system $AX = F$ by a first Gauss transformation gives

$$X = X_0 + \sum_{i=1}^{n} a_i s_i \, ,$$

$$a_i = \frac{(\bar{r}_{i-1}, \bar{r}_{i-1})}{(s_i, A'As_i)} \, ,$$

$$s_1 = \bar{r}_0 \, ,$$

$$\bar{r}_i = \bar{r}_{i-1} - a_i A'As_i \, ,$$

$$s_{i+1} = \bar{r}_i + b_i s_i \, ,$$

$$b_i = \frac{(\bar{r}_i, \bar{r}_i)}{(\bar{r}_{i-1}, \bar{r}_{i-1})} \, .$$

Here $\bar{r}_i$ is the residual of the transformed system. It is clear that

$$\bar{r}_i = A'r_i$$

where $r_i$ is the residual of the initial system. Keeping this in mind, and transforming scalar products, we arrive at the following computational formulas:

$$X = X_0 + \sum_{i=1}^{n} a_i s_i \, ,$$

$$a_i = \frac{(A'r_{i-1}, A'r_{i-1})}{(As_i, As_i)} \, ,$$

$$s_1 = A'r_0 \, ,$$

$$r_i = r_{i-1} - a_i As_i \, ,$$      (22)

$$s_{i+1} = A'r_i + b_i s_i \, ,$$

$$b_i = \frac{(A'r_i, A'r_i)}{(A'r_{i-1}, A'r_{i-1})} \, .$$

The method is controlled by satisfying the orthogonality conditions

$$(A'r_i, A'r_j) = 0 \, .$$

In Table VI.22 is given an illustrative example with the data of Table II.1.

9. **The method of conjugate gradients after a second Gauss transformation.** Let $A$ be a non-singular matrix. Applying the method of conjugate gradients to the auxiliary system

$$AA'Y = F$$

obtained from the initial system $AX = F$ by a second Gauss transformation, we obtain

$$Y = Y_0 + \sum_{i=1}^{n} a_i s_i \, ,$$

$$a_i = \frac{(r_{i-1}, r_{i-1})}{(s_i, AA's_i)} \, ,$$

**TABLE VI.23　The Solution of a Linear System by Craig's Method**

| | $r_0$ | $g_1$ | $Ag_1$ | $r_1$ | $A'r_1$ | $g_2$ | $Ag_2$ | $r_2$ | $A'r_2$ |
|---|---|---|---|---|---|---|---|---|---|
| I | 0.3 | 0.953 | 1.04047 | -0.11580161 | -0.26309916 | -0.10839731 | -0.08891071 | 0.15303758 | 0.04080381 |
| | 0.5 | 1.183 | 2.36831 | -0.44644450 | -0.34013073 | -0.14809265 | -0.13742156 | -0.03092312 | -0.01716715 |
| | 0.7 | 1.284 | 1.30906 | 0.17686213 | -0.03964159 | 0.16879197 | 0.00985241 | 0.14707141 | 0.03473364 |
| | 0.9 | 0.384 | 1.87906 | 0.14906582 | 0.09851721 | 0.16085248 | 0.09831923 | -0.14822190 | -0.16451905 |
| II | | 3.804 | 6.59692 | $0.4 \cdot 10^{-8}$ | -0.54435427 | 0.07315449 | -0.11816063 | $-0.9 \cdot 10^{-8}$ | -0.10614875 |
| | | | | | | | | $0.2 \cdot 10^{-8}$ | |
| III | 1.64 | 4.103810 | | 0.26622354 | | 0.088045659 | | 0.067976472 | |
| IV | | 0.16233143 | 0.39962864 | | | 0.25533607 | 3.0236986 | | |

| | $g_3$ | $Ag_3$ | $r_3$ | $A'r_3$ | $g_4$ | $Ag_4$ | $X$ |
|---|---|---|---|---|---|---|---|
| I | 0.01312607 | -0.08234041 | 0.38156664 | 0.42952349 | 0.48061945 | 0.52191943 | 0.4408885 |
| | -0.05498055 | 0.04283959 | -0.14982089 | -0.10797849 | -0.32200171 | -0.20492996 | -0.3630310 |
| | 0.07783232 | 0.14849649 | -0.26506842 | -0.40244471 | -0.09946623 | -0.36256932 | 1.1667983 |
| | -0.12344761 | -0.11185027 | 0.16220935 | 0.61234865 | 0.13180318 | 0.22187530 | 0.3935672 |
| II | -0.08746977 | -0.00285460 | $0.68 \cdot 10^{-7}$ | 0.53144894 | 0.19095469 | 0.17629545 | |
| | | | $-0.14 \cdot 10^{-7}$ | | | | |
| | | | $0.22 \cdot 10^{-9}$ | | | | |
| III | 0.024492337 | | 0.26461254 | | 0.36194577 | | |
| IV | 3.8927078 | 2.7754180 | | | | 0.73108339 | |

$$s_1 = r_0 \ ,$$
$$r_i = r_{i-1} - a_i A A' s_i \ ,$$
$$s_{i+1} = r_i + b_i s_i \ ,$$
$$b_i = \frac{(r_i, r_i)}{(r_{i-1}, r_{i-1})} \ .$$

Here $r_i$ are the residuals of the transformed system, which obviously coincide with the residuals of the initial system.

We write

$$A' s_i = g_i \ .$$

Since $X = A' Y$, we obtain the following computational formulas after simple transformations:

$$X = X_0 + \sum_{i=1}^{n} a_i g_i \ ,$$
$$a_i = \frac{(r_{i-1}, r_{i-1})}{(g_i, g_i)} \ ,$$
$$g_1 = A' r_0 \ ,$$
$$r_i = r_{i-1} - a_i A g_i \ , \qquad (23)$$
$$g_{i+1} = A' r_i + b_i g_i \ ,$$
$$b_i = \frac{(r_i, r_i)}{(r_{i-1}, r_{i-1})} \ .$$

It is obvious that $(r_i, r_j) = 0$ for $i \neq j$.

The last method is equivalent to a method described by Craig (1) but differs slightly from it in its computational scheme.

In Table VI.23 is given a numerical illustration of the method for the system of Table II.1

# Chapter VII

# GRADIENT ITERATIVE METHODS

In the present chapter we shall present iterative methods useful in solving both linear systems and the special eigenvalue problem in the case of a positive definite matrix. These methods are based primarily on the idea of relaxation which has already been dealt with in Chapters III and V. In contrast with the methods presented in these chapters, the vectors in whose directions the minimization of the chosen functional is accomplished will not be coordinate vectors but vectors related to the functional itself; in fact these will be the functional gradients. Such a choice of minimization-direction is bound up with the fact that (cf. paragraph 1, Section 14, Chapter I) the direction opposite to the direction of the functional gradient at a given point guarantees the most rapid decrease of the functional in a neighborhood of this point. Because of this, certain gradient methods are called *methods of steepest descent*.

An ideal gradient method would be to construct the line of steepest descent starting from an initial approximation and running to the point giving a minimum for the functional, i.e. a line whose direction at each point is opposite to the direction of the gradient of the functional at this point. The differential equation for the line of steepest descent is

$$\frac{dX}{dt} = - \rho(t)\,\mathrm{grad}F(X) \qquad (1)$$

where $\rho(t)$ is any positive function in the parameter $t$. The choice of the function $\rho(t)$ affects only the parametrization of the line of steepest descent.

For example, if in solving the system $AX = F$ with a positive definite matrix $A$ we take as the functional the error function, then equation (1) for $\rho(t) = 1$ will be

$$\frac{dX}{dt} = F - AX. \tag{2}$$

Consequently the line of steepest descent will be determined as the solution of a system of linear differential equations with constant coefficients.

For a selected parametrization the desired solution of the algebraic system is obtained as $\lim\limits_{t\to\infty} X(t)$ independent of the choice of initial approximation.

In fact it is easy to see that the general solution of system (2) is $X = = X^* + e^{-At}C$ where $X^*$ is the exact solution of the system $AX = F$ and $C$ is an arbitrary constant vectors.

In one-step methods of steepest descent the trajectory of the steepest descent, starting from the given initial approximation, is replaced by a broken line formed from segments tangent to the trajectories of steepest descent which pass through the peaks of this broken line (fig. 6).

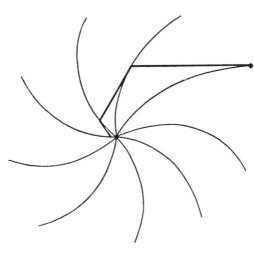

FIGURE 6

The first gradient method to appear was the method of steepest descent proposed by L. V. Kantorovich (1) and independently by Temple (1).

Detailed study of this method has indicated slow convergence of the process in many cases; this fact has led to the development of other gradient methods, methods with incomplete relaxation and $s$-step processes. These latter processes have turned out to be connected naturally (for $s = n$) with the method of conjugate gradients (for solving a system) and with the method of minimal iterations (in the case of solving an eigenvalue problem). Thus these methods may be considered as limit cases for multi-step descent methods.

In Sections 70-73 we consider iterative gradient proceesses for solving linear systems; in Sections 74-76, for solving the special eigenvalue problem. As a rule the matrix $A$ is assumed positive definite in this chapter.

## 70. THE METHOD OF STEEPEST DESCENT FOR SOLVING LINEAR SYSTEMS

Let $A$ be a positive definite matrix and let

$$AX = F \tag{1}$$

be a given linear system. In the work of L. V. Kantorovich (1)-(3) the solution of this system is connected with the problem of finding the vector which gives a minimum for the functional

$$H(X) = (AX, X) - 2(F, X) \tag{2}$$

that differs only by a constant (but unknown beforehand) element $(AX^*, X^*)$ from the error function $f(X) = (AY, Y)$. Here $X^*$ is the exact solution of the system and coincides with the vector that realizes the minimum for $H(X)$; $Y = X^* - X$ is the error vector.

The problem we have posed is solved in the following manner. An arbitrary vector $X_0$ is chosen. We compute the direction opposite to the gradient of the functional $H(X)$ (or, what is the same thing, to the gradient of the error function) at this point; this direction coincides with the direction of the residual $r_0 = F - AX_0$ for the chosen initial approximation. From the point $X_0$ we move in the selected direction to the point $X_1$ at which the functional $H(X)$ becomes minimal.

Since

$$
\begin{aligned}
H(X_0 + \alpha r_0) &= (AX_0 + \alpha A r_0, X_0 + \alpha r_0) - 2(F, X_0 + \alpha r_0) \\
&= (AX_0, X_0) + 2\alpha(AX_0, r_0) + \alpha^2(Ar_0, r_0) - 2(F, X_0) - 2\alpha(F, r_0) \\
&= H(X_0) - 2\alpha(r_0, r_0) + \alpha^2(Ar_0, r_0) \\
&= H(X_0) - \frac{(r_0, r_0)^2}{(Ar_0, r_0)} + (Ar_0, r_0)\left[\alpha - \frac{(r_0, r_0)}{(Ar_0, r_0)}\right]^2,
\end{aligned}
$$

then this expression attains a minimum for

$$\alpha = \alpha_0 = \frac{(r_0, r_0)}{(Ar_0, r_0)} \tag{3}$$

and this minimum equal to

$$H(X_0) - \frac{(r_0, r_0)^2}{(Ar_0, r_0)}. \tag{4}$$

Thus

$$X_1 = X_0 + \alpha_0 r_0$$

where

$$r_0 = F - AX_0 ,$$

$$\alpha_0 = \frac{(r_0 , r_0)}{(Ar_0 , r_0)} ,$$

$$H(X_1) = H(X_0) - \frac{(r_0 , r_0)^2}{(Ar_0 , r_0)} ,$$

$$f(X_1) = f(X_0) - \frac{(r_0 , r_0)^2}{(Ar_0 , r_0)} .$$

We then determine $X_2 = X_1 + \alpha_1 r_1$ where $r_1 = F - AX_1$ and

$$\alpha_1 = \frac{(r_1 , r_1)}{(Ar_1 , r_1)}$$

and the process is continued further by the formulas

$$X_{k+1} = X_k + \alpha_k r_k ,$$

$$r_k = F - AX_k = r_{k-1} - \alpha_{k-1} Ar_{k-1}, \tag{5}$$

where

$$\alpha_k = \frac{(r_k , r_k)}{(Ar_k , r_k)} . \tag{6}$$

Therefore

$$f(X_{k+1}) - f(X_k) = \frac{(r_k , r_k)^2}{(Ar_k , r_k)} . \tag{7}$$

We observe at once that in actually carrying out the process it is more convenient to compute the vectors $r_k$, especially for a matrix of a high order, from the formula $r_k = r_{k-1} - \alpha_{k-1} Ar_{k-1}$. However, as a consequence of rounding-off errors the vectors $r_k$ so computed may begin to diverge from the true residuals $F - AX_k$ after several steps of the process. Therefore one should compute the vectors $r_k$ from time to time by the formula $r_k = F - AX_k$.

THEOREM 70.1. *The successive approximations* $X_0, X_1, X_2, \cdots$ *converge to the solution of the system* $AX = F$ *with the rate of a geometric progression.*[†]

*Proof.* We show first of all that the sequence of values for the error function approaches zero for $k \to \infty$. We have

$$f(X_{k+1}) - f(X_k) = - \frac{(r_k , r_k)^2}{(Ar_k , r_k)} .$$

On the other hand

$$f(X_k) = (A^{-1} r_k , r_k) .$$

Therefore

---

[†] L. V. Kantorovich (2).

$$\frac{f(X_{k+1})}{f(X_k)} = 1 - \frac{(r_k, r_k)^2}{f(X_k)(Ar_k, r_k)}$$

and

$$\frac{f(X_{k+1})}{f(X_k)} = 1 - \frac{(r_k, r_k)^2}{(A^{-1}r_k, r_k)(Ar_k, r_k)} \ .$$

We shall estimate from below the subtrahend in the right member of the last equation.

Let $\lambda_1 \leq \lambda_2 \leq \cdots \leq \lambda_n$ be eigenvalues of matrix $A$; let $U_1, U_2, \cdots, U_n$ be their associated eigenvectors, orthogonal to each other and normalized so that $(U_i, U_i) = 1$ for $i = 1, \cdots, n$. Since $A$ is positive definite, all $\lambda_i > 0$. Let $m \leq \lambda_1$, $\lambda_n \leq M$. Furthermore let

$$r_k = c_1 U_1 + \cdots + c_n U_n$$

where not all $c_i$ are equal to zero. Then

$$Ar_k = c_1 \lambda_1 U_1 + \cdots + c_n \lambda_n U_n \ ,$$
$$A^{-1} r_k = c_1 \lambda_1^{-1} U_1 + \cdots + c_n \lambda_n^{-1} U_n \ .$$

Consequently

$$(r_k, r_k) = \sum_{i=1}^{n} c_i^2 \ ,$$

$$(Ar_k, r_k) = \sum_{i=1}^{n} \lambda_i c_i^2 \ ,$$

$$(A^{-1} r_k, r_k) = \sum_{i=1}^{n} \frac{1}{\lambda_i} c_i^2$$

and therefore

$$\frac{(r_k, r_k)^2}{(A^{-1} r_k, r_k)(Ar_k, r_k)} = \frac{\left( \sum\limits_{i=1}^{n} c_i^2 \right)^2}{\sum\limits_{i=1}^{n} \lambda_i c_i^2 \sum\limits_{i=1}^{n} \frac{1}{\lambda_i} c_i^2} \ .$$

To estimate the last quotient from below we apply the inequality

$$\frac{\sum\limits_{i=1}^{n} \gamma_i a_i \sum\limits_{i=1}^{n} \frac{1}{\gamma_i} a_i}{\left( \sum\limits_{i=1}^{n} a_i \right)^2} \leq \frac{1}{4} \left[ \sqrt{\frac{M}{m}} + \sqrt{\frac{m}{M}} \right]^2 , \tag{8}$$

which holds under the condition that all numbers $a_i > 0$ and the numbers $\gamma_i$ satisfy the inequalities $0 < m \leq \gamma_i \leq M$.

This inequality is verified as follows. The expression

$$\Gamma = \sum_{i=1}^{n} \gamma_i a_i \sum_{i=1}^{n} \frac{1}{\gamma_i} a_i \ ,$$

considered as a function of one of the parameters $\gamma_i$, with the remaining ones fixed, has the form

$$A\gamma_i + \frac{B}{\gamma_i} + C$$

where $A > 0$ and $B > 0$. The last function obviously does not have a maximum for positive values of $\gamma_i$ and therefore for varying $\gamma_i$ in the interval $(m, M)$, it will have a maximum at one of the endpoints. Thus the expression $\Gamma$ for a change in $\gamma_i$ in the interval $(m, M)$ takes on a maximal value when certain $\gamma_i$ are equal to $m$ and the remainder to $M$.

Without loss of generality we may assume that $\gamma_i = m$ for $i = 1, \cdots, k + 1, \gamma_i = M$ for $i = k + 1, \cdots, n$.

We introduce the notation

$$S_1 = \sum_{i=1}^{k} a_i \, ,$$

$$S_2 = \sum_{i=k+1}^{n} a_i \, .$$

Then

$$\Gamma \le (S_1 m + S_2 M)\left(\frac{S_1}{m} + \frac{S_2}{M}\right) = S_1^2 + S_2^2 + \left(\frac{m}{M} + \frac{M}{m}\right) S_1 S_2$$

$$= (S_1 + S_2)^2 + \frac{(m - M)^2}{Mm} S_1 S_2 \le (S_1 + S_2)^2 \left[1 + \frac{(m - M)^2}{4Mm}\right],$$

since

$$S_1 S_2 \le \frac{(S_1 + S_2)^2}{4} \, .$$

Furthermore,

$$1 + \frac{(m - M)^2}{4Mm} = \frac{1}{4}\left[\frac{m}{M} + 2 + \frac{M}{m}\right] = \frac{1}{4}\left[\sqrt{\frac{m}{M}} + \sqrt{\frac{M}{m}}\right]^2 \, .$$

Thus

$$\Gamma \le (S_1 + S_2)^2 \frac{1}{4}\left[\sqrt{\frac{m}{M}} + \sqrt{\frac{M}{m}}\right]^2 = \left(\sum_{i=1}^{n} a_i\right)^2 \frac{1}{4}\left(\sqrt{\frac{m}{M}} + \sqrt{\frac{M}{m}}\right)^2,$$

which also proves inequality (8).

On the basis of inequality (8)

$$\frac{(r_k, r_k)^2}{(A^{-1}r, r_k)(Ar_k, r_k)} \ge \frac{4}{\left[\sqrt{\frac{m}{M}} + \sqrt{\frac{M}{m}}\right]^2} \tag{9}$$

and consequently

$$\frac{f(X_{k+1})}{f(X_k)} \le 1 - \frac{4}{\left[\sqrt{\frac{M}{m}} + \sqrt{\frac{m}{M}}\right]^2} = \left[\frac{M - m}{M + m}\right]^2 < 1 \, .$$

Thus

$$f(X_{k+1}) \le \left[\frac{M - m}{M + m}\right]^2 f(X_k) \tag{10}$$

and consequently

$$f(X_{k+1}) \leq \left[\frac{M-m}{M+m}\right]^{2(k+1)} f(X_0) . \qquad (11)$$

Thus $f(X_{k+1}) \to 0$ for $k \to \infty$ and therefore $X_{k+1} \to X^*$ where $X^*$ is the exact solution of the system $AX = F$.

We now estimate the length of the error vector, i.e. the vector $Y_k = = X^* - X_k$. Since

$$f(X_k) = (AY_k, Y_k) \geq m \, |Y_k|^2 ,$$

then

$$|Y_k| \leq \sqrt{\frac{f(X_0)}{m}} \left(\frac{M-m}{M+m}\right)^k . \qquad (12)$$

We have thus proved that $|Y_k|$ approaches zero with the rapidity of a geometric progression. The theorem has been proved.

It is possible to give the estimate for $f(X_k)$ a slightly different form if we consider the $P$-number of condition, i.e. the number $\rho = \lambda_n/\lambda_1$. That is, we may take $m = \lambda_1$, $M = \lambda_n$. Then

$$f(X_k) \leq \left(\frac{\lambda_n - \lambda_1}{\lambda_n + \lambda_1}\right)^{2k} f(X_0) = \left(\frac{\rho - 1}{\rho + 1}\right)^{2k} f(X_0) . \qquad (11')$$

We observe two properties of the approximations $X_k$ for the method of steepest descent.

1. The residuals of two successive approximations are orthogonal to each other.

Actually $r_{k+1} = r_k - \alpha_k Ar_k$, from which $(r_{k+1}, r_k) = (r_k, r_k) - \alpha_k(Ar_k, r_k) = 0$, from the definition of $\alpha_k$.

2. Each successive approximation is closer to the exact solution than the preceding one, i.e.

$$|X^* - X_{k+1}| < |X^* - X_k| .$$

In other words, the length of the error vector in passing to a new approximation is strictly decreasing.

In fact

$$Y_{k+1} = Y_k - \alpha_k r_k$$

and consequently

$$(Y_{k+1}, Y_{k+1}) = (Y_k, Y_k) - 2\alpha_k(Y_k, r_k) + \alpha_k^2(r_k, r_k)$$

$$= (Y_k, Y_k) - \alpha_k(Y_k, r_k) - \frac{\alpha_k^2}{(r_k, r_k)}\left[\frac{(Y_k, r_k)(r_k, r_k)}{\alpha_k} - (r_k, r_k)^2\right]$$

$$= (Y_k, Y_k) - \alpha_k(Y_k, r_k) - \frac{\alpha_k^2}{(r_k, r_k)}[(Y_k, r_k)(Ar_k, r_k) - (r_k, r_k)^2] .$$

We shall show that

TABLE VII.1  The Solution of a System of Linear Equations by the Method of Steepest Descent

| | $X_0$ | $r_0$ | $Ar_0$ | $X_1$ | $r_1 = r_0 - \alpha_0 Ar_0$ | $r_1 = F - AX_1$ | $Ar_1$ | $X_2$ | $X_8$ | $X_9$ | $X_{10}$ |
|---|---|---|---|---|---|---|---|---|---|---|---|
| I | 0 | 0.76 | 0.3616 | 1.4245790 | 0.08220033 | 0.08220030 | 0.06456777 | 1.5280471 | 1.5349633 | 1.5349634 | 1.5349650 |
| | 0 | 0.08 | 0.0496 | 0.1499557 | -0.01297252 | -0.01297254 | -0.00258459 | 0.1336268 | 0.1220118 | 0.1220090 | 0.1220097 |
| | 0 | 1.12 | 0.6576 | 2.0993795 | -0.11263569 | -0.11263567 | -0.09805664 | 1.9576015 | 1.9751502 | 1.9751560 | 1.9751560 |
| | 0 | 0.68 | 0.3120 | 1.2746233 | 0.09517285 | 0.09517284 | 0.06715236 | 1.3944203 | 1.4129515 | 1.4129545 | 1.4129552 |
| II | | | 1.3808 | | | | 0.03107890 | | | | |
| III | | 2.3008 | 1.227456 | | | 0.02866984 | 0.02277678 | | | | |
| IV | | 1.8744460 | | | | 1.2587313 | | | | | |

$$(Y_k, r_k)(Ar_k, r_k) - (r_k, r_k)^2 \geq 0 .$$

We assume $A = B^2$ where $B$ is a positive definite matrix. Then

$$(Y_k, r_k)(Ar_k, r_k) - (r_k, r_k)^2 = (A^{-1}r_k, r_k)(Ar_k, r_k) - (r_k, r_k)^2$$
$$= (B^{-1}r_k, B^{-1}r_k)(Br_k, Br_k) - (Br_k, B^{-1}r_k)^2 \geq 0$$

on the basis of the Cauchy-Bunyakovskii inequality. Thus

$$(Y_{k+1}, Y_{k+1}) \leq (Y_k, Y_k) - \alpha_k(Y_k, r_k) < (Y_k, Y_k) ,$$

since $\alpha_k > 0$ and $(Y_k, r_k) = (AY_k, Y_k) > 0$ .

Finally it follows from the comparison of corresponding formulas that the approximation $X_{k+1}$ coincides with the first approximation for the method of conjugate gradients, carried out with $X_k$ as initial approximation.
We give examples of applications of the method of steepest descent.

*Example 1.* We shall solve a system with matrix

$$\begin{bmatrix} 0.78 & -0.02 & -0.12 & -0.14 \\ -0.02 & 0.86 & -0.04 & 0.06 \\ -0.12 & -0.04 & 0.72 & -0.08 \\ -0.14 & 0.06 & -0.08 & 0.74 \end{bmatrix}$$

and with constant column $(0.76, 0.08, 1.12, 0.68)'$. In Table VII.1 is given the beginning of the computational process (to illustrate the computational scheme of the method) and the results of the last three steps.

In the first part of the table are written in sequence the vectors $X_i$, $r_i$ and $Ar_i$, in the second—the result of control computation by column sums for $Ar_i$, in the third—corresponding scalar products $(r_i, r_i)$ and $(Ar_i, r_i)$, in the fourth—the coefficients $\alpha_i$. For comparison purposes the vector $r_1$ is computed by two methods.

Comparison of the course of the steepest descent process with the results of computing by the method of successive approximations and by the cyclic one-step process shows that in the example given the method of steepest descent converges more rapidly. That is, the eighth step of the method of steepest descent gives better results than the tenth step in both methods. The tenth step gives the solution accurately to within $1 \cdot 10^{-6}$.

As a second example we consider the solution of system (9) in Section

**TABLE VII.2   The Solution of a System of Linear Equations by the Method of Steepest Descent**

| $X_0$ | 0 | 0 | 0 | 0 |
|---|---|---|---|---|
| $X_1$ | 0.1515525 | 0.2525875 | 0.3536225 | 0.4546575 |
| $X_{18}$ | $-1.2546235$ | 0.0434210 | 1.0365064 | 1.4786412 |
| $X_{25}$ | $-1.2573084$ | 0.0435634 | 1.0389273 | 1.4820379 |
| $X_{30}$ | $-1.2577348$ | 0.0434859 | 1.0391170 | 1.4823232 |
| $X_{40}$ | $-1.2577912$ | 0.0434873 | 1.0391642 | 1.4823900 |
| $X_{52}$ | $-1.2577937$ | 0.0434873 | 1.0391662 | 1.4823928 |

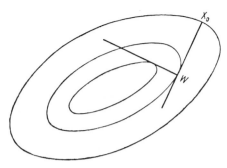

**FIGURE 7**

23.   The results obtained by the method of steepest descent are given in Table VII.2.

Theoretical estimates for the rate of convergence of the method of steepest descent show that for an increase in the condition number of the matrix of coefficients the convergence of the method soon slows down. It turns out that theoretical estimates are hardly overstated and actually the process converges very slowly even for not very large condition numbers. Thus in the last example the $P$-number of condition is about 10.

The method of steepest descent may be given the following geometric interpretation. We consider in an $n$-dimensional space the family of surfaces

$$f(X) = c$$

where $f(X)$ is the error function. This family is a family of similar ellipsoids. The process of steepest descent may be explained geometrically as follows. Through the approximation $X_0$ passes the ellipsoid

$$f(X) = X_0 ,$$

at the point $X_0$ is constructed the normal to this ellipsoid and then the ellipsoid of the family which is tangent to this normal is found. The point of tangency $W$ will be the next approximation. For $n = 2$ a geometric picture is given in Fig. 7.

It is obvious geometrically that if the ellipsoids of the family are stretched

**FIGURE 8**

in one direction and some approximation falls close enough to the major axis, then subsequent approximations will also be close to the endpoints of major axes of similar ellipsoids. Thus for $n = 2$ if the initial approximation is found at a point whose normal forms an angle of $45°$ with the major axis of a corresponding ellipse, then the angle between the normal and the major axis will equal $\pm 45°$ for all subsequent approximations as well (Fig. 8). It is not hard to convince oneself (for $n = 2$) that for just such a choice of the initial approximation the convergence of the process will be very slow and the rate of convergence will coincide with the theoretical estimate.

## 71. THE GRADIENT METHOD WITH MINIMAL RESIDUALS

This method has been described in the work of M. A. Kransnosel'skii and S. G. Krein (1). Let $A$ be a positive definite matrix and let $X_0$ be the initial approximation to the solution of the system $AX = F$. The next approximation $X_1$ is sought, just as in the method of steepest descent, in the form $X_0 + \beta r_0$ but the parameter $\beta$ is chosen so that the *length* $|r|$ *of the residual vector*, or, what is the same thing, $(r, r) = |r|^2$ is minimized. Thus the functional $(r, r)$ is minimized here in the direction opposite to the gradient of *another* functional, namely, the error function. After completing the first step the process is repeated.

Let us derive the formulas relating adjacent approximations. Let

$$X_{k+1} = X_k + \beta_k r_k \tag{1}$$

and

$$r_{k+1} = r_k - \beta_k A r_k . \tag{2}$$

Consequently,

$$(r_{k+1}, r_{k+1}) = (r_k, r_k) - 2\beta_k(Ar_k, r_k) + \beta_k^2(Ar_k, Ar_k)$$
$$= (r_k, r_k) - \frac{(Ar_k, r_k)^2}{(Ar_k, Ar_k)} + (Ar_k, Ar_k)\left[\beta_k - \frac{(Ar_k, r_k)}{(Ar_k, Ar_k)}\right]^2 .$$

from which it follows that $(r_{k+1}, r_{k+1})$ takes on a minimal value equal to

$$(r_k, r_k) - \frac{(Ar_k, r_k)^2}{(Ar_k, Ar_k)} \quad \text{for} \quad \beta_k = \frac{(Ar_k, r_k)}{(Ar_k, Ar_k)} .$$

Thus the working formulas of the method will be

$$X_{k+1} = X_k + \beta_k r_k ,$$
$$r_k = F - AX_k = r_{k-1} - \beta_{k-1} A r_{k-1} , \tag{3}$$
$$\beta_k = \frac{(Ar_k, r_k)}{(Ar_k, Ar_k)} .$$

THEOREM 71.1. *The successive approximations* $X_0, X_1, \cdots$ *converge to the solution of the system* $AX = F$ *with the rate of a geometric progression.*

The truth of the theorem will form a more general theorem which will be proved in the next paragraph.

In Table VII.3 is given the solution of the first example in Section 70. The rate of convergence for the process in the example given is the same as in the method of steepest descent.

## 72. GRADIENT METHODS WITH INCOMPLETE RELAXATION

Let $A$ be a positive definite matrix as before and let us seek the solution for the system $AX = F$.

Consider the iterative process in which each successive approximation is obtained from the previous one by a change in the direction opposite to the gradient of the error function so that at each step the error function decreases. Formulas for obtaining the successive approximations must obviously have the form

$$X_{k+1} = X_k + \gamma_k r_k. \tag{1}$$

For more extensive investigation it is convenient to let

$$\gamma_k = q_k \alpha_k \tag{2}$$

where $\alpha_k$ is the corresponding coefficient in the method of steepest descent. We have

$$
\begin{aligned}
f(X_{k+1}) &= f(X_k) - 2\gamma_k(r_k, r_k) + \gamma_k^2(Ar_k, r_k) \\
&= f(X_k) - 2q_k\alpha_k(r_k, r_k) + q_k^2\alpha_k^2(Ar_k, r_k) \\
&= f(X_k) - 2q_k\frac{(r_k, r_k)^2}{(Ar_k, r_k)} + q_k^2\frac{(r_k, r_k)^2}{(Ar_k, r_k)} \\
&= f(X_k) - (2q_k - q_k^2)\frac{(r_k, r_k)^2}{(Ar_k, r_k)} .
\end{aligned}
\tag{3}
$$

From this formula it is clear that for $f(X_{k+1})$ to be less than $f(X_k)$ it is necessary and sufficient that for the relation factors $q_k$ the inequalities

$$0 < q_k < 2 \tag{4}$$

are satisfied.

We shall call the group of methods in which not all $q_k$ are equal to 1 methods of *incomplete gradient relaxation* (one-step). If all the relaxation factors $q_k \leq 1$ but not all are equal to one, the method is called the method of *under-relaxation*; if all $q_k \geq 1$ but not all $q_k = 1$, it is called the method of *over-relaxation*.

Thus the method with minimal residuals is the method of under-relaxation, since in it

$$q_k = \frac{\beta_k}{\alpha_k} = \frac{(Ar_k, r_k)^2}{(Ar_k, Ar_k)(r_k, r_k)} \leq 1 \tag{5}$$

according to the Cauchy-Bunyakovskii inequality. Here the equality sign is possible only if $r_k$ is an eigenvector of matrix $A$.

The gradient method with constant multiplier $\gamma_k = \gamma$ is included in the

**TABLE VII.3. The Solution of a System of Linear Equations by the Gradient Method with Minimal Residuals**

|  | $X_0$ | $r_0$ | $AX_0$ | $X_1$ | $r_1 = F - \cdot AX_1$ | $r_1 = r_0 - \beta_0 Ar_0$ | $Ar_1$ | $X_2$ | $X_9$ | $X_{10}$ |
|---|---|---|---|---|---|---|---|---|---|---|
| I | 0 | 0.76 | 0.3616 | 1.4070460 | 0.09054233 | 0.09054233 | 0.06822351 | 1.5213114 | 1.5349632 | 1.5349648 |
|  | 0 | 0.08 | 0.0496 | 0.1481101 | −0.01182826 | −0.01182826 | −0.00194231 | 0.1331827 | 0.1220091 | 0.1220099 |
|  | 0 | 1.12 | 0.6576 | 2.0735415 | −0.09746508 | −0.09746505 | −0.08875643 | 1.9505396 | 1.9751557 | 1.9751558 |
|  | 0 | 0.68 | 0.3120 | 1.2589359 | 0.10237059 | 0.10237059 | 0.07016581 | 1.3381287 | 1.4129543 | 1.4129551 |
| II |  |  | 1.3808 |  |  |  | 0.04769058 |  |  |  |
| III |  | 1.227456 | 0.66299648 |  | $0.22033655 \cdot 10^{-1}$ |  | $0.17459165 \cdot 10^{-1}$ |  |  |  |
| IV |  | 1.8513763 |  |  | 1.2620108 |  |  |  |  |  |

group of incomplete relaxation methods if this multiplier satisfies the inequality

$$0 < \gamma < \frac{1}{M} \tag{6}$$

where $M$ is the largest eigenvalue of matrix $A$.

In fact under this assumption

$$0 < q_k = \frac{\gamma}{\alpha_k} = \frac{\gamma(Ar_k, r_k)}{(r_k, r_k)} \le \gamma M < 2 .$$

The condition $0 < \gamma < 2/M$ is also necessary for the error function to decrease for any initial vector. In fact,

$$0 < q_k < 2$$

gives

$$0 < \gamma \frac{(Ar_k, r_k)}{(r_k, r_k)} < 2 ,$$

from which

$$0 < \gamma < \frac{(2r_k, r_k)}{(Ar_k, r_k)} .$$

Since the last inequality must be fulfilled for all $r_k$, we must also satisfy the inequality

$$0 < \gamma < \min \frac{2(z, z)}{(Az, z)} = \frac{2}{M} .$$

We observe that the gradient method with constant multiplier is none other than the process of successive approximations applied to a system prepared in the following way:

$$X = (E - \gamma A)X + \gamma F .$$

The necessary and sufficient convergence conditions for the method coincide with the condition $0 < \gamma < 2/M$. In fact the largest eigenvalue of the matrix $E - \gamma A$ will be the larger of the numbers $1 - \gamma M$ and $1 - \gamma m$. Thus for the method of successive approximations to converge it is necessary and sufficient to satisfy the inequalities

$$-1 < 1 - \gamma m < 1$$
$$-1 < 1 - \gamma M < 1 ,$$

from which it follows that $\gamma > 0$, $\gamma < 2/m$ and $\gamma < 2/M$. Satisfying the third inequality guarantees satisfying the second.

The largest eigenvalue of the matrix $E - \gamma A$ will be the smallest of those possible if $1 - \gamma m = -(1 - \gamma M)$, i.e. if $\gamma = 2/(m + M)$. In this case $1 - \gamma m = (M - m)/(M + m)$. Consequently the rate of convergence with such a choice of the multiplier $\gamma$ will be estimated by the inequality

$$| Y_k | \leq \left(\frac{M - m}{M + m}\right)^k | Y_0 | ,$$

i.e. it has the same order as in the method of steepest descent.[†]

The whole group of incomplete relaxation methods (including even the case of complete relaxation when all $q_k = 1$, which corresponds to the method of steepest descent) is contained naturally in the general scheme of iterative methods described in Chapter III. That is, they are obtained from the general iterative formula

$$X_{k+1} = X_k + H^{(k+1)}(F - AX_k)$$

for

$$H^{(k+1)} = \gamma_k E .$$

THEOREM 72.1. *If in the process of incomplete gradient relaxation the relaxation factors satisfy the condition $\varepsilon < q_k < 2 - \varepsilon, 0 < \varepsilon < 1$, then the process converges to the solution with the speed of a geometric progression.*

*Proof.* Let $X_k$ be the $k$th approximation for an incomplete relaxation method satisfying the theorem's conditions. We denote by $\bar{X}_{k+1}$ the approximation obtained from $X_k$ by one step of the method of steepest descent, and by $Y_{k+1}$ and $\bar{Y}_{k+1}$ the corresponding error vectors.

As we saw above

$$f(\bar{X}_{k+1}) \leq \tau f(X_k)$$

where

$$\tau = \left(\frac{M - m}{M + m}\right)^2 .$$

Consequently

$$f(X_k) - f(\bar{X}_{k+1}) \geq (1 - \tau)f(X_k) .$$

Furthermore

$$f(X_{k+1}) = f(X_k) - (2q_k - q_k^2)\frac{(r_k, r_k)^2}{(Ar_k, r_k)}$$

from which

$$f(X_k) - f(X_{k+1}) = (2q_k - q_k^2)\frac{(r_k, r_k)^2}{(Ar_k, r_k)} = (2q_k - q_k^2)[f(X_k) - f(\bar{X}_{k+1})]$$

since

$$f(X_k) - f(\bar{X}_{k+1}) = \frac{(r_k, r_k)^2}{(Ar_k, r_k)} .$$

Thus

$$f(X_k) - f(X_{k+1}) \geq (2q_k - q_k^2)(1 - \tau)f(X_k)$$

---

[†] I. P. Natanson (1).

and therefore

$$f(X_{k+1}) \le [1 - (2q_k - q_k^2)(1 - \tau)]f(X_k) \le [1 - (2\varepsilon - \varepsilon^2)(1 - \tau)]f(X_k) \,.$$

We let

$$\tau_1 = 1 - (2\varepsilon - \varepsilon^2)(1 - \tau) \,.$$

It is clear that $0 < \tau_1 < 1$, since $0 < 2\varepsilon - \varepsilon^2 < 1, 0 < 1 - \tau < 1$. It follows from the last inequality that

$$f(X_{k+1}) \le \tau_1^k f(X_0) \,. \tag{7}$$

Thus for $k \to \infty$, $f(X_k) \to 0$ with the speed of a geometric progression, $Y_k \to 0$ and $X_k \to X^*$. The theorem has been proved.

It follows from the theorem, in particular, that the method with minimal residuals converges, since in this case the factors $q_k$ satisfy the inequality

$$\varepsilon \le q_k \le 1 \,,$$

where

$$\varepsilon = \frac{4}{\left[ \sqrt{\dfrac{M}{n}} + \sqrt{\dfrac{m}{M}} \right]^2} \,.$$

In fact,

$$q_k = \frac{\beta_k}{\alpha_k} = \frac{(Ar_k, r_k)^2}{(Ar_k, Ar_k)(r_k, r_k)} = \frac{(A^{1/2}r_k, A^{1/2}r_k)^2}{(A^{3/2}r_k, A^{1/2}r_k)(A^{-1/2}r_k, A^{1/2}r_k)}$$

$$= \frac{(z, z)^2}{(Az, z)(A^{-1}z, z)} \,.$$

Here $z = A^{1/2}r_k$. Therefore from inequality (9) in Section 70 we obtain

$$q_k > \frac{4}{[\sqrt{M/m} + \sqrt{m/M}]^2} \,.$$

The inequality $q_k \le 1$ has already been noted.

The rate of convergence of the method, with such a method of procedure, is greatly reduced. In the above mentioned work of M. A. Krasnosel'skii and S. G. Krein it was established that the rate of convergence has the same order as in the method of steepest descent.

In estimate (7) it was not considered whether, at given step, $q_k$ would

TABLE VII.4   The Solution of a System of Linear Equations by the Gradient Method
with Incomplete Relaxation for $q = 0.8$

| $X_0$ | 0 | 0 | 0 | 0 |
|---|---|---|---|---|
| $X_1$ | 0.1212420 | 0.2020700 | 0.2828980 | 0.3637260 |
| $X_{18}$ | −1.2574919 | 0.0435581 | 1.0390689 | 1.4822318 |
| $X_{30}$ | −1.2577926 | 0.0434876 | 1.0391658 | 1.4823922 |
| $X_{35}$ | −1.2577936 | 0.0434873 | 1.0391662 | 1.4823928 |

be more or less than one. There are grounds for assuming that if over-relaxation is applied, then these estimates are not appreciably raised. For under-relaxation they are in all probability increased, since with under-relaxation the broken line with peaks at successive approximations will adhere more closely to the line of steepest descent than the analogous broken lines for complete and over-relaxation.

From the point of view of computing, it is convenient to take the relaxation factors $q_k$ independent of $k$.

**TABLE VII.5   The Solution of a System of Linear Equations by the Gradient Method with Incomplete Relaxation for $q = 1.2$**

| $X_0$ | 0 | 0 | 0 | 0 |
|---|---|---|---|---|
| $X_1$ | 0.1818630 | 0.3031050 | 0.4243470 | 0.5455890 |
| $X_{18}$ | $-1.2251212$ | 0.0452191 | 1.0164571 | 1.4498670 |
| $X_{30}$ | $-1.2551442$ | 0.0436275 | 1.0373248 | 1.4797554 |
| $X_{35}$ | $-1.2566377$ | 0.0437154 | 1.0386884 | 1.4816669 |
| $X_{52}$ | $-1.2577672$ | 0.0434887 | 1.0391478 | 1.4823665 |
| $X_{62}$ | $-1.2577904$ | 0.0434875 | 1.0391640 | 1.4823796 |

In Tables VII.4 and VII.5 are given the results of computing the solution of system (9), Section 23, by the gradient method with application of incomplete relaxation for $q = 0.8$ and $q = 1.2$. It is apparent from the tables that in the given example under-relaxation leads to the solution more rapidly than complete (Table VII.2), and over-relaxation less rapidly.

In concluding this paragraph we consider a method which is the limiting case for the over-relaxation method. The formulas for the method[†] are as follows:

$$X_{k+1} = X_k + 2\alpha_k r_k \, ,$$

where $\alpha_k$ is the coefficient for the method of steepest descent. In this case the error function does not decrease from step to step and the successive approximations do not approach the solution. However, if the initial approximation is not orthogonal to the eigenvector belonging to the largest eigenvalue of matrix $A$, then the sequences $X_0, X_2, \cdots$ and $X_1, X_3, \cdots$ converge. The half-sum of the limits of these sequences gives the system's solution and the half-difference is the eigenvector belonging to the largest eigenvalue.

The reader will find a proof of the corresponding theorem in the specified work of V. N. Kostarchuk (1).

## 73.  S-STEP GRADIENT METHODS OF STEEPEST DESCENT

In the previous three paragraphs we have considered one-step gradient methods connected with complete or incomplete relaxation of the error

[†] V. N. Kostarchuk (1).

function and established that the best result for one-step is given by the method of steepest descent. This naturally raises the question of how to determine the multipliers $\gamma_1, \cdots, \gamma_s$ so as to obtain, starting from an initial approximation $X^{(0)}$, the best result after $s$-steps of the process and how to combine these $s$-steps into one-step of a new computational process. In other words, how do we construct a computational process, one step of which is equivalent to $s$-steps of the one-step gradient method chosen according to the best strategy?

Let

$$
\begin{aligned}
X_0 &= X^{(0)}, \\
X_1 &= X_0 + \gamma_0(F - AX_0), \\
X_2 &= X_1 + \gamma_1(F - AX_1), \\
&\cdots\cdots\cdots\cdots, \\
X^{(1)} = X_s &= X_{s-1} + \gamma_{s-1}(F - AX_{s-1}).
\end{aligned} \tag{1}
$$

Then passing over to the error vectors we obtain

$$
\begin{aligned}
Y_1 &= Y_0 - \gamma_0 A Y_0 = (E - \gamma_0 A)Y_0, \\
Y_2 &= Y_1 - \gamma_1 A Y_1 = (E - \gamma_1 A)Y_1, \\
&\cdots\cdots\cdots\cdots\cdots\cdots\cdots, \\
Y^{(1)} = Y_s &= Y_{s-1} - \gamma_{s-1} A Y_{s-1} = (E - \gamma_{s-1}A)Y_{s-1},
\end{aligned} \tag{2}
$$

from which

$$
\begin{aligned}
Y^{(1)} &= (E - \gamma_0 A)(E - \gamma_1 A) \cdots (E - \gamma_{s-1}A)Y_0 \\
&= (E + c_1 A + c_2 A^2 + \cdots + c_s A^s)Y_0 \\
&= Y_0 + c_1 r_0 + \cdots + c_s A^{s-1} r_0,
\end{aligned} \tag{3}
$$

where

$$
r_0 = F - AX^{(0)}
$$

and

$$
X^{(1)} = X^{(0)} - c_1 r_0 - \cdots - c_s A^{s-1} r_0. \tag{4}
$$

Thus the problem reduces to determining the coefficients $c_1, \cdots, c_s$ so that the value $f(X^{(1)})$ of the error function will be the least value.

This problem has already been solved in Section 69 while studying the method of conjugate gradients. In fact we established that the $s$th approximation of the method of conjugate gradients minimizes the error function among the vectors $X^{(0)} + V$ where the vector $V$ belongs to the subspace spanned by $r_0, Ar_0, \cdots, A^{s-1}r_0$.

Thus the result of one-step of the $s$-step method of steepest descent coincides with the result of the $s$th approximation for the method of conjugate gradients.

Namely,

$$
X^{(1)} = X^{(0)} + \sum_{j=1}^{s} a_j s_j \tag{5}
$$

where

$$s_1 = r_0,$$
$$r_i = r_{i-1} - a_i A s_i,$$
$$a_i = \frac{(s_i, r_{i-1})}{(s_i, As_i)} = \frac{(r_{i-1}, r_{i-1})}{(s_i, As_i)},$$
$$s_{i+1} = r_i + b_i s_i,$$
$$b_i = \frac{(r_i\ r_i)}{(r_{i-1}, r_{i-1})} = -\frac{(r_i, As_i)}{(s_i, As_i)},$$
$$(i = 1, 2, \cdots, s - 1).$$

(6)

Therefore in actually carrying out the $s$-step method of steepest descent we need neither the coefficients $c_1, \cdots, c_s$ nor, what is more, the coefficients $\gamma_1, \cdots, \gamma_s$.

However, it is not hard to realize what these and other coefficients represent. In considering the method of conjugate gradients it was pointed out that

$$r_s = r_s(A)r_0$$

(7)

and consequently

$$Y^{(1)} = r_s(A)Y^{(0)}$$

where $r_s(t)$ is the polynomial of Section 69, paragraph 3.

But on the other hand,

$$Y^{(1)} = (E + c_1 A + \cdots + c_s A^s)Y_0.$$

Consequently,

$$1 + c_1 t + \cdots + c_s t^s = r_s(t),$$

i.e. the coefficients $c_1, \cdots, c_s$ are none other than the coefficients of the polynomial $r_s(t)$. Moreover from the equation

$$(1 - \gamma_0 t)(1 - \gamma_1 t) \cdots (1 - \gamma_{s-1} t) = 1 + c_1 t + c_2 t^2 + \cdots + c_s t^s = r_s(t)$$

if follows that the numbers $\gamma_0, \cdots, \gamma_{s-1}$ are the numbers inverse to the roots of the polynomial $r_s(t)$.

We observe that if instead of formula (5) we compute the vector $X^{(1)}$ recurrently by formulas (1), then at certain steps of this process an increase in the error function may even come about, since not all the numbers $\gamma_0, \cdots, \gamma_{s-1}$ guarantee relaxation.

Actually if $s = n$ and we take for $\gamma_0$ the number $1/\lambda_1$ and for $r_0$ the vector $U_n$, then $\alpha_0 = (r_0, r_0)/(Ar_0, r_0) = 1/\lambda_n$, $\gamma_0 = q_0 \alpha_0 = q_0 \lambda_n^{-1}$, so that $q_0 = \lambda_n/\lambda_1 = \rho$ and therefore $q_0 > 2$ if $\rho > 2$.

The approximation $X^{(1)}$ may be computed by using the method of $A$-minimal iterations as well. In fact, as we saw in Section 69, successive approximations obtained by this method for $q_0 = r_0$ coincide with the corresponding approximations for the method of conjugate gradients.

The multi-step method of steepest descent was described for the first time

**TABLE VII.6** The Two-step Method of Steepest Descent. The Conjugate Gradients Scheme

| $X^{(0)}$ | $r_0 = s_1$ | $As_1$ | $r_1$ | $s_2$ | $As_2$ | $X^{(1)}$ | $X^{(2)}$ | $X^{(3)}$ | $X^{(4)}$ |
|---|---|---|---|---|---|---|---|---|---|
| 0 | 0.3 | 1.482 | −0.44866929 | −0.40302542 | −0.12915358 | −1.2648181 | −1.2572773 | −1.2577994 | −1.2577932 |
| 0 | 0.5 | 1.246 | −0.12944800 | −0.05337489 | −0.03272692 | 0.0650097 | 0.0434888 | 0.0435032 | 0.0434873 |
| 0 | 0.7 | 1.220 | 0.08368655 | 0.19018891 | 0.02000431 | 1.0220120 | 1.0387864 | 1.0391674 | 1.0391658 |
| 0 | 0.9 | 1.472 | 0.15658246 | 0.29331407 | 0.04567390 | 1.4854645 | 1.4818603 | 1.482840 | 1.4823923 |
| | | 5.420 | | | −0.09620229 | | | | |
| | 1.64 | 3.2464 | −0.49392752 | 0.24951983 | 0.071000367 | | | | |
| | −0.15214623 | 0.50517496 | | | 3.5143456 | | | | |

**TABLE VII.7** The Two-step Method of Steepest Descent. The Minimal Iterations Scheme

| $X^{(0)}$ | $r_0 = q_0$ | $Aq_0$ | $\bar{q}_1$ | $Aq_1$ | $X^{(1)}$ | $X^{(2)}$ | $X^{(3)}$ | $X^{(4)}$ |
|---|---|---|---|---|---|---|---|---|
| 0 | 0.3 | 1.482 | 0.79779375 | 0.25566113 | −1.2648180 | −1.2572775 | −1.2577994 | −1.2577932 |
| 0 | 0.5 | 1.246 | 0.10565625 | 0.06478338 | 0.0650097 | 0.0434887 | 0.0435032 | 0.0434873 |
| 0 | 0.7 | 1.220 | −0.37648125 | −0.03959875 | 1.0220119 | 1.0387863 | 1.0391673 | 1.0391659 |
| 0 | 0.9 | 1.472 | −0.58061875 | −0.09041200 | 1.4854643 | 1.4818599 | 1.4823841 | 1.4823923 |
| | | 5.420 | | 0.19043375 | | | | |
| | 0.2464 | 7.404024 | 0.27821271 | | | | | |
| | 2.2806875 | | | | | | | |
| | 1.64 | | −0.49392750 | | | | | |
| | 0.50517496 | | −1.7753592 | | | | | |

in the work of L. V. Kantorovich (2), (3). In these works the successive approximations were formed by actually finding the coefficients $c_1, \cdots, c_s$ in solving a system of $s$ linear equations. This system had the following form for the first step:

$$
\begin{aligned}
(r_0, r_0) - (r_0, Ar_0)c_1 - \cdots - (r_0, A^s r_0)c_s &= 0 , \\
(r_0, Ar_0) - (r_0, A^2 r_0)c_1 - \cdots - (r_0, A^{s+1} r_0)c_s &= 0 , \\
\cdots\cdots\cdots\cdots\cdots\cdots\cdots\cdots\cdots\cdots\cdots\cdots\cdots , & \\
(r_0, A^{s-1} r_0) - (r_0, A^s r_0)c_1 - \cdots - (r_0, A^{2s-1} r_0)c_s &= 0 .
\end{aligned}
\tag{8}
$$

For each subsequent step the initial residual $r_0$ had to be replaced by a residual obtained as a result of the previous step.

We shall consider application of the two-step method for finding the solution of system (9), Section 23.

The computations are given for different schemes in Tables VII.6, VII.7 and VII.8.

**TABLE VII.8   The Two-step Method of Steepest Descent.   Scheme with Solution of the System**

| $X^{(0)}$ | $r_0$ | $Ar_0$ | $A^2 r_0$ | $X^{(1)}$ | $X^{(2)}$ | $X^{(3)}$ | $X^{(4)}$ |
|---|---|---|---|---|---|---|---|
| 0 | 0.3 | 1.482 | 3.63564 | $-1.2648177$ | $-1.2572769$ | $-1.2577994$ | $-1.2577931$ |
| 0 | 0.5 | 1.246 | 2.90652 | 0.0650098 | 0.0434889 | 0.0435034 | 0.0434873 |
| 0 | 0.7 | 1.220 | 2.74284 | 1.0220119 | 1.0387865 | 1.0391674 | 1.0391657 |
| 0 | 0.9 | 1.472 | 3.26676 | 1.4854642 | 1.4818605 | 1.4823838 | 1.4823923 |
| | | 5.420 | 12.55176 | | | | |

To find $X^{(1)}$ in Table VII.8 we compute the coefficients $c_1$ and $c_2$ from the system of equations

$$3.2464c_1 + 7.404024c_2 = 1.64 ,$$
$$7.404024c_1 + 17.164478c_2 = 3.2464 .$$

This gives

$$c_1 = 4.5542139 ,$$
$$c_2 = 1.7753589 .$$

For $s > 2$ the first and second schemes are more preferable, since in the last scheme we need to solve an auxiliary system of linear equations at each step of the process.

As in the case of one-step gradient processes the $s$-step method of steepest descent may be included in the general scheme of iterative processes

$$X^{(k)} = X^{(k-1)} + H^{(k)}(F - AX^{(k-1)}) ,$$

letting

$$H^{(k)} = H_s^{(k)}(A)$$

for

$$H_s^{(k)}(t) = \frac{1 - r_s^{(k)}(t)}{t} . \tag{9}$$

In fact,

$$Y_s^{(k)} = r_s^{(k)}(A)Y^{(k-1)} = Y^{(k-1)} + (r_s^{(k)}(A) - E)Y^{(k-1)} .$$

From here,

$$X^{(k)} = X^{(k-1)} + (E - r_s^{(k)}(A))Y^{(k-1)} = X^{(k-1)} + H_s^{(k)}(A)AY^{(k-1)}$$
$$= X^{(k-1)} + H_s^{(k)}(A)(F - AX^{(k-1)}) .$$

It is clear that the polynomials $H_s^{(k)}(t)$ change here from step to step.

We shall now establish the convergence of the $s$-step process of steepest descent and shall estimate the rate of convergence.

We observe first of all that one-step of the $s$-step process of steepest descent does not give a worse result, in the sense of a decreasing error function, than $s$-steps of the one-step method of steepest descent. From this it follows directly that the $s$-step method of steepest descent converges, and for the error function the estimate

$$f(X^{(k)}) \le \left(\frac{M - m}{M + m}\right)^{2s} f(X^{(k-1)}) \tag{10}$$

holds; consequently,

$$f(X^{(k)}) \le \left(\frac{M - m}{M + m}\right)^{2sk} f(X_0) . \tag{11}$$

However, for the $s$-step method of steepest descent it is possible to give even better estimates,[†] comparing it with the stationary iterative process

$$X^{(k)} = X^{(k-1)} + H_s(A)(F - AX^{(k-1)}) \tag{12}$$

where $H_s(t)$ is a certain polynomial of degree $s - 1$.

It is clear that whatever the polynomial $H_s(t)$, each step of the $s$-step method of steepest descent will guarantee no worse a decrease in the error function than one-step carried out according to formula (12), starting from the preceding approximation in the method of steepest descent. Let $X^{(k-1)}$ be this approximation and let $X^{(k)}$ be the next approximation of the $s$-step method of steepest descent

$$\bar{X}^{(k)} = X^{(k-1)} + H_s(A)(F - AX^{(k-1)}) . \tag{13}$$

Furthermore let $Y^{(k-1)}$, $Y^{(k)}$ and $\bar{Y}^{(k)}$ be the corresponding error vectors. From formula (13) it follows that

$$\bar{Y}^{(k)} = \Phi_s(A)Y^{(k-1)}$$

where

$$\Phi_s(t) = 1 - tH_s(t) .$$

To estimate the error function we assume, as has been done repeatedly,

† M. Sh. Birman (1).

that $A = B^2$ where $B$ is a positive definite matrix. Then

$$f(\bar{X}^{(k)}) = (A\bar{Y}^{(k)}, \bar{Y}^{(k)}) = (B\bar{Y}^{(k)}, B\bar{Y}^{(k)}) = |B(\bar{Y}^{(k)})|^2 \, .$$

But

$$B\bar{Y}^{(k)} = B\Phi_s(A)Y^{(k-1)} = \Phi_s(A)BY^{(k-1)}$$

and therefore

$$|B\bar{Y}^{(k)}| = |\Phi_s(A)BY^{(k-1)}| \le ||\Phi_s(A)|| \, |BY^{(k-1)}| \, .$$

Let

$$Q_s = ||\Phi_s(A)|| \, . \tag{14}$$

Then

$$f(\bar{X}^{(k)}) \le Q_s^2 f(X^{(k-1)}) \tag{15}$$

and moreover

$$f(X^{(k)}) \le Q_s^2 f(X^{(k-1)}) \, . \tag{16}$$

The matrix $\Phi_s(A)$ is symmetric so that its norm $Q_s$ is equal to the largest modulus among its eigenvalues, which are obviously equal to $\Phi_s(\lambda_1), \cdots, \Phi_s(\lambda_n)$.

The polynomial $\Phi_s(t)$ obviously possesses the property $\Phi_s(0) = 1$; its degree is equal to $s$—beyond this it is arbitrary. We obtain the best estimate (16) if among polynomials of the type indicated we choose the polynomial $\Phi_s(t)$ such that $Q_s = \max_i \Phi_s(\lambda_i)$ will be least. This best estimate naturally depends on $\lambda_1, \cdots, \lambda_n$.

For the class of matrices whose eigenvalues are included in the interval $(m, M)$ we may give an estimate depending on the numbrs $m$ and $M$ which is best for this class as a whole. For this we should take, as the polynomial $\Phi_s(t)$, a polynomial which diverges least from zero in the interval $(m, M)$ and which is normalized so that $\Phi(0) = 1$.

By the linear transformation $t = (M - m)\tau/2 - (M + m)/2$ we reduce the problem to constructing a polynomial diverging least from zero in the interval $-1 \le \tau \le 1$ and taking on the value 1 at the point $\tau_0 = (M + m)/(M - m)$. The solution of the last problem[†] is given by the polynomial

$$\tilde{T}(\tau) = \frac{\cos s \, \text{arc} \cos \tau}{\cos s \, \text{arc} \cos \tau_0} \, , \tag{17}$$

where the maximum divergence is equal to

$$L_s = \max_{-1 \le \tau \le 1} |\tilde{T}_s(\tau)| = \frac{1}{|\cos s \, \text{arc} \cos \tau_0|} = \frac{1}{T_s(\tau_0)}$$

where $T_s(\tau) = \cos s \, \text{arc} \cos \tau$ is a Chebyshev polynomial.

It is well-known[‡] that

---

[†] V. L. Goncharov. *The theory of interpolating and approximating functions*,, GTTI, 1934, p. 281.

[‡] V. L. Goncharov, *op. cit.*, p. 27.

$$T_s(\tau) = \frac{(\tau + \sqrt{\tau^2 - 1})^s + (\tau - \sqrt{\tau^2 - 1})^s}{2}.$$

Therefore

$$L_s = \frac{2}{\left(\dfrac{M+m}{M-m} + \sqrt{\left(\dfrac{M+m}{M-m}\right)^2 - 1}\right)^s + \left(\dfrac{M+m}{M-m} - \sqrt{\left(\dfrac{M+m}{M-m}\right)^2 - 1}\right)^s}. \qquad (18)$$

In particular,

$$L_1 = \frac{M-m}{M+m},$$

$$L_2 = \frac{(M-m)^2}{(M+m)^2 + 4mM}, \qquad (19)$$

$$L_3 = \frac{(M-m)^3}{(M+m)[(M+m)^2 + 12mM]}, \quad \cdots$$

It is easy to see that

$$1 > L_1 > \sqrt{L_2} > \sqrt[3]{L_3} > \cdots \qquad (20)$$

The last inequalities mean that for sufficiently large $N$ the result of applying $[N/s]$ steps of the $s$-step process gives a better approximation than $[N/(s-1)]$ steps of the $(s-1)$-step process.

We observe that in the above arguments we did not use the results of Kantorovich on estimating an error function in the one-step method of steepest descent and therefore the equality

$$L_1 = \frac{M-m}{M+m}$$

gives a different conclusion than estimate (10), Section 70.

It is also possible to give other more exact estimates for the rate of convergence of the $s$-step method of steepest descent if we have more exact information on the distribution of eigenvalues for matrix $A$.

Thus in the work of B. A. Samokish (1) the situation is considered where the largest eigenvalue $\lambda_1$ is known, as is the interval $(m, M_1)$ in which all remaining eigenvalues are placed.

In this case we may take as $\Phi_s(t)$ the polynomial which diverges least from zero on the set consisting of the point $\lambda_1$ and the interval $(m, M_1)$. We denote the divergence from zero of the selected polynomial by $\bar{L}_s$. Then

$$\bar{L}_s < L_s \quad \text{if} \quad a = \frac{2\lambda_1 - M_1 - m}{M_1 - m} > \frac{3 - \cos(\pi/s)}{1 + \cos(\pi/s)}.$$

Moreover $\bar{L}_s$ is monotonically increasing, approaching $L_{s-1}$ for $a \to \infty$ where $L_{s-1}$ is constructed for the interval $(m, M_1)$.

It may be calculated that for the two-step method of steepest descent

$$\bar{L}_2 = \frac{M_1 - m}{M_1 + m + [2mM_1/(\lambda_1 - M_1)]} \quad \text{if} \quad M_1 > \frac{\lambda_1 + m}{2}.$$

For arbitrary $s$ the corresponding estimates are obtained by using polynomials studied by E. I. Zolotarev[†] but these turn out to be unwieldy. B. A. Samokish (1) has applied the approximate formula

$$\bar{L}_s \approx \frac{1}{R_s(\tau, a)} \, ,$$

$$\tau = \frac{M_1 + m}{M_1 - m} \, ,$$

$$R_s(\tau, a) = \frac{1}{2} \left[ v^s \frac{v - \alpha}{1 + \alpha v} + v^{-s} \frac{1 - \alpha v}{v - \alpha} \right] ,$$

$$v = \tau + \sqrt{\tau^2 - 1} \, , \quad \alpha = a + \sqrt{a^2 - 1} \, .$$

## 74. DETERMINATION OF THE ALGEBRAICALLY LARGEST EIGENVALUE AND ITS ASSOCIATED EIGENVECTOR FOR A SYMMETRIC MATRIX USING GRADIENT METHODS

The extremal theory of eigenvalues allows us to apply relaxation gradient methods in determining extreme eigenvalues (i.e. the algebraically largest $\lambda_1$ and the algebraically smallest $\lambda_n$) for a symmetric matrix $A$ as well as their associated eigenvectors. In fact,

$$\lambda_1 = \max \frac{(AX, X)}{(X, X)} \, ,$$

$$\lambda_n = \min \frac{(AX, X)}{(X, X)}$$

while the eigenvectors belonging to these eigenvalues will be vectors realizing an extremum. Thus the problem of finding $\lambda_1$ or $\lambda_n$ is connected with the problem of maximizing or minimizing the functional

$$\mu(X) = \frac{(AX, X)}{(X, X)} \, .$$

In view of theory's complete analogy we shall consider the problem of finding the algebraically largest eigenvalue and its associated eigenvector.

As was explained in Section 14, the gradient of the functional $\mu(X)$ will be the vector

$$\frac{2}{(X, X)} [AX - \mu(X)X] = \frac{2}{(X, X)} \xi$$

where

$$\xi = AX - \mu(X)X \, . \tag{1}$$

The direction of the gradient coincides with the direction of the vector $\xi$, since $(X, X) > 0$. If $X$ is not an eigenvector of matrix $A$, then $\xi \neq 0$. In

[†] E. I. Zolotarev. *The Application of Elliptic Functions to the Problem of Functions Least and Most Divergent from Zero*, (Collected Works, 2nd ed.).

what follows when we speak about the gradient of the functional $\mu(X)$ we shall mean by this the vector $\xi$.

Let $X_0$ be an arbitrary vector which is not an eigenvector of matrix $A$. Let

$$X_1 = X_0 + \gamma \xi_0 . \tag{2}$$

It follows from the properties of the gradient for $\gamma$ small enough in modulus the inequalities $\mu(X_1) > \mu(X_0)$ for $\gamma > 0$ and $\mu(X_1) < \mu(X_0)$ for $\gamma < 0$ are valid.

We shall explain in more detail how the difference $\mu(X_1) - \mu(X_0)$ behaves for a change in $\gamma$ along the whole real axis. We have

$$(AX_1, X_1) = (AX_0, X_0) + 2\gamma(AX_0, \xi_0) + \gamma^2(A\xi_0, \xi_0)$$
$$= (AX_0, X_0) + 2\gamma(\xi_0, \xi_0) + \gamma^2(A\xi_0, \xi_0) ,$$

since

$$(AX_0, \xi_0) = (\xi_0, \xi_0) .$$

Furthermore,

$$(X_1, X_1) = (X_0, X_0) + \gamma^2(\xi_0, \xi_0) ,$$

since $(X_0, \xi_0) = 0$.

Thus

$$\mu(X_1) = \mu(X_0 + \gamma \xi_0) = \frac{(AX_0, X_0) + 2\gamma(\xi_0, \xi_0) + \gamma^2(A\xi_0, \xi_0)}{(X_0, X_0) + \gamma^2(\xi_0, \xi_0)}$$
$$= \frac{\mu(X_0) + 2\gamma t_0^2 + \gamma^2 t_0^2 \mu(\xi_0)}{1 + \gamma^2 t_0^2} ,$$

where

$$t_0^2 = \frac{(\xi_0, \xi_0)}{(X_0, X_0)} . \tag{3}$$

Therefore

$$\mu(X_1) - \mu(X_0) = \frac{\mu(X_0) + 2\gamma t_0^2 + \gamma^2 t_0^2 \mu(\xi_0) - \mu(X_0) - \gamma^2 t_0^2 \mu(X_0)}{1 + \gamma^2 t_0^2}$$
$$= \frac{2\gamma t_0^2 - \gamma^2 t_0^2 [\mu(X_0) - \mu(\xi_0)]}{1 + \gamma^2 t_0^2} = \frac{2\gamma - \gamma^2 s_0}{1 + \gamma^2 t_0^2} t_0^2 , \tag{4}$$

where

$$s_0 = \mu(X_0) - \mu(\xi_0) . \tag{5}$$

From equality (4) it is clear that $\mu(X_1) = \mu(X_0)$ for $\gamma = 0$ and for $\gamma = 2/s_0$. Moreover it is clear that $\mu(X_1) - \mu(X_0)$ is a continuous function in $\gamma$ for all real $\gamma$ including $\gamma = \infty$, since $\lim_{\gamma \to +\infty} [\mu(X_1) - \mu(X_0)] = \lim_{\gamma \to -\infty} [\mu(X_1) - \mu(X_0)] = -s_0$. We observe that $\mu(X_1) - \mu(X_0) = -s_0$ at one other point, namely, for $\gamma = -s_0/2t_0^2$. Thus the graph of $\mu(X_1) - \mu(X_0)$, dependent on the sign of $s_0$, has the form given in Figs. 9, 10, and 11.

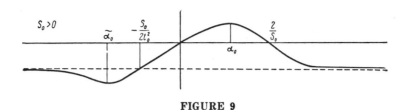

$$S_0 > 0$$

**FIGURE 9**

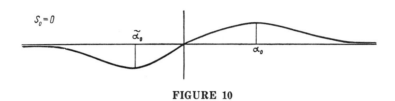

$$S_0 = 0$$

**FIGURE 10**

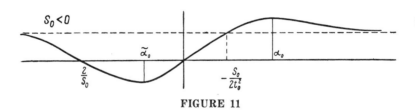

$$S_0 < 0$$

**FIGURE 11**

It is apparent from the graphs that the inequality $\mu(X_1) - \mu(X_0) > 0$ is satisfied in the region $0 < \gamma < 2/s_0$ if $s_0 > 0$, in the region $\gamma > 0$ if $s_0 = 0$ and in the region $\gamma > 0$ or $\gamma < 2/s_0$ if $s_0 < 0$.

We shall show that for a given matrix $A$ it is possible to prescribe an interval of change in $\gamma$ such that $\mu(X_1) - \mu(X_0) > 0$ independent of the choice of $X_0$, i.e. independent of the value $s_0$. Such an interval is $0 < \gamma < 2/(M - m)$. In fact for $s_0 \leq 0$, $\mu(X_1) - \mu(X_0) > 0$ for any positive $\gamma$; if $s_0 > 0$, then $\mu(X_1) - \mu(X_0) > 0$ for $0 < \gamma < 2/(M - m)$, since

$$\frac{2}{s_0} = \frac{2}{\mu(X_0) - \mu(\xi_0)} \geq \frac{2}{M - m}.$$

We shall now find values of the parameter $\gamma$ for which $\mu(X_1) - \mu(X_0)$ will take on extremal values. It is clear from the above graphs that the maximum lies to the right and the minimum to the left. The derivative of $\mu(X_1) - \mu(X_0)$ with respect to $\gamma$ (accurate to within a positive multiplier) will be

$$(2 - 2\gamma s_0)(1 + \gamma^2 t_0^2) - 2\gamma t_0^2(2\gamma - \gamma^2 s_0) = -2\gamma^2 t_0^2 - 2\gamma s_0 + 2$$

and therefore the critical values $\alpha$ and $\tilde{\alpha}$ of the parameter $\gamma$ are determined from the quadratic equation

$$\gamma^2 t_0^2 + \gamma s_0 - 1 = 0 .$$

The positive root of this equation

$$\alpha_0 = \frac{-s_0 + \sqrt{s_0^2 + 4t_0^2}}{2t_0^2} = \frac{2}{\sqrt{s_0^2 + 4t_0^2} + s_0} \tag{6}$$

realizes the maximum and the negative root

$$\tilde{\alpha}_0 = \frac{-s_0 - \sqrt{s_0^2 + 4t_0^2}}{2t_0^2} = -\frac{2}{\sqrt{s_0^2 + 4t_0^2} - s_0} \tag{7}$$

realizes the minimum.   For this

$$\mu(X_0 + \alpha_0 \xi_0) - \mu(X_0) = \frac{2\alpha_0 - \alpha_0^2 s_0}{1 + \alpha_0^2 t_0^2} t_0^2 = \frac{(2 - \alpha_0 s_0)\alpha_0}{1 + 1 - \alpha_0 s_0} t_0^2 = \alpha_0 t_0^2 , \tag{8}$$

$$\mu(X_0 + \alpha_0 \xi_0) - \mu(X_0) = \alpha_0 t_0^2 . \tag{9}$$

We observe two properties of the root $\alpha_0$.

1.  $\alpha_0 = \dfrac{1}{\mu(X_1) - \mu(\xi_0)} \geq \dfrac{1}{M - m} .$ \hfill (10)

In fact

$$\mu(X_1) - \mu(\xi_0) = \mu(X_0 + \alpha_0 \xi_0) - \mu(X_0) + \mu(X_0) - \mu(\xi_0) = \alpha_0 t_0^2 + s_0 = \frac{1}{\alpha_0}$$

on the basis of the equation for $\alpha_0$.

2.  $\alpha_0 t_0^2 = \mu(X_0 + \alpha_0 \xi_0) - \mu(X_0) \leq M - m .$ \hfill (*11*)

We shall call the coefficient $\alpha_0$ the optimal coefficient.

We now consider the following group of iterative processes, which it is natural to call gradient processes.   Let $X_0$ be a certain initial vector different from the null one.   We construct the vector sequence

$$X_k = X_{k-1} + \gamma_{k-1} \xi_{k-1} \qquad (k = 1, 2, \cdots) \tag{12}$$

where

$$\xi_{k-1} = AX_{k-1} - \mu_{k-1} X_{k-1} , \tag{13}$$

$\mu_{k-1} = \mu(X_{k-1})$, and $\gamma_{k-1}$ is a certain positive number chosen so that in every case $\mu_k$ will be less than $\mu_{k-1}$.

We observe that if $X_{k-1} \neq 0$, then $X_k \neq 0$ as well.   In fact,

$$(X_k, X_k) = (X_{k-1} + \gamma_{k-1} \xi_{k-1}, X_{k-1} + \gamma_{k-1} \xi_{k-1})$$
$$= (X_{k-1}, X_{k-1}) + \gamma_{k-1}^2 (\xi_{k-1}, \xi_{k-1}) > 0 .$$

Thus we continue the process indefinitely.

The next two theorems† give sufficient conditions for the convergence of gradient methods.

THEOREM 74.1.  *If at all steps of a gradient process*

$$\mu(X_{k+1}) - \mu(X_k) \geq \delta \frac{(\xi_k, \xi_k)}{(X_k, X_k)} \qquad (\delta > 0),$$

*then the sequence* $\mu(X_k)$ *converges to the largest eigenvalue of matrix A in the invariant subspace generated by the vector* $X_0$ *and the sequence of vectors* $X_k$ *converges directionally to the corresponding eigenvector.*

*Proof.*  Let $U_1, \cdots, U_r$ be the normalized eigenvectors forming a basis of the cyclic subspace $P_0$ generated by the vector $X_0$ and let $\lambda_1 > \lambda_2 > \cdots > \lambda_n$ be their corresponding eigenvalues.  Let

$$X_0 = a_1^{(0)} U_1 + \cdots + a_r^{(0)} U_r.$$

Then (theorem 11.8) $a_1^{(0)} \neq 0, \cdots, a_r^{(0)} \neq 0$.  We may assume without loss of generality that $a_1^{(0)} > 0$.

It is clear that all the vectors $X_1, \cdots, X_k, \cdots$ are contained in the subspace $P_0$ and therefore

$$X_k = a_1^{(k)} U_1 + \cdots + a_r^{(k)} U_r.$$

We shall show that $a_1^{(k)} > 0$.  We have

$$
\begin{aligned}
a_1^{(k)} &= (X_k, U_1) = (X_{k-1}, U_1) + \gamma_{k-1}(\xi_{k-1}, U_1) \\
&= (X_{k-1}, U_1) + \gamma_{k-1}(AX_{k-1} - \mu_{k-1}X_{k-1}, U_1) \\
&= (X_{k-1}, U_1) + \gamma_{k-1}\lambda_1(X_{k-1}, U_1) - \gamma_{k-1}\mu_{k-1}(X_{k-1}, U_1) \\
&= [1 + \gamma_{k-1}(\lambda_1 - \mu_{k-1})](X_{k-1}, U_1) \\
&= [1 + \gamma_{k-1}(\lambda_1 - \mu_{k-1})]a_1^{(k-1)} \\
&\phantom{=} \cdot\ \cdot\ \cdot\ \cdot\ \cdot\ \cdot\ \cdot\ \cdot\ \cdot\ \cdot\ \cdot\ \cdot\ \cdot\ \cdot\ \cdot\ \cdot\ \cdot\ \cdot \\
&= [1 + \gamma_{k-1}(\lambda_1 - \mu_{k-1})] \cdots [1 + \gamma_0(\lambda_1 - \mu_0)]a_1^{(0)} > 0
\end{aligned}
$$

since all factors of the last product are positive.

We observe first of all that the process may be stabilized if it turns out at some step that $\xi_k = 0$.  In this case the vector $X_k$ will be an eigenvector of matrix $A$ and since $(X_k, U_1) = a_1^{(k)} > 0$, then this eigenvector will be proportional to $U_1$.  Thus the theorem is proved for the case of a stabilized process.

We turn to consideration of the general case when $\xi_k \neq 0$.  We introduce the notation

$$b_i^{(k)} = \frac{a_i^{(k)}}{|X_k|}.$$

Then

$$X_k = |X_k|(b_1^{(k)} U_1 + \cdots + b_r^{(k)} U_r),$$

---

† Hestenes and Karush (1).

where $b_1^{(k)} > 0$.

Furthermore,

$$\xi_k = a_1^{(k)}(\lambda_1 - \mu_k)U_1 + \cdots + a_r^{(k)}(\lambda_r - \mu_k)U_r \, .$$

Consequently,

$$t_k^2 = \frac{(\xi_k, \xi_k)}{(X_k, X_k)} = \frac{\sum\limits_{i=1}^{r} a_i^{(k)\,2}(\lambda_i - \mu_k)^2}{|X_k|^2} = \sum_{i=1}^{r} b_i^{(k)\,2}(\lambda_i - \mu_k)^2$$

Since $\mu_1 < \mu_2 < \cdots < \mu_k < M = \lambda_1$, then $\lim\limits_{k \to \infty} \mu_k$ exists. We denote it by $\mu$.
By the theorem's condition,

$$t_k^2 \leq \frac{1}{\delta}(\mu_{k+1} - \mu_k) \to 0 \qquad (k \to \infty) \, .$$

Therefore $\sum\limits_{i=1}^{r} b_i^{(k)\,2}(\lambda_i - \mu_k)^2 \to 0$ for $k \to \infty$, i.e.

$$b_i^{(k)\,2}(\lambda_i - \mu_k)^2 \to 0 \quad \text{for all} \quad i = 1, 2, \cdots, r \, .$$

If $\lambda_i - \mu \neq 0$, then $b_i^{(k)} \to 0$ for $k \to \infty$. However, this may not occur for all $i = 1, 2, \cdots, r$, since $\sum\limits_{i=1}^{r} b_i^{(k)\,2} = 1$. Thus $j$ is found such that $\mu = \lambda_j$. Then $\lim\limits_{k \to \infty} b_i^{(k)\,2} = 0$ for $i \neq j$ and $\lim\limits_{k \to \infty} b_j^{(k)\,2} = 1$.

We shall show that $j = 1$. If we assume $j > 1$, then

$$\left| \frac{b_1^{(k)}}{b_j^{(k)}} \right| = \left| \frac{a_1^{(k)}}{a_j^{(k)}} \right| \to 0 \, .$$

On the other hand, we have

$$\left| \frac{a_1^{(k)}}{a_j^{(k)}} \right| = \left| \frac{1 + \gamma_{k-1}(\lambda_1 - \mu_{k-1})}{1 + \gamma_{k-1}(\lambda_j - \mu_{k-1})} \right| \left| \frac{a_1^{(k-1)}}{a_j^{(k-1)}} \right|$$

$$= \left| 1 + \frac{\gamma_{k-1}(\lambda_1 - \lambda_j)}{1 + \gamma_{k-1}(\lambda_j - \mu_{k-1})} \right| \left| \frac{a_1^{(k-1)}}{a_j^{(k-1)}} \right| > \left| \frac{a_1^{(k-1)}}{a_j^{(k-1)}} \right|$$

since

$$\frac{\gamma_{k-1}(\lambda_1 - \lambda_j)}{1 + \gamma_{k-1}(\lambda_j - \mu_{k-1})} > 0 \, ,$$

because $\gamma_{k-1} > 0$ by construction and $\mu_{k-1} < \mu = \lambda_j < \lambda_1$. The resulting contradiction shows that $j = 1$. Thus we have proved that $\mu_k \to \lambda_1$ for $k \to \infty$. In the same way it has been established that $b_1^{(k)\,2} \to 1$ and since $b_1^{(k)} > 0$, then $b_1^{(k)} \to 1$. Therefore $b_i^{(k)} \to 0$ for $i = 2, \cdots, r$. But

$$\frac{X_k}{|X_k|} = b_1^{(k)}U_1 + \cdots + b_r^{(k)}U_r \, .$$

Consequently

$$\frac{X_k}{|X_k|} \to U_1 \, ,$$

i.e. the sequence $X_k$ converges to $U_1$ directionally.

THEOREM 74.2 *If, besides the condition of theorem 74.1, all numbers $\gamma_k$ are bounded above as a set, then* $\lim\limits_{k\to\infty} X_k = LU_1$, *where L is some positive number.*

*Proof.* On the basis of theorem 74.1 we need to prove that $|X_k| \to L$ for $k \to \infty$. But

$$(X_k, X_k) = (X_{k-1}, X_{k-1}) + \gamma_{k-1}^2(\xi_{k-1}, \xi_{k-1}) = (1 + \gamma_{k-1}^2 t_{k-1}^2)(X_{k-1}, X_{k-1})$$

Consequently,

$$(X_k, X_k) = (1 + \gamma_0^2 t_0^2)(1 + \gamma_1^2 t_1^2) \cdots (1 + \gamma_{k-1}^2 t_{k-1}^2)(X_0, X_0) .$$

The infinite product

$$\prod_{k=0}^{\infty}(1 + \gamma_k^2 t_k^2)$$

converges, since $\sum\limits_{k=0}^{\infty} t_k^2$ converges (since it is majorized by the convergent series $\sum\limits_{k=0}^{\infty} \dfrac{1}{\delta}(\mu_{k+1} - \mu_k)$ and $\gamma_k$ are bounded above by the condition of the theorem.

We now consider several special gradient methods.

1. $\gamma_k = \gamma = $ **const.: the constant multiplier method.** In this case for increasing $\mu_k$ at each step of the process it is necessary (for any initial vector) to satisfy the inequality

$$0 < \gamma < \frac{2}{M - m} .$$

We shall show that in satisfying this inequality we also satisfy the conditions of theorems 74.1 and 74.2. In fact let $\gamma = \beta/(M - m)$ for $0 < \beta < 2$. Then

$$\frac{\mu(X_k) - \mu(X_{k-1})}{t_{k-1}^2} = \frac{2\gamma - \gamma^2 s_{k-1}}{1 + \gamma^2 t_{k-1}^2}$$

$$= \frac{\gamma\left(2 - \dfrac{\beta}{M - m} s_{k-1}\right)}{1 + \gamma^2 t_{k-1}^2} \geq \frac{\gamma(2 - \beta)}{1 + \gamma^2 l^2} = \delta > 0 ,$$

since $t_{k-1}^2 \leq l^2$ where $l$ is the spherical norm of matrix $A$. In fact,

$$t_k^2 = \frac{(\xi_k, \xi_k)}{(X_k, X_k)} = \frac{(AX_k, \xi_k)}{(X_k, X_k)} = \frac{(AX_k, AX_k)}{(X_k, X_k)}$$

$$\frac{(AX_k, X_k)^2}{(X_k, X_k)^2} \leq \frac{(AX_k, AX_k)}{(X_k, X_k)} = \frac{\|AX_k\|^2}{|X_k|^2} \leq \|A\|^2 .$$

The theorem's remaining conditions are obviously satisfied.

2. $\gamma_k = \alpha_k$ **where $\alpha_k$ is the optimal coefficient of the $k$th step: the method of steepest descent.** In this case

$$\frac{\mu(X_k) - \mu(X_{k-1})}{t_{k-1}^2} = \alpha_k$$

and therefore to check the conditions for satisfying theorem 74.1 we need to convince ourselves that the numbers $\alpha_k$ are bounded from below. But we have already established that $\alpha_k \geq 1/(M - m)$ so that boundedness from below actually holds.

In order to be convinced that the condition of theorem 74.2 is also satisfied we need to prove that the numbers $\alpha_k$ are bounded from above. It turns out here, as opposed to the previous estimates, that an upper bound exists but depends on the initial approximation. That is, it is not hard to see that all values $\alpha_k$ for large enough $k$ satisfy the condition

$$\alpha_k < \frac{2}{\lambda_1 - \lambda_2} .$$

In fact,

$$\alpha_k = \frac{1}{\mu(X_k) - \mu(\xi_{k-1})} .$$

For large enough $k$, $\mu(X_k)$ becomes as close as we wish to $\lambda_1$, i.e. $\mu(X_k) > \lambda_1 - \varepsilon$ for $\varepsilon > 0$. On the other hand, $\xi_{k-1}$ is orthogonal to $X_{k-1}$ and consequently $\xi_{k-1}$ after normalization (to unit length) approaches as close as we wish to the subspace orthogonal to $U_1$. In this subspace $\mu(X)$ does not exceed $\lambda_2$. Therefore $\mu(\xi_{k-1}) \leq \lambda_2 + \varepsilon$ for large enough $k$. Consequently, for sufficiently large $k$,

$$\alpha_k \leq \frac{1}{\lambda_1 - \varepsilon - \lambda_2 - \varepsilon} = \frac{1}{\lambda_1 - \lambda_2 - 2\varepsilon} < \frac{2}{\lambda_1 - \lambda_2} ,$$

if we take $\varepsilon < (\lambda_1 - \lambda_2)/4$. Thus theorem 74.2 also turns out to be true for the method of steepest descent.

3. $\gamma_k = \beta\alpha_k$ **where $\alpha_k$ is the optimal coefficient of the $k$th step, $0 < \beta < 1$: the method of incomplete steepest descent.** To substantiate theorem 74.1 we need to prove that the sequence $[\mu(X_{k+1}) - \mu(X_k)]/t_k^2$ is bounded from below. We have

$$\begin{aligned}
\frac{\mu(X_{k+1}) - \mu(X_k)}{t_k^2} &= \frac{2\alpha_k\beta - \alpha_k^2\beta^2 s_k}{1 + a_k^2\beta^2 t_k^2} \\
&= \frac{2\alpha_k\beta - \alpha_k\beta^2(1 - t_k^2\alpha_k^2)}{1 + \alpha_k^2\beta^2 t_k^2} = \alpha_k\beta\frac{2 - \beta + \beta t_k^2\alpha_k^2}{1 + \alpha_k^2\beta^2 t_k^2} \\
&= \alpha_k\beta\left(1 + \frac{1 - \beta + \beta(1 - \beta)\alpha_k^2 t_k^2}{1 + \alpha_k^2\beta^2 t_k^2}\right) \\
&= \alpha_k\beta\left(1 + (1 - \beta)\frac{1 + \beta\alpha_k^2 t_k^2}{1 + \alpha_k^2\beta^2 t_k^2}\right) \geq \alpha_k\beta .
\end{aligned}$$

**TABLE VII.9  Determining the Largest Eigenvalue by the Gradient Method with Constant Multiplier $\gamma = 0.6$**

|   | $X_0$ | $AX_0$ | $\xi_0$ | $X_1$ | $X_2$ | $X_3$ | $X_4$ | $\tilde{X}_4$ | $\tilde{X}_7$ |
|---|---|---|---|---|---|---|---|---|---|
| I | 0.34 | 1.4050 | 0.78278286 | 0.80966972 | 0.81117064 | 0.80557137 | 0.80682530 | 1.0000 | 1.00000 |
|   | 0.20 | 0.9608 | 0.59478992 | 0.55687395 | 0.63814979 | 0.63972334 | 0.64009468 | 0.7934 | 0.79359 |
|   | 0.90 | 1.3126 | −0.33444536 | 0.69933278 | 0.60957215 | 0.60444577 | 0.60307521 | 0.7475 | 0.74781 |
|   | 0.75 | 1.2604 | −0.11213780 | 0.68271732 | 0.70658793 | 0.71601739 | 0.71543204 | 0.8867 | 0.88732 |
| II |  | 4.9388 |  |  |  |  |  |  |  |
| III | 1.5281 | 2.7965 |  |  |  |  |  |  |  |
| IV | 1.8300504 |  |  | 2.3087048 | 2.3226218 | 2.3227458 | 2.3227486 |  | 2.3227488 |

**TABLE VII.10  Determining the Largest Eigenvalue by the Method of Steepest Descent**

| $X_0$ | $AX_0$ | $\xi_0$ | $A\xi_0$ | $X_1$ | $X_2$ | $X_3$ | $X_4$ | $X_5$ | $\tilde{X}_4$ | $\tilde{X}_5$ |
|---|---|---|---|---|---|---|---|---|---|---|
| 0.34 | 1.4050 | 0.78278286 | 0.77798318 | 0.87034015 | 0.82112664 | 0.82844787 | 0.82774590 | 0.82787672 | 1.0000 | 1.00000 |
| 0.20 | 0.9608 | 0.59478992 | 0.76719557 | 0.60297379 | 0.65447856 | 0.65670073 | 0.65693195 | 0.65697730 | 0.7936 | 0.79357 |
| 0.90 | 1.3126 | −0.33444536 | 0.25391984 | 0.67341123 | 0.62913176 | 0.62014585 | 0.61926655 | 0.61910096 | 0.7481 | 0.74783 |
| 0.75 | 1.2604 | −0.11213780 | 0.59262847 | 0.67402596 | 0.73573673 | 0.73327297 | 0.73460252 | 0.73455418 | 0.8875 | 0.88727 |
|  | 4.9388 | $-0.18 \cdot 10^{-7}$ | 2.39172706 |  |  |  |  |  |  |  |
| 1.5281 | 2.7965 | 1.09095264 | 0.91393373 |  |  |  |  |  |  |  |
| 1.8300504 |  |  |  | 2.3137407 | 2.3226083 | 2.3227463 | 2.3227487 | 2.3227488 |  |  |

Consequently,

$$\frac{\mu(X_{k+1}) - \mu(X_k)}{t_k^2} \geq \frac{1}{M-m} \ .$$

Theorem 74.2 is proved in the same way as in the method of steepest descent.

4.   $\gamma_k = \dfrac{1}{\mu(X_k)} = \dfrac{1}{\mu_k}$: **the power method.** In this case

$$X_{k+1} = \frac{1}{\mu_k} AX_k \ .$$

We have

$$\frac{\mu(X_{k+1}) - \mu(X_k)}{t_k^2} = \frac{2/\mu_k - s_k/\mu_k^2}{1 + t_k^2/\mu_k^2} = \frac{2\mu_k - s_k}{\mu_k^2 + t_k^2} = \frac{\mu(X_k) + \mu(\xi_k)}{\mu^2(X_k) + t_k^2} \ .$$

If $A$ is a positive definite matrix, then

$$\frac{\mu(X_{k+1}) - \mu(X_k)}{t_k^2} \geq \frac{2m}{M^2 + l^2} > 0$$

where $l$ is the spherical norm of matrix $A$. It is obvious likewise that $\gamma_k = 1/\mu_k \leq 1/m$. Thus in the case of a positive definite matrix the power method is not only convergent but even produces relaxation, since at each step $\mu(X_k)$ increases.

In the case of an arbitrary symmetric matrix this may not happen in the first steps of the process. However, if $\lambda_1 > -\lambda_n$, i.e. the algebraically largest eigenvalue is the one largest in modulus, then from the directional convergence of the vectors $X_k$ and $U_1$ it follows that $\mu(k) \to \lambda_1$ and $\mu(\xi_k) \geq m$; consequently

$$\frac{\mu(X_{k+1}) - \mu(X_k)}{t_k^2} \geq \frac{(\lambda_1 - \varepsilon) + m}{\lambda_1^2 + l^2}$$

beginning with a certain place. Thus, in this case, the power method preserves the relaxation character starting with a certain step of the process.

In Table VII.9 the largest eigenvalue of matrix (4), Section 51, is deter-

**TABLE VII.11**   **Determining the Smallest Eigenvalue by the Method of Steepest Descent**

| $X_0$ | $AX_0$ | $\xi_0$ | $A\xi_0$ | $X_1$ | $X_8$ | $X_{13}$ | $\tilde{X}_{13}$ |
|---|---|---|---|---|---|---|---|
| 0.34 | 1.4050 | 0.78278286 | 0.77798318 | −1.2783558 | −2.0012022 | −2.0088782 | 1.0000 |
| 0.20 | 0.9608 | 0.59478992 | 0.76719557 | −1.0296919 | −0.2776540 | −0.2691368 | 0.3140 |
| 0.90 | 1.3126 | −0.33444536 | 0.25391984 | 1.5914454 | 1.0838913 | 1.0828515 | −0.5390 |
| 0.75 | 1.2604 | −0.11213780 | 0.59262847 | 0.9818381 | 1.5991528 | 1.5917026 | −0.7923 |
| | 4.9388 | −0.18·10⁻⁷ | 2.39172706 | | | | |
| 1.5281 | 2.7965 | 1.09095264 | 0.91393373 | | | | |
| 1.8300504 | | 0.83773917 | | | 0.24227723 | 0.24226089 | |

mined by the gradient method with constant $\gamma$; in Table VII.10 this is done using the method of steepest descent. In Table VII.11 the smallest eigenvalue is determined for the same matrix.

In the last row of the tables are written the values of $\mu(X)$.

At each step of the process in Table VII.10 is found the optimal coefficient using formula (6). For the first step we have

$$s_0 = 0.99231126 ,$$
$$t_0^2 = 0.71392752$$

so that

$$\alpha_0 = \frac{2}{\sqrt{3.8403817} + 0.99231126} = 0.67750609 ,$$

We also give the values

$$\alpha_1 = 0.60271425 ,$$
$$\alpha_2 = 0.51669127 ,$$
$$\alpha_3 = 0.59837767 ,$$
$$\alpha_4 = 0.51670397 .$$

At each step of the process in Table VII.11 the optimal coefficient $\tilde{\alpha}_k$ is found using formula (7). For the first step we have

$$s_0 = 0.99231126 ,$$
$$t_0^2 = 0.71392752$$

$$\alpha_0 = \frac{-2}{\sqrt{3.8403917} - 0.99231126} = -2.0674390 .$$

A comparison of Tables VII.9 and VII.10 shows that in the given example the method of a constant multiplier with $\gamma = 0.6$ converges only a little slower than the method of steepest descent. The volume of computations for one step of the constant multiplier method is half as much as in the method of steepest descent.

The question of choosing the constant multiplier's value has hardly been investigated. For a positive definite matrix it is possible in any case to take as $\gamma$ a number not exceeding $2/\|A\|_I$.

The computations carried out for matrix (4) in Section 51 for $\gamma = 0.5$, $\gamma = 0.4$, $\gamma = 0.3$ and $\gamma = 0.25$ show that decreasing $\gamma$ slows down the convergence of the process.

The application of incomplete steepest descent for $\beta = 0.8$ and $\beta = 0.9$ in the given example turned out to be inadvisable, since convergence of the process was not improved.

We shall now return to the constant multiplier method where we shall consider this time that

$$0 < \gamma < \frac{1}{M - m} .$$

We observe the following important property of successive approximations.

LEMMA.  *If in the constant multiplier method* $\gamma = \dfrac{\beta}{M-m}$ *and* $0 < \beta < 1,$
*then* $\lim\limits_{k\to\infty} \dfrac{a_j^{(k)}}{a_i^{(k)}} = 0$ *for* $i < j.$

Here $a_1^{(k)},\ a_2^{(k)},\ \cdots,\ a_r^{(k)}$ are the coefficients in the expansion of the approximation $X_k$ in terms of eigenvectors $U_1, \cdots, U_r$ contained in the invariant subspace $P_0$ generated by the initial approximation.

*Proof.*  Let

$$X_0 = a_1^{(0)} U_1 + a_2^{(0)} U_2 + \cdots + a_r^{(0)} U_r\,.$$

Then

$$X_k = a_1^{(k)} U_1 + a_2^{(k)} U_2 + \cdots + a_r^{(k)} U_r\,,$$

where

$$a_i^{(k)} = [1 + \gamma(\lambda_i - \mu_{k-1})]a_i^{(k-1)}\,.$$

From this it follows that $a_i^{(k)} > 0$, since

$$1 + \gamma(\lambda_i - \mu_{k-1}) = 1 + \beta\frac{\lambda_i - \mu_{k-1}}{M - m} > 0\,.$$

We write

$$\delta_i = 1 + \gamma(\lambda_i - \lambda_1) = 1 + \beta\frac{\lambda_i - \lambda_1}{M - m}\,.$$

From the choice of $\beta$ it is obvious that

$$0 < \delta_r < \delta_{r-1} < \cdots < \delta_2 < \delta_1 = 1\,.$$

Since $\mu_k \to \lambda_1$, then $a_i^{(k)}/a_i^{(k-1)} \to \delta_i$.  Therefore

$$\frac{a_j^{(k)}}{a_i^{(k)}} : \frac{a_j^{(k-1)}}{a_i^{(k-1)}} \to \frac{\delta_j}{\delta_i}$$

and $0 < \delta_j/\delta_i < 1$, if $i < j$.  Consequently after a certain point

$$\frac{a_j^{(k)}}{a_j^{(k)}} \le L\left(\frac{\delta_j}{\delta_i} + \varepsilon\right)^k \qquad (k > k_0)\,.$$

Choosing $\varepsilon$ so that $\delta_j/\delta_i + \varepsilon < 1$ we obtain that $a_j^{(k)}/a_i^{(k)} \to 0$ for $i < j$.

In this constant multiplier method the behavior of the sequence $\xi_k$ is also interesting.

THEOREM 74.3.[†]  *If in the constant multiplier method* $\gamma = \beta/(M-m)$ *and* $0 < \beta < 1,$ *then* $\lim\limits_{k\to\infty} \xi_k/|\xi_k| = U_2$ *and* $\lim\limits_{k\to\infty} \mu(\xi_k) = \lambda_2.$

*Proof.*  We have

$$X_k = a_1^{(k)} U_1 + a_2^{(k)} U_2 + \cdots + a_r^{(k)} U_r,\ a_i^{(k)} > 0\,,$$

$$\xi_k = (\lambda_1 - \mu_k)a_1^{(k)} U_1 + (\lambda_2 - \mu_k)a_2^{(k)} U_2 + \cdots + (\lambda_r - \mu_k)a_r^{(k)} U_r\,,$$

† Hestenes and Karush (1).

where

$$(\xi_k, \xi_k) = (\lambda_1 - \mu_k)^2 a_1^{(k)2} + (\lambda_2 - \mu_k)^2 a_2^{(k)2} + \cdots + (\lambda_r - \mu_k)^2 a_r^{(k)2} \neq 0$$

since $r \geq 2$. Moreover we have

$$(X_k, \xi_k) = 0 = (\lambda_1 - \mu_k) a_1^{(k)2} + (\lambda_2 - \mu_k) a_2^{(k)2} + \cdots + (\lambda_r - \mu_k) a_r^{(k)2}.$$

From here, dividing by $a_2^{(k)2}$ and passing to the limit we obtain, on the basis of the previous lemma

$$\frac{\lambda_1 - \mu_k}{a_2^{(k)2}} \cdot a_1^{(k)2} \to \lambda_1 - \lambda_2 \,.$$

But

$$\frac{1}{a_2^{(k)}} \xi_k = \frac{(\lambda_1 - \mu_k) a_1^{(k)2}}{a_2^{(k)2}} \frac{a_2^{(k)}}{a_1^{(k)}} U_1 + (\lambda_2 - \mu_k) U_2 + \cdots + \frac{(\lambda_r - \mu_k) a_r^{(k)}}{a_2^{(k)}} U_r$$

and consequently

$$\frac{1}{a_2^{(k)}} \xi_k \to (\lambda_2 - \lambda_1) U_2 \,.$$

On the basis of Section 13 we conclude that the sequence $\xi_k$ approaches $U_2$ directionally.

Furthermore,

$$\mu(\xi_k) = \mu\left(\frac{\xi_k}{|\xi_k|}\right) \to \mu(U_2) = \lambda_2 \,.$$

As an example we shall carry out the described process for the matrix

$$\begin{bmatrix} 0.22 & 0.02 & 0.12 & 0.14 \\ 0.02 & 0.14 & 0.04 & -0.06 \\ 0.12 & 0.04 & 0.28 & 0.08 \\ 0.14 & -0.06 & 0.08 & 0.26 \end{bmatrix}.$$

Here the exact values are $\lambda_1 = 0.48$, $\lambda_2 = 0.24$, $\lambda_3 = 0.12$, $\lambda_4 = 0.06$, $U_1 = (1, 0, 1, 1)'$, $U_2 = (0, -1, -1, 1)'$.

In applying the constant multiplier method we may take $\gamma = 1$. We have

| $X_0$ | $X_{25}$ | $\xi_{25}$ | $X_{43}$ | $\xi_{43}$ | $\tilde{\xi}_{43}$ |
|---|---|---|---|---|---|
| 0.34 | 0.72458928 | $0.008256 \cdot 10^{-4}$ | 0.724591532513 | $0.000237 \cdot 10^{-6}$ | 0.001 |
| 0.20 | 0.00013557 | $-0.323536 \cdot 10^{-4}$ | 0.000000981434 | $-0.235470 \cdot 10^{-6}$ | -0.998 |
| 0.90 | 0.72473074 | $-0.336452 \cdot 10^{-4}$ | 0.724592515860 | $-0.235923 \cdot 10^{-6}$ | -1.000 |
| 0.75 | 0.72445458 | $0.328216 \cdot 10^{-4}$ | 0.724590551128 | $0.235687 \cdot 10^{-6}$ | 0.999 |
| $\mu$ | 0.48000000 | 0.239942 | 0.48000000 | 0.23999989 | |

We see that for $k = 43$ the second eigenvalue is determined accurately to within $1 \cdot 10^{-7}$. Its associated eigenvector is determined with significantly less accuracy.

We get quite another picture in the method of steepest descent. Here the sequence $\xi_k$ does not converge. More than this, the following theorem is valid.

THEOREM 74.4. *Gradients of adjacent approximations in the method of steepest descent are orthogonal.*

*Proof.* We have

$$\xi_k = AX_k - \mu_k X_k$$

$$\xi_{k+1} = AX_{k+1} - \mu_{k+1}X_{k+1} = A(X_k + \alpha_k\xi_k) - \mu_{k+1}(X_k + \alpha_k\xi_k) .$$

Consequently,

$$(\xi_{k+1}, \xi_k) = (AX_k, \xi_k) + \alpha_k(A\xi_k, \xi_k) - \mu_{k+1}(X_k, \xi_k) - \alpha_k\mu_{k+1}(\xi_k, \xi_k)$$
$$= [1 + \alpha_k\mu(\xi_k) - \alpha_k\mu_{k+1}](\xi_k, \xi_k) = 0 ,$$

since

$$\alpha_k = \frac{1}{\mu(X_{k+1}) - \mu(\xi_k)} = \frac{1}{\mu_{k+1} - \mu(\xi_k)} .$$

## 75. SOLVING THE SPECIAL EIGENVALUE PROBLEM USING LANCZOS POLYNOMIALS

The property established in theorem 74.3 for a sequence of approximation gradients in the constant multiplier method admits of generalization, allowing us to find several eigenvalues and their corresponding eigenvectors with a number $m$ (fixed beforehand) of eigenvalues subject to determination.

We introduce the following notation. We denote by $p_0^{(k)}, p_1^{(k)}, \cdots, p_{m-1}^{(k)}$ the first $m$ Lanczos vectors constructed from an initial vector $X_k$ and by $p_0^{(k)}(t), p_1^{(k)}(t), \cdots, p_{m-1}^{(k)}(t)$ the polynomials corresponding to them. It is clear that $p_0^{(k)} = X_k$ and $p_1^{(k)} = \xi_k$.

THEOREM 75.1. *Let* $X_k = p_0^{(k)} = a_1^{(k)}U_1 + a_2^{(k)}U_2 + \cdots + a_r^{(k)}U_r$ *be a sequence of vectors in the subspace spanned by the normalized eigenvectors* $U_1, U_2, \cdots, U_r$ *of a symmetric matrix* $A$, *and which correspond to the eigenvalues* $\lambda_1 > \lambda_2 > \cdots > \lambda_r$. *Let*

$$\frac{a_2^{(k)}}{a_1^{(k)}} \to 0, \frac{a_3^{(k)}}{a_2^{(k)}} \to 0, \cdots, \frac{a_r^{(k)}}{a_{r-1}^{(k)}} \to 0 \quad \text{for } k \to \infty .$$

*Then if* $p_i^{(k)} = p_i^{(k)}(A)p_0^{(k)}$ *is the Lanczos vector, then the corresponding polynomial* $p_i^{(k)}(t)$ *converge to the polynomial* $\bar{p}_i(t) = (t - \lambda_1) \cdots (t - \lambda_i)$ *and the vectors* $p_i^{(k)}$ *converge directionally to the vector* $U_{i+1}$.

*Proof.* We shall prove the theorem by induction. For $i = 0$ the statement is correct, since

$$\frac{1}{a_1^{(k)}} p_0^{(k)} = U_1 + \frac{a_2^{(k)}}{a_1^{(k)}} U_2 + \cdots + \frac{a_r^{(k)}}{a_1^{(k)}} U_r \to U_1 .$$

We assume that the statement of the theorem is correct for the indices 0, 1, $\cdots$, $i-1$ and under this assumption we shall prove it for the index $i$. We have

$$p_j^{(k)} = a_1^{(k)} p_j^{(k)}(\lambda_1) U_1 + \cdots + a_{j+1}^{(k)} p_j^{(k)}(\lambda_{j+1}) U_{j+1} + \cdots + a_r^{(k)} p_j^{(k)}(\lambda_r) U_r$$
$$= a_{j+1}^{(k)} p_j^{(k)}(\lambda_{j+1}) [b_{1j}^{(k)} U_1 + \cdots + b_{jj}^{(k)} U_j + U_{j+1}$$
$$+ b_{j+2,\,j}^{(k)} U_{j+2} + \cdots + b_{rj}^{(k)} U_r] = a_{j+1}^{(k)} p_j^{(k)}(\lambda_{j+1}) V_j^{(k)} ,$$

where

$$b_{sj}^{(k)} = \frac{a_s^{(k)}}{a_{j+1}^{(k)}} \frac{p_j^{(k)}(\lambda_s)}{p_j^{(k)}(\lambda_{j+1})} ,$$

$$V_j^{(k)} = b_{1j}^{(k)} U_1 + \cdots + b_{jj}^{(k)} U_j + U_{j+1} + b_{j+2,\,j}^{(k)} U_{j+2} + \cdots + b_{rj}^{(k)} U_r .$$

From the induction hypothesis:

1. $p_j^{(k)}(t) \to \bar{p}_j(t)$, for $j < i$,
2. $V_j^{(k)} \to U_{j+1}$ for $j < i$.

We shall prove the same thing for $j = i$. We have

$$p_i^{(k)}(t) = (t - \alpha_{i-1}^{(k)}) p_{i-1}^{(k)}(t) - \beta_{i-1}^{(k)} p_{i-2}^{(k)}(t)$$

where

$$\alpha_{i-1}^{(k)} = \frac{(A p_{i-1}^{(k)}, p_{i-1}^{(k)})}{(p_{i-1}^{(k)}, p_{i-1}^{(k)})} = \frac{(A V_{i-1}^{(k)}, V_{i-1}^{(k)})}{(V_{i-1}^{(k)}, V_{i-1}^{(k)})} ,$$

$$\beta_{i-1}^{(k)} = \frac{(p_{i-1}^{(k)}, p_{i-1}^{(k)})}{(p_{i-2}^{(k)}, p_{i-2}^{(k)})} = \left[\frac{a_i^{(k)}}{a_{i-1}^{(k)}}\right]^2 \left[\frac{p_{i-1}^{(k)}(\lambda_i)}{p_{i-2}^{(k)}(\lambda_i)}\right]^2 \frac{(V_{i-1}^{(k)}, V_{i-1}^{(k)})}{(V_{i-2}^{(k)}, V_{i-2}^{(k)})} .$$

On the strength of the induction hypothesis

$$\alpha_{i-1}^{(k)} \to \frac{(A U_i, U_i)}{(U_i, U_i)} = \lambda_i .$$

Moreover,

$$\frac{(V_{i-1}^{(k)}, V_{i-1}^{(k)})}{(V_{i-2}^{(k)}, V_{i-2}^{(k)})} \to \frac{(U_i, U_i)}{(U_{i-1}, U_{i-1})} = 1 ,$$

$$\frac{p_{i-1}^{(k)}(\lambda_i)}{p_{i-2}^{(k)}(\lambda_i)} \to \frac{\bar{p}_{i-1}(\lambda_i)}{\bar{p}_{i-2}(\lambda_i)} ,$$

and $a_i^{(k)}/a_{i-1}^{(k)} \to 0$ from the condition of the theorem. Consequently $\beta_{i-1}^{(k)} \to 0$ for $k \to \infty$ and therefore

$$p_i^{(k)}(t) \to (t - \lambda_i) \bar{p}_{i-1}(t) = \bar{p}_i(t) .$$

To prove the second statement we need to establish that the coefficients $b_{si}^{(k)} \to 0$ for $k \to \infty$ if $s \ne i+1$.

For $s \ge i+2$ this is established quite simply. Namely,

$$b_{si}^{(k)} = \frac{a_s^{(k)}}{a_{i+1}^{(k)}} \frac{p_i^{(k)}(\lambda_s)}{p_i^{(k)}(\lambda_{i+1})} .$$

From the theorem's condition $a_s^{(k)}/a_{i+1}^{(k)} \to 0$ for $s \geq i+2$ and, in view of what has already been proved, $p_i^{(k)}(\lambda_s) \to \bar{p}_i(\lambda_s)$ and

$$p_i^{(k)}(\lambda_{i+1}) \to \bar{p}_i(\lambda_{i+1}) = (\lambda_{i+1} - \lambda_1) \cdots (\lambda_{i+1} - \lambda_i) \neq 0 .$$

It is more difficult to establish the limit relationships $b_{si}^{(k)} \to 0$ if $s \leq i$. For the proof we assume

$$b_{si}^{(k)} = c_{si}^{(k)} \frac{a_{i+1}^{(k)}}{a_s^{(k)}} \qquad (s \leq i) , \tag{1}$$

where

$$c_{si}^{(k)} = \left[ \frac{a_s^{(k)}}{a_{i+1}^{(k)}} \right]^2 \frac{p_i^{(k)}(\lambda_s)}{p_i^{(k)}(\lambda_{i+1})} . \tag{2}$$

From the theorem's condition $a_{i+1}^{(k)}/a_s^{(k)} \to 0$ for $s \leq i$. We shall prove that $c_{si}^{(k)}$ approach finite limits as $k \to \infty$. In this way it will be established that $b_{si}^{(k)} \to 0$ for $s \leq i$.

For the proof we use orthogonality of the vector $p_i^{(k)}$ to the vectors $p_0^{(k)}$, $\cdots$, $p_{i-1}^{(k)}$. We have

$$\begin{aligned}
0 = (p_i^{(k)}, p_s^{(k)}) &= a_1^{(k)2} p_i^{(k)}(\lambda_1) p_s^{(k)}(\lambda_1) \\
&+ \cdots + a_i^{(k)2} p_i^{(k)}(\lambda_i) p_s^{(k)}(\lambda_i) + a_{i+1}^{(k)2} p_i^{(k)}(\lambda_{i+1}) p_s^{(k)}(\lambda_{i+1}) \\
&+ a_{i+2}^{(k)2} p_i^{(k)}(\lambda_{i+2}) p_s^{(k)}(\lambda_{i+2}) + \cdots + a_r^{(k)2} p_i^{(k)}(\lambda_r) p_s^{(k)}(\lambda_r) .
\end{aligned}$$

We divide this equality by $a_{i+1}^{(k)2} p_i^{(k)}(\lambda_{i+1})$ and transfer the elements beginning with the $(i+1)$th to the other part of the equality. We obtain

$$c_{1i}^{(k)} p_s^{(k)}(\lambda_1) + c_{2i}^{(k)} p_s^{(k)}(\lambda_2) + \cdots + c_{ii}^{(k)} p_s^{(k)}(\lambda_i) = d_{is}^{(k)} , \tag{3}$$

where

$$\begin{aligned}
d_{is}^{(k)} = &- p_s^{(k)}(\lambda_{i+1}) - \left[ \frac{a_{i+2}^{(k)}}{a_{i+1}^{(k)}} \right]^2 \frac{p_i^{(k)}(\lambda_{i+2}) p_s^{(k)}(\lambda_{i+2})}{p_i^{(k)}(\lambda_{i+1})} \\
&- \cdots - \left[ \frac{a_r^{(k)}}{a_{i+1}^{(k)}} \right]^2 \frac{p_i^{(k)}(\lambda_r) p_s^{(k)}(\lambda_r)}{p_i^{(k)}(\lambda_{i+1})} .
\end{aligned}$$

The system of equalities (3) for $s = 0, \cdots, i-1$ may be considered as a system of linear equations relative to the coefficients $c_{1i}^{(k)}, \cdots, c_{ii}^{(k)}$.

As $k$ becomes infinite the coefficients of this system $p_s^{(k)}(\lambda_j)$ approach finite limits $\bar{p}_s(\lambda_j)$. In turn the constants of the system also approach finite limits. Namely,

$$d_{is}^{(k)} \to - \bar{p}_s(\lambda_{i+1}) ,$$

since

$$\frac{p_i^{(k)}(\lambda_j) p_s^{(k)}(\lambda_j)}{p_i^{(k)}(\lambda_{i+1})} \to \frac{\bar{p}_i(\lambda_j) \bar{p}_s(\lambda_j)}{\bar{p}_i(\lambda_{i+1})} ,$$

and

$$\frac{a_j^{(k)}}{a_{i+1}^{(k)}} \to 0 \quad \text{for} \quad j > i+1 .$$

We consider the limit system

$$c_{1i}\,\bar{p}_0(\lambda_1) + c_{2i}\,\bar{p}_0(\lambda_2) + \cdots + c_{ii}\,\bar{p}_0(\lambda_i) = -\,\bar{p}_0(\lambda_{i+1})\,,$$
$$c_{1i}\,\bar{p}_1(\lambda_1) + c_{2i}\,\bar{p}_1(\lambda_2) + \cdots + c_{ii}\,\bar{p}_1(\lambda_i) = -\,\bar{p}_1(\lambda_{i+1})\,,$$
$$\cdots \cdots \cdots \cdots \cdots \cdots \cdots \cdots \cdots \cdots \cdots \cdots \cdots,$$
$$c_{1i}\,\bar{p}_{i-1}(\lambda_1) + c_{2i}\,\bar{p}_{i-1}(\lambda_2) + \cdots + c_{ii}\,\bar{p}_{i-1}(\lambda_i) = -\,\bar{p}_{i-1}(\lambda_{i+1})\,.$$

This system has a triangular matrix, since $\bar{p}_1(\lambda_1) = \bar{p}_2(\lambda_1) = \bar{p}_2(\lambda_2) = \cdots = \bar{p}_{i-1}(\lambda_{i-1}) = 0$ and its determinant

$$\varDelta = \bar{p}_0(\lambda_1)\,\bar{p}_1(\lambda_2) \cdots \bar{p}_{i-1}(\lambda_i)$$

is not equal to zero.

On the basis of theorem 13.1 its solution $c_{1i}, \cdots, c_{ii}$ will be the limit for $c_{1i}^{(k)}, \cdots, c_{ii}^{(k)}$.

We have thus established that the sequence $c_{1i}^{(k)}, \cdots, c_{ii}^{(k)}$ has finite limits. Consequently,

$$b_{si}^{(k)} = c_{si}^{(k)} \frac{a_{i+1}^{(k)}}{a_s^{(k)}} \to 0 \qquad \text{(for } s \le i)\,.$$

Thus

$$V_i^{(k)} \to U_{i+1}$$

and therefore the vectors $p_i^{(k)}$ converge to $U_{i+1}$ directionally. The theorem has been completely proved.

We observe that during the proof we obtained an estimate for the rate of convergence. Namely,

$$b_{si}^{(k)} = O\!\left(\frac{a_{i+1}^{(k)}}{a_s^{(k)}}\right) \qquad \text{(for } s \le i)\,,$$

$$b_{si}^{(k)} = O\!\left(\frac{a_s^{(k)}}{a_{i+1}^{(k)}}\right) \qquad \text{(for } s \ge i+2)\,.$$

In the lemma of Section 74 it was proved that the successive approximations $X_k$ computed by the constant multiplier method for $0 < \gamma < 1/(M-m)$ satisfy the conditions of theorem 75.1.

In this way it has been proved that if

$$X_{k+1} = X_k - \gamma \xi_k\,,$$
$$\xi_k = AX_k - \mu(X_k)X_k\,,$$
$$0 < \gamma < \frac{1}{M-m}\,,$$

and $p_1^{(k)}, \cdots, p_{m-1}^{(k)}$ are Lanczos vectors constructed starting with $p_0^{(k)} = X_k$, then $p_0^{(k)}, \cdots, p_{m-1}^{(k)}$ converge directionally to the first (in decreasing order of eigenvalues) eigenvectors located in the subspace generated by the initial vector $X_0$.

We observe that in computing the vectors $p_0^{(k)}, \cdots, p_{m-1}^{(k)}$, starting with the vector $X_k$, it is necessary to subtract close values so that the process indicated is not very useful in actual computing.

We have already seen this in the last example of Section 74.

In conclusion we observe that the power method also satisfies the conditions of theorem 75.1. Therefore if

$$X_{k+1} = AX_k \qquad (k = 0, 1, \cdots),$$

and $p_1^{(k)}, \cdots, p_{m-1}^{(k)}$ are Lanczos vectors constructed starting with $p_0^{(k)} = X_k$, then $p_0^{(k)}, \cdots, p_{m-1}^{(k)}$ converge directionally to the corresponding eigenvectors in the subspace generated by the initial vector $X_0$.

## 76. THE S-STEP METHOD OF STEEPEST DESCENT

As a matter of fact, the first approximation $X_1$ of any gradient method lies in the subspace $P^{(2)}$ spanned by the vectors $X_0$ and $AX_0$; in the method of steepest descent it produces maximization of the functional $\mu(X)$ in this subspace. The second approximation $X_2$ of any gradient method lies in the subspace $P^{(3)}$ spanned by $X_0$, $AX_0$, $A^2X_0$, the third in the subspace $P^{(4)}$ spanned by $X_0$, $AX_0$, $A^2X_0$, $A^3X_0$, etc. However the second approximation, even in the method of steepest descent, no longer maximizes $\mu(X)$ in the subspace $P^{(3)}$. It is natural to raise the question of finding the vector which maximizes the functional $\mu(X)$ in the subspace $P^{(s+1)}$ spanned by the vectors $X_0$, $AX_0$, $\cdots$, $A^sX_0$ for some number $s$ fixed beforehand. Solving this problem allows us to construct an iterative process for determining the algebraically largest eigenvalue of a symmetric matrix and its associated eigenvector—the so-called s-step method of steepest descent, which consists of the following.

We take an initial approximation $X_0$, construct the vector $X_1$ which maximizes $\mu(X)$ in the subspace $P_0^{(s+1)}$ spanned by the vectors $X_0$, $AX_0$, $\cdots$, $A^sX_0$, and then look for the vector $X_2$ which maximizes $\mu(X)$ in the subspace $P_1^{(s+1)}$ spanned by the vector $X_1$, $AX_1$, $\cdots$, $A^sX_1$ and so forth.

Before explaining the convergence of the process we shall show how the problem of maximizing $\mu(X)$ in the subspace $P_0^{(s+1)}$ is solved. In this subspace we choose any basis $V_0$, $V_1$, $\cdots$, $V_s$. Let $X$ be any vector of the subspace and let

$$X = b_0 V_0 + b_1 V_1 + \cdots + b_s V_s \qquad (1)$$

where not all $b_i$ are equal to zero.

We associate with the vector $X$ the vector $\bar{X}$ with components $b_0, b_1, \cdots,$ $b_s$ of the arithmetic space $R^{(s+1)}$. Then

$$\begin{aligned}
(X, X) &= \sum_{i=0, j=0}^{s} c_{ij} b_i b_j = (C\bar{X}, \bar{X}), \\
(AX, X) &= \sum_{i=0, j=0}^{s} d_{ij} b_i b_j = (D\bar{X}, \bar{X}),
\end{aligned} \qquad (2)$$

where

$$c_{ij} = (V_i, V_j), \qquad C = (c_{ij}),$$
$$d_{ij} = (AV_i, V_j), \qquad D = (d_{ij}). \tag{3}$$

It is obvious that the matrix $C$ is positive definite. Thus our problem reduces to maximizing the functional

$$\frac{(D\bar{X}, \bar{X})}{(C\bar{X}, \bar{X})} \tag{4}$$

in the space $R^{(s+1)}$.

The desired maximum $\mu^{(0)}$ is obtained as the major root of the equation $|D - Ct| = 0$ and the vector realizing it is determined from the system of linear homogeneous equations $(D - \mu^{(0)}C)\bar{X} = 0$. The vector $X_1$ corresponding to it is then obtained from expansion (1). Thus our problem reduces to the solution of the special generalized eigenvalue problem for a matrix of order $s + 1$. If we take the vectors $X_0, AX_0, \cdots, A^sX_0$ for the basis, then

$$c_{ij} = (A^iX_0, A^jX_0) = (A^{i+j}X_0, X_0),$$
$$d_{ij} = (A^{i+1}X_0, A^jX_0) = (A^{i+j+1}X_0, X_0). \tag{5}$$

The problem's solution is greatly simplified if we take some orthogonal basis in the subspace $P_0^{(s+1)}$. The basis $p_0, p_1, \cdots, p_s$ consisting of the first $(s + 1)$ Lanczos vectors turns out to be very convenient in this respect.[†] In this case

$$c_{ij} = (p_i, p_j)$$

and therefore

$$c_{ij} = 0,$$
$$i \neq j; \tag{6}$$
$$c_{ii} = (p_i, p_i).$$

Moreover $d_{ij} = (Ap_i, p_j)$ so that

$$d_{ij} = 0 \quad \text{for} \quad |i - j| > 1;$$
$$d_{ii} = (Ap_i, p_i) = \alpha_i(p_i, p_i); \tag{7}$$
$$d_{ii-1} = d_{i-1i} = \beta_i(p_{i-1}, p_{i-1}).$$

Thus,

$|D - Ct|$

$$= \begin{vmatrix} \alpha_0(p_0, p_0) - t(p_0, p_0) & \beta_1(p_0, p_0) & 0 & \cdots & 0 \\ \beta_1(p_0, p_0) & \alpha_1(p_1, p_1) - t(p_1, p_1) & \beta_2(p_1, p_1) & \cdots & 0 \\ \cdots & \cdots & \cdots & \cdots & \cdots \\ 0 & 0 & 0 & \cdots & \alpha_s(p_s, p_s) - t(p_s, p_s) \end{vmatrix}$$

$$= \begin{vmatrix} \alpha_0(p_0, p_0) - t(p_0, p_0) & \beta_1(p_0, p_0) & 0 & \cdots & 0 \\ (p_1, p_1) & \alpha_1(p_1, p_1) - t(p_1, p_1) & \beta_2(p_1, p_1) & \cdots & 0 \\ \cdots & \cdots & \cdots & \cdots & \cdots \\ 0 & 0 & 0 & \cdots & \alpha_s(p_s, p_s) - t(p_s, p_s) \end{vmatrix} \tag{8}$$

[since $\beta_i(p_{i-1}, p_{i-1}) = (p_i, p_i)$]

$$= (p_0, p_0)(p_1, p_1)\cdots(p_s, p_s) \begin{vmatrix} \alpha_0 - t & \beta_1 & 0 \cdots 0 & 0 \\ 1 & \alpha_1 - t & \beta_2 \cdots 0 & 0 \\ \cdots & \cdots & \cdots \\ 0 & 0 & 0 \cdots 1 & \alpha_s - t \end{vmatrix} = l_s p_{s+1}(t).$$

---

† Karush (1).

where $l_s = (-1)^{s+1}(p_0, p_0) \cdots (p_s, p_s)$ and $p_{s+1}(t)$ is the $s + 1$th Lanczos polynomial.

Thus the value $\mu^{(0)}$ for the maximum of $\mu(X)$ in the subspace $P_0^{(s+1)}$ is the largest root of the $(s + 1)$th Lanczos polynomial $p_{s+1}(t)$.

We now determine the vector which realizes this maximum. The coordinates $b_0, b_1, \cdots, b_s$ of this vector (relative to the basis $p_0, p_1, \cdots, p_s$) are determined from the system of linear homogeneous equations

$$[\alpha_0(p_0, p_0) - \mu^{(0)}(p_0, p_0)]b_0 + \beta_1(p_0, p_0)b_1 = 0 \,,$$
$$\beta_1(p_0, p_0)b_0 + [\alpha_1(p_1, p_1) - \mu^{(0)}(p_1, p_1)]b_1 + \beta_2(p_1, p_1)b_2 = 0 \,,$$
$$\cdots \cdots \cdots \cdots \cdots \cdots \cdots \cdots \cdots \cdots \,, \qquad (9)$$
$$\beta_s(p_{s-1}, p_{s-1})b_{s-1} + (\alpha_s - \mu^{(0)})(p_s, p_s)b_s = 0 \,.$$

Making the substitution

$$b_i = \frac{b_i'}{(p_i, p_i)} \,,$$

and remembering that $\beta_i = (p_i, p_i)/(p_{i-1}, p_{i-1})$ we obtain for determining $b_i'$ the system

$$(\alpha_0 - \mu^{(0)})b_0' + b_1' = 0 \,,$$
$$\beta_1 b_0' + (\alpha_1 - \mu^{(0)})b_1' + b_2' = 0 \,,$$
$$\cdots \cdots \cdots \cdots \cdots \cdots \,,$$
$$\beta_s b_{s-1}' + (\alpha_s - \mu^{(0)})b_s' = 0 \,.$$

We assume

$$b_0' = 1 = p_0(\mu^{(0)}) \,.$$

Then

$$b_1' = \mu^{(0)} - \alpha_0 = p_1(\mu^{(0)}) \,,$$
$$b_2' = (\mu^{(0)} - \alpha_1)b_1' - \beta_1 b_0' = (\mu^{(0)} - \alpha_1)p_1(\mu^{(0)}) - \beta_1 p_0(\mu^{(0)}) = p_2(\mu^{(0)})$$

and so on. From the last equation we obtain

$$b_s' = p_s(\mu^{(n)}) \,.$$

The last equation turns out to be satisfied, since

$$p_{s+1}(\mu^{(0)}) = 0 \,.$$

Thus

$$b_i = \frac{p_i(u^{(0)})}{(p_i, p_i)} \qquad (i = 0, \cdots, s)$$

and the vector $X_1$ which realizes the maximum is given by the formula

$$X_1 = \sum_{i=0}^{s} \frac{p_i(\mu^{(0)})}{(p_i, p_i)} p_i \,. \qquad (10)$$

The derived formulas allow us to give the following form to the $s$-step method of steepest descent. For the initial approximation we take an

arbitrary vector $X_0$. After constructing the vectors $X_k$ we may construct
the vector $X_{k+1}$ by the formula

$$X_{k+1} = \sum_{i=0}^{s} \frac{p_i^{(k)}(\mu^{(k)})}{(p_i^{(k)}, p_i^{(k)})} \, p_i^{(k)} , \qquad (11)$$

where the vectors $p_0^{(k)}, p_1^{(k)}, \cdots, p_s^{(k)}$ are Lanczos vectors constructed starting
with the vector $p_0^{(k)} = X_k$ according to the recurrent relationships for the
method of minimal iterations; $\mu^{(k)}$ is the largest root of the Lanczos polynomial $p_{s+1}^{(k)}(t)$.

THEOREM 76.1. *Let the sequence of vectors $X_0, X_1, \cdots, X_k, \cdots$ be constructed so that $X_k$ realizes the maximum of $(AX, X)/(X, X)$ in the subspace
spanned by the vectors $X_{k-1}, AX_{k-1}, \cdots, A^s X_{k-1}$. Then the quotient $(AX_k,
X_k)/(X_k, X_k)$ converges to the largest eigenvalue of the matrix $A$ in the invariant
subspace generated by the vector $X_0$ and the vectors $X_k$ converge directionally
to the eigenvector belonging to this eigenvalue.*

*Proof.* Let $r = r_0$ be the dimension of the invariant subspace generated
by the vector $X_0$, let $\lambda_1 > \lambda_2 > \cdots > \lambda_r$ be the eigenvalues of matrix $A$ in
this subspace, and let $U_1, U_2, \cdots, U_r$ be the corresponding normalized
eigenvectors. Then

$$X_0 = a_1^{(0)} U_1 + a_2^{(0)} U_2 + \cdots + a_r^{(0)} U_r .$$

All the coefficients $a_1^{(0)}, a_2^{(0)}, \cdots, a_r^{(0)}$ differ from zero and may be considered
positive without loss of generality.

Moreover, we denote by $r_k$ the dimension of the invariant subspace generated by the vector $X_k$. Since each successive vector $X_k$ is contained in
the invariant subspace generated by the preceding vector $X_{k-1}$, then $r_0 \geq
\geq r_1 \geq \cdots \geq r_k \geq \cdots$

First of all we consider the degenerate case where for some $k$ it turns
out that $r_k \leq s + 1$. Then the subspace spanned by the vectors $X_k, AX_k,
\cdots, A^s X_k$ will be itself invariant (if $r_k < s + 1$, and then among the enu-
merated vectors these will be some linearly dependent ones) and the vector
$X_{k+1}$ for which the maximum of $(AX, X)/(X, X)$ is realized in this subspace
turns out to be an eigenvector of matrix $A$. At the next step the subspace
spanned by the vectors $X_{k+1}, AX_{k+1}, \cdots, A^s X_{k+1}$ will be one-dimensional so
that the process is stabilized. Later on we shall show that the eigenvector
so obtained will be proportional to $U_1$ and in this very way the theorem
will be proved for the degenerate case.

We shall now consider one non-degenerate step of the process. Let

$$X_k = a_1^{(k)} U_1 + a_2^{(k)} U_2 + \cdots + a_r^{(k)} U_r ,$$

$p_i^{(k)}$, $(i = 0, \cdots, s, s + 1)$, be Lanczos vectors constructed starting with the
vector $X_k$, let $p_i^{(k)}(t)$ be their corresponding polynomials, and let $\mu_i^{(k)}$ be
their largest roots. As we saw above, $\mu_{s+1}^{(k)}$ is the maximum of $(AX, X)/(X,X)$
in the subspace spanned by the vectors $X_k, AX_k, \cdots, A^s X_k$, and was denoted

earlier by $\mu^{(k)}$, so that

$$\mu_{s+1}^{(k)} = \mu(X_{k+1}) \ . \tag{12}$$

In accordance with formula (11)

$$X_{k+1} = \sum_{i=0}^{s} \frac{p_i^{(k)}(\mu_{s+1}^{(k)})}{(p_i^{(k)}, p_i^{(k)})} \, p_i^{(k)} \ .$$

In turn,

$$p_i^{(k)} = a_1^{(k)} p_i^{(k)}(\lambda_1) U_1 + \cdots + a_r^{(k)} p_i^{(k)}(\lambda_r) U_r \ . \tag{13}$$

Consequently

$$X_{k+1} = a_1^{(k+1)} U_1 + \cdots + a_r^{(k+1)} U_r$$

where

$$a_1^{(k+1)} = a_1^{(k)} \sum_{i=0}^{s} \frac{p_i^{(k)}(\mu_{s+1}^{(k)}) p_i^{(k)}(\lambda_1)}{(p_i^{(k)}, p_i^{(k)})} = a_1^{(k)} M_1^{(k)} \ ,$$

$$\cdots \cdots \cdots \cdots \cdots \cdots \cdots \cdots ,$$

$$a_r^{(k+1)} = a_r^{(k)} \sum_{i=0}^{s} \frac{p_i^{(k)}(\mu_{s+1}^{(k)}) p_i^{(k)}(\lambda_r)}{(p_i^{(k)}, p_i^{(k)})} = a_r^{(k)} M_r^{(k)} \ .$$

We shall draw certain conclusions from the formulas constructed. Since roots of Lanczos polynomials separate each other we have $\mu_1^{(k)} < \mu_2^{(k)} < \cdots < \mu_s^{(k)} < \mu_{s+1}^{(k)} \leq \lambda_1$ so that $\mu_{s+1}^{(k)}$ and $\lambda_1$ are strictly greater than all roots of the polynomials $p_i^{(k)}(t)$. Consequently $M_1^{(k)} > 0$. We assume that $a_1^{(0)} > 0$. Therefore $a_1^{(1)} > 0, \cdots, a_1^{(k+1)} > 0$ so that the invariant subspace generated by the vector $X_{k+1}$ includes the eigenvector $U_1$ belonging to the eigenvalue $\lambda_1$. Consequently if the degenerate case holds for $X_{k+1}$, then in the next step in the subspace spanned by the vectors $X_{k+1}, AX_{k+1}, \cdots A^s X_{k+1}$ (it will be invariant!) the maximum of the quotient $(AX, X)/(X, X)$ will be attained precisely for the eigenvector $U_1$.

It now remains to consider the process without degeneracies. We observe first of all that

$$\mu(X_0) < \mu(X_1) < \cdots < \lambda_1,$$

so that there exists

$$\lim_{k \to \infty} \mu(X_k) = \mu \ .$$

We assume

$$X_k = |X_k| (b_1^{(k)} U_1 + \cdots + b_r^{(k)} U_r). \tag{14}$$

Just as in proving theorem 74.1 we shall prove first of all that one of the coefficients $b_j^{(k)} \to 1$ and the remaining coefficients $b_i^{(k)} \to 0$, $i \neq j$. With this in mind we again consider the quotient

$$t_k^2 = \frac{(\xi_k, \xi_k)}{(X_k, X_k)} = b_1^{(k)2}[\lambda_1 - \mu(X_k)]^2 + \cdots + b_r^{(k)2}[\lambda_r - \mu(X_k)]^2 \ . \tag{15}$$

In our notation

$$t_k^2 = \frac{(p_1^{(k)}, p_1^{(k)})}{(p_0^{(k)}, p_0^{(k)})} = \beta_1^{(k)} .$$

But

$$\beta_1^{(k)} = (t - \alpha_0^{(k)})(t - \alpha_1^{(k)}) - p_2^{(k)}(t) . \tag{16}$$

Here $\alpha_0^{(k)} = \mu(X_k)$, $\alpha_1^{(k)} = \mu(p_1^{(k)})$, and $p_2^{(k)}(t)$ is the second Lanczos polynomial. We substitute into (16) $t = \mu(X_{k+1}) = \mu_{s+1}^{(k)}$. It is clear that $p_2^{(k)}(\mu_{s+1}^{(k)}) > 0$, since $\mu_{s+1}^{(k)}$ for $s > 1$ is larger than both roots of the polynomial $p_2^{(k)}(t)$. Therefore

$$\beta_1^{(k)} < (\mu_{s+1}^{(k)} - \alpha_0^{(k)})(\mu_{s+1}^{(k)} - \alpha_1^{(k)}) = [\mu(X_{k+1}) - \mu(X_k)][\mu(X_{k+1}) - \mu(p_1^{(k)})] . \tag{17}$$

It follows that $\beta_1^{(k)} \to 0$, since the first factor in the right member of inequality (17) approaches zero and the second is bounded.

From equality (15) we conclude that $b_i^{(k)2}[\lambda_i - \mu(X_k)]^2 \to 0$ for all $i = 1, 2, \cdots, r$. Of the factors $(\lambda_i - \mu(X_k))$ no more than one of them may approach zero; $b_i^{(k)}$ can not approach zero in any case, since $\sum_{i=1}^{r} \beta_i^{(k)2} = 1$. Therefore $\mu(X_k) \to \lambda_j$ for some defined $j$, and all $b_i^{(k)}$ approach zero for $i \neq j$; $b_j^{(k)2} \to 1$.

Thus the vectors $X_k$ converge directionally to $U_j$. It remains to prove that $j = 1$.

$$\frac{b_1^{(k)}}{b_j^{(k)}} = \frac{a_1^{(k)}}{a_j^{(k)}} \to 0 \qquad (\text{for } k \to \infty) .$$

But on the other hand,

$$\frac{a_1^{(k+1)}}{a_j^{(k+1)}} = \frac{a_1^{(k)}}{a_j^{(k)}} \frac{M_1^{(k)}}{M_j^{(k)}} .$$

We shall show that $M_1^{(k)}/M_j^{(k)} > 1$. In fact,

$$\frac{M_1^{(k)}}{M_j^{(k)}} = \frac{\sum_{i=0}^{s} \dfrac{p_i^{(k)}(\mu_{s+1}^{(k)})p_i^{(k)}(\lambda_1)}{(p_i^{(k)}, p_i^{(k)})}}{\sum_{i=0}^{s} \dfrac{p_i^{(k)}(\mu_{s+1}^{(k)})p_i^{(k)}(\lambda_j)}{(p_i^{(k)}, p_i^{(k)})}} .$$

But $\mu_{s+1}^{(k)} = \mu(X_{k+1}) < \lambda_j$ so that $\lambda_j$ is larger than all the roots of the $(s+1)$th Lanczos polynomial for any $k$, and all the more larger the polynomials $p_i^{(k)}(t)$. Since, besides this, $\lambda_1 > \lambda_j$, then $p_i^{(k)}(\lambda_1) > p_i^{(k)}(\lambda_j)$ for $i = 1, 2, \cdots, s$ and consequently,

$$M_1^{(k)} > M_j^{(k)} > 0 .$$

Thus,

$$\left| \frac{a_1^{(k+1)}}{a_j^{(k+1)}} \right| > \left| \frac{a_1^{(k)}}{a_j^{(k)}} \right| .$$

This inequality is satisfied for all $k$ and contradicts the limit relationship

$$\frac{a_1^{(k)}}{a_j^{(k)}} \to 0 .$$

Therefore the inequality $j > 1$ is impossible. The theorem has been proved.

*Note.* If the process of steepest descent is carried out according to the formulas

$$X_{k+1} = X_k + \sum_{i=1}^{s} \frac{(p_0^{(k)}, p_0^{(k)}) p_i^{(k)}(\mu_{s+1}^{(k)})}{(p_i^{(k)}, p_i^{(k)})} p_i^{(k)} ,$$

then the vector $X_{k+1} - X_k$ will be orthogonal to the vector $X_k$. In this case the vector sequence $X_k$ will converge to $U_1$ (and not only directionally). This fact, proved by Karush (1), follows easily from estimates of the rate of convergence for the process.

THEOREM 76.2. *In the s-steepest descent the inequality*

$$\lambda_1 - \mu_{k+1} \leq (1 + \varepsilon) Q_s^2 (\lambda_1 - \mu_k)$$

*holds for sufficiently large k. Here $Q_s$ is the least divergence from zero at points of the set $\lambda_2, \cdots, \lambda_r$ of the polynomials $F_s(t)$ of degree s, normalized by the condition $F_s(\lambda_1) = 1$.*

*Proof.* Let $\Phi_s(t)$ be the polynomial which produces the least divergence from zero in the point set $\lambda_2, \cdots, \lambda_r$ under the normalization $\Phi_s(\lambda_1) = 1$.

We assume that the approximation $X_k$ has been constructed. Along with the next approximation $X_{k+1}$ we consider the vector $\tilde{X}_{k+1} = \Phi_s(A)X_k$. We denote by $\mu_k$, $\mu_{k+1}$, $\tilde{\mu}_{k+1}$ the corresponding values of the functional $\mu(X)$. It is clear that

$$\mu_{k+1} \geq \tilde{\mu}_{k+1} ,$$

since $\mu_{k+1}$ is the maximum for $\mu(X)$ in the subspace $P_k^{(s+1)}$ and $\tilde{\mu}_{k+1}$ is one of the values for $\mu(X)$ in the same subspace.

We now compare $\lambda_1 - \mu_{k+1}$ and $\lambda_1 - \mu_k$. Let

$$X_k = a_1^{(k)} U_1 + a_2^{(k)} U_2 + \cdots + a_r^{(k)} U_r$$

be the expansion of the vector $X_k$ in terms of the eigenvectors contained in the invariant subspace generated by the vector $X_0$. Then

$$\tilde{X}_{k+1} = a_1^{(k)} U_1 + \Phi_s(\lambda_2) a_2^{(k)} U_2 + \cdots + \Phi_s(\lambda_r) a_r^{(k)} U_r .$$

Futhermore,

$$\lambda_1 - \mu_k = \lambda_1 - \frac{(AX_k, X_k)}{(X_k, X_k)}$$

$$= \lambda_1 - \frac{\lambda_1 a_1^{(k)2} + \lambda_2 a_2^{(k)2} + \cdots + \lambda_r a_r^{(k)2}}{a_1^{(k)2} + a_2^{(k)2} + \cdots + a_r^{(k)2}}$$

$$= \frac{(\lambda_1 - \lambda_2) a_2^{(k)2} + \cdots + (\lambda_1 - \lambda_r) a_r^{(k)2}}{a_1^{(k)2} + a_2^{(k)2} + \cdots + a_r^{(k)2}} .$$

Thus,

$$\lambda_1 - \tilde{\mu}_{k+1} = \frac{(\lambda_1 - \lambda_2) \Phi_s^2(\lambda_2) a_2^{(k)2} + \cdots + (\lambda_1 - \lambda_r) \Phi_s^2(\lambda_r) a_r^{(k)2}}{a_1^{(k)2} + \Phi_s^2(\lambda_2) a_2^{(k)2} + \cdots + \Phi_s^2(\lambda_r) a_r^{(k)2}} .$$

Since

$$|\Phi_s(\lambda_i)| \le Q_s \qquad (\text{for } i = 2, \cdots, r),$$

then

$$\lambda_1 - \tilde{\mu}_{k+1} \le Q_s^2 \frac{(\lambda_1 - \lambda_2)a_2^{(k)2} + \cdots + (\lambda_1 - \lambda_r)a_r^{(k)2}}{a_1^{(k)2}}.$$

Therefore

$$\frac{\lambda_1 - \tilde{\mu}_{k+1}}{\lambda_1 - \mu_k} \le Q_s^2 \frac{a_1^{(k)2} + \cdots + a_r^{(k)2}}{a_1^{(k)2}} = Q_s^2 \frac{1}{b_1^{(k)2}},$$

where

$$b_1^{(k)2} = \frac{a_1^{(k)2}}{a_1^{(k)2} + \cdots + a_r^{(k)2}}.$$

Thus

$$\frac{\lambda_1 - \mu_{k+1}}{\lambda_1 - \mu_k} \le \frac{\lambda_1 - \tilde{\mu}_{k+1}}{\lambda_1 - \mu_k} \le Q_s^2 \frac{1}{b_1^{(k)2}} = Q_s^2(1 + \sigma_k),$$

where $\sigma_k = [1 - b_1^{(k)2}]/b_1^{(k)2}$. On the basis of theorem 75.1, $b_1^{(k)2} \to 1$ for $k \to \infty$ so that $\sigma_k$ becomes less than $\varepsilon$ for sufficiently large $k$, and therefore

$$\lambda_1 - \mu_{k+1} \le Q_s^2(1 + \varepsilon)(\lambda_1 - \mu_k) \qquad \text{for } k > k_0.$$

The estimates obtained show that the rate of convergence of the sequence $\mu_k$ to $\lambda_1$ is not less than the rate of convergence of a geometric progression with denominator $Q_s^2(1 + \varepsilon)$.

It is clear that $Q_s$ does not exceed the divergence from zero of the polynomial that diverges least from zero in the interval $(\lambda_2, \lambda_r)$ with the previous normalization at the point $t = \lambda_1$. This least divergence, as is well-known, is equal to

$$E_s = \frac{1}{T_s(l)} = \frac{2}{(l + \sqrt{l^2 - 1})^s + (1 - \sqrt{l^2 - 1})^s}$$

$$l = \frac{2\lambda_2 - \lambda_2 - \lambda_r}{\lambda_2 - \lambda_r}.$$

Applying the two-step method of steepest descent to matrix (4) of Section 51 for $X = (0.34, 0.20, 0.90, 0.75)'$ gives

$$\mu^{(0)} = 1.8300504,$$
$$\mu^{(1)} = 2.3227312,$$
$$\mu^{(2)} = 2.3227487.$$

For this,

$$\tilde{X}_1 = (1.000, 0.799, 0.752, 0.887)',$$
$$\tilde{X}_2 = (1.0000, 0.7931, 0.7479, 0.8867)'.$$

Thus applying the two-step method here turns out to be sufficient to obtain the first eigenvalue with high accuracy. The eigenvector corresponding to it is determined with significantly less accuracy.

# Chapter VIII

# ITERATIVE METHODS FOR SOLVING
# THE COMPLETE EIGENVALUE PROBLEM

Iterative methods for solving the complete eigenvalue problem have appeared only recently. They are naturally more laborious than iterative methods for solving the special problem. As a rule their practical realization, even for matrices of not very high order, requires the use of high-speed computers.

However, their undoubted advantage over exact methods (Chapter IV) is the possibility of computing all eigenvalues without the necessity of computing the characteristic polynomial. As was already observed above, rounding-off errors in the coefficients of the characteristic may strongly influence accuracy in computing its roots.

## 77. THE DIVISION AND SUBTRACTION ALGORITHM

The algorithm (Rutishauser (2)) which we are about to describe is built upon the biorthogonal algorithm. This algorithm allows us to determine the eigenvalues of a matrix $A$ as limits of numerical sequences which are constructed recurrently from simple formulas. The coefficients $\rho_k$ and $\sigma_k$ of the binomial form of the biorthogonal algorithm serve as the initial data for the algorithm.

First of all we present the theory of the method for the simplest case, which assumes matrix $A$ (for which the algorithm is constructed) to be positive definite.

The algorithm is based on the properties of several infinite vector sequences $p_i^{(k)}$ where the vectors lie in the cyclic invariant subspace generated by the initial vector $X_0$.

This allows us, in presenting the method's theory, to consider all the eigenvalues of matrix $A$ to be pairwise distinct and all components in the expansion of the initial vector $X_0$ in terms of eigenvalues to be different from zero.

For fixed $k$ the vectors $p_i^{(k)}$, $i = 0, 1, \cdots, n-1$, are constructed by $A$-orthogonalizing the vector sequence $X_0, AX_0, \cdots, A^{n-1}X_0$. For all $k$ we have $p_0^{(k)} = X_0$, $p_i^{(k)} = p_i^{(k)}(A)X_0$ where $p_i^{(k)}(t) = t^i + \cdots$ are polynomials of degree $i$.

We shall show that the vectors $p_i^{(k)}$ are related by simple recurrent relationships

$$
\begin{aligned}
Ap_{i-1}^{(k+1)} - p_i^{(k)} &= \rho_{i-1}^{(k+1)} p_{i-1}^{(k)} &&(i = 1, 2, \cdots, n), \\
p_i^{(k)} - p_i^{(k+1)} &= \sigma_i^{(k+1)} p_{i-1}^{(k+1)} &&(i = 1, 2, \cdots, n-1).
\end{aligned}
\tag{1}
$$

For $i = n$ we assume $p_n^{(k)} = (0)$.

In fact the vector $Ap_{i-1}^{(k+1)} - p_i^{(k)}$ belongs to the subspace $P^{(i)}$ spanned by the vectors $X_0, AX_0, \cdots, A^{i-1}X_0$ and is $A^k$-orthogonal to the vectors $X_0$, $AX_0, \cdots, A^{i-2}X_0$. On the basis of the uniqueness of the normalized vector satisfying these conditions, the vector $Ap_{i-1}^{(k+1)} - p_i^{(k)}$ differs from the vector $p_{i-1}^{(k)}$ only by a numerical factor.

Analogously, the vector $p_i^{(k)} - p_i^{(k+1)}$ belongs to the subspace $P^{(i)}$, is $A^{k+1}$-orthogonal to the vectors $X_0, AX_0, \cdots, A^{i-2}X_0$, and consequently differs from the vector $p_{i-1}^{(k+1)}$ only by a numerical factor.

Among the polynomials $p_i^{(k)}(t)$ the relationships

$$
\begin{aligned}
tp_{i-1}^{(k+1)}(t) - p_i^{(k)}(t) &= \rho_{i-1}^{(k+1)} p_{i-1}^{(k)}(t), \\
p_i^{(k)}(t) - p_i^{(k+1)}(t) &= \sigma_i^{(k+1)} p_{i-1}^{(k+1)}(t).
\end{aligned}
\tag{2}
$$

are also obviously satisfied.

We now derive the dependence between the numbers $\rho_i^{(k)}$ and $\sigma_i^{(k)}$. With this in mind we pass to the trinomial relationships which relate the vectors $p_{i+1}^{(k)}$, $p_i^{(k)}$, and $p_{i-1}^{(k)}$. Such relationships may be constructed in two ways. First, eliminating the vectors $Ap_i^{(k+1)}$ and $Ap_{i-1}^{(k+1)}$ from the relationships

$$
\begin{aligned}
Ap_i^{(k+1)} - p_{i+1}^{(k)} &= \rho_i^{(k+1)} p_i^{(k)}, \\
Ap_{i-1}^{(k+1)} - p_i^{(k)} &= \rho_{i-1}^{(k+1)} p_{i-1}^{(k)}, \\
Ap_i^{(k)} - Ap_i^{(k+1)} &= \sigma_i^{(k+1)} Ap_{i-1}^{(k+1)}
\end{aligned}
$$

we obtain

$$
p_{i+1}^{(k)} + (\sigma_i^{(k+1)} + \rho_i^{(k+1)})p_i^{(k)} - Ap_i^{(k)} + \rho_{i-1}^{(k+1)}\sigma_i^{(k+1)}p_{i-1}^{(k)} = 0.
\tag{3}
$$

Second, eliminating the vectors $p_{i+1}^{(k-1)}$ and $p_i^{(k-1)}$ from the relationships

$$
\begin{aligned}
Ap_i^{(k)} - p_{i+1}^{(k-1)} &= \rho_i^{(k)} p_i^{(k-1)}, \\
p_{i+1}^{(k-1)} - p_{i+1}^{(k)} &= \sigma_{i+1}^{(k)} p_i^{(k)}, \\
p_i^{(k-1)} - p_i^{(k)} &= \sigma_i^{(k)} p_{i-1}^{(k)}
\end{aligned}
$$

we obtain

$$p_{i+1}^{(k)} + (\sigma_{i+1}^{(k)} + \rho_i^{(k)})p_i^{(k)} - A p_i^{(k)} + \sigma_i^{(k)} \rho_i^{(k)} p_{i-1}^{(k)} = 0 . \tag{4}$$

Comparing trinomial relationships (3) and (4) we conclude that

$$\sigma_i^{(k+1)} + \rho_i^{(k+1)} = \sigma_{i+1}^{(k)} + \rho_i^{(k)} ,$$
$$\sigma_i^{(k+1)} \rho_{i-1}^{(k+1)} = \sigma_i^{(k)} \rho_i^{(k)} . \tag{5}$$

These last relationships allow us to construct successively the numbers $\rho_i^{(k+1)}$, $\sigma_i^{(k+1)}$ in the order $\rho_0^{(k+1)}$, $\sigma_1^{(k+1)}$, $\rho_1^{(k+1)}$, $\cdots$, $\rho_{n-2}^{(k+1)}$, $\sigma_{n-1}^{(k+1)}$, $\rho_{n-1}^{(k+1)}$ as soon as the numbers $\rho_i^{(k)}$ and $\sigma_i^{(k)}$ have been constructed.

The computational formulas of the algorithm (the division and subtraction scheme is $QD$) are, for $k \geq 1$,

$$\rho_i^{(k+1)} = \sigma_{i+1}^{(k)} + \rho_i^{(k)} - \sigma_i^{(k+1)} \qquad (i = 0, 1, \cdots, n-1) ,$$
$$\sigma_i^{(k+1)} = \frac{\sigma_i^{(k)} \rho_i^{(k)}}{\rho_{i-1}^{(k+1)}} \qquad (i = 1, 2, \cdots, n-1) . \tag{6}$$

In this it is advisable to consider $\sigma_0^{(k)} = \sigma_n^{(k)} = 0$. The initial row of the algorithm

$$\rho_0^{(1)}, \sigma_1^{(1)}, \cdots, \sigma_{n-1}^{(1)}, \rho_{n-1}^{(1)}$$

is determined by coefficients of the binomial formulas for the method of minimal iterations or is obtained from the coefficients $\alpha_i$ and $\beta_i$ of the trinomial formulas of this method in the form

$$\rho_0^{(1)} = \alpha_0 , \quad \sigma_i^{(1)} = \frac{\beta_i}{\rho_{i-1}^{(1)}} , \quad \rho_i^{(1)} = \alpha_i - \sigma_i^{(1)} . \tag{7}$$

For computing the eigenvalues of a matrix (in the invariant subspace generated by the vector $X_0$) we need to form only the sequence of numbers $\rho_i^{(k)}$, $\sigma_i^{(k)}$. The vectors $p_i^{(k)}$ do not need to be constructed, so that they need be introduced only to clarify the theory of the method. Moreover, the following theorem holds.

THEOREM 77.1. *Let $A$ be a positive definite matrix, let $X_0$ be the given vector, and let $\lambda_1 > \lambda_2 > \cdots > \lambda_r$ be the eigenvalues of matrix $A$ in the invariant subspace generated by the vector $X_0$. Let $\rho_i^{(k)}(i = 0, 1, \cdots, r-1)$ and $\sigma_i^{(k)}(i = 1, \cdots, r-1)$ be numbers constructed according to the division and subtraction scheme. Then*

$$\lim_{k \to \infty} \rho_i^{(k)} = \lambda_{i+1}, \lim_{k \to \infty} \sigma_i^{(k)} = 0 . \tag{8}$$

*Proof.* The polynomials $p_i^{(k)}(t) = t^i + \cdots$ are characterized by the orthogonality conditions $(A^k p_i^{(k)}(A)X_0, p_j^{(k)}(A)X_0) = 0$ for $i \neq j$. These conditions may be rewritten in the form $(p_i^{(k)}(A)A^{k/2}X_0, p_j^{(k)}(A)A^{k/2}X_0) = 0$; this means that $p_i^{(k)}(t)$ are Lanczos polynomials constructed for the initial vector $Y_0^{(k)} = A^{k/2}X_0$.

The sequence of vectors $A^{k/2}X_0$ obviously satisfies the conditions of theorem 75.1. Consequently the limit relationships

$$\lim_{k \to \infty} p_i^{(k)}(t) = \bar{p}_i(t) = (t - \lambda_1) \cdots (t - \lambda_i)$$

holds.

Passing to the limit in the equality

$$\rho_i^{(k)} = \frac{t p_i^{(k)}(t) - p_{i+1}^{(k-1)}(t)}{p_i^{(k-1)}(t)},$$

arising from (2), we obtain

$$\lim_{k \to \infty} \rho_i^{(k)} = \frac{t(t - \lambda_1) \cdots (t - \lambda_i) - (t - \lambda_1) \cdots (t - \lambda_i)(t - \lambda_{i+1})}{(t - \lambda_1) \cdots (t - \lambda_i)} = \lambda_{i+1} .$$

For $\sigma_i^{(k)}$ we obtain, analogously,

$$\lim_{k \to \infty} \sigma_i^{(k)} = \frac{\bar{p}_i(t) - \bar{p}_i(t)}{\bar{p}_{i-1}(t)} = 0 .$$

The theorem has been proved.

We shall now generalize the algorithm in two respects. First, we shall extend it to any matrices with real distinct eigenvalues and, second, we shall consider the broader class of weight functions governing orthogonalization. As we shall see below, a rational choice of the sequence of weight functions may improve the convergence of the process significantly.

Let

$$\varphi_0(t) = 1, \ \varphi_1(t) = t, \ \varphi_k(t) = t(t - t_2) \cdots (t - t_k) \qquad (k = 2, 3, \cdots) \qquad (9)$$

be a sequence of polynomials. Each polynomial is obtained from the preceding one by multiplying by a linear binomial.

Furthermore let $A$ be a given matrix whose eigenvalues are real and distinct and let $X_0$ and $Y_0$ be certain initial vectors.

Starting from the vector system $X_0$, $AX_0$, $\cdots$, $A^{n-1}X_0$ and the vector system $Y_0$, $A'Y_0$, $\cdots$, $A'^{n-1}Y_0$ we construct the vector systems $p_0^{(k)}$, $p_1^{(k)}$, $\cdots$, $p_{n-1}^{(k)}$ and $\tilde{p}_0^{(k)}$, $\tilde{p}_1^{(k)}$, $\cdots$, $\tilde{p}_{n-1}^{(k)}$ of vectors biorthogonal by weight $\varphi_k(A)$. We assume that the initial vectors $X_0$ and $Y_0$ are selected so that, for all $k = 0, 1, \cdots$, the construction of the vectors $p_0^{(k)}$, $\cdots$, $p_{n-1}^{(k)}$ and $\tilde{p}_0^{(k)}$, $\cdots$, $\tilde{p}_{n-1}^{(k)}$ is possible, i.e. so that the orthogonalization process is carried out without degeneracy, for all $k$.

The vectors $p_i^{(k)}$ are completely characterized by satisfaction of the following two requirements:

1.  $p_i^{(k)} = p_i^{(k)}(A)X_0$, where $p_i^{(k)}(t) = t^i + c_1^{(k)}t^{i-1} + \cdots$;

2.  $(\varphi_k(A)p_i^{(k)}, A'^j Y_0) = 0$ for $j = 0, 1, \cdots, i - 1$ . $\qquad (10)$

It is not hard to give the concise formulas for the vectors $p_i^{(k)}$ and the polynomials $p_i^{(k)}(t)$. Namely, letting

$$\tilde{p}_i^{(k)}(t) = \begin{vmatrix} m_0^{(k)} & \cdots & m_{i-1}^{(k)} & 1 \\ m_1^{(k)} & \cdots & m_i^{(k)} & t \\ \cdot & \cdots & \cdots & \cdot \\ m_i^{(k)} & \cdots & m_{2i-1}^{(k)} & t^i \end{vmatrix}, \qquad (11)$$

where

$$m_j^{(k)} = (\varphi_k(A)A^j X_0, Y_0),$$

we have

$$(\varphi_k(A)\tilde{p}_i^{(k)}(A)X_0, A'^j Y_0) = (\varphi_k(A)A^j \tilde{p}_i^{(k)}(A)X_0, Y_0)$$

$$= \begin{vmatrix} m_0^{(k)} & \cdots & m_{i-1}^{(k)} & m_j^{(k)} \\ m_1^{(k)} & \cdots & m_i^{(k)} & m_{j+1}^{(k)} \\ \cdot & \cdot & \cdot & \cdot \\ m_i^{(k)} & \cdots & m_{2i-1}^{(k)} & m_{j+i}^{(k)} \end{vmatrix} = 0 \qquad (j = 0, 1, \cdots, i-1).$$

Therefore the vector $\tilde{p}_i^{(k)}(A)X_0$ differs from the vector $p_i^{(k)}$ only by a numerical factor. This factor is equal to $1/\Delta_i^{(k)}$ where

$$\Delta_i^{(k)} = \begin{vmatrix} m_0^{(k)} & \cdots & m_{i-1}^{(k)} \\ \cdot & \cdot & \cdot & \cdot \\ m_{i-1}^{(k)} & \cdots & m_{2i-2}^{(k)} \end{vmatrix}. \tag{12}$$

Thus, to guarantee a non-degenerate course of the orthogonalization process at each step we need to assume that all the determinants $\Delta_i^{(k)}$ are different from zero.

Then

$$p_i^{(k)} = p_i^{(k)}(A)X_0$$

where

$$p_i^{(k)}(t) = \frac{1}{\Delta_i^{(k)}} \tilde{p}_i^{(k)}(t). \tag{13}$$

We observe one property of the polynomials $\tilde{p}_i^{(k)}(t)$ which will be important later on. Namely, we shall prove the identity

$$\tilde{p}_i^{(k)}(t_{k+1}) = (-1)^i \Delta_i^{(k+1)}. \tag{14}$$

Actually, from the equality $\varphi_{k+1}(t) = \varphi_k(t)(t - t_{k+1})$ we have

$$m_j^{(k+1)} = (\varphi_k(A)(A - t_{k+1}E)A^j X_0, Y_0)$$
$$= (\varphi_k(A)A^{j+1}X_0, Y_0) - t_{k+1}(\varphi_k(A)A^j X_0, Y_0) = m_{j+1}^{(k)} - t_{k+1}m_j^{(k)}.$$

If in the determinant

$$\tilde{p}_i^{(k)}(t_{k+1}) = \begin{vmatrix} m_0^{(k)} & \cdots & m_{i-1}^{(k)} & 1 \\ m_1^{(k)} & \cdots & m_i^{(k)} & t_{k+1} \\ \cdot & \cdot & \cdot & \cdot \\ m_i^{(k)} & \cdots & m_{2i-1}^{(k)} & t_{k+1}^i \end{vmatrix}$$

we subtract from each row the preceding row multiplied by $t_{k+1}$, then we obtain

$$\tilde{p}_i^{(k)}(t_{k+1}) = \begin{vmatrix} m_0^{(k)} & \cdots & m_{i-1}^{(k)} & 1 \\ m_0^{(k+1)} & \cdots & m_{i-1}^{(k+1)} & 0 \\ \cdot & \cdot & \cdot & \cdot \\ m_{i-1}^{(k+1)} & \cdots & m_{2i-2}^{(k+1)} & 0 \end{vmatrix} = (-1)^i \Delta_i^{(k+1)}.$$

It is easily established that the vectors $p_i^{(k)}$ are connected by the recurrent relationships

$$(A - t_{k+1}E)p_{i-1}^{(k+1)} - p_i^{(k)} = \rho_{i-1}^{(k+1)}p_{i-1}^{(k)}$$
$$p_i^{(k)} - p_i^{(k+1)} = \sigma_i^{(k+1)}p_{i-1}^{(k+1)} \,, \tag{15}$$

and the polynomials $p_i^{(k)}(t)$ by the relationships

$$(t - t_{k+1})p_{i-1}^{(k+1)}(t) - p_i^{(k)}(t) = \rho_{i-1}^{(k+1)}p_{i-1}^{(k)}(t) \,,$$
$$p_i^{(k)}(t) - p_i^{(k+1)}(t) = \sigma_i^{(k+1)}p_{i-1}^{(k+1)}(t) \,. \tag{16}$$

The coefficients $\rho_i^{(k)}$ and $\sigma_i^{(k)}$ in turn satisfy the relationships

$$\sigma_i^{(k+1)} + \rho_i^{(k+1)} + t_{k+1} = \sigma_{i+1}^{(k)} + \rho_i^{(k)} + t_k \,,$$
$$\sigma_i^{(k+1)}\rho_{i-1}^{(k+1)} = \sigma_i^{(k)}\rho_i^{(k)} \,. \tag{17}$$

These relationships are derived by comparing the trinomial relationships connecting the vectors $p_{i+1}^{(k)}$, $p_i^{(k)}$, and $p_{i-1}^{(k)}$ which are obtained in two ways from the binomial relationships, just as was done above.

Relationships (17) permit us to compute successively the coefficients $\rho_i^{(k+1)}$ and $\sigma_i^{(k+1)}$ in the order $\rho_0^{(k+1)}$, $\sigma_1^{(k+1)}$, $\rho_1^{(k+1)}$, $\sigma_2^{(k+1)}$, $\cdots$, $\rho_{n-2}^{(k+1)}$, $\sigma_{n-1}^{(k+1)}$, $\rho_{n-1}^{(k+1)}$, as soon as the numbers of the previous row have been computed.

The computational formulas (division and subtraction scheme with displacement) are

$$\rho_i^{(k+1)} = \sigma_{i+1}^{(k)} + \rho_i^{(k)} - \sigma_i^{(k+1)} - t_{k+1} + t_k \qquad (i = 0, 1, \cdots, n-1) \,,$$
$$\sigma_i^{(k+1)} = \frac{\sigma_i^{(k)}\rho_i^{(k)}}{\rho_{i-1}^{(k+1)}} \qquad (i = 1, \cdots, n-1) \,. \tag{18}$$

We need to assume here that $\sigma_0^{(k)} = \sigma_n^{(k)} = 0$.

The initial row is formed from the coefficients of the biorthogonal algorithm's binomial formulas, which may be computed directly or found using the coefficients of the trinomial formulas.

If we consider that all $t_k = 0$, then formulas (18) coincide exactly with formulas of the $QD$ scheme derived above for a positive definite matrix $A$.

The following formula for the coefficients $\rho_i^{(k)}$ will turn out to be useful later on:

$$\rho_{i-1}^{(k+1)} = \frac{\Delta_{i-1}^{(k)}\Delta_i^{(k+1)}}{\Delta_i^{(k)}\Delta_{i-1}^{(k+1)}} \,. \tag{19}$$

To derive this formula we let $t = t_{k+1}$ in the first of relationships (16). Then

$$\rho_{i-1}^{(k+1)} = -\frac{p_i^{(k)}(t_{k+1})}{p_{i-1}^{(k)}(t_{k+1})} = -\frac{\Delta_{i-1}^{(k)}\tilde{p}_i^{(k)}(t_{k+1})}{\Delta_i^{(k)}\tilde{p}_{i-1}^{(k)}(t_{k+1})} = \frac{\Delta_{i-1}^{(k)}\Delta_i^{(k+1)}}{\Delta_i^{(k)}\Delta_{i-1}^{(k+1)}} \,.$$

THEOREM 77.2. *If the weight polynomials $\varphi_k(t)$ and the initial vectors $X_0$ and $Y_0$ are chosen so that*

1.  *all $\Delta_i^{(k)} \neq 0$,*
2.  *the sequence $t_k$ converges to a finite limit $\tau$,*

3. *for a certain numbering of the eigenvalues*

$$|\lambda_1 - \tau| > |\lambda_2 - \tau| > \cdots > |\lambda_n - \tau|,$$

*then the sequences* $\rho_i^{(k)}$ *converge to the limits* $\lambda_{i+1} - \tau$ *and the sequences* $\sigma_i^{(k)}$ *converge to zero.*

*Proof.* We observe first of all that in satisfying the conditions of the theorem $\tau \neq \lambda_i$ for $i = 1, \cdots, n-1$. The equality $\tau = \lambda_n$ is not excluded.

We shall derive asymptotic formulas for $\rho_i^{(k+1)}$. For this we estimate first of all the determinant $\Delta_i^{(k)}$. Let

$$X_0 = c_1 U_1 + \cdots + c_n U_n,$$
$$Y_0 = d_1 V_1 + \cdots + d_n V_n$$

where $U_1, \cdots, U_n$ are eigenvectors of matrix $A$ belonging to the eigenvalues $\lambda_1, \cdots, \lambda_n$ numbered according to condition 3) of the theorem. Correspondingly, $V_1, \cdots, V_n$ are eigenvectors of matrix $A'$. Then

$$m_j^{(k)} = b_1 \varphi_k(\lambda_1)\lambda_1^j + \cdots + b_n \varphi_k(\lambda_n)\lambda_n^j$$

where

$$b_i = c_i d_i (U_i, V_i).$$

Consequently,

$$\Delta_i^{(k)} = \begin{vmatrix} \sum b_s \varphi_k(\lambda_s) & \sum b_s \varphi_k(\lambda_s)\lambda_s & \cdots & \sum b_s \varphi_k(\lambda_s)\lambda_s^{i-1} \\ \sum b_s \varphi_k(\lambda_s)\lambda_s & \sum b_s \varphi_k(\lambda_s)\lambda_s^2 & \cdots & \sum b_s \varphi_k(\lambda_s)\lambda_s^i \\ \cdots & \cdots & \cdots & \cdots \\ \sum b_s \varphi_k(\lambda_s)\lambda_s^{i-1} & \sum b_s \varphi_k(\lambda_s)\lambda_s^i & \cdots & \sum b_s \varphi_k(\lambda_s)\lambda_s^{2i-2} \end{vmatrix}.$$

Here all sums are extended to $s = 1, 2, \cdots, n$.

The matrix found under the determinant sign for $\Delta_i^{(k)}$ may be represented as the product of the following two rectangular matrices:

$$\begin{bmatrix} b_1 \varphi_k(\lambda_1) & b_2 \varphi_k(\lambda_2) & \cdots & b_n \varphi_k(\lambda_n) \\ \lambda_1 b_1 \varphi_k(\lambda_1) & \lambda_2 b_2 \varphi_k(\lambda_2) & \cdots & \lambda_n b_n \varphi_k(\lambda_n) \\ \cdots & \cdots & \cdots & \cdots \\ \lambda_1^{i-1} b_1 \varphi_k(\lambda_1) & \lambda_2^{i-1} b_2 \varphi_k(\lambda_2) & \cdots & \lambda_n^{i-1} b_n \varphi_k(\lambda_n) \end{bmatrix}$$

and

$$\begin{bmatrix} 1 & \lambda_1 & \cdots & \lambda_1^{i-1} \\ 1 & \lambda_2 & \cdots & \lambda_2^{i-1} \\ \cdots & \cdots & \cdots & \cdots \\ 1 & \lambda_n & \cdots & \lambda_n^{i-1} \end{bmatrix}.$$

Using the well-known theorem on the determinant of the product of two rectangular matrices, we obtain

$$\Delta_i^{(k)} = \sum_{s_1 < s_2 < \cdots < s_i} b_{s_1} \varphi_k(\lambda_{s_1}) \cdots b_{s_i} \varphi_k(\lambda_{s_i}) \begin{vmatrix} 1 & \lambda_{s_1} & \cdots & \lambda_{s_1}^{i-1} \\ \cdots & \cdots & \cdots & \cdots \\ 1 & \lambda_{s_i} & \cdots & \lambda_{s_i}^{i-1} \end{vmatrix}^2.$$

It is not hard to see that for sufficiently large $k$

$$|\varphi_k(\lambda_1)| > |\varphi_k(\lambda_2)| > \cdots > |\varphi_k(\lambda_n)|$$

and moreover

$$\lim_{k \to \infty} \frac{\varphi_k(\lambda_i)}{\varphi_k(\lambda_{i-1})} = 0 .$$

In fact,

$$\frac{\varphi_k(\lambda_i)}{\varphi_k(\lambda_{i-1})} = \frac{(\lambda_i - t_1) \cdots (\lambda_i - t_k)}{(\lambda_{i-1} - t_1) \cdots (\lambda_{i-1} - t_k)} = \frac{\varphi_{k0}(\lambda_i)}{\varphi_{k0}(\lambda_{i-1})} \prod_{s=k_0+1}^{k} \frac{\lambda_i - t_s}{\lambda_{i-1} - t_s} .$$

We choose $k_0$ large enough so that, for $s > k_0$,

$$\left| \frac{\lambda_i - t_s}{\lambda_{i-1} - t_s} \right| < \left| \frac{\lambda_i - \tau}{\lambda_{i-1} - \tau} \right| + \varepsilon ,$$

where $\varepsilon$ is a small number such that

$$\left| \frac{\lambda_i - \tau}{\lambda_{i-1} - \tau} \right| + \varepsilon = q < 1 .$$

Then

$$\left| \frac{\varphi_k(\lambda_i)}{\varphi_k(\lambda_{i-1})} \right| < \left| \frac{\varphi_{k0}(\lambda_i)}{\varphi_{k0}(\lambda_{i-1})} \right| q^{k-k_0} \to 0$$

for $k \to \infty$.

Therefore the predominant element in $\Delta_i^{(k)}$ will be the one in which $s_1 = 1$, $s_2 = 2, \cdots, s_i = i$. Next in value will be the one in which $s_1 = 1$, $s_2 = 2, \cdots$, $s_{i-1} = i - 1$, $s_i = i + 1$. Consequently,

$$\Delta_i^{(k)} = b_1 \cdots b_i \varphi_k(\lambda_1) \cdots \varphi_k(\lambda_i) \begin{vmatrix} 1 & \lambda_1 & \cdots & \lambda_1^{i-1} \\ \cdot & \cdot & \cdot & \cdot \\ 1 & \lambda_i & \cdots & \lambda_i^{i-1} \end{vmatrix}^2 \left[ 1 + O\left( \frac{\varphi_k(\lambda_{i+1})}{\varphi_k(\lambda_i)} \right) \right]. \qquad (20)$$

This formula is true for $i = 1, 2, \cdots, n - 1$. For $i = 0$ we must consider $\Delta_0^{(k)} = 1$; for $i = n$ we have the exact formula

$$\Delta_i^{(k)} = b_1 \cdots b_n \varphi_k(\lambda_1) \cdots \varphi_k(\lambda_n) \begin{vmatrix} 1 & \lambda_1 & \cdots & \lambda_1^{n-1} \\ \cdot & \cdot & \cdot & \cdot \\ 1 & \lambda_n & \cdots & \lambda_n^{n-1} \end{vmatrix}^2 . \qquad (21)$$

Furthermore,

$$\frac{\Delta_i^{(k+1)}}{\Delta_i^{(k)}} = \frac{b_1 \cdots b_i \varphi_{k+1}(\lambda_1) \cdots \varphi_{k+1}(\lambda_i) \begin{vmatrix} 1 & \lambda_1 & \cdots & \lambda_1^{i-1} \\ \cdot & \cdot & \cdot & \cdot \\ 1 & \lambda_i & \cdots & \lambda_i^{i-1} \end{vmatrix}^2 \left[ 1 + O\left( \frac{\varphi_{k+1}(\lambda_{i+1})}{\varphi_{k+1}(\lambda_i)} \right) \right]}{b_1 \cdots b_i \varphi_k(\lambda_1) \cdots \varphi_k(\lambda_i) \begin{vmatrix} 1 & \lambda_1 & \cdots & \lambda_1^{i-1} \\ \cdot & \cdot & \cdot & \cdot \\ 1 & \lambda_i & \cdots & \lambda_i^{i-1} \end{vmatrix}^2 \left[ 1 + O\left( \frac{\varphi_k(\lambda_{i+1})}{\varphi_k(\lambda_i)} \right) \right]}$$

$$= (\lambda_1 - t_{k+1}) \cdots (\lambda_i - t_{k+1}) \frac{1 + O\left( \dfrac{\varphi_{k+1}(\lambda_{i+1})}{\varphi_{k+1}(\lambda_i)} \right)}{1 + O\left( \dfrac{\varphi_k(\lambda_{i+1})}{\varphi_k(\lambda_i)} \right)} .$$

For $i < n - 1$ the values $\varphi_k(\lambda_{i+1})/\varphi_k(\lambda_i)$ and $\varphi_{k+1}(\lambda_{i+1})/\varphi_{k+1}(\lambda_i)$ have the same order of smallness, since

$$\frac{\varphi_{k+1}(\lambda_{i+1})}{\varphi_{k+1}(\lambda_i)} = \frac{\varphi_k(\lambda_{i+1})}{\varphi_k(\lambda_i)} \frac{(\lambda_{i+1} - t_{k+1})}{(\lambda_i - t_k)} \approx \frac{\varphi_k(\lambda_{i+1})}{\varphi_k(\lambda_i)} \frac{\lambda_{i+1} - \tau}{\lambda_i - \tau} .$$

For $i = n - 1$ the order of smallness for $\varphi_{k+1}(\lambda_n)/\varphi_{k+1}(\lambda_{n-1})$ may be higher than the order for $\varphi_k(\lambda_n)/\varphi_k(\lambda_{n-1})$ if $\tau = \lambda_n$.

Therefore for $i \le n - 1$ we have the formula

$$\frac{\Delta_i^{(k+1)}}{\Delta_i^{(k)}} = (\lambda_1 - t_{k+1}) \cdots (\lambda_i - t_{k+1}) \left[ 1 + O\left( \frac{\varphi_k(\lambda_{i+1})}{\varphi_k(\lambda_i)} \right) \right]. \tag{22}$$

For $i = n$ we have the exact formula

$$\frac{\Delta_n^{(k+1)}}{\Delta_n^{(k)}} = (\lambda_1 - t_{k+1}) \cdots (\lambda_n - t_{k+1}) . \tag{23}$$

From here we obtain asymptotic formulas for $\rho_i^{(k+1)}$. Namely, for $i = 0, 1, \cdots, n - 2$,

$$\rho_i^{(k+1)} = \frac{\Delta_{i+1}^{(k+1)}}{\Delta_{i+1}^{(k)}} : \frac{\Delta_i^{(k+1)}}{\Delta_i^{(k)}}$$

$$= \frac{(\lambda_1 - t_{k+1}) \cdots (\lambda_{i+1} - t_{k+1}) \left[ 1 + O\left( \frac{\varphi_k(\lambda_{i+2})}{\varphi_k(\lambda_{i+1})} \right) \right]}{(\lambda_1 - t_{k+1}) \cdots (\lambda_i - t_{k+1}) \left[ 1 + O\left( \frac{\varphi_k(\lambda_{i+1})}{\varphi_k(\lambda_i)} \right) \right]} \tag{24}$$

$$= (\lambda_{i+1} - t_{k+1}) \left[ 1 + O\left( \frac{\varphi_k(\lambda_{i+2})}{\varphi_k(\lambda_{i+1})} \right) + O\left( \frac{\varphi_k(\lambda_{i+1})}{\varphi_k(\lambda_i)} \right) \right].$$

For $\rho_{n-1}^{(k+1)}$ the asymptotic formula has the simpler form

$$\rho_{n-1}^{(k+1)} = (\lambda_n - t_{k+1}) \left[ 1 + O\left( \frac{\varphi_k(\lambda_n)}{\varphi_k(\lambda_{n-1})} \right) \right]. \tag{25}$$

Passing to the limit, we obtain

$$\lim_{k \to \infty} \rho_i^{(k+1)} = \lambda_{i+1} - \tau \qquad (i = 0, 1, \cdots, n - 1)$$

which is what we were to prove.

In proving the theorem we have also obtained an estimate for the rate of convergence of the sequences $\rho_i^{(k)}$.

It remains to prove that the sequences $\sigma_i^{(k)}$ approach zero. We have

$$\frac{\sigma_i^{(k+1)}}{\sigma_i^{(k)}} = \frac{\rho_{i+1}^{(k)}}{\rho_i^{(k+1)}} .$$

We have already established that

$$\frac{\rho_{i+1}^{(k)}}{\rho_i^{(k+1)}} \to \frac{\lambda_{i+1} - \tau}{\lambda_i - \tau} ,$$

and consequently, for $k > k_0$,

$$\left|\frac{\sigma_i^{(k+1)}}{\sigma_i^{(k)}}\right| < \left|\frac{\lambda_{i+1} - \tau}{\lambda_i - \tau}\right| + \varepsilon = q < 1.$$

Therefore $|\sigma_i^{(k)}| < |\sigma_i^{(k_0)}| \, q^{k-k_0} \to 0$.

It follows from the theorem proved, in particular, that if we take $t_2 = t_3 = \cdots = 0$ (*QD* scheme), then for $|\lambda_1| > |\lambda_2| > \cdots > |\lambda_n|$ we obtain

$$\lim_{k\to\infty} \rho_i^{(k)} = \lambda_{i+1}.$$

More precisely,

$$\rho_i^{(k)} = \lambda_{i+1} + O\left(\frac{\lambda_{i+2}}{\lambda_{i+1}}\right)^k + O\left(\frac{\lambda_{i+1}}{\lambda_i}\right)^k. \tag{26}$$

Thus the result of theorem 77.1 is extended to any matrices with real eigenvalues satisfying the inequalities $|\lambda_1| > |\lambda_2| > \cdots > |\lambda_n|$ only if we assume that all $\varDelta_i^{(k)}$ are unequal to zero.

In the work of Rutishauser (2) the case of complex eigenvalues is also considered as well as possible degeneracies in the process. The problem of computing the eigenvectors is considered in his work (4).

Convergence of the division and subtraction algorithm, especially for eigenvalues close together, is quite slow. Since this process itself does not correct this, there arises the danger during its prolonged course of some accumulation of rounding-off errors. What is more interesting is the possibility of using displacements for speeding up the convergence of the process. That is, for a certain defined choice of translations a process is obtained with second-order convergence in turn for each eigenvalue.

THEOREM 77.3. *Assume that at a certain step of the QD process (with or without translation) under the conditions of the previous theorem already obtained*

$$|\lambda_n - p_{n+1}^{(k)} - t_k| < \varepsilon.$$

*Then taking*

$$t_{k+1} = \rho_{n-1}^{(k)} + t_k$$

*we obtain*

$$|\lambda_n - \rho_{n-1}^{(k+1)} - t_{k+1}| < \mu\varepsilon^2$$

*where $\mu$ is a certain constant.*

*Proof.* It is possible to prove that in the asymptotic formula (25) the order of the residual element, under the conditions of theorem 77.2, turns out to be exact. Namely,

$$\rho_{n-1}^{(k)} = (\lambda_n - t_k)\left[1 + M\frac{\varphi_{k-1}(\lambda_n)}{\varphi_{k-1}(\lambda_{n-1})}(1 + \varepsilon_k)\right],$$

where $M$ is a certain constant and $\varepsilon_k \to 0$. Therefore

$$\frac{\rho_{n-1}^{(k+1)} - \lambda_n + t_{k+1}}{\rho_{n-1}^{(k)} - \lambda_n + t_k} = \frac{\varphi_{k+1}(\lambda_n)\varphi_{k-1}(\lambda_{n-1})}{\varphi_k(\lambda_{n-1})\varphi_k(\lambda_n)}(1 + \varepsilon_k') = \frac{\lambda_n - t_{k+1}}{\lambda_{n-1} - t_k}(1 + \varepsilon_k'),$$

where $\varepsilon_k' \to 0$. For

$$t_{k+1} = \rho_{n-1}^{(k)} + t_k$$

we have

$$\frac{\lambda_n - \rho_{n-1}^{(k+1)} - t_{k+1}}{\lambda_n - \rho_{n-1}^{(k)} - t_k} = \frac{\lambda_n - \rho_{n-1}^{(k)} - t_k}{\lambda_{n-1} - t_k}(1 + \varepsilon_k'),$$

from which

$$|\lambda_n - \rho_{n-1}^{(k+1)} - t_{k+1}| = \frac{(\lambda_n - \rho_{n-1}^{(k)} - t_k)^2}{|\lambda_{n-1} - t_k|}(1 + \varepsilon_k'), \tag{27}$$

that is,

$$|\lambda_n - \rho_{n-1}^{(k+1)} - t_{k+1}| < \mu\varepsilon^2$$

for

$$\mu < \frac{2}{|\lambda_{n-1} - \tau|}$$

We now describe the process with acceleration. We assume for simplicity that all eigenvalues are positive and $\lambda_1 > \lambda_2 > \cdots > \lambda_n$.

We assume that several steps of the algorithm $QD$ without translation have been carried out so that the last column has begun to stabilize as a crude approximation. Let this occur at the $k_1$-th step. Then we take $t_{k_1+1} = \rho_{n-1}^{(k_1)}$, $t_{k_1+2} = \rho_{n-1}^{(k_1+1)} + t_{k_1+1}, \cdots$ In the new process we obtain second-order convergence of the sequence $t_k$ to $\lambda_n$; the sequences $\sigma_{n-1}^{(k)}$ and $\rho_{n-1}^{(k)}$ converge to zero with the same speed.

For $k = k_2$ let the numbers $\sigma_{n-1}^{(k)}$ and $\rho_{n-1}^{(k)}$ actually become equal to zero. Then we may take $\lambda_n \approx t_{k_2}$ with the same degree of accuracy and may pass to the process with acceleration to determine $\lambda_{n-1}$. Namely, we assume $t_{k_2+1} = \rho_{n-2}^{(k_2)}$, $t_{k_2+2} = \rho_{n-2}^{(k_2+1)} + t_{k_2+1} \cdots$ For this we strike from the scheme the column consisting of the values $\rho_{n-1}^{(k)}$, since the latter are no longer needed either for determining $\lambda_n$ or for continuing the scheme, since $\sigma_{n-1}^{(k)}$ has become and remains equal to zero. After determining $\lambda_{n-1}$ we pass in turn to the determination of $\lambda_{n-2}, \lambda_{n-3}, \cdots, \lambda_1$.

The fact that in this process second-order convergence will hold for each eigenvalue follows from the fact that during the process the vectors $p_0^{(k)}$, $p_1^{(k)}, \cdots, p_{n-1}^{(k)}$ will fall into invariant subspaces of decreasing dimensions and $\lambda_{n-1}, \lambda_{n-2}, \cdots, \lambda_1$ will in turn play the role of the smallest eigenvalues.

The indicated process may be applied in passing to the second row, letting $t_2 = \rho_{n-1}^{(1)}$, $t_3 - t_2 = \rho_{n-1}^{(2)} \cdots$ However, in this, several of the first translations will have a random character and the process will begin to converge rapidly only when one of the translations turns out to be close to some eigenvalue. That is, this eigenvalue will be determined by the first one. Generally the eigenvalue next determined will be the one closest to the one obtained.

The application of translation is possible at the first step of the process as well, i.e. we may take the weight polynomial $\varphi_k(t)$ equal to $(t - t_1)(t - t_2)$ $\cdots (t - t_k)$ for $t_1 \neq 0$. This will involve changing the initial row of the process. It is easy to see that in this case it must be constructed according to the formulas

$$\bar{\rho}_0^{(1)} = \alpha_0 - t_1 \,,$$

$$\bar{\sigma}_i^{(1)} = \frac{\beta_i}{\rho_{i-1}^{(1)}} \qquad (i = 1, 2, \cdots, n - 1) \,,$$

$$\bar{\rho}_i^{(1)} = -\bar{\sigma}_i^{(1)} + \alpha_i - t_1 \qquad (i = 1, 2, \cdots, n - 1) \,.$$

In this form the $QD$ process is used for making the crude approximation to one of the eigenvalues obtained by some other means more accurate. We should take for $t_1$ this known rough approximation and should then apply the $QD$ process with translations, as was described above.

In Tables VIII. 1-VIII. 6 are given numerical illustrations for the course of the $QD$ process.

In Table VIII. 1 is given the course of the $QD$ process without acceleration for the positive definite matrix (4) of Section 51. The first row of the table is filled in from the data of Table VI. 16.

In Table VIII. 2 we give for the same matrix the $QD$ process with the translations.

$$t_1 = 0, \quad t_{k+1} = t_k + \rho_3^{(k)} \,.$$

In this case all four eigenvalues are determined after twelve steps of the process.

In Table VIII. 3 is given the course of the $QD$ process for a non-positive definite matrix (and even non-symmetric) of Leverrier. The first row of the scheme is taken from Table VI. 17.

In Table VIII. 4 the $QD$ process is carried out for Leverrier's matrix with the translations

$$t_1 = t_2 = t_3 = t_4 = t_5 = t_6 = 0 \,,$$

$$t_{k+1} = t_k + \rho_3^{(k)} \qquad \text{(for } k \geq 6) \,.$$

The larger roots were obtained with low accuracy because of a certain loss of accuracy in the seventh and eighth steps.

In Table VIII. 5 is given the accuracy increase in the larger roots of Leverrier's matrix using the $QD$ scheme with constant translations.

Finally in Table VIII. 6 is given the accuracy increase for the first eigenvalue with a change in the initial row of the process by $t_1 = -17.863248$ and with successive constant translations.

## 78. THE TRIANGULAR POWER METHOD

The triangular power method (Bauer (7)) is a generalization of the stepped power method suitable for finding all the eigenvalues of a matrix $A$. We

**TABLE VIII.1   The QD Scheme without Translations**

| k | | $\rho_0^{(k)}$ | $\sigma_1^{(k)}$ | $\rho_1^{(k)}$ | $\sigma_2^{(k)}$ | $\rho_2^{(k)}$ | $\sigma_3^{(k)}$ | $\rho_3^{(k)}$ | | $\Sigma$ |
|---|---|---|---|---|---|---|---|---|---|---|
| 1 | 0 | 2.2 | 0.018 | 0.45668599 | 0.11169690 | 0.39432359 | 0.02841890 | 0.69087470 | 0 | 4.00000008 |
| 2 | 0 | 2.318 | 0.00354631 | 0.56483658 | 0.07797782 | 0.34476467 | 0.05694870 | 0.63392600 | 0 | 4.00000008 |
| 3 | 0 | 2.32154631 | 0.00086282 | 0.64195158 | 0.04187854 | 0.35983483 | 0.10032731 | 0.53359869 | 0 | 4.00000008 |
| 13 | 0 | 2.32274877 | 0.00000001 | 0.77144752 | 0.00436073 | 0.65910897 | 0.00004616 | 0.24228792 | 0 | 4.00000008 |
| 14 | 0 | 2.32274878 | 0 | 0.77580825 | 0.00370478 | 0.65545035 | 0.00001706 | 0.24227086 | 0 | 4.00000008 |
| 21 | 0 | | | 0.79176600 | 0.00095776 | 0.64226676 | 0.00000002 | 0.24226076 | 0 | |
| 22 | | | | 0.79272376 | 0.00077598 | 0.64149080 | 0.00000001 | 0.24226075 | | |
| 69 | | | | 0.79670669 | 0.00000002 | 0.63828384 | | | | |

**TABLE VIII.2   The QD Scheme with Translations**

| k | | $\rho_0^{(k)}$ | $\sigma_1^{(k)}$ | $\rho_1^{(k)}$ | $\sigma_2^{(k)}$ | $\rho_2^{(k)}$ | $\sigma_3^{(k)}$ | $\rho_3^{(k)}$ | $t_k$ | $\lambda_i$ |
|---|---|---|---|---|---|---|---|---|---|---|
| 1 | 0 | 2.2 | 0.018 | 0.45668599 | 0.11169690 | 0.39432359 | 0.02841890 | 0.69087470 | 0 | |
| 2 | 0 | 1.62712530 | 0.00505207 | -0.12754388 | -0.34532995 | 0.07719774 | 0.25433256 | -0.25433256 | 0.69087470 | |
| 3 | 0 | 1.88650993 | -0.00034156 | -0.21819971 | 0.12217565 | 0.46368721 | -0.13950148 | 0.13950148 | 0.43654214 | |
| 4 | 0 | 1.74666689 | 0.00004267 | -0.23556821 | -0.24048782 | 0.42517207 | -0.04577126 | 0.04577126 | 0.57604362 | |
| 5 | 0 | 1.70093830 | -0.00000591 | -0.52182133 | 0.19594579 | 0.13768376 | -0.01521609 | 0.01521609 | 0.62181488 | |
| 6 | 0 | 1.68571630 | 0.00000183 | -0.34109351 | -0.07909430 | 0.18634588 | -0.00124247 | 0.00124247 | 0.63703097 | |
| 7 | 0 | 1.68447566 | -0.00000037 | -0.42142991 | 0.03497354 | 0.14888740 | -0.00001037 | 0.00001037 | 0.63827344 | |
| 8 | 0 | 1.68446492 | 0.00000009 | -0.38646683 | -0.01347365 | 0.16234031 | 0 | 0 | 0.63828381 | 0.63828381 |
| 9 | 0 | 1.52212470 | -0.00000002 | -0.56228077 | 0.00389008 | -0.00389008 | | | 0.80062412 | |
| 10 | 0 | 1.52601476 | 0.00000001 | -0.55450062 | 0.00002729 | -0.00002729 | | | 0.79673404 | 2.32274880 |
| 11 | 0 | | | -0.55444604 | 0 | 0 | | | 0.79670675 | 0.79670675 |
| 12 | 0 | | | 0 | | | | | 0.24226071 | 0.24226071 |

**TABLE VIII.3   The *QD* Scheme without Translations**

| k | $\rho_0^{(k)}$ | $\sigma_1^{(k)}$ | $\rho_1^{(k)}$ | $\sigma_2^{(k)}$ | $\rho_2^{(k)}$ | $\sigma_3^{(k)}$ | $\rho_3^{(k)}$ | $\Sigma$ |
|---|---|---|---|---|---|---|---|---|
| 1 | − 7.5036210 | − 0.23490436 | − 13.224755 | − 1.5827643 | − 5.9951147 | 1.3221017 | − 20.669372 | − 47.888430 |
| 2 | − 7.7385254 | − 0.40143988 | − 14.406079 | − 0.65867010 | − 4.0143429 | 6.8073437 | − 27.476717 | − 47.888431 |
| 3 | − 8.1399653 | − 0.71046673 | − 14.354282 | − 0.18420479 | − 2.9772056 | − 62.825173 | − 35.348456 | − 47.888430 |
| 4 | − 8.8504320 | − 1.1522872 | − 13.386200 | 0.04096872 | − 59.888936 | 37.081521 | − 1.7330650 | − 47.888430 |
| 5 | − 10.002719 | − 1.5420554 | − 11.803176 | 0.20787398 | − 23.015289 | 2.7922606 | − 4.5253256 | − 47.888430 |
| 6 | − 11.544774 | − 1.5765705 | − 10.018732 | 0.47753345 | − 20.700562 | 0.61041324 | − 5.1357388 | − 47.888431 |
| 35 | − 18.653538 | 0.06359643 | − 16.425749 | − 0.00000001 | − 7.5740424 | 0.00000061 | − 5.2986996 | |
| 137 | − 17.869245 | 0.00024064 | − 17.146687 | | | | | |

**TABLE VIII.4   The *QD* Scheme with Translations**

| k | $\rho_0^{(k)}$ | $\sigma_1^{(k)}$ | $\rho_1^{(k)}$ | $\sigma_2^{(k)}$ | $\rho_2^{(k)}$ | $\sigma_3^{(k)}$ | $\rho_3^{(k)}$ | $t_k$ |
|---|---|---|---|---|---|---|---|---|
| 6 | − 11.544774 | − 1.5765705 | − 10.018732 | 0.47753345 | − 20.700562 | 0.61041324 | − 5.1357388 | 0 |
| 7 | − 7.9856057 | − 1.9779636 | − 2.4474962 | 4.0721838 | − 19.026594 | 0.16476533 | − 0.16476533 | − 5.1357388 |
| 8 | − 9.7988040 | − 0.49000869 | 2.2994616 | − 33.694752 | 14.997689 | − 0.00181012 | 0.00181012 | − 5.3005041 |
| 9 | − 10.290623 | 0.10949348 | − 31.506594 | 16.039290 | − 1.0452212 | 0.00000313 | − 0.00000313 | − 5.2986940 |
| 10 | − 10.181126 | 0.33883939 | − 15.806140 | 1.0606388 | − 2.1058537 | 0 | 0 | − 5.2986971 |
| 11 | − 7.7364329 | 0.69227548 | − 13.331923 | 0.16753398 | − 0.16753398 | | | − 7.4045508 |
| 12 | − 6.8766234 | 1.3421359 | − 14.338991 | 0.00195743 | − 0.00195743 | | | − 7.5720848 |
| 13 | − 5.5325301 | 3.484943 | − 17.813570 | 0.00000022 | − 0.00000022 | | | − 7.5740422 |
| 14 | − 2.0540356 | 30.167151 | − 47.980721 | 0 | 0 | | | − 7.5740424 |
| 15 | 76.093836 | − 19.021799 | − 19.021799 | | | | | − 55.554763 |
| 16 | 38.050238 | 9.5092398 | 9.5092398 | | | | | − 36.532964 |
| 17 | 19.031758 | 4.7513026 | 4.7513026 | | | | | − 27.023724 |
| 18 | 9.5291528 | 2.3690329 | 2.3690329 | | | | | − 22.272421 |
| 19 | 4.7910870 | 1.1714078 | 1.1714078 | | | | | − 19.903388 |
| 20 | 2.4482714 | 0.56047554 | 0.56047554 | | | | | − 18.731980 |
| 21 | 1.3273203 | 0.23666694 | 0.23666694 | | | | | − 18.171504 |
| 22 | 0.85398642 | 0.06558798 | 0.06558798 | | | | | − 17.934837 |
| 23 | 0.72281046 | 0.00595147 | 0.00595147 | | | | | − 17.869249 |
| 24 | 0.71090752 | 0.00004982 | 0.00004982 | | | | | − 17.863298 |
| 25 | 0.71080788 | 0 | 0 | | | | | − 17.863248 |
| | | | | | | | | − 17.152440 |

**TABLE VIII.5  Improving Eigenvalue Accuracy using the QD Scheme with Constant Translations**

| $k$ | $\rho_0^{(k)}$ | $\sigma_1^{(k)}$ | $\rho_1^{(k)}$ | $\sigma_2^{(k)}$ | $\rho_2^{(k)}$ | $\sigma_3^{(k)}$ | $\rho_3^{(k)}$ | $t_k$ | $\lambda_i$ |
|---|---|---|---|---|---|---|---|---|---|
| 1 | $-7.5036210$ | $-0.23490436$ | $-13.224755$ | $-1.5827643$ | $-5.9951147$ | $1.322017$ | $-20.669372$ | $0$ | |
| 2 | $10.124723$ | $0.30682839$ | $2.7489003$ | $3.4518725$ | $9.7383625$ | $-2.806198$ | $-0.00000420$ | $-17.863248$ | |
| 3 | $10.431551$ | $0.08085477$ | $6.1199180$ | $5.4928163$ | $1.4394264$ | $0.0000819$ | $-0.00001239$ | $-17.863248$ | |
| 4 | $10.512406$ | $0.04707054$ | $11.565664$ | $0.68361875$ | $0.75581584$ | $0$ | $-0.00001239$ | | $-17.863260$ |
| 1 | $-7.5036210$ | $-0.23490436$ | $-13.224755$ | $-1.5827643$ | $-5.9951147$ | $1.322017$ | $-20.669372$ | $0$ | |
| 2 | $9.4139146$ | $0.32999583$ | $2.0149249$ | $4.7092839$ | $7.7701431$ | $-3.5169251$ | $-0.00000690$ | $-17.152440$ | |
| 3 | $9.7439104$ | $0.06823921$ | $6.6559696$ | $5.4975926$ | $-1.243746$ | $-0.00001950$ | $0.00001260$ | $-17.152440$ | |
| 4 | $9.8121496$ | $0.04628936$ | $12.107273$ | $-0.56503756$ | $-0.67935654$ | $0$ | $0.00001260$ | | $-17.152427$ |

**TABLE VIII.6  Improving Eigenvalue Accuracy using the QD Scheme with Altered First Row**

| $k$ | $\bar\rho_0^{(k)}$ | $\bar\sigma_1^{(k)}$ | $\bar\rho_1^{(k)}$ | $\bar\sigma_2^{(k)}$ | $\bar\rho_2^{(k)}$ | $\bar\sigma_3^{(k)}$ | $\bar\rho_3^{(k)}$ | $t_k$ | $\lambda_i$ |
|---|---|---|---|---|---|---|---|---|---|
| 1 | $10.359627$ | $0.17014447$ | $4.2334445$ | $4.9443593$ | $5.3410097$ | $-1.480175$ | $-0.00000550$ | $-17.863248$ | |
| 2 | $10.529771$ | $0.68405768$ | $8.4493461$ | $3.1090958$ | $0.74789640$ | $0.00001091$ | $-0.00001641$ | $-17.863248$ | |
| 3 | $11.213829$ | $0.51812919$ | $11.084713$ | $0.20977372$ | $0.53813359$ | $0$ | $-0.00001641$ | | $-17.863264$ |

shall assume that all eigenvalues of matrix $A$ are real and distinct in absolute value. We denote them by $\lambda_1, \cdots, \lambda_n$, numbering them in order of decreasing absolute value.

As a basis for the triangular power method we carry out the following discussion. Let $K$ and $M$ be arbitrary matrices. Consider the sequence of matrices $A^{(k)} = KA^k M$. We shall represent each of these matrices as the product of a left triangular matrix $C_k$ with unit diagonal elements, a diagonal matrix $D_k$ and a right triangular matrix $B_k$ with unit diagonal elements:

$$A^{(k)} = C_k D_k B_k .$$

It is assumed that such an expansion is possible for all $k$. Then under certain additional limitations the matrices $D_k = [d_1^{(k)}, \cdots, d_n^{(k)}]$ are such that

$$\lim_{k \to \infty} \frac{d_i^{(k)}}{d_i^{(k-1)}} = \lambda_i \qquad (1)$$

and the matrices $C_k$ and $B_k$ approach the limit matrices.

We shall prove this. First of all we observe (Section 1, paragraph 12) that

$$d_i^{(k)} = \frac{\varDelta_i^{(k)}}{\varDelta_{i-1}^{(k)}} \qquad (2)$$

where $\varDelta_i^{(k)}$ is the upper principal minor of order $i$ for the matrix $A^{(k)}$. We shall estimate $\varDelta_i^{(k)}$. Let $A = P^{-1} \varLambda P$ where $\varLambda = [\lambda_1, \cdots, \lambda_n]$. Then

$$A^{(k)} = KA^k M = KP^{-1} \varLambda^k PM .$$

We let

$$KP^{-1} = \begin{bmatrix} q_{11} & \cdots & q_{1n} \\ \cdot & \cdot \cdot \cdot \cdot & \cdot \\ q_{n1} & \cdots & q_{nn} \end{bmatrix}.$$

$$PM = \begin{bmatrix} p_{11} & \cdots & p_{1n} \\ \cdot & \cdot \cdot \cdot \cdot & \cdot \\ p_{n1} & \cdots & p_{nn} \end{bmatrix}. \qquad (3)$$

Then

$$\varLambda^k PM = \begin{bmatrix} \lambda_1^k p_{11} & \lambda_1^k p_{12} & \cdots & \lambda_1^k p_{1n} \\ \lambda_2^k p_{21} & \lambda_2^k p_{22} & \cdots & \lambda_2^k p_{2n} \\ \cdot & \cdot \cdot \cdot & \cdot \cdot \cdot & \cdot \\ \lambda_n^k p_{n1} & \lambda_n^k p_{n2} & \cdots & \lambda_n^k p_{nn} \end{bmatrix}.$$

The matrix of the minor $\varDelta_i^{(k)}$ is the product of two rectangular matrices, the first of which is formed from the first $i$ rows of the matrix $KP^{-1}$ and the second of which from the first $i$ columns of the matrix $\varLambda^k PM$. Therefore

$$\varDelta_i^{(k)} = \sum_{j_1 < \cdots < j_i} \begin{bmatrix} q_{1j_1} & \cdots & q_{1j_i} \\ \cdot & \cdot \cdot \cdot \cdot & \cdot \\ q_{ij_1} & \cdots & q_{ij_i} \end{bmatrix} \cdot \begin{bmatrix} p_{j_1 1} & \cdots & p_{j_1 i} \\ \cdot & \cdot \cdot \cdot \cdot & \cdot \\ p_{j_i 1} & \cdots & p_{j_i i} \end{bmatrix} \lambda_{j_1}^k \cdots \lambda_{j_i}^k . \qquad (4)$$

For sufficiently large $k$ the predominant element in $\Delta_i^{(k)}$ will generally be the element corresponding to $j_1 = 1, \cdots, j_i = i$. The element second in magnitude will be the one corresponding to $j_1 = 1, \cdots, j_{i-1} = i-1, j_i = i+1$. This will be true if

$$Q_{ii} = \begin{bmatrix} q_{11} \cdots q_{1i} \\ \cdot \quad \cdots \quad \cdot \\ q_{i1} \cdots q_{ii} \end{bmatrix} \neq 0 \quad \text{and} \quad P_{ii} = \begin{bmatrix} p_{11} \cdots p_{1i} \\ \cdot \quad \cdots \quad \cdot \\ p_{i1} \cdots p_{ii} \end{bmatrix} \neq 0 . \tag{5}$$

We shall assume that the latter conditions are satisfied for all $i$. Under this assumption

$$\Delta_i^{(k)} = Q_{ii} P_{ii} \lambda_1^k \lambda_2^k \cdots \lambda_i^k \left[ 1 + O\left( \frac{\lambda_{i+1}}{\lambda_i} \right)^k \right]. \tag{6}$$

Consequently,

$$d_i^{(k)} = \frac{\Delta_i^{(k)}}{\Delta_{i-1}^{(k)}} = \frac{Q_{ii} P_{ii}}{Q_{i-1\,i-1} P_{i-1\,i-1}} \lambda_i^k \left[ 1 + O\left( \frac{\lambda_{i+1}}{\lambda_i} \right)^k + O\left( \frac{\lambda_i}{\lambda_{i-1}} \right)^k \right]. \tag{7}$$

Similarly,

$$d_i^{(k-1)} = \frac{Q_{ii} P_{ii}}{Q_{i-1\,i-1} P_{i-1\,i-1}} \lambda_i^{k-1} \left[ 1 + O\left( \frac{\lambda_{i+1}}{\lambda_i} \right)^k + O\left( \frac{\lambda_i}{\lambda_{i-1}} \right)^{k-} \right].$$

Therefore

$$\frac{d_i^{(k)}}{d_i^{(k-1)}} = \lambda_i + O\left( \frac{\lambda_{i+1}}{\lambda_i} \right)^k + O\left( \frac{\lambda_i}{\lambda_{i\,1}} \right)^k \quad (i = 1, \cdots, n-1) . \tag{8}$$

As the smallest eigenvalue in modulus we obtain

$$\frac{d_n^{(k)}}{d_n^{(k-1)}} = \lambda_n + O\left( \frac{\lambda_n}{\lambda_{n-1}} \right)^k . \tag{9}$$

We shall now investigate the matrices $B_k$ and $C_k$. We recall the formulas for elements of $B_k$ and $C_k$ in Section 1, paragraph 12. Namely,

$$b_{ij}^{(k)} = \frac{\beta_{ij}^{(k)}}{\beta_{ii}^{(k)}} ,$$

$$c_{ji}^{(k)} = \frac{\gamma_{ji}^{(k)}}{\gamma_{ii}^{(k)}} ,$$

where $\beta_{ii}^{(k)} = \gamma_{ii}^{(k)} = \Delta_i^{(k)}$ and $\beta_{ij}^{(k)}$ and $\gamma_{ji}^{(k)}$ are certain minors of order $i$ for the matrix $A^{(k)}$.

It is obvious that the arguments we applied in estimating the principal minors $\Delta_i^{(k)}$ remain in force for estimating any minors so that

$$\beta_{ij}^{(k)} = Q_{ij} P_{ij} \lambda_1^k \cdots \lambda_i^k \left[ 1 + O\left( \frac{\lambda_{i+1}}{\lambda_i} \right)^k \right],$$

$$\gamma_{ji}^{(k)} = Q_{ji} P_{ji} \lambda_1^k \cdots \lambda_i^k \left[ 1 + O\left( \frac{\lambda_{i+1}}{\lambda_i} \right)^k \right]. \tag{10}$$

Here $Q_{ij}, P_{ij}, Q_{ji},$ and $P_{ji}$ are certain minors of order $i$ formed from elements of the matrices $KP^{-1}$ and $PM$. Therefore

$$b_{ij}^{(k)} = \frac{Q_{ij}P_{ij}}{Q_{ii}P_{ii}}\left[1 + O\left(\frac{\lambda_{i+1}}{\lambda_i}\right)^k\right],$$

$$c_{ji}^{(k)} = \frac{Q_{ji}P_{ji}}{Q_{ii}P_{ii}}\left[1 + O\left(\frac{\lambda_{i+1}}{\lambda_i}\right)^k\right].$$

(11)

Thus, under the assumption made above that all determinants $Q_{ii}$ and $P_{ii}$ are non-zero, all elements $b_{ij}^{(k)}$ and $c_{ji}^{(k)}$ have limits for $k \to \infty$ while equalities (11) give an estimate for the rate of convergence.

We now describe the method's computational scheme.

Let $C_0$ be an arbitrary matrix and let $A$ be a matrix with real eigenvalues distinct in absolute value for which we must solve the complete eigenvalue problem.

We construct a sequence of left triangular matrices $C_1, C_2, \cdots, C_k, \cdots$ with unit diagonal elements using the recurrent relationships

$$AC_0 = C_1 R_1,$$
$$AC_1 = C_2 R_2,$$
$$\cdots \cdots \cdots,$$
$$AC_{k-1} = C_k R_k,$$
$$\cdots \cdots \cdots$$

(12)

Here $R_1, R_2, \cdots, R_k, \cdots$ are right triangular matrices.

The process should be carried out until the matrices $C_k$ are stabilized with sufficient accuracy. It is easy to prove that $\lim\limits_{k\to\infty} C_k$ exists. Namely, eliminating the matrices $C_1, C_2, \cdots, C_{k-1}$ from relationships (12) we obtain

$$A^k C_0 = C_k R_k R_{k-1} \cdots R_1.$$

(13)

The matrix $R_k R_{k-1} \cdots R_1$ is right triangular. Therefore the matrices $C_k$ and $R_k R_{k-1} \cdots R_1$ coincide with the matrices $C_k$ and $D_k B_k$ in the previous notation, set up for the matrix

$$A^{(k)} = A^k C_0.$$

Consequently $\lim\limits_{k\to\infty} C_k = C$ exists.

The matrices $R_k$ are also stabilized simultaneously with stabilization of the matrices $C_k$, since for $k \to \infty$

$$R_k = C_k^{-1} A C_{k-1} \to C^{-1} A C = R.$$

(14)

Knowledge of the limit matrices $C$ and $R$ makes it possible to solve the complete eigenvalue problem for matrix $A$.

In fact it follows from equality (14) that the eigenvalues of matrices $A$ and $R$ coincide and the eigenvectors $U_1, \cdots, U_n$ of matrix $A$ are equal respectively to $CV_1, \cdots, CV_n$ where $V_1, \cdots, V_n$ are eigenvectors of matrix $R$. The matrix $R$ is triangular so that its eigenvalues are equal to its diagonal elements and the eigenvectors $V_1, \cdots, V_n$ are easily determined from the solution of the triangular system.

The rate at which the matrix diagonal elements converge to the desired eigenvalues may be estimated from the following considerations.

First of all, from equation

$$R_k R_{k-1} \cdots R_1 = D_k B_k$$

it follows that

$$R_k = D_k B_k B_{k-1}^{-1} D_{k-1}^{-1} .$$

Therefore the diagonal elements of the matrix $R_k$ coincide with the diagonal elements of the matrix $D_k D_{k-1}^{-1}$ whose rate of convergence to the eigenvalues is estimated in formulas (8) and (9).

The matrices $C_k$ and $R_k$ may be computed at each step using the compact scheme of Gauss's method. However, there is no need to compute the matrices $R_k$ until the process is stabilized, since only the limit matrix $R$ is necessary to solve the given problem.

This means we may limit ourselves to computing only the matrices $C_k$; for this we may use, for example, the single division scheme (with elimination by columns) applied in Section 17 to compute a determinant. The same scheme gives us the diagonal elements of matrix $R_k$ so that approximate determination of matrix $A$'s eigenvalues does not require additional computations. If it is also necessary to compute the eigenvectors, then at the last step of the process we need to compute the whole matrix $R_k$ which gives an approximate value for the limit matrix $R$.

The triangular power method is a self-corrective process, since each approximation $C_k$ may be considered an initial approximation.

In Section 57 we mentioned the connection between the stepped power method and the triangular power method. It is now possible to say something about this in more detail. Namely, the $r$-stepped power method may be presented as follows. Starting with a rectangular matrix with $r$ columns $C_0^{(r)}$ we need to form the rectangular matrix $A C_0^{(r)}$, which we then represent as the product of a left stepped rectangular matrix $C_1^{(r)}$ with a unit principal diagonal and a right triangular matrix $R_1^{(r)}$ of order $r$. The process is then repeated. It is easy to see that if we add the matrix $C_0^{(r)}$ to the square matrix $C_0$ in some way and apply the triangular power method to it, then the matrices $C_k^{(r)}$ will coincide with the matrices formed from the first $r$ columns of the matrices $C_k$ and the matrices $R_k^{(r)}$ will be left upper cells of order $r$ of the matrices $R_k$.

Therefore the evaluation of the triangular power method's convergence is transferred to the $r$-stepped method without change.

The triangular power method allows for modification using translations. For a suitable choice of translations we may obtain in turn second-order convergence to each eigenvalue.

The basis of such a modification consists of the following. Let

$$\varphi_k(t) = (t - t_1) \cdots (t - t_k) \tag{15}$$

be a sequence of polynomials such that each successive polynomial is obtained from the preceding one by multiplication by a linear binomial. It is assumed that $t_k \to \tau$ and the eigenvalues of matrix $A$ under a certain enumeration satisfy the conditions

$$|\lambda_1 - \tau| > |\lambda_2 - \tau| > \cdots > |\lambda_n - \tau|.$$

Then, as we have seen in Section 77,

$$\frac{\varphi_k(\lambda_i)}{\varphi_k(\lambda_{i-1})} \to 0 \qquad (\text{for } k \to \infty).$$

We consider the matrix sequence

$$A^{(k)} = K\varphi_k(A)M \tag{16}$$

where $K$ and $M$ are certain fixed matrices. It is not hard to give estimates for minors of any order formed from elements of the matrices $A^{(k)}$. Namely, any minor of order $i$ is equal to

$$\hat{Q}\hat{P}\varphi_k(\lambda_1) \cdots \varphi_k(\lambda_i)\left[1 + O\left(\frac{\varphi_k(\lambda_{i+1})}{\varphi_k(\lambda_i)}\right)\right],$$

where $\hat{Q}$ and $\hat{P}$ are certain numbers dependent on the distribution of the minor inside the matrix $A^k$. In particular, for the upper principal minors we have

$$\Delta_i^{(k)} = Q_{ii}P_{ii}\varphi_k(\lambda_1) \cdots \varphi_k(\lambda_i)\left[1 + O\left(\frac{\varphi_k(\lambda_{i+1})}{\varphi_k(\lambda_i)}\right)\right], \tag{17}$$

where $Q_{ii}$ and $P_{ii}$ have the same meaning as in the previous paragraph. The result of these asymptotic formulas does not differ in any way from the calculation just made for the case $\varphi_k(t) = t^k$.

Writing the matrices $A^{(k)}$ as

$$A^{(k)} = C_k D_k B_k,$$

where the matrices $C_k$, $D_k$, and $B_k$ have their former structure, we obtain

$$\lim_{k \to \infty} \frac{d_i^{(k)}}{d_i^{(k-1)}} = \lambda_i - \tau \tag{18}$$

and the matrices $C_k$ and $B_k$ approach certain limit matrices as $k \to \infty$.

The computational scheme for the triangular power method with translation is as follows. An arbitrary matrix $C_0$ is chosen and a matrix $(A - t_1E)C_0$ is formed which is then set up as the product of a left triangular matrix $C_1$ with unit principal diagonal and a right triangular matrix $R_1$. The matrix $(A - t_2E)C_1$ is then formed and set up as the product of triangular matrices $C_2$ and $R_2$ with the previous structure. The process is continued in this way. It is clear that

$$\varphi_k(A)C_0 = C_k R_k R_{k-1} \cdots R_1. \tag{19}$$

From here it follows that there exist limit matrices

$$C = \lim_{k \to \infty} C_k \quad \text{and} \quad R = \lim_{k \to \infty} R_k$$

and that

$$R_k = D_k B_k B_{k-1}^{-1} D_{k-1}^{-1} \qquad (20)$$

so that the diagonal elements $r_i^{(k)}$ of the matrices $R_k$ coincide with the diagonal elements of the matrices $D_k D_{k-1}^{-1}$ which, as we saw, approach the numbers $\lambda_i - \tau$. More precisely,

$$r_i^{(k)} = (\lambda_i - t_k)\left[1 + O\left(\frac{\varphi_k(\lambda_i)}{\varphi_k(\lambda_{i+1})}\right) + O\left(\frac{\varphi_k(\lambda_{i+1})}{\varphi_k(\lambda_i)}\right)\right] \qquad (21)$$

for $i = 1, 2, \cdots, n - 1$.

For the last diagonal element $r_n^{(k)}$ the asymptotic formula will be simpler. Namely,

$$r_n^{(k)} = (\lambda_n - t_k)\left[1 + O\left(\frac{\varphi_k(\lambda_n)}{\varphi_k(\lambda_{n-1})}\right)\right] \qquad (22)$$

To guarantee second-order convergence we should take $t_k + r_n^{(k)}$ for $t_{k+1}$ until the point where, for a certain $k = k_1$, the number $r_n^{(k_1)}$ will not be equal to zero, given the degree of accuracy required. It is then necessary to strike out the last column from matrix $C_{k_1}$ and to pass to the stepped algorithm for $n - 1$ columns, applying the numbers $t_{k+1} = t_k + r_{n-1}^{(k)}$, $k > k_1$, as translations. After $r_{n-1}^{(k_2)}$ turns out to be practically equal to zero the last column is struck from the matrix $C_{k_2}$ and the process is continued as the stepped one with $n - 2$ columns, etc.

The approximate values for the eigenvalues will be the numbers $t_{k_1}$, $t_{k_2}$, $\cdots$, $t_{k_n}$.

## 79. THE *LR*-ALGORITHM

The *LR*-algorithm (Rutishauser (5)) consists of the following. The matrix $A$ is arranged as the product of two triangular matrices (left and right)

$$A = L_1 R_1 \qquad (1)$$

where the left matrix is taken with unit diagonal elements. The matrix $R_1 L_1$ is then formed and the analogous expansion

$$R_1 L_1 = L_2 R_2 \qquad (2)$$

is constructed for it.

The process is then repeated. Thus as a result of the process two matrix sequences $L_1, L_2, \cdots$, and $R_1, R_2, \cdots$ are constructed, connected by the relationships

$$\begin{aligned} A &= L_1 R_1 , \\ R_1 L_1 &= L_2 R_2 , \\ &\cdots\cdots\cdots , \\ R_{k-1} L_{k-1} &= L_k R_k , \\ &\cdots\cdots\cdots \end{aligned} \qquad (3)$$

We shall establish the connection between the $LR$-algorithm and the tri-angular power method for $C_0 = E$. With this in mind we let

$$L_1 L_2 \cdots L_k = C_k . \tag{4}$$

It is clear that $C_k$ is a left triangular matrix with a unit principal diago-nal. We shall now prove that

$$A C_{k-1} = C_k R_k . \tag{5}$$

For $k = 1$ we have

$$A C_0 = A = L_1 R_1 = C_1 R_1 .$$

We assume that equation (5) is valid for indices less than $k$ and we shall prove that it is also true for the index $k$. We have

$$A C_{k-1} = A C_{k-2} L_{k-1} = C_{k-1} R_{k-1} L_{k-1} = C_{k-1} L_k R_k = C_k R_k .$$

Thus equality (5) is proved.

From equality (5) it follows that the matrices $C_k$ and $L_k$ of the $LR$-algorithm coincide with corresponding matrices of the triangular power method for $C_0 = E$. In this way it is proved that under the conditions of convergence for the triangular power method the diagonal elements of the matrices $R_k$ converge to the eigenvalues of matrix $A$.

The $LR$-algorithm is slightly simpler in its computational scheme than is the triangular power method. However, it is not a self-corrective algorithm. Moreover it is less applicable in determining the eigenvectors of a matrix, since to solve this problem we must reconstruct matrix $C_k$ which reduces to multiplying a large number of triangular matrices together and may be accompanied by an increase in rounding-off errors.

It is worth noting the connection between the $LR$-algorithm and the $QD$-algorithm. That is, the relationships

$$\sigma_i^{(k+1)} + \rho_i^{(k+1)} = \sigma_{i+1}^{(k)} + \rho_i^{(k)} ,$$
$$\sigma_i^{(k+1)} \rho_{i-1}^{(k+1)} = \sigma_i^{(k)} \rho_i^{(k)} ,$$

which lie at the basis of the $QD$-algorithm are equivalent to the following matrix equalities:

$$
\begin{bmatrix}
\rho_0^{(k)} & 1 & & & 0 \\
\rho_1^{(k)} & 1 & & & \\
& \cdot & \cdot & & \\
& & \cdot & \cdot & \\
& & & \cdot & 1 \\
0 & & & & \rho_{n-1}^{(k)}
\end{bmatrix}
\begin{bmatrix}
1 & & & & 0 \\
\sigma_1^{(k)} & 1 & & & \\
& \cdot & \cdot & & \\
& & \cdot & \cdot & \\
& & & \cdot & \\
0 & & & \sigma_{n-1}^{(k)} & 1
\end{bmatrix}
$$
$$
=
\begin{bmatrix}
1 & & & & 0 \\
\sigma_1^{(k+1)} & 1 & & & \\
& \cdot & \cdot & & \\
& & \cdot & \cdot & \\
& & & \cdot & \\
0 & & & \sigma_{n-1}^{(k)} & 1
\end{bmatrix}
\begin{bmatrix}
\rho_0^{(k+1)} & 1 & & & 0 \\
\rho_1^{(k+1)} & 1 & & & \\
& \cdot & \cdot & & \\
& & \cdot & \cdot & \\
& & & \cdot & 1 \\
0 & & & & \rho_{n-1}^{(k+1)}
\end{bmatrix} ,
$$

so that the $QD$-algorithm may be considered as the $LR$-algorithm applied
to a certain initial matrix

$$
J = L_1 R_1 =
\begin{bmatrix}
1 & & & & 0 \\
\sigma_1^{(1)} & 1 & & & \\
 & \cdot & \cdot & \cdot & \\
 & & \cdot & \cdot & \\
0 & & & \sigma_{n-1}^{(1)} & 1
\end{bmatrix}
\begin{bmatrix}
\rho_0^{(1)} & 1 & & & 0 \\
 & \rho_1^{(1)} & 1 & & \\
 & & \cdot & \cdot & \\
 & & & \cdot & 1 \\
0 & & & & \rho_{n-1}^{(1)}
\end{bmatrix}
$$

$$
=
\begin{bmatrix}
\alpha_0 & 1 & & & \\
\beta_1 & \alpha_1 & 1 & & \\
 & \cdot & \cdot & \cdot & \\
 & & \cdot & \cdot & 1 \\
 & & & \beta_{n-1} & \alpha_{n-1}
\end{bmatrix},
$$

where $\alpha_i = \rho_i^{(1)} + \sigma_i^{(1)}$, $\beta_i = \rho_{i-1}^{(1)} \sigma_i^{(1)}$.

We recall that in the $QD$-algorithm we took as initial values for $\rho_i^{(1)}$ and
$\sigma_i^{(1)}$ the coefficients of the binomial relationships for the biorthogonal
algorithm. Thus the numbers $\alpha_i$ and $\beta_i$ which determine the matrix $J$ are
coefficients of the trinomial relationships for the biorthogonal algorithm.
As we have seen, the matrix $J'$ is obtained from matrix $A$ by a similarity
transformation produced by passing to the basis consisting of the vectors
$p_0, \cdots, p_{n-1}$. Therefore the eigenvalues of the matrix $J'$, and consequently
of matrix $J$ as well, coincide with the eigenvalues of matrix $A$.

Applying the $LR$-algorithm to matrices of a general type requires a very
large number of computational operations. The number of operations is
reduced significantly if the initial matrix is a *band matrix*, i.e. a matrix
such that the equalities $a_{ij} = 0$ are satisfied for its elements $a_{ij}$ where $|i - j|$
$> m$, and $m$ is some number significantly less than the order $n$ of the
matrix. In other words, a band matrix has the form

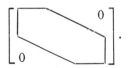

A decrease in the number of operations occurs for this case since all
successive matrices $L_k R_k$ will remain band matrices of the same structure.

The $LR$-algorithm allows for *modification with translations* equivalent to
the corresponding modification of the triangular power method for $C_0 = E$.
In this modification the process is carried out according to the following
rule:

$$
\begin{aligned}
(A - t_1 E) &= L_1 R_1, \\
R_1 L_1 - (t_2 - t_1)E &= L_2 R_2, \\
\cdots \cdots \cdots &\cdots, \\
R_{k-1} L_{k-1} - (t_k - t_{k-1})E &= L_k R_k, \\
\cdots \cdots \cdots &\cdots,
\end{aligned}
\tag{6}
$$

where $L_k$ and $R_k$ are matrices of the former structure.

We write

$$L_1 L_2 \cdots L_k = C_k$$

and we wish to show that

$$(A - t_k E) C_{k-1} = C_k R_k .$$

For $k = 1$ this is true, assuming $C_0 = E$. Assume that the statement is true for indices less than $k$. We shall prove that it is also true for the index $k$. In fact,

$$
\begin{aligned}
(A - t_k E) C_{k-1} &= (A - t_{k-1} E) C_{k-1} - (t_k - t_{k-1}) C_{k-1} \\
&= (A - t_{k-1} E) C_{k-2} L_{k-1} - (t_k - t_{k-1}) C_{k-1} \\
&= C_{k-1} R_{k-1} L_{k-1} - (t_k - t_{k-1}) C_{k-1} \\
&= C_{k-1} (R_{k-1} L_{k-1} - (t_k - t_{k-1}) E) \\
&= C_{k-1} L_k R_k = C_k R_k .
\end{aligned}
$$

We have thus shown that the matrices $C_k$ and $R_k$ coincide with corresponding matrices of the triangular power method with translations $t_1, t_2, \cdots$

Just as in Section 78, we may take the translations so that the corresponding $LR$-method with translations has second-order convergence. That is, we should take

$$t_{k+1} - t_k = r_n^{(k)}$$

where $r_n^{(k)}$ is the last diagonal element of the matrix $R_k$.

## 80.  THE $\Lambda P$-ALGORITHM

The computational scheme for the $\Lambda P$-algorithm[†] approximates that for the $LR$-algorithm but is slightly more laborious at each step. The convergence of the process under the convergence conditions of the triangular power method is second-order.

The algorithm permits us to compute successively the matrices $C_k$ of the triangular power method with the numbers $k$ being powers of two (for $C_0 = E$).

We let

$$A^{2^m} = \Lambda_m \Sigma_m \tag{1}$$

where $\Lambda_m$ is a left triangular matrix with a unit principal diagonal and $\Sigma_m$ is a right triangular matrix. It is clear that $\Lambda_m = C_{2^m}$ in the notation of the triangular power method for $C_0 = E$.

Squaring equation (1) we obtain

$$A^{2^{m+1}} = \Lambda_m \Sigma_m \Lambda_m \Sigma_m . \tag{2}$$

We now expand the matrix $\Sigma_m \Lambda_m$ as the product of a left triangular

---

[†] Rutishauser and Bauer (1).

matrix $\tilde{L}_m$ with a unit principal diagonal and a right triangular matrix $\tilde{R}_m$:

$$\Sigma_m \Lambda_m = \tilde{L}_m \tilde{R}_m . \tag{3}$$

Then we obtain

$$A^{2^{m+1}} = \Lambda_m \tilde{L}_m \tilde{R}_m \Sigma_m$$

from which

$$\Lambda_{m+1} = \Lambda_m \tilde{L}_m ,$$
$$\Sigma_{m+1} = \tilde{R}_m \Sigma_m . \tag{4}$$

These formulas allow us to compute successively the matrices $\Lambda_m$ and $\Sigma_m$ for $m = 0, 1, 2, \cdots$, beginning with matrices $\Lambda_0$ and $\Sigma_0$ which are found by expanding the initial matrix $A$ as the product of two triangular matrices.

The matrices $\Lambda_m$, under the convergence conditions for the triangular power method, will converge to a limit matrix $C$ while convergence will be of order $O(|\lambda_i/\lambda_{i-1}|^{2^m})$, i.e. convergence will be second-order.

Having found matrix $C$ we then construct the matrix $R = C^{-1}AC$, which is not difficult since $C$ is a triangular matrix with a unit principal diagonal. The matrix $R$ will be a right triangular matrix, the same as the one which appeared as the limit for the matrices $R_k$ in the triangular power method. The eigenvalues of matrix $A$ are equal to the diagonal elements of matrix $R$; the eigenvectors are determined using matrices $C$ and $R$ as in the triangular power method.

An insufficiency in the computational scheme just described is the rapid growth (or rapid disappearance) of elements in the matrices $\Sigma_m$ with an increase in $m$. This phenomenon is partly removed by normalizing the right triangular matrices $\Sigma_m$ to a unit principal diagonal at each step. A process with such normalization is also included under the ΛP-algorithm. We present the computational formulas for this algorithm.

We let

$$\Sigma_m = \Delta_m P_m , \tag{5}$$

where $\Delta_m$ is a diagonal matrix and $P_m$ is a right triangular matrix with a unit principal diagonal. Then

$$A^{2^m} = \Lambda_m \Delta_m P_m .$$

Squaring this equation we get

$$A^{2^{m+1}} = \Lambda_m \Delta_m P_m \Lambda_m \Delta_m P_m .$$

We now expand the matrix $P_m \Lambda_m$ as the product of a left triangular matrix with a unit principal diagonal $L_m$, a diagonal matrix $D_m$ and a right triangular matrix with a unit principal diagonal $R_m$:

$$P_m \Lambda_m = L_m D_m R_m . \tag{6}$$

Then

$$A^{2^{m+1}} = \Lambda_m \Delta_m L_m D_m R_m \Delta_m P_m = \Lambda_m (\Delta_m L_m \Delta_m^{-1}) \Delta_m D_m \Delta_m (\Delta_m^{-1} R_m \Delta_m) P_m .$$

It is clear that $\varDelta_m L_m \varDelta_m^{-1}$ is a left triangular matrix with a unit principal diagonal, the matrix $\varDelta_m^{-1} R_m \varDelta_m$ is a right triangular matrix with a unit principal diagonal and the matrix $\varDelta_m D_m \varDelta_m = \varDelta_m^2 D_m$ is diagonal.

Therefore

$$\varLambda_{m+1} = \varLambda_m (\varDelta_m L_m \varDelta_m^{-1}) ,$$
$$P_{m+1} = (\varDelta_m^{-1} R_m \varDelta_m) P_m , \tag{7}$$
$$\varDelta_{m+1} = \varDelta_m^2 D_m ,$$

where

$$P_m \varLambda_m = L_m D_m R_m .$$

The latter formulas are also a characterization of the $\varLambda P$-algorithm. The beginning of the process is determined by expanding the given matrix $A$ as the product $\varLambda_0 \varDelta_0 P_0$.

We observe that each of the factors $\varDelta_m L_m \varDelta_m^{-1}$ and $\varDelta_m^{-1} R_m \varDelta_m$ approaches the unit matrix.

The $\varLambda P$-algorithm ends with the stabilization of the matrices $\varLambda_m$. The matrix $R = C^{-1} A C$ is determined according to the limit matrix $C$. Its diagonal elements give the eigenvalues of matrix $A$; the eigenvectors for matrix $A$ are

$$U_1 = C V_1 , \cdots , U_n = C V_n$$

where $V_1 , \cdots , V_n$ are eigenvectors of matrix $R$.

## 81. ITERATIVE PROCESSES BASED ON THE APPLICATION OF ROTATIONS

Rotations, already considered by us in Section 51 in connection with the transformation of a symmetric matrix to a tridiagonal one, may be used to construct iterative processes which solve the complete eigenvalue problem.

These processes for symmetric matrices consist of a sequence of similarity transformations which produce a diagonal matrix as a limit so that its eigenvalues are determined directly. Such a process was proposed for the first time by Jacobi (1) in 1846. However its actual application became possible only with the development of high-speed computing devices. At the present time a whole series of modifications of Jacobi's method exist.

The elementary step for each Jacobian process consists of a similarity transformation using the matrix

$$T_{ij} = \begin{bmatrix} 1 & & & & & & & \\ & \ddots & & & & & & \\ & & \ddots & & & & & \\ & & & c & \cdots & -s & & \cdots i \\ & & & \ddots & & \ddots & & \\ & & & & 1 & & & \\ & & & \ddots & & \ddots & & \\ & & & s & \cdots & c & & \cdots j \\ & & & & & & \ddots & \\ & & & & & & & 1 \end{bmatrix} \qquad (i < j)$$

where $c^2 + s^2 = 1$.    As we have seen in Section 51 the matrix $T_{ij}$ is the matrix rotating the plane spanned by the $i$th and $j$th coordinate vectors by the angle $\theta$ such that $\cos \theta = c$, $\sin \theta = s$. The matrix $T_{ij}$ is orthogonal so that $T'_{ij} = T_{ij}^{-1}$.

The process as a whole consists of constructing the matrix sequence $A = A^{(0)}$, $A^{(1)}$, $A^{(2)}$, $\cdots$, each of which is obtained from the preceding one using an elementary step. These elementary steps must be chosen so that the matrices $A^{(k)}$ are infinitely close to a diagonal matrix for $k \to \infty$.

The closeness of a symmetric matrix $A$ to diagonal form will be characterized by the number $t^2(A)$ equal to the sum of the squares of all non-diagonal elements of matrix $A$. This closeness may be also characterized by any norm of the matrix $A - D$ where $D$ is the diagonal matrix formed from diagonal elements of matrix $A$.

The Jacobian process will be called *relaxational* or *monotonic* if $t^2(A^{(k)})$ decreases at each step.

We shall explain how we must take the matrix $T_{ij}$ for fixed indices $i$ and $j$ so that $t^2(T'_{ij} A T_{ij})$ will be less than $t^2(A)$.

We write

$$T'_{ij} A T_{ij} = C = (c_{kl}) .$$

We recall (Section 51) that elements of matrix $C$ coincide with elements of matrix $A$ with the exception of the elements lying in rows with the numbers $i$ and $j$ or in columns with the numbers $i$ and $j$. In particular $c_{kk} = a_{kk}$ for $k \neq i$, $k \neq j$ .

Let

$$n^2(A) = \sum_{i,j=1}^{n} a_{ij}^2 = \operatorname{Sp} A'A = \operatorname{Sp} A^2 .$$

It is easy to see that $n^2(A) = n^2(C)$. In fact,

$$n^2(C) = \operatorname{Sp} C^2 = \operatorname{Sp} (T_{ij}^{-1} A T_{ij})^2 = \operatorname{Sp} A^2 = n^2(A) .$$

Moreover let

$$\tilde{C} = \begin{bmatrix} c_{ii} & c_{ij} \\ c_{ji} & c_{jj} \end{bmatrix} ,$$

$$\tilde{A} = \begin{bmatrix} a_{ii} & a_{ij} \\ a_{ji} & a_{jj} \end{bmatrix} .$$

Then $\tilde{C} = \tilde{T}' \tilde{A} \tilde{T}$ where $\tilde{T} = \begin{bmatrix} c & -s \\ s & c \end{bmatrix}$ and consequently $n^2(\tilde{C}) = n^2(\tilde{A})$.

It is clear that

$$t^2(C) - t^2(A) = n^2(C) - \sum_{k=1}^{n} c_{kk}^2 - n^2(A) + \sum_{k=1}^{n} a_{kk}^2 = a_{ii}^2 + a_{jj}^2 - c_{ii}^2 - c_{jj}^2$$

since $n^2(C) = n^2(A)$ and $c_{kk} = a_{kk}$ for $k \neq i$, $k \neq j$. Consequently,

$$t^2(C) - t^2(A) = n^2(\tilde{A}) - 2a_{ij}^2 - n^2(\tilde{C}) + 2c_{ij}^2 = 2(c_{ij}^2 - a_{ij}^2)$$

since $n^2(\tilde{A}) = n^2(\tilde{C})$.

Thus for the process to be relaxational at a given step it is necessary that $|c_{ij}| < |a_{ij}|$. This will happen only if $a_{ij} \neq 0$.

It is easily verified that

$$c_{ij} = a_{ij}(c^2 - s^2) + (a_{jj} - a_{ii})cs = a_{ij}\cos 2\theta + \tfrac{1}{2}(a_{jj} - a_{ii})\sin 2\theta = \alpha_{ij}\sin(2\theta - 2\theta_0)$$

where $\alpha_{ij} = \pm \sqrt{a_{ij}^2 + (a_{jj} - a_{ii})^2/4}$ and $\tan 2\theta_0 = 2a_{ij}/(a_{ii} - a_{jj})$. The angle $\theta_0$ is determined accurately to within a whole multiple of $\pi/2$. We shall assume that $-\pi/4 < \theta_0 \leq \pi/4$. Under such a condition for choosing $\theta_0$

$$\alpha_{ij} = \text{sign}\,(a_{jj} - a_{ii})\sqrt{a_{ij}^2 + (a_{jj} - a_{ii})^2/4}\;.$$

In Fig. 12 is given a graph of $c_{ij}$ as a function of $\theta$ under the assumption $a_{ij} > 0$ and $\theta_0 > 0$. It is clear that for other signs of $a_{ij}$ and $\theta_0$ the corresponding graphs for $c_{ij}$ will have the same form (to within symmetry) relative to the abscissa (for a change in the sign of $a_{ij}$) or relative to the ordinate (for a change in the sign of $\theta_0$).

We may conclude from the graphs that the values of $\theta$ which guarantee the inequality $t^2(C) < t^2(A)$ occupy four intervals in the segment $(-\pi, \pi)$, each of which touches with one of its ends the points $0, \pi/2, -\pi/2, -\pi$ (or $+\pi$). We shall later consider that the angle of rotation $\theta$ is taken at each step of the Jacobian process from the interval $(0, 2\theta_0)$ or $(2\theta_0, 0)$ touching the point $0$ so that $\theta = q\theta_0$ where $0 < q < 2$ and $|\theta_0| \leq \pi/4$.

The greatest drop in value for $t^2(A)$ in one step is obtained for $c_{ij} = 0$, which will be guaranteed when $\theta = \theta_0$. By analogy with the relaxation processes for solving systems of linear equations we shall call Jacobian processes in which $\theta = \theta_0$ at each step processes with *complete relaxation*. Analogously, processes in which $0 < q < 1$ will be called processes with *under-relaxation*; processes for which $1 < q < 2$ will be called processes with *over-relaxation*.

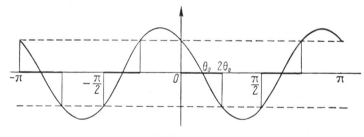

**FIGURE 12**

The choice of index pairs $(i_1, j_1), (i_2, j_2), \cdots, (i_k, j_k)$ for successive steps of the process may be accomplished *a priori* or with control during the process. The most natural *a priori* choice is a cyclic alternation of pairs numbered in one way or another.

We shall turn to a consideration of several Jacobian processes recently investigated.

**1. The classical method of Jacobi.** The rotational matrix $T_{ij}$ is chosen at each step so that the element of the $i$th row and $j$th column of the transformed matrix becomes zero. Index pairs $(i, j)$ are thus chosen at each step so that the non-diagonal element largest in absolute value of the matrix obtained from the previous step of the process is annihilated. Thus Jacobi's method is the method of complete relaxation with direction.

The convergence of the method is easily proved. Let $A^{(k)}$ be the matrix obtained at the $k$th step of the process and let $a_{i_k j_k}^{(k)}$ be its non-diagonal element largest in modulus. Then

$$t^2(A^{(k+1)}) = t^2(A^{(k)}) - 2(a_{i_k j_k}^{(k)})^2 \ .$$

On the other hand, $t^2(A^{(k)}) \leq n(n-1)(a_{i_k j_k}^{(k)})^2$, from which $(a_{i_k j_k}^{(k)})^2 \geq 1/n(n-1)$ $t^2(A^{(k)})$. Consequently,

$$t^2(A^{(k+1)}) \leq t^2(A^{(k)}) \left( 1 - \frac{2}{n(n-1)} \right)$$

and therefore

$$t^2(A^{(k+1)}) \leq t^2(A) \left( 1 - \frac{2}{n(n-1)} \right)^{k+1} \ .$$

Thus $t(A^{(k)}) \to 0$ for $k \to \infty$ which proves the convergence for the Jacobian process.

We shall give the computational formulas for one step of the method, again denoting the initial matrix of the step by $A$ and letting $C = T'_{ij} A T_{ij}$.

As has already been said above, the pair $(i, j)$ is chosen so that $a_{ij}$ is the largest element in modulus of matrix $A$. Then

$$
\begin{aligned}
c_{kl} &= a_{kl} \quad &&\text{(for } k \neq i, \, k \neq j, \, l \neq i, \, l \neq j)\ , \\
c_{ki} &= c_{ik} = ca_{ki} + sa_{kj}\ , \\
c_{kj} &= c_{jk} = -sa_{ki} + ca_{kj} \quad &&\text{(for } k \neq i, \, k \neq j)\ .
\end{aligned}
\tag{1}
$$

Finally,

$$
\begin{aligned}
c_{ii} &= c^2 a_{ii} + 2cs a_{ij} + s^2 a_{jj}\ , \\
c_{jj} &= s^2 a_{ii} - 2cs a_{ij} + c^2 a_{jj}\ , \\
c_{ij} &= (c^2 - s^2) a_{ij} + cs(a_{jj} - a_{ii}) = 0\ .
\end{aligned}
\tag{2}
$$

The numbers $c$ and $s$ are determined from the formulas

$$
\begin{aligned}
c &= \cos \theta\ , \\
s &= \sin \theta\ ,
\end{aligned}
\tag{3}
$$

where

$$\tan 2\theta = \frac{2a_{ij}}{a_{ii} - a_{jj}}\ ,$$

$$|\theta| \leq \frac{\pi}{4}\ . \tag{4}$$

We may replace formulas (3) and (4) by

$$c = \sqrt{\frac{1}{2}\left(1 + \frac{|a_{ii} - a_{jj}|}{d}\right)},$$

$$s = \text{sign}\,[a_{ij}(a_{ii} - a_{jj})]\,\sqrt{\frac{1}{2}\left(1 - \frac{|a_{ii} - a_{jj}|}{d}\right)},\qquad(5)$$

where

$$d = \sqrt{(a_{ii} - a_{jj})^2 + 4a_{ij}^2}\,.$$

Formulas (2) may be replaced by

$$c_{ii} = \frac{a_{ii} - a_{jj}}{2} + \text{sign}\,(a_{ii} - a_{jj})\frac{d}{2}\,,$$

$$c_{jj} = \frac{a_{ii} + a_{jj}}{2} - \text{sign}\,(a_{ii} - a_{jj})\frac{d}{2}\,,$$

$$c_{ij} = c_{ji} = 0\,.$$

After the numbers $c$ and $s$ have been computed matrix $C$ may be constructed directly as well from formulas in Section 51.

If the matrix does not have multiple eigenvalues, then by annihilating the non-diagonal elements to within $\varepsilon$ we obtain that the diagonal elements already approximate the eigenvalues to within $\varepsilon^2$. In fact if the non-diagonal elements of the symmetric matrix

$$A = \begin{bmatrix} a_{11} & a_{12} & \cdots & a_{1n} \\ a_{21} & a_{22} & \cdots & a_{2n} \\ \cdot & \cdot & \cdots & \cdot \\ a_{n1} & a_{n2} & \cdots & a_{nn} \end{bmatrix}$$

are small numbers, then the approximate formula

$$\lambda_i \approx a_{ii} + \sum_{\substack{j=1 \\ j \neq i}}^{n} \frac{a_{ij}^2}{a_{ii} - a_{jj}} \qquad (6)$$

is correct to within small third-order numbers.

In fact, let $a_{ij} = \varepsilon \alpha_{ij}$ where $\varepsilon$ is a small number. We let

$$\lambda_1 = a_{11} + k\varepsilon^2\,.$$

Then to determine $k$ we obtain the equation

$$\begin{vmatrix} -k\varepsilon^2 & \varepsilon\alpha_{12} & \cdots & \varepsilon\alpha_{1n} \\ \varepsilon\alpha_{21} & a_{22} - a_{11} - k\varepsilon^2 & \cdots & \varepsilon\alpha_{2n} \\ \cdot & \cdots\cdots\cdots\cdots & \cdots & \cdot \\ \varepsilon\alpha_{n1} & \varepsilon\alpha_{n2} & \cdots & a_{nn} - a_{11} - k\varepsilon^2 \end{vmatrix} = 0\,.$$

We shall remove $\varepsilon$ from the first row and first column of the determinant and replace $\varepsilon$ by zero in the resultant equation. We obtain

$$\begin{vmatrix} -k & \alpha_{12} & \cdots & \alpha_{1n} \\ \alpha_{21} & a_{22} - a_{11} & \cdots & 0 \\ \cdot & \cdots\cdots\cdots & \cdots & \cdot \\ \alpha_{n1} & 0 & \cdots & a_{nn} - a_{11} \end{vmatrix} \approx 0\,,$$

from which

$$k \approx \frac{\alpha_{12}\alpha_{21}}{a_{11} - a_{22}} + \frac{\alpha_{13}\alpha_{31}}{a_{11} - a_{33}} + \cdots + \frac{\alpha_{1n}\alpha_{n1}}{a_{11} - a_{nn}}$$

accurately to within small numbers of order $\varepsilon$.  Consequently,

$$\lambda_1 = a_{11} + \frac{a_{12}a_{21}}{a_{11} - a_{22}} + \frac{a_{13}a_{31}}{a_{11} - a_{33}} + \cdots + \frac{a_{1n}a_{n1}}{a_{11} - a_{nn}} + O(\varepsilon^3)$$

$$= a_{11} + \sum_{j=2}^{n} \frac{a_{1j}^2}{a_{11} - a_{jj}} + O(\varepsilon^3) \ .$$

Analogously,

$$\lambda_i = a_{ii} + \sum_{\substack{j=1 \\ j \neq i}}^{n} \frac{a_{ij}^2}{a_{ii} - a_{jj}} + O(\varepsilon^3) \ .$$

In cruder form the last formula reduces to

$$\lambda_i = a_{ii} + O(\varepsilon^2) \ .$$

We shall apply Jacobi's method to determine the eigenvalues of matrix (4) in Section 51.

The first step of the process consists in a rotational transformation, using the matrix $T_{14}$.

Computing by formulas (4) we obtain

$$c = s = 0.70710678 \ .$$

Furthermore,

$$T'_{14} A T_{14} = \begin{bmatrix} 1.66 & 0.60811183 & 0.53740115 & 0 \\ 0.60811183 & 1 & 0.32 & 0.01414214 \\ 0.53740115 & 0.32 & 1 & -0.22627417 \\ 0 & 0.01414214 & -0.22627417 & 0.34 \end{bmatrix} \ .$$

Subsequent steps require transformations with the matrices $T_{12}$, $T_{13}$, $T_{34}$, $T_{14}$, $T_{13}$, $T_{24}$, $T_{23}$.  As a result we obtain the matrix

$$\begin{bmatrix} 2.3227487 & -0.00048637 & 0.00001483 & 0.00004994 \\ -0.00048637 & 0.63828393 & 0 & 0.00004930 \\ 0.00001483 & 0 & 0.79670201 & 0.00161630 \\ 0.00004994 & 0.00004930 & 0.00161630 & 0.24226544 \end{bmatrix} \ .$$

Its non-diagonal elements are already so small that we may apply formula (6) in determining the eigenvalues.  After computing we obtain

$$\lambda_1 = 2.3227487 \ + 0.00000014 = 2.32274884$$
$$\lambda_2 = 0.63828393 - 0.00000014 = 0.63828379$$
$$\lambda_3 = 0.79670201 + 0.00000471 = 0.79670672$$
$$\lambda_4 = 0.24226544 - 0.00000471 = 0.24226073.$$

These results were obtained accurately to within $4 \cdot 10^{-8}$ for all eigenvalues.

It is readily shown that when the initial matrix has no multiple eigenvalues Jacobi's method possesses second-order convergence.

We assume that the process is carried out far enough so that all non-diagonal elements of the matrix become less in modulus than the small number $\varepsilon$. Then, as we saw above, the diagonal elements approximate the eigenvalues accurately to within $\varepsilon^2$. Therefore at each step of Jacobi's process the angle of rotation $\theta$ will have order $\varepsilon$, since $\tan 2\theta = 2a_{ij}/(a_{ii} - a_{jj})$, the numerator has order $\varepsilon$, and the denominator—equal to $\lambda_i - \lambda_j + O(\varepsilon^2)$—is bounded from below.

All non-diagonal elements being altered at this step, except for $a_{ij}$, change by a value of order $\varepsilon^2$. In fact,

$$c_{ki} = c_{ik} = a_{ki} \cos \theta + a_{kj} \sin \theta \, ,$$

from which

$$c_{ki} - a_{ki} = a_{ki}(\cos \theta - 1) + a_{kj} \sin \theta = O(\varepsilon^2) \, .$$

Analogously,

$$c_{kj} - a_{kj} = O(\varepsilon^2) \, .$$

In particular, the element annihilated at the previous step becomes at most of the order of magnitude of $\varepsilon^2$. Therefore after at most $n(n-1)/2$ steps of the process all non-diagonal elements have the order of magnitude of $\varepsilon^2$. This also proves the second-order convergence of the process.

The constants for the given estimates are compiled in the work of Henrici (1).

We shall say a few words about finding the eigenvectors. Let the process be carried out up to the point where the matrix

$$A^{(k)} = [\prod_m T'_{i_m j_m}] A [\prod_m T_{i_m j_m}]$$

turns out to be practically diagonal. Then the columns of the matrix $\prod T_{i_m j_m}$ will be the eigenvectors of the initial matrix $A$.

In the case where the eigenvalues of matrix $A$ are pairwise distinct and the non-diagonal elements of the matrix have been made zero (to within $\varepsilon$) it is easy to give a method for computing components of the eigenvectors $U_i$ to within $\varepsilon^2$

Namely,

$$U_i = \prod T_{i_m j_m} V_i$$

where

$$V_i = \left( \frac{a_{1i}}{a_{ii} - a_{11}} , \; \cdots , \; \frac{a_{i-1\,i}}{a_{ii} - a_{i-1\,i-1}} , \; 1, \; \frac{a_{i+1\,i}}{a_{ii} - a_{i+1\,i+1}} , \; \cdots , \; \frac{a_{ni}}{a_{ii} - a_{nn}} \right)' .$$

Here the $a_{ij}$ are elements of the matrix $A^{(k)}$ for which we stopped the process.

As an example we shall determine the eigenvector belonging to the smallest eigenvalue for the matrix considered above in connection with determination of the eigenvalues.

We have

$$a_{44} - a_{11} = -2.08048326,$$
$$a_{44} - a_{22} = -0.39601849,$$
$$a_{44} - a_{33} = -0.55445657$$

so that

$$V_4 = (-0.00002400,\ -0.00012449,\ -0.00291521,\ 1)'\,.$$

We then find successively from formulas (8), Section 51, the vectors

$$T_{23}V_4,\ T_{24}T_{23}V_4,\ T_{13}T_{24}T_{23}V_4,\ \cdots,$$
$$T_{14}T_{12}T_{13}T_{34}T_{14}T_{13}T_{24}T_{23}V_4 = U_4$$

placing their components successively in columns I-IV of Table VIII. 7.

In columns V and VI are given the coefficients $c$ and $s$ respectively. In the last row is given the eigenvector $U_4$ normalized to a unit first component.

The classical method of Jacobi requires at each step of the process a choice of the largest non-diagonal element. To carry out this operation on a high-speed computer requires a considerable expenditure of machine

**TABLE VIII.7   Determination of an Eigenvector by Jacobi's Method**

| $i$ | $j$ | I | II | III | IV | V | VI |
|---|---|---|---|---|---|---|---|
| 2 | 3 | −0.00002400 | −0.00003556 | −0.00291765 | 1 | 0.99953524 | 0.03048475 |
| 2 | 4 | −0.00002400 | −0.02812808 | −0.00291765 | 0.99960433 | 0.99960533 | 0.02809253 |
| 1 | 3 | 0.00004740 | −0.02812808 | −0.00291736 | 0.99960433 | 0.99970055 | 0.02447059 |
| 1 | 4 | 0.03967421 | −0.02812808 | −0.00291736 | 0.99881670 | 0.99921394 | −0.03964253 |
| 3 | 4 | 0.03967421 | −0.02812808 | 0.40999603 | 0.91079448 | 0.91066679 | −0.41314164 |
| 1 | 3 | −0.13986866 | −0.02812808 | 0.38743715 | 0.91079448 | 0.90350564 | 0.42857618 |
| 1 | 2 | 0.10581153 | −0.09569928 | 0.38743715 | 0.91079448 | 0.85934868 | 0.51139013 |
| 1 | 4 | −0.71884900 | −0.09569928 | 0.38743715 | 0.56920890 | 0.70710678 | 0.70710678 |
| | | 1 | 0.133128 | −0.538969 | −0.791834 | | |

time.   Thus cyclic Jacobian processes and particularly cyclic processes with barriers turn out to be more convenient.

**2.   Cyclic Jacobian processes.**[†]   In carrying out a cyclic process a specific numbering of the pairs $(i, j)$ is chosen and annihilation of non-diagonal elements occurs in cycles; during each cycle the elements $a_{ij}$ are annihilated in turn according to the numbering of pairs of indices.

An elementary step of the process differs in no way from an elementary step of Jacobi's classical method, so that the formulas introduced above are still the working formulas.   As was pointed out in an article by Gregory (3), the method's convergence has been established in an unpublished work of Forsythe and Henrici.

After the non-diagonal elements become small enough the convergence

† Gregory (3).

of the process becomes second-order; we may easily convince ourselves of this using the arguments introduced for Jacobi's method.

The most natural way to number the pairs is according to rows, from left to right and from top to bottom, namely

$$(1, 2), (1, 3), \cdots, (1, n), (2, 3), \cdots, (2, n), \cdots, (n - 1, n)$$

or according to columns from top to bottom and from left to right

$$(1, 2), (1, 3), (2, 3), (1, 4), (2, 4), (3, 4), \cdots, (1, n), (2, n), \cdots, (n - 1, n) .$$

**3. Cyclic processes with barriers.**[†]     The cyclic method with barriers occupies an intermediate position between Jacobi's classical method and the cyclic method. An inadequacy in the cyclic method is the fact that during the process it is necessary to annihilate small non-diagonal elements, although large ones are still present in the matrix. This inadequacy is removed by introducing a "barrier". That is, we introduce the sequence of numbers $\alpha_1, \alpha_2, \cdots$, which is monotonically decreasing to zero and in annihilating successively the non-diagonal elements we omit those steps for which we would have to annihilate elements less than $\alpha_1$. After all the non-diagonal elements become no larger than $\alpha_1$ in modulus, the "barrier" is moved—the number $\alpha_1$ is replaced by the number $\alpha_2$, and so forth.

It is easy to establish that the process with barriers converges.

Actually the "barrier" $\alpha_1$ will pass through no more than $t^2(A)/\alpha_1^2$ elementary steps, since for each step until the barrier is overcome the number $t^2(A)$ decreases by no less than $\alpha_1^2$. In an analogous way the subsequent barriers $\alpha_2, \alpha_3$, etc. will be overcome in a finite number of steps. Since $\alpha_k \to 0$, then after a sufficiently large number of elementary steps all non-diagonal elements of the matrix become as small as we wish. This also proves that the process converges.

As before, the convergence of the process will be second-order after a certain point when the eigenvalues are pairwise distinct.

**4. Jacobian processes with rational formulas.**     An elementary step of these processes differs from one of a process with complete relaxation in the choice of angle for the rotation. That is, instead of the formula $\tan 2\theta_0 = 2a_{ij}/(a_{ii} - a_{jj})$ we use the formula

$$\tan \frac{\theta}{2} = \frac{a_{ij}}{2(a_{ii} - a_{jj})} = \alpha, \text{ if } \left| \frac{a_{ij}}{2(a_{ii} - a_{jj})} \right| < \sqrt{2} - 1 \approx 0.414 ,$$

$$\theta = \frac{\pi}{4}, \text{ if } \left| \frac{a_{ij}}{2(a_{ii} - a_{jj})} \right| \geq \sqrt{2} - 1 .$$

For small angles $\theta \approx \theta_0$ the angle $\theta$ is always a little larger than $\theta_0$. It is not hard to verify that $| \theta_0 | < | \theta | < | 2\theta_0 |$ so that the indicated choice for the angle of rotation guarantees over-relaxation.

For computing the numbers $c$ and $s$ we use the rational formulas

---

[†] Pope, Tompkins (1).

$$c = \frac{1-\alpha^2}{1+\alpha^2}, \; s = \frac{2\alpha}{1+\alpha^2} \quad (\alpha < \sqrt{2}-1) \, ,$$

$$c = \frac{\sqrt{2}}{2}, \quad s = \pm\frac{\sqrt{2}}{2} \quad (\alpha \geq \sqrt{2}-1) \, .$$

The rotational transformation itself should be carried out according to formulas in Section 51.

The pairs may either be chosen cyclically or according to the indices of the non-diagonal element largest in modulus.

Certain other Jacobian processes have also been described in the literature of mathematics.

The methods described are transformed with hardly a change by diagonalization of Hermitian matrices. Instead of elementary rotations one makes transformations with unitary matrices of the form

$$T_{ij} = \begin{bmatrix} 1 & & & & & & & \\ & \ddots & & & & & & \\ & & \ddots & & & & & \\ & & & c_1 & \cdot & \cdot & \cdot & -s_1 & & \cdots i \\ & & & \cdot & \ddots & & & \cdot \\ & & & \cdot & & 1 & & \cdot \\ & & & \cdot & & & \ddots & \cdot \\ & & & s_2 & \cdot & \cdot & \cdot & c_2 & & \cdots j \\ & & & & & & & & \ddots \\ & & & & & & & & & 1 \end{bmatrix}$$

(unitary rotations),

The conditions that a matrix be unitary

$$|c_1|^2 + |s_2|^2 = 1 \, ,$$
$$|s_1|^2 + |c_2|^2 = 1 \, ,$$
$$-\bar{c}_1 s_1 + \bar{s}_1 c_1 = 0$$

permit us to choose the numbers $c_1$, $s_1$, $c_2$ and $s_2$ as

$$c_1 = \cos\theta e^{i\alpha} \, , \qquad s_1 = \sin\theta e^{i\beta} \, ,$$
$$s_2 = \sin\theta e^{i\gamma} \, , \qquad c_2 = \cos\theta e^{i\delta} \, .$$

Here $\theta$, $\alpha$, $\beta$, $\gamma$ and $\delta$ are real numbers where

$$\alpha - \beta - \gamma + \delta \equiv 0 (\text{mod } 2\pi) \, .$$

We shall not give the corresponding working formulas here.

In the case of an arbitrary matrix it is no longer possible in general to bring the matrix to diagonal form with similarity transformations using rotations.

However, the following theorem of J. Schur is true. For any complex matrix $A$ there exists a unitary matrix $U$ such that $U^{-1}AU$ is an upper triangular matrix.

For proof we consider the matrix $A$ as the matrix of an operator in a certain orthonormal basis and we pass to a new orthonormal basis which includes one of the eigenvectors of the matrix. In this new basis there will correspond to the operator the matrix

$$U_1^{-1}AU_1 = \begin{bmatrix} \lambda_1 & b_{12} & \cdots & b_{1n} \\ 0 & b_{22} & \cdots & b_{2n} \\ \cdot & \cdot & \cdots & \cdot \\ 0 & b_{n2} & \cdots & b_{nn} \end{bmatrix}.$$

Applying the same argument to the matrix of order $n - 1$

$$\begin{bmatrix} b_{22} & \cdots & b_{2n} \\ \cdot & \cdots & \cdot \\ b_{n2} & \cdots & b_{nn} \end{bmatrix}$$

we obtain

$$U_2^{-1}U_1^{-1}AU_1U_2 = \begin{bmatrix} -\lambda_1 & c_{12} & c_{13} & \cdots & c_{1n} \\ 0 & \lambda_2 & c_{23} & \cdots & c_{2n} \\ 0 & 0 & c_{33} & \cdots & c_{3n} \\ \cdot & \cdot & \cdot & \cdots & \cdot \\ 0 & 0 & c_{n3} & \cdots & c_{nn} \end{bmatrix}.$$

Continuing the process further, and remembering that the product of unitary matrices is a unitary matrix, we arrive at the required result.

Schur's theorem lets us assume that by choosing a sequence of unitary rotation matrices in a suitable way we may change a given matrix to triangular form using similarity transformations.

Two such processes are described in the work of Greenstadt (1) and Lotkin (4). In the first of these it is proposed that sub-diagonal elements be annihilated in turn by appropriate unitary rotations. A convergence proof for the process is lacking; only an indication is given via experimental verification of the method. In the second work the rotations are selected so that the sum of the squares of moduli for sub-diagonal elements decreases at each step. The process turns out to be quite involved; one must solve an auxiliary cubic equation at each step.

We shall not describe these processes in detail.

## 82. SOLUTION OF THE COMPLETE EIGENVALUE PROBLEM USING SPECTRAL ANALYSIS OF SUCCESSIVE ITERATIONS

This method, belonging to Lanczos (7), consists of the following. Let a matrix $A$ be symmetric[†] with its eigenvalues contained in the interval $(m, M)$. We introduce the matrix

---

[†] These arguments remain in force even in the case where matrix A is not symmetric, if it has pairwise distinct real eigenvalues.

$$B = \frac{M+m}{M-m}\left(E - \frac{2}{M+m}A\right), \tag{1}$$

whose eigenvalues $\mu_1, \cdots, \mu_n$ already belong to the interval $(-1, 1)$. If matrix $A$ is positive definite $m$ may be taken, for example, as zero.

We compute the vector sequence $X_0$, $X_1 = BX_0$,

$$X_k = 2BX_{k-1} - X_{k-2} \qquad (k = 2, 3, \cdots, N).$$

Let

$$X_0 = U_1 + \cdots + U_r \tag{2}$$

where $U_1, \cdots, U_r$ are eigenvectors of matrix $B$ and consequently of matrix $A$ as well.

Then

$$X_k = T_k(\mu_1)U_1 + \cdots + T_k(\mu_r)U_r \tag{3}$$

where $T_k(t) = \cos k \arccos t$.

We let $\mu_i = \cos \theta_i$ for $0 < \theta_i < \pi$. It is clear that $T_k(\mu_i) = \cos k \arccos \mu_i = \cos k\theta_i$ so that

$$X_k = \cos k\theta_1 U_1 + \cdots + \cos k\theta_r U_r. \tag{4}$$

From here any component of the vector $X_k$, denoted here by $x_k$, will have the form

$$x_k = a_1 \cos k\theta_1 + \cdots + a_r \cos k\theta_r \tag{5}$$

where $a_1, \cdots, a_r$ are certain defined (but unknown to us beforehand) numbers equal to the chosen components of the vectors $U_1, \cdots, U_r$.

Lanczos proposes to determine the angles $\theta_1, \cdots, \theta_r$ (and consequently the eigenvalues $\mu_1, \cdots, \mu_r$ as well) by subjecting the sequence $x_0, x_1, \cdots, x_N$ to harmonic analysis (as a check we must take likewise a second sequence consisting of other components of the iterations $X_k$). Thereupon the amplitudes $a_1, \cdots, a_r$ are also determined in passing; these are the selected components of the eigenvectors $U_1, \cdots, U_r$. After the eigenvalues have been determined, the remaining components of the eigenvectors are found, as we shall see below, without carrying out a complete harmonic analysis of the sequence of corresponding vector components $x_0, x_1, \cdots, x_N$.

We consider first the case where

$$x_k = a \cos k\theta, \quad k = 0, \cdots, N. \tag{6}$$

Let

$$y_m = \frac{x_0}{2} + x_1 \cos \frac{\pi m}{N} + x_2 \cos \frac{2\pi m}{N}$$

$$+ \cdots + x_{N-1} \cos \frac{(N-1)\pi m}{N} + x_N \cos \frac{N\pi m}{N} \qquad (m = 0, \cdots, N). \tag{7}$$

The sequence $y_0, y_1, \cdots, y_N$ is the "Fourier transform" of the sequence

$x_0, x_1, \cdots, x_N$. It is not hard to verify that

$$y_m = (-1)^m \frac{a}{2} \sin N\theta \sin \theta \frac{1}{\cos \pi m/(N) - \cos \theta} . \tag{8}$$

We shall investigate the behavior of $y_m$ with respect to changes in $m$ for a fixed $\theta$. We assume first that

$$\theta = \frac{\pi}{N} p , \tag{9}$$

where $p$ is a whole number.

In this case the denominator of expression (8) becomes zero for $m = p$ and $\sin N\theta = 0$. Consequently $y_m = 0$ for $m \neq p$ and, as is easy to compute, $y_p = aN/2$. Thus among the values $y_m$ is encountered a unique non-zero value $y_p$.

Analogously, if

$$x_k = a_1 \cos \theta_1 + \cdots + a_r \cos \theta_r \tag{10}$$

and

$$\theta_i = \frac{\pi}{N} p_i \tag{11}$$

where $p_i$ are whole numbers, then among the values $y_m$ are $r$ non-zero values $y_{p1} = a_1 N/2, \cdots, y_{pr} = a_r N/2$ and all remaining values will be zero. This allows us to determine both the angles $\theta_1, \cdots, \theta_r$ (in terms of the numbers $p_1, \cdots, p_r$) and the amplitudes $a_1, \cdots, a_r$ (in terms of the values $y_{p1}, \cdots, y_{pr}$).

We now turn to an investigation of the case

$$x_k = a \cos k\theta$$

under the assumption that $\theta \neq (\pi/N)p$ for integral $p$. Let

$$\theta = \frac{\pi}{N} (p + \tau) \qquad (0 < \tau < 1) . \tag{12}$$

In this case $\cos \pi m/N - \cos \theta > 0$ for $0 \leq m \leq p$ and $\cos \pi m/N - \cos \theta < 0$ for $p + 1 \leq m \leq N$. Therefore the values $y_m$ will have alternating signs as $m$ goes from 0 to $p$. The signs of $y_p$ and $y_{p+1}$ coincide and then alternate again. The greatest absolute values for the $y_m$ will be for $m = p$ and $m = p + 1$ and will decrease as $m$ moves away from $p$. (cf. fig. 13)

Such behavior of the sequence $y_m$ makes it possible to determine the number $p$ and also, as a consequence of (13), the value $\theta$ accurately to within $\pi/N$.

After determining $p$ we may obtain a more accurate value for $\theta$ from the easily verified formula

$$\cos \theta = \frac{y_{p+1} \cos \pi(p + 1)/N + y_p \cos \pi p/N}{y_{p+1} + y_p} . \tag{13}$$

In passing to the general case

$$x_k = a_1 \cos k\theta_1 + \cdots + a_r \cos k\theta_r \tag{14}$$

the situation becomes slightly more involved.   In this case

$$y_m = y_m^{(1)} + y_m^{(2)} + \cdots + y_m^{(r)}$$

where

$$y_m^{(i)} = (-1)^m \frac{a_i}{2} \sin N\theta_i \sin \theta_i \frac{1}{\cos \pi m/N - \cos \theta_i}$$

$$\theta_i = \frac{\pi}{N}(p_i + \tau_i), \qquad (0 \le \tau_i < 1) . \tag{15}$$

For $m$ close to $p_i$ the element $y_m^{(i)}$ will generally predominate over the others.   In particular, $y_{p_i} \approx y_{p_i}^{(i)}$ and $y_{p_i+1} \approx y_{p_i+1}^{(i)}$.   Therefore $y_{p_i}$ and $y_{p_i+1}$ will be of one sign and this allows us to determine $p_i$ and thus $\theta_i$ accurately to within $\pi/N$.

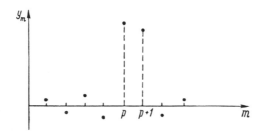

FIGURE 13

However in improving the accuracy of $\theta_i$ by the method indicated above, the influence of other elements $y_m^{(j)}$, $j \ne i$, may still be too significant and it is impossible to ignore them.

Lanczos proposed considering, along with the numbers $y_m$, the numbers

$$z_m = y_{m-1} - 2y_m + y_{m+1} . \tag{16}$$

It is clear that

$$z_m = z_m^{(1)} + \cdots + z_m^{(i)} + \cdots + z_m^{(r)} . \tag{17}$$

As $m$ moves away from $p_i$ the value of $z_m^{(i)}$ decreases much more quickly than $y_m^{(i)}$ so that peaks for $z_m$ are expressed more distinctly than for $y_m$.

We shall compute an approximate value of $z_m^{(i)}$ for $m$ close to $p_i$.   In doing this we first replace the exact formula for $y_m^{(i)}$ by the more convenient approximate formula which is valid for $m$ close to $p_i$.

We let $m = p_i + q$ where $q$ is a small whole number.   We have

$$y_m^{(i)} = (-1)^{p_i+q} \frac{a_i}{2} \sin N\frac{\pi(p_i + \tau_i)}{N} \sin \theta_i \; \frac{1}{\cos \dfrac{(p_i + q)\pi}{N} - \cos \theta_i}$$

$$\approx (-1)^{p_i+q} \frac{a_i}{2} (-1)^{p_i} \sin \pi\tau_i \sin \theta_i \; \frac{1}{\sin \theta_i \left( \theta_i - \dfrac{(p_i + q)\pi}{N} \right)} \qquad (18)$$

$$= (-1)^q \frac{a_i N \sin \pi\tau_i}{2\pi} \frac{1}{\tau_i - q} \,.$$

Correspondingly,

$$z_m^{(i)} = \frac{(-1)^q a_i N \sin \pi\tau_i}{2\pi} \left( \frac{1}{\tau_i - q + 1} - \frac{2}{\tau_i - q} + \frac{1}{\tau_i - q - 1} \right)$$

$$= \frac{(-1)^q a_i N \sin \pi\tau_i}{\pi} \frac{1}{(\tau_i - q)(\tau_i - q + 1)(\tau_i - q - 1)} \,, \qquad (19)$$

from which it follows that $z_m^{(i)}$ decreases with $1/|q|^3$ for increasing $|q|$. On this basis we shall consider that

$$z_{p_i} \approx z_{p_i}^{(i)} \approx \frac{a_i N \sin \pi\tau_i}{\pi} \frac{1}{\tau_i(\tau_i + 1)(\tau_i - 1)}$$

and

$$z_{p_i+1} \approx z_{p_i+1}^{(i)} \approx - \frac{a_i N \sin \pi\tau_i}{\pi} \frac{1}{(\tau_i - 1)\tau_i(\tau_i - 2)} \,.$$

From here

$$\frac{z_{p_i}}{z_{p_i+1}} \approx - \frac{\tau_i - 2}{\tau_i + 1} \,,$$

and consequently,

$$\tau_i \approx \frac{2 - z_{p_i}/(z_{p_i+1})}{1 + z_{p_i}/(z_{p_i+1})} \,. \qquad (20)$$

The accuracy of this approximate equality will be satisfactory if the angles $\theta_i$ are not very close to each other.

Having determined $\tau_i$ we find the amplitudes $a_i$ by the formula

$$a_i \approx \frac{\pi}{N \sin \pi\tau_i} \tau_i(\tau_i + 1)(\tau_i - 1) z_{p_i} \,. \qquad (21)$$

However the amplitudes $a_i$ may also be computed from the formulas

$$a_i \approx \frac{2\pi}{N \sin \pi\tau_i} \tau_i(1 - \tau_i) [y_{p_i} + y_{p_i+1}] \,. \qquad (22)$$

This last formula, slightly more precise than the preceding one, requires a knowledge of only two adjacent values of $y_m$. It makes sense to use it to find all the components of the eigenvector $U_i$ after the corresponding eigenvalue $\mu_i$ has been computed.

That is, it is necessary to compute the vectors $y_{p_i}$, $y_{p_i+1}$ for $i = 1, \cdots, r$ and to apply formula (22) to all their components.

When using the method in practice one should take fairly large numbers for $N$, in any case exceeding the order of the matrix by as many times as (10, 20). For convenience in computing one should also take the number $N$ as a multiple of 180.

The method requires a very large number of operations (several million multiplications for $N = 1080$) and it may be used only on high-speed computing devices. At the present time it has been applied only for matrices of low order.

In the work of Lanczos the question of the influence of a pair of close eigenvalues and of eigenvalues close in modulus on the course of the process has been considered.

# Chapter IX

# UNIVERSAL ALGORITHMS

The present chapter is devoted to the theory and description of computational schemes known as universal algorithms as applied to the problem of solving a system of linear equations $AX = F$ or, in prepared form, $X = BX + G$.

By a universal algorithm we mean an iterative process, generally not stationary, which is carried out according to formulas of the form

$$X^{(k+1)} = X^{(k)} + h^{(k)}(A)[F - AX^{(k)}]$$

or in the case of a prepared system according to the formulas

$$X^{(k+1)} = X^{(k)} + f^{(k)}(B)[BX^{(k)} + G - X^{(k)}],$$

in which the sequence of polynomials $h^{(k)}(t)$ (or $f^{(k)}(t)$) is constructed once and for all for a broad class of matrices with a given distribution of eigenvalues, for example, for the class of symmetric matrices whose eigenvalues are contained in a known interval.

A characteristic peculiarity of universal algorithms is the fact that their rate of convergence does not depend on the order of the matrix but is determined only by the matrix condition.

The simplest universal algorithm is the algorithm for which

$$h^{(k)}(t) = h = \text{const.}$$

Such an algorithm is nothing less than the process of successive approximations applied to the system $AX = F$, prepared in the form

$$X = (E - hA)X + hF.$$

The basic principle for constructing universal algorithms is the idea of "suppressing components". In the next paragraph we explain it for the simplest case.

## 83. THE GENERAL IDEA OF COMPONENT SUPPRESSION

Let $AX = F$ be a system of linear equations where it is known that all eigenvalues of matrix $A$ are real, positive, and distinct. Let $X$ be a certain initial vector and let

$$X' = X + h(A)[F - AX] \qquad (1)$$

where $h(t)$ is a certain polynomial (it may have degree zero).

As we saw in Chapter VII, the successive approximations for applying gradient iterative processes in the case of a positive definite matrix $A$ are determined from formula (1) where the polynomial $h(t)$ may change from step to step. In the gradient methods considered above the coefficients of the polynomials depend essentially on both matrix $A$ and on the initial approximation, so that the methods described in Chapter VII are not universal in the sense of the above determination.

To construct universal algorithms we look at formula (1) from the following point of view. We denote by $Y$ and $Y'$ the corresponding error vectors for the approximations $X$ and $X'$. Then

$$Y' = (E - Ah(A))Y = g(A)Y,$$

where

$$g(t) = 1 - th(t). \qquad (2)$$

Let $\lambda_1, \cdots, \lambda_n$ be pairwise distinct eigenvalues of matrix $A$ and let $U_1, \cdots, U_n$ be their corresponding eigenvectors. Let

$$Y = a_1 U_1 + \cdots + a_n U_n.$$

Then

$$Y' = a_1 g(\lambda_1) U_1 + \cdots + a_n g(\lambda_n) U_n. \qquad (3)$$

Thus in passing from approximation $X$ to approximation $X'$, the error vector components, expanded in terms of eigenvectors of matrix $A$, are multiplied respectively by the values $g(\lambda_1), \cdots, g(\lambda_n)$, which are called quite naturally *damping factors*. Passing from the vector $X$ to the vector $X'$ may be considered "successful" if one or several of the factors $g(\lambda_1), \cdots, g(\lambda_n)$ are very small and the remainder not too large. Having "suppressed" one or several of the error vector components in this way, we may go on to the next step of the process, ordering them so that other components are suppressed in this step.

The choice of the polynomial $g(t)$ for one step of the process is quite arbitrary. A natural condition associated with the choice of $g(t)$ is the requirement $g(0) = 1$.

The ideal choice of the polynomial $g(t)$ for a given matrix $A$ is one for which $g(\lambda_1) = \cdots = g(\lambda_n) = 0$, since with such a choice one step of the process already leads to the exact solution. We arrive at this essentially by applying the method of conjugate gradients.

We now give criteria which should guide us in the choice of universal polynomial $g(t)$, starting with the natural assumption that we know nothing about the eigenvalue distribution for matrix $A$ except for the interval $(0, M)$ in which these values are contained.

Let the graph of $g(t)$ be given by Fig. 14.

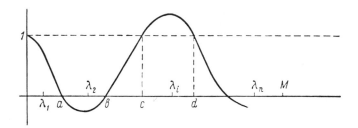

**FIGURE 14**

Then the damping factors $g(\lambda_1), \cdots, g(\lambda_n)$ will be the ordinates of the graph with abscissas $\lambda_1, \cdots, \lambda_n$.

It is clear that damping will be most effective in the neighborhood of the roots for the polynomial $g(t)$, less effective in the neighborhood of points where $g(t)$ is close to $\pm 1$, and instead of damping, the components will increase at points where $|g(t)|$ is greater than 1. Thus in Fig. 14 damping is strong for eigenvalues close to the points $a$ and $b$ but does not hold at all for eigenvalues in the interval $(c, d)$. Thus at each step of the process it is advisable to choose the polynomial $g(t)$ so that it runs as close as possible to the abscissa axis and so that $|g(t)|$ does not exceed 1 anywhere in the interval $(0, M)$. It is clear that the greatest difficulty in this will be to suppress components corresponding to eigenvalues close to zero, since $g(0) = 1$. This is quite natural, since the presence of eigenvalues close to zero for matrix $A$ indicates ill-condition of the system.

The choice of the polynomial $h(t)$ (and of $g(t)$ as well) may be made by applying another criterion based on a comparison of components of the error vector $Y'$ with components of the residual $r = F - AX = AY$ for the approximation $X$. Let

$$r = b_1 U_1 + \cdots + b_n U_n .$$

Then

$$Y = \frac{b_1}{\lambda_1} U_1 + \cdots + \frac{b_n}{\lambda_n} U_n$$

and

$$Y' = \frac{b_1}{\lambda_1} g(\lambda_1) U_1 + \cdots + \frac{b_n}{\lambda_n} g(\lambda_n) U_n$$

$$= b_1 \left( \frac{1}{\lambda_1} - h(\lambda_1) \right) U_1 + \cdots + b_n \left( \frac{1}{\lambda_n} - h(\lambda_n) \right) U_n . \tag{4}$$

With such an approach, this choice of the polynomial $h(t)$ becomes natural in order for its values in the interval $(0, M)$ to diverge as little as possible from values of the function $1/t$.

Both of the criteria indicated are actually close to each other. That is, fulfilling one of them involves fulfilling the other to a greater or lesser degree. However the second criterion gives stronger requirements to suppress components corresponding to small eigenvalues. In fact if

$$\left| \frac{1}{\lambda_i} - h(\lambda_i) \right| < \delta \qquad (\delta > 0) ,$$

then

$$| g(\lambda_i) | = | 1 - \lambda_i h(\lambda_i) | < \delta \lambda_i .$$

It is advisable to be guided by the first criterion in those cases where using formula (1) is an elementary step in an iterative process, since in carrying out several steps of the iterative process the damping factors are simply multiplied. If formula (1) is applied only once in all, then it makes sense to use the second criterion if there is a basis for assuming that components of the initial residual (for example, the constants $F$) have coefficients of the same order in their expansion in eigenvectors.

In the case where some additional information on the eigenvalue distribution for matrix $A$ exists, we may pose the question of a best choice of the polynomial $g(t)$ of a given degree, in the sense of the first or second criteria. Thus if it is known that all the eigenvalues are contained in the interval $(m, M)$, then the best polynomial in the sense of the first criterion will be the polynomial $g(t)$ which diverges least from zero on the interval $(m, M)$, normalized by the condition $g(0) = 1$; the best polynomial in the sense of the second criterion will be the polynomial $h(t)$ which diverges least from $1/t$ in the same interval. We shall consider below some universal algorithms based on the realization of such a choice.

For a system prepared in the form

$$X = BX + G \tag{5}$$

with the condition that all eigenvalues of matrix $B$ lie in the interval $(-1, 1)$ one step of the universal algorithm is carried out according to the formula

$$X' = X + f(B)[BX + G - X] . \tag{6}$$

The error vectors for two adjacent approximations are connected in this case by the relationship

$$Y = Y + f(B)(BY - Y) = [E - (E - B)f(B)]Y = e(B)Y ,$$

where $e(t) = 1 - (1 - t)f(t)$. Thus the polynomial $e(t)$ must here satisfy the requirement $e(1) = 1$. The damping factors will be the values $e(\mu_1), \cdots, e(\mu_n)$, where $\mu_1, \cdots, \mu_n$ are eigenvalues of matrix $B$. Criteria for the choice of polynomial $e(t)$ or of the polynomial $f(t)$ are, first, minimal divergence from zero of values for $e(t)$ in the interval $(-1, 1)$ normalized by the condition $e(1) = 1$ or, second, minimal divergence of $f(t)$ from the function $1/(1 - t)$ in the same interval.

### 84. L. A. LUSTERNIK'S METHOD FOR ACCELERATING THE CONVERGENCE OF THE SUCCESSIVE APPROXIMATIONS METHOD IN SOLVING A SYSTEM OF LINEAR EQUATIONS[†]

In solving the system $X = BX + G$ by the successive approximations method certain information on the eigenvalue distribution for matrix $B$ is obtained right in the course of the process. Namely, if the eigenvalue $\mu_1$ largest in modulus for matrix $B$ is isolated from the remaining eigenvalues, then it may be determined from the ratios of corresponding components of the vectors $X^{(k+1)} - X^{(k)}$ and $X^{(k)} - X^{(k-1)}$. In fact $X^{(k+1)} - X^{(k)} = B^k(X_1 - X_0)$ and $X^{(k)} - X^{(k-1)} = B^{k-1}(X_1 - X_0)$ where $X_0$ is the initial approximation.

Knowing $\mu_1$ we may annihilate the coefficient for the eigenvector $U_1$ in the error vector, starting from the approximation $X^{(k)}$ which has already been constructed by the successive approximations method. For this it is sufficient, for passing from $X^{(k)}$ to the next approximation $X^{(k+1)}$, to take as the polynomial $f^{(k)}(t)$ the constant $1/(1 - \mu_1)$. Actually the corresponding polynomial $e^{(k)}(t)$ is equal to $1 - (1 - t)/(1 - \mu_1)$ so that $e^{(k)}(\mu_1) = 0$. The last step will then be carried out according to the formula

$$\bar{X}^{(k+1)} = X^{(k)} + \frac{1}{1 - \mu_1}(BX^{(k)} + G - X^{(k)})$$

$$= X^{(k)} + \frac{1}{1 - \mu_1}(X^{(k+1)} - X^{(k)}) \tag{1}$$

where $X^{(k+1)} = BX^{(k)} + G$ is the next approximation after $X^{(k)}$ in the successive approximations method.

If the successive approximations method is carried out from the formula

$$X^{(k)} = \sum_{i=0}^{k-1} B^i G ,$$

then formula (1) acquires the still simpler form

$$\bar{X}^{(k+1)} = X^{(k)} + \frac{1}{1 - \mu_1} B^k G . \tag{2}$$

The error vector components for approximations found by formulas (1) and (2) will obviously have order $O(|\mu_2|^k)$ where $\mu_2$ is the eigenvalue of

† L. A. Lusternik (1).

matrix $B$ which is next after $\mu_1$ in modulus.

We shall illustrate the method described with an example from Section 30. On the basis of Table III.1 in Section 53 (example 3) it was determined that $\mu_1 = 0.4800$ from the component ratios of the 14th and 13th iterations of the vector $G = (0.76, 0.08, 1.12, 0.68)'$ with matrix $B$. From the same table

$$X^{(13)} = (1.53490847, 0.12200958, 1.97509985, 1.41289889)'.$$

We compute

$$\frac{B^{14}G}{1 - 0.4800} = (0.00005656, 0, 0.00005656, 0.00005656)'.$$

Thus

$$\bar{X}^{(14)} = (1.534965, 0.122010, 1.975156, 1.412955)'.$$

The given solution coincides to within $1 \cdot 10^{-6}$ with the one found by Gauss's method.

The ratios of the 7th and 6th iterations give (Section 53) for $\mu_1$ the value $\mu_1 = 0.4792$.

Since

$$X^{(6)} = (1.52533, 0.12201, 1.96551, 1.40333)'$$

and

$$\frac{B^7G}{0.5208} = (0.00962, 0.00002, 0.00963, 0.00960)',$$

then

$$\bar{X}^{(7)} = (1.53495, 0.12203, 1.97514, 1.41293)'.$$

From Table III.1 it is apparent that $\bar{X}^{(7)}$ is closer to the exact value than is $X^{(14)}$.

Lusternik's method may be applied to a cyclic one-step process as well, since this method, applied to the system $X = BX + G$, is equivalent to the successive approximations method for a certain equivalent system.

## 85. COMPONENT SUPPRESSION USING LOWER-DEGREE POLYNOMIALS

We shall consider, in the light of the component suppression idea, some methods for preparing the system $AX = F$ with a positive definite matrix $A$ for application of the successive approximations method.

For a system prepared in the form $X = X + h(F - AX) = (E - hA)X + hF$ the formula of the successive approximation method will be

$$X^{(k)} = X^{(k-1)} + h(F - AX^{(k-1)}). \tag{1}$$

In this case the damping coefficients will be the values $1 - h\lambda_i$ for $i = 1, \cdots, n$.

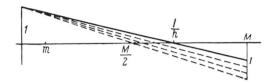

**FIGURE 15**

All damping coefficients will obviously be less than one and consequently the process will converge if $1/h > M/2$, i.e. $h < 2/M$ (Fig. 15).

The damping rate for components in different parts of the interval $(m, M)$ will be different and the fastest rate will obviously be in the zone touching the point $1/h$ (if $h > 1/M$) or at the right end of the interval (if $h \leq 1/M$). Slowest damping will occur at the right end of the interval if $h < 1/(M + m)$ and at the left end if $h > 1/(M + m)$. It is clear that in order to accelerate the successive approximations process it is advisable from time to time to take $h = 1/M$ or $h = 1/m$, respectively, if one of these numbers is known.

Lusternik's method described above also uses these facts as applied to the prepared system $X = BX + G$ for $B = E - hA$, $G = hF$.

Actually the eigenvalues $\mu_i$ of matrix $B$ are just the damping coefficients $\mu_i = 1 - h\lambda_i$. The largest of them in modulus will be $\mu_1 = 1 - hM$ or $\mu_2 = 1 - hm$. For $h = 1/M$ one step gives

$$\bar{X}^{(k+1)} = X^{(k)} + \frac{1}{M}(F - AX^{(k)})$$

or, passing to matrix $B$,

$$\bar{X}^{(k+1)} = X^{(k)} + \frac{h}{1 - \mu_1}\left[F - \frac{1}{h}(E - B)X^{(k)}\right]$$

$$= X^{(k)} + \frac{1}{1 - \mu_1}[G - (E - B)X^{(k)}]$$

$$= X^{(k)} + \frac{BX^{(k)} + G - X^{(k)}}{1 - \mu_1} = X^{(k)} + \frac{1}{1 - \mu_1}(X^{(k+1)} - X^{(k)}).$$

We arrive at such a result if the largest eigenvalue of matrix $B$ turns out to be $1 - hm$ and we assume $h = 1/M$ at the $(k + 1)$th step.

We obtain a rather effective method for suppressing components if we change the constant $h$ from step to step so that the root $1/h_k$ of the polynomial $g^{(k)}(t) = 1 - h_k t$ moves along the interval $(M/2, M)$ from right to left. The zone of effective component suppression for the error vector will thereupon be moved over, finally covering the whole interval.

If we want to suppress components of eigenvectors which belong to eigenvalues lying in the interval $(0, M/2)$ quickly, then we need to take

$h > 2/M$; but this will lead to an increase in components belonging to eigenvalues close to $M$.

If we want to construct the process so that an increase in any components does not occur in one step, then we must turn to polynomials of higher degrees for further displacement of the zone of most effective component suppression to the left.

We shall consider a crude scheme for choosing the polynomials $g_s^{(k)}(t)$ of low degree $s$ serving a significant part of the interval $(0, M)$ with a gradual displacement of the zone of most effective component suppression from right to left (the "flatiron" method, cf. Fig. 16). As we have already seen, the interval $(M/2, M)$ is well covered by the polynomials of the first system.

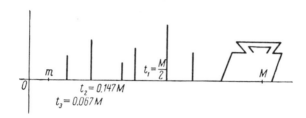

**FIGURE 16**

It is well-known that of all polynomials of a given degree $s$ which satisfy the requirements $g_s(0) = 1$ and $|g_s(t)| \leq 1$ for $0 \leq t \leq M$ a Chebyshev polynomial possesses a least root closest to zero for the interval $(0, M)$, i.e. the polynomial

$$\cos s \arccos \frac{M - 2t}{M}.$$

The least root of this polynomial is equal to $t_s = M \sin^2 \pi/4s$ so that for successive Chebyshev polynomials the least roots will be

$$t_1 = \frac{M}{2},$$

$$t_2 = \frac{2 - \sqrt{2}}{4} M \approx 0.147M,$$

$$t_3 = \frac{2 - \sqrt{3}}{4} M \approx 0.067M,$$

$$\cdots \cdots \cdots \cdots,$$

$$t_s \approx \frac{\pi^2}{16s^2} M.$$

Therefore by using second-degree polynomials we may widen the zone of operation for the method from $M/2$ to $0.147M$, by using third-degree polynomials we may widen it to $0.067M$, and so on. Thus for matrices with condition less than 15 we may confine ourselves to polynomials up through the third degree.

First-degree polynomials, for example, may be chosen as follows. In the interval $(M/2, M)$ we take the points $a_0 = M > a_1 > \cdots > a_{p_1} \geq M/2$ and let $g_1^{(k)}(t) = 1 - t/a_k$, $k = 0, \cdots, p_1$ (Fig. 17).

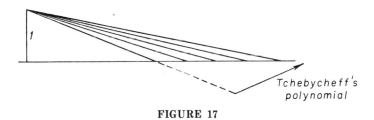

Tchebycheff's polynomial

**FIGURE 17**

To construct second-degree polynomials we then choose the point sequence $b_0 = M/2 > b_1 > \cdots > b_{p_2} = 0.147M$ and let

$$g_2^{(k)}(t) = 1 - \frac{t^2 - Mt}{b_{k-p_1}^2 - Mb_{k-p_1}} \qquad (k = p_1 + 1, \cdots, p_1 + p_2)$$

(cf. Fig. 18). These polynomials obviously satisfy the first two requirements given for the polynomials $g_2^{(k)}(t)$ and are reduced to zero for $t = b_{k-p_1}$.

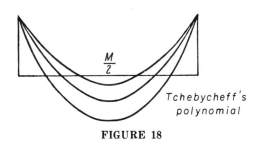

$\frac{M}{2}$

Tchebycheff's polynomial

**FIGURE 18**

For covering the zone from $0.147M$ to $0.067M$ we have the third-degree polynomials

$$g_3^{(k)}(t) = \frac{(t^2 - Mt)(2t - M)}{(c_j^2 - Mc_j)(2c_j - M)},$$

where $c_j$ are points between $t_2$ and $t_3$.

We shall not enter into details regarding the choice of polynomials of higher degree, since below we shall consider other methods (fulfilled more simply) for suppressing components corresponding to eigenvalues close to zero. We observe only that in everything we have presented, knowledge of the number $M$ plays an important role. An error in estimating this number (on the side of a decrease) holds a certain danger. An error which produces an increase may raise the volume of work only slightly.

The choice of dividing points is conditioned by the accuracy requirement, on the one hand, and by the desire to have as few of them as possible (to cover the zone more thoroughly) on the other.

We shall give the result of using lower-degree polynomials to find the solution of system (9), Section 23.

Here $M = 2.62$. Letting

$$a_1 = 2.62, \quad a_2 = 2.30, \quad a_3 = 2.00, \quad a_4 = 1.70, \quad a_5 = 1.40$$
$$b_1 = 1.20, \quad b_2 = 0.9, \quad b_3 = 0.6, \quad b_4 = 0.3$$

we obtain

$$X' = (-1.2284428, \quad 0.0473912, \quad 1.0233490, \quad 1.4591532)' .$$

The approximation $X'$ is obtained as a result of applying thirteen iterations. The length of the error vector for the approximation constructed is 1.85% of the length of the error vector for the initial approximation $X_0 = (0, 0, 0, 0)'$.

The method described for choosing the suppression polynomials is quite crude and, as we shall see below, with a more careful choice of polynomials we may obtain a better result using the same number of iterations.

## 86. DIFFERENT FORMS FOR CARRYING OUT UNIVERSAL ALGORITHMS

Before describing concrete universal algorithms we shall touch on the question of their numerical realization, limiting ourselves to considering one step of a process for different ways of assigning the polynomial which determines this step.

**1. The coefficients of the polynomial $h(t)$ are known or, what is the same thing, the coefficients of $g(t)$ are known.** Let

$$h(t) = c_0 t^{s-1} + c_1 t^{s-2} + \cdots + c_{s-1} . \tag{1}$$

In this case $X'$ may be computed using the two following methods.

1. First of all we compute the residual $r = F - AX$. The vectors $Ar$, $A^2 r$, $\cdots$, $A^{s-1} r$ are then computed successively, and finally $X'$ is found as a known linear combination of already known vectors

$$X' = X + c_{s-1} r + c_{s-2} Ar + \cdots + c_0 A^{s-1} r . \tag{2}$$

With such a construction of the vector $X'$, the accretion of rounding-off errors is possible, on the one hand, and also the loss of significant figures on the other, since as a rule the coefficients $c_0, c_1, \cdots, c_{s-1}$ have different signs.

2. Having computed $r = F - AX$ we find successively the vectors

$$Z_0 = c_0 r \,,$$
$$Z_1 = AZ_0 + c_1 r \,,$$
$$\cdots \cdots \cdots \,, \tag{3}$$
$$Z_{s-1} = AZ_{s-2} + c_{s-1} r \,.$$

It is clear that $h(A)r = Z_{s-1}$, and consequently,

$$X' = X + Z_{s-1} \,.$$

This method essentially is nothing less than an application of Horner's well known scheme. With such a construction of the vector $X'$, rounding-off errors are slightly less than in the previous case.

2. **The roots $\varepsilon_1, \cdots, \varepsilon_s$ of the polynomial $g(t)$ are known.** Then

$$g(t) = \left(1 - \frac{t}{\varepsilon_1}\right) \cdots \left(1 - \frac{t}{\varepsilon_s}\right) \,. \tag{4}$$

In this case the vector $X'$ may be found as the last elements $Z_s$ of the following vector sequence:

$$Z_1 = X + \frac{1}{\varepsilon_1}(F - AX)$$
$$Z_i = Z_{i-1} + \frac{1}{\varepsilon_i}(F - AZ_{i-1}) \qquad (i = 2, \cdots, s) \,. \tag{5}$$

In fact for all $i$,

$$X^* = X^* + \frac{1}{\varepsilon_i}(F - AX^*)$$

where $X^*$ is the exact solution of the system and therefore

$$X^* - Z_i = (X^* - Z_{i-1}) - \frac{1}{\varepsilon_i} A(X^* - Z_{i-1}) = \left(E - \frac{1}{\varepsilon_i} A\right)(X^* - Z_{i-1}) \,.$$

Consequently,

$$X^* - Z_s = \left(E - \frac{1}{\varepsilon_s} A\right) \cdots \left(E - \frac{1}{\varepsilon_1} A\right)(X^* - X) = g(A)Y = Y - h(A)AY$$
$$= Y - h(A)(F - AX)$$

and

$$Z_s = X^* - Y + h(A)(F - AX) = X + h(A)(F - AX) = X' \,.$$

This scheme was described for the first time by Richardson (1).

In applying this scheme the steps most dangerous in terms of the influence of rounding-off errors are those which correspond to small roots $\varepsilon_i$. It is advisable to arrange the roots in decreasing order.

3. **Recurrent relationships for determining the polynomial $g(t)$ are known.** We shall consider the case of the trinomial relations most often encountered. Let $g(t) = g_s(t)$ where

$$g_0(t) = 1 , \quad g_1(t) = \alpha t + 1 ,$$
$$g_i(t) = [\alpha_i t + (\beta_i + 1)]g_{i-1}(t) - \beta_i g_{i-2}(t) \qquad (i = 2, \cdots, s) . \tag{6}$$

The form of the recurrence relationships is chosen so that all the polynomials $g_i(t)$ are normalized by the condition $g_i(0) = 1$.

We construct the vector sequence

$$X_i = X_{i-1} - \alpha_i(F - AX_{i-1}) + \beta_i(X_{i-1} - X_{i-2}) \tag{7}$$

for $X_0 = X$ and $X_1 = X_0 - (F - AX)$.

Then

$$X' = X_s .$$

Actually from (7) it follows that

$$Y_i = Y_{i-1} + \alpha_i A Y_{i-1} + \beta_i(Y_{i-1} - Y_{i-2}) .$$

From here we conclude by induction that

$$Y_i = g_i(A)Y_0 .$$

In particular

$$Y_s = g_s(A)Y_0$$

and therefore

$$X_s = X' .$$

The following methods are related to a system prepared in the form

$$X = BX + G .$$

**4. The coefficients of the polynomial $f(t)$ are known,** namely,

$$f(t) = b_0 t^{s-1} + b_1 t^{s-2} + \cdots + b_{s-1} . \tag{8}$$

1. Having computed the residual of the initial approximation $r = BX + G - X$, we compute the vectors $Br, \cdots, B^{s-1}r$ and find $X'$ as a known linear combination of already known vectors:

$$X' = X + b_{s-1}r + \cdots + b_0 B^{s-1}r . \tag{9}$$

2. Having computed the residual $r$, we construct the vector sequence

$$Z_0 = b_0 r$$
$$Z_i = BZ_{i-1} + b_i r \qquad (i = 1, \cdots, s - 1) . \tag{10}$$

Then

$$X' = X + Z_{s-1} .$$

**5. The coefficients of the polynomial $e(t)$ are known,** namely,

$$e(t) = a_0 t^s + a_1 t^{s-1} + \cdots + a_s . \tag{11}$$

1. We compute the successive approximations $X_0 = X$, $X_1 = BX_0 + G$, $\cdots$, $X_s = BX_{s-1} + G$ and we form their linear combination

$$a_0 X_s + a_1 X_{s-1} + \cdots + a_s X_0 . \tag{12}$$

It will also be the desired approximation $X'$.
   In fact let

$$Y_0 = X^* - X_0, \ Y_1 = X^* - X_1, \ \cdots, \ Y_s = X^* - X_s .$$

Then

$$Y_1 = BY_0, \ \cdots, \ Y_s = B^s Y_0 \quad \text{and} \quad a_0 X_s + a_1 X_{s-1} + \cdots + a_s X_0$$
$$= (a_0 + a_1 + \cdots + a_s) X^* - a_0 B^s Y_0 - a_1 B^{s-1} Y_0 - \cdots - a_s Y_0$$
$$= e(1) X^* - e(B) Y_0 = X^* - Y_0 + f(B)(E - B) Y_0$$
$$= X_0 + f(B)(BX_0 + G - X_0) = X' .$$

   2.   We compute the vector sequence from the formulas

$$Z_0 = a_0 X$$
$$Z_{i+1} = BZ_i + a_{i-1}X + (a_0 + \cdots + a_i)G \qquad (i = 0, \ \cdots, \ s-1) . \tag{13}$$

Then the last vector $Z_s$ will be equal to $X'$.
   **6.  The roots $\varepsilon_1, \cdots, \varepsilon_s$ of the polynomial $e(t)$ are known.**  Then

$$e(t) = \frac{(t - \varepsilon_1) \cdots (t - \varepsilon_s)}{(1 - \varepsilon_1) \cdots (1 - \varepsilon_s)} \tag{14}$$

since $e(1) = 1$.
   We may pass from the vector $X$ to the vector $X'$ in $s$-steps, assuming at the $i$th step that $e_i(t) = (t - \varepsilon_i)/(1 - \varepsilon_i)$, since in applying several steps the polynomials $e_i(t)$ are multiplied.   For this $f_i(t) = 1/(1 - \varepsilon_i)$, since

$$e_i(t) = 1 - \frac{1}{1 - \varepsilon_i}(1 - t) = 1 - f_i(t)(1 - t) .$$

Consequently the vector $X'$ is found as the $s$th element $Z_s$ of the sequence

$$Z_i = Z_{i-1} + \frac{1}{1 - \varepsilon_i}(BZ_{i-1} + G - Z_{i-1}) \tag{15}$$

starting with $Z_0 = X$.
   **7.  Recurrence relationships for determining the polynomial $f(t)$ are known.**
Let $f(t) = f_{s-1}(t)$ where

$$f_0(t) = \beta_0 , \quad f_1(t) = \alpha_1 t + \beta_1 ,$$
$$f_i(t) = (\alpha_i t + \beta_i) f_{i-1}(t) - \gamma_i f_{i-2}(t) . \tag{16}$$

We compute the vector sequence

$$r = BX + G - X , \quad Z_0 = \beta_0 r , \quad Z_1 = \alpha_1 Br + \beta_1 r ,$$
$$Z_i = \alpha_i BZ_{i-1} + \beta_i Z_{i-1} - \gamma_i Z_{i-2} \qquad (i = 1, 2, \cdots, s-1) . \tag{17}$$

It is clear that

$$Z_{s-1} = f(B)(BX + G - X) ,$$
$$X' = X + Z_{s-1} .$$

**8.   Recurrence relationships for determining the polynomial $e(t)$ are known.**
Let $e(t) = e_s(t)$ where

$$e_0(t) = 1 , \quad e_1(t) = \alpha_1 t + \beta_1 ,$$
$$e_i(t) = (\alpha_i t + \beta_i)e_{i-1}(t) - \gamma_i e_{i-2}(t) . \tag{18}$$

It is assumed, besides this, that all polynomials $e_i(t)$ satisfy the condition
$e_i(1) = 1$ so that $\alpha_1 + \beta_1 = 1$, $\alpha_i + \beta_i - \gamma_i = 1$, $i = 2, \cdots, s$. In this case $X'$
will be equal to the vector $X_s$ in the sequence

$$X_0 = X , \quad X_1 = \alpha_1(BX_0 + G) + \beta_1 X_0 ,$$
$$X_i = \alpha_i(BX_{i-1} + G) + \beta_i X_{i-1} - \gamma_i X_{i-2} . \tag{19}$$

In fact from the condition $\alpha_1 + \beta_1 = 1$, $\alpha_i + \beta_i - \gamma_i = 1$ we have

$$X^* = \alpha_1(BX^* + G) + \beta_1 X^* ,$$
$$X^* = \alpha_i(BX^* + G) + \beta_i X^* - \gamma_i X^* .$$

Subtracting, we obtain

$$Y_1 = \alpha_1 BY_0 + \beta_1 Y_0 = e_1(B)Y_0 ,$$
$$Y_i = (\alpha_i B + \beta_i E)Y_{i-1} - \gamma_i Y_{i-2} ,$$

from which, by induction,

$$Y_i = e_i(B)Y_0 .$$

In particular, $Y_s = e(B)Y_0$ and consequently

$$X' = X_s .$$

As a rule schemes using recurrence relationships turn out to be most convenient.

## 87.   THE BEST UNIVERSAL ALGORITHM IN THE SENSE OF THE FIRST CRITERION[†]

Let it be known that all eigenvalues of a matrix $A$ are located in the
interval $(m, M)$, $0 < m < M$, and are distinct.  We shall prepare the system
$AX = F$ in the form $X = BX + G$ so that the eigenvalues of matrix $B$ are
located in a symmetric interval with its center at the coordinate origin.
It is obvious that we then need to take $h = 2/(M + m)$, $B = E - hA$,
$G = hF$.  Then all eigenvalues of matrix $B$ lie in the interval $(-1/\gamma, 1/\gamma)$
where $\gamma = (M + m)/(M - m) > 1$.
We construct the universal algorithm

$$X_s = X_0 + f_{s-1}(B)(BX_0 + G - X_0) , \tag{1}$$

choosing the polynomial $f_{s-1}(t)$ so that for its given degree $(s - 1)$ there will
be guaranteed a maximal component suppression for the whole class of
matrices $B$ with eigenvalues contained in the interval $(-1/\gamma, 1/\gamma)$.

† M. Sh. Birman (1).

For this we obviously need to take as the polynomial $e_s(t) = 1 - (1-t)f_{s-1}(t)$ the polynomial which diverges least from zero in the interval $(-1/\gamma,\ 1/\gamma)$ and which is normalized by the condition $e_s(1) = 1$.

Such a polynomial will be

$$\tilde{T}_s(t) = \frac{T_s(\gamma t)}{T_s(\gamma)}\,, \qquad (2)$$

where

$$T_s(t) = \cos s \operatorname{arc\,cos} t\,.$$

In Fig. 19 is given the graph of $\tilde{T}_6(t)$ for $\gamma = 5/4$.

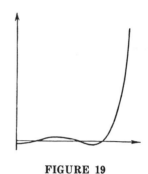

**FIGURE 19**

The polynomials $\tilde{T}_i(t)$ are connected by simple recurrence relationships. In fact for Chebyshev polynomials we have the relationships

$$T_i(t) = 2t\,T_{i-1}(t) - T_{i-2}(t)$$

for $T_0(t) = 1$, $T_1(t) = t$. Therefore

$$\tilde{T}_i(t) = \left[1 + \frac{T_{i-2}(\gamma)}{T_i(\gamma)}\right]t\tilde{T}_{i-1}(t) - \frac{T_{i-2}(\gamma)}{T_i(\gamma)}\,\tilde{T}_{i-2}(t)\,,$$

$$\tilde{T}_0(t) = 1\,, \qquad\qquad\qquad\qquad\qquad (3)$$

$$\tilde{T}_1(t) = t\,.$$

In accordance with recurrence relationships (3), the universal algorithm (Section 86, par. 8) is constructed from the formulas

$$X_1 = BX_0 + G\,,$$

$$X_i = \left[1 + \frac{T_{i-2}(\gamma)}{T_i(\gamma)}\right](BX_{i-1} + G) \qquad\qquad (4)$$

$$\qquad - \frac{T_{i-2}(\gamma)}{T_i(\gamma)}X_{i-2} \qquad (i = 2,\ \cdots,\ s)$$

In computing with formula (4) we need to prepare the values $[T_{i-2}(\gamma)]/T_i(\gamma)$ beforehand according to the recurrence relationships. For increasing $i$ these values approach the limit

$$\alpha = (\gamma - \sqrt{\gamma^2 - 1})^2$$

quite rapidly.

The rate of convergence for the process described for $s$ will have order

$$\frac{1}{T_s(\gamma)} = \frac{2}{(\gamma + \sqrt{\gamma^2 - 1})^s + (\gamma - \sqrt{\gamma^2 - 1})^s}\,,$$

so that the process converges significantly more rapidly than the method of successive approximations, for which the rate of convergence will be

$\gamma^{-s}$.

Thus, for example, for $\gamma = 25/24$ ($\gamma^{-1} = 0.96$) it will be

$$\frac{1}{T_s(\gamma)} = \frac{2}{\left(\dfrac{4}{3}\right)^s + \left(\dfrac{3}{4}\right)^s} \approx 2\left(\frac{3}{4}\right)^s,$$

when $\gamma^{-s} = (24/25)$.

When using formulas (4) it is possible, instead of taking all increasing values for $s$, to take not very large $s$ but to repeat the process several times, applying the approximation obtained at the end of a cycle as the initial approximation of a new process.

It is not easy to prepare the given matrix $A$ in a form which allows us to use the process described. In fact if in estimating the number $M$ we may use at least Gershgorin's estimate, then it is still much more laborious to determine the number $m$. The process itself is determined by the choice of $\gamma = (M + m)/(M - m)$ which soon approaches 1 for $m \to 0$; too crude a choice of $\gamma$ may greatly decelerate the rate of convergence for the process (if the true value for $\gamma$ is significantly larger than the one taken on the basis of the crude estimates for $M$ and $m$).

The process described is essentially identical to the one with which the method of steepest descent was compared in deriving estimates for the rate of convergence.

For the polynomial $\tilde{T}_s(t)$ it is easy to compute the coefficients which make it possible for us to use the scheme of paragraph 5, Section 86. However in using this scheme a considerable loss of significant figures occurs.

A slightly better scheme is the one of paragraph 6, Section 86, which uses roots of the polynomial $T_s(t)$. It is obvious that these roots are equal to

$$\varepsilon_i - \frac{1}{\gamma} \cos \frac{(2i - 1)\pi}{2s} \qquad (i = 1, \cdots, s).$$

When using this scheme the approximation $X_s$ is obtained as the last element in the sequence

$$Z_i = Z_{i-1} + \frac{1}{1 - \varepsilon_i}(BZ_{i-1} + G - Z_{i-2}) \qquad (i = 1, \cdots, s),$$

starting with $Z_0 = X_0$.

It is clear that for such a construction of $X_s$ we need to fix the value $s$ beforehand. Convergence of the process is assured by cyclic repetition.

To lessen the influence of rounding-off errors we should arrange the roots $\varepsilon$ in decreasing order, since the process is most sensitive to rounding-off errors at those steps where $\varepsilon$ is close to one.

We shall show the course of the process in the example of system (9), Section 23, taking for $M$ and $m$ three-digit approximations to the largest and smallest eigenvalues of the system's matrix, that is, $M = 2.322$ and

$m = 0.242$. This gives $h = 0.78003120$ so that

$$B = \begin{bmatrix} 0.21996880 & -0.32761310 & -0.42121685 & -0.51482059 \\ -0.32761310 & 0.21996880 & -0.24960998 & -0.34321373 \\ -0.42121685 & -0.24960998 & 0.21996880 & -0.17160686 \\ -0.51482059 & -0.34321373 & -0.17160686 & 0.21996880 \end{bmatrix}$$

$$G = (0.23400936, \quad 0.39001560, \quad 0.50602184, \quad 0.70202808)' ,$$

Moreover $\gamma = 1.2326923$, $\alpha = 0.26204959$.

We have

| | | | | |
|---|---|---|---|---|
| $X_0$ | 0 | 0 | 0 | 0 |
| $X_1$ | $-0.23400936$ | 0.39001560 | 0.54602184 | 0.70202808 |
| $X_2$ | $-0.64639927$ | 0.03264644 | 0.52125464 | 0.75775989 |
| $X_8$ | $-1.2463278$ | 0.0430531 | 1.0292117 | 1.4685821 |
| $X_{13}$ | $-1.2571934$ | 0.0436358 | 1.0389835 | 1.4821027 |
| $X_{18}$ | $-1.2577816$ | 0.0434862 | 1.0391543 | 1.4823759 |
| $X_{22}$ | $-1.2577930$ | 0.0434873 | 1.0391655 | 1.4823918 |
| $X_{24}$ | $-1.2577936$ | 0.0434873 | 1.0391661 | 1.4823926 |

## 88. THE BEST UNIVERSAL ALGORITHM IN THE SENSE OF THE SECOND CRITERION

Let it be known for the system $X = BX + G$ that all eigenvalues of matrix $B$ lie in the interval $(-1/\gamma, 1/\gamma)$ for $\gamma > 1$. A universal algorithm will be best in the sense of the second criterion if the polynomial $f(t)$ determining it is the polynomial which diverges least from $1/(1-t)$ in the interval $(-1/\gamma, 1/\gamma)$. It is obvious that $f(t) = \gamma F(\gamma t)$ where $F(t)$ is the polynomial which diverges least from the function $1/(\gamma - t)$ in the interval $(-1, 1)$.

The polynomial of degree $s-1$ which satisfies the last requirement was known to Chebyshev. Namely,

$$F_{s-1}(t) = -\frac{2\alpha^{s/2}}{(1-\alpha)^2} \frac{1}{\gamma - t} [T_s(t) - 2\sqrt{\alpha}\, T_{s-1}(t) + \alpha T_{s-2}(t)] + \frac{1}{\gamma - t} ,$$

where

$$\alpha = (\gamma - \sqrt{\gamma^2 - 1})^2 ,$$
$$T_s(t) = \cos s \arccos t .$$

Thus

$$f_{s-1}(t) = \frac{1}{1-t} - \frac{2\alpha^{s/2}}{(1-\alpha)^2} \frac{1}{1-t} [T_s(\gamma t) - 2\sqrt{\alpha}\, T_{s-1}(\gamma t) + \alpha T_{s-2}(\gamma t)] .$$

In the work of M. K. Gavurin (1), which proposed such a choice of the polynomial $f(t)$ for the first time, it is recommended to compute the coefficients of the polynomial $f(t)$ and then to look for a solution as a linear combination of vectors $B^i r_0$ in accordance with the scheme of paragraph 4, a), Section 86. For large powers $s$ a great loss of significant figures occurs.

A much more convenient computational scheme is obtained if we switch to the polynomial $e_s(t)$. We have

$$e_s(t) = 1 - (1 - t)f_{s-1}(t)$$

$$= \frac{2\alpha^{s/2}}{(1 - \alpha)^2} [T_s(\gamma t) - 2\sqrt{\alpha}\, T_{s-1}(\gamma t) + \alpha T_{s-2}(\gamma t)].$$

We let

$$\tilde{e}_s(t) = \alpha^{-s/2} e_s(t) = \frac{2}{(1 - \alpha)^2} [T_s(\gamma t) - 2\sqrt{\alpha}\, T_{s-1}(\gamma t) + \alpha T_{s-2}(\gamma t)].$$

The polynomials $T_i(\gamma t)$ are connected by the recurrence relationship

$$T_i(\gamma t) = 2\gamma t\, T_{i-1}(\gamma t) - T_{i-2}(\gamma t)$$

with coefficients which do not depend on the polynomial's number. Therefore any combination of several adjacent Chebyshev polynomials is connected by relationships of the same type. In particular,

$$\tilde{e}_i(t) = 2\gamma t \tilde{e}_{i-1}(t) - \tilde{e}_{i-2}(t).$$

Multiplying by $\alpha^{i/2}$ and passing to the polynomials $e_i(t)$ we obtain

$$e_i(t) = 2\gamma\sqrt{\alpha}\, t e_{i-1}(t) - \alpha e_{i-2}(t) = (1 + \alpha)t e_{i-1}(t) - \alpha e_{i-2}(t)$$

since

$$2\gamma\sqrt{\alpha} = 2\sqrt{\alpha}\,\frac{\sqrt{\alpha} + \sqrt{\alpha^{-1}}}{2} = 1 + \alpha.$$

For this

$$e_1(t) = \left(\frac{1 + \alpha}{1 - \alpha}\right)^2 (t - 1) + 1,$$

$$e_2(t) = t e_1(t) + \frac{2\alpha}{1 - \alpha}(t - 1).$$

Thus the successive approximation may be computed by the recurrence relationships

$$X_i = (1 + \alpha)(BX_{i-1} + G) - \alpha X_{i-2}$$

(according to paragraph 8, Section 86) starting with the initial approximations $X_1$ and $X_2$ which are computed from the formulas

$$X_1 = X + \left(\frac{1 + \alpha}{1 - \alpha}\right)^2 (BX + G - X),$$

$$X_2 = BX_1 + G + \frac{2\alpha}{1 - \alpha}(BX + G - X).$$

In numerical application of the best algorithm in the sense of the second criterion, we need to assume the same information about the eigenvalue distribution for the matrix of coefficients as in using the algorithm best in the sense of the first criterion. We shall show the course of the process in the example of system (9), Section 23, prepared just as in Section 87.

Here $\alpha = 0.26204959$. We have

| | | | | |
|---|---|---|---|---|
| $X_0$ | 0 | 0 | 0 | 0 |
| $X_1$ | 0.6844342 | 1.1407237 | 1.5970131 | 2.0533026 |
| $X_2$ | $-1.5527282$ | $-0.4096498$ | 0.3597126 | 6.6343461 |
| $X_3$ | $-0.7490792$ | 0.3334810 | 1.1875589 | 1.6324174 |
| $X_8$ | $-1.2636344$ | 0.0348866 | 1.0266917 | 1.4668835 |
| $X_{13}$ | $-1.2571287$ | 0.0438702 | 1.0393853 | 1.4826318 |
| $X_{18}$ | $-1.2578043$ | 0.0434744 | 1.0391495 | 1.4823719 |
| $X_{24}$ | $-1.2577941$ | 0.0434870 | 1.0391660 | 1.4823925 |
| $X_{25}$ | $-1.2577935$ | 0.0434875 | 1.0391665 | 1.4823930 |

## 89. A. A. ABRAMOV'S METHOD FOR ACCELERATING THE CONVERGENCE OF THE SUCCESSIVE APPROXIMATIONS METHOD IN SOLVING A SYSTEM OF LINEAR EQUATIONS

A. A. Abramov (1) has proposed the following method for accelerating the convergence of the successive approximations process for a system written in the form $X = BX + G$ where $B$ is a matrix whose eigenvalues are real and lie in the interval $(-1, 1)$.

From time to time the normal course of the successive approximations process, carried out according to the formula

$$X^{(k)} = BX^{(k-1)} + G, \tag{1}$$

whose application will be called a $B$-step for brevity, is interrupted by one or several more complex $B_2$-steps or $B_4$-steps which will now be described.

A $B_2$-step consists in constructing an approximation $\bar{X}^{(k+2)}$ according to the formula

$$\bar{X}^{(k+2)} = 2X^{(k+2)} - X^{(k)} \tag{2}$$

where $X^{(k+2)}$ is the second successive approximation constructed from $X^{(k)}$, i.e.

$$X^{(k+2)} = BX^{(k+1)} + G,$$
$$X^{(k+1)} = BX^{(k)} + G. \tag{3}$$

Thus a $B_2$-step requires the application of two $B$-steps and the forming of a linear combination.

Furthermore, a $B_4$-step consists in constructing the approximation $X^{(k+4)}$ according to the formula

$$\bar{\bar{X}}^{(k+4)} = 2\bar{X}^{(k+4)} - X^{(k)} \tag{4}$$

where $\bar{X}^{(k+4)}$ is obtained from $X^{(k)}$ by double application of the $B_2$-step.

A $B_8$-step, $B_{16}$-step, etc. are determined in an analogous way.

As a rule one should limit oneself in practical computations to using only $B$, $B_2$, $B_4$ and $B_8$-steps, alternating them with each other in a certain sequence.

It follows from the description that the computational scheme of Abramov's process is hardly more complex than the computational scheme for the classical successive approximations method. The volume of computations for the process $B^{k_0}B_2^{k_1}B_4^{k_2}B_8^{k_3}$ (i.e. the process consisting of $k_0$ $B$-steps, $k_1$ $B_2$-steps, etc.) is only a little larger than the volume of computations for $k_0 + 2k_1 + 4k_2 + 8k_3$ steps of the successive approximations process.

We shall now explain why applying Abramov's process gives a better result in comparison with the result, equivalent in volume of computations, obtained by the successive approximations method.

With this in mind we compute the damping factors in the error vector components for separate steps of Abramov's process.

It is clear that

$$Y^{(k+1)} = BY^{(k)} ,$$
$$\bar{Y}^{(k+2)} = 2Y^{(k+2)} - Y^{(k)} = (2B^2 - E)Y^{(k)} = B_2 Y^{(k)} = T_2(B)Y^{(k)} ,$$
$$\bar{Y}^{(k+4)} = (2B_2^2 - E)Y^{(k)} = B_4 Y^{(k)} = T_2(B_2)Y^{(k)} = T_4(B)Y^{(k)} .$$

Here the polynomials $T_2(t)$, $T_4(t)$, $\cdots$ are nothing less than the Chebyshev polynomials

$$T_s(t) = \cos s \arccos t .$$

This is obvious for $s = 2$; for $s = 4, 8, \cdots$, this is true on the basis of the well-known relationship

$$T_{s_1 s_2}(t) = T_{s_1}(T_{s_2}(t)) .$$

In Fig. 20 are given the graphs of the functions $t^2$ and $|T_2(t)| = |2t^2 - 1|$, in Fig. 21 the graphs of the functions $t^4$ and $|T_4(t)|$, in Fig. 22 the graphs of the functions

$$|T_2^2(t)| \quad \text{and} \quad |T_4(t)| .$$

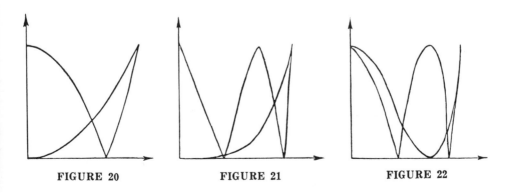

FIGURE 20         FIGURE 21         FIGURE 22

Examination of these graphs allows us to compare damping factors for components in an expansion of the error vector for steps which are almost equivalent in volume of computations: two $B$-steps and one $B_2$-step (Fig. 20), four $B$-steps and one $B_4$-step (Fig. 21), and finally two $B_2$-steps and one $B_4$-step (Fig. 22).

From Fig. 20 we see that two $B$-steps are better than one $B_2$-step for $0 < t < 1/\sqrt{3} \approx 0.58$. But one $B_2$-step is significantly better than two $B$-steps for $t > 1/\sqrt{3}$. Therefore having "suppressed" sufficiently the error vector components for eigenvalues from the interval $0 \le t \le 1/\sqrt{3}$ by the successive approximations method, we should switch to $B_2$-steps.

Moreover a $B_4$-step turns out to be better than two $B_2$-steps for the interval $0.89 < t < 1$ (Fig. 22). From Fig. 21 it likewise follows that one $B_4$-step is better than four $B$-steps for the interval $0.86 < t < 1$. Therefore we should begin to apply $B_4$-steps after the $B$ and $B_2$-steps have already suppressed the error vector components for the interval $0 < t < 0.89$.

From time to time one should return to the lower steps to suppress the accumulating rounding-off errors in components corresponding to small eigenvalues.

We shall not give a numerical example here, since Abramov's method belongs to the group of $BT$-processes which have a more uniform computational scheme as a special case.

## 90. *BT*-PROCESSES

Abramov's method may be generalized slightly without complicating the computational scheme. That is, Abramov's process may be considered as a special case of "$BT$-processes" which are defined as follows. Let the system

$$X = BX + G \tag{1}$$

be given with a matrix $B$ which has real eigenvalues located in the interval $(-1, 1)$. Let there then be given a sequence of the letters $B$ and $T$, starting with the letter $B$. For example,

$$BBTTTBTBTBBTT \cdots \tag{2}$$

In accordance with this sequence we construct the successive approximations $X^{(0)}, X^{(1)}, X^{(2)}, \cdots$, having been given the initial approximation $X^{(0)}$ arbitrarily. We construct the approximation $X^{(k)}$ in the following manner: if the letter $B$ is found in the $k$th place in sequence (2), then we let

$$X^{(k)} = BX^{(k-1)} + G . \tag{3}$$

If the letter $T$ is found at the $k$th place, then we let

$$X^{(k)} = 2[BX^{(k-1)} + G] - X^{(k-2)} \tag{4}$$

We shall consider how the error vector components change in the $BT$-process. With this in mind we break up sequence (2) which determines the process into "words" so that each word contains the letter $B$ once and begins with it. For example,

$$BB\ TTT\ BT\ BT\ BB\ TT \cdots = B(BTTT)(BT)(BT)B(BTT) \cdots$$
$$= B_1 B_4 B_2 B_2 B_1 B_3 \cdots$$

Here

$$B_s = BT \cdots T = BT^{s-1}, \quad B_1 = B.$$

It is clear that

$$Y^{(k)} = BY^{(k-1)},$$

if the letter $B$ is found at the $k$th step, and

$$Y^{(k)} = 2BY^{(k-1)} - Y^{(k-2)},$$

if the letter $T$ is found at the $k$th step.

We shall consider how applying the words $B_s$ influences the error vector. We have

$$Y^{(k+1)} = BY^{(k)} = T_1(B)Y^{(k)},$$
$$Y^{(k+2)} = 2BY^{(k+1)} - Y^{(k)} - (2B^2 - E)Y^{(k)} = T_2(B)Y^{(k)},$$
$$Y^{(k+3)} = 2BY^{(k+2)} - Y^{(k+1)} = (2BT_2(B) - T_1(B))Y^{(k)} = T_3(B)Y^{(k)},$$
$$\cdots\cdots\cdots\cdots\cdots\cdots\cdots\cdots\cdots\cdots\cdots,$$
$$Y^{(k+s)} = [2BT_{s-1}(B) - T_{s-2}(B)]Y^{(k)} = T_s(B)Y^{(k)}.$$

Here the polynomials $T_i(t)$ are determined by the recurrence relationships

$$T_i(t) = 2tT_{i-1}(t) - T_{i-2}(t)$$

under the initial conditions $T_0(t) = 1$, $T_1(t) = t$ and consequently they coincide with the Chebyshev polynomials $\cos i$ arc $\cos t$. Thus applying the word $B_s$ involves multiplying error vector components by the factors $T_s(\mu_1), \cdots,$ $T_s(\mu_i)$. Applying the sequence of letters $B$ and $T$ involves multiplying each error vector component by the product of the values at a corresponding point of Chebyshev polynomials corresponding to the separate words $B_s$ of which the determining sequence (?) consists.

Abramov processes are special cases of $BT$-processes in which the determining sequences consist of the words $B, B_2, B_4, B_8, \cdots$. The successive approximations method is written as a $BBBB\cdots$-process.

Just at in Abramov processes, the application of words with high numbers hastens the decrease of components corresponding to just those eigenvalues of matrix $B$ which are close to 1 in modulus. Therefore it is advisable to work only with words $B$ at first, introducing longer words at subsequent stages and returning to the shorter words for clearing away rounding-off errors.

We shall apply the $BT$-process to find the solution of system (9), Section 23, prepared with $h = 2/2.62$ (Section 31, (3)).

Applying the word

$$B^{10}(BT)^8 B^2 (BTT)^2 (BT)^2$$

gives

$$X = (-1.2577936, \quad 0.0434873, \quad 1.0391661, \quad 1.4823926)' .$$

The chosen word consists of 38 letters so that 38 iterations had to be computed to obtain the approximation.

As we saw in Section 31, the successive approximations method gave the same accuracy with 75 iterations.

The word

$$B^{10}(BT)^4(BTT)^3(BTTT)^2 B^2(BT)^3 ,$$

consisting of 43 letters, gives

$$X = (-1.2577945, \quad 0.0434870, \quad 1.0391667, \quad 1.4823935)' .$$

This approximation is worse than the previous one.

Supplementing the given word with the letters $TBBT$ we obtain

$$X = (-1.2577936, \quad 0.0434874, \quad 1.0391662, \quad 1.4823927)' .$$

The more successful choice of the first word as compared with the second is explained by the fact that the condition of the given system is such that there is no need to fall back on words longer than $BTT$.

A good choice of the word which guides the $BT$-process may be realized by the following "$BT$-process with direction". During this process we must compute along with the successive approximations $X_k$ three systems of auxiliary vectors: the residuals $r_k$, the vectors $s_k = Br_k$ and the vectors $\bar{s}_k = 2s_k - r_{k-1}$. We let

$$Z_{k+1} = BX_k + G ,$$
$$\bar{Z}_{k+1} = 2(BX_k + G) - X_{k-1} .$$

These vectors are obtained from the vectors $X_k$ by applying a $B$-step and a $T$-step. Computation of the vectors $Z_{k+1}$ and $\bar{Z}_{k+1}$ does not require carrying out the iterations, since

$$Z_{k+1} = X_k + r_k ,$$
$$\bar{Z}_{k+1} = 2(X_k + r_k) - X_{k-1} .$$

We shall compute their residuals $w_{k+1}$ and $\bar{w}_{k+1}$. It is easy to see that

$$w_{k+1} = Br_k = s_k ,$$
$$\bar{w}_{k+1} = 2Br_k - r_{k-1} = 2s_k - r_{k-1} = \bar{s}_k .$$

We compare norms of the vectors $s_k$ and $\bar{s}_k$ and we let

$$X_{k+1} = Z_{k+1} = X_k + r_k ,$$
$$r_{k+1} = s_k \tag{5}$$

if $\| s_k \| \leq \| \bar{s}_k \|$ and

$$X_{k+1} = \bar{Z}_{k+1} = 2(X_k + r_k) - X_{k-1} ,$$
$$r_{k+1} = \bar{s}_k \tag{6}$$

if $\|s_k\| > \|\bar{s}_k\|$.

The vectors $s_{k+1} = Br_{k+1}$ and $\bar{s}_{k+1} = 2s_{k+1} - r_k$ are then computed and the process is continued further. From time to time, in order to decrease rounding-off errors, we should compute the the residual $r_{k+1}$ directly from the formula $r_{k+1} = BX_{k+1} + G - X_{k+1}$ and not from formulas (5) and (6).

Applying the $BT$-process with direction to the previous example led to the word $BTBTBTBTBTBTBBTBBBTBTBBTBBTBBTBBBTBT = (BT)^5 B(BT)B^2(BT)^2 B(BT)B(BT)B(BT)B^2(BT)^2(34$ letters); as a result we obtained the approximation.

$$(-1.2577936, \quad 0.0434874, \quad 1.0391661, \quad 1.4823928)'$$

which differs from the exact solution by no more than $2 \cdot 10^{-7}$ in each component. The residuals were computed directly at the 11th, 16th, 21st, 26th and 31st steps.

## 91.  GENERAL TRINOMIAL ITERATIVE PROCESSES

The $BT$-processes presented above and the algorithm best in the sense of the first criterion are contained in the following general scheme of trinomial universal processes.[†]

To solve the system

$$X = BX + G \tag{1}$$

with a matrix $B$ whose eigenvalues are contained in the interval $(-1, 1)$ a sequence of approximations according to the formulas

$$X^{(1)} = BX^{(0)} + G, \tag{2}$$

$$X^{(k)} = (1 + \alpha_k)(BX^{(k-1)} + G) - \alpha_k X^{(k-2)} \tag{3}$$

is constructed, starting with a certain initial approximation $X^{(0)}$. The first step may be formally considered to arise from the general formula (3), letting $\alpha_1 = 0$.

It is clear that if $\alpha_{s+1} = 0$, then the process goes along, beginning with the $s + 1$th step, as if the $s$th approximation had been taken as the initial one and the process is then applied for $\alpha'_2 = \alpha_{s+2}$, $\alpha'_3 = \alpha_{s+3}$, $\cdots$. In particular, the cyclic repetition of the process with certain $\alpha_2, \cdots, \alpha_s$ is equivalent to applying a single process in which the sequence $\alpha_1, \alpha_2, \cdots, \alpha_s$, $\alpha_{s+1}, \cdots$ consists of cyclically recurring segments $0, \alpha_2, \cdots, \alpha_s$.

In the successive approximations method all $\alpha_k = 0$; in the $BT$-processes the sequence $\alpha_k$ consists of zeros and ones; in the optimal process $\alpha_k = [T_{k-2}(\gamma)]/T_k(\gamma)$.

The error vectors in the trinomial process satisfy the relationships

$$Y^{(1)} = BY^{(0)},$$

$$Y^{(k)} = (1 + \alpha_k)BY^{(k-1)} - \alpha_k Y^{(k-2)} \qquad (k = 2, 3, \cdots)$$

---

† D. K. Faddeev (2).

so that

$$Y^{(k)} = P_k(B)Y^{(0)}$$

where $P_k(t)$ is a sequence of polynomials constructed from the recurrence relationships

$$P_0(t) = 1 ,$$
$$P_1(t) = t ,$$
$$P_k(t) = (1 + \alpha_k)tP_{k-1}(t) - \alpha_k P_{k-2}(t) .$$

THEOREM 91.1.  *The polynomials $P_k(t)$ satisfy the inequalities $P_k(t) \leq 1$ for* $-1 \leq t \leq 1$, $-1 \leq \alpha_k \leq 1$ $(k = 2, 3, \cdots)$.

*Proof.* We write $P_k(t) = \Phi \ (t, \alpha_2, \cdots, \alpha_k)$. For fixed $t$ the polynomial $\Phi \ (t, \alpha_2, \cdots, \alpha_k)$ is a linear function of each of the arguments $\alpha_2, \cdots, \alpha_k$. Therefore in changing the parameters $\alpha_2, \cdots, \alpha_k$ in the cube $-1 \leq \alpha_2 \leq 1$, $\cdots, -1 \leq \alpha_k \leq 1$ the function $\Phi \ (t, \alpha_2, \cdots, \alpha_k)$ takes on extremal values at one of the cube's vertices, i.e.

$$|\Phi \ (t, \alpha_2, \cdots, \alpha_k)| \leq |\Phi \ (t_1, \varepsilon_2, \cdots, \varepsilon_k)|$$

where

$$\varepsilon_2 = \pm 1, \cdots. \varepsilon_k = \pm 1 .$$

We shall show that

$$\bar{P}_k(t) = \Phi \ (t, \varepsilon_2, \cdots, \varepsilon_k) = T_{s_k}(t) = \cos s_k \arccos t$$

where the Chebyshev number $s_{k+1}$ of each successive polynomial is greater or less by one than the number $s_k$ of the preceding one.

In fact this is true for $k = 0$, $k = 1$, since

$$\bar{P}_0(t) = 1 = T_0(t) ,$$
$$\bar{P}_1(t) = t = T_1(t) .$$

We assume that this is true for polynomials $\bar{P}_{k-1}(t)$ and $\bar{P}_k(t)$, i.e. we assume that $\bar{P}_k(t) = T_{s_k}(t)$ and $\bar{P}_{k-1}(t) = T_{s_k \mp 1}(t)$.  Then

$$\bar{P}_{k+1}(t) = (1 + \varepsilon_{k+1})t\bar{P}_k(t) - \varepsilon_{k+1}\bar{P}_{k-1} = (1 + \varepsilon_{k+1})tT_{s_k}(t) - \varepsilon_{k+1}T_{s_k \mp 1}(t) .$$

For $\varepsilon_{k+1} = -1$ we obtain

$$\bar{P}_{k+1}(t) = T_{s_k \mp 1}(t) .$$

For $\varepsilon_{k+1} = 1$ we obtain

$$\bar{P}_{k+1}(t) = 2tT_{s_k}(t) - T_{s_k \mp 1}(t) = T_{s_k \mp 1}(t) .$$

In both cases the polynomials $\bar{P}_{k+1}(t)$ and $\bar{P}_k(t)$ turned out to be Chebyshev polynomials with adjacent numbers.

In the same way

$$|P_k(t)| = |\Phi \ (t, \alpha_2, \cdots, \alpha_k)| \leq |\bar{P}_k(t)| = |T_{s_k}(t)| \leq 1 ,$$

which is what we were to prove.

It is possible to prove that if $0 < \alpha_k < \alpha < 1$, then the polynomials $P_k(t)$ approach zero uniformly in any interval $-\gamma \le t \le \gamma \le 1$ so that under these conditions the trinomial iterative process will be convergent.

**1. The trinomial algorithm with constant $a$.** Let $\alpha_k = \alpha$, $0 < \alpha < 1$ for all $k \ge 2$. Then, as is easy to see,

$$P_k(t) = \alpha^{k/2} T_k(\gamma t) + \frac{1-\alpha}{2\sqrt{\alpha}} \alpha^{k/2} t U_{k-1}(\gamma t) ,$$

where

$$\gamma = \frac{1+\alpha}{2\sqrt{\alpha}} .$$

$$\alpha = (\gamma - \sqrt{\gamma^2 - 1})^2 ,$$

$$T_k(t) = \cos k \arccos t ,$$

$$U_k(t) = \frac{\sin (k + 1) \arccos t}{\sin \arccos t} .$$

Therefore for $-1/\gamma \le t \le 1/\gamma$ we have

$$|P_k(t)| \le \alpha^{k/2}\left(1 + \frac{1-\alpha}{1+\alpha}k\right) = (\gamma - \sqrt{\gamma^2 - 1})^k\left(1 + \frac{1-\alpha}{1+\alpha}k\right) .$$

Thus if all eigenvalues of matrix $B$ are contained in the interval $(-1/\gamma, 1/\gamma)$, then the process with constant $\alpha_k = \alpha = (\gamma - \sqrt{\gamma^2 - 1})^2$ converges almost as rapidly as the optimal process.

**2. The universal algorithm with Chebyshev polynomials of the second type.** Let $\alpha_k = (k-1)/(k+1)$. In this case

$$P_k(t) = \frac{1}{k+1} \frac{\sin (k+1) \arccos t}{\sin \arccos t} = \frac{1}{k+1} U_k(t)$$

is a Chebyshev polynomial of the second type, normalized by the condition $P_k(1) = 1$.

It is obvious that the sequence $P_k(t)$ approaches zero uniformly in any

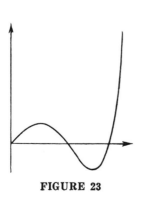

**FIGURE 23**

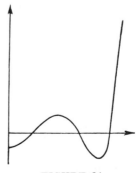

**FIGURE 24**

segment inside the interval $(-1, 1)$, although the sufficient condition for convergence formulated above is not fulfilled here. The process will converge quite slowly; however its merit consists in the fact that, even for small $k$, a perceptible reduction of components in the wider interval occurs.

In fact the maximum closest to one for the modulus $\sigma_k$ of the polynomial $P_k(t)$ is asymptotically equal to $|\cos \rho_1| = 0.217$ where $\rho_1$ is the least positive root of the transcendental equation $\tan t = t$. This maximum is attained at the point $t_1 = \cos \rho_1/(k+1) \approx 1 - \rho_1^2/[2(k+1)^2]$, close to one.

In Figs. 23 and 24 are given the graphs of the polynomials $P_5(t)$ and $P_6(t)$ and in Table 91.1 are given the values of the interior maximum of the modulus $\sigma_k$ of the polynomial $P_k(t)$ for $k = 2, 3, \cdots, 6$ and the bounds of the interval $(-\gamma_k, \gamma_k)$ in which $|P_k(t)| \leq \sigma_k$.

**TABLE 91.1  Values of $\sigma_k$ and $\gamma_k$**

| $k$ | $\sigma_k$ | $\gamma_k$ |
|-----|------------|------------|
| 2 | 0.333 | 0.707 |
| 3 | 0.272 | 0.816 |
| 4 | 0.250 | 0.878 |
| 5 | 0.239 | 0.913 |
| 6 | 0.233 | 0.935 |

| | $\alpha=1/2$ | | | |
|------|------------|------------|------------|------------|
| $X_0$ | 0.22900763 | 0.38167939 | 0.53435115 | 0.68702290 |
| $X_1$ | $-0.4055708$ | 0.0372933 | 0.3577880 | 0.5162869 |
| $X_2$ | $-0.5442885$ | 0.1987509 | 0.7684031 | 1.0678660 |
| $X_{18}$ | $-1.2593757$ | 0.0436874 | 1.0408131 | 1.4846543 |
| $X_{25}$ | $-1.2574775$ | 0.0435276 | 1.0389918 | 1.4821385 |
| $X_{30}$ | $-1.2578296$ | 0.0434832 | 1.0391862 | 1.4824220 |
| $X_{40}$ | $-1.2577930$ | 0.0434875 | 1.0391659 | 1.4823924 |
| $X_{41}$ | $-1.2577939$ | 0.0434874 | 1.0391664 | 1.4823932 |

| | $\alpha=1/3$ | | | |
|------|------------|------------|------------|------------|
| $X_0$ | 0.22900763 | 0.38167939 | 0.53435115 | 0.68702290 |
| $X_1$ | $-0.4055708$ | 0.0372939 | 0.3577880 | 0.5162869 |
| $X_2$ | $-0.4583667$ | 0.2190763 | 0.7423973 | 1.0255501 |
| $X_{18}$ | $-1.2577739$ | 0.0434788 | 1.0391330 | 1.4823466 |
| $X_{25}$ | $-1.2577920$ | 0.0434873 | 1.0391649 | 1.4823909 |
| $X_{30}$ | $-1.2577939$ | 0.0434873 | 1.0391662 | 1.4823929 |

It is advisable to apply the described universal algorithm cyclically for not very large $k$. Applying $m$ cycles suppresses the error vector components of the interval $(-\gamma_k, \gamma_k)$, with intensity $\sigma_k^m$.

We shall carry out the results of solving system (9), Section 23, prepared with $h = 2/2.62$ (Section 31, (3)) according to the trinomial algorithm with constant $\alpha$.

We shall also give the results of applying to the same system the algorithm with Chebyshev polynomials of the second type.

The cyclic passage of the process through four steps is contained in the general scheme for $\alpha_1 = 0$, $\alpha_2 = 1/3$, $\alpha_3 = 1/2$, $\alpha_4 = 3/5$, $\alpha_5 = 0$, $\alpha_6 = 1/3$, $\alpha_7 = 1/2$, $\alpha_8 = 3/5$, $\alpha_9 = 0$, $\cdots$. We obtain

| | | | | |
|---|---|---|---|---|
| $X_0$ | 0 | 0 | 0 | 0 |
| $X_4$ | $-1.2048821$ | 0.0715392 | 1.0432993 | 1.4836599 |
| $X_8$ | $-1.2594577$ | 0.0409823 | 1.0371510 | 1.4800347 |
| $X_{12}$ | $-1.2576202$ | 0.0437954 | 1.0393241 | 1.4825350 |
| $X_{16}$ | $-1.2578018$ | 0.0434441 | 1.0391507 | 1.4823855 |
| $X_{20}$ | $-1.2577945$ | 0.0434944 | 1.0391677 | 1.4823922 |
| $X_{24}$ | $-1.2577936$ | 0.0434859 | 1.0391659 | 1.4823933 |
| $X_{28}$ | $-1.2577938$ | 0.0434876 | 1.0391663 | 1.4823927 |

Applying polynomials of higher degrees in the given case gives a worse result. Namely, six steps of the process with $\alpha_1 = 0$, $\alpha_2 = 1/3$, $\alpha_3 = 1/2$, $\alpha_4 = 3/5$, $\alpha_5 = 2/3$, $\alpha_6 = 5/7$, $\alpha_7 = \alpha_1$, $\cdots$, gives

| | | | | |
|---|---|---|---|---|
| $X_0$ | 0 | 0 | 0 | 0 |
| $X_6$ | $-1.5437461$ | 0.0461754 | 1.2796197 | 1.8191787 |
| $X_{12}$ | $-1.1941027$ | 0.0504107 | 0.9848809 | 1.4059616 |
| $X_{18}$ | $-1.2734516$ | 0.0443237 | 1.0509812 | 1.4992437 |
| $X_{24}$ | $-1.2543485$ | 0.0436402 | 1.0363966 | 1.4784862 |
| $X_{30}$ | $-1.2586028$ | 0.0434955 | 1.0397846 | 1.4832740 |

## 92. THE UNIVERSAL ALGORITHM OF LANCZOS

A convenient universal algorithm has been proposed by Lanczos (6). We present it with certain insignificant changes which concern the preliminary preparation of the system. Namely, we shall consider that the system has been prepared in the from

$$X = BX + G \qquad (1)$$

(instead of $Y = \frac{1}{2}BY + 2c_0$ in the author's notation) and that all eigenvalues of matrix $B$ are located in the interval $(-1, 1)$.

As the polynomial $e_s(t)$ we take

$$e_s(t) = \frac{1 - T_{s+1}(t)}{(s+1)^2(1-t)}, \qquad (2)$$

where $T_{s+1}$ is a Chebyshev polynomial. It is obvious that $e_s(t)$ is really a polynomial, since $T_{s+1}(1) = 1$, and consequently $1 - T_{s+1}(t)$ is divisible by $1 - t$. It is easy to verify that $e_s(1) = 1$. In fact letting $t = \cos \theta$ we obtain that

$$e_s(t) = \frac{1 - \cos(s+1)\theta}{(s+1)^2(1 - \cos \theta)} = \frac{\sin^2 \dfrac{s+1}{2}\theta}{(s+1)^2 \sin^2 \dfrac{\theta}{2}} \qquad (3)$$

and consequently

$$e_s(1) = \lim_{\theta \to 0} \frac{\sin^2 \frac{s+1}{2}\theta}{(s+1)^2 \sin^2 \frac{\theta}{2}} = 1 \,.$$

From formula (3) it is apparent that the polynomial $e_s(t)$ is transformed by the substitution $t = \cos\theta$ into the Fejer kernel $K_s(\theta)$, normalized by the condition $K_s(0) = 1$. We give the graphs of $e_5(t)$ and $e_6(t)$ in Figs. 25 and 26.

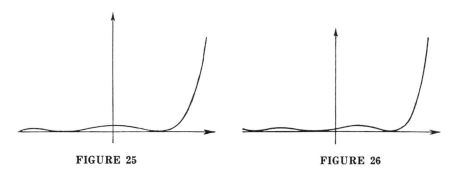

FIGURE 25             FIGURE 26

The sequential maxima of the polynomial $e_s(t)$ decrease as they move along the abscissa from right to left. The largest interior maximum $\sigma_s$ approaches the limit 0.0472 for $s \to \infty$. In Table 92.1 are given the values of the interior maximum $\sigma_s$ of the polynomials $e_s(t)$ for $s = 3, \cdots, 7$ and the boundary $\gamma_s$ of the interval $[-1, \gamma_s]$ in which $|e_s(t)| \leq \sigma_s$.

Comparison with Table 91.1 shows that for the same number $s$ the values of the maximum for the modulus in the Lanczos algorithm are significantly lower than in the algorithm with Chebyshev polynomials of the second type. The numbers $\gamma_s$ approximate 1 earlier in the latter algorithm.

**TABLE 92.1  Values of $\sigma_s$ and $\gamma_s$**

| $s$ | $\sigma_s$ | $\gamma_s$ |
|-----|--------|--------|
| 3 | 0.0741 | 0.334 |
| 4 | 0.0625 | 0.541 |
| 5 | 0.0572 | 0.683 |
| 6 | 0.0543 | 0.750 |
| 7 | 0.0525 | 0.806 |

The universal algorithm of Lanczos is carried out according to the scheme which uses recurrence relationships for the polynomial

$$f_{s-1}(t) = \frac{1 - e_s(t)}{1 - t} \,.$$

We shall derive these relationships. We have

$$f_{s-1}(t) = \frac{1 - e_s(t)}{1 - t} = \frac{1 - \dfrac{1 - T_{s+1}(t)}{(s+1)^2(1-t)}}{1 - t} = \frac{(s+1)^2(1-t) - 1 + T_{s+1}(t)}{(s+1)^2(1-t)^2} \ .$$

Therefore

$$(s+1)^2(1-t)^2 f_{s-1}(t) - (s+1)^2(1-t) + 1 = T_{s+1}(t) \ .$$

On the basis of the relationship

$$T_{s+2}(t) = 2t\,T_{s+1}(t) - T_s(t)$$

we obtain

$$(s+2)^2(1-t)^2 f_s(t) - (s+2)^2(1-t) + 1$$
$$= 2t(s+1)^2(1-t)^2 f_{s-1}(t) - 2t(s+1)^2(1-t)$$
$$+ 2t - s^2(1-t)^2 f_{s-2}(t) + s^2(1-t) - 1 \ .$$

After bringing together similar elements and cancelling $(1-t)^2$ we arrive at the relationship

$$(s+2)^2 f_s(t) = 2t(s+1)^2 f_{s-1}(t) - s^2 f_{s-2}(t) + 2(s+1)^2 \tag{4}$$

with initial polynomials $f_{-1} = 0$, $f_0 = 1/2$.

We now introduce for consideration the polynomials

$$F_s(t) = \frac{(s+2)^2}{4} f_s(t) \ . \tag{5}$$

These polynomials satisfy the still simpler recurrence relationship

$$F_s(t) = 2t F_{s-1}(t) - F_{s-2}(t) + \tfrac{1}{2}(s+1)^2 \tag{6}$$

under the initial conditions $F_0(t) = \tfrac{1}{2}$, $F_1(t) = t + 2$.   These last relationships permit us to carry out the universal algorithm in the following way.

Having computed the residual of the initial approximation $r = BX + G - X$, we form the vector sequence

$$Z_0 = \tfrac{1}{2}r \ ,$$
$$Z_1 = 2BZ_0 + 4Z_0 \ ,$$
$$Z_i = 2BZ_{i-1} - Z_{i-2} + (1+i)^2 Z_0 \qquad (i = 2, 3, \cdots, s-1) \ .$$

Then

$$X' = X + \frac{4}{(s+1)^2} Z_{s-1} \ .$$

In the notation of Lanczos $(Y = \tfrac{1}{2}BY + 2c_0)$ the formulas acquire the still simpler form

$$Z_0 = \tfrac{1}{2}r \ , \quad r = \tfrac{1}{2}BY + 2c_0 - Y \ ,$$
$$Z_1 = BZ_0 + 4Z_0 \ .$$
$$Z_i = BZ_{i-1} - Z_{i-2} + (i+1)^2 Z_0$$

and

$$Y' = Y + \frac{4}{(s+1)^2} Z_{s-1} .$$

The universal process we have been considering converges, with infinitely increasing $s$, very slowly, since the damping factors $e_{s+1}(\mu_i)$ at each fixed interval contained in $(-1, 1)$ decrease only with speed $1/s^2$.

However for large values of $s$ it gives significant component suppression for the error vector over a fairly broad interval $(-1, \gamma_s)$ for the eigenvalues.

Cyclic repetition of the process guarantees good convergence for systems with a $P$-number of condition which does not exceed $(1 + \gamma_s)/(1 - \gamma_s)$.

Lanczos recommends that we take $s = 7$. In this case (as applied to the system $Y = \frac{1}{2}BY + 2c_0$) we have

$$Z_6 = \frac{1}{2} (B^6 + 4B^5 + 4B^4 + 4B^2 + 16B + 32) ,$$

$$Y' = Y + \frac{1}{16} Z_6 .$$

It is recommended that the vector $Z_6$ not be computed from the recurrence relationships but from the formulas of Horner's scheme (Section 86, par. 4, b).

Lanczos advises that we use the described universal algorithm for high-order systems with the aim of preparing the initial approximation beforehand in applying the minimal iterations method, since with such a choice of the initial approximation the latter becomes degenerate quite rapidly. As an example we shall give the result of cyclically applying the Lanczos process for $s = 9$ to system (9) of Section 23, prepared with $h = 2/2.62$.

The first cycle gives

$$X = (-1.2574119, \quad 0.0428473, \quad 1.0387672, \quad 1.4829441)' ,$$

the second cycle:

$$X = (-1.2577870, \quad 0.0434717, \quad 1.0391658, \quad 1.4823988)' ,$$

and the third cycle:

$$X = (-1.2577935, \quad 0.0434868, \quad 1.0391663, \quad 1.4823931)' .$$

As a result of the third cycle (27 iterations) we obtain the components of the solution accurately to within $3 \cdot 10^{-7}$.

In the given example four cycles with $s = 7$ (28 iterations) give a worse result.

$$X = (-1.2577659, \quad 0.0434909, \quad 1.0391514, \quad 1.4823708)' .$$

### 93. UNIVERSAL ALGORITHMS BEST IN THE MEAN

As we have seen, in choosing polynomials which suppress the error vector components for the solution of the system

$$X = BX + G \tag{1}$$

under the condition that the eigenvalues of matrix $B$ are contained in the interval $(-1, 1)$ we are necessarily guided by two poorly compatible conditions, namely, the smallness of the polynomial's deviation from zero inside the interval $(-1, 1)$ and the conversion into 1 at its right end.

Since it is impossible to satisfy these two requirements literally, it is natural to try to satisfy the first of them by making the mean squared deviation small.

That is, we shall construct the polynomial $e_s(t)$ that minimizes

$$\int_{-1}^{+1} \rho(t)e_s^2(t)dt$$

in the class of polynomials of degree $s$ which satisfy the condition $e_s(1) = 1$. The weight function $\rho(t) > 0$ may be expressed in a different way on the basis of information about the eigenvalue distribution density for the given matrix $B$ in the interval $(-1, 1)$ and information about the character of the distribution of components for the initial error vector. If such information is lacking it is natural to take $\rho(t) = 1$ for $-1 \le t \le 1$.

It is easily established that the polynomials $e_s(t)$ form an orthogonal system in terms of the weight $(1 - t)\rho(t)$. In fact let

$$e_s(t) = 1 + a_1(1 - t) + a_2(1 - t)^2 + \cdots + a_s(1 - t)^s ,$$

$$J = \int_{-1}^{1} \rho(t)e_s^2(t)dt . \tag{2}$$

Then it is obvious that

$$\frac{\partial J}{\partial a_i} = 2 \int_{-1}^{1} \rho(t)(1 - t)^i e_s(t)dt .$$

For an extremal polynomial, all partial derivatives are equal to zero, so that $e_s(t)$ will be orthogonal by weight $\rho(t)$ to the polynomials $(1 - t)^i$ for $i = 1, \cdots, s$ and by weight $(1 - t)\rho(t)$ to the polynomials $(1 - t)^{i-1}$ for $i = 1, \cdots, s$, and consequently to any polynomial whose degree is less than $s$. In particular,

$$\int_{-1}^{1} (1 - t)\rho(t)e_s(t)e_i(t)dt = 0 \quad \text{for} \quad i = 0, 1, \cdots, s - 1 . \tag{3}$$

It follows from the theory of orthogonal polynomials that for any weight function the polynomials $e_i(t)$ are connected by the trinomial recurrence relationships

$$e_i(t) = (\alpha_i t + \beta_i)e_{i-1}(t) - \gamma_i e_{i-2}(t) . \tag{4}$$

This lets us construct successive approximations to the solution of the system in the universal algorithm best in the mean for a given choice of weight function according to the formulas of paragraph 8, Section 86, as soon as the coefficients $\alpha_i$, $\beta_i$ and $\gamma_i$ have been computed.

For $\rho(t) = 1$ relationships (4) have the form

$$e_0(t) = 1,$$

$$e_1(t) = \frac{3}{4} t + \frac{1}{4},$$

$$e_i(t) = \left[\frac{i(2i+1)}{(i+1)^2} t + \frac{i}{(2i-1)(i+1)^2}\right] e_{i-1}(t) - \frac{(2i+1)(i-1)^2}{(2i-1)(i+1)^2} e_{i-2}(t).$$

Therefore the successive approximations are computed from the formulas

$$X_1 = \frac{3}{4}(BX_0 + G) + \frac{1}{4} X_0,$$

$$X_i = \frac{i(2i+1)}{(i+1)^2}(BX_{i-1} + G) + \frac{i}{(2i-1)(i+1)^2} X_{i-1} - \frac{(2i+1)(i-1)^2}{(2i-1)(i+1)^2} X_{i-2}.$$

In Figs. 27 and 28 are given graphs of

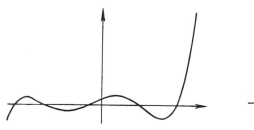 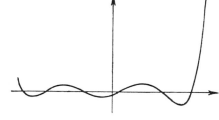

**FIGURE 27**                                    **FIGURE 28**

$$e_5(t) = \frac{77}{32} t^5 + \frac{35}{32} t^4 - \frac{35}{32} t^3 - \frac{35}{48} t^2 + \frac{35}{96} t + \frac{5}{96}$$

and

$$e_6(t) = \frac{429}{112} t^6 + \frac{99}{56} t^5 - \frac{495}{112} t^4 - \frac{45}{28} t^3 + \frac{135}{112} t^2 + \frac{15}{56} t - \frac{5}{112}.$$

In Table 93.1 are given values for $\sigma_s$ and $\gamma_s$ where $\sigma_s$ and $\gamma_s$ have the same meaning as in Sections 91 and 92.

**TABLE 93.1   Values for $\sigma_s$ and $\gamma_s$**

| $s$ | $\sigma_s$ | $\gamma_s$ |
|-----|------------|------------|
| 2 | 1/3 | 0.6 |
| 3 | 1/4 | 0.735 |
| 4 | 1/5 | 0.804 |
| 5 | 1/6 | 0.860 |
| 6 | 1/7 | 0.894 |

The polynomials $e_s(t)$ are easily related to the Legendre polynomials

$$L_s(t) = \frac{1}{2^s \cdot s!} \frac{d^s (t^2 - 1)^s}{dt^s}.$$

Namely,

$$e_s(t) = \frac{L_{s+1}(t) - L_s(t)}{(s+1)(t-1)}.$$

For

$$\rho(t) = (1 - t)^{\alpha}(1 + t)^{\beta} \qquad (\alpha > -1, \ \beta > -1)$$

we obtain polynomials orthogonal by weight $(1 - t)^{\alpha+1}(1 + t)^{\beta}$, i.e. the so-called Jacobian polynomials (hypergeometric polynomials). The application of such polynomials is considered in the work of Stiefel (5).

The universal algorithm best in the mean for $\rho(t) = 1$ was applied cyclically to find the solution of system (9), Section 23, prepared with $h = 2/2.62$. We obtained:

| | | | | |
|---|---|---|---|---|
| $X_0$ | 0 | 0 | 0 | 0 |
| $X_6$ | $-1.3813506$ | 0.0338070 | 1.1209397 | 1.5985210 |
| $X_{12}$ | $-1.2464295$ | 0.0446982 | 1.0325678 | 1.4726554 |
| $X_{18}$ | $-1.2588065$ | 0.0433614 | 1.0397375 | 1.4832186 |
| $X_{24}$ | $-1.2577045$ | 0.0434987 | 1.0391177 | 1.4823210 |
| $X_{30}$ | $-1.2578015$ | 0.0434862 | 1.0391704 | 1.4823990 |
| $X_{35}$ | $-1.2577939$ | 0.0434873 | 1.0391663 | 1.4823930 |
| $X_{36}$ | $-1.2577930$ | 0.0434874 | 1.0391658 | 1.4823923 |

Here $X_{35}$ turned out to be closer to the exact solution than $X_{36}$. This indicates that one of the roots for the polynomial $e_5(t)$ turned out to be closer to the eigenvalue whose component was still insufficiently suppressed by the previous cycles.

In using special polynomials for suppressing error vector components it is generally advisable to use likewise the "anomalous" suppression that originates from the proximity of one of the roots to an eigenvalue, as is done in the $BT$-process "with direction"

## 94. A METHOD FOR SUPPRESSING COMPONENTS IN A COMPLEX REGION

Until now all the universal algorithms we have considered have been constructed for systems with real eigenvalues in the matrix of coefficients.

It is easy to extend theoretically the idea of component suppression to systems whose matrices have complex eigenvalues.

In the complex plane let there be given a bounded closed set $\Sigma$ whose complement is the connected region $\Delta$ which contains the point $z = 0$.

We consider the class of systems

$$AX = F$$

such that all eigenvalues of matrices $A$ lie in the set $\Sigma$. For such a class of systems we may construct (on a theoretical level) a universal algorithm best in the sense of the first criterion.

With this in mind it is sufficient to construct polynomials $g_s(z)$ which satisfy the requirement $g_s(0) = 1$ and which diverge least from zero in the set $\Sigma$ and then to construct approximate solutions of the system according to the formula

$$X_s = X_0 + h_{s-1}(A)(F - AX_0)$$

where $h_{s-1}(t) = [1 - g_s(t)]/t$.

We shall give an estimate for the convergence of such a universal algorithm for $s \to \infty$.

Let

$$\tau_s(z) = z^s + \cdots$$

be the polynomial which diverges least from zero in the set $\Sigma$ and such that the roots of the polynomial $\tau_s(z)$ lie in the set $\Sigma$. Then the divergence of the polynomial $g_s(z)$ from zero in the set $\Sigma$ is no more than the divergence of the polynomial $\tau_s(z)/\tau_s(0)$ from zero. We denote the divergence of the polynomial $\tau_s(z)$ from zero by $\tau_s$. Then, as is well-known,[†]

$$\lim \sqrt[s]{\frac{\tau_s}{|\tau_s(0)|}} = e^{-G(0)}$$

where $G(z)$ is Green's function for the region $\Delta$ with a logarithmic singularity at the point $z = \infty$. Therefore for $z \in \Sigma$

$$|g_s(z)| \le \frac{\tau_s}{|\tau_s(0)|} \le C(\varepsilon)e^{-s(G(0)-\varepsilon)}$$

where $\varepsilon$ is an arbitrarily small positive number and $C(\varepsilon)$ is a constant depending on $\varepsilon$.

On the other hand, let $R = \max_{z \in \Sigma} g_s(z)$ and let $\Sigma^*$ be an aggregate of points such that $|g_s(z)| \le R$; $\Sigma^*$ is a bounded closed set containing $\Sigma$ and its complement $\Delta^*$ is a connected region contained in $\Delta$. It is clear that $1/s \log |g_s(z)|/R$ is Green's function for the region $\Delta^*$. Since $\Delta^* \subset \Delta$, then the inequality $1/s \log |g_s(z)|/R \le G(z)$ will be satisfied for all $z \in \Delta^*$. This inequality will also hold for all points $z \in \Delta$, since if $z \in \Delta$ and $z \notin \Delta^*$, then $G(z) \ge 0$ and $|g_s(z)| \le R$, so that $1/s \log |g_s(z)|/R \le 0$. Letting $z = 0$ we obtain

$$\frac{1}{s} \log \frac{1}{R} \le G(0)$$

from which

$$R = \max_{z \in \Sigma} g_s(z) \ge e^{-sG(0)}.$$

In this way,

$$e^{-sG(0)} \le R \le C(\varepsilon)e^{-s(G(0)-\varepsilon)}.$$

Thus the universal algorithm best in the sense of the first criterion converges with the rate of a geometric progression with denominator $e^{-(G(0)-\varepsilon)}$.

General methods for constructing the polynomials $g(z)$ are unknown, so that the described algorithm may be applied only to certain special regions. In Section 97 the possibility of constructing $\Sigma$-universal algorithms will be

[†] G. M. Goluzin, Geometricheskaya teoriya funktsii kompleksnogo peremennogo (Geometric theory of functions of a complex variable), G.I.T.T.L., 1952, Ch. VII.

established using conformal mapping of the unit circle into the region $\varDelta$ (or into its simply connected covering); for simply-connected regions $\varDelta$ these possess almost the same rate of convergence.

Technically it is easier to construct the universal algorithm best in the mean. This comes down to constructing the polynomials $g_s(t)$ which minimize $\int |g_s(z)|^2 d\mu$ where $\mu$ is a certain non-negative measure concentrated in the set $\Sigma$.

It is not hard to give formulas for constructing the polynomials $g_s(z)$ as soon as the system of orthogonal polynomials in measure $\mu$ is known. Let $P_i(z)$ be the orthogonal system of polynomials in measure $\mu$. We shall look for a polynomial $g_s(z)$ in the form

$$g_s(z) = \sum_{i=0}^{s} c_i P_i(z) \, .$$

Then

$$I = \int |g_s(z)|^2 \, d\mu = |c_0|^2 + |c_1|^2 + \cdots + |c_s|^2$$

The condition $g_s(0) = 1$ takes the form

$$\sum_{i=0}^{s} c_i P_i(0) = 1 \, .$$

According to the Cauchy-Bunyakovskii inequality

$$\sum_{i=0}^{s} |c_i|^2 \sum_{i=0}^{s} |P_i(0)|^2 \geq \left| \sum_{i=0}^{s} c_i P_i(0) \right|^2 = 1$$

from which

$$I \geq \frac{1}{\displaystyle\sum_{i=0}^{s} |P_i(0)|^2} \, .$$

Equality is attained for

$$c_i = \frac{\overline{P_i(0)}}{\displaystyle\sum_{i=0}^{s} |P_i(0)|^2}$$

where $\overline{P_i(0)}$ is the complex number conjugate to $P_i(0)$. Consequently,

$$g_s(z) = \frac{1}{\displaystyle\sum_{i=0}^{s} |P_i(0)|^2} \sum_{i=0}^{s} \overline{P_i(0)} P_i(z) \, .$$

Orthogonal polynomials of a given measure for different sets $\Sigma$ have been studied in the classic works of Szegö, Bochner and V. I. Smirnov.

## 95. THE APPLICATION OF CONFORMAL MAPPING IN SOLVING LINEAR SYSTEMS

Until now we have considered universal algorithms for the system $X = BX + G$ under the condition that all eigenvalues of matrix $B$ were real and contained in the interval $(-1, 1)$. V. N. Kublanovskaya[†] has worked out a method which uses the theory of functions of a complex variable to construct universal algorithms for systems $X = BX + G$, with other assumptions concerning the eigenvalue distribution for matrix $B$.

Instead of the system

$$X = BX + G \tag{1}$$

we study the system

$$X = zBX + G \tag{2}$$

which contains the complex parameter $z$. The solution of the latter system is obviously $X(z) = (E - zB)^{-1}G$. All components of the solution are analytic and even rational functions of $z$, with poles only at the points $1/\mu_1, \cdots, 1/\mu_n$ where $\mu_1, \cdots, \mu_n$ are the eigenvalues of matrix $B$. In fact,

$$X(z) = (E - zB)^{-1}G = \frac{1}{|E - zB|}C(z)G \tag{3}$$

where $C(z)$ is the matrix adjoint to $E - zB$. The elements of $C(z)$ are obviously polynomials in $z$ and the roots of the denominator are numbers reciprocal to the eigenvalues of matrix $B$. Moreover we have

$$(E - zB)^{-1}G = G + zBG + z^2B^2G + \cdots \tag{4}$$

The radius of convergence for this series is equal to $1/\max|\mu_i|$ and if there exist among the eigenvalues of matrix $B$ some which are greater than one in modulus, then the interesting point $z = 1$ turns out to be beyond the limits of the convergence circle.

We now let

$$z = z(w) = a_1 w + a_2 w^2 + \cdots \tag{5}$$

where $z(w)$ is a function meromorphic in the unit circle $|w| < 1$ which takes on the value $z = 1$ at a certain point $\theta$ of this circle and which does not take on the values $1/\mu_1, \cdots, 1/\mu_n$ inside the circle (likewise, if among the numbers $\mu_i$ there exists a zero, then the function $z(w)$ must be regular).

All components of the vector $X(z)$ will be regular functions of $w$ in the circle $|w| < 1$. In fact the denominator $|E - zB|$ will not be reduced to zero for $|w| < 1$. At the poles of the function $z(w)$, if they exist (which is possible only for $|B| \neq 0$), all components of $X(z)$ will be equal to zero, since $(E - zB)^{-1}G = 1/z(1/z E - B)^{-1}G \to 0$ for $z \to \infty$. Therefore the solution $X(z)$ of system (2) may be expanded in a series according to powers of $w$ with a convergence radius which is not less than one.

[†] V. N. Kublanovskaya (1), (2).

To construct this series we expand the function $1/(1 - zt)$ in terms of powers of $w$. We have

$$\frac{1}{1 - zt} = 1 + tz + t^2z^2 + \cdots$$

$$= 1 + t(a_1w + a_2w^2 + \cdots) + t^2(a_1w + a_2w^2 + \cdots)^2 + \cdots$$
$$= 1 + a_1tw + (a_2t + a_1^2t^2)w^2 + \cdots \tag{6}$$
$$= 1 + b_1(t)w + b_2(t)w^2 + \cdots + b_i(t)w^i + \cdots ,$$

where $b_i(t)$ are certain polynomials in $t$ of degree $i$. The coefficients of the polynomials $b_i(t)$ may be computed as soon as the expansion of the function $z(w)$ is known.

For $w = \theta$ the series (6) is transformed into the expansion

$$\frac{1}{1 - t} = 1 + b_1(t)\theta + \cdots + b_i(t)\theta^i + \cdots \tag{7}$$

We now use equality (6) to expand the solution $X(z)$ in a series in powers of $w$ and obtain

$$X(z) = (E - zB)^{-1}G = G + wb_1(B)G + w^2b_2(B)G + \cdots$$

The solution of the initial system $X = X(1)$ is found by the formula

$$X = (E - B)^{-1}G = G + \theta b_1(B)G + \theta^2 b_2(B)G + \cdots \tag{8}$$

This series will always converge, since $0 < |\theta| < 1$, and the convergence will be faster as $|\theta|$ decreases.

We now assume that, regarding matrix $B$, it is known that all its eigenvalues belong to a certain bounded set $S$ whose complement is a simply-connected region containing the point 1. Under these conditions it is natural to raise the question of an $S$-universal algorithm, i.e. an algorithm applied to all systems $X = BX + G$ such that the eigenvalues of matrix $B$ belong to $S$.

$S$-universal algorithms may be constructed using conformal mapping. We denote by $D$ the region which is obtained from the complement of $S$ by mapping the function $1/z$. The region $D$ so constructed will be simply-connected and will contain the points 0 and 1; it will not contain numbers reciprocal to the eigenvalues of all matrices $B$ of the class considered. The point $\infty$ may or may not belong to the region $D$.

In applying formula (8) to the whole class of systems considered we may take as $z(w)$ any function meromorphic in the unit circle whose values belong to the region $D$ and which takes on the value 1 at a certain point $\theta$ of this circle. In particular, such a function is the one which realizes a conformal mapping of the unit circle into the region $D$. From the theory of conformal mapping it follows that such a choice of the function $z(w)$ is the most propitious one, since the smallest possible value for $|\theta|$ is achieved for just this function.

In fact let $\tilde{z}(w) = \tilde{a}_1w + \tilde{a}_2w^2 + \cdots$ be any permissible function where

$\tilde{z}(\tilde{\theta}) = 1$.  Furthermore, $z = z(w) = a_1 w + a_2 w^2 + \cdots$ is a function which realizes a conformal mapping of the unit circle into region $D$ and $w = F(z)$ is the functional inverse to $z(w)$ where $\theta = F(1)$.  We shall show that $|\theta| < |\tilde{\theta}|$.

With this in mind we shall consider the function $W(w) = (F(\tilde{z}(w))$.  It is clear that $W(0) = 0$, $W(w)$ is regular in the unit circle, and $|W(w)| < 1$ for $|w| < 1$.  According to a lemma of Schwartz[†] for all points of the unit circle we have

$$|W(w)| \le |w| .$$

Letting $w = \tilde{\theta}$ we have

$$|W(\tilde{\theta})| = |F(1)| = |\theta| \le |\tilde{\theta}| .$$

The equality sign may hold only if $\tilde{z}(w) = z(w\varepsilon)$, $|\varepsilon| = 1$, i.e. if the function $\tilde{z}(w)$ itself realizes a conformal mapping of the unit circle into region $D$.

The conformal mapping method may be applied in the case where the complement of $S$, and along with it $D$, are multiply connected regions.  In this case the best $z(w)$ is a function which realizes a conformal mapping of the unit circle into a universal Riemann surface covering $D$.

We shall now go on to describe computational schemes for the method. In the work of V. N. Kublanovskaya (2) are given auxiliary tables of coefficients for the polynomials

$$L_s(t) = 1 + \theta b_1(t) + \theta^2 b_2(t) + \cdots + \theta^s b_s(t)$$
$$= l_{s0} + l_{s1} t + \cdots + l_{ss} t^s$$

for a series of mapping functions $z(w)$ for $s = 5$ and $s = 10$.

This lets us compute the approximations $L_s(B)G$ to the solution $X$ as a linear combination of successive iterations $G$, $BG$, $B^2 G, \cdots$ or by applying Horner's scheme.  For this we must fix beforehand the number of elements taken in formula (8).

Restricting ourselves to small $s$ makes it hard to count on obtaining sufficient accuracy.  However a single application of the approximate formula $X = L_s(B)G$ may be looked on as an elementary step in the iterative process $X^{(k)} = X^{(k-1)} + L_s(B)r_{k-1}$ where $r_{k-1} = BX^{(k-1)} + G - X^{(k-1)}$.

It is easy to construct computational schemes for the method which use only the coefficients $a_j$ of the mapping function $z(w)$.  Namely, from the identity

$$[1 - t(a_1 w + a_2 w^2 + \cdots)][1 + b_1(t)w + b_2(t)w^2 + \cdots] = 1 ,$$

equivalent to formula (6), we obtain recurrent relationships for the polynomials $b_i(t)$.  Namely,

$$b_1(t) = a_1 t ,$$
$$b_i(t) = a_1 t b_{i-1}(t) + a_2 t b_{i-2}(t) + \cdots + a_{i-1} t b_1(t) + a_i t . \qquad (9)$$

[†] G. M. Goluzin, *op. cit.*, p. 29.

Writing $b_i(B)G = G_i$, $G_0 = G$, we obtain

$$G_i = B(a_1 G_{i-1} + \cdots + a_i G_0) \qquad (i = 1, 2, \cdots)$$

$$X = \sum_{i=0}^{\infty} \theta^i G_i . \tag{10}$$

The components of the vectors $G_i$ will either not increase or will increase very slowly, since the radius of convergence for the series $\sum G_i w^i$ is equal to one.

Recurrent relationships analogous to (9) may also be established for the polynomials $L_i(t)$. Namely, it is easy to verify that

$$L_i(t) = 1 + a_1 \theta t L_{i-1}(t) + a_2 \theta^2 t L_{i-2}(t) + \cdots + a_i \theta^i t L_0(t) . \tag{11}$$

This makes it possible to construct the successive approximations $X_i = L_{i-1}(B)G$ themselves by recurrence formulas. Namely,

$$X_1 = G , \quad X_2 = G + a_1 \theta BG$$

$$X_i = G + a_1 \theta B X_{i-1} + a_2 \theta^2 B X_{i-2} + \cdots + a_{i-1} \theta^{i-1} B X_1 . \tag{12}$$

It is obvious that formulas (12) may have the form

$$X_1 = G , \quad X_2 = (1 - a_1 \theta)G + a_1 \theta (B X_1 + G)$$

$$X_i = (1 - a_1 \theta - a_2 \theta^2 - \cdots - a_{i-1} \theta^{i-1})G + a_1 \theta (B X_{i-1} + G) \tag{13}$$

$$+ \cdots + a_{i-1} \theta^{i-1}(B X_1 + G) .$$

Computations by recurrence formulas (10), (12) or (13) require a slightly larger number of computational operations than does computation using co-efficients $l_{ij}$ computed beforehand; however they have the advantage that it is not necessary to fix the number of series elements beforehand.

Formulas (13) correspond to par. 7, Section 86. To obtain formulas an-alogous to par. 8, Section 86, we consider the polynomial

$$l_i(t) = 1 - (1 - t)L_{i-1}(t) . \tag{14}$$

It is easy to verify that the polynomials $l_i(t)$ are connected by the recur-rence relationships

$$l_i(t) = a_1 \theta t l_{i-1}(t) + \cdots + a_{i-1} \theta^{i-1} t l_1(t)$$

$$+ (1 - a_1 \theta - \cdots - a_{i-1} \theta^{i-1})t . \tag{15}$$

From here we obtain for the $i$th approximation $X_i = l_i(B)X_0$

$$X_i = a_1 \theta (B X_{i-1} + G) + \cdots + a_{i-1} \theta^{i-1}(B X_1 + G)$$

$$+ (1 - a_1 \theta - \cdots - a_{i-1} \theta^{i-1})(B X_0 + G) . \tag{16}$$

Formula (16) is transformed into formula (13) for $X_0 = 0$.

In many cases recurrence formulas constructed starting with the coeffici-ents $d_j$ of the expansion of the function $w/z$ as a series in powers of $w$ turn out to be more convenient. This function will be regular, since $z = 0$ only for $w = 0$ (if the region $D$ is multiply connected, then the function $w/z$ will be meromorphic).

Let

$$\frac{w}{z} = \frac{w}{a_1 w + a_2 w^2 + \cdots} = d_0 + d_1 w + d_2 w^2 + \cdots \qquad (d_0 \neq 0) . \qquad (17)$$

Then

$$\frac{1}{1 - zt} = \frac{w/z}{w/z - tw} = \frac{d_0 + d_1 w + d_2 w^2 + \cdots}{d_0 + d_1 w + d_2 w^2 + \cdots - tw}$$

$$= \frac{d_0 + d_1 w + d_2 w^2 + \cdots}{d_0 + (d_1 - t)w + d_2 w^2 + \cdots} = 1 + b_1(t)w + b_2(t)w^2 + \cdots \qquad (18)$$

From here, equating coefficients of like powers, we obtain

$$d_0 b_1(t) + (d_1 - t) = d_1 ,$$
$$d_0 b_2(t) + (d_1 - t)b_1(t) = 0 ,$$
$$d_0 b_i(t) + (d_1 - t)b_{i-1}(t) + d_2 b_{i-2}(t) + \cdots + d_{i-1} b_1(t) = 0 , \qquad (19)$$

so that

$$b_1(t) = \frac{1}{d_0} t; \quad b_2(t) = \frac{t - d_1}{d_0} b_1(t) ,$$

$$b_i(t) = \frac{t - d_1}{d_0} b_{i-1}(t) - \frac{d_2}{d_0} b_{i-2}(t) - \cdots - \frac{d_{i-1}}{d_0} b_1(t) \qquad (i \geq 3) . \qquad (20)$$

Formulas (20) permit us to compute the vectors $G_i$ in formula (10) according to the recurrence relationships

$$G_1 = \frac{1}{d_0} BG , \quad G_2 = \frac{1}{d_0} BG_1 - \frac{d_1}{d_0} G_1 ,$$

$$G_i = \frac{1}{d_0} BG_{i-1} - \frac{d_1}{d_0} G_{i-1} - \cdots - \frac{d_{i-1}}{d_0} G_1 \qquad (i \geq 3) . \qquad (21)$$

It is likewise easy to derive the recurrence formulas for the polynomials $L_i(t)$ and $l_i(t)$. Namely,

$$L_0(t) = 1; \quad L_1(t) = 1 + \frac{\theta}{d_0} t; \quad L_2(t) = \frac{\theta(t - d_1)}{d_0} L_1(t) + \frac{d_0 + d_1 \theta}{d_0};$$

$$L_i(t) = \frac{\theta(t - d_1)}{d_0} L_{i-1}(t) - \frac{\theta^2 d_2}{d_0} L_{i-2}(t) - \cdots - \frac{\theta^{i-1} d_{i-1}}{d_0} L_1(t) \qquad (22)$$

$$+ \frac{d_0 + d_1 \theta + \cdots + d_{i-1}\theta^{i-1}}{d_0} .$$

Furthermore,

$$l_i(t) = \frac{\theta(t - d_1)}{d_0} l_{i-1}(t) - \frac{\theta^2 d_2}{d_0} l_{i-2}(t) - \cdots - \frac{\theta^{i-2} d_{i-2}}{d_0} l_2(t)$$

$$+ \frac{d_0 + d_1 \theta + \cdots + d_{i-2}\theta^{i-2} - \theta}{d_0} l_1(t) \qquad (i \geq 3) \qquad (23)$$

for $l_1(t) = t$, $l_2(t) = t - \frac{\theta}{d_0} t + \frac{\theta}{d_0} t^2$ .

Therefore the successive approximations may be computed by the formulas

$$X_1 = G; \quad X_2 = \frac{\theta}{d_0} BX_1 + X_1;$$

$$X_i = \frac{\theta}{d_0} BX_{i-1} - \frac{\theta d_1}{d_0} X_{i-1} - \frac{\theta^2 d_2}{d_0} X_{i-2} - \cdots - \frac{\theta^{i-2} d_{i-2}}{d_0} X_2$$

$$+ \frac{d_0 + d_1\theta + \cdots + d_{i-2}\theta^{i-2}}{d_0} X_1 \quad (i \geq 3) \tag{24}$$

or by the formulas

$$X_1 = BX_0 + G; \quad X_2 = \frac{\theta}{d_0}(BX_1 + G) + \frac{d_0 - \theta}{d_0} X_1;$$

$$X_i = \frac{\theta}{d_0}(BX_{i-1} + G) - \frac{\theta d_1}{d_0} X_{i-1} - \frac{\theta^2 d_2}{d_0} X_{i-2} - \cdots - \frac{\theta^{i-2} d_{i-2}}{d_0} X_2$$

$$+ \frac{d_0 + d_1\theta + \cdots + d_{i-2}\theta^{i-2} - \theta}{d_0} X_1 \quad (i \geq 3) . \tag{25}$$

We shall now consider the conformal mapping method from the point of view of component suppression. Let $X_0$ be the initial approximation and let $X' = X_s = X_0 + L_{s-1}(B)r_0$ where $r_0 = BX_0 + G - X_0$. Then corresponding error vectors are connected by the relationship

$$Y' = Y_0 - L_{s-1}(B)r_0 = l_s(B)Y_0$$

where $l_s(t) = 1 - (1 - t)L_{s-1}(t)$.

We shall estimate $|l_s(t)|$ for any point $t$ belonging to the set $S$.

Let $0 < \rho_0 < 1$ and let $C_{\rho_0}$ be the image of the circle $|w| = \rho_0$ for the mapping $z = z(w)$ of the unit circle into the region $D$. We have

$$|b_j(t)| - \left| \frac{1}{2\pi i} \int_{|w|=\rho_0} \frac{dw}{w^{j+1}[1 - tz(w)]} \right|.$$

While $|w| = \rho_0$ the function $z(w)$ runs along the curve $C_{\rho_0}$. If $t \in S$, then $1/t$ does not belong to the region $D$. Therefore $1 - tz$ is not equal to zero for $t \in S$, $z \in D$ and $|1 - tz|$ is uniformly bounded from below by the constant $d_{\rho_0}$, while $z \in C_{\rho_0}$, $t \in S$. Therefore

$$|b_j(t)| \leq \frac{1}{2\pi} \frac{2\pi\rho_0}{\rho_0^{j+1}d_{\rho_0}} = \frac{1}{\rho_0^j d_{\rho_0}} .$$

holds as a uniform estimate for $t \in S$.

Thus, on the basis of formula (7),

$$\left| \frac{1}{1-t} - L_{s-1}(t) \right| \leq \left| \sum_{j=s}^{\infty} b_j(t)\theta^j \right| \leq \sum_{j=s}^{\infty} \left( \frac{|\theta|}{\rho_0} \right)^j \frac{1}{d_{\rho_0}} = \frac{1}{d_{\rho_0}} \frac{\left( \frac{|\theta|}{\rho_0} \right)^s}{1 - \frac{|\theta|}{\rho_0}},$$

from which

$$| l_s(t) | \leq | 1 - t | \frac{1}{d\rho_0} \frac{\left( \frac{|\theta|}{\rho_0} \right)^s}{1 - \frac{|\theta|}{\rho_0}} \leq C(\rho_0)[(1 + \varepsilon)| \theta |]^s .$$

Here $\varepsilon = \dfrac{1}{\rho_0} - 1$ is a positive number which may be made as small as we wish and

$$C(\rho_0) = \max | 1 - t | \frac{1}{d_{\rho_0} \left( 1 - \dfrac{|\theta|}{\rho_0} \right)} .$$

Thus the error vector components are "damped" with at least speed $[(1 + \varepsilon)| \theta |]^s$.

We make the following observation regarding application of the conformal mapping method.

If we introduce the parameter $z$ into the system $X = BX + G$ in the form

$$X = zBX + \rho(z)G$$

where $\rho(z)$ is a regular function in the region $D$ which reduces to one for $z = 1$, then by the method of conformal mapping we obtain the solution to the initial system in the form

$$X = c_0 G + \theta c_1(B)G + \theta^2 c_2(B)G + \cdots$$

where $c_i(t)$ are polynomials which are the coefficients in the expansion of the function $\sigma(w)/[1 - tz(w)] = c_0 + c_1(t)w + c_2(t)w^2 + \cdots$. Here $\sigma(w) = \rho(z(w))$ is a function which is regular in the unit circle. For any choice of the function $\sigma(z)$ (or, what is the same thing, of $\rho(z)$) the order of the rate of convergence will be one and the same.

## 96.  EXAMPLES OF $S$-UNIVERSAL ALGORITHMS

We shall now consider several concrete $S$-algorithms. As we have seen above, each such algorithm is determined by a bounded closed set $S$ which contains the eigenvalues of matrices $B$.

1.  $S$ is a circle of radius $1/\gamma$, $\gamma > 1$, which center at the coordinate origin (Fig. 29). In this case region $D$ is the interior of a circle of radius $\gamma$ with center at the coordinate origin. The mapping function is

$$z(w) = \gamma w ,$$

$$\theta = \frac{1}{\gamma} .$$

Therefore the successive approximations are computed by the formula

$$X_i = BX_{i-1} + G ,$$

i.e. in this case we arrive at the classical method of successive approximations.

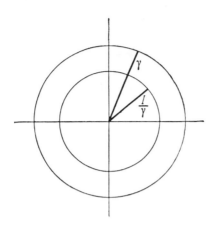

**FIGURE 29**

2. $S$ is a segment of the real axis $(-1/\gamma, 1/\gamma)$, $\gamma > 1$ (Fig. 30). In this case region $D$ is the plane with two cuts along the real axis going from the points $-\gamma$ and $\gamma$ to infinity. The mapping function is

$$z(w) = \frac{2\gamma w}{1 + w^2},$$

$$\theta = \gamma - \sqrt{\gamma^2 - 1}.$$

However in this case the function $w/z(w)$ will be even simpler. That is $w/z(w) = (1 + w^2)/2\gamma$, so that $d_0 = d_2 = 1/2\gamma$, $d_1 = d_3 = d_4 = \cdots = 0$. Therefore the successive approximations are conveniently constructed according to recurrence formulas (25), which become the formulas

$$X_i = 2\gamma\theta(BX_{i-1} + G) - \theta^2 X_{i-2}$$
$$= (1 + \theta^2)(BX_{i-1} + G) - \theta^2 X_{i-2} \qquad (\text{for } i \geq 3)$$

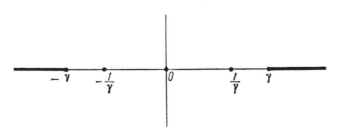

**FIGURE 30**

$$X_1 = BX_0 + G,$$
$$X_2 = 2\gamma\theta(BX_1 + G) + (1 - 2\gamma\theta)X_1$$
$$= (1 + \theta^2)(BX_1 + G) - \theta^2 X_1 .$$

The process almost coincides with the universal trinomial algorithm with constant multiplier $\alpha = \theta^2$ (Section 91), differing from it only at the beginning of the process.

3.  $S$ is an ellipse with foci at the points $-1/\gamma$ and $1/\gamma$ and with vertices at the points $-1/\alpha, 1/\alpha, \gamma > \alpha > 1$ (Fig. 31). In this case the region $D$ will be the interior of a certain oval. The mapping function is

$$z(w) = \frac{2\rho\gamma w}{1 + \rho^2 w^2} ,$$

$$\rho = \frac{\gamma}{\alpha} - \sqrt{\frac{\gamma^2}{\alpha^2} - 1} ,$$

$$\theta = \frac{\gamma - \sqrt{\gamma^2 - 1}}{\rho} = \frac{\theta_1}{\rho} ,$$

where $\theta_1 = \gamma - \sqrt{\gamma^2 - 1}$.

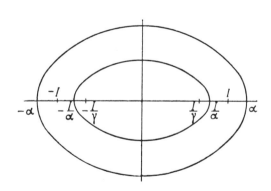

**FIGURE 31**

Again the function $w/z(w)$ turns out to be the quadratic polynomial

$$\frac{1}{2\rho\gamma} + \frac{\rho}{2\gamma} w^2 .$$

Therefore

$$d_0 = \frac{1}{2\rho\gamma} , \quad d_2 = \frac{\rho}{2\gamma} , \quad d_1 = d_3 = d_4 = \cdots = 0 .$$

From formula (25) we obtain

$$X_1 = BX_0 + G\,,$$
$$X_2 = 2\gamma\theta_1(BX_1 + G) + (1 - 2\gamma\theta_1)X_1 = (1 + \theta_1^2)(BX_1 + G) - \theta_1^2 X_1\,,$$
$$X_i = (1 + \theta_1^2)(BX_{i-1} + G) - \theta_1^2 X_{i-2}\,.$$

As we shall see, these formulas coincide with the formulas obtained above for a segment of the real axis.

4. $S$ is a segment of the imaginary axis $(-i/\beta,\ i/\beta)$ (Fig. 32). In this case region $D$ is the plane with two cuts along the imaginary axis running from the points $-\beta i$ and $\beta i$ to infinity. The mapping function is

$$z(w) = \frac{2\beta w}{1 - w^2}\,,$$

$$\theta = \sqrt{\beta^2 + 1} - \beta\,.$$

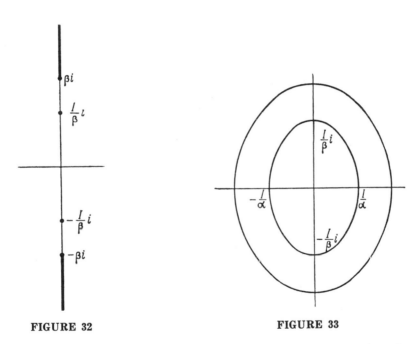

**FIGURE 32**                              **FIGURE 33**

Just as in the previous case, the function $w/z(w)$ will be a quadratic polynomial, namely, $w/z(w) = (1 - w^2)/2\beta$, and therefore $d_0 = 1/2\beta$, $d_2 = -1/2\beta$, $d_1 = d_3 = d_4 = \cdots = 0$. For successive approximations of formula (25) we obtain the recurrence formulas

$$X_i = 2\beta\theta(BX_{i-1} + G) + \theta^2 X_{i-2}$$
$$= (1 - \theta^2)(BX_{i-1} + G) + \theta^2 X_{i-2} \qquad (i \geq 3)$$
$$X_1 = BX_0 + G\,,$$
$$X_2 = (1 - \theta^2)(BX_1 + G) + \theta^2 X_1\,.$$

5. $S$ is an ellipse with foci at the points $-1/\beta i$ and $1/\beta i$ and with an imaginary pole $1/\alpha$, $\alpha > 1$ (Fig. 33).

In this case region $D$ will be the interior of a certain oval.

The mapping function is

$$z(w) = \frac{2\rho\beta w}{1 - \rho^2 w^2} \, ,$$

where

$$\rho = \frac{\sqrt{\alpha^2 + \beta^2} - \beta}{\alpha} \, ,$$

$$\theta = \frac{\sqrt{\beta^2 + 1} - \beta}{\rho} = \frac{\theta_1}{\rho} \, ,$$

$$\theta_1 = \sqrt{\beta^2 + 1} - \beta \, .$$

Since $w/z(w) = (1/2\rho)\beta - (\rho/2)\beta w^2$ we obtain from formulas (25), after simple transformations,

$$X_1 = BX_0 + G \, ,$$
$$X_2 = (1 - \theta_1^2)(BX_1 + G) + \theta_1^2 X_1 \, ,$$
$$X_i = (1 - \theta_1^2)(BX_{i-1} + G) + \theta_1^2 X_{i-2}$$

The computational formulas for the approximations coincide with the formulas of ex. 4 after replacing $\theta$ by $\theta_1$ in them.

6. $S$ is two tangent circles having the segments $(0, 1/\gamma)$ and $(0, -1/\gamma)$ as their diameters, $\gamma > 1$ (Fig. 34). In this case region $D$ is the pole $-\gamma < \operatorname{Re} z < \gamma$ and the mapping function is

$$z(w) = \frac{4\gamma}{\pi} \operatorname{arctg} w = \frac{2\gamma}{\pi i} \lg \frac{1 + wi}{1 - wi} = \frac{4\gamma}{\pi} \left[ w - \frac{w^3}{3} + \frac{w^5}{5} - \cdots \right], \quad \theta = \operatorname{tg} \frac{\pi}{4\gamma} \, .$$

*Note.* When actually using the conformal mapping method, it is not essential to use the smallest set $S$ known to us which contains the eigenvalues of the given matrix, in view of existing information. The set $S$

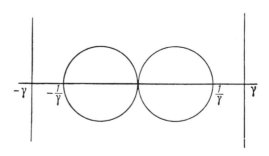

FIGURE 34

should be chosen so that the mapping function has the simplest possible form. Small changes in the set $S$ involve only a small increase in $\theta$.

## 97. THE CONFORMAL MAPPING METHOD AS APPLIED TO AN UNPREPARED SYSTEM

The conformal mapping method may be applied directly in solving the system

$$AX = F. \tag{1}$$

Let $\Sigma$ be a bounded closed set whose complement $\Delta$ (including infinitely remote points) is a simply-connected region which contains the point 0. We assume that all eigenvalues of matrix $A$ belong to the set $\Sigma$. We shall consider the system

$$AX - uX = F, \tag{2}$$

dependent on the complex parameter $u$. The initial system is obtained from (2) for $u = 0$.

We assume that the function $u = u(w)$ maps the unit circle $|w| < 1$ into the region $\Delta$ so that $u(0) = \infty$. We denote by $\theta$ the pre-image of 0. The solution of system (2) is written as

$$X(u) = (A - uE)^{-1}F. \tag{3}$$

We shall investigate the function $1/(t - u)$. This function is regular in the region $\Delta$, including the point $u = \infty$, for any $t$ belonging to the set $\Sigma$. It is clear that the function $u(w)$ has the following series expansion:

$$u(w) = -\frac{c_{-1}}{w} - c_0 - c_1 w - c_2 w^2 - \cdots, \tag{4}$$

where $c_{-1} \neq 0$. Consequently,

$$\frac{1}{t - u(w)} = \frac{w}{c_{-1} + (t + c_0)w + c_1 w^2 + \cdots}$$

$$= \frac{w}{c_{-1}}\left(1 + \frac{c_0 + t}{c_{-1}}w + \frac{c_1}{c_{-1}}w^2 + \cdots\right)^{-1}$$

$$= w[d_0(t) + d_1(t)w + d_2(t)w^2 + \cdots], \tag{5}$$

where $d_i(t)$ are polynomials of degree $i$. The radius of convergence for this series will not be less than one.

In accordance with expansion (5), the solution of system (2) will be

$$X(u) = w[d_0(A)F + wd_1(A)F + w^2 d_2(A)F + \cdots] \tag{6}$$

The solution of the initial system is represented as the convergent series

$$X = \theta[d_0(A)F + \theta d_1(A)F + \theta^2 d_2(A)F + \cdots], \tag{7}$$

whose segments give approximate solutions for the system.

The method permits the following variation. Instead of the function $1/[t - u(w)]$ we may consider the function $\rho(w)/w[t - u(w)]$ where $\rho(w)$ is any function, regular in the unit circle, which satisfies the condition $\rho(\theta) = \theta$. Then

$$\frac{\rho(w)}{w[t - u(w)]} = \alpha_0(t) + \alpha_1(t)w + \alpha_2(t)w^2 + \cdots \tag{8}$$

where $\alpha_i(t)$ are certain polynomials in $t$. The radius of convergence for the last series is equal to one as before. Correspondingly, the solution of system (1) is represented as

$$X = \alpha_0(A)F + \theta\alpha_1(A)F + \theta^2\alpha_2(A)F + \cdots \tag{9}$$

We shall now compare the solution of the system $AX = F$ found from series (7) with the solution of the same system found from series (8), Section 95, after preliminary preparation of system (1) to the form $X = BX + G$ with $B = E - hA$ and $G = Fh$.

The eigenvalues of matrix $B$ are connected with the eigenvalues of matrix $A$ by the relationship $\mu_i = 1 - h\lambda_i$. Therefore, as the set $S$ for matrix $B$ we may take the point set $1 - ht$ for $t \in \Sigma$. Then region $D$ will be the complement of the set $1/(1 - th)$, $t \in \Sigma$, so that (in view of the single-valuedness of the function $1/(1 - \xi h)$ in the whole complex plane, including the point at infinity) $D$ is obtained from $\Delta$ by mapping, using the function $1/(1 - \xi h)$.

The function $u(w)$ maps the unit circle into the region $\Delta$ and consequently $z(w) = 1/[1 - hu(w)]$ maps the unit circle into $D$, while $z(0) = 0$, $z(\theta) = 1$. Thus the function $z(w)$ satisfies the requirements of Section 95.

Letting $t_1 = 1 - ht$ (here $t$ is the "representative" of matrix $A$ and $t_1$ is the "representative" of matrix $B$) we have

$$X = b_0(B)G + \theta b_1(B)G + \theta^2 b_2(B)G + \cdots$$

where $b_i(t)$ are coefficients in the expansion of the function $1/[1 - t_1 z(w)]$ in powers of $w$. Substituting $B = E - hA$ and $G = hF$ we obtain

$$X = h\beta_0(A)F + \theta h\beta_1(A)F + \theta^2 h\beta_2(A)F + \cdots$$

where $\beta_i(t) = b_i(1 - ht)$.

It is clear that the polynomials $\beta_i(t)$ will be coefficients in the expansion of the function

$$\frac{1}{1 - (1 - ht)z(w)} = \frac{1}{1 - \dfrac{1 - ht}{1 - hu(w)}} = \frac{1 - hu(w)}{h(t - u(w))},$$

and the polynomials $h\beta_i(t)$ will be coefficients in the expansion of the function

$$\frac{1 - hu(w)}{t - u(w)}.$$

Thus, comparing this with formula (8), we obtain

$$h\beta_i(t) = \alpha_i(t)$$

for

$$\rho(w) = w(1 - hu(w)) .$$

From this we draw the conclusion that the order of the rate of convergence for series giving the solution of the system $AX = F$ for its different
preparations does not depend on the number $h$ that determines the given
preparation, since this order is determined only by the number $\theta$. It is
interesting to note that the solution given by series (7) is obtained as the
limiting case of the solution to the preparared system for $h \to 0$.

Green's function $G(u)$ for the region $\varDelta$ and the function $u(w)$ are connected, as is well-known, in the following manner:

$$G(u) = - \log | w(u) |$$

where $w(u)$ is the inverse function for $u(w)$.

Therefore

$$| \theta | = | w(0) | = e^{-G(0)} .$$

Thus the conformal mapping method has the same order for the rate of
convergence as the component suppression method with polynomials that
diverge least from zero (cf. Section 94).

We shall not stop to analyze concrete $\Sigma$-algorithms which may be constructed in the same way as concrete $S$-algorithms.

We now consider the case where the region $\varDelta$ is multiply connected.

The method of conformal mapping is extended to this case with hardly
a change. A function $z(w)$ is chosen which is meromorphic in the unit
circle and which has a simple pole at the point $w = 0$; it takes on the value
0 at a certain point $\theta$ and does not take on values from the set $\Sigma$. Then

$$\frac{1}{t - z(w)} = d_0 + d_1(t)w + d_2(t)w^2 + \cdots$$

is a convergent series in the circle $|w| < 1$ for any $t \in \Sigma$. The solution
of the system $AX = F$ is written as

$$X = d_0F + \theta d_1(A)F + \theta^2 d_2(A)F + \cdots \tag{10}$$

Series (10) will converge more rapidly as $|\theta|$ decreases, so that we should
take for $\theta$ the pre-image least in modulus of the point $z = 0$, if there are
several them.

The best function $z(w)$ is the function $z = \varphi(w)$ which maps the unit circle
into a simply-connected Reimann surface covering the region $\varDelta$.

In fact, let $z(w)$ be any function of the class considered. Each branch
of the function $\varphi^{-1}(z(w))$ will be regular inside the unit circle so that the
multi-valued function $\varphi^{-1}(z(w))$ really breaks down on single-valued regular
branches. Of them we choose the one, $\varPhi(w)$, for which $\varPhi(0) = 0$. It is

clear that $|\Phi(w)| < 1$ for $|w| < 1$. From the well-known lemma of Schwartz

$$|\Phi(w)| \leq |w| .\tag{11}$$

Let $\theta$ be the pre-image smallest in modulus of the point $z = 0$ under the mapping $z = z(w)$ and let $\theta_0 = \Phi(\theta)$. It is clear that $\varphi(\theta_0) = 0$ so that $\theta_0$ is one of the pre-images of the point $z = 0$ under the mapping $z = \varphi(w)$. Letting $w = \theta$ in inequality (11) we obtain

$$|\theta_0| \leq |\theta| .$$

If we denote by $\theta^*$ the pre-image of the point $z = 0$ least in modulus under the mapping $z = \varphi(w)$, then it is all the more true that

$$|\theta^*| \leq |\theta| .$$

It follows from the same lemma of Schwartz that the equality sign is possible only if $\Phi(w) = \nu w$ for $|\nu| = 1$, i.e. if $z(w) = \varphi(\nu w)$. In this very way we prove that the function $z = \varphi(w)$ comes from the number $\theta$ least in modulus.

We shall now show that in the case of a multiply connected region the conformal mapping method gives a worse result than the method of suppressing components using polynomials that diverge least from zero.

Let $G(z)$ be Green's function for the region $\Delta$ with a logarithmic singularity at the point $z = \infty$. We denote by $\psi(z)$ the value smallest in modulus for the function $\varphi^{-1}(z)$ (if there are several, then we choose any one) and we consider the function $H(z) = -\log|\psi(z)|$. It is clear that $H(z)$ is a harmonic function in the neighborhood of the point $z = \infty$, excluding the point itself at which $H(z)$ has a logarithmic singularity, since $\psi(z)$ in the neighborhood of the infinitely distant point coincides with that branch of the function $\varphi^{-1}(z)$ for which $\varphi^{-1}(\infty) = 0$.

Let $z_0$ be an arbitrary point of the region $\Delta$ and let $\delta$ be a circle with center at the point $z_0$ which is contained with its boundary $\gamma$ inside the region $\Delta$. We take that branch of the function $\varphi^{-1}(z)$ for which $\varphi^{-1}(z_0) = \psi(z_0)$. Then $|\psi(z)| \leq |\varphi^{-1}(z)|$ at all points of the circle $\delta$, including its boundary $\gamma$. It is clear that

$$H(z_0) = -\log|\varphi^{-1}(z_0)| = \frac{1}{2\pi r}\int_\gamma -\log|\varphi^{-1}(z)|\,d\sigma ,$$

since $-\log|\varphi^{-1}(z)|$ is a harmonic function in the circle $\delta$ (let $r$ designate the radius of this circle). Therefore

$$H(z_0) \leq \frac{1}{2\pi r}\int_\gamma -\log|\psi(z)|\,d\sigma = \frac{1}{2\pi r}\int_\gamma H(z)\,d\sigma .$$

Thus $H(z)$ is a subharmonic function in region $\Delta$. Furthermore, $H(z) - G(z)$ is a subharmonic function bounded in the neighborhood of the point $z = \infty$; its largest value may not be attained inside the region $\Delta$. We assume that $\Delta$ is bounded by a finite number of analytic curves. Then, at the boundary of $\Delta$, both $G(z)$ and $H(z)$ are defined and equal to zero. Consequently either $H(z) - G(z) < 0$ at all points of the region $\varDelta$ or $H(z) = G(z)$ in $\varDelta$. The second possibility disappears, since for a multiply connected region $H(z)$ is not a harmonic function at points for which there exists more than one pre-image with smallest modulus. Thus $H(z) - G(z) < 0$ in $\varDelta$. In particular, $H(0) = -\log|\theta| < G(0)$, from which

$$|\theta| > e^{-G(0)} . \tag{12}$$

The assumption about the boundary of the region $\varDelta$, from which inequality (12) was derived, is reduced by a limit process from regions $\varDelta_\varepsilon$ bounded by $\varepsilon$-lines for Green's function to the initial region $\varDelta$.

Inequality (12) means that the rate of convergence for the method of conformal mapping is inferior to the rate of convergence for the component suppression method using polynomials which diverge least from zero.

We shall give an example which illustrates this fact.

Let $\Sigma$ be the set of two segments $(-b, -a)$ and $(a, b)$ on the real axis (Fig. 35). In this case the region $\varDelta$ is doubly connected. The function which maps the unit circle into the universal covering is given by the equation

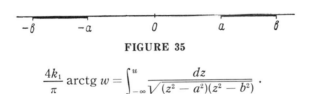

**FIGURE 35**

$$\frac{4k_1}{\pi} \operatorname{arctg} w = \int_{-\infty}^{u} \frac{dz}{\sqrt{(z^2 - a^2)(z^2 - b^2)}} .$$

Here

$$k_1 = \int_b^\infty \frac{dz}{\sqrt{(z^2 - a^2)(z^2 - b^2)}} = \int_0^{\pi/2} \frac{d\varphi}{\sqrt{b^2 - a^2 \sin^2 \varphi}} .$$

and

$$|\theta| = \frac{1 - e^{-(\pi/2)(k_2/k_2)}}{1 + e^{-(\pi/2)(k_2/k_2)}} ,$$

where

$$k_2 = \int_a^b \frac{dz}{\sqrt{(z^2 - a^2)(b^2 - z^2)}} = \int_0^{\pi/2} \frac{d\varphi}{\sqrt{b^2 - (b^2 - a^2)\sin^2 \varphi}} .$$

If we take $a = 1$, $b = \sqrt{2}$, then $k_1 = k_2$ and therefore

$$|\theta| = \frac{1 - e^{-\pi/2}}{1 + e^{-\pi/2}} \approx 0.6558 .$$

We shall show that the method of suppressing components leads to a process with faster convergence.

Consider the polynomial

$$g_{2s}(t) = \frac{T_s(3 - 2t^2)}{T_s(3)}$$

where $T_s(t) = \cos s \arccos t$.  For $t \in \Sigma$

$$|g_{2s}(t)| \leq \frac{1}{T_s(3)} \approx \frac{2}{(3 + \sqrt{8})^s} = \frac{2}{(\sqrt{2} + 1)^{2s}} = 2(\sqrt{2} - 1)^{2s}$$

Moreover $g_{2s}(0) = 1$.  We let

$$h_{2s-1}(t) = \frac{1 - g_{2s}(t)}{t}$$

and we shall find the approximation

$$X_{2s} = X_0 + h_{2s-1}(A)(F - AX_0) .$$

Then

$$Y_{2s} = X^* - X_{2s} = Y_0 - h_{2s-1}(A)AY_0 = g_{2s}(A)Y_0 .$$

Consequently the components of the vector $Y_{2s}$, in the expansion in eigen-vectors, are multiplied by factors $g_{2s}(\lambda_i)$.  But

$$|g_{2s}(\lambda_i)| \leq \frac{1}{T_s(3)} \approx 2(\sqrt{2} - 1)^{2s} .$$

Thus, using $2s$ iterations of the initial vector, we obtained an error estimate of order $(\sqrt{2} - 1)^{2s} \approx (0.4142)^{2s}$, while the conformal mapping method with the same number of iterations gives a rate of convergence or order $\theta^{2s} \approx (0.6558)^{2s}$.

## 98. APPLICATION OF THE COMPONENT SUPPRESSION IDEA IN SOLVING THE SPECIAL EIGENVALUE PROBLEM

The power method for determining the eigenvector belonging to the eigenvalue greatest in modulus was based on the fact that the vector sequence $A^k Y_0$ with arbitrary initial vector $Y_0$ converged directionally to the indicated eigenvector.

Actually the components of the eigenvectors $U_1, \cdots, U_n$, in the expansion of the initial vector

$$Y_0 = c_1 U_1 + \cdots + c_n U_n$$

as a result of $n$-tuple application of iteration with matrix $A$, obtain the multipliers $\lambda_1^k, \cdots, \lambda_n^k$, among which $\lambda_1^k$ predominates over the others.  If we normalize the process so that the coefficient for $U_1$ remains equal to one, then the remaining components are "suppressed" by the factors $(\lambda_2/\lambda_1)^k, \cdots, (\lambda_n/\lambda_1)^k$, which are approaching zero.  This idea of "suppressing" components may be generalized as follows.

Let it be known that all eigenvalues of matrix $A$ except $\lambda_i$, subject to determination, lie in a certain bounded set $\Sigma$ whose complement $\Delta$ is a connected region of the complex plane for the variable $z$. Let $\tau_z(t) = t^k + \cdots$ be the polynomial of degree $k$ which diverges least from zero on the set $\Sigma$ such that its roots, in turn, lie in the set $\Sigma$.

We construct the vector $1/\tau_k(\lambda_i)\,\tau_k(A)Y_0$. We shall assume that $Y_0$ has a non-zero $i$th component in the eigenvector expansion. It is clear that the $i$th component of the constructed vector will be as before and the remaining components will acquire the factors $\tau_k(\lambda_j)/\tau_k(\lambda_i)$ $(j \neq i)$. The moduli of these "suppression factors" satisfy the inequality

$$\frac{\tau_k(\lambda_j)}{\tau_k(\lambda_i)} \leq \frac{\tau_k}{|\tau_k(\lambda_i)|}.$$

Here $\tau_k = \max_{t \in s} |\tau_k(t)|$. It is known[†] that

$$\lim \sqrt[k]{\frac{\tau_k}{|\tau_k(\lambda_i)|}} = e^{-G(\lambda_i)} < 1$$

where $G(t)$ is Green's function for the region $\Delta$. Therefore the vector sequence $\tau_0(A)Y_0$, $\tau_1(A)Y_0$, $\cdots$ converges directionally to the eigenvector belonging to the eigenvalue $\lambda_i$ with the rate $e^{-k[G(\lambda_i)-\varepsilon]}$, where $\varepsilon$ is as small a number as we wish. After determining the eigenvector, the eigenvalue $\lambda_i$ is easily determined.

## 99. APPLICATION OF CONFORMAL MAPPING IN SOLVING THE SPECIAL EIGENVALUE PROBLEM

The power method may be interpreted in the following way. Consider the vector $Y(w) = (E - wA)^{-1}Y_0$. Its components are analytic functions of the complex variable $w$ which have poles at the points $1/\lambda_1, \cdots, 1/\lambda_n$ where $|\lambda_1| > |\lambda_2| \geq \cdots \geq |\lambda_n|$ are eigenvalues of matrix $A$. The smallest pole in modulus will be a number which is the reciprocal of the largest eigenvalue in modulus. We shall expand the vector $Y(w)$ as a series in powers of $w$

$$Y(w) = (E - wA)^{-1}Y_0 = Y_0 + wAY_0 + w^2A^2Y_0 + \cdots \qquad (1)$$

The radius of convergence for this series will obviously be equal to $1/|\lambda_i|$ and in its circle of convergence will exist a unique simple pole $1/\lambda_1$. The components of the vector $Y(w)$ will be analytic functions which generally have a simple pole $1/\lambda_1$ at the boundary of convergence for series (1). For any such component $y(w)$ we have

$$y(w) = y_0 + y_1 w + y_2 w^2 + \cdots. \qquad (2)$$

Here $y_k$ is the chosen component of the vector $A^k Y_0$.

The limit formula

$$\lambda_1 = \lim_{k \to \infty} \frac{y_{k+1}}{y_k} \qquad (3)$$

---

† G. M. Goluzin, *op. cit.*, Ch. VII.

may be interpreted as the result of applying Koenig's theorem on the expansion coefficients for a function which has a unique simple pole on the boundary of the circle of convergence.

An estimate of the rate of convergence for the process (of order $|\lambda_2/\lambda_1|$ where $\lambda_2$ is the eigenvalue next in modulus for matrix $A$) also follows from Koenig's theorem.

Such an examination of the power method lets us generalize it in the following direction. Let

$$u(w) = -\frac{c_{-1}}{w} - c_0 - c_1 w - c_2 w^2 - \cdots$$

be a function meromorphic in the unit circle, where $u(0) = \infty$ and $u(\theta) = \lambda_1$ for $0 < |\theta| < 1$, while for $|w| \leq |\theta|$, $w \neq \theta$, $u(w)$ does not take on values equal to the eigenvalues of matrix $A$.

Consider the vector $(A - u(w)E)^{-1}Y_0$. According to formula (6), Section 97, $(A - u(w)E)^{-1}Y_0 = w[d_0(A)Y_0 + d_1(A)Y_0 w + d_2(A)Y_0 w^2 + \cdots]$ where $d_i(t)$ are certain polynomials of degree $i$. Any component of the vector $(A - u(w)E)^{-1}$ will be meromorphic in the unit circle and regular in the circle $|w| \leq |\theta|$, except for the point $w = \theta$ at which it will generally have a simple pole.

Consequently, according to Koenig's theorem, the ratios of the chosen components of the vectors $d_{k+1}(A)Y_0$ and $d_k(A)Y_0$ will approach $\theta$ and the vectors themselves will converge directionally to the eigenvector belonging to the eigenvalue $\lambda_i$. The eigenvalue is found as $u(\theta)$. The rate of convergence for the process will have order $|\theta/\theta^*|^k$ where $|\theta^*| \geq 1$ if there are no pre-images of eigenvalues except $\theta$ inside the unit circle. If there are also pre-images of eigenvalues of matrix $A$ in the unit circle, then $\theta^*$ is the smallest of them in modulus.

We now assume that, regarding the eigenvalues of matrix $A$, it is known that all of them except the one subject to determination lie in the bounded closed set $\Sigma$ whose complement $\Delta$ is a simply-connected region. In this case we should take as the function $u(w)$ a function which realizes a conformal mapping of the unit circle into the region $\Delta$. According to the arguments presented in Section 95, such a choice of the function will be best in the sense of the rate of convergence for the process obtained.

In carrying out the computations it is possible to use auxiliary tables for coefficients of the successive polynomials $d_k(t)$. The vectors $d_k(A)Y_0$ may also be constructed from recurrent relationships.

We also observe that since the vectors $d_k(A)Y_0$ converge directionally to the eigenvector of matrix $A$ which belongs to the eigenvalue $\lambda_i$ being determined, then the latter may be found as the ratio of components of the vectors $Ad_k(A)Y_0$ and $d_k(A)Y_0$ (and not as the image $\theta$).

The use of conformal mapping to find eigenvalues was first proposed by V. N. Kublanovskaya (1), (2), who considered instead of the function $u(w)$ the function $z(w) = 1/u(w)$ which maps the unit circle into the region $\tilde{\Delta}$

obtained from the region $\varDelta$ by mapping, using $z = 1/u$. In these works the function

$$(E - zA)^{-1} = - u(A - uE)^{-1}$$

is investigated; the poles of this function coincide with the poles of the function $A - uE$ considered above. In work (2) are given the tables of coefficients of corresponding polynomials for a series of regions $\tilde{\varDelta}$.

# CONCLUSION

The considerable number of methods considered by us in the present book hardly exhausts the increasing catalogue of methods proposed for the numerical solution of basic problems in linear algebra. We have completely neglected "Monte Carlo" methods whose character is more theoretic-probabilistic than algebraic. Of the many elimination schemes, we have considered only a few of the most useful ones, a far cry from describing all iterative processes. Schemes capable of solving special problems which have a narrow area of application were hardly examined. Finally, methods which have been published very recently are not described in the book.

The material we have considered gives us grounds for raising the question as to which of the described methods must be recommended for practical calculations as being preferable to others. It is difficult, and even imposssible, to give a definite answer to this question, since under different conditions various requirements are imposed on the methods.

The most important criterion for the qualitative evaluation of a numerical method is its *reliability*, i.e. its ability to use almost all information in a problem's conditions to solve it. However there are other criteria, besides that of reliability, which are quite essential.

One is the *simplicity of the computational scheme*. Another is the *minimality of the number of computational operations*. There is also the *minimal work load for a machine's memory* with program guidance and the *compactness* of notation when using such machines. The possibility of using *individual peculiarities of the problem* which facilitate its solution (domination of diagonal elements in a matrix, the presence of a large number of zero elements, etc.) often turns out to be important. Finally, the suitability of a method for serial solution of single-type problems is sometimes important.

Each numerical method worthy of attention must satisfy all the indicated requirements to some degree. But under different concrete conditions the particular weight of each requirement may be different. It might be that even such a necessary requirement as reliability may be relegated to a minor role, as, for example, when solving a long series of single-type problems which do not require much accuracy in the result. Of two methods, the first of which has a simpler computational scheme but requires a larger number of computational operations than the other, one would ordinarily prefer the first. However, in problems connected with higher-order matrices the criterion that the number of computational operations be minimal may turn out to be decisive.

Comparison of methods according to criteria of computational scheme simplicity, memory work load, applicability of the problem's individual peculiarities, and serial usefulness, is not difficult. Figuring out the quantity of computational operations is not complicated either in exact methods or in iterative ones, if the rate of convergence is known beforehand.

It is a lot more complicated to evaluate methods as to criteria of reliability. Actually in the approximate statement of computational problems there must always exist in their solution some measure of indeterminacy arising from the initial data. This indeterminacy may be significant, for example, in solving an ill-conditioned system. A numerical method must be considered reliable if the solution gotten from its application is obtained with error not exceeding the inevitable error imposed by the indeterminacy of the initial data, as was indicated above.

Basic factors which reduce the reliability of a method are rounding-off errors in intermediate computations. Thus, a strict account of the reliability must be based on an estimate of the influence of rounding-off errors. In practice, probability estimates are interesting, since estimates "at the maximum" almost always are employed in practice, but there is small probability that they are actually attained.

In the literature there are quite a few works devoted to investigation of the influence of rounding-off errors in the basic arithmetic operations and in certain numerical methods of linear algebra. Relating to this are the works: Neumann and Goldstine (1), Turing (1), Goldstine and Neumann (1), Dwyer and Waugh (1), Abramov (2), (3), Householder (3), (11), Goldstine, Murray and Neumann (1), Carr (1), and others. In particular, Gauss's scheme for principal elements has been subjected to a detailed analysis by Neumann and Goldstine. However, at the present time few of the newer basic methods have been subjected to such investigation.

Moreover, sometimes such investigation turns out to be impossible in principle because of the dependence of rounding-off error on the influence of factors which cannot be accounted for a priori.

Thus, in carrying out the single-division sheme for solving a system of linear equations with the order for eliminating the unknowns fixed beforehand, the reliability of the result is determined to a significant degree by whether a loss of significant figures occurs during the process or not; this in turn depends on whether or not small principal minors exist for the coefficient matrix. It may turn out for one and the same system that, with a different choice of order for eliminating the unknowns, the reliability of the result will also be affected. The most reliable choice will be an order of elimination which coincides with the order in the principal elements scheme. Thus the single-division scheme with a fixed order for eliminating the unknowns does not turn out to be an unconditionally reliable method; at any rate, it is less reliable than the scheme of principal elements. However it does not yet follow from this that the scheme of principal elements is always preferable, since to realize it is a much more involved process than to realize the scheme with a fixed order of elimination. Many other numerical methods of linear algebra turn out to be reliable only conditionally. Such methods include the biorthogonal algorithm and the method of minimal iterations. Their reliability depends a great deal on a successful choice of the initial vector. Among methods for determining eigenvalues the most reliable are those in which the eigenvalues are determined without the necessity of computing the coefficients of the characteristic polynomial, since even insignificant errors in determining these coefficients may involve significant errors in computing the roots. But as against this, iterative methods for solving the complete eigenvalue problem turn out to be much more laborious than the exact ones connected with computation of the characteristic polynomial's coefficients.

It seems to us that in estimating the applicability of conditionally reliable methods, the experience of using them has a basic value, and only by generalizing and extending this experience will it be possible to make more definite judgements on this question.

We shall say more about certain means for increasing the reliability of the methods. Theoretically speaking, each "exact" method of linear algebra offers an unlimited depth as to the reliability which may be realized through the accuracy in carring out the intermediate computations. But significant increase in accuracy over the whole computational process often turns out to be impossible, and the accuracy demand should be refined only in the most "vulnerable" situations. A source of significant rounding-off errors arises from the computation of product sums $\sum_{k=1}^{n} a_k b_k$, when rounding-off to a given number of digits in each element. It is often advisable to perform this operation with double accuracy, computing each element without prior rounding-off and eliminating excess only after completing the addition. Of course it is sometimes superfluous to apply double accuracy, as, for example, in carrying out a self-correcting iterative process.

In using conditionally reliable methods it may often be advisable to carry out the algorithm in two non-equivalent variations, for example, to carry out the single-division scheme for two choices of order in eliminating the unknowns, or to apply the biorthogonal algorithm starting from two initial vectors, etc.

Applying control during the course of a process may serve as a fairly good indirect check on reliability.

# SUPPLEMENT

We present one more iterative process for solving the complete eigenvalue problem, communicated to the authors by V. N. Kublanovskaya. The process consists of the following.

Let the matrix $A$ be symmetric, and assume that its square has only simple eigenvalues. We construct the matrix sequence

$$AP_1 = \Lambda_1;\ P_1'\Lambda_1 = P_1^{-1}AP_1 = A_1;$$
$$A_1P_2 = \Lambda_2;\ P_2'\Lambda_2 = P_2^{-1}A_1P_2 = A_2;$$
$$\cdots\cdots\cdots\cdots\cdots\cdots\cdots;$$
$$A_{k-1}P_k = \Lambda_k;\ P_k'\Lambda_k = P_k^{-1}A_{k-1}P_k = A_k;$$
$$\cdots\cdots\cdots\cdots\cdots\cdots\cdots$$

Here $P_i$ are orthogonal matrices and $\Lambda_i$ are left triangular matrices. The matrices $P_i$ may be computed, for example, as products of rotational or reflection matrices, as was done in Section 16 in solving linear systems.

We shall show that the matrix sequence $A_k$ converges to a diagonal matrix formed from eigenvalues of matrix $A$ arranged in order of decreasing moduli, and that the matrix $Q_k = P_1 \cdots P_k$, for large enough $k$, has columns as close as we wish to the normalized eigenvectors of matrix $A$.

For proof, we shall establish the connection of the process with the $LR$-algorithm applied to the matrix $A^2$.

We have

$$\Lambda_1\Lambda_1' = AP_1P_1'A = A^2,$$
$$\Lambda_1'\Lambda_1 = P_1'A^2P_1 = A_1^2 = \Lambda_2\Lambda_2',$$
$$\cdots\cdots\cdots\cdots\cdots\cdots\cdots,$$
$$\Lambda_k'\Lambda_k = P_k'A_{k-1}^2P_k = A_k^2 = \Lambda_{k+1}\Lambda_{k+1}',$$
$$\cdots\cdots\cdots\cdots\cdots\cdots\cdots,$$

Let $\Delta_i$ be the diagonal matrix formed from the diagonal elements of $\Lambda_i$. We let

$$L_k = \Delta_{k-1} \cdots \Delta_1 \Lambda_k \Delta_1^{-1} \cdots \Delta_{k-1}^{-1} \Delta_k^{-1} \, ,$$
$$R_k = \Delta_k \Delta_{k-1} \cdots \Delta_1 \Lambda_k' \Delta_1^{-1} \cdots \Delta_{k-1}^{-1} \, .$$

It is clear that $L_k$ is a left triangular matrix with unit diagonal and $R_k$ is a right triangular matrix with diagonal $\Delta_k^2$. It is easy to verify that $L_1 R_1 = A^2$ and $R_{k-1} L_{k-1} = L_k R_k$, i.e. the matrices $L_i$ and $R_i$ coincide with the corresponding matrices of the $LR$-algorithm applied to matrix $A^2$.

In view of the fact that the matrix $A^2$ is positive definite and its eigenvalues are pairwise distinct, the $LR$-algorithm converges for it; in particular, diagonal elements of the matrices $R_k$ converge to the squares of eigenvalues of matrix $A$. Consequently,

$$\mathrm{Sp}\, R_k = \mathrm{Sp}\, \Delta_k^2 \to \mathrm{Sp}\, A^2 \, .$$

Moreover, letting $\Lambda_k = (l_{ij,\,k})$, we have

$$\mathrm{Sp}\, A^2 = \mathrm{Sp}\, A_k^2 = \mathrm{Sp}\, \Lambda_k' \Lambda_k = \sum_{i,j} l_{ij,\,k}^2$$
$$= \sum_i l_{ii,\,k}^2 + \sum_{i>j} l_{ij,\,k}^2 = \mathrm{Sp}\, \Delta_k^2 + \sum_{i>j} l_{ij,\,k}^2 \, .$$

Therefore for $k \to \infty$, $\sum_{i>j} l_{ij,\,k}^2 \to 0$, so that all non-diagonal elements of matrix $\Lambda_k$ approach zero. Consequently,

$$A_k^2 = \Lambda_k' \Lambda_k \to [\lambda_1^2, \cdots, \lambda_n^2] \text{ for } k \to \infty \, .$$

The matrices $A_k$ become as close to the diagonal matrices $[\pm \lambda_1, \cdots, \pm \lambda_n]$ as we wish for large enough $k$. But since the matrices $A_k$ are similar to matrix $A$ for all $k$, it follows that

$$A_k \to [\lambda_1, \cdots, \lambda_n] \, .$$

Furthermore, $A_k = Q_k' A Q_k$ and consequently for large enough $k$ the columns of matrix $Q_k$ are as close as we wish to the eigenvectors of matrix $A$, normalized in view of the orthogonality of $Q_k$. Since the choice of matrices $P_k$ is not unique, the sequence of matrices $Q_k$ may not converge. It will converge only with accuracy to within column signs. Convergence will hold if, starting at a certain place, we take the matrix $P_k$ as close to the unit matrix as possible at each step.

For accelerating the convergence, the process allows us to use both translations (similar to the $LR$-algorithm) and the formulas of Jacobi's method for improving accuracy.

# LITERATURE

Here and in the Supplementary Literature the following abbreviations indicate where reviews are to be found: R. Zh. M., *Referativnyi Zhurnal Matematiki* (Mathematics Reference Journal, Russian); M. R., *Mathematical Reviews* (United States).

ABRAMOV, A. A.

(1) Uskorenie skhodimosti v iterativnykh protsessakh (The acceleration of convergence in iterative processes), *Dokl. AN SSSR*, 1950, **74**, 1051-1052; M. R., 12, 861.

(2) O vliyanii oshibok okrugleniya pri reshenii uravneniya Laplasa (On the influence of rounding-off errors in solving the Laplace equation), *Vychisl. matem. i vychisl. tekhnika*, 1953, **1**, No. 1, 37-40; M. R., 16, 1156.

(3) Ob oshibke okruglenii pri reshenii sistem Lineinykh uravnenii (On rounding-off error in solving systems of linear equations), *Dokl. AN SSSR*, 1954, **97** No. 2, 189-191; R. Zh. M., 1955, 6107.

(4) Ob oshibke okruglenii pri reshenii sistem lineinykh uravnenii, (On rounding-off error in solving systems of linear equations), *Ber. Internat. Math.-Kolloq.*, 1955 (1957), Nov., 151-153; R. Zh. M., 1958, 6232.

AFRIAT, S. N.

(1) An iterative process for the numerical determination of characteristic values of cartain matrices, *Quart. J. Math., Oxford.* Ser. (**2**), 1951, **2**, 121-122; M. R., 12, 861-862.

AITKEN, A.

(1) On Bernoulli's numerical solution of algebraic equations. *Proc. Roy. Soc. Edinburgh*, 1926, **46**, 289.

(2) Further numerical studies in algebraic equations and matrices. *Proc. Roy. Soc. Edinburgh*, 1931, **51**, 80.

(3) On the evaluation of determinants, the formation of their adjugates, and the partial solution of simultaneous linear equations. *Proc. Edinburgh Math. Soc.*, II, 1933, **3**, 207-219.

(4) Studies in practical mathematics. I. The evaluation with applications, of a certain triple product matrix. *Proc. Roy. Soc. Edinburgh*, 1937, **57**, 172-181.

(5) Studies in Practical Mathematics. II. The evaluation of the latent roots and latent vectors of a matrix. *Proc. Roy. Soc. Edinburgh*, Ser. A., 1936, 1937, **57**, 269-304.

(6) Studies in practical mathematics. V. On the iterative solution of a system of linear equations. *Proc. Roy. Soc. Edinburgh*, Ser. A, 1950, **63**, 52-60; M. R., 12, 56.

(7) *Determinants and matrices*. 9th ed. Edinburgh—London, Oliver and Boyd; New York, Interscience, 1956, vii, 144 pp.; R. Zh. M. 1957, 6159.


ALBERT, A. A.
(1) A rule for computing the inverse of a matrix, *Amer. Math Monthly*, 1941, 48, 198-199; M.R., 2, 100.


ALESKEROV, S. S.
(1) K voprosu resheniya sistemy lineinykh chislennykh uravnenii (On solving a system of linear numerical equations), *Tr. Azerbaidzh. industr. in-ta*, 1957, **16**, 5-10,; R. Zh. M., 1958, 4254.


ALLEN, D. N. DE G.
(1) *Relaxation methods*, McGraw-Hill, New York, Toronto, London, 1954, 257 pp.; M. R., 15, 831.


ALLEN, D. W.
(1) Numerical solution of "n" linear equations in "n" unknowns, and the evaluation of "n"th order determinant (complex coefficients), *J. Roy. Aeronaut Soc.*, 1956, **60**, 350-353; M. R., 17, 1137.


Alt, F. L.
(1) Machine methods for finding characteristic roots of a matrix. *Proc. Comput. Sem. Dec.*, 1949, *N.Y. IBM. Corp.*, 1951, **49 53**; M. R., 13, 496.


ALTMAN, M.
(1) On the solution of linear algebraic equations, *Bull. Acad. Polon. Sci.*, cl. 3, 3, 1957, **5**, No. 2, 93-97, IX; R. Zh. M., 1959, 7478.

(2) On the approximate solution of linear algebraic equations, *Bull. Acad. Polon. Sci.*, cl. 3, 1957, **5**, No. 4, 365-370, XXIX; R. Zh. M., 1959, 8560.


ANDERSEN, E.
(1) Solution of great systems of normal equations together with an investigation of Andrae's dotfigure. An arithmeticaltechnical investigation, *Geodaetisk Inst. Skriften* (*Mem. Inst. Geod. Danemark*), 1947, **3**, 11, 65 pp.; M. R., 9, 622.

(2) Solution of great systems of normal equations, *Bull. Geod.*, 1950, **15**, 19-29; M. R., 11, 693.


ANDREE, R. V.
(1) Computation of the inverse of a matrix, *Amer. Math. Monthly*, 1951, **58**, 87-92; M. R., 12, 639.

ANGELITCH, T. P.

(1) Resolutions des systemes d'equations lineaires algebriques par la methode de Banachiewicz, *Srpska Akad. Nauka. Zbornik Radova*, 1952, **18**, Mat. Inst. 2, 71-92; M. R., 14, 501.

APARO, ENZO.

(1) Sulle equazioni algebriche matriciali, *Atti Accad. naz. Lincei. Rend. Cl. Sci. fis., mat. e natur.*, Ser. 8, 1957, **22**, 20-23; M. R., 19, 685.

ARMS, R. J., GATES, L. D., and ZONDEK, B.

(1) A method of block iteration, *J. Soc. Industr. and Appl. Math.*, 1956, **4**, No. 4, 220-229; R. Zh. M., 1958, 2434.

ARNOLDI, W. E.

(1) The principle of minimized iteration in the solution of the matrix eigenvalue problem, *Quart. Appl. Math.*, 1951, **9**, 17-29; M. R., 13, 163.

ARZHANYKH, I. S.

(1) Rasprostranenie metoda A. N. Krylova na polinomial'nye matritsy (An extension of A. N. Krylov's method to polynomial matrices), *Dokl. AN SSSR*, 1951, **81**, No. 5, 749-752; M. R., 14, 92.

ATTA, SUSIE A.

(1) Effect of propagated error on inverse of Hilbert matrix, *J. Assoc. Comput. Machinery*, 1957, **4**, No. 1, 36-40; R. Zh. M., 1959, 3244.

AZBELEV, N., and VINOGRAD, P.

(1) Protsess posledovatel'nykh priblizhenii dlya otyskaniya sobstvennykh chisel i sobstvennykh vektorov (The successive approximations process for finding eigenvalues and eigenvectors), *Dokl. AN SSSR*, 1952, **83**, No. 2, 173-174; M. R., 14, 126.

AZPEITIA, A. G.

(1) Un metodo para el calculo de la matriz inversa, *Rev. Real acad. cienc. exact., fis. y natur.*, Madrid, 1956, **50**, No. 4, 463-470; R. Zh. M., 1957, 8962.

BABUSKA, IVO.

(1) O jednom numerickem reseni uplne regularnich systemu linearnich rovnic a o jeho aplikaci na staticke reseni patrovych ramu, *Casop. pestov. mat.*, 1955, **80**, No. 1, 60-88; R Zh. M., 1956, 4107.

BACKMAN, G.

(1) Rekursionsformeln zur Losung der Normalgleichungen auf Grund der Krakovianenmethodik, *Ark. Mat., Astr. Fys.*, 1946, **33A**, No. 1, 1-14; M. R., 8, 287.

BAETSLÉ, P. L.

(1) Sur les méthodes itératives de calcul numérique des vecteurs propres d'une matrice. *III-e Congrès National des Sciences*, Bruxelles, 1950, **2**, 104-106; M. R., 17, 666.

(2) Systématisation des calculs numériques de matrices. *Bull. géod.*, 1951, 22-41; M. R., 12, 861.

BALLANTINE, J. P.

(1) Numerical Solutions of linear equations by vectors, *Amer. Math. Monthly*, 1931, **38**, 275-277.

BANACHIEWICZ, T.

(1) Zur Berechnung der Determinanten, wie auch der Inversen, und zur darauf basierten Auflosung der Systeme linearer Gleichungen, *Acta Astron.*, Ser. C, 1937, **3**, 42-67.

(2) Calcul des determinants par la methode des cracoviens, *Bull. intern. Acad. Polon. Sci. A.*, 1937, 109-120.

(3) Sur la resolution numerique d'un systeme d'equations lineaires, *Bull. intern. Acad. Polon. Sci. A.*, 1937, 350-354.

(4) Principes d'une nouvelle technique de la methode des moindres carres, *Bull. intern. Acad. Polon. Sci. A* , 1938, 134-135.

(5) Methode de resolution numerique des equations lineaires, du calcul des determinants et des inverses et de reduction des formes quadratiques, *Bull. intern. Acad. Polon. Sci. A.*, 1938, 393-401.

(6) La regle de Chio, cracoviens et matrices, *Bull. intern. Acad. Polon. Sci. A.*, 1939, 405-412.

(7) An outline of the Cracovian algorithm of the method of least squares, *Astr. J.*, 1942, **50**, 38-41; M. R., 4, 90-91.

(8) Fragmentos de novo algorithmo de methodo de minimo quadratos, *Rocznik Astr. Observ. Krakov. Suppl. Internat.*, 1949, **20**, 87-98; M. R., 11, 403.

(9) Sur la resolution des equations normales de la methode des moindres carres, *Soc. Sci. Let. Varsovie. C. R. Cl. III, Sci. Math., Phys.*, 1948, **41**, 63-68 (1950); M. R. 13, 285.

(10) Resolution d'un systeme d'equations lineaires algebriques par division, *Ensignement Math.*, 1951, **39** (1942-1950), 34-45; M. R., 12, 861.

BANDEMER, HANS.

(1) Berechnung der reellen Eigenwerte einer reellen Matrix mit dem Verfahren von Rutishauser, *Wiss. Z. Martin-Luther Univ. Halle-Wittenberg. Math.-naturwiss. Reihe*, 1957, **6**, No. 5, 807-814; R. Zh. M., 1959, 2015.

BANDYOPADHYAY, G., and NARASHIMHAN, R. K.

(1) Special types of group relexation for simultaneous linear equations, *Quart. J. Mech. and Appl. Math.*, 1956, **9**, No. 1, 122-128; R. Zh. M., 1957, 2683.

BARANKIN, EDWARD W.

(1) Bounds for the characteristic roots of a matrix, *Bull. Amer. Math. Soc.*, 1945, **51**, 767-770; M. R., 7, 107.

(2) Bounds on characteristic values, *Bull. Amer. Math. Soc.*, 1948, **54**, 728-735.

BARTLETT, M. S.

(1) An inverse matrix adjustment arising in discriminant analysis, *Ann. Math. Statistics*, 1951, **22**, 107-111; M. R., 12, 639.

BARTSCH, HELMUT.

(1) Ein Einschliessungssatz fur die charakteristischen Zahlen allgemeiner Matrizen-Eigenwertaufgaben, *Arch. Math.*, 1953, **4**, No. 2, 133-136; R. Zh. M., 1954, 3235.

(2) Abschatzungen fur die kleinste charakteristische Zahl einer positiv-definiten hermiteschen Matrix *Z. angew. Math. und Mech.*, 1954, **34**, No. 1-2, 72-74; R. Zh. M., 1955, 4710.

BASILE, R.

(1) *Resolution de systemes d'equations lineaires algebriques et inversions de matrices au moyen des machines de mecanographie comptable, Complement pratique par R. Janin.* Office National d'Etudes et de Recherches Aeronautiques, Paris, 1949, **28**; M. R., 11, 692.

BASILE, R., et JANIN, R.

(1) *Resolution de systemes d'equations lineaires algebriques et inversions de matrices au moyen des machines de mecanographie comptable,* Office National d'Etudes et de Recherches Aeronautiques, Paris, 1949, **28**; M.R., 12, 208.

BAUER, FRIEDRICH L.

(1) Der Newton-Prozess als quadratisch konvergente Abkurzung des allgemeinen linearen stationaren Iterationsverfahrens. 1. ORDNUNG (Wittmeyer-Prozess), *Z. Angew. Math. und Mech.*, 1955, **35**, No. 12, 469-470; R. Zh. M., 1957, 912.

(2) Das Verfahren der abgekurzten Iteration fur algebraische Eigenwertprobleme, insbesondere zur Nullstellenbestimmung eines Polynoms, *Z. angew Math und Phys.*, 1956, **7**, No. 1, 17-32; R. Zh. M., 1958, 752.

(3) Zur numerischen Behandlung von algebraischen Eigenwertproblemen hoherer Ordnung, *Z. angew. Math. und Mech.*, 1956, **36**; M. R., 18, 766.

(4) Iterationsverfahren der linearen Algebra vom bernoullischen Konvergenztyp, *Nachrichtentechn. Fachber.*, 1956, **4**, 171-175, 221. Diskuss; R. Zh. M., 1957, 8266.

(5) Zusammenhange zwischen einigen Iterationsverfahren der linearen Algebra, *Ber. Internat. Math. Kolloq.*, 1955 (1957), Nov., 99-111; R. Zh. M., 1958, 10300.

(6) Beitrage zum Daniewski-Verfahren, *Ber. Internat. Math. Kolloq.*, 1955 (1957), Nov., 133-139; R. Zh. M., 1959, 5229.

(7) Das Verfahren der Treppeniteration und verwandte Verfahren zur Losung algebraischer Eigenwertprobleme, *Z. angew. Math. und Phys.*, 1957, **8**, No. 3, 214-235; R. Zh. M., 1958, 10299.

(8) On modern matrix iteration process of Bernoulli and Greffe type, *J. Assoc. Comput. Machinery*, 1958, **5**, No. 3, 246-258; R. Zh. M., 1959, 8573.

BAWIE, O. L.

(1) Practical solution of simultaneous linear equations, *Quart. Appl. Math.*, 1951, **8**. 369-373: M. R.. 12. 538.

BEJARANO, GABRIEL G., and ROSENBLATT, BRUCE R.

(1) A solution of simultaneous linear equations and matrix inversion with high speed Computing devices, *Math. Tables and other Aids Comput.*, 1953, **7**, No. 42, 77-81; R. Zh. M., 1954, 3837.

BELL, W. D.

(1) Punched card techniques for the solution of simultaneous equations and other matrix operations, *Proc. Scient. Comput. Forum*, 1948, N. Y., I. B. M. Corp., 1950, 28-31; M. R., 13, 387.

BENDIXSON, I.

(1) Sur les racines d'une equation fondamentale, *Acta Math.*, 1902, **25**, 359-365.

DI BERARDINO, V.

(1) Risoluzione dei sistemi di equazioni algebriche lineari per incrementi successivi delle incognite. Nuovo metodo di calculo., *Riv. catasto e serv. tecn. erariarli*, 1956,

**11**, No. 5, 6, 334-338; R. Zh. M., 1958, 7200.

(2) Il metodo di Hardy Cross e la sua giustificazione mediante un nuovo procedimento di risoluzione dei sistemi di equazioni algebriche lineari, *Ingegneria ferroviaria*, 1957, **12**, No. 10, 821-831; R. Zh. M., 1959, 876.

DI BERARDINO, V., E GIRARDELLI, L.

(1) Nuovo metodo di risoluzione di particolari sistemi di equazioni algebriche lineari. sistemi a catena., *Riv. catasto e serv. tech. erariali*, 1957, **12**, No. 1, 46-50; R. Zh. M., 1958, 7201.

DI BERARDINO, V., E FRANDI, P.

(1) Formule ricorrenti per la risoluzione graduale dei sistemi di equazioni algebriche lineari, *Archimede*, 1950, **2**, 108-113; M. R., 13, 586.

(2) Formule ricorrenti per la risoluzione graduale dei sistemi di equazioni algebriche lineari, *Ricerca Sci.*, 1950, **20**, 662-666; M. R., 13, 587.

BERGER, E. J., and SAIBEL, EDWARD.

(1) On the inversion of continuant matrices, *J. Franklin Inst.*, 1953, **256**, No. 3, 249-253; R. Zh. M., 1954, 5067.

BERRY, C. E.

(1) A criterion of convergence for the classical itrative method of solving linear simultaneous equations, *Ann. Math. Statistics*, 1945, **16**, 398-400; M. R., 7, 338.

BIEDENHARN, L. C., AND BLATT, J. M.

(1) A variation principle for eigenfunctions, *Physical Rev.*, 1954, (2) **93**, 230-232; M. R., 15, 745.

BIEZENO, C. B., AND BOTTEMA, O.

(1) The convergence of a specialized iterative process in use in structural analysis, *Proc. Koninkl. nederl. akad. wetensch.*, 1946, **49**, 489-499 (Also *Indagationes math.*, **8**); M. R., 9, 104.

BILY, J.

(1) Solution of a system of linear equations with large coefficients in the diagonal, *Aktuarske Vedy*, 1949, **5**, No. 3, 114-127; M. R., 11, 403.

BINGHAM, M. D.

(1) A new method for obtaining the inverse matrix, *J. Amer. Statist. Assoc.*, 1941, **36**, 530-534; M. R., 3, 154.

BIRMAN, M. SH.

(1) Nekotorye otsenki dlya metoda naiskoreishego spuska (Some estimates for the method of steepest descent), *Uspekhi matem. nauk*, 1950, **5**, No. 3, 152-155.

(2) Ob odnom variante metoda posledovatel'nykh priblizhenii (A variant of the successive approximations method), *Vestn. Leningr. in-ta, seriya matem., fiz. i khim.*, 1952, **9**, 69-76.

(3) O vychislenii sobstvennykh chisel metodom naiskoreishego spuska (Computing eigenvalues by the method of steepest descent), *L. Zapiski Gorn. in-ta*, 1952, **27**, No. 1, 209-216.

BISSHOPP, K. E.
(1)  The inverse of a stiffness matrix, *Quart. Appl. Math.*, 1945, **3**, 82-84; M. R., 6, 218.

Bjerhammar, A.
(1)  Rectangular reciprocal matrices with special reference to geodetic calculations. *Bull. géod.*, 1951, 188-220; M. R., 13, 312.
(2)  Triangular matrices for adjustment of triangular networks. *Kungl. Tekn. Högsk. Handl.*, Stockholm, 1956, **105**, 82.

BLANC, CH., et LINIGER, W.
(1)  Erreurs de chute dans la resolution de systemes algebriques lineaires, *Comment. math. helv.*, 1956, **30**, No. 4, 257-264; R. Zh. M., 1957, 1839.

BLACK, A. N.
(1)  Further notes on the solutions of algebraic linear simultaneous equations, *Quart. J. Mech. and Appl. Math.*, 1949, **2**, 321-324; M. R., 11, 743.

BLUMENTHAL, O.
(1)  Uber die Genauigkeit der Wurzeln linearer Gleichungen, *Z. Math. und Phys.*, 1914, **62**, 539-362.

BODEWIG, E.
(1)  Comparison of some direct methods for computing determinants and inverse matrices, *Proc. Koninkl nederl. akad. wetensch.*, 1947, A, **50**, 49-57 (Also *Indagationes math.*, **9**); M. R., 8, 407.
(2)  Bericht uber die verschiedenen Methoden zur Losung eines Systems linearer Gleichungen mit reellen Koeffizienten I, II, III, IV, V, *Proc. Koninkl. nederl. akad. wetensch.*, 1947, **50**, 930-941; 1104-1116; 1285-1295; 1948, **51**, 53-64; 211-219 (Also *Indagationes math.*, **9**; **10**); M. R., 9, 250, 382, 621.
(3)  Bericht uber die Methoden zur numerischen Losung von algebraischen Eigenwertproblemen, I, II, *Atti., Sem. Mat. Fis. Univ. Modena*, 1951, No. 3, 3-39; No. 4, 133-193; M. R., 13, 991.
(4)  A practical refutation of the iteration method for the algebraic eigenproblem, *Math. Tables and Other Aids Comput.*, 1954, **8**, No. 48, 237-240; R. Zh. M., 1956, 762.
(5)  Zum Matrizenkalkul. IV. *Proc. Koninkl. nederl. akad. wetensch.*, 1955, A **58**, No. 1, 95-106; 1956, A **59**, No. 3, 301-304; 1956, A **59**, No. 3, 305-312; 1957, A **60**, No. 1, 82-87; 1957, A **60**, No. 3, 242-247. (Also *Indagationes Math.*, **17**; **18**; **19**); R. Zh. M., 1955, 4876; 1957, 4614; 1958, 3298, 3299; 1959, 881.
(6)  *Matrix calculus*, Amsterdam, North-Holl and, 1956, xii, 334; R. Zh. M., 1957, 4623.
(7)  Zu Stiefels Berechnung der Eigenwerte aus den Schwarzschen Konstanten, *Z. angew. Math. und Mech.*, 1958, **38**, No. 1-2, 72-73; R. Zh. M., 1959, 880.

BODEWIG, E., und ZURMÜHL, R.
(1)  Zu R. Zurmühl. Zur numerischen Auflosung linearer Gleichungssysteme nach dem Matrizenverfahren von Banachiewicz, *Z. angew. Math. und Mech.*, 1950, **30**, 130-132; M. R., 11, 743.

BOLIE, VICTOR W.
(1)  Minimum-storage matrix inversion, *Z. angew. Math. und Mech.*, 1958, **38**, No. 9-10, 369-372; R. Zh. M., 1959, 5223.

BOLTZ, H.

(1) Entwicklungsverfahren zur Ansgleichung Geodatischer Netze nach der Methode der kleinsten Quadrate, *Veröff. Preussischen Geod. Inst.*, 1923, **90**.

BONDAR', N. G.

(1) O tochnosti nekotorykh priblizhennykh metodov vychisleniya sobstvennykh chisel kvadratnykh matrits, (On the accuracy of certain approximating methods for computing eigenvalues of quadratic matrices) *Tr. Dnepropetrovskogo in-ta inzh. zh.-d. transporta,* 1953, **23**, 61–69; R. Zh. M., 1954, 3468.

BORODYANSKII, M. YA.

(1) Privedenie nekotorogo tipa matrits k diagonal'nomu vidu (Transforming a certain type of matrix to diagonal form), *Tr. Kievsk. Tekhnol. in-ta pishch. prom-ti*, 1953, **13**, 195–196.

BORSCH-SUPAN, W., UND BOTTENBRUCH, H.

(1) Eine Methode zur Eingrenzung samtlicher Eigenwerte einer hermiteschen Matrix mit uberwiegender Hauptdiagonale, *Z. angew. Math. und. Mech.*, 1958, **38**, No. 5–6, 169–171; R. Zh. M., 1958, 10301.

BOTTEMA, O.

(1) A geometrical interpretation of the relaxation method, *Quart. Appl. Math.*, 1950, **7**, 422–423; M. R., 11, 403.

BOSCHAN, PAUL.

(1) The consolidated Doolittle technique, *Ann. Math. Statistics*, 1946, **17**, 503.

BOWKER, A. H.

(1) On the norm of matrix, *Ann Math. Statistics*, 1947, **18**, 285–288; M. R., 9, 75.

BRANSTETTER, R. D.

(1) A round-off theory for scalar products, *Iowa State Coll. J. Sci.*, 1954, **28**, No. 3, 283–284; R. Zh. M., 1955, 4715.

BRAUER, ALFRED.

(1) Limits for the characteristic roots of a matrix. I—VI (VI along with La Borde, H. T.), *Duke Math. J.*, 1946, **13**, 387–395; 1947, **14**, 21–26; 1948, **15**, 871–877; 1952, **19**, 75–91; 1952, **19**, 553–562; 1955, **22**, 253–261; M. R., 8, 192; 8, 192; 8, 559; 10, 231; 13, 813; 14, 836; 17, 1044.

(2) On the characteristic equations of certain matrices, *Bull. Amer. Math. Soc.*, 1947, **53**, 605–607; M. R., 8, 559.

(3) Matrices with all their characteristic roots in the interior of the unit circle, *J. Elisha Mitchell Sci. Soc.*, 1952, **68**, 188–193; M. R., 14, 836.

(4) Uber die Lage der charakteristischen Wurzeln einer Matrix, *J. reine und angew. Math.*, 1953, **192**, No. 2, 113–116; R. Zh. M., 1954, 5451.

(5) Bounds for characteristic roots of matrices, *Nat. Bur. Standards. Appl. Math. Ser.*, 1953, **29**, 101–106.

(6) Bounds for the ratios of the coordinates of the characteristic vectors of a matrix. *Proc. Nat. Acad. U. S. A.*, 1955, **41**, No. 3, 162–164; R. Zh. M., 1956, 2780

(7) The theorem of Ledermann and Ostrowski on positive matrices. *Duke Math. J.*, 1957, **24**, No. 2, 265–274; R. Zh. M., 1958, 2733.

(8) A new-proof of theorems of Perron and Frobenius on nonnegative matrices. I. Positive matrices. *Duke Math. J.*, 1957, **24**, No. 3, 367–378.

(9)  A method for the computation on the greatest root of a positive matrix. *J. Soc. Indust. Appl. Math,*. 1957, **5**, 250–253; M. R., 19, 1197.

BRENNER, J. L., and REITWIESNER, G. W.
(1)  Remark on determination of characteristic roots by iteration.  *Math. Tables and Other Aids Comput.*, 1955, **9**, No. 51, 117–118; R. Zh. M., 1957, 903.

BRODSKII, M. L.
(1)  Veroyatnostnye otsenki pogreshnostei pri opredelenii sobstvennykh znachenii i· sobstvennykh vektorov var'iruyushcheisya matritsy (Probability estimates of the error in determining the eigenvalues and eigenvectors of a variable matrix), *Uspekhi matem. nauk*, 1952, **7**, No. 5, 205–214; M. R., 14, 692.

BROCK, JOHN E.
(1)  Variation of coefficients of simultaneous linear equations.  *Quart. Appl. Math.*, 1953, **11**, No. 2, 234–240; R. Zh. M., 1954, 2351.

BROOKER, R. A., and SUMNER, F. H.
(1)  The method of Lanczos for calculating the characteristic roots and vectors of a real symmetric matrix.  *Proc. Inst. Electr. Engrs.*, 1956, B **103** Suppl. No. 1, 114–119.  Discuss., 120–122; R. Zh. M., 1958, 2439.

BROWN, R. D., and BASSETT, J. M.
(1)  A method for calculating the first order perturbation of an eigenvector of a finite matrix, with applications to molecularorbital theory, *Proc. Phys. Soc.*, 1958, **71**, No. 5, 724–732; R. Zh. M., 1959, 4249.

BROWNE, E. T.
(1)  The characteristic equation of a matrix, *Bull. Amer. Math. Soc.*, 1928, **34**, 363–368.
(2)  The characteristic of a matrix, *Bull. Amer. Math. Soc.*, 1930, **36**, 705–710.
(3)  Limits to the characteristic roots of a matrix, *Amer. Math. Monthly*, 1939, **46**, 252–265.

BRUNER, N.
(1)  Note on the Doolittle solution. *Econometrica*, 1947, **5**, 43–44; M. R., 8, 407.

BURDINA, V. I.
(1)  K odnomu metodu resheniya sistem lineinykh algebraicheskikh uravnenii (Toward a method for solving systems of linear algebraic equations), *Dokl. AN SSSR*, 1958, **120**, No. 2, 235–238; R. Zh. M., 1959, 5218.

BURGER, A. P.
(1)  Inversion of matrices with the aid of punched card machines, *Statistica, Rijswijk*, 1952, **6**, 121–133; M. R., 14, 1128.

BURGESS, H. T.
(1)  On the matrix equation BX C, *Amer. Math. Monthly*, 1916, **23**, 152–155.

BURKHARDT, FELIX
(1)  Über spezielle lineare Gleichungs-systeme mit der Eigenschaft $\lim_{\nu \to \infty} (\mathfrak{F} - \mathfrak{A})^{\nu}$ Null-matrix.  *Wiss. Z. Univ. Leipzig*, 1952/53, No. 5, 187–192; R. Zh. M., 1946, 4106.

BÜCKNER, H.
(1) Über ein unbeschrankt anwendbares Iterationsverfahren für Systeme linearer Gleichungen. *Arch. March.*, 1950, **2**, 172-177; M. R., 11, 743.

CAIRONI, MARIO
(1) Osservazioni sui procedimenti di approssima1ioni successive nel metodo delle forze. *Ingegnere*, 1957, **31**, No. 5, 410-420; R. Zh. M., 1958, 2436.

CAPRIOLI, LUIGI
(1) Sulla risoluzione dei sistemi di equazioni lineari con il metodo di Cimmino. *Boll. Unione mat.*, 1953, **8**, No. 3, 260-265; R. Zh. M., 1955, 6110.

CARR, JOHN W.
(1) Error analysis in floating point arithmetic. *Comm. Assoc. Comput. Machinery*, 1959, **2**, No. 5, 10-15.

CASSINA, U.
(1) Sul numero delle operazioni elementari necessarie per risoluzione dei sistemi di equazioni lineari. *Boll. Unione mat. ital.*, ser. *III*, 1948, **3**, 142-147; M. R., 10, 405.

CASSINIS, GINO
(1) I metodi di H. Boltz per risoluzione dei sistemi di equazioni lineari e il loro impiego nelle compensazioni della triangolazione. *Riv. catasto e ser. tecn. erariali*, 1944, 1.
(2) Risoluzione dei sistemi di equazioni algebriche lineari. *Rend. Sem. mat. e fis.*, *Milano*, 1946, **17**, 62-78; M. R., 9, 622.

CAUCHY, A. L.
(1) Méthode générale pour la résolution des systèmes d'équations simultanées. *C. r. Acad. sci.*, 1847, **25**, 536-538.

CAUSEY, ROBERT L.
(1) Computing eigenvalues of non-hermitian matrices by methods of Jacobi type. *J. Soc. Industr. and Appl. Math.*, 1958, **6**, No. 2, 172-181.
(2) On some error bounds of Givens. *J. Assoc. Comput. Machinery*, 1958, **5**, No. 2, 127-131; R. Zh. M., 1959, 5228.

CESARI, LAMBERTO
(1) Sulla risoluzione dei sistemi di equazioni lineari, per approssimazioni sucessivi. *Atti. Acad. naz. Lincei. Rend. Ser.* 6, 1937, **25**, 422-428.
(2) Sulla risoluzione dei sistemi di equazioni ineari per approssimazioni successive. *Estratta della Rass. Paste, Telegr., e Telef.*, 1937, **4**, 37.

CHANCELLOR, JUSTUS, SHELDON, J. W. and TATUM, G. L.
(1) The solution of simultaneous linear equations using the IBM Card-programmed Electronic Calculator. *Proc. Indus. Comput. Sem., N. Y., IBM Corp.*, 1951, 57-61; M. R., 13, 587.

CHAO, F. H.
(1) A gradient method for solving simultaneous equations. *Acta math. sinica*, 1953, **3**, 328-342; M. R. 17, 194.
(2) Konechno-raznostnyi metod dlya resheniya sistem lineinykh algebraicheskikh uravnenii (Finite-difference method for solving systems of linear algebraic equations),

*Acta math. sinica*, 1955, **5**, No. 2, 149–159; R. Zh. M., 1957, 8264.

(3) Metod tabulirovaniya dlya rescheniya sistemy lineinykh algebraicheskikh uravnenii (A tabulation method for solving a system of linear algebraic equations), *Chinese
J. Civil Engng.*, 1956, **3**, No. 4, 463–474; R. Zh. M., 1957, 8267.

(4) Sravnenie gradientnykh metodov (A comparison of gradient methods), *Acta math.
sinica*, 1957, **7**, No. 1, 63–78; R. Zh. M., 1958, 10295.

CHEN, T. C., and WILLOUGHBY, R. A.

(1) A note on the computation of eigenvalues and vectors of Hermitean matrices.
*IBM J. Res. and Developm.*, 1958, **2**, No. 2, 169–170; R. Zh. M., 1959, 5230.

CHOW, C.

(1) Gradual developing method. *Bull Géod.*, 1951, 221–229; M. R., 13, 496.

CHERENKOV, F. S.

(1) O reschenii sistem lineinykh uravnenii metodom iteratsii (On solving systems of
linear equations by an iterative method), *Matem. sb.*, 1936, **1** (43), No. 6, 953–958.

CHERUBINO, SALVATORE

(1) *Calcolo delle matrici.* Montr. mat. Consigilo naz. ricerche, 4, Roma. Ed. Cremonese,
1957, VII, 322 p.; R. Zh. M., 1958, 9554.

CHIO, F.

(1) *Mémoire sur les Fonctions Connues sous le Nom de Résultants ou de Déterminants*, Turin, 1853.

CICALA, P.

(1) Determination of modes and frequencies above the fundamental by matrix iteration. *J. Aeronaut. Sci.*, 1952, **19**, 719–720; M. R., 14, 587.

CIMMINO, GIANFRANCO

(1) Calcolo approssimato per resoluzioni dei sistemi di equazioni lineari. *Ricerca
Scien. Roma*, (2), 1938, **9**, 326–333.

CLASEN, B. I.

(1) Sur une nouvelle méthode de résolution des équations linéaires et sur l'application
de cette méthode au calcul des déterminants. *Ann. Soc. scient., Bruxelles*, 1888, **12**,
A 50–59, B 251–281.

CLERC, D.

(1) Sur le calcul par itération des modes propres d'ordre supérieur. I. Part. *Rech.
aèronaut.*, 1956, No. 54, 39–48; R. Zh. M., 1957, 7400.

COLLAR, A. R.

(1) On the reciprocation of certain matrices. *Proc. Roy-Soc. Edinburgh*, 1939, **59**,
95–206.

(2) Some notes on Jahn's method for the improvement of approximate latent roots
and vectors of a square matrix. *Quart. J. Mech. and Appl. Math.*, 1948, **1**, 145–148;
M. R., 10, 152.

COLLATZ, L.

(1) Fehlerabschätzung für das Iterationsverfahren zur Auflösung linearer Gleichungssysteme. *Z. angew. Math. und Mech.*, 1942, **22**, 357–361; M. R., 5, 50.

(2) Einschliessungssatz für die charakterischen Zahlen von Matrizen. *Math. Z.*, 1942, **48**, 221–226; M. R., 5, 30.

(3) *Eigenwertaufgaben mit technischen Anwendungen.* Mathematik und ihre Anwendungen in Physik un Technik, Reihe A, 1949, **19**, Akademische Verlagsgesellschaft, Leipzig, XVII + 466 pp; M. R., 11, 137.

(4) Über die Konvergenzkriterien bei Iterationsverfahren für lineare Gleichungssysteme *Math. Z.* 1950, **53**, 149–161; M. R., 12, 361.

(5) Zur Herleitung von Konvergenzkriterien für Iterationsverfahren bei linearen Gleichungssystemen. *Z. angew. Math. and Mech.*, 1950, **30**, 278–280.

(6) Zur Fehlerabschätzung bei linearen Gleichungssystemen. *Z. angew. Math. und Mech.*, 1954, **34**, No. 1–2, 71–72; R. Zh. M., 1955, 439.

Conference on matrix computations (Wayne State Univ. Sept. 3rd–6th, 1957). *J. Assoc. Comput. Machinery*, 1957, **4**, No. 4, 520–523.

COOPER, J. L. B.
(1) The solution of natural frequency equations by relaxation methods. *Quart. Appl. Math.*, 1948, **6**, 179–183; M. R., 10, 70.

COÜARD, A.
(1) Résolution d'un système d'équations linéaires par approximations successives. *Génie civil*, 1953, **130**, No. 6, 114; R. Zh. M., 1953, 454.

COUFFIGNAL, LOUIS
(1) Recherches de mathématiques utilisables. La résolution numérique des systèmes d'équations linéaires. I. L'opération fondamentale de réduction d'un tableau. *Revue. sci.*, 1944, **82**, 67–78; M. R., 8, 128.

(2) Sur la précision des solutions approchées d'un système d'équations linéaires. *C. r. Acad. sci.*, 1948, **227**, 30–32, M. R., 10, 212.

(3) Sur la résolution numérique des systèmes d'équations linéaires. II. *Rev. sci.* 1951, **89**, 3–10; M. R., 13, 284.

(4) Méthodes pratiques de réalisation des calculs matriciels. *Rend mat. e appl.*, 1954, **14**, No. 1–2, 85–97; R. Zh. M., 1956, 6867.

COWDEN, D. J.
(1) Correlation concepts and the Doolittle method. *J. Amer. Statist. Assoc.*, 1943, **38**, 327–334; M. R., 5, 42.

CRAIG, EDWARD J.
(1) The N-step iteration procedures. *J. Math. and Phys.*, 1955, **34**, No. 1, 64–73; R. Zh. M., 1956, 6866.

CRANDALL, S. H.
(1) *Engineering analysis; A survey of numerical procedures.* McGraw-Hill, New York, Toronto, London, 1956; M. R., 18, 674–675.

CROCKETT, JEAN BRONFENBRENNER, and CHERNOFF, HERMAN
(1) Gradient methods of maximization. *Pacif. J. Math.*, 1955, **5**, No. 1, 33–50; R. Zh. M., 1956, 4925.

CROUT, P. D.
(1) A short method for evaluating determinants and solving systems of linear equa-

tions with real or complex coefficients. *Trans. Amer. Inst. Elec. Engng*, 1941, **60**, 1235-1240.

CURRIE, J. C.

(1) Cassini ovals associated with a second order matrix. *Amer. Math. Monthly*, 1948, **55**, 487-489; M. R., 10, 177.

CURRY, H. B.

(1) The method of steepest descent for non-linear minimization problems. *Quart. Appl. Math.*, 1944, **2**, 258-261; M. R., 6, 52.

CURTISS, J. H.

(1) A generalization of the method of conjugate gradients for solving systems of linear algebraic equations. *Math. Tobles and Other Aids Comput.*, 1954, **8**, No. 48, 189-293; R. Zh. M., 1956, 4105.

(2) A theoretical comparison of the efficiency of two classical methods and a Monte Carlo method for computing one component of the solution of a set of linear algebraic equations. *Sympos. Monte Carlo Methods*, New York Wiley, 1956, 191-233; R. Zh. M., 1958, 8272.

DAVYDOV, V. V.

(1) Reschenie trekhchlennykh uravenii, vstrechayushchiknsya v stroitel'noi mekhanike korablya (The solution of trinominal equations encountered in the mechanics of ship construction), *Tr. Gor'kovsk. in-ta inzh. vodn. transp.*, 1957, **14**, 10-24; R. Zh.M., 1958, 3297.

DANILEVSKII, A. M.

(1) O chislennom reschenii vekovogo uravneniya (The numerical solution of a secular equation), *Matem. sb.*, 1937, **2**, (44), 169-171.

DANTZIG, GEORGE B., and ORCHARD-HAYS, WM.

(1) The product from for the inverse in the simplex method. *Math. Tables and Other Aids Comput.*, 1954, **8**, No. 46, 64-67; R. Zh. M., 1956, 6149.

DEMING, H. G.

(1) A systematic method for the solution of simultaneous linear equations. *Amer. Math. Mouthly.* 1928, **35**, 360-363.

DENIS-PAPIN, et KAUFMANN, A.

(1) *Cours de calcul matriciel appliqué.* Electro-techn., hydraul., radio, 1958, **237**, Suppl., 150-157; R. Zh. M., 1959 5555.

DERWIDUÉ, L.

(1) La méthode de L. Coeffignal pour la résolution numérique des systèmes d'équations linéaires. *Mathesis*, 1954, **63**, No. 1-2, 9-12, R. Zh. M., 1955, 4712.

(2) Une méthode mécanique de calcul des vecteurs d'une matrice quelconque. *Bull. Soc. Roy. Sci.*, Liège, 1955, **24**, No. 5, 150-171; R. Zh. M., 1956, 9115.

DIAS, AGUDO ROLDAO FERNANDO

(1) On the characteristic equation of a matrix. *Univ, Lisboa Rev. Fac. Ci. A. Ci. Mat.*, 1954, (2) **3**, 87-136; R. Zh. M., 1957, 6854.

DIMSDALE, B.
(1) The non-convergence of a characteristic root method. *J. Soc. Industr. and Appl. Math.*, 1958, **6**, No. 1, 23–25, R. Zh. M., 1959, 5227.

DMITRIEV, N., and DYNKIN, E.
(1) Kharakteristicheskie korni stakhasticheskikh matrits (Characteristic roots of Stochastic Matrices), *Izd. AN SSSR, ser. matem.*, 1946, **10**, No. 2, 167–184; M. R., 8, 129.

DOWNING, A. C., Jr., and HOUSEHOLDER, A. S.
(1) Some inverse characteristic value problems. *J. Assoc. Comput. Machinery.* 1956, **3**, No. 3, 203–207; R. Zh. M., 1957, 8963.

DOYLE, THOMAS C.
(1) Inversion of symmetric coefficient matrix of positive-definite quadratic form. *Math. Tables and Other Aids Comput.*, 1957, **11**, No. 58, 55–58; R. Zh. M., 1958, 6214,

DULEAU, JACQUES
(1) Résolution numérique de certains systemes d'équations linéiares vectorielles. *C. r. Acad. sci.*, 1956, **242**, No. 7, 870–873; R. Zh. M., 1956, 9130.

DUNCAN, W. J.
(1) Some devices for the solution of large sets of simultaneous linear equations. *Philos. Mag.*, 1944, (7) **35**, 660–670; M. R., 7, 84.

DURAND, DAVID
(1) A note on matrix inversion by the square root method. *J. Amer. Statist. Assoc.*, 1956, **51**, No. 274, 288–292; R. Zh. M., 1958, 6215.

DWYER, P. S.
(1) The solution of simultaneous equations. *Psychometrika*, 1941, **6**, 101–129; M. R., 2, 367.
(2) The evaluation of the determinants. *Psychometrika*, 1941, **6**, 191–204; M. R., 2, 367.
(3) The evaluation of linear forms. *Psychometrika*, 1941, **6**, 355–365, M. R., 2, 154.
(4) The Doolittle technique. *Ann. Math. Statistics*, 1941, 449–458; M. R., 3, 276.
(5) Recent developments in correlation techniques. *J. Amer. Statist. Assoc.*, 1942, **37**, 441–460; M. R., 4, 164.
(6) A matrix presentation of least squares and correlation theory with matrix justification of improved methods of solution. *Ann Math. Statistics.*, 1944, **15**, 82–89; M. R., 5, 245.
(7) The square root method and its use in correlation and regression. *J. Amer. Statist. Assoc.*, 1945. **40**, 493–503; M. R., 7, 338.
(8) *Linear Computations*, 1951, pp. 344, New York, Wiley; R. Zh. M., 1954, 1829.
(9) Errors of matrix computations. *Nat. Bur. Standards. Appl. Math. Ser.*, 1953, **29**, 49–58; R. Zh. M., 1956, 6868.

DWYER, PAUL S. and WAUGH, FREDERICK V.
(1) On errors in matrix inversion. *J. Amer. Statist. Assoc.*, 1953, **48**, No. 262, 289–319; R. Zh. M., 1954, 5785.

EGERWÁRY, JENÖ
(1) Über die Faktorisation von Matrizen und ihre Anwendung auf die Lösung von

linearen Gleichungssystemen. *Z. angew. Math. und Mech.*, 1955, **35**, No. 3, 111–118; R. Zh. M., 1956, 6389.

(2) Az inverz matrix általánosítása. *Megyar tud. akad. Mat. kutató int. közl.*, 1956, **1**, No. 3, 315–324; R. Zh. M., 1959, 3513.

(3) Regi és új módszerek lineáris egyenletrendszerek megoldására. *Magyar tud. akad. Mat. kutató int. közl.*, 1956, **1**, 1–2, 109–123; R. Zh. M., 1959, 7479.

(4) Uber eine Veralgemeinerung der Purcellschen Methode zür Auflösung linearer Gleichungssysteme. *Osterr. ingr.-Arch.*, 1957, **11**, 4, 249–251; R. Zh. M., 1959, 5216.

EISEMANN, KURT

(1) Removal of ill-conditioning for matrices. *Quart. Appl. Math.*, 1957, **15**, No. 3, 225–230; R. Zh. M., 1958, 6216.

EISEN, AXEL.

(1) Beitrag zur Lösung linearer Gleichungen. *Internat. Vereinig. Brücken. und Hochban.*, 1935, **3**, 56–66.

ERSHOV, A. P.

(1) Ob odnom metode obrashcheniya matrits (On a method for inverting matrices), *Dokl. AN SSSR*, 1955, **100**, No. 2, 209–211; R, Zh. M., 1955, 4242.

FABIAN, VACLAV

(1) Zufälliges Abrunden und die Konvergenz des linearen (seidelschen) Iterationsverfahren. *Math. Nachr.*, 1957, **16**, No. 5-6, 265–270; R. Zh. M., 1958, 7203.

FADDEEV, D. K.

(1) O preobrazovanii kharakteristicheskogo uravneniya matritsy (On transforming the characteristic equation of a matrix), *Tr. in-ta inzh. prom. stroit.*, 1937, N. 4, 78–96.

(2) O nekotorykh posledovatel'nostyakh polinomov, poleznykh dlya postroeniya iteratsionnykh metodov resheniya sistem lineinykh algebraicheskikh uravnenii (On some successive polynomials used in constructing iterative methods for solving systems of linear algebraic equations), *Vestn. Leningr. un-ta*, 1958, No. 7, 155–159; R. Zh. M., 1959, 875.

(3) K voprosu o verkhnei relaksatsii pri reshenii sistem lineinykh uravnenii (On upper relaxation in solving systems of linear equations), *Izd. vysshikh uchebn. zavedenii. Matematika*, 1958, No. 5, 122–125, R. Zh. M., 1959, 5219.

(4) Ob obuslovlennosti matrits (On matrix condition), *Tr. Matem. in-ta AN SSSR*, 1959, **53**, 387–391.

FADDEEV, D. K., and FADDEEVA, V. N.

(1) Vychislitel'nye metody lineinoi algebry (Computational methods of linear algebra), *Tr. 3rd Vsesoyuznogo matem. s''ezda*, 1958, **3**, 434–445.

FADDEEVA, V. N.

(1) *Vychislitel'nye metody lineinoi algebry* (Computational methods of linear algebra), Gostekhnizdat, 1950; M. R., 13, 872.

FALK, SIGURD

(1) Ein übersichtliches Schema für die Matrizenmultiplikation. *Z. angew. Math. und Mech.*, 1951, **31**, 152–153; M. R., 12, 751.

(2) Neue Verfahren zur direkten Lösung des allgemeinen Matrizeneigenwertproblemes.

*Z. angew. Math. und Mech.*, 1954., **34**, No. 8/9, 289-291; R. Zh. M., 1956, 761.

(3) Das Ersatzwertverfahren als Hilfsmittel bei der iterativen Bestimmung von Matrizen- Eigenwerten. *Abhandl. Braunschweig. wiss. Ges.*, 1956, **8**, 99-110; R. Zh., M., 1958, 6219.

FAN, Ky.
(1) Note on circular disks containing the eigenvalues of a matrix. *Duke Math. J.*, 1958, **25**, 3, 441-445; R. Zh. M., 1959, 7485.

FAN, Ky, and HOFFMAN, A. J.
(1) Lower bounds for the rank and location of the eigenvalues of a matrix. *Nat. Bur. Standards. Appl. Math. Ser.*, 1954, **39**, 117-130; R. Zh. M. 1958, 5190.

FARNELL, A. B.,
(1) Limits for the characteristic roots of a matrix. *Bull. Amer. Math. Soc.*. 1944, **50**, 789-794; M. R., 6, 113.
(2) Limits for the field of a matrix. *Amer. Math. Monthly*, 1945, **52**, 488-493.

FAURE, G. SIMON-SUISSE, J., et RONA, TH.
(1) Deux circuits analogiques pour l'inversion des matrices symétriqus et la recherche de la vitesse critique de flutter. *Proc. Seventh Internat. Congress Appl. Mech.*, 1948, **4**, 81-95; M. R., 11, 403.

FEHLBERG E.
(1) Bemerkungen zur Konvergenz des Iterationsverfahrens bei linearen Gleichungs-systemen. *Z. angew. Math. und Mech.*, 1951, **31**, 387-389; M. R., 13, 990.

FELLER, W. and FORSYTHE, G. E.
(1) New matrix transformation for obtaining characteristic vectors. *Quart. Appl. Math.*, 1951, **8**, 325-331; M. R., 12, 538.

FETTIS, H.
(1) A method for obtaining the characteristic equation of a matrix and computing the associated modal columns. *Quart. Appl. Math.*, 1950, **8**, 206-212; M. R., 12, 209.

FIEDLER, M., und PTAK, V.
(1) Über die Konvergenz des verallgemeinerten Seidelschen Verfahrens zur Lösung von Systemen linearer Gleichungen. *Math. Nachr.*, 1956, **15**, 31-38; R. Zh. M., 1959, 104, 84.

FILIPOWSKY, R.
(1) Numerical calculations in electrical engineering and electronics. I. Calculation of higher order and the solution of simultaneous algebraic equations. *J. Madras Inst. Tech.*, 1952, **1**, 64-88; M. R., 14, 692.

FISCHBACH, JOSEPH W.
(1) Some applications of gradient methods, *Proc. Sympos. Appl. Math.*, 6 New York, Toronto, London, 1956, 52-72; R. Zh. M., 1957, 5962.

FISHER, MICHAEL E., and FULLER, A. T.
(1) On the stabilization of matrices and the convergence of linear iterative processes. *Proc. Cambridge Philos. Soc.*, 1958, **54**, 417-425; R. Zh. M., 1959, 8561.

FLANDERS, D., and SHORTLEY, G.
(1) Numerical determination of fundamental modes. *J. Appl. Phys.*, 1950, **21**, 1326–1332; M. R., 12, 640.

FLOMENHOFT, H. I.
(1) A method for determining mode shapes and frequencies above the fundamental by matrix iteration. *J. Appl. Mech.*, 1950, **17**, 249–256; M. R., 12, 287.

FORSYTHE, A. I., and FORSYTHE, G. E.
(1) Punched-card experiments with accelerated gradient methods for linear equations. *Nat. Bur. Standards. Appl. Math. Ser.*, 1954, **39**, 55–69; R. Zh. M., 1958, 10296.

FORSYTHE, G. E.
(1) Gauss to Gerling on relaxation. *Math. Tables and Other Aids Comput.*, 1951, **5**, 255–258.
(2) Alternative derivations of Fox's escalator formulae for latent roots. *Quart. J. Mech. and Appl. Math.*, 1952, **5**, 191–195; M. R., 14, 92.
(3) Tentative classification of methods and bibliography on solving systems of linear equations. *Nat. Bur. Standards. Appl. Math. Ser.*, 1953, **29**, 1–28; M. R., 15, 164.
(4) Solving linear algebraic equations can be interesting. *Bull. Amer. Math. Soc.* 1953, **59**, No. 4, 299–329; R. Zh. M., 1954, 3840.

FORSYTHE, G. E., and LEIBLER, R. A.
(1) Matrix inversion by a Monte Carlo method. *Math. Tables and Other Aids Comput.*, 1950, **4**, 127–129; M. R., 12, 361.

FORSYTHE, G. E., and Motzkin, T. S.
(1) Asymptotic properties of the optimum gradient method. *Bul. Amer. Math. Soc.*, 1951, **57**, 183.
(2) Acceleration of the optimum gradient method. *Bul. Amer. Math. Soc.*, 1951, **57**, No. 4, 304.
(3) An extension of Gauss' transformation for improving the condition of systems of linear equations. *Math. Tables and Other Aids Comput.*, 1952, **6**, 9–17; M. R., 13, 991.

FORSYTHE, G. E., and STRAUS, ERNST G.
(1) On best conditioned matrices. *Proc. Internat. Congr. Math.*, 1954, 2, *Amsterdam*, 1954, 102–103; R. Zh. M., 1956, 2791.
(2) On best conditioned matrices. *Proc. Amer. Math. Soc.*, 1955, **6**, No. 3, 340–345; R. Zh. M., 1956, 7882.

FORSYTHE, G. E., and STRAUS, LOUISE W.
(1) The Souriau-Frame characteristic equation algorithm on a digital computer. *J. Math. and Phys.*, 1955, **34**, No. 3, 152–156; R. Zh. M., 1957, 900.

FORSYTHE, G. E., HESTENES, M. R., and ROSSER, J. B.
(1) Iterative methods for solving linear equations. *Bul. Amer. Math. Soc.* 1951, **57**, 480.

FOX, L.
(1) Some improvements in the use of relaxation methods for the solution of ordinary

and partial differential equations. *Proc. Roy. Soc. Ser. A.* 1947, **190**, 31–59.

(2) A short account of relaxation methods. *Quart. J. Mech. and Appl. Math.*, 1948, **1**, 253–280; M. R., 10, 574.

(3) Practical methods for the solution of linear equations and the inversion of matrices. *J. Roy. Statist. Soc. Ser. B.*, 1950, **12**, 120–136; M. R., 12, 538.

(4) Escalator methods for latent roots. *Quart. J. Mech. and Appl. Math.*, 1952, **5**, 178–190; M. R., 14, 92.

(5) Practical solution of linear equations and inversion of matrices. *Nat. Bur. Standards. Appl. Math. Ser.*, 1954, **39**, 1–54; R, Zh. M., 1957, 8956.

FOX, L., HUSKEY, H. D., and WILKINSON, F. H.

(1) Notes on the solution of algebraic linear simultaneous equations. *Quart. J. Mech. and Appl Math.*, 1948, **7**, 149–173; M. R., 10, 152.

FOX, L., and HAYES, J. G.

(1) More practical methods for the inversion of matrices. *J. Roy. Statistics Soc. Ser. B.*, 1951, **13**, 83–91; M. R., 13, 990.

FRAME, J. S.

(1) A simple recurrent formula for inverting a matrix. *Bull. Amer. Math. Soc.*, 1949, **55**, 1045.

FRANK, W. L.

(1) Computing eigenvalues of complex matrices by determinant evaluation and by methods of Danilewski and Wielandt. *J. Soc. Indust.and Appl. Math.*, 1958, **6**, No. 4, 378–392.

FRANCKX, ED.

(1) Résolution pratique des systèmes linéaires par la méthode des matrices de relaxation. *Bull. Soc. roy. scy. Liège*, 1957, **26**, No. 7–12, 390–395; R. Zh. M., 1959, 3242.

FRAZER, R. A.

(1) Note on the Morris escalator process for the solution of linear simultaneous equation. *Philos. Mag.*, 1947, (**7**) **38**, 287–289; M. R., 9, 250.

(2) Some problems in aerodynamics and structural engineering related to eigenvalues. *Nat. Bur. Standards. Appl. Math. Ser.*, 1953, **29**, 65–74; M. R., 15, 164.

FRAZER, R. A., DUNCAN, W. J., and COLLAR, A. R.

(1) *Elementary matrices and some application to dynamics and differential equations.* 1946; M. R., **8**, 365.

(2) *Elementary Matrices.* Oxford Univ. Press, 1951.

FREEMAN, G. F.

(1) On the iterative solution of linear simultaneous equations. *Philos. Mag.*, 1943, (**7**), **34**, 409–416; M. R., 5, 50.

FREIRE, RÉMY

(1) A matricial method for the solution of certain systems of linear equations. *Soc. Parana Mat. Anuario*, 1956, **3**, 54–59.

FRIEDRICH, K., und JENNE, W.

(1) Geometrischanschauliche Auflösung lineardr mit Nullkoeffizienten ausgestatteter

Gleichungssysteme. *Deutsche Akad. Wiss. Berlin. Veröff. Geodät. Inst. Potsdam*, 1951, **5**, 68; M. R., 13, 387.

FRÖBERG, CARL-ERIC
(1) Solutions of linear systems of equations on a relay machine. *Nat. Bur. Standards. Appl. Math. Ser.*, 1953, **29**, 39-42; R. Zh. M., 1956, 3331.

FRUCHTER, B.
(1) Note on the computation of the inverse of a triangular matrix. *Psychometrika*, 1949, **14**, 89-93; M. R., 11, 403.

FURUYA, SHIGERU
(1) Methods of numerical calculation for simultaneous linear equations and inverse matrices. *Sugaku*, 1957/1958, **9**, 240-249; M. R., 20, 1406.

GANTMAKHER, F. P.
(1) K algebraicheskomu analizu metoda ak. A. N. Krylova preobrazovaniya vekovogo uravneniya (Toward an algebraic analysis of A. N. Krylov's method for transforming a secular equation), *Tr. 2nd Vsesoyuznogo matem. s"ezda*, 1937, 45-48.
(2) *Teoriya matrits* (Matrix theory), Gostekhizdat, 1953.

GARZA, A. DE LA.
(1) Error bounds on approximate solutions to systems of linear algebraic equations. *Math. Tables and Other Aids Comput.*, 1953, **7**, No. 42, 81-84; R. Zh. M., 1954, 3838.
(2) Error bounds for a numerical solution of a recurring linear system. *Quart. Appl. Math.*, 1956, **13**, No. 4, 453-456, R. Zh. M. 1958, 4239.

GASTINEL, NOËL
(1) Procédé itératif pour la résolution numérique d'un système d'équations linéaires. *C. r. Acad. sci.*, 1958, **246**, No. 18, 2571-2574; R. Zh. M., 1959, 4244.

GATTO, F.
(1) Sulla risoluzione numerica dei sistemi di equazioni lineari. *Riserca Sci.*, 1949, **19**, 1935-1388; M. R., 11, 743.

GAUSS, C. F.
(1) Supplementum theoriae combinationis observationum erroribus minimis obnoxiae. *Werke*, Göttingen, 1873, **4**, 55-93.

GAUTSCHI, WERNER
(1) The asymptotic behaviour of powers of matrices. *Duke Math. J.* 1953, **20**, 127-140; M. R., 15, 94.
(2) The asymptotic behaviour of powers of matrices. II. *Duke Math. J.* 1953, **20**, No. 3, 375-379; R. Zh. M., 1954, 3620.
(3) Bounds of matrices with regard to an Hermitian metric. *Compositio Math.*, 1954, **12**, 1-16.
(4) On norms of matrices and some relations between norms and eigenvalues. *Philosoph.-Naturwiss. Fak. der Universität Basel*, Basel, Buchdruckerei Birkhäuser, AG, 1954, **39** S.; R. Zh. M., 1959, 5559.

GAVRILOV, YU. M.
(1) Pro zbizhnist' prostikh iteratsii ta kriterii znakoviznachenosti kvadratichnikh

form, *Dopovidi AN URSR*, 1953, No. 6, 389-393; R. Zh. M., 1955, 436.

(2) O skhodimosti iteratsionnykh protsessov i kriteriyakh znakoopredelennosti kvadratichnykh form, *Izv. AN SSSR, Ser. mat.*, 1954, **18**, No. 1, 87-94; R. Zh. M., 1955, 437.

(3) Pro zbizhnist' prostikh, a takozh grupovikh iteratsii pri rozv'yazuvanni sistem normal'nikh rivnyan', *Nauch. zap. L'vovsk. politekhn in-ta*, 1955, **29**, 114-120; R. Zh. M., 1956, 8348.

GAVURIN, M. K.

(1) Primenenie polinomov nailuchshego priblizheniya dlya uluchsheniya skhodimosti iterativnykh protsessov (Applying polynomials of best approximation to improve the convergence of iterative processes), *Uspekhi matem. nauk*, 1950, **5**, No. 3, 156-160; M. R., 12, 209.

GEIRINGER, H.

(1) Zur Praxis der Lösung linearer Gleichungen in der Statik. *Z. angew. Math., und Mech.*, 1928, **8**, 446-447.

(2) On the numerical solution of linear problems by group iteration. *Bull. Amer. Math. Soc.*, 1942, **48**, p. 370.

(3) On the solution of systems of linear equations by certain methods. *Reissner Anniversary Volume, Contrib. Appl. Mech.*, 1948, 365-393, Ann. Arbor. Mich.; M. R., 10, 574.

GEL'FAND, I. M.

(1) *Lektsii po lineinoi algebre* (Lectures on linear algebra), Izd. 2-e, M. L., 1951.

GERI, GERO

(1) Il simbolismo delle matrici nella soluzione del sistema normale gaussiano. *Riv. calusto e serw. tecn. erariali*, 1954, 9, No. 2, 117-120; R. Zh. M., 1955, 6111.

GERMANSKY, BORIS

(1) Zur angenäherten Auflösung linearer Gleichungssysteme mittels Iteration. *Z. angew. Math. und Mech.*, 1936, **16**, 57-58.

GERSIIGORIN, S. A.

(1) Ueber die Abgrenzung der Eigenwerte einer Matrix. *AH CCCP, Ser. matem*, 1931, **7**, 749-754.

GIVENS, WALLACE

(1) A method of computing eigenvalues and eigenvectors suggested by classical results on symmetric matrices. *Nat. Bur. Standards. Appl. Math. Ser.*, 1953, **29**, 117-122; R. Zh. M., 1956, 6861.

(2) Numerical computation of the characteristic values of a real symmetric matrix. *U. S. Atomic Energy Comm. Repts.*, 1954, ORNL-1574, 107 pp.; R. Zh. M., 1956, 8347.

(3) The characteristic value-vector problem. *J. Assoc. Comput. Machinery*, 1957, **4**, No. 3, 298-307; R. Zh. M., 1959, 2014.

(4) Computation of plane rotations transforming a general matrix to triangular form. *J. Soc. Industr. and Appl. Math.*, 1958, **6**, No. 1, 26-50; M. R., 19, 1081.

GLODEN, A.

(1) La méthode du Luxembourgeois B. I. Clasen pour la résolution d'un systeme d'équations linéaires (méthode des coefficients égaux). *Rev. histoire sci.*. 1953, **6**, No.

2, 168-170; R. Zh. M., 1954, 2353.

GOFMAN, SH. M.

(1) K voprosu o reshenii sistemy lineinykh algebraicheskikh uravenii (On solving a system of linear algebraic equations), *Dokl. AN Uz. SSR*, 1949, No. 2, 7-10.

(2) Ob otsenke odnogo opredelitelya i o nekotorykh neravenstvakh, svyazannykh s iteratsionnymi Zeidelya (On estimating a determinant and on some inequalities connected with Seidel iterative processes), *Tr. Tashkentsk. in-ta zh.-d. transp.*, 1956, vyp. **5**, 178-186; R. Zh. M., 1957, 902.

(3) Ob odnom variante rescheniya sistemy trekhchlennykh matrichnykh uravnenii (A variant solution of a system of trinomial matrix equations), *Tr. Tashkentsk. in-ta zh.-d. transp.*, 1956, vyp. **5**, 204-206; R. Zh. M., 1957, 2684.

GOLDSTINE, H., and NEUMANN, J.

(1) Numerical inverting of matrices of high order, II. *Proc. Amer. Math., Soc.*, 1951, **2**, 188-202; M. R., 12, 861.

GOLDSTINE, H. H., MURAY, F. I., and VON NEUMANN

(1) The Jacobi method for real symmetric matrices. *J. Assoc. Comput. Machinery*, 1960, **6**, No. 1, 59-97.

GOLDSTINE, H. H., and HORWITZ, L. P.

(1) A procedure for the diagonalization of normal matrices. *J. Assoc. Comput. Machinery*, 1959, **6**, No. 2, 176-195.

GOODEY, W. J.

(1) Note on the improvement of approximate latent roots and modal columns of a symmetrical matrix. *Quart. J. Mech. and Appl. Math.*, 1955, **8**, No. 4, 452-453; R. Zh., M., 1956, 6865.

GOPSHTEIN, N. M.

(1) O reshenii odnorodnykh lineinykh uravnenii metodom iteratsii (On solving homogeneous linear equations by an iterative method), *Dokl. AN SSSR*, 1944, **43**, 332-395, M. R., 6, 218.

GOTTHARDT, E.

(1) Bolzsches Entwicklungsverfahren und Gaußscher Algorithmus. *Z. Vermessungswesen*, 1953, **78**, No. 4, 97-104.

GOUARNÉ, REUÉ

(1) *Méthodes algébriques de la physique et de la chimie, Résolution rapide des systèmes d'équations linéires.* These Doc. Sci. math. Ann. Univ. Paris, 1956, **26**, No. 4, 588-591; R. Zh. M., 1958, 1599.

(2) Calcul automatique des déterminants. *C. r. Acad. sci.*, 1957, **245**, No. 8, 824-826; R. Zh. M., 1959, 3246.

(3) Remarques sur le calcul automatique des déterminants et polynomes caractéristiques par la méthode des cycles. *C. r. Acad. sci.*, 1957, **245**, No. 23, 1998-2000, R. Zh. M., 1959, 3247.

(4) Calcul automatique des polynomes caractéristiques. *C. r. Acad. sci.*, 1957, **245**, No. 14, 1114-1117.

GRANDALL, S. H.
(1) On a relaxation method for eigenvalue problems. *J. Math. and Phys.*, 1951, **30**, 140–145; M. R., 13, 496.
(2) Iterative procedures related to relaxation methods for eigenvalue problems. *Proc. Roy. Soc.*, Ser. A. 1951, **207**, 416–423; M. R., 13, 163.

GRAY, H. J. JR.
(1) Numerical methods in digital real-time simulation. *Quart. Appl. Math.*, 1954, **12**, 133–140; M. R., 15, 991.

GREENSPAN, DONALD
(1) Methods of matrix inversion. *Amer. Math. Monthly*, 1955, **62**, No. 5, 303–318, R. Zh. M., 1956, 6870.

GREENSTADT, J.
(1) A method for finding roots of arbitrary matrices. *Math. Tables and Other Aids Comput.*, 1955, **9**, No. 50, 47–52.

GREGORY, ROBERT T.
(1) Computing eigenvalues and cigenvectors of a symmetric matrix on the ILLIAC. *Math. Tables and Other Aids Comput.*, 1953, **7**, 44, 215–221.
(2) On the convergence rate of an iterative process. *Math. Mag.*, 1955, **29**, No. 2, 63–68; R. Zh. M., 1955, 7646.
(3) Results using Lanczos method for finding eigenvalues of arbitrary matrices. *J. Soc., Industr. and Appl. Math.*, 1958, **6**, No. 2, 182–188.

GROHNE, D.
(1) Rechenverfahren zur Auflösung von Gleichungssystemen. *Veröffentlichungen Math. Inst. Tech. Hochschule Braunschweing*, 1946, No. **4**, i + 28 p.

GROSSMAN, D. P.
(1) K probleme chislennogo rescheniya sistemy odnorodnykh lineinykh algebraichc-skikh uravenenii (On the problem of solving numerically a system of homogeneous linear algebraic equations), *Uspekhi matem. nauk*, 1950, **5**, No. 3, 87–103; M. R., 13, 586.

GUBERMAN, I. O.
(1) Sproschena skhema rozv'yazannya sistem liniinikh algebraichnikh rivnyan', *Prikl. mekhanika*, 1957, **3**, No. 1, 108–112; R. Zh. M., 1957, 8268.

GUEST, I.
(1) The solution of linear simultaneous equations by matrix iteration. *Austral. J. Phys.*, 1955, **8**, No. 4, 425–439; R. Zh. M., 1957, 6710.

GUTSHALL, W. D.
(1) Practical inversion of matrices of high order. *Proc. Comput. Sem. Dec. 1949*, N. Y., IBM Corp., 1951, 171–173; M. R., 13, 387.

GUTTMAN, L.
(1) Enlargement methods for computing the inverse matrix. *Ann. Math. Statistics* 1946, **17**, 336–343; M. R., 8, 171.

HALLERT, B.

(1) Über einige Verfahren zur Lösung von Normalgleichungen. *Z. Vermessungswesen*, 1943, **72**, 238–244; M. R., 8, 171.

HAMMERSLEY, J. M.

(1) The numerical reduction of nonsingular matrix pencils. *Philos. Mag.*, 1949, (**7**) **40**, 783–807; M. R., 11, 464.

HARMAN, HARRY H.

(1) The square root method and multiple group methods of factor analysis. *Psychometrika*, 1954, **19**, 39–55; M. R., 16, 177.

HAYES, J. G., and VICKERS, T.

(1) The fitting of polynomials to unequally-spaced data. *Philos. Mag.*, 1951, (**7**) **42**, 1387–1400; M. R., 13, 990.

HEAD, Y. W., and OULTON, G. M.

(1) The solution of ill-conditioned linear simultaneous equations. *Aircraft Engng*, 1958, **30**, 356, 309–312; M. R., 20, 343.

HECHT, JOSEF

(1) Poznámka k řešeni soustav algebraickych lineárnich rovnic. *Aplikace mat.*, 1958, **3**, 233–237; R. Zh. M., 1959, 5217.

HEINRICH, HELMUT

(1) Bemerkungen zu den Verfahren von Hossenberg und Voetter. *Z. angew. Math. und Mech.*, 1956, **36**, 250–252; R. Zh. M., 1959, 7481.

(2) Zur Eingrenzung der charakteristischen Zahlen einer beliebigen matrix. *Technik*, 1958, **13**, No. 2, 82–86; R. Zh. M., 1958, 9264.

HELLER, J.

(1) Ordering properties of linear successive iteration schemes applied to multi-diagonal type linear systems, *J. Soc. Industr. and Appl. Math.*, 1957, **5**, 238–243; M. R., 19, 1080.

HENRICI, P.

(1) On the speed of convergence of cyclic and quasi-cyclic Jacobi methods for computing eigenvalues of Hermitian matrices, *J. Soc. Industr. and Appl. Math.*, 1958, **6**, No. 2, 144–162; M. R., 20, 343.

(2) The quotient-difference algorithm. *Nat. Bur. Standards. Appl. Math. Ser.*, 1958, 49, 23–46; M. R., 20, 233.

HERRMANN, A.

(1) Bestimmung der höheren Eigenwerte einer Matrix durch Iteration. *Ann. Univ. Saraviensis*, 1952, **1**, 220–223; M. R., 14, 1129.

HERZBERGER, M.

(1) The normal equations of the method of least squares and their solution. *Quart. Appl. Math.*, 1949, 7, 217–223; M. R., 11, 57.

HERZBERGER, M., and MORRIS, R. II.

(1) A contribution to the method of least squares. *Quart Appl. Math.*, 1947, **5**, 354–357; M. R., 9, 210.

HESTENES, MAGNUS R.
(1) Determination of eigenvalues and eigenvectors of matrices. *Nat. Bur. Standards. Appl. Math. Ser.*, 1953, **29**, 89–94; R. Zh. M., 1956, 6863.
(2) Iterative computational methods. *Communic. Pure and Appl. Math.*, 1955, **8**, 8595; M. R., 16, 863.
(3) The conjugate-gradient method for solving linear systems. *Proc. Sympos. Appl. Math.*, 6., New York, Toronto, London, 1956, 83–102; R. Zh. M., 1959, 878.
(4) Inversion of matrices by biorthogonalization and related results. *J. Soc. Industr. and Appl. Math.*, 1958, **6**, 51–90.

HESTENES, MAGNUS R., and KARUSH, W.
(1) A method of gradients for the calculation of the characteristic roots and vectors of a real symmetric matrix. *J. Res. Nat. Bur. Standards*, 1951, **47**, 45–61; M. R., 13, 283.

HESTENES, MAGNUS R., and STIEFEL, EDUARD
(1) Methods of conjugate gradients for solving linear systems. *J. Res. Nat. Bur. Standards*, 1952 (1953), **49**, 409–436; R. Zh. M., 1956, 7645.

HOEL, P. G.
(1) The errors involved in evaluating correlation determinants. *Ann. Math. Statistics*, 1940, **11**, 58–65.
(2) On methods of solving normal equations. *Ann. Math. Statistics*, 1941, **12**, 354–359; M. R., 3, 154.

HOFFMAN, A. J.
(1) Lower bounds for the rank and location of the eigenvalues of a matrix. *Nat. Bur. Standards. Appl. Math. Ser.*, 1954, **39**, 117–130.

HOFFMAN, A. J., and WIELANDT, H. W.
(1) The variation of the spectrum of a normal matrix. *Duke Math. J.*, 1953, **20**, No. 1, 37–39; R. Zh. M., 1953, 587.

HOLDT, RICHARD ELTON
(1) An iterative procedure for the calculation of the eigenvalues and eigenvectors of a real symmetric matrix. *J. Assoc. Comput. Machinery*, 1956, **3**, No. 3, 223–238; R. Zh. M., 1957, 8966.

HOLZER, L., und MELAN, E.,
(1) Ein Beitrag zur Auflösung linearer Gleichungssysteme mit positiv definiter Matrix mittels Iteration. *Akad. Wiss. Wien*, S.-B. IIa, 1942, **151**, 249–254; M. R., 8, 407.

HORVAY, G.
(1) Solution of large equation systems and eigenvalue problems by Lanczós' matrix iteration method. *Gen. Electr. Co., Knolls Atomic Power Lab. Schenectady, N. Y., Rept.*, 1953, No. KAPL-1004, 113 pp.; R. Zh. M., 1957, 8958.

HORST, PAUL
(1) A method for determining the coefficients of a characteristic equation. *Ann. Math. Statistics*, 1935, **6**, 83–84.

HOTELLING, H.

(1) Analysis of a complex of statistical variables into principal components. *J. Educ. Phys.*, 1933, **24**, 417–441, 498–520.

(2) Simplified calculation of principal components. *Psychometrika*, 1936, **1**, 27–35.

(3) Some new methods in matrix calculation. *Ann. Math. Statistics*, 1943, **14**, 1–34; M. R., 4, 202.

(4) Further points on matrix calculation and simultaneous equations. *Ann. Math. Statistics*, 1943, **14**, 440–441; M. R., **10**, 245.

(5) Practical problems of matrix calculation. *Proc. Berkeley Symp. Math. Stat. Prob.*, 1945, 1946, 275–293; M. R., **10**, 574.

HOUSEHOLDER, ALSTON S.

(1) Some numerical methods for solving systems of linear equations. *Amer. Math. Monthly*, 1950, **57**, 453–459; M. R., 12, 538.

(2) Errors in iterative solution of linear systems. *Proc. Assoc. Comput. Mach. Meeting at Toronto, Ont.*, 1952, Sept. 1953, 30–33; R. Zh. M., 1954, 5263.

(3) *Principles of numerical analysis.* McGraw–Hill, 1953; M. R., 15, 470.

(4) The geometry of some iterative methods of solving linear systems. *Nat. Bur. Standards. Appl. Math. Ser.*, 1953, **29**, 35–37; R. Zh. M., 1956, 3157.

(5) On norms of vectors and matrices. *Oak Ridge Nat. Lab. Oak. Ridge Tenn. Rep. ORNL* 1756, 1954, 18 pp.; R. Zh. M., 1957, 6855.

(6) Terminating and nonterminating iterations for solving linear systems. *J. Soc. Industr. and Appl. Math.*, 1955, **3**, No. 2, 67–72; R. Zh. M., 1958, 1593.

(7) On the convergence of matrix iterations. *J. Assoc. Comput Machinery*, 1956, **3**, No. 4, 314–324; R. Zh. M., 1958, 6211.

(8) A survey of some closed methods for inverting matrices. *J. Soc. Industr. and Appl. Math.*, 1957, **5**, 155–169; M. R., 19, 982.

(9) A class of methods for inverting matrices. *J. Soc. Indust. and Appl. Math.*, 1958, **6**, No. 2, 189–195.

(10) The approximate solution of matrix problems. *J. Assoc. Comput. Machinery*, 1958, **5**, No. 3, 205–243.

(11) Generated error in rotational tridiagonalization. *J. Assoc. Comput. Machinery*, 1958, **5**, No. 4, 335–338.

(12) Unitary triangularization of a nonsymmetric matrix. *J. Assoc. Comput. Machinery*, 1958, **5**, No. 4, 339–342.

HRUSKA, V.

(1) Lösung von Gleichungssystemen durch das Iterationsverfahren. *Acad. Tchèque Sci. Bull. Cl. Sci. Math. Nat.*, 1943, **44**, 239–304, 399–422.

IDEL'SON, N. I.

(1) Vychislenie vesov neizvestnykh v metode naimen'shikh kvadratov (Computing the weights of unknowns in the least squares method), *Astr. zhurnal*, 1943, **20**, 11–13; M. R., 6, 51.

ISHAQ, M.

(1) Sur les spectres des matrices. *Sèmin. P. Dubreil et Ch. Pisot. Fac. sci. Paris*, 1955–1956, **9**, No. 14, 1–14; R. Zh. M., 1958, 2730.

ITO, M.

(1) A geometrical study of the characteristic equation. *Tôhoku Math. J.*, 1932, **35**, 294–303.

IVANOV, V. K.

(1) O skhodimosti protsessov iteratsii pri reshenii sistem lineinykh algebraicheskikh uravnenii (On the convergence of iterative processes in solving systems of linear algebraic equations), *Izv. AN SSSR, ser. matem.* 1939, No. 4, 477-483; M. R., 2, 118.

JACOBI, C. G. J.

(1) Ueber eine neue Auflösungsart der bei der Methode der kleinsten Quadrate vorkommenden lineären Gleichungen. *Astr. Nachr.*, 1845, **22**, No. 523, 297-306; Jacobis Werké 3, 467.

JAHN, H. A.

(1) Improvement of an approximate set of latent roots and modal columns of a matrix by methods akin to those of classical perturbation theory. *Quart. J. Mech. and Appl. Math.*, 1948, **1**, 131-144; M. R., 10, 152.

JANIN, R.

(1) Résolution de systèmes d'équations algébriques linéaire d'ordre élevé, à l'aide des méthodes mécanographiques (Emploi du calculateur électronique). *Rech. aéronaut.*, 1955, No. 44, 47-50; R. Zh. M., 1956, 4914.

JENNE, W.

(1) Zur Auflösung linearer Gleichungssysteme. *Astr. Nachr.*, 1949, 278, 79-95.

JENSEN, H.

(1) An attempt at a systematic classification of some methods for the solution of normal equations. *Geod. Inst. Medd.*, 1944, No. 18, 45 pp.; M. R., **7**, 488.

JOSSA, F.

(1) Risoluzione progressiva di un sistema di equazioni lineari. Analogia con un problema meccanico. *Rend. Accad. Sci. fis. e mat.*, Napoli, Ser. 4, 1940, **10**, 346-352; M. R., 8, 535.

JÜRGENS, E.

(1) *Zur Auflösung linearer Gleichungssysteme und numerischen Berehnung von Determinanten.* Festgabe. Aachen Palm., 1886.

KACZMARZ, S.

(1) Angenäherte Auflösung von Systemen linearer Gleichungen. *Bull. intern. Acad. Polon. Sci. A*, 1937, 355-357.

KALINOVSKAYA, S. S.

(1) Otsinka shvidkosti zbizhnosti deyakikh iteratsiinikh protsesiv, *Nauk. zap. Luts'k. derzh. ped. in-tu*, 1955, **3**, No. 2, 11-17; R. Zh. M., 1956, 7649.

KAMELA, C.

(1) Die Lösung der Normalgleichungen nach der Methode von Prof. Dr. T. Banachiewicz (sogenannte ,, Krakovianenmethode'') *Schweiz. Z. vermessungswesen Kulturtech.*, 1943, **41**, 225-232, 265-275; M. R., 7, 488.

KANTOROVICH, L. V.

(1) Ob odnom effektivnom metode resheniya ekstremal'nykh zadach dlya kvadratich-

nykh funktsionalov (An effective method for solving extremal problems for quadratic Functionals), *Dokl. AN SSSR*, 1945, **48**, No. 7, 455–460; M. R., 8, 30.

(2) O metode naiskoreishego spuska (On the method of steepest descent), *Dokl, AN SSSR*, 1947, **56**, No. 3, 233–236; M. R., 9, 308.

(3) Funktsional'nyi analiz i prikladnaya matematika (Functional analysis and applied mathematics), *Uspekhi metem. nauk*, 1948, **3**, No. 6, 89–185; M. R., 10, 380.

(4) Metod N'yutona (Newton's Method), *Tr. Matem. in-ta AN SSSR*, 1949, **28**, 104–144; M. R., 12, 419.

KAPPUS, R.

(1) L'algorithme de Gauss modernisé et son application à des systèmes d'équations linéaires dégénérés ou mal ordonnés. *Note techn. O. N. E. R. A.*, 1953, No. 11, 133 p. ill.; R. Zh. M., 1959, 5215.

KARPELEVICH, F. I.

(1) Kharakteristicheskie korni matrits s neotritsatel'nymi koeffitsientami (Characteristic roots of matrices with non-negative coefficients), *Uspekhi matem. nauk*, 1949, **4**, No. 5 (33), 177–178; M. R., 11, 154.

(2) O kharakteristicheskikh kornyakh matrits s neotritsatel'nymi elementami (On charocteristic roots of matrices with non-negative elements), *Izv. AN SSSR, ser. matem.*, 1951, **15**, 361–383; M. R., 13, 201.

KARUSH, W.

(1) An iterative method for finding characteristic vectors of a symmetric matrix. *Pacif. J. Math.*, 1951, **1**, No. 2, 233–247; M. R., 13, 388.

(2) Convergence of a method of solving linear problems. *Proc. Amer. Math. Soc.*, 1952, **3**, 839–851; M. R., 14, 1127.

KASAJIMA, TOMOMI

(1) A note on a theorem of Rutishauser. *Comment. math. Univ. St. Pauli.* 1957, **6**, No. 1, 89–91; R. Zh. M. 1959, 7488.

KAŠANIN, R.

(1) Interprétation géométrique du schéma de Banchiewicz. *Srpska Akad. Nauka. Zbornik Padowa*, 1952, **18**, *Mat. Inst.*, **2**, 93–96; M. R., 14, 501.

KATO, T.

(1) On the upper and lower bounds of eigenvalues. *J. Phys. Soc. Japan*, 1949, **4**, 334–339; M. R., 12, 447.

KELLY, J.

(1) Matrix multiplication on the IBM Card–Programmed Electronic Calculator. *Proc. Comput. Sem. Dec. 1949, N. Y. IBM Corp.*, 1951, 47–48; M. R., 13, 387.

KERKHOFS, W.

(1) Résolution de systèmes d'équations simultanées à un grand nombre d'inconnues. *Ossature Métallique*, 1947, **12**, 187–195; M. R., 10, 70.

KHAN, N. A.

(1) A theorem on the characteristic roots of matrices. *J. Univ. Bombay, sect. A.*, 1955, **38**, 13–18.

KHLODOVSKII, I. N.

(1) K teorii obshchego sluchaya preobrazovaniya vekovogo uravneniya metodom akad. A. N. Krylova (Toward a theory of the general case for transforming a secular equation by the method of A. N. Krylov), *IAN OMEN*, 1933, **8**, 1077–1102.

KHUBLAROVA, S. L.

(1) K voprosu o mekhanizatsii resheniya bol'shikh sistem normal'nykh uravnenii (On mechanizing the solution of great systems of normal equations), *Sb. ref. Tsentr. n.-i. in-ta. geod. aeros''emki i kartogr.*, 1954, No. 1, 15–16; R. Zh. M., 1958, 8286.

KIKUKAWA, M.

(1) A numerical method for multiplication, reciprocation and division of matrices and for the solution of simultaneous linear algebraic equations. *Tech. Rep. Osaka Univ.*, 1952, **2**, 11–30; M. R., 14, 501.

KIMBALL, BRADFORD F.

(1) Note on computation of orthogonal predictors. *Ann. Math. Statistics*, 1953, **24**, 299–303; M. R., 14, 1019.

KINCAID, W. M.

(1) Numerical methods for finding characteristic roots and vectors of matrices. *Quart. Appl. Math.*, 1947, **5**, 320–345; M. R., 9, 210.

KNÖDEL, W.

(1) Lineare Gleichungen. Beispiele aus den Anwendungen, Lösungsmethoden, Lochkartenverfahren. *MTW-Mitt.*, 1956, 138–140; R. Zh. M., 1957, 1838.

KOGBETLIANTZ, ERVAND GEORGE

(1) Diagonalization of general complex matrices as a new method for solution of linear equations. *Proc. Internat. Congr. Math.*, 1954, **2**, *Amsterdam*, 1954, 356–357; R. Zh. M., 1955, 6109.

(2) Solution of linear equations by diagonalization of coefficients matrix. *Quart. Appl. Math.*, 1955, **13**, No. 1, 123–132; R. Zh. M., 1958, 1595.

KOHN, W.

(1) A variational iteration method for solving secular equations. *Chem. Phys.*, 1949, **17**, 670; M. R., 11, 136.

KOSKO, E.

(1) Reciprocation of triply partitioned matrices. *J. Roy. Aeronaut. Soc.*, 1956, 60, 490–491; M. R., 18, 418.

(2) Matrix inversion by partitioning. *Aeronaut. Quart.*, 1957, **8**, No. 2, 157–184; R. Zh. M., 1958, 751.

KOSTARCHUK, V. N.

(1) Ob odnom metode resheniya sistem lineinykh uravnenii i otyskaniya sobstvennykh vectorov matritsy (A method of solving systems of linear equations and of finding the eigenvectors of a matrix), *Dokl. AN SSSR*, 1954, **98**, No. 4, 531–534; R. Zh. M., 1955, 4709.

(2) Zastosuvannya metodu minimal'nikh nev'yazok do znakhozhdeniya vlasnikh chisel matritsi, *Nauk. zap. Zhitomirs'k. derzh. ped. in-t, ser. fiz.-mat.*, 1957, **3**, 63–76; R. Zh. M., 1958, 1598.

KOSTARCHUK, V. N., and PUGACHEV, B. P.

(1) Tochnaya otsenka umen'sheniya pogreshnosti na odnom shage metoda naiskorei-shego spuska (An exact estimate of the error decrease in one step of the steepest descent method), *Tr. semin. po funktsion. analizu. Voronezhsk. un-t.* 1956, **2**, 25-30,

KOSTRO, J., I BORRO, L.

(1) Resoluçao de sistemas de equaçoes lineares pelo método de Cross. *Técnica*, 1953, **28**, No. 230, 365-370; R. Zh., M. 1953, 1407.

KOTELYANSKII, D. M.

(1) O raspolozhenii tochek matrichnogo spectra (On the point distribution of a matrix spectrum), *Ukrain. matem. zhurn.*, 1955, **7**, No. 2, 131-133; M. R., 17, 228.

(2) O nekotorykh dostatochnykh priznakakh veschestvennosti i prostoty matrichnogo spektra (Some sufficient indicators of whether a matrix spectrum is simple or real), *Matem., sb.*, 1955, **36** (78), 163-168; M. R., 16, 894.

(3) O vliyanii preobrazovaniya Gaussa na spektry matrits (The influence of a Gauss transformation on matrix spectra), *Uspekhi matem. nauk*, 1955, **10**, No. 1, 117-121; R. Zh. M., 1956, 1055.

KRARUP, T., and SVEJGARD, B.

(1) A method for matrix multiplication, matrix inversion, and problems of adjustment by punched card equipment. *Geodoet. Inst. Kobenhavn. Medd.*, 1956, **31**, 31; M. R., 18, 337.

KRASNOSEL'SKII, M. A.

(1) O nekotorykh priemakh priblizhennogo vychisleniya sobstvennykh znachenii i sobstvennykh vektorov polozhitel'no-opredelennoi matritsy (Some methods for computing approximately the eigenvalues and eigenvectors of a positive definite matrix), *Uspekhi matem. nauk*, 1956, **11**, No. 3, 151-158; R. Zh. M., 1957, 1841.

KRASNOSEL'SKII, M. A., and KREIN, S. G.

(1) Iterativnyi protsess s minimal'nymi nevyazkami (Iterative processes with minimal residuals), *Matem. sb.*, 1952, **31** (73), 315-334; M. R., 14, 692.

(2) Zamechanie o raspredelenii oshibok pri reshenii sistemy lineinykh uravnenii pri pomoshchi iteratsionnogo protsessa (Note on error distribution in solving a system of linear equations by an iterative process), *Uspekhi matem, nauk*, 1952, **7**, No. 4, 157-161; M. R., 14, 501.

KREYSZIG, E.

(1) Die Einschliessung von Eigenwerten hermitescher Matrizen beim Iterationsverfahren. *Z. angew. Math. and Mech.*, 1954, **34**, No. 12, 459-469; R. Zh. M., 1956, 4919.

(2) Die Ausnutzung zusätzlicher Vorkenntnisse für die Einschliessung von Eigenwerten beim Iterationsverfahren. *Z. angew. Math. and Mech.*, 1955, **35**, No. 3, 89-95; R. Zh. M., 1956, 4920.

(3) Einschliessung von Eigenwerten und Mohrsches Spannungs diagramm. *Z. angew. Math und Phys.*, 1958, **9a**, No. 2, 202-206; R. Zh. M., 1959, 7483.

KRON, GABRIEL.

(1) Detailed example of interconnecting piecewise solutions. *J. Franklin Inst.*, 1955, **259**, No. 4, 307-333; R. Zh. M,, 1956, 6872.

(2) Inverting a $256 \times 256$ matrix, *Engineering*, 1955, **179**, No. 4 650, 309-312; R.

Zh. M., 1956, 4915.

(3) Improved procedure for interconnecting piece-wise solutions. *J. Franklin Inst.*, 1956, **262**, 385–392; M. R., 19, 64.

(4) Factorized inverse of partitioned matrices. *Matrix Tensor Quart.*, 1957, **8**, 39–41; M. R,, 19, 1198.

KRYLOV, A. N.

(1) O chislennom reshenii uravneniya, kotorym v tekhnicheskikh voprosakh opredel-yayutsya chastoty malykh kolebanii material'nykh sistem (On numerical solutions which determine the frequencies of small oscillations of material systems in technical problems), *IAN OMEN*, 1931, **4**, 491–539.

KRYLOV, N. M., and BOGOLYUBOV, N. N.

(1) Novye metody dlya resheniya nekotorykh matematicheskikh problem, vstre-chaemykh v tekhnike (New methods for solving some mathematical problems encountered in engineering), *Ukr. nauchn. in-t sooruzhenii*, 1933, **78**, 78–95.

KUBLANOVSKAYA, V. N.

(1) *Primenenie analiticheskogo prodolzheniya v chislennykh metodakh analiza* (Application of analytic extension in numerical methods of analysis), Avtoreferat diss. kand. fiz. matem. n., LGU, L. 1955; R. Zh. M., 1956, 1688.

(2) Primenenie analiticheskogo prodolzheniya metodom zameny peremennykh v chislennom analize (Application of analytic extension by the replacement of variables method in numerical analysis), *Tr. Matem. in-ta AN SSSR*, 1959, **53**, 145–185.

KUNZ, K. S.

(1) Matrix methods. *Proc. Comput. Sem. Dec.*, 1949, No. 9, IBM Corp., 1951, 37–42; M. R., 13, 496.

LADERMAN, J.

(1) The square root method for solving simultaneous linear equations. *Math. Tables and Other Aids Comput.*, 1948, **3**, 13–16; M. R., 9, 622.

LAMINE, T.

(1) Diagonalisation des matrices. *Bull. Assoc. ingrs issus Ecole applic., artill et génie*, 1958, **36**, No. 1, 49–74; R. Zh. M., 1959, 1252.

LANCZOS, CORNELIUS

(1) A simple recurrence method for solving a set of linear equations. *Bull. Amer. Math. Soc.*, 1936, **42**, 325; M. R., 1, 97,

(2) An iteration method for the solution of the eigenvalue problem of linear differential and integral operators. *J. Res. Nat. Bur. Standards*, 1950, **45**, No. 4, 255–288; M. R., 13, 163.

(3) An iteration method for the solution of the eigenvalue problem of linear differential and integral operators. *Proc. Second. Symp. Large-Scale Digital Calcul. Mach.*, (1949), 1951, 164–206; M. R., 13, 589.

(4) An iteration method for the solution of the eigenvalue problem of linear differential and integral operators. *Proc. Symp. Spectral Theory and Differential Problems. Oklahoma Agricultural Mech. College*, 1951, 301–316; M. R., 13, 497.

(5) Solution of systems of linear equations by minimized iterations. *J. Res. Nat. Bur. Standards.* 1952, **49**, No. 1, 33–53; M. R.. 14, 501.

(6) Chebyshev polynomials in the solution of large-scale linear systems. *Proc. Assoc. Comput Mach., Meeting at Toronto Ont.*, 1952, *Sept.* 1953, 124-133; R. Zh. M., 1954, 5262.

(7) Spectroscopic eigenvalue analysis. *J. Wash. Acad. Sci.*, 1955, **45**, No. 10, 315-323; R. Zh. M., 1956, 9114.

(8) *Applied analysis.* Prentice Hall, Englewood Cliffs., N. Y., 1956, xx+539 pp.; R. Zh. M., 1958, 4265.

(9) Iterative solution of large-scale linear systems. *J. Soc. Indust. and Appl. Math.*, 1958, **6**, 1, 91-109; M. R., 20, 70.

LANGEFORS, B.

(1) Approximate solution of simultaneous equations by means of transformation of variables. Applications to aeronautical problems. *SAAB TN*, 1953, No. 7, 26 pp; R. Zh. M., 1959, 3241.

(2) Ill-conditioned matrices. *SAAB TN*, 1953, No. 22; R. Zh. M., 1956, 8349.

(3) On the practical solution of linear equations. *SAAB TN*, 1955, No. 35, 24 pp; R. Zh. M., 1959, 9505.

LAVRENT'EV, M. M.

(1) K voprosu ob uluchshenii tochnosti resheniya sistemy lineinykh uravnenii (Improving the accuracy for a solution of a system of linear equations), *Dokl. AN SSSR*, 1953, **92**, No. 5, 885-886; R. Zh. M., 1954, 3839.

(2) O tochnosti resheniya lineinykh uravnenii (On the accuracy of a solution of linear equations), *Matem. sb.*, 1954, **34** (76), No. 2, 259-268; R. Zh. M., 1954, 5784.

(3) Ob otsenke tochnosti resheniya sistem lineinykh uravnenii (Estimating the accuracy of the solution of a system of linear equations), *Dokl. AN SSSR*, 1954, **95**, No. 3, 447-448; R. Zh. M., 1955, 4714.

LAVUT, A. P.

(1) Raspolozhenie sobstvennykh chisel preobrazovanii Zeidelya dlya sistem normal'nykh uravnenii (Eigenvalue distribution of a Seidel transformation for systems of normal equations), *Uspekhi matem. nauk*, 1952, **7**, No. 6, 197-202; M. R., 14, 1128.

LEAVENS, D. H.

(1) Accuracy in the Doolittle solution. *Econometrica*, 1947, **15**, 45-50; M. R., 8, 407.

LEDERMANN, WALTER

(1) Bounds for the greatest latent-roots of a positive matrix. *J. London Math. Soc.*, 1950, **25**, 265-268; M. R., 12, 312.

LEGRAS, JEAN

(1) La méthode de relaxation. *Age nucléaire*, 1957, No. 6, 65-66; R. Zh. M., 1958, 6208.

LEIDERMAN, YU. R.

(1) Ob odnom sposobe resheniya sistemy lineinykh algebraicheskikh uravnenii, kogda obychnye metody posledovatel'nogo priblizheniya neprimenimy (A method for solving a system of linear algebraic equations when the usual successive approximation methods are not applicable), *Dokl. AN Uzb. SSR*, 1953, No. 1, 8-11; R. Zh. M., 1953, 902.

(2) K voprosu o reshenii sistemy lineinykh algebraicheskikh uravnenii metodom posledovatel'nykh priblizhenii (On solving a system of linear algebraic equations by the successive approximations method), *Dokl. AN Uzb. SSR*, 1953, No. 10, 6-9; R. Zh. M., 1954, 4919.

(3) Ob odnom metode resheniya sovmestnykh lineinykh algebraicheskikh uravnenii (A method for solving combined linear algebraic equations), *Tr. In-ta matem. i mekhan. AN Uzb. SSR*, 1954, v. 13, 153–158; R. Zh. M., 1957, 6711.

LENG, SEN-MING.
(1) The characteristic roots of a matrix. *Duke Math.* J., 1952, **19**, No. 1, 139–154; M. R., 14, 7.

LENTI, RAUMO.
(1) Eine Methode von sukzessiven Projektionen zur Lözung der linearen algebraischen Vektorgleichung und ihre Anwendungen fur Inversion von Matrizen. *Soc. Sci. Fennica. Jr. Phys.-Math.*, 1958, XXIs.

LEPPERT, E. L.
(1) A fraction series solution for characteristic values useful in some problems of airplane dynamics. *J. Aeronaut. Sci.*, 1955, **22**, No. 5, 326–328; R. Zh. M., 1956, 5510.

LEVERRIER, U. J. J.
(1) Sur les variations séculaires des éléments des orbites. *J. Math.*, 1840.
(2) Recherches Astronomiques. *Ann. l'Obser.*, *Paris*, 1856, **11**, 128.

LIDSKII, V. B.
(1) O sobstvennykh znacheniyakh summy i proizvedeniya simmetrichnykh matrits (Eigenvalues of the sum and product of symmetric matrices), *Dokl. AN SSSR*, 1950, **75**, 769–772.

LINETSKII, V. D.
(1) Reshenie sistemy trekhchlennykh uravnenii s pomoshch'yu fokusnykh otnoshenii (Solving a system of trinomial equations with the aid of focal ratios), *Nauchn. tr. Leningr. inzh.-stroit. in-ta*, 1954, **17**, 185–190; R. Zh. M., 1956, 765.

LITVINOV, M. V.
(1) Obchislennya koefitsientiv vplivu dlya skladenikh sitkovikh oblastei za dopomogoyu matrichnikh peretvoren', *Dopovidi AN URSR*, 1955, No. 3, 222–226; R. Zh. M., 1956, 6871.

LOHMAN, J. B.
(1) An iterative method for finding the smallest eigenvalue of a matrix. *Quart. Appl. Math.*, 1949, **7**, 234; M. R., 10, 743.

LONSETH, A. T.
(1) Systems of linear equations with coefficients subject to error. *Ann. Math. Statistics*, 1942, No. 3, 332–337; M. R., 4, 90.
(2) On relative errors in systems of linear equations. *Ann. Math. Statistics*, 1944, **15**, 323–325; M. R., 6, 51.
(3) An extension of an algorithm of Hotelling. *Berkeley Symp. Math. Stat. Prob.*, 1945, 1946, 353–357; M. R., 10, 627.
(4) The propagation of error in linear problems. *Trans. Amer. Math. Soc.*, 1947, **62**, 193–212; M. R., 9, 192.

LOO, W., et KWAN, CHAO-CHIH.
(1) La méthode de col dans le problème de relaxation. *Acta Math. sinica*, 1955, **5**, 497–504; M. R., 17, 791.

590                                                                           *Literature*

LOPSHITS, A. M.
   (1) Chislennyi metod nakhozhdeniya sobstvennykh znachenii i sobstvennykh ploskostei lineinogo operatora (A numerical method for finding eigenvalues and eigensurfaces of a linear operator), *Tr. sem. po vektorn. i tenzorn. analizu*, 1949, **7**, 233–259; M. R., 13, 991.
   (2) Ekstremal'naya teorema dlya giperellipsoida i ee primenenie k resheniyu sistemy lineinykh algebraicheskikh uravnenii (An extremal theorem for a hyperellipsoid and its application in solving a system of linear algebraic equations), *Tr. sem. po vektorn. i tenzorn. analizu*, 1952, **9**, 183–197; M. R., 14, 1127.

LOTKIN, MARK.
   (1) A set of test matrices. *Math. Tables and Other Aids Comput.*, 1955, **9**, No. 52, 153–161; R. Zh. M., 1957, 2685.
   (2) Characteristic values of arbitrary matrices. *Quart. Appl. Math.*, 1956, **14**, No. 3, 267–275; R. Zh. M., 1957, 8964.
   (3) The diagonalization of skew-Hermitian matrices. *Duke Math. J.*, 1957, **24**, 9–14; M. R., 19, 685.

LOTKIN, MARK, and REMAGE, RUSSELL.
   (1) Scaling and error analysis for matrix inversion by partitioning. *Ann. Math. Statistics*, 1953, **24**, 428–439; R. Zh. M., 1954, 5265.
   (2) Matrix inversion by partitioning. *Proc. Assoc. Comput. Math., Meeting at Toronto, Ont.*, 1952, *Sept.* 1953, 36–41; R. Zh. M., 1954, 5264.

LOWASS-NAGY, VIKTOR.
   (1) *Mátrixzámitás* Müszaki matematikai gyakorlatok. C. IV. Budapest, Tankönyvkiadó, 1956; R. Zh. M., 1958, 6487.

LOWE, J.
   (1) Solution of simultaneous linear algebraic equations using the IBM Type 604 Electronic Calculating Punch. *Proc. Comput. Sem.*, *Dec.* 1949, *N. Y. IBM Corp.*, 1951, 54–56; M. R., 13, 388.

LUKASZEWICZ, JÓZEF, I WARMUS, MIECZYSLAW.
   (1) *Metody numeryczne i graficzne*. I. Państwowe Wydaw. Nauk., Warszawa, 1956, 429 pp.; M. R., 18, 235.

LUZIN, N. N.
   (1) O metode ak. A. N. Krylova sostavleniya vekovogo uravneniya (On A. N. Krylov's method of setting up the secular equation), *IAN OMEN*, 1931, 903–958.
   (2) O nekotorykh svoistvakh peremeshchayushchegosya mnozhitelya v metode A. N. Krylova (Some properties of the transforming factor in A. N. Krylov's method), I, II, III. *IAN OMEN*, 1932, 596–638, 735–762, 1065–1102.

LYUSTERNIK, L. A.
   (1) Zamechaniya k chislennomu resheniyu kraevykh zadach uravneniya Laplasa i vychisleniyu sobstvennykh znachenii metodom setok (Notes toward a numerical solution of extreme problems for the Laplace equation and computation of eigenvalues by the method of nets), *Tr. Matem. in-ta AN SSSR*, 1947, **20**, 49–64; M. R., 10, 71.
   (2) O skhodimosti pri sluchainykh nachal'nykh dannykh i nakoplenii oshibok iteratsionnogo protsessa resheniya sistemy algebraicheskikh uravnenii (On convergence for random initial data and error accumulation of an iterative process in solving a system

of algebraic equations), *Vychisl. matem. i vychisl. tekhnika*, 1953, N. 1, 41–45; R. Zh. M., 1955, 435.

(3) Reshenie zadach lineinoi algebry metodom nepreryvnikh drobei (Solving problems of linear algebra by the method of continuous fractions), *Tr. semin. po funktsion. analizu. Voronezhsk. un-t*, 1956, **2**, 85–90; R. Zh. M., 1957, 5971.

MacMILLAN, R. H.

(1) A new method for the numerical evaluation of determinants. *J. Roy Aeronaut. Soc.*, 1955, **59**, No. 539, 772–773; R. Zh M., 1956, 6151.

MACKENZIE, J. K.

(1) A least squares solution of linear equations with coefficients subject to a special type of error. *Austral. J. Phys.*, 1957, **10**, 1, 103–109; R. Zh. M., 1958, 7204.

MADIP, PETAR.

(1) L'étude de solubilité des systèmes des équations algébriques linéaires. *Bull. Isnt. Nuclear Sci. ,,Boris Kidrich"*, 1952, **18**, 13–15; M. R., 17, 1008.

(2) Domen greshke u reshenima sistema linearnikh algebarskikh jednachina, *Vesn. Drusht. matem. i fiz. N. R. Srbije*, 1956, **8**, Nos. 3–4, 191–194; R. Zh. M., 1958, 1594.

(3) Sur une méthode de résolution des systèmes d'équations algébriques linéaires. *C. r. Acad. sci.*, 1956, **242**, No. 4, 439–441; R. Zh. M., 1956, 9129.

MAGNIER, A.

(1) Sur le calcul numérique des matrices. *C. r. Acad. sci.*, 1948, **226**, 464–465; M. R., 9, 471.

MAL'TSEV, A. I.

(1) *Osnovy lineinoi algebry* (Foundations of linear algebra), Izd. 2-e, Gostekhizdat, 1956.

MANLEY, R. G.

(1) Roots of frequency equations. Nature and distribution of roots. *Aircraft Engrg.*, 1944, **16**, 203; M. R., 6, 74.

MARCUS, M.

(1) An eigenvalue inequality for the product of normal matrices. *Amer. Math. Monthly*, 1956, **63**, 3, 173–174; R. Zh. M., 1957, 2063.

(2) On the optimum gradient method for systems of linear equations. *Proc. Amer. Math. Soc.*, 1956, **7**, No. 1, 77–81; R. Zh. M., 1957, 2681.

MARKOV, O. O.

(1) Do pitannya pro metod relaksatsii, *Nauk. zap. Kherson'sk. derzh. ped. in-t*, 1956, **7**, 45–48; R. Zh. M., 1958, 9265.

MASLOV, P. G.

(1) K opredeleniyu obratnykh matrits potentsial'noi energii mnogoatomnykh molekul (Metod opredelitelei); [On determining inverse matrices for the potential energy of multi-atomic molecules (method of determinants)], *Dokl. AN SSSR*, 1949, **67**, 819–822.

(2) K opredeleniyu obratnykh matrits potentsial'noi energii mnogoatomnykh molekul (On determining inverse matrices for the potential energy of multi-atomic molecules), *L., Zap. Gorn. in-ta*, 1949, **24**, 185.

(2) Inequalities for normal and Hermitian matrices. *Duke Math. J.*, 1957, **24**, No. 4, 591-599; R. Zh. M., 1958, 8578.

MISES, R., and GERINGER, H.
(1) Praktische Verfahren der Gleichungsauflösung. *Z. angew. und Mech.*, 1929, **9**, 58-77, 152-164.

MITANI
(1) Chislennoe reshenie sistemy lineinykh algebraicheskikh uravnenii na vychislitel'-noi mashine (Numerical solution of a system of linear algebraic equations on a computer), *Bull. Electrotechn. Lab.*, 1955, **19**, No. 8, 576-581; R. Zh. M., 1957, 5959.

MITRA, S. K.
(1) On an orthogonalization method of evaluating the reciprocal and the determinant of a matrix and its Gaussian transform. *Proc. Second Congr. Theor. and Appl. Mech.*, *New Delhi*, Octob., 1956, 261-268; M. R., 19, 1080.

MITCHELL, H. F., Sr.
(1) Inversion of a matrix of order 38. *Math. Tables and Othr Aids Comput.*, 1948, **3**, 161-166; M. R., 10, 152.

MITCHELL, A. R., and RUTHERFORD, D. E.
(1) On the theory of relaxation. *Proc. Glasgow Math. Assoc.*, 1953, **1**, 101-110; M. R., 15, 353.

MORRIS, J.
(1) On a simple method for solving simultaneous linear equations by means of successive approximations. *J. Roy. Aeronaut. Soc.*, 1935, **39**, 349.
(2) A succesive approximation process for solving simultaneous linear equations. *Aeronaut. Res. Comm.*, 1936, Rep. 1711, 1-12.
(3) Frequency equations. *Aircraft Engrg*, 1942, **14**, 108-110; M. R., 4, 90.
(4) An escalator process for the solution of linear simultaneous equations. *Philos. Mag.*, 1946, (**7**)**37**, 106-120; M. R., 8, 287.
(5) The escalator process for the solution of damped Lagrangian frequency equations. *Poilos. Mag.*, 1947, (**7**)**38**, 275-287; M. R., 9, 210.
(6) An application of the escalator process. Solution thereby of quasi-Hermitian frequency equations encountered in specific practical problems. *Aircraft Engrg*, 1951, **23**, 136-137; M. R., 12, 862.

MORRIS, J., and HEAD, J. W.
(1) Lagrangian frequency equations. An ,,escalator" method for numerical solution. *Aircraft Engrg*, 1942, **14**, 312-314, 316; M. R., 4, 148.
(2) The ,,escalator" process for the solution of Lagrangian frequency equations. *Philos. Mag.*, 1944, (**7**)**35**, 735-759; M. R., 7, 84.

MORRISON, J. F.
(1) The solution of three-term simultaneous linear equation by the use of submatrices. *Engineering J.*, 1946, **29**, 80-83; M. R., 8, 128.

MOULTON, F. R.
(1) On the solutions of linear equations having small determinants. *Amer. Math. Monthly*, 1913, **20**, 242-249.

MUNKELT, KARL.
(1) Formeln zur maschinellen Berechnung von Kehrmatrizen. *Dtsch. hydrogr. Z.*, 1956, **9**, No. 3, 143–146; R. Zh. M., 1959, 4248.

MUHLIG, F., and KOPPERMANN, E.
(1) Die Rechenvorschrift des „modernisierten" Gaußschen Algorithmus in ihrer einfachsten Form. *Z. Vermessungswesen*, 1953, **78**, No. 12, 389–393; R. Zh. M., 1955, 4713.

MÜLLER, E.
(1) Genauigkeit der bei Reduktion von Fehlergleichungen eliminierten Unbekannten. *Z. Vermessungswesen*, 1942, **71**, 186–190; M. R., 5. 161.

MURRAY, F. J.
(1) Simultaneous linear equations. *Proc. Scient. Comput. Forum.*, 1948, IBM Corp., 1950, 105–106; M. R., 13, 496.

NAGLER, H.
(1) On the simultaneous numerical inversion of a matrix and all its leading submatrices. *Math. Tables and Other Aids Comput.*, 1956, **10**, No. 56; 225–226; R. Zh. M., 1957, 8271.

NATANSON, I. P.
(1) K teorii priblizhennogo resheniya uravnenii (Toward a theory for solving equations approximately), *Uch. zap. Leningr. gos. ped. in-ta im. A. I. Gertsena*, 1948, **64**, 3–8.

NEKRASOV, P. A.
(1) Opredelenie neizvestnykh po sposobu naimen'shikh kvadratov pri ves'ma bol'shom chisle neizvestnykh (Determining unknowns by the least squares method for a very large number of unknowns), *Matem. Sb.*, 1885, **12**, 189–204.
(2) K voprosu o reshenii lineinoi sistemy uravnenii s bol'shim chislom neizvestnykh posredstvom posledovatel'nykh priblizhenii (On solving a linear system of equations with a large number of unknowns by successive approximations), *Pril. k t. LXIX. Zap. Ak. Nauk*, 1892, No. 5, 1–18.

NEVILLE, E. H.
(1) Ill-conditioned sets of linear equations. *Philos. Mag.*, 1948, (7)**39**, 35–48; M. R., 9, 382.

von NEUMAN, J.
(1) Some matrix inequalities and metrization of matrix-space. *Izv. in-ta matem. i mekhanika. tomskii un-t*, 1937,

von NEUMAN, J., and GOLDSTINE, H. H.
(1) Numerical inverting of matrices of high order. *Bull. Amer. Math. Soc.*, 1947, **53**, 1021–1099; M. R., 9, 471

NEWMARK, N. M.
(1) Bounds and convergence of relaxation and iteration procedures. *Proc. First U.S. Nat. Congr. Appl. Mech. Chicago, 1951, N. Y. Amer, Soc. Mech. Eng.*, 1952; M. R., 15, 353.

NIKOLAEVA, M. V.

(1) O metode relaksatsii (On a relaxation metood), *Tr. Matem. in-ta AN SSSR,* 1949, **28**, 160–182; M. R., 12, 539.

OLDENBURGER, RUFUS.

(1) Infinite powers of matrices and characteristic roots. *Duke Math. J.,* 1940, **6**, 357–361; M. R. 1, 324.

ORLOFF, CONSTANTIN.

(1) Méthode spectrale pratique d'évaluation numérique des déterminants et de résolution du systéme d'équation linéaires. *Vest. Drushva matem. i fiz. n.r. srbije,* H. P. Cpouje, 1953, **5**, No. 1–2, 17–30; R. Zh. M., 1956, 7651.

OSBORNE, ELMER E.

(1) On acceleration and matrix deflation processes used with the power method. *J. Soc. Indust. and Appl. Math.,* 1958, **6**, No. 3, 279–287.

OSTROWSKI, A. M.

(1) Sur la variation de la matrice inverse d'une matrice donnée. *C. r. Acad. sci.,* 1950, **231**, 1019–1021; M. R., 12, 396.

(2) Un nouveau théorème d'existence pour les systèmes d'équations. *C. r. Acad. sci.,* 1951, **232**, 786–788; M. R., 12, 596.

(3) Sur les matrices peu differentes d'une matrice triangulaire. *C. r. Acad. sci.,* 1951, **233**, 1558–1560; M. R., 13, 900.

(4) Sur les conditions générales pour la régularité des matrices. *Univ. Roma Ist. Naz. Alta. Mat. Rend. Mat. e Appl.,* 1951, (5) **10**, 156–168; M. R., 14, 125.

(5) Ueber das Nichtverschwinden einer Klasse von Determinanten und die Lokalisierung der charakteristischen Wurzeln von Matrizen. *Compositio Math.,* 1951, **9**, 209–226; M. R., 13, 524.

(6) Bounds for the greatest latent root of a positive matrix. *J. London Math. Soc.,* 1952, **27**, No. 106, 253–256; M. R., 14, 126.

(7) On over and under relaxation in the theory of the cyclic single step iteration. *Math. Tables and Other Aids Comput.* 1953, **7**, 152–159; R. Zh. M., 1954, 5783.

(8) On the linear iteration procedures for symmetric matrices. *Rend. mat. e applic.,* 1954, **14**, No. 1–2, 140–163; R. Zh. M., 1956, 8346.

(9) On nearly triangular matrices. *J. Res. Nat. Bur. Standards,* 1954, **52**, No. 6, 319–345; R. Zh. M., 1955, 4232.

(10) Über Normen von Matrizen. *Math. Z.,* 1955, **63**, No. 1, 2–18; R. Zh. M., 1956, 6395.

(11) Über Verfahren von Steffensen und Householder zur Konvergenzverbesserung von Iterationen. *Z. angew. Math. und Phys.,* 1956, **7**, No. 3, 218–229; R. Zh. M., 1957, 911.

(12) Determinanten mit überwiegender Hauptdiagonale und die absolute Konvergenz von linearen Iterationsprozessen. *Comment. math. helv.,* 1956, **30**, No. 3, 175–210; R. Zh. M., 1957, 5960.

(13) On Gauss'speeding up device in the theory of single step iteration. *Math. Tables and Other Aids Comput.,* 1958, **18**, No. 62, 116–132.

PANC, VLADIMIR

(1) Upravená relaxační methoda. *Aplikace mat.,* 1957, **2**, No. 3, 184–201; R. Zh. M., 1958, 7206.

PANOV, D. Yu.
(1) Reshenie sistem lineinykh uravnenii (Solving systems of linear equations); Addendum to the book of D. Scarborough. *Chislennye metody matematicheskogo analiza* (Numerical methods of mathematical analysis), M.-L., 1934.

PARKER, W. V.
(1) The characteristic roots of a matrix. *Duke Math. J.*, 1937, **3**, 484-487.
(2) Limits to the characteristic roots of a matrix. *Duke Math. J.*, 1943, **10**, 479-482; M. R., 5, 30.
(3) The characteristic roots of matrices. *Duke Math. J.*, 1945, **12**, 519-526, M. R., 7, 107.
(4) Characteristic roots and the field of values of a matrix. *Duke Math. J.*, 1948, **15**, 439-442; M. R., 10, 4.
(5) Sets of complex numbers associated with a matrix. *Duke Math. J.*, 1948, **15**, 711-715; M. R., 10, 230.
(6) Characteristic roots and field of values of a matrix. *Bull. Amer. Math. Soc.*, 1951, **57**, 103-108; M. R., 12, 581.

PARKES, E. W.
(1) Linear simultaneous equations. Some practical aspects of their solution in respect to the time involved with a series and the relative accuracy of the results. *Aircraft Engrg*, 1950, **22**, 48, 56; M. R., 11, 618.

PARODI, MAURICE
(1) Remarque sur la stabilité. *C. r. Acad. sci.*, 1949, **228**, 51-52; M. R., 10, 501.
(2) Complément à un travail sur la stabilité. *C. r. Acad. sci.*, 1949, **228**, 1198-1200; M. R., 10, 501.
(3) Sur les limites des modules des racines des équations algébriques. *Bull. Sci. Math.*, (2), 1949, **73**, 135-144; M. R.. 11, 307.
(4) Quelques propriétés des matrices H. *Ann. Soc. sci. Bruxelles.*, ser. I, 1950, **64**, 22-25; M. R., 12, 234.
(5) Sur une limite supérieure du rapport des valeurs caractéristiques de deux matrices symétriques définies positives, à éléments réels, dont les éléments correspondants différent peu. *C. r. Acad. sci.*, 1950, **230**, 705 707; M. R., 11, 413.
(6) Sur des familles de matrices auxquelles est applicable une méthode d'itération. *C. r. Acad. sci.*, 1951, **232**, 1053-1054; M. R., 12, 639.
(7) Sur un théorème de M. Ostrowski. *C. r. Acad. sci.*, 1952, **234**, 282-284; M. R., 14, 126.
(8) Sur quelques proprietes des valeurs caractéristiques des matrices carrées *Mem. sci. Math.*, 1952, **118**, 64 pp.; M. R., 14, 236.
(9) Sur la localisation des valeurs caractéristiques des matrices dans la plan complexe. *C. r. Acad. sci.*, 1956, **242**, 2617-2618; M. R., 18, 4.
(10) Sur une méthode de localisation des valeurs caractéristiques de certaines matrices. *C. r. Acac. sci.*, 1957, **244**, 1597-1598; M. R., 19, 379.

PELTIER, JEAN
(1) Détermination de vecteurs propres de certaines matrices à déterminant faible. *C. r. Acad. sci.*, 1955, **240**, No. 23, 2201-2203; R. Zh. M., 1957, 901.
(2) Mécanisation des problèmes linéaires sur machines électroniques. *C. r. Acad. sci.*, 1957, **244**, No. 8, 1003-1005; R. Zh. M., 1959, 3243.

PENROSE, R.

(1) On best approximation solutions of linear matrix equations. *Proc. Cambridge Philos. Soc.*, 1956, **52**, 17-19; M. R., 17, 536.

PERES, M.

(1) On solution of systems of simultaneous linear equations. *Las Ciencias*. Marid, 1952, **17**, 443-449; M. R., 17, 1137.

PETRIE, GEORGE W.

(1) Matrix inversion and solution of simultaneous linear algebraic equations with the IBM 604 electronic calculating punch. *Nat. Bur. Standards. Appl. Math. Ser.*, 1953, **29**, 107-112; R. Zh. M., 1956, 3332.

PLUNKETT, R.

(1) On the convergence of matrix iteration processes. *Quart. Appl. Math.*, 1950, **7**, 419-421; M. R., 11, 464.

POKOPNÁ, O.

(1) Řešení soustav linearnich algebraických rovnic minimisací součtu čtverců residuí. *Sbor. Českosl. akad. věd. Lab. mat. strojů*, 1954, No. 2, 111-116; R. Zh. M., 1956, 3329.

(2) Řešení soustav lineárních algebraickch rovniýc prěhled a srovnani metod. *Stroje zpracov. inform.*, 1955, **3**, 139-196; R. Zh. M., 1957, 6151.

(3) Schema pro řešení soustav linearních algebraických rovnic eliminací. *Aplikace mat.*, 1957, **2**, No. 3, 235-241; R. Zh. M., 1958, 6209.

POLLACZEK-GEIRINGER, H.

(1) Zur Praxis der Lösung linearer Gleichungen in der Statik. *Z. angew. Math. und Mech.*, 1928, **8**, 446-447.

POPE, DAVID A., and TOMPKINS, C.

(1) Maximizing functions of rotations-experiments concerning speed of diagonalization of symmetric matrices using Jacobi's method. *J. Assoc. Comput. Machinery*, 1957, **4**, No. 4, 459-466; R. Zh. M., 1959, 2013.

POPOVICI, CONST. C.

(1) Asupra metodei iteratiei, aplicată la un sistem de ecuatii lineare. *Studii si cercetari mat.*, 1953, **4**, No. 1-2, 233-247; R. Zh. M., 1955, 3987.

PORTER, R. E.

(1) Single order reduction of a complex matrix. *Proc. Comput. Sem. Dec., 1949. N. Y., IBM Corp.*, 1951, 138-140; M. R., 13, 387.

POTRON, l'ABBÉ

(1) Sur les matrices non négatives, et les solutions positives de certains systèmes linéaires. *Bull. Soc. math. France*, 1939, **67**, 56-61; M. R., 1, 97.

PUGACHEV, B. P.

(1) O dvukh priemakh priblizhennogo vychisleniya sobstvennykh znachenii i sobstvennykh vektorov (Two methods for approximate computation of eigenvalues and eigenvectors), *Dokl. AN SSSR*, 1956, **110**, No. 3, 334-337; R. Zh. M., 1957, 5181.

(2) Ob odnom metode priblizhennogo otyskaniya sobstvennykh znachenii (An appro-

ximate method for finding eigenvalues), *Tr. 3-go Vsesoyuznogo matem. s"ezda*, 1956, **2**, 153–154; R. Zh. M., 1956, 9113.

PURCELL, EVERETT W.
(1) The vector method of solving simultaneous linear equations. *J. Math. and Phys.*, 1953, **32**, 180–183; R. Zh. M., 1954, 5787.

QUADE, W.
(1) Auflösung linearer Gleichungen durch Matrizen-iteration. *Ber. Math.-Tagung Tübingen*, 1946, 1947, 123–124; M. R., 9, 104.

QUENOUILLE, M. H.
(1) A further note on discriminatory analysis. *Ann. Eugenics*, 1949, **15**, 11–14; M. R., 11, 743.

RAHMAN, A.
(1) Numerical evaluation of determinants. *Bull. cl. sci. Acad. roy. Belgique*, 1954, **40**, No. 8, 798–801; R. Zh. M., 1956, 764.

RALL, L. B.
(1) Error bounds for iterative solutions of Fredholm integral equations. *Pacif. J. Math.*, 1955, **5**, Suppl. No. 2, 977–986; R. Zh. M., 1957, 5178.

RAYMONDI, C.
(1) Contributo allo studio dei sistemi elastici staticamenta indeterminanti. *Atti. Ist. Costruzioni Univ. Pisa*, 1949, **13**, 1–15; M. R., 13, 587.

REGGINI, HORACIO C.
(1) Resolucion de sistemas de ecuaciones lineales. *Cienc. y tecn.*, 1952, **125**, No. 628, 158–167; R. Zh. M., 1959, 4245.

REDHEFFER, R.
(1) Errors in simultaneous linear equations. *Quart. Appl. Math.*, 1948, **6**, 342–343; M. R., 10, 152.

REICH, E.
(1) On the convergence of the classical iterative method of solving linear simultaneous equations. *Ann. Math. Statistics*, 1949, **20**, 448–451; M. R., 11, 136.

REIERSOL, O.
(1) A method for recurrent computation of all the principal minors of a determinant, and its application in confluence analysis. *Ann, Math. Statistics*, 1940, **11**, 193–198; M. R., 2, 61.

RICCI, L.
(1) Confronto fra i metodi di Banachiewicz, Roma e Volta per la risoluzione dei sistemi di equazioni algebriche lineari. *Atti. Accad. naz. Lincei. Rend. Cl. sci. fis., mat. e natur.* ser. 8, 1949, **7**, 72–76; M. R., 11, 743.

RICE, LEPINE HALL.
(1) The rank of a matrix, the value of a determinant, and the solution of a system of linear equations. *J. Math. and Phys.*, 1932, **11**, 146–149.

RICHARDSON, L. E.
(1) A purification method for computing the latent columns of numerical matrices

and some integrals of differential equations. *Phil. Trans. Roy. Soc. London, Ser. A*, 1950, **242**, 439–491; M. R., 12, 133.

RICHTER, HANS.
(1) Bemerkung zur Norm der Inversen einer Matrix. *Arch. Math.*, 1954, **5**, No. 4–6, 447–448; R. Zh. M., 1956, 1054.

RILEY, JAMES D.
(1) Solving systems of linear equations with a positive definite, symmetric, but possibly ill-conditioned matrix. *Math. Tables and Other Aids Comput.*, 1955, **9**, No. 51, 96–101; R. Zh. M., 1957, 5958.

ROMA, M. S.
(1) Il metodo dell'ortogonalizzazione per la risoluzione numerica dei sistemi di equazioni lineari algebriche. *Ricerca Sci.*, 1946, **16**, 309–312; M. R., 8, 171.
(2) Il metodo dell'ortogonalizzazione per la risoluzione numerica dei sistemi di equazioni algebriche. *Pubblicazioni 1st. Appl. Calcolo*, 1947, **189**, 12 pp; M. R., 10, 574.
(3) Sulla risoluzione numerica dei sistemi di equazioni algebriche lineari col metodo della ortogonalizzazione. *Pubblicazioni 1st. Appl. Calcolo*, 1950, **283**, M. R., 13, 691.

ROSSER, J. BARKLEY.
(1) A general iteration scheme for solving simultaneous equations. *Bull. Amer. Math. Soc.*, 1950, **56**, 176–177.
(2) A method of computing exact inverses of matrices with integer coefficients. *J. Res. Nat. Bur. Standards*, 1952, **49**, 349–353; M. R., **14**, 1128.
(3) Rapidly converging iterative methods for solving linear equations. *Nat. Bur. Standards. Appl. Math. Ser.*, 1953, **29**, 59–64; M. R., 15, 651.

ROSSER, J. B., HESTENES, M. R., KARUCH, W., and LANCZOS, C.
(1) Separation of close eigenvalues of a real symmetric matrix. *J. Res. Nat. Bur. Standards*, 1951, **47**, No. 4, 291–297; M. R., 14, 92.

ROTH, J. P., and SCOTT, D. S.
(1) A vector method for solving linear equations and inverting matrices. *J. Math. and Phys.*, 1956, **35**, No. 3, 312–317; R. Zh. M., 1958, 749.

ROY, S. W.
(1) A useful theorem in matrix theory. *Proc. Amer. Math. Soc.*, 1954, **5**, 635–638; M. R., 16, 4.

RUBINSTEIN, H., and RUTLEDGE, Y. D.
(1) High order matrix computations on the Univac. *Proc. Assoc. Comput. Mach.*, 1952, 181–186; M. R., 14, 1019.

RUGGIERO, R. J.
(1) Investigation of three methods for solving the flutter equations and their relative merits. *J. Aeronaut. Sci.*, 1946, **13**, 3–22; M. R., 7, 338.

RUSHTON, S.
(1) On least squares fitting by orthonormal polynomials using the Choleski method. *J. Roy. Statist. Soc., Ser. B*, 1951, **13**, 92–99; M. R., 13, 990.

RUTHERFORD, D. E.
(1) On the rational solution of the matrix equations $sx = xt$. *Proc. Koninkl. nederl.*

*akad. wetensch.*, 1933, **36**, 432–442.

RUTISHAUSER, HEINZ.
(1) Beiträge zur Kenntnis des Biorthogonalisierungs–Algorithmus von Lanczos. *Z. angew. Math. und Phys.*, 1953, **4**, No. 1, 35–56; R. Zh. M., 1953, 453.
(2) Der Quotienten-Differenzen-Algorithmus. *Z. angew. Math. und Phys.*, 1954, **5**, No. 3, 233–251; R. Zh. M., 1955, 5316.
(3) Anwendungen des Quotienten-Differenzen-Algorithmus. *Z. angew. Math. und Phys.*, 1954, **5**, No. 6, 496–503; R. Zh. M., 1956, 4926.
(4) Bestimmung der Eigenwerte und Eigenvektoren einer Matrix mit Hilfe des Quotienten-Differenzen-Algorithmus. *Z. angew. Math. und Phys.*, 1955, **6**, No. 5, 387–401; R. Zh. M., 1956, 6862.
(5) Une méthode pour la détermination des valeurs propres d'une matrice. *C. r. Acad. sci.*, 1955, **240**, No. 1, 34–36; R. Zh. M., 1956, 4916.
(6) *Der Quotienten-Differenzen-Algorithmus.* Mitt. Inst. angew. Math. Eidgenoss. techn. Hochschule Zürich, 1957, No. 7, 745; R. Zh. M., 1958, 760.
(7) Solution of eigenvalue problems with the LR-transformation. *Nat. Bur. Standards. Appl. Math. Ser.*, 1958, **49**, 47–81; M. R., 19, 770.
(8) Zur Bestimmung der Eigenwerte schiefsymmetrischer Matrizen. *Z. Angew. Math. und Phys.*, 1958, **9b**, No. 5-6, 586–590; R. Zh. M., 1959, 7482.

RUTISHAUSER, HEINZ, et BAUER, FRIEDRICH L.
(1) Détermination des vecteurs propres d'une matrice par une méthode itérative avec convergence quadratique. *C. r. Acad. sci.*, 1955, **240**, No. 17, 1680–1681; R. Zh. M., 1956, 4917.

RŮŽIČKA, MIROSLAV.
(1) Zkrácení iteračního způsobu výpočtu systému lineárních rovnic. *Inžen. stavby*, 1957, **5**, No. 9, 490–492; R. Zh. M., 1958, 8288.

SABROFF, R. R., and HIGGINS, T. J.
(1) A critical study of Kron's method of "tearing". *Matrix Tensor Quart.*, 1957, **7**, 107–113; M. R., 19, 64.

SAIBEL, EDWARD.
(1) A modified treatment of the iterative method. *J. Franklin Inst.*, 1943, 235, 163–166; M. R., 4, 148.
(2) A rapid method of inversion of certain types of matrices. *J. Franklin Inst.*, 1944, **237**, 197–201; M. R., 5, 245.

SAIBEL, EDWARD, and BERGER, W. J.
(1) On finding the characteristic equation of a square matrix. *Math. Tables and Other Aids Comput.*, 1953, **7**, No. 49, 228–236; R. Zh. M., 1954, 5786.

SAMOKISH, B. A.
(1) Issledovanie bystroty skhodimosti metoda naiskoreishego spuska (An investigation of the rate of convergence for the method of steepest descent), *Uspekhi matem. nauk*, 1957, **12**, No. 1, 238–240.

SAMSONOV, K. V.
(1) Pribor dlya resheniya sistemy lineinykh algebraicheskikh uravnenii metodom iteratsii (A device for solving a system of linear algebraic equations by an iterative method),

*Prikl. matem. i mekhanika*, 1935, **2**, 309–313.

SAMUELSON, P. A.

(1) A method of determining explicitly the coefficients of the characteristic equation. *Ann. Math. Statistics*, 1942, **13**, 424–429; M. R., 4, 148.

(2) Efficient computation of the latent vectors of a matrix. *Proc. Nat. Acad. Sci. U.S.A.*, 1943, **29**, 393–397; M. R., 5, 161.

(3) A convergent iterative process. *J. Math. and Phys.*, 1945, **24**, 131–134.

(4) Solving linear equations by continuous substitution. *Bull. Amer. Math. Soc.*, 1950, **56**, 159.

SASSENFELD, H.

(1) Ein hinreichendes Konvergenzkriterium und eine Fehlerabschätzung fur die Iteration in Einzelschritten bei linearen Gleichungen. *Z. angew. Math. und Mech.*, 1951, **31**, 92–94; M. R., 14, 692.

SATTERTHWAITE, F. E.

(1) Error control in matrix calculation. *Ann. Math. Statistics*, 1944, **15**, 373–389; M. R., 6, 218.

SCHERMAN, JACK.

(1) Computations relating to inverse matrices. *Nat. Bur. Standards. Appl. Math. Ser.*, 1953, **29**, 123–124; R. Zh. M., 1956, 6150.

SCHMEIDLER, W.

(1) *Vorträge über Determinanten und Matrizen mit Anwendungen in Physik und Technik.* Berlin, 1949.

SCHMIDT, R. J.

(1) On the numerical solution of linear simultaneous equations by an iterative method. *Philos. Mag.*, 1941, (**7**) **32**, 369–383; M. R., 3, 276.

SCHMIDTMAYER, JOSEF.

(1) Über die Auflösung des Systems linearer algebraischer Gleichungen mit komplexen Koeffizienten. *Z. angew. Math. und Mech.*, 1958, **38**, 74–77; M. R., 19, 1080.

SCHMIDTMAYER, JOSEF, a MAYER, DANIEL.

(1) Výhodné řešení lineárnich problemů v oboru komplexnich čísei. *Slaboproudý obzor*, 1958, **19**, No. 7, 472–477; R. Zh. M., 1959, 4246.

SCHNEIDER, HANS.

(1) Regions of exclusion for the latent roots of a matrix. *Proc. Amer. Math. Soc.*, 1954, **5**, No. 2, 320–322; R. Zh. M., 1956, 760.

SCHRÖDER, JOHANN.

(1) Eine Bemerkung zur Konvergenz der Iterationsverfahren für lineare Gleichungssysteme. *Arch. Math.*, 1953, **4**, No. 4, 322–326; R. Zh. M., 1954, 4191.

(2) Neue Fehlerabschätzungen für verschiedene Iterationsverfahren. *Z. angew. Math. und Mech.*, 1956, **36**, No. 5-6, 168–181; R. Zh. M., 1957, 1827.

SCHUR, J.

(1) Über Potenzreihen, die im Inneren des Einheitskreises beschränkt sind. *J. reine und angew. Math.*, 1917, **147**, 205–232.

SCHWARZ, H. R.
(1) Ein Verfahren zur Stabilitätsfrage bei Matrizen-Eigenwertproblemen. *Z. angew. Math. und Phys.*, 1956, **7**, 473–500; M. R., 18, 676.

SCOTTO, L. G.
(1) Sul calcolo ed affinamento delle caratteristiche delle vibrazioni dei sistemi elastici ad n gradi di libertà. *Rend. Ist. Lombrardo sci. e lettere Cl. sci. mat. e natur. Ser.* 3, 1956, **21** (**90**), 89–106; M. R., 18, 676.

SEIDEL, L.
(1) Über ein Verfanren, die Gleichungen, auf welche die Methode der kleinsten Quadrate führt, sowie lineäre Gleichungen überhaupt, durch successive Annäherung aufzulösen. *Abh. math.-Phys. Kl., Bayrische Akad. Wiss., München*, 1874, **11**, No. 3, 81–108.

SEMENDYAEV, K. A.
(1) O nakhozhdenii sobstvennykh znachenii i invariantnykh mnogoobrazii matrits posredstvom iteratsii (On finding eigenvalues and invariant type of matrices using iteration), *Prikl. matem. i mekhanika*, 1943, **7**, 193–222; M. R., 6, 51.

SEREBRYANNIKOV, S. V.
(1) O reshenii chislennykh uravnenii metodom iteratsii (On solving numerical equations by an iterative method), *Novocherkassk. Tr. inzh. meliorat. in-ta*, 1939, **3**, 168–172.

SHANKS, DANIEL.
(1) On analogous theorems of Fredholm and Frame and on the inverse of a matrix. *Quart. Appl. Math.*, 1955, **13**, 95–98; R. Zh. M., 1957, 8270.

SHAW, T. S.
(1) *An introduction to relaxation methods*. N.Y., Dover, 1953, 396 pp.; M. R. 15, 353.

SHERMAN, J., and MORRISON, W. S.
(1) Adjustment of an inverse matrix corresponding to a change in one element of a given matrix. *Ann. Math. Statistics*, 1950, **21**, 124–127; M. R., 11, 693.

SHINGO, TAKAICHI.
(1) An exact method of solving the linear simultaneous equations with the principal diagonal coefficients and those adjacent to them only. *Trans. Japan soc. Civil. Engrs*, 1954, No. 19, 1–7; R. Zh. M., 1956, 1681.

SHINOMIYA, TETSURO
(1) Obrashchenie matrits metodom posledovatel'nykh priblizhenii (Matrix inversion by the successive approximations method), *Res. Repts Fac. Engng. Gifu Prefect. Univ.*, 1955, No. 5, 5–10; R. Zh. M., 1958, 4252.
(2) Prakticheskii metod obrashcheniya matrits pri pomoshchi vychisleniya opredelitelei (A practical method for inverting matrices by computing determinants), *Res. Repts Fac. Engng. Gifu Prefect. Univ.*, 1956, No. 6, 1–6; R. Zh. M., 1958, 4253.
(3) Some notes on the process of reversing a matrix. *Proc. 5th Japan Nat. Congr. Appl. Mech.*, 1955, Tokyo, 1956, 481–484; R. Zh. M., 1958, 10302.

SHMUL'YAN, YU. L.
(1) Zamechanie po povodu stat'i Yu. M. Gavrilova "O skhodimosti iteratsionnykh protsessov" (A note regarding Yu. M. Gavrilov's article "On convergence of iterative

processes"), *Izv. AN SSSR, ser. matem.*, 1955, 19, No. 2, 191; R. Zh. M., 1956, 5507.

SHREIDER, YU. A.

(1) Reshenie sistemy lineinykh sovmestnykh algebraicheskikh uravnenii (Solving a system of linear combined algebraic equations), *Dokl. AN SSSR*, 1951, **76**, 651–654; M. R., 12, 639.

SHURA-BURA, M. R.

(1) Otsenka pogreshnostei pri vychislenii obratnoi matritsy dlya matritsy vysokogo poryadka (An error estimate for computing the inverse matrix of a high-order matrix), *Uspekhi matem. nauk*, 1951, **6**, No. 4, 121–150; M. R., 13, 284.

SINDEN, FRANK W.

(1) An oscillation theorem for algebraic eigenvalue problems and its applications. *Mitt. Inst. angew. Math.*, Zürich, 1954, **4**, 57 pp.; M. R., 16, 666.

SINOMIYA and OTI

(1) Neskol'ko formul dlya obratnykh matrits (Some formulas for inverse matrices), *Trans. Japan. Soc. Civil. Engrs*, 1955, **24**, 78–82; R. Zh. M., 1956, 7109.

SNYDER, JAMES N.

(1) On the improvement of the solutions to a set of simultaneous linear equations using the ILLIAC. *Math. Tables and Other Aids Comput.*, 1955, **9**, No. 52, 177–184; R. Zh. M., 1957, 2682.

SOKOLOFF, N. P.

(1) Sur l'application des déterminants supérièurs a la résolution de certains systèmes d'équations linéaires. *Ann. Soc. Scient. Bruxelles, Sér. I*, 1937, **57**, 60–66.

SOURIAU, J. M.

(1) Une méthode pour la décomposition spectrale et l'inversion des matrices. *C. r. Acad. sci.*, 1948, **227**, 1010–1011; M. R., 10, 348.

SOURIAU, J. M., et BONNARD, R.

(1) Théorie des erreurs en calcul matriciel. *Rech. aéronaut.*, 1951, No. 19, 41–48; M. R., 12, 638.

SOUTHWELL, R. V.

(1) *Relaxation Methods in engineering Science, a Treatise on Approximate Computation.* Oxford Univ. Press, 1940.
(2) *Relaxation Methods in Theoretical Physics.* Oxford Univ. Press., 1946.

SPOERL, Ch. A.

(1) A fundamental proposition in the solution of simultaneous linear equations. *Trans Actuar. Soc. Amer.*, 1943, **44**, 276–288; M. R., 5, 161.
(2) On solving simultaneous linear equations. *Trans. Actuar. Soc. Amer.*, 1944, **45**, 18–32, 67–69; M. R., 6, 50.

STANKIEWICZ, L.

(1) Sur les opérations arithmétiques dans le calcul des inverses d'après la méthode de M. T. Banachiewicz. *Bull. intern. Acad. Polon. Sci. A.*, 1937, 363–376.
(2) Sur les méthodes de Césari et Kaczmarz relatives a la résolution de systèmes d'équations linéaires a l'aide d'approximations successives. *Bull. intern. Acad. Polon.*

*Sci. A*, 1937, 521-529.

STEARN, J. L.
(1) Iterative solutions of normal equations. *Bull. Géod.*, 1951, 331-339; M. R., 13, 990.

STEIN, M.
(1) Gradient methods in the solution of systems of linear equations. *J. Res. Nat. Bur. Standards*, 1952, **48**, No. 6, 407-413; M. R., 14, 322.
(2) Determining the mode shapes and frequencies of low aspect ratio multispar winks. *J. Aeronaut Sci.*, 1955, **22**, No. 2, 137-138, R. Zh. M., 1958, 8287.

STEIN, P.
(1) The convergence of Seidel iterants of nearly symmetric matrices. *Math. Tables and Other Aids Comput.*, 1951, **5**, 237-239.
(2) A note on inequalities for the norm of a matrix. *Amer. Math. Monthly*, 1951, **58**, 558-559; M. R., 13, 717.
(3) A note on bounds of multiple characteristic roots of a matrix. *J. Res. Nat. Bur. Standards*, 1952, **48**, 59-60; M. R., 13, 813.
(4) A note on the bounds of the real parts of the characteristic roots of a matrix. *J. Res. Nat. Bur. Standarns*, 1952, **48**, 106-108; M. R., 14, 8.

STEIN, P., and ROSENBERG, R. L.
(1) On the solution of linear simultaneous equations by iteration. *J. London Math. Soc.*, 1948, **23**, 111-118; M. R., 10, 485.

STESIN, I. M.
(1) Vychislenie sobstvennykh znachenii pri pomoshchi nepreryvnykh drobei (Computing eigenvalues with non-continued fractions), *Uspekhi matem. nauk*, 1954, **9**, 191-198; R. Zh. M., 1957, 4384.

STIEFEL, E.
(1) Über einige Methoden der Relaxationsrechnung. *Z. angew. Math. und Phys.*, 1952, **3**, 1-33; M. R., 13, 874.
(2) Zur Interpolation von tabellierten Funktionen durch Exponentialsummen und zur Berechnung von Eigenwerten aus den schwarzschen Konstanten. *Z. angew. Math. und Mech.*, 1953, **33**, No. 8-9, 260-262; R. Zh. M., 1954, 4195.
(3) Some special methods of relaxation technique. *Nat. Bur. Standards. Appl. Math. Ser.* 1953, **29**, 43-48; R. Zh. M., 1956, 7643.
(4) Ausgleichung ohne Aufstellung der Gausschen Normalgleichungen. *Wiss. Z. Techn. Hochschule Dresden*, 1953, **2**, 441-442; M. R., 16, 1155.
(5) Relaxationsmethoden bester Strategie zur Lösung linearer Gleichungssysteme. *Comment. math. helv.*, 1955, **29**, 157-179; R. Zh. M., 1956, 7644.
(6) *Kernel polynomials in linear algebra and their numerical applications.* Four lectures on solving linear equations and determining eigenvalues. Nat. Bur. Standards, Washington, 1955, 52 pp.; M. R., 17, 790.
(7) Kernel polynomials in linear algebra and their numerical applications. *Nat. Bur. Standards. Appl. Math. Ser.*, 1958, **49**, 1-22; M. R., 19, 1080.

STOJAKOVICH, MIRKO
(1) O nekim postuptsima za reshavan'e sistema linearnikh algebarskikh jednachina, *Zb. Mash. fak. Un-t Beogradu*, 1954-1955 (1956), 19-27; R. Zh. M., 1957, 8957.

SWINDLCHURST, BEVERLY
(1) On the solution of simultaneous linear equations. *Proc. Montana Acad. Sci*, 1956, **16**, 59–60; R. Zh. M., 1959, 5220.

SYNGE, J. L.
(1) A geometrical interpretation of the relaxation method. *Quart. Appl. Math.*, 1944, **2**, 87–89; M. R., 6, 50.

TAGA, Y.
(1) O lineinykh vychisleniyakh na avtomaticheskoi releinoi vychislitel'noi mashine Instituta matematicheskoi statistiki (On linear computations with an automatic relay computer at the Institute of Mathematical Statistics), *Proc. Inst. Statist. Math.*, 1957, **5**, No. 1, 32–48; R. Zh. M., 1958, 4251.

TAUSSKY, OLGA
(1) Bounds for characteristic roots of matrices. *Duke Math. J.*, 1948, **15**, 1043–1044; M. R., 10, 501.
(2) A recurring theorem on determinants. *Amer. Math. Monthly*, 1949, **56**, 672–676; M. R., 11, 307.
(3) Notes on numerical analysis. II. Notes on the condition of matrices. *Math. Tables and Other Aids Comput.*, 1950, **4**, 111–112; M. R., 12, 361.
(4) Bounds for characteristic roots of matrices. II. *J., Res. Nat. Bur. Standards*, 1951, **46**, 124–125; M. R., 13, 311.

TAUSSKY, OLGA, and TODD, JOHN
(1) Systems of equations, matrices and determinants. *Math. Mag.*, 1952, **25**, 9–20, 71–78; M. R., 14, 715.

TEMPLE, G.
(1) The general theory of relaxation methods applied to linear systems. *Proc. Roy. Soc. Ser. A.*, 1939, **169**, 476–500.
(2) The accuracy of Rayleigh's method of calculating the natural frequencies of vibrating systems. *Proc. Roy. Soc. Ser. A*, 1952, **211**, 204–224; M. R., 13, 691.

TERNER, L. R.
(1) Improvement in the convergence of methods of successive approximation. *Proc. Comput. Sem.*, Dec. 1949, *N. Y. IBM Corp.*, 1951, 135–137; M. R., 13, 586.

TERRACINI, ALESSANDRO
(1) Un procedimento per la risoluzione numerica dei sistemi di equazioni lineari. *Ric. ingegn.*, 1935, **3**, 40–48.

TODD, J.
(1) The condition of certain matrices. I. *Quart. J. Mech. and Appl. Math.*, 1949, **2**, 469–472; M. R., 11, 619.
(2) The condition of a certain matrix. *Proc. CambridgePhilos. Soc.*, 1950, **46**, 116–118; M. R., 11, 403.
(3) Experiments on the inversion of a $16 \times 16$ matrix. *Nat. Bur. Standards. Appl. Math. Ser.*, 1953, **29**, 113–115; R. Zh. M., 1956, 3333.
(4) The condition of certain matrices. II. *Arch. Math.*, 1954, **5**, No. 4–6, 249–257; R. Zh. M., 1956, 763.
(5) The condition of matrices. *Proc. Internat. Congr. Math.*, 1954, **2**, Amsterdam, 1954, 385–386; R. Zh. M., 1955, 4244.

(6) The condition of the finite segments of the Hilbert matrix. *Nat. Bur. Standards. Appl. Math. Ser.*, 1954, **39**, 109–116; M. R., **16**, 861.

(7) The condition of certain matrices. III. *J. Res. Nat. Bur. Standards*, 1958, **60**, No. 1, 1–7; R. Zh. M., 1958, 6217.

TODOROW, MARKO
(1) Über die iterative Behandlung linearer Gleichungssysteme. *Bautechnik*, 1958, **35**, No. 4, 136–138; R. Zh. M., 1959, 877.

TUCKER, L. R.
(1) The determination of successive principal components without computation of tables of residual correlation coefficients. *Psychometrika*, 1944, **9**, 149–153; M. R., 6, 51.

TUCKERMAN, L. B.
(1) On the mathematically significant figures in the solution of simultaneous linear equations. *Ann. Math. Statistics*, 1941, **12**, 307–316; M. R., 3, 154.

TURETSKY, R.
(1) The least square solution for a set of complex linear equations. *Quart. Appl. Math.*, 1951, **9**, 108–110; M. R., 12, 641.

TURING, A. M.
(1) Rounding-off errors in matrix processes. *Quart. J. Mech., and Appl. Math.*, 1948, **1**, 287–308; M. R., 10, 405.

TURNBULL, H. W., and AITKEN, A. C.
(1) *An introduction to the theory of canonical matrices*. London, Blackie, 1932.

TURTON, F. J.
(1) On the solution of the numerical simultaneous equations arising in the analysis of redundant structures. *J. Roy. Aeronaut. Soc.*, 1945, **49**, 104–111; M. R., 6, 218.

UHLIG, J.
(1) Untersuchung der Genauigkeit einer Reihe von Verfahren zur Bestimmung von charakteristischen Zahlen und der Eigenvektoren einer Matrix. *Z. angew. Math. und Mech.*, 1957, **37**, 7–8, 265; R. Zh. M., 1959, 7486.

(2) Über ein inverses Eigenwertproblem. *Z. angew. Math. und Mech.*, 1958, **38**, 7–8, 284; R. Zh. M., 1959, 5226.

ULLMAN, J.
(1) The probability of convergence of an iterative process of inverting a matrix. *Ann. Math. Statistics*, 1944, **15**, 205–213; M. R., 6, 51.

UMANSKII, A. A.
(1) *Kurs stroitel'noi mekhaniki* (A course in structural mechanics), 1935.

UNGER, H.
(1) Orthogonalisierung von Matrizen nach E. Schmidt und ihre praktische Durchführung. *Z. angew. Math. und Mech.*, 1951, **31**, 53–54; M. R., 14, 692.

(2) Zur Auflösung umfangreicher linearer Gleichungssysteme. *Z. angew. Math. und Mech.*, 1952, **32**, 1–9; M. R., 14, 92.

(3) Zur Praxis der Biorthonormierung von Eigen-und-Hauptvektoren. *Z. angew. Math. und Mech.*, 1953, **33**, 319–331; M. R., 15, 560.

(4) Über direkte Verfahren bei Matrizeneigenwertproblemen. *Wiss. Z. Techn. Hochschule Dresden*, 1952, 1953, **2**, No. 3, 449–456; R. Zh. M., 1958, 10298.

VARGA, RICHARD

(1) Eigenvalues of circulant matrices. *Pacif. J. Math.*, 1954, **4**, No. 1, 151–160; R. Zh. M., 1956, 4918.

(2) A comparison of the successive overrelaxation method and semi-iterative methods using Chebyshev polynomials. *J. Soc. Industr. and Appl. Math.*, 1957, **5**, 39–46; R. Zh. M., 1958, 9266.

VASILEVSKII, S.

(1) Skhema dlya resheniya normal'nykh uravnenii na vychislitel'nykh mashinakh (A scheme for solving normal equations on computers), *Tr. Latviiskogo un-ta*, 1940, **3**, No. 11, 3–12; M. R., 3, 154.

VERZUH, F. M.

(1) The solution of simultaneous linear equations with the aid of the 60-r calculating punch. *Math. Tables and Other Aids Comput.*, 1949, **3**, 453–462; M. R., 11, 57.

VINOGRADE, B.

(1) Note on the escalator method. *Proc. Amer. Math. Soc.*, 1950, **1**, 162–164; M. R., 11, 618.

VOETTER, H.

(1) Über die numerishe Berechung der Eigenwerte von Säkulargleichungen. *Z. angew. Math. und Phys.*, 1952, **3**, 314–316; M. R., 14, 501.

VOLTA, E.

(1) Un nuovo metodo per la risoluzione rapida di sistemi di equazioni lineari. *Atti Accad. naz. Lincei. Rend. Cl. sci. fis., mat. e natur. ser.* 8, 1949, **7**, No. 50, 203–207; M. R., 11, 743.

VRIES, D. de

(1) Eigenwaarden van matrices. *Tijdschr. kadaster en landmeetkunde*. 1953, **69**, No. 5, 316-322; R. Zh. M., 1956, 4104.

VUCHKOVICH, MILORAD

(1) Sistem linearnikh jednachina i negova primena y reshavany statichki neodrenenikh konstruktsija, *Izgradna*, 1956, **10**, No. 7, 8, 3–13; R. Zh. M., 1958, 2437.

VUJAKLIJA, G.

(1) Sur le calcul des déterminants. *Godišnjak Techn. Fak. Univ. Beograd*, 1946–47, No. 1-4, 1949; M. R., 11, 154.

VZOROVA, A. I.

(1) O reshenii sistemy lineinykh algebraicheskikh uravnenii sposobom Yu. A. Shreidera (On solving a system of linear algebraic equations using Yu. A. Schroeder's method), *Vychisl. matem. i vychisl. tekhika*, 1953, No. 1, 90–94; R. Zh. M., 1954, 5266.

WAGNER, HARVEY M.

(1) A partitioning method of inverting symmetric definite matrices on a card-program-

med calculator. *Math. Tables and Other Aids Comput.*, 1954, **8**, No. 47, 132–139; R. Zh. M., 1956, 3334.

WALKER, A. G., and WESTON, J. D.
(1) Inclusion theorems for the eigenvalues of a normal matrix. *J. London Math. Soc.*, 1949, **24**, 28–31; M. R., 10, 501.

WASHIZU, K.
(1) On the bounds of eigenvalues. *Quart. J. Mech. and Appl. Math.*, 1955, **8**, No. 3, 311–325; R. Zh. M., 1956, 6864.

WASOW, W. R.
(1) A note on the inversion of matrices by random values. *Math. Tables and Other Aids Comput.*, 1952, **6**. 78–81.

WAUGH, F. V.
(1) A note concerning Hotelling's method of inverting a partitioned matrix. *Ann. Math. Statistics*, 1945, **16**, 216–217; M. R., 7, 84.
(2) Inversion of the Leontief matrix by power series. *Econometrica*, 1950, **18**, 142–154; M. R., 12, 133.

WAUGH, F. V., and DWYER, P. S.
(1) Compact computation of the inverse of a matrix. *Ann. Math. Statistics*, 1945, **16**, 259–271; M. R., 7, 218.

WAYLAND, H.
(1) Expansion of determinantal equations into polynomial form. *Quart. Appl. Math.*, 1945, **2**, 277–306; M. R., 6, 218, 1947, **2**, No. 4, 128–158.

WAŻEWSKI, TADEUSZ
(1) Sur l'algorithmisation des méthodes d'éliminations successives. *Ann. Soc. polon.*, 1953, **24**, No. 2, 157–164; R. Zh. M., 1956, 4108.
(2) Einfluss der Änderung einer Matrix auf die Lösung des Z ugehörigen Gleichungssystems, sowie auf die charakteristischen Zahlen und die Eigenvektoren. *Z. angew. Math. und Mech.*, 1936, **16**, No. 5, 287–300.
(3) Über die Lösung von linearen Gleichungssystemen durch Iteration. *Z. angew. Math. und Mech.*, 1936, **16**. No. 5, 301–310.
(4) Berechnung einzelner Eigenwerte eines algebraischen linearen Eigenwertproblems durch „Störiteration". *Z. angew. Math. und Mech.*, 1955, **35**, No. 12, 441–452; R. Zh. M., 1957, 2686.

WEBB, JOHN
(1) Matrices. I. *Electr. Rev.*, 1956, **159**, No. 6, 237–240.
(2) Matrices. II. Solution of simultaneous equations. *Electr. Rev.*, 1956, **159**, No. 9, 397–400; R. Zh. M., 1958, 750.

WEBER, R.
(1) Sur les méthodes de calcul employées pour la recherche des valeurset vecteurs propres d'une matrice. *Rech. aéronaut.*, 1949, No. 10, 57–60; M. R., 11, 266.

WEGNER, UDO
(1) Bemerkungen zur Matrizentheorie. *Z. angew. Math. und Mech.*, 1953, **33**, 262–264; M. R., 15, 388.

(2) Contributi alla teoria dei procedimenti iterativi per la risoluzione numerica dei sistemi di equazioni lineari algebriche. *Atti Accad. naz. Lincei Mem.*, cl. sci. fis., mat. e natur., 1953, **4**, No. 1, 1–48; R. Zh. M., 1954, 5781.

WEINER, B. L.
(1) Variations of coefficients of simultaneous linear equations. (With Discussion). *Trans. Amer. Soc. Civil Engrs*, 1948, **113**, 1349–1390.

WEISSINGER, JOHANNES
(1) Uber das Iterationsverfahren. *Z. angew. Math. und Mech.*, 1951, **31**, 245–246.
(2) Zur Theorie und Anwendung des Iterationsverfahrens. *Math. Nachr.*, 1952, **8**, 193–212.
(3) Verallgemeinerung des Seidelschen Iterationsverfahrens. *Z. angew. Math. und Mech.*, 1953, **33**, 155–163.

WENKE, KLAUS
(1) Erfahrungen und Probleme bei der Iochkartenmässigen Berechnung von Kehr-matrizen. *Nachrichtentechn. Fachber.*, 1956, **4**, 198–201, 227; R. Zh. M., 1958, 1596.

WHEELER, D. J., and NASH, J. P.
(1) Digital computer methods for solving linear algebraic equations and finding eigenvalues and eigenvectors. *Digital and Analog Computers and Computing Methods. Symposium at the 18th Appl. Mech. Div. Conf. of the Asme held at the Univ. of Minnesota, June 18–20, 1953*, N. Y. 1953, 21–35; R. Zh. M., 1955, 1519.

WIELANDT, HELMUT
(1) Ein Einschliessungssatz fur charakteristische Wurzeln normaler Matrizen. *Arch. Math.*, 1949, **1**, 348–352; M. R., 11, 4.
(2) Die Einschliessung von Eigenwerten normaler Matrizen. *Math. Ann.*, 1949, **121**, 234–241, M. R., 11, 307.
(3) Inclusion theorems for eigenvalues. *Nat. Bur. Standards. Appl. Math. Ser.*, 1953, **29**, 75–78; R. Zh. M. 1956, 1975.
(4) Einschliessung von Eigenwerten hermitescher Matrizen nach dem Abschnittsver-fahren. *Arch. Math.*, 1954, **5**, No. 1–3, 108–114; R. Zh. M., 1956, 4103.
(5) An extremum property of sums of eigenvalues. *Proc. Amer. Math. Soc.*, 1955, **6**, No. 1, 106–110; R. Zh. M., 1956, 6394.

WILKINSON, J. N.
(1) The calculation of the latent roots and vectors of matrices on the pilot model of the A. C. E. *Proc. Cambridge Philos. Soc.*, 1954, **50**, No. 4, 536–566; R. Zh. M., 1955, 6108.
(2) The use of iterative methods for finding the latent roots and vectors of matrices. *Math. Tables and Other Aids Comput.*, 1955, **9**, No. 52, 184–191; R. Zh. M., 1957, 8965.

WILKES, M. V.
(1) Solution of linear algebraic and differential equations by the long-division algorithm. *Proc. Cambridge Philos. Soc.*, 1956, **52**, No. 4, 758–763; R. Zh. M., 1957, 3524.

WITTMEYER, HELMUT
(1) *Einfluss der Anderung einer Matrix auf die Lösung des zugehörigen Gleichungs-systems, sowie auf die charakteristischen Zahlen und die Eigenvektoren.* Dissertation. Darmstadt, 1934.

WORCH, G.
(1) Uber die zweckmässigste Art, lineare Gleichungen durch Elimination aufzulösen. *Z. angew. Math. und Mech.*, 1932, **12**, 175–181.

WRIGHT, L. T., Jr.
(1) The solution of simultaneous linear equations by an approximation method. *Cornell Univ., Engrg. Exper. Station Bull.*, 1943, **31**, 6 pp; M. R., 5, 110.

YOUNG, DAVID
(1) On Richardson's method for solving linear systems with positive definite matrices. *J. Math. and Phys.*, 1954, **32**, 243–255; R. Zh. M., 1954, 5782.
(2) Iterative methods for solving partial difference equations of elliptic type. *Trans. Amer. Math. Soc.*, 1954, **76**, No. 1, 92–111; R. Zh. M., 1955, **4**, 1953.
(3) On the solution of linear systems by iteration. *Proc Sympos. Appl. Math.*, 6, New York, Toronto, London, 1956, 283–298; R. Zh. M., 1957, 5961.

ZADUNAISKY, PEDRO E.
(1) Un metodo de iteration para la resolucion de sistemas de ecuaciones lineales algebraicas. *Rev. Union mat. argent. y Asoc. fis. argent.*, 1955, **17**, 335–343; R. Zh. M., 1957, 8265.

ZHIVOGLYADOV, V. G.
(1) *O nekotorykh chislennykh metodakh resheniya sistemy lineinykh algebraicheskikh uravnenii, ikh prisposoblenii dlya drugikh vychislenii i ob oshibkakh okrugleniya v etikh protsessakh* (Some numerical methods for solving a system of linear algebraic equations, their adaptability for other computations with remarks on rounding-off errors in these processes), Avtoreferat diss. kand. fiz. matem. n. Kazanskii un-t, Kazan', 1933; R. Zh. M., 1956, 1689.

ZURMÜHL, R.
(1) Das Eliminationsverfahren von Gauss zur Auflösung linearer Gleichungssysteme. *Ber. Inst. Prakt. Math.*, T. H. Darmstadt, Prof. Dr. A. Walther, Z. W. B. Unters. u. Mitt., 1944, **774**, 11–14.
(2) Zur numerischen Auflösung linearer Gleichungssysteme nach dem Matrizenverfahren von Banachiewicz. *Z. angew. Math. und Mech.*, 1949, **29**, 76–84; M. R., 10, 743.
(3) *Matrizen*. Berlin, Gottingen, Heidelberg, 1950; M. R., **12**, 73.
(4) Zur Iteration einzelner Eigenwerte von Matrizen. *Z. angew. Math. und Mech.*, 1957, **37**, No. 5-6, 228; R. Zh. M., 1958, 1597.

ZYL'EV, V. P.
(1) Priznaki skhodimosti i otsenki pogreshnostei reshenii sistemy lineinykh algebraicheskikh uravnenii sposobom iteratsii v matrichnom izlozhenii (Convergence indicators and error estimates for solving a system of linear algebraic equations by iterations in a matrix expansion), *M.-L., Inzhen. stroit. in-t im. Kuibysheva, Sb. trudov*, 1939, **2**, 232–245.

# SUPPLEMENTARY LITERATURE

ANSORGE, R.
(1) Uber ein Iterationsverfahren von G. Schulz zur Ermittlung der Reziproken einer Matrix. *Z. angew. Math. und Mech.*, 1959, **39**, No. 3-4, 164–165.
(2) Bemerkungen zu einem Iterationsverfahren von Bodewig zur Auflösung linearer Gleichungsysteme. *Z. angew. Math. und Mech.*, 1959, **39**, No. 3-4, 165.
(3) Das Hertwigsche Iterationsverfahren zur Auflösung linearer Gleichungssysteme als Gesamt-und Einzelschrittverfahren. *Z. angew. Math. und Mech.*, 1959, **39**, No. 5-6, 248–249.

BAKER, GEORGE A.
(1) A new derivation of Newton's identities and their application to the calculation of the eigenvalues of a matrix. *J. Soc. Industr. and Appl. Math.*, 1959, **7**, No. 2, 143–148.

BAUER, FRIEDRICH L.
(9) Sequential reduction to tridiagonal form. *J. Soc. Industr. and Appl. Math.* 1959, **7**, No. 1, 107–113.

BESSMERTNYKH, G. A.
(1) Ob odnovremennom otyskanii dvukh sobstvennykh chisel samocopryazhennogo operatora (On finding simultaneously two eigenvalues of a self-conjugate operator), *Dokl. AN SSSR*, 1959, 128, No. 6, 1106–1109.

BROEDER, GEORGE G., and SMITH, HARRY J.
(1) A property of semi-definite Hermitian matrices. *J. Assoc. Comput. Machinery*, 1958, **5**, No. 3, 244–245.

BROWN, J.
(1) Propagation in coupled transmission line systems. *Quart. J. Mech. and Appl. Math.*, 1958, **11**, 236–243.

CUTHILL, ELIZABETH H., and VARGA, RICHARD S.
(1) A method of normalized block iteration. *J. Assoc. Comput. Machinery*, 1959, **6**, No. 2, 236–244.

DÜCK, W.
(1) Eine Fehlerabschätzung zum Einzelschrittverfahren bei linearen Gleichungssystemen. *Numerische Math.*, 1959, **1**, No. 1, 73–77.

FORSYTHE, G. E.
(5) Singularity and near singularity in numerical analysis. *Amer. Math. Monthly*, 1959, **65**, No. 4, 229–240.

FRANK, W. L.
(2) Finding zeros of arbitrary functions. *J. Assoc. Comput. Machinery*, 1958, **5**, No. 2, 154–160.

FRÖBERG, C. E.
(2) Diagonalization of Hermitian matrices. *Math. Tables and Other Aids Comput.*, 1958, **12**, 219–220.

GOL'DBAUM, YA. S.
(1) K preobrazovaniyu vekovogo uravneniya (On transforming the secular equation), *Prikl. matem. i mekhanika*, 1958, **22**, No. 4, 539–541.

GREENSPAN, DONALD
(2) On popular methods and extent problems in the solution of polynomial equations. *Math. Mag.*, 1958, **31**, No. 5, 239–253; R. Zh. M., 1959, 8572.

GUN, G.
(1) Limits for the characteristic roots of a matrix. I. *Advancement in Math.*, 1958, **4**, 450–456; M. R., 20, 6, 3893.

HENRICI, P.
(3) On the speed of convergence of cyclic and quasi-cyclic Jacobi methods for computing eigenvalues of Hermitian matrices. *Abstr. Short communs Internat. Congress Math. in Edinburgh*. Edinburg, Univ. Edinburgh, 1958, 160; R. Zh. M., 1959, 8565.

HORNICK, S. D.
(1) IBM 709 Tape Matrix Compiler. *Comm. Assoc. Comput. Machinery*, 1959, **2**, No. 9, 31–32.

HOUSEHOLDER, ALSTON S., and BAUER, FRIEDRICH L.
(1) On certain methods for expanding the characteristic polynomial. *Numerische Math.*, 1959, **1**, No. 1, 29–37.

LANCZOS, C.
(10) Linear systems in self-adjoint form *Amer. Math. Monthly*, 1958, **65**, No. 9, 665–679.

LOTKIN, MARK
(4) Note on the method of contractants *Amer. Math. Monthly*, 1959, **66**, No. 6, 476–479.
(5) Determination of characteristic values. *Quart. Appl. Math.*, 1959, **17**, No. 3.

McGINN, LAURENCE C.
(1) The matrix math compiler. *J. Franklin Inst.*, 1957, **264**, No. 5, 415–416; R. Zh. M., 1959, 11619.

MANAIRA, MARIO
(1) L'inversione delle matrici con L'UNIVAC. *Idee e sist.*, 1958, No. 24–25, 9–11; R. Zh. M., 1959, 8562.

MARATHE, C. R.
(1) Note on some semimoduli of a rectangular matrix. *Amer. Math. Monthly*, 1958, **65**, No. 4, 259–263.

MIRSKY, L.
(3) On the minimization of matrix norms. *Amer Math. Monthly*, 1958, **65**, No. 2, 106–107; R. Zh. M., 1959, 8828.

(4) Diagonal elements of orthogonal matrices. *Amer. Math. Monthly*, 1959, **66**, No. 1, 19–22.

MONJALLON, ALBERT
(1) *Initiation au calcul matriciel. Matices. Déterminants. Applications à l'algèbre et à la géométrie analytique*. Paris, Librairie Vuibert, 1955, 131 pp; R. Zh. M., 1959, 8834.

MOSTOWSKI, ANDRZEJ, and STARK, MARCELI
(1) *Algebra liniowa* (Bibliot. mat., 19), Warszawa, PWN, 1958, 188 s.; R. Zh. M., 1959, 9789.

NEWMAN, MORRIS, and TODD, JOHN
(1) The evaluation of matrix inversion programs. *J. Soc. Industr. and Appl. Math.*, 1958, **6**, No. 4, 466–476.

NITSCHE, J.
(1) Einfache Fehlerschranken beim Eigenwertproblem symmetrischer Matrizen. *Z. angew. Math. und Mech.*, 1959, **39**, Nos. 7–8, 322–325.

NOBLE, B.
(1) The numerical solution of an infinite set of linear simultaneous equations. *Quart. Appl. Math.*, 1959, **17**, No. 1, 98–102.

PARODI, MAURICE
(1) Sur une méthode de localisation des valeurs caractèristiques de certaines matrices. *C. r. Acad. sci.*, 1958, **247**, No. 5, 571–573; R. Zh. M., 1959, 10834.

PUGACHEV, B. P.
(3) Ob odnom sposobe odnovremennogo vychisleniya dvukh granits spektra (On a method for computing simultaneously the two boundaries of a spectrum), *Tr. semin. po funktsion. analizu, Voronezhsk. un-t*, 1957, **5**, 52–70.
(4) K voprosu o bystrote skhodimosti metoda normal'nykh khord (On the rate of convergence of the normal chords method), *Tr. semin. po funktsion. analizu, Rostovsk. n/D i Voronezhsk. gos. un-ty*, 1960, **3–4**, 77–80.
(5) Issledovanie odnogo metoda priblizhennogo vychisleniya sobstvennykh chisel i vektorov (Investigation of an approximate method for computing eigenvalues and eigen-

vectors), *Tr. semin. po funktsion. analizu, Rostovsk. n/D i Voronezhsk. gos. un-ty*, 1960, **3-4**, 81-97.

RAICHL, JIRI
(1) The economical coding of high-order matrices for automatic computers. *Stroje na zpracovàni informaci*, 1956, **4**, 257-271; M. R., 20, 6, 4345.

ROSENBLUM, MARVIN
(1) On the Hilbert matrix I. *Proc. Amer. Math. Soc.*, 1958, **9**, No. 1, 137-140.
(2) On the Hilbert Matrix II. *Proc. Amer. Math. Soc.*, 1958, **9**, No. 4, 581-585.

RUTISHAUSER, HEINZ
(9) Zur Matrizeninversion nach Gauss-Jordan. *Z. angew. Math. und Phys.*, 1959, **10**, No. 3, 281-291.
(10) Deflation bei Bandmatrizen. *Z. angew. Math. und Phys.*, 1959, **10**, No. 3, 314-319.

SARHAN, A. E., and GREENBERG, B. G.
(1) Inverting patterned matrices. *Abstr. Short communs Internat. Congress Math. in Edinburgh*, Edinburgh, Univ. Edinburgh, 1958, 128; R. Zh. M., 1959, 8563.

SCHECHTER, S.
(1) On the inversion of certain matrices. *Math. Tables and Other Aids Comput.*, 1959, **13**, 73-77.

SHELDON, J. W.
(1) On the spectral norms of several iterative processes. *J. Assoc. Comput. Machinery*, 1959, **6**, No. 4, 494-505.

SCHMIDTMAYER, JOSEF
(2) Linear computations over a complex field. *J. Roy. Aeronaut. Soc.*, 1958, **62**, No. 570, 451-455; R. Zh. M., 1959, 10483.

SHREIDER, Yu. A.
(2) Reshenie sistem lineinykh algebraicheskikh uravnenii po metodu Monte-Karlo (Solving systems of linear algebraic equations by a Monte-Carlo method), *Vopr. teorii matem. mashin. 1. M., Fizmatgiz.* 1958, 167-171; R. Zh. M., 1959, 8559.

TUPPER, S. J.
(1) Ill-conditioned linear equations. *Math. Gaz.*, 1958, **42**, No. 342, 299-300; R. Zh. M., 1959, 9507.

WHITE, PAUL A.
(1) The computation of eigenvalues and eigenvectors of a matrix. *J. Soc. Industr. and Appl. Math.*, 1958, **6**, No. 4, 393-437.

WILF, HERBERT S.
(1) Matrix inversion by the annihilation of rank. *J. Soc. Industr. and Appl. Math.*, 1959, **7**, No. 2, 149-151.

WILKINSON, J. H.
(3) Linear algebra on the Pilot A.C.E. Automatic Digital Comput. at N.P.L., 1955, 129-137.

(4) The calculation of the eigenvectors of codiagonal matrices. *Comput. J.*, 1958, **1**, 90–96.

(5) The calculation of eigenvectors by the method of Lanczos. *Comput. J.*, 1958, **1**, No. 3, 148–152.

(6) The evaluation of the zeros of ill-conditioned polynomials. Part I. *Numerische Math.*, 1959, **1**, No. 3, 150–166.

(7) The evaluation of the zeros of ill-conditioned polynomial. Part II. *Numerische Math.*, 1959, **1**, No. 3, 167–180.

(8) Stability of the reduction of a matrix to almost triangular and triangular forms by elementary similarity transformations. *J. Assoc. Comput. Machinery*, 1959, **5**, No. 3, 336–359.

WILSON, L. B.

(1) Solution of certain large sets of equations on Pegasus using matrix methods. *Comput. J.*, 1959, **2**, No. 3, 130–133.

WINDLEY, P. F.

(1) Transposing matrices in a digital computer. *Comput. J.*, 1959, **2**, No. 1, c. 47–48.

WYNN, P.

(1) A sufficient condition for the instability of the $q$-$d$-algorithm. *Numerische Math.*, 1959, **1**, No. 4, 203–207.

(2) On the propagation of error in certain non-linear algorithms. *Numerische Math.*, 1959, **1**, No. 3, 142–149.

# INDEX

26313

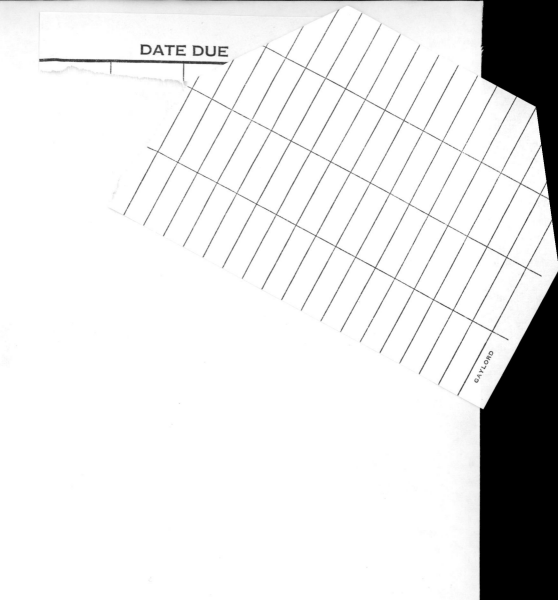

**DATE DUE**

GAYLORD